Clifford (Geometric) Algebras

with applications
to physics, mathematics, and engineering

William E. Baylis
Editor

Birkhäuser
Boston • Basel • Berlin

William E. Baylis
Department of Physics
University of Windsor
Windsor, Ontario
Canada N9B 3P4

Library of Congress Cataloging-In-Publication Data

Clifford (geometric) algebras with applications to physics,
mathematics, and engineering / William E. Baylis, editor.
p. cm.
Proceedings of the 1995 Summer School on Theoretical Physics of
the Canadian Association of Physicists, held in Banff, Alberta, July
30-Aug. 12, 1995.
Includes bibliographical references and index.
ISBN 0-8176-3868-7 (alk. paper). -- ISBN 3-7643-3868-7 (alk.
paper)
1. Clifford algebras--Congresses. 2. Mathematical physics-
-Congresses. I. Baylis, William E. (William Eric), 1939- .
II. Summer School on Theoretical Physics of the Canadian Association
of Physicists (1995 : Banff, Alta.)
QC20.7.C55C58 1996
530.1'5257--dc20 96-25142
CIP

Printed on acid-free paper

ISBN 0-8176-3868-7
ISBN 3-7643-3868-7

Camera-ready text prepared by the Editor in LaTeX.
Printed and bound by Hamilton Printing Company, Rensselaer, NY.
Printed in the United States of America.

9 8 7 6 5 4 3 2

Contents

Preface

This volume is an outgrowth of the 1995 Summer School on Theoretical Physics of the Canadian Association of Physicists (CAP), held in Banff, Alberta, in the Canadian Rockies, from July 30 to August 12, 1995. The chapters, based on lectures given at the School, are designed to be tutorial in nature, and many include exercises to assist the learning process. Most lecturers gave three or four fifty-minute lectures aimed at relative novices in the field. More emphasis is therefore placed on pedagogy and establishing comprehension than on erudition and superior scholarship. Of course, new and exciting results are presented in applications of Clifford algebras, but in a coherent and user-friendly way to the nonspecialist.

The subject area of the volume is Clifford algebra and its applications. Through the geometric language of the Clifford-algebra approach, many concepts in physics are clarified, united, and extended in new and sometimes surprising directions. In particular, the approach eliminates the formal gaps that traditionally separate classical, quantum, and relativistic physics. It thereby makes the study of physics more efficient and the research more penetrating, and it suggests resolutions to a major physics problem of the twentieth century, namely how to unite quantum theory and gravity. The term "geometric algebra" was used by Clifford himself, and David Hestenes has suggested its use in order to emphasize its wide applicability, and because the developments by Clifford were themselves based heavily on previous work by Grassmann, Hamilton, Rodrigues, Gauss, and others.

The summer school concentrated on applications to physics, but important developments in mathematics as well as applications to other areas such as engineering were also presented. Some diversity of approach will be evident to the reader and is only natural in an actively growing field of research with contributions from the different cultures of mathematics and physics. The reader will benefit from the degree of uniformity that has been established in notation by mathematical physicists who bridge the cultures, especially by Pertti Lounesto, who has been called the policeman of Clifford algebras for his efforts to ensure accuracy and consistency in the field. Nevertheless, as indicated in the introduction, some further work is still needed along these lines.

Lectures at the School generally built on material familiar to undergraduate physics majors, and the same accessibility is evident in this volume. Anyone new to Clifford algebras may wish to start by reading the Introduction, but after that, practically any set of chapters can be read independently of the others.

This volume contains material from most but not all of the lectures given at the Summer School. Unfortunately, a couple of speakers did not find time to prepare their lectures for publication. Notably absent are those of David Hestenes, who provided crucial organizational help to the school.

There is a wide range of material in this volume. The reader can sample the history of Clifford algebras in Lounesto's chapters, learn how many features of Clifford algebras can be introduced through unipodal numbers as discussed by Sobczyk, or study a generalization of Grassmann algebras to both vectors and forms in the chapter by Jancewicz. The Cambridge group (Doran, Gull, and Lasenby) present a coherent package of lectures covering applications of Hestenes' spacetime algebra (STA) to electromagnetic theory, quantum theory, and general relativity. Generalizations of traditional symmetries in relativity theory is the topic of Crawford's chapters. The lectures by Chisholm and Farwell discuss the application of higher-dimensional Clifford algebras to theories of fundamental particles. Several single chapters are also presented on such applications as computer vision and robotics (Joan Lasenby), projective geometry (Maks), Hamiltonian mechanics (Pappas), and symbolic computer calculations (Abłamowicz).

Most lecture series have bibliographies for further reading. Lounesto's is a particularly rich source of references to primary material. An index is provided for the entire volume.

Many thanks are in order for the help I received in preparing this volume. In particular, I want to thank all the contributing authors for their well-prepared latex files, my Windsor colleague John Huschilt for his careful proof-reading of several chapters, co-op student Anthony Mandarino for his help with the initial assembling of chapters, and Ann Kostant and her associates at Birkhäuser Boston for their professional but friendly assistance.

May the reader enjoy and be enriched!

W.E. Baylis Windsor, Ont., June 1996

Clifford (Geometric) Algebras

with applications to physics, mathematics, and engineering

Chapter 1

Introduction

Clifford (geometric) algebras extend the real number system to include *vectors* $\mathbf{u}, \mathbf{v}, \mathbf{w}, \ldots$, and their products $\mathbf{uv}$, $\mathbf{uvw}, \ldots$. They are useful for modeling geometry, and their vector products, representing surfaces and higher-dimensional objects, allow simple but rigorous descriptions of rotations, reflections, and other geometric transformations. The name Clifford algebra honors the English mathematician William Kingdon Clifford (1845-79), who recognized the importance of ideas set forth by the German high-school mathematics teacher Hermann Günther Grassmann (1809-77). Clifford developed Grassmann's ideas into what he called *geometric algebras.* Complex numbers and quaternions ("hypercomplex numbers"), form two particularly simple geometric algebras. The aim of this introduction is to present a brief overview of Clifford algebras and their applications in physics.

Clifford algebras are closed under addition and multiplication, and products of three or more vectors are associative: $\mathbf{u}(\mathbf{vw}) = (\mathbf{uv})\mathbf{w} \equiv \mathbf{uvw}$. The basic axiom of Clifford algebras is that the product of any vector $\mathbf{v}$ with itself is its square length:

$$\mathbf{v}^2 = \mathbf{v}\,\mathbf{v} = \mathbf{v} \cdot \mathbf{v}. \tag{1.1}$$

Suppose $\mathbf{v}$ is the sum of two other vectors: $\mathbf{v} = \mathbf{a} + \mathbf{b}$. Relation (1.1) becomes $(\mathbf{a}+\mathbf{b})^2 = (\mathbf{a}+\mathbf{b}) \cdot (\mathbf{a}+\mathbf{b})$, whose expansion, followed by the elimination of $\mathbf{a}^2 = \mathbf{a} \cdot \mathbf{a}$ and $\mathbf{b}^2 = \mathbf{b} \cdot \mathbf{b}$, leaves the basic result

$$\mathbf{ab} + \mathbf{ba} = 2\mathbf{a} \cdot \mathbf{b}. \tag{1.2}$$

If $\mathbf{a}$ and $\mathbf{b}$ are not collinear, $\mathbf{ab}$ cannot be a scalar since $\mathbf{aab} = \mathbf{a}(\mathbf{ab}) = (\mathbf{aa})\mathbf{b}$ would then equate vectors pointing in different directions. Consequently,

$$\mathbf{ab} - \mathbf{ba} = 2(\mathbf{ab} - \mathbf{a} \cdot \mathbf{b}) \tag{1.3}$$

vanishes only if $\mathbf{a}$ and $\mathbf{b}$ are aligned. Thus, unlike scalars, vector products generally do not commute: $\mathbf{ab} \neq \mathbf{ba}$. Otherwise however, vectors multiply like scalars. In particular, products are linear and distributive over addition. The multiplication rules for vectors are the same as for square matrices.

A Clifford algebra common in physics is $\mathcal{C}\ell_3$, the Pauli algebra, based on vectors in three-dimensional Euclidean space $\mathbb{R}^3$. A conventional representation uses the

2×2 Pauli spin matrices

$$\sigma_1 = \begin{pmatrix} 0 & 1 \\ 1 & 0 \end{pmatrix}, \ \sigma_2 = \begin{pmatrix} 0 & -i \\ i & 0 \end{pmatrix}, \ \sigma_3 = \begin{pmatrix} 1 & 0 \\ 0 & -1 \end{pmatrix}, \tag{1.4}$$

to represent unit vectors along the Cartesian axes. A vector $\mathbf{v}$ with Cartesian components v_x, v_y, v_z, is thus represented by the matrix $v_x\sigma_1 + v_y\sigma_2 + v_z\sigma_3$.[1] An infinite number of different matrix representations may be used, but only the algebra of their products (which they have in common) is physically significant.

From Eq. (1.3), products of any two real n-dimensional vectors are sums

$$\mathbf{ab} = \mathbf{a} \cdot \mathbf{b} + \mathbf{a} \wedge \mathbf{b} \tag{1.5}$$

of a dot product and a wedge (or exterior) product defined by

$$\mathbf{a} \wedge \mathbf{b} := \frac{1}{2} (\mathbf{ab} - \mathbf{ba}) \,. \tag{1.6}$$

The dot product $\mathbf{a} \cdot \mathbf{b}$ is a scalar, but $\mathbf{a} \wedge \mathbf{b}$ is neither a scalar nor a vector; it is a new element called a *bivector*. In three dimensions, $\mathbf{a} \wedge \mathbf{b}$ is related to the vector cross product $\mathbf{a} \times \mathbf{b}$ (see below), but unlike cross products, wedge products are associative and well defined in spaces of higher dimension.

Just as n-dimensional vectors $\mathbf{a}$ and $\mathbf{b}$ can be expanded in a Cartesian basis $\{\mathbf{e}_1, \mathbf{e}_2, \cdots, \mathbf{e}_n\}$, so substitution into (1.6) shows that any bivector $\mathbf{a} \wedge \mathbf{b}$ can be expanded in a basis of $n(n-1)/2$ unit bivectors. In three dimensions,

$$\mathbf{a} \wedge \mathbf{b} = (a_x b_y - a_y b_x)\, \mathbf{e}_1\mathbf{e}_2 + (a_y b_z - b_y a_z)\, \mathbf{e}_2\mathbf{e}_3 + (a_z b_x - a_x b_z)\, \mathbf{e}_3\mathbf{e}_1 \tag{1.7}$$

Whereas the vector $\mathbf{e}_1$ corresponds to a directed line segment of unit length, the bivector $\mathbf{e}_1\mathbf{e}_2 = \mathbf{e}_1 \wedge \mathbf{e}_2 = -\mathbf{e}_2\mathbf{e}_1$ represents an oriented plane patch of unit area that contains both $\mathbf{e}_1$ and $\mathbf{e}_2$. Its orientation is the twist that rotates $\mathbf{e}_1$ most directly into $\mathbf{e}_2$. In three dimensions, the bivectors $\mathbf{e}_1\mathbf{e}_2, \mathbf{e}_2\mathbf{e}_3, \mathbf{e}_3\mathbf{e}_1$ span a linear space distinct from the original vector space spanned by $\mathbf{e}_1, \mathbf{e}_2, \mathbf{e}_3$. Unit vectors in Euclidean space square to 1 [see Eq. (1.2)], but the basis bivectors square to -1, for example, $(\mathbf{e}_1\mathbf{e}_2)^2 = \mathbf{e}_1\mathbf{e}_2\mathbf{e}_1\mathbf{e}_2 = -\mathbf{e}_1\mathbf{e}_1\mathbf{e}_2\mathbf{e}_2 = -1$.[2]

The geometrical significance of $\mathbf{e}_1\mathbf{e}_2$ is further revealed by noting from Eq. (1.2) that it is an operator that rotates both $\mathbf{e}_1$ and $\mathbf{e}_2$ by the same right angle:

$$\mathbf{e}_1 (\mathbf{e}_1\mathbf{e}_2) = \mathbf{e}_2 \,, \ \mathbf{e}_2 (\mathbf{e}_1\mathbf{e}_2) = -\mathbf{e}_1 \,. \tag{1.8}$$

Any vector $\mathbf{v} = v_x\mathbf{e}_1 + v_y\mathbf{e}_2$ in the xy plane is similarly rotated. Linear combinations of the identity and $\mathbf{e}_1\mathbf{e}_2$ can produce rotations by an arbitrary angle θ:

$$\mathbf{v}' = \mathbf{v} (\cos\theta + \mathbf{e}_1\mathbf{e}_2 \sin\theta) = \mathbf{v} \exp(\mathbf{e}_1\mathbf{e}_2\theta) \ , \tag{1.9}$$

[1] In physics literature, this matrix is often written $\mathbf{v} \cdot \boldsymbol{\sigma}$, but one should understand that it represents a vector, not a scalar.

[2] One can identify Hamilton's hypercomplex (or quaternion) units $\mathbf{i}, \mathbf{j}, \mathbf{k}$ as unit bivectors: $\mathbf{i} = \mathbf{e}_3\mathbf{e}_2$, $\mathbf{j} = \mathbf{e}_1\mathbf{e}_3$, $\mathbf{k} = \mathbf{e}_2\mathbf{e}_1$, but Hamilton also used them as basis *vectors* in a notation still common today. The confusion of vectors and bivectors persists in the vector cross product.

as one verifies by expanding in powers of θ. To rotate a vector $\mathbf{r} = x\mathbf{e}_1 + y\mathbf{e}_2 + z\mathbf{e}_3$ with a component z normal to the rotation plane, one can use[3]

$$\mathbf{r}' = \exp(\mathbf{e}_2\mathbf{e}_1\theta/2)\,\mathbf{r}\exp(\mathbf{e}_1\mathbf{e}_2\theta/2)\ . \tag{1.10}$$

If $\theta = \omega t$, Eqs. (1.9–1.10) describe rotations at constant angular velocity ω.

Bivectors can also be used for reflections. The reflection of $\mathbf{r}$ in the xy plane is

$$(\mathbf{e}_1\mathbf{e}_2)\,\mathbf{r}\,(\mathbf{e}_1\mathbf{e}_2) = x\mathbf{e}_1 + y\mathbf{e}_2 - z\mathbf{e}_3\,. \tag{1.11}$$

Two successive reflections in intersecting planes can be seen to be equivalent to a rotation by twice the angular opening between the planes.

The product $\mathbf{e}_1\mathbf{e}_2\mathbf{e}_3$ is a *trivector* representing an oriented unit volume. There is only one linearly independent trivector in $\mathbb{R}^3$ since $\mathbf{e}_1\mathbf{e}_2\mathbf{e}_3 = -\mathbf{e}_2\mathbf{e}_1\mathbf{e}_3 = \mathbf{e}_2\mathbf{e}_3\mathbf{e}_1 = \cdots$. It squares to -1 and commutes with all the basis vectors, and hence with all elements of the algebra. It can be identified with the unit imaginary:

$$\mathbf{e}_1\mathbf{e}_2\mathbf{e}_3 = i\,. \tag{1.12}$$

The identification (1.12) associates bivectors with imaginary vectors directed normal to the plane, for example, $\mathbf{e}_1\mathbf{e}_2 = \mathbf{e}_1\mathbf{e}_2\mathbf{e}_3\mathbf{e}_3 = i\mathbf{e}_3$. The vector $\mathbf{e}_3$ is called the vector dual to the bivector $\mathbf{e}_1\mathbf{e}_2$; it is the axis of the rotation plane. More generally [see Eq. (1.7)], the vector cross product $\mathbf{a} \times \mathbf{b}$ is the vector dual to the bivector $\mathbf{a} \wedge \mathbf{b}$:

$$\mathbf{a} \wedge \mathbf{b} = i\mathbf{a} \times \mathbf{b}. \tag{1.13}$$

Like the cross product itself, the dual relationship between bivectors and vectors is meaningful only in three dimensions. Bivectors, however, are used for rotations and reflections in spaces of any dimension $n \geq 2$.

A general element p of $C\ell_3$ is a real sum of scalar S, vector $\mathbf{V}$, bivector $\mathbf{B}$, and trivector T parts or *grades*:

$$p = S + \mathbf{V} + \mathbf{B} + T\,. \tag{1.14}$$

Two conjugations are introduced that have the effect of changing the signs on some of the parts. Both *reversion*[4]

$$p^\dagger = S + \mathbf{V} - \mathbf{B} - T \tag{1.15}$$

and *Clifford conjugation*[5]

$$\bar{p} = S - \mathbf{V} - \mathbf{B} + T \tag{1.16}$$

[3] Elements of the matrix representation of $\exp(\mathbf{e}_1\mathbf{e}_2\theta/2)$ are the Caley-Klein parameters.

[4] So called because it results when all factors are expressed as products of vectors and then the order of the vector factors is reversed. Other notation (in $C\ell_3$) is $\tilde{p}$ (ALS, P) and p^* (M), where the letters stand for authors: ALS = Abłamowicz, Lounesto, and Sobczyk, P = Pappas, and M = Maks. The tilde notation is also often used in other Clifford algebras.

[5] Alternative notation (by DGL and M) is $\tilde{p}$, where DGL = Doran, Gull, and Lasenby.

are *antiautomorphic involutions*: the conjugate of a product is the product of the conjugates in reverse order: $(pq)^\dagger = q^\dagger p^\dagger$. Their combination (applied in either order) gives the automorphic *grade involution*:[6]

$$\bar{p}^\dagger = S - \mathbf{V} + \mathbf{B} - T\,, \tag{1.17}$$

which changes the signs only of the odd grades.

Because of Eq. (1.12), every element p in the Pauli algebra can be expressed as the sum of a complex scalar and a complex vector in a four-dimensional space and thus as a complex linear combination of what is called the *paravector basis* $\{\mathbf{e}_0, \mathbf{e}_1, \mathbf{e}_2, \mathbf{e}_3\}$, where for convenience one defines $\mathbf{e}_0 = 1$:

$$p = p_0\mathbf{e}_0 + p_x\mathbf{e}_1 + p_y\mathbf{e}_2 + p_z\mathbf{e}_3\,. \tag{1.18}$$

If the coefficients $p_0\,, p_x, p_y, p_z$ are real, p is a real paravector, and such elements can represent spacetime vectors in relativity. Their boosts (velocity transformations) are simply rotations [Eq. (1.10)] in spacetime planes containing the time axis $\mathbf{e}_0$. The spacetime metric is built into the scalar norm

$$p\bar{p} = p_0^2 - p_x^2 - p_y^2 - p_z^2 \tag{1.19}$$

where $\bar{p} = p_0\mathbf{e}_0 - (p_x\mathbf{e}_1 + p_y\mathbf{e}_2 + p_z\mathbf{e}_3)$. The momenta of light signals are spacetime vectors p with null norms ($p\bar{p} = 0$); such p are non-zero elements whose inverse $p^{-1} = \bar{p}/(p\bar{p})$ does not exist. They have no counterpart in the fields $\mathbb{R}$ and $\mathbb{C}$.

The center (commuting part) of the Pauli algebra is the complex field, and the even subalgebra, containing only products of zero and two real vectors (real scalars and bivectors), is the quaternion algebra. Clifford algebras $\mathcal{C}\ell_{p,q}$ are also readily found for n-dimensional vector spaces with basis vectors $\mathbf{e}_j$ satisfying $\mathbf{e}_j \cdot \mathbf{e}_k = \frac{1}{2}(\mathbf{e}_j\mathbf{e}_k + \mathbf{e}_k\mathbf{e}_j) = 0$, $j \neq k$ and

$$\mathbf{e}_j^2 = \begin{cases} +1, & j = 1, \ldots, p \\ -1, & j = p+1, \ldots, n \end{cases}\,. \tag{1.20}$$

Clifford himself concentrated on cases with $p = n$ and $p = 0$. Elements of the algebra, sometimes called cliffors reside in a 2^n-dimensional real vector space with parts comprising a scalar, an n-dimensional vector, an $n(n-1)/2$-dimensional bivector, and higher-order multivectors.

The field of complex numbers is the Clifford algebra with $n = 1$ and $p = 0$, and the quaternion algebra is that with either $n = 2$ or $n = 3$ (the two algebras are equivalent) and $p = 0$. The Pauli algebra discussed above has $n = p = 3$. In Dirac's theory of the electron, the algebra of the Dirac matrices is the Clifford algebra with $n = 4$ and $p = 1$. Only Clifford algebras with $2p - n = 3 + 4N$ for any integer N share the Pauli-algebra property that they possess a natural complex structure in a space of 2^{n-1} dimensions. Many applications to physics follows in subsequent chapters. Further introductory material on Clifford algebras is developed especially in section 2.2 and in chapters 3, 6, and 17.

[6] A common notation for the grade involution is $\hat{p}$ (ALS, M). An alternative name is the *main involution* (M).

Chapter 2

Clifford Algebras and Spinor Operators

Pertti Lounesto
Institute of Mathematics
Helsinki University of Technology
FIN-02150 Espoo, Finland

This paper begins with a historical survey on Clifford algebras and a model on how to start an undergraduate course on Clifford algebras. The Dirac equation and the bilinear covariants are discussed. The Fierz identities are sufficient to reconstruct a Dirac spinor from its bilinear covariants, up to a phase. However, the Weyl and Majorana spinors cannot be reconstructed using the Fierz identities alone. This paper introduces a new concept, the boomerang, for the reconstruction of the Weyl and Majorana spinors. This method reveals a new class of spinors residing in between the Weyl, Majorana and Dirac spinors, namely the flag-dipole spinors.

2.1 A History of Clifford Algebras

Clifford algebras were created and classified by William K. Clifford 1878/1882, when he presented a new multiplication rule for vectors in Grassmann's exterior algebra $\bigwedge \mathbb{R}^n$. In the special case of $\mathbb{R}^3$ this construction embodied Hamilton's quaternions. Clifford algebras were independently rediscovered by Lipschitz 1880/1886, who also gave their first application to geometry, namely the representation of rotations in $\mathbb{R}^n$.

Spinor representations of the rotation group $SO(n)$, or more precisely of the spin group $\mathbf{Spin}(n)$, were introduced by E. Cartan 1913 and later by Brauer and Weyl 1935. Cartan 1908 also discovered the periodicity of 8 in matrix representations of real Clifford algebras (rediscovered by Atiyah, Bott, and Shapiro 1964), and Brauer and Weyl also presented multiplication of Clifford numbers by binary indices (this was discovered earlier by Vahlen 1897). Vahlen 1902 introduced a representation of Möbius transformations of $\mathbb{R}^n$ by 2×2-matrices in $C\ell_n(2)$ with entries in the

Clifford algebra $C\ell_n$ of $\mathbb{R}^n$ (or more precisely in $C\ell_{0,n}$ of $\mathbb{R}^{0,n}$).

The first one to associate Clifford algebras with quadratic forms was Ernst Witt 1937, who determined Clifford algebras of non-degenerate quadratic forms over arbitrary fields of characteristic $\neq 2$. The generalization to the exceptional characteristic 2 was given by Chevalley 1954 in his construction $C\ell(Q) \subset \mathrm{End}(\bigwedge V)$. Chevalley went further and gave the most general definition, $C\ell(Q) = \otimes V/\mathcal{I}_Q$, valid not only for fields, but also for commutative rings. Marcel Riesz 1958 reconstructed Grassmann's exterior algebra from the Clifford algebra, in any characteristic $\neq 2$, by

$$\mathbf{x} \wedge u = \frac{1}{2}(\mathbf{x}u + (-1)^k u\mathbf{x})$$

where $\mathbf{x} \in V$ and $u \in \bigwedge^k V$ [Chevalley 1946 related exterior products of vectors to antisymmetric Clifford products of vectors, but his relationship was valid only in characteristic 0].

Cartan's periodicity of 8 of real Clifford algebras, with an involution, was enhanced by Porteous 1969 (and later by Harvey 1990), who found that the real graded Clifford algebras, with an anti-involution, have a periodicity $(8 \times 8)/2$ of a chessboard. This periodicity classified all possible scalar products of spinors in the real Clifford algebras $C\ell_{p,q}$.

On the physical side, Pauli 1927 and Dirac 1928 presented their spinor equations for the description of the electron spin. Juvet 1930 and Sauter 1930 replaced column spinors by square matrix spinors, where only the first column was non-zero. Marcel Riesz 1947 was the first one to consider spinors as elements in a minimal left ideal of a Clifford algebra (although the special case of pure spinors had been considered earlier by Cartan 1938). Riesz 1958 also condensed the Maxwell equations, in the special case of a homogeneous and isotropic medium, into a single equation by bivectors in the Clifford algebra $C\ell_{1,3}$ (this had been discovered by Juvet and Schidlof 1932).

Gürsey 1956-58 rewrote the Dirac equation with 2×2 quaternion matrices in $\mathbb{H}(2)$ (see also Gsponer and Hurni 1993). Kustaanheimo 1964 presented the spinor regularization of the Kepler motion, the KS-transformation, which emphasized the operator aspect of spinors. This led David Hestenes 1966-74 to a reformulation of the Dirac theory, where the role of spinors [in columns $\mathbb{C}^4$ or in minimal left ideals of the complex Clifford algebra $\mathbb{C} \otimes C\ell_{1,3} \simeq \mathbb{C}(4)$] was taken over by operators in the even subalgebra $C\ell_{1,3}^+$ of the real Clifford algebra $C\ell_{1,3} \simeq \mathbb{H}(2)$.

Spinors were reconstructed from their bilinear covariants by Y. Takahashi 1983 and J. Crawford 1985.

2.2 Teaching Clifford algebras

In this section I will tell how I have started my undergraduate courses on Clifford algebras. An undergraduate course on vectors presents them as geometric objects, directed line segments, depicts their computation rules by diagrams, and emphasizes the distinction between vectors and scalars, the real numbers. After introducing the linear structure on the set $\mathbb{R}^2 = \mathbb{R} \times \mathbb{R}$, which makes it a vector plane $\mathbb{R}^2$, one usually discusses the metric, the scalar product,

$$\mathbf{a} \cdot \mathbf{b} = a_1 b_1 + a_2 b_2$$

of $\mathbf{a} = a_1 \mathbf{e}_1 + a_2 \mathbf{e}_2$ and $\mathbf{b} = b_1 \mathbf{e}_1 + b_2 \mathbf{e}_2$ in an orthonormal basis $\{\mathbf{e}_1, \mathbf{e}_2\}$. At this point one is ready for Clifford algebras.

2.2.1 The Clifford Product of Vectors.

Take two orthogonal unit vectors $\mathbf{e}_1$ and $\mathbf{e}_2$ in the vector plane $\mathbb{R}^2$. The length of the vector $\mathbf{r} = x\mathbf{e}_1 + y\mathbf{e}_2$ is $|\mathbf{r}| = \sqrt{x^2 + y^2}$. If the vector $\mathbf{r}$ is multiplied with itself, $\mathbf{r}\mathbf{r} = \mathbf{r}^2$, it seems natural to require that the product equals the length of $\mathbf{r}$ squared

$$\mathbf{r}^2 = |\mathbf{r}|^2.$$

In coordinate form we have introduced a product for vectors in such a way that

$$(x\mathbf{e}_1 + y\mathbf{e}_2)^2 = x^2 + y^2.$$

Use the distributive rule without assuming commutativity to obtain

$$x^2\mathbf{e}_1^2 + y^2\mathbf{e}_2^2 + xy(\mathbf{e}_1\mathbf{e}_2 + \mathbf{e}_2\mathbf{e}_1) = x^2 + y^2.$$

This is satisfied if the orthogonal unit vectors $\mathbf{e}_1$, $\mathbf{e}_2$ obey the multiplication rules

$$\boxed{\begin{array}{l} \mathbf{e}_1^2 = \mathbf{e}_2^2 = 1 \\ \mathbf{e}_1\mathbf{e}_2 = -\mathbf{e}_2\mathbf{e}_1 \end{array}} \quad \text{which correspond to} \quad \boxed{\begin{array}{l} |\mathbf{e}_1| = |\mathbf{e}_2| = 1 \\ \mathbf{e}_1 \perp \mathbf{e}_2 \end{array}}$$

Use associativity to calculate the square $(\mathbf{e}_1\mathbf{e}_2)^2 = -\mathbf{e}_1^2\mathbf{e}_2^2 = -1$. Since the square of the product $\mathbf{e}_1\mathbf{e}_2$ is negative, it follows that $\mathbf{e}_1\mathbf{e}_2$ is neither a scalar nor a vector. The product is a new kind of unit, called a **bivector**, representing the oriented plane area of the square with sides $\mathbf{e}_1$ and $\mathbf{e}_2$. Denote for short $\mathbf{e}_{12} = \mathbf{e}_1\mathbf{e}_2$.

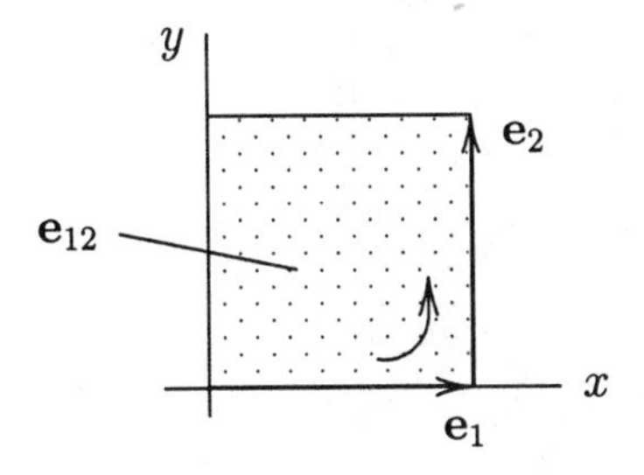

Example. Compute $\mathbf{e}_1\mathbf{e}_{12} = \mathbf{e}_1\mathbf{e}_1\mathbf{e}_2 = \mathbf{e}_2$, $\mathbf{e}_{12}\mathbf{e}_1 = \mathbf{e}_1\mathbf{e}_2\mathbf{e}_1 = -\mathbf{e}_1^2\mathbf{e}_2 = -\mathbf{e}_2$, $\mathbf{e}_2\mathbf{e}_{12} = \mathbf{e}_2\mathbf{e}_1\mathbf{e}_2 = -\mathbf{e}_1\mathbf{e}_2^2 = -\mathbf{e}_1$ and $\mathbf{e}_{12}\mathbf{e}_2 = \mathbf{e}_1\mathbf{e}_2^2 = \mathbf{e}_1$. Note in particular that $\mathbf{e}_{12}$ anticommutes with both $\mathbf{e}_1$ and $\mathbf{e}_2$. ∎

The four elements

1	scalar
$\mathbf{e}_1$, $\mathbf{e}_2$	vectors
$\mathbf{e}_{12}$	bivector

form a basis of the **Clifford algebra** $C\ell_2$ of the vector plane $\mathbb{R}^2$, that is, an arbitrary element

$$u = u_0 + u_1\mathbf{e}_1 + u_2\mathbf{e}_2 + u_{12}\mathbf{e}_{12} \quad \text{in} \quad C\ell_2$$

is a linear combination of a scalar u_0, a vector $u_1\mathbf{e}_1 + u_2\mathbf{e}_2$ and a bivector $u_{12}\mathbf{e}_{12}$. [1] The Clifford algebra $C\ell_2$ is a 4-dimensional real linear space with basis elements 1, $\mathbf{e}_1$, $\mathbf{e}_2$, $\mathbf{e}_{12}$ which have the multiplication table

	$\mathbf{e}_1$	$\mathbf{e}_2$	$\mathbf{e}_{12}$
$\mathbf{e}_1$	1	$\mathbf{e}_{12}$	$\mathbf{e}_2$
$\mathbf{e}_2$	$-\mathbf{e}_{12}$	1	$-\mathbf{e}_1$
$\mathbf{e}_{12}$	$-\mathbf{e}_2$	$\mathbf{e}_1$	-1

In particular, the *Clifford product* of two vectors $\mathbf{a} = a_1\mathbf{e}_1 + a_2\mathbf{e}_2$ and $\mathbf{b} = b_1\mathbf{e}_1 + b_2\mathbf{e}_2$ is seen to be $\mathbf{ab} = a_1b_1 + a_2b_2 + (a_1b_2 - a_2b_1)\mathbf{e}_{12}$.

2.2.2 The Exterior Product.

Extracting the scalar and bivector parts of the Clifford product we have as products of two vectors $\mathbf{a} = a_1\mathbf{e}_1 + a_2\mathbf{e}_2$ and $\mathbf{b} = b_1\mathbf{e}_1 + b_2\mathbf{e}_2$

$$\mathbf{a} \cdot \mathbf{b} = a_1b_1 + a_2b_2 \qquad \text{the scalar product “}\mathbf{a}\text{ dot }\mathbf{b}\text{”}$$
$$\mathbf{a} \wedge \mathbf{b} = (a_1b_2 - a_2b_1)\mathbf{e}_{12} \quad \text{the exterior product “}\mathbf{a}\text{ wedge }\mathbf{b}\text{”.}$$

The bivector $\mathbf{a} \wedge \mathbf{b}$ represents the oriented plane segment of the parallelogram with sides $\mathbf{a}$ and $\mathbf{b}$. The area of this parallelogram is $|a_1b_2 - a_2b_1|$, and we will take the *magnitude* of the bivector $\mathbf{a} \wedge \mathbf{b}$ to be this area $|\mathbf{a} \wedge \mathbf{b}| = |a_1b_2 - a_2b_1|$.

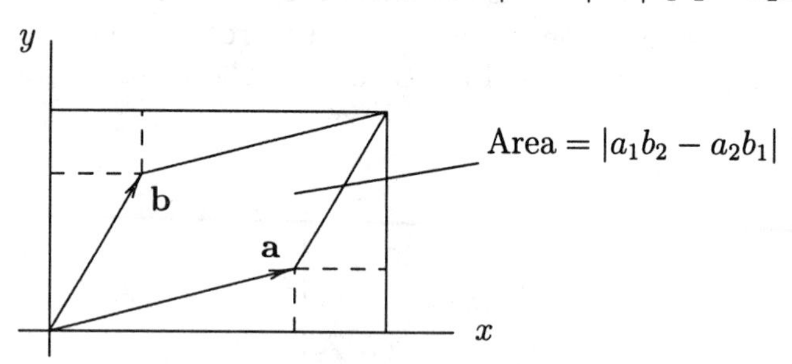

The parallelogram can be regarded as a kind of geometrical product of its sides:

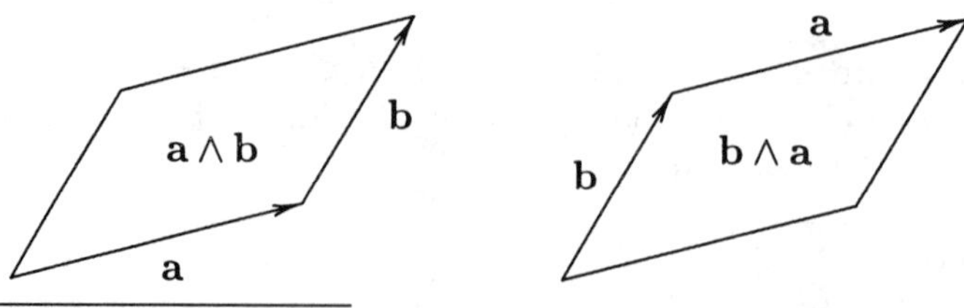

[1] An element $\mathbf{u} = u_1\mathbf{e}_1 + u_2\mathbf{e}_2 \in \mathbb{R}^2$ is called a vector or a 1-vector, but an arbitrary element $u = u_0 + u_1\mathbf{e}_1 + u_2\mathbf{e}_2 + u_{12}\mathbf{e}_{12} \in C\ell_2$ is called just an element (and not a vector although it is an element of a linear space).

The bivectors $\mathbf{a} \wedge \mathbf{b}$ and $\mathbf{b} \wedge \mathbf{a}$ have the same magnitude but opposite senses of rotation. This can be expressed simply by writing

$$\mathbf{a} \wedge \mathbf{b} = -\mathbf{b} \wedge \mathbf{a}.$$

Using the multiplication table of the Clifford algebra $C\ell_2$ we notice that the Clifford product

$$(a_1\mathbf{e}_1 + a_2\mathbf{e}_2)(b_1\mathbf{e}_1 + b_2\mathbf{e}_2) = a_1b_1 + a_2b_2 + (a_1b_2 - a_2b_1)\mathbf{e}_{12}$$

of two vectors $\mathbf{a} = a_1\mathbf{e}_1 + a_2\mathbf{e}_2$ and $\mathbf{b} = b_1\mathbf{e}_1 + b_2\mathbf{e}_2$ is a sum of a scalar $\mathbf{a} \cdot \mathbf{b} = a_1b_1 + a_2b_2$ and a bivector $\mathbf{a} \wedge \mathbf{b} = (a_1b_2 - a_2b_1)\mathbf{e}_{12}$. As an equation,

$$\mathbf{ab} = \mathbf{a} \cdot \mathbf{b} + \mathbf{a} \wedge \mathbf{b}. \qquad (a)$$

The commutative rule $\mathbf{a} \cdot \mathbf{b} = \mathbf{b} \cdot \mathbf{a}$ together with the anticommutative rule $\mathbf{a} \wedge \mathbf{b} = -\mathbf{b} \wedge \mathbf{a}$ imply a relation between $\mathbf{ab}$ and $\mathbf{ba}$. Thus,

$$\mathbf{ba} = \mathbf{a} \cdot \mathbf{b} - \mathbf{a} \wedge \mathbf{b}. \qquad (b)$$

Adding and subtracting equations (a) and (b), we find

$$\mathbf{a} \cdot \mathbf{b} = \frac{1}{2}(\mathbf{ab} + \mathbf{ba}) \quad \text{and} \quad \mathbf{a} \wedge \mathbf{b} = \frac{1}{2}(\mathbf{ab} - \mathbf{ba}).$$

Two vectors $\mathbf{a}$ and $\mathbf{b}$ are parallel $\mathbf{a} \parallel \mathbf{b}$ when they commute $\mathbf{ab} = \mathbf{ba}$, that is, $\mathbf{a} \wedge \mathbf{b} = 0$ or $a_1b_2 = a_2b_1$, and orthogonal $\mathbf{a} \perp \mathbf{b}$ when they anticommute $\mathbf{ab} = -\mathbf{ba}$, that is, $\mathbf{a} \cdot \mathbf{b} = 0$. Thus,

$$\begin{aligned} \mathbf{ab} = \mathbf{ba} \quad &\Longleftrightarrow \quad \mathbf{a} \parallel \mathbf{b} \quad \Longleftrightarrow \quad \mathbf{a} \wedge \mathbf{b} = 0 \quad \Longleftrightarrow \quad \mathbf{ab} = \mathbf{a} \cdot \mathbf{b}, \\ \mathbf{ab} = -\mathbf{ba} \quad &\Longleftrightarrow \quad \mathbf{a} \perp \mathbf{b} \quad \Longleftrightarrow \quad \mathbf{a} \cdot \mathbf{b} = 0 \quad \Longleftrightarrow \quad \mathbf{ab} = \mathbf{a} \wedge \mathbf{b}. \end{aligned}$$

2.2.3 Components of a Vector in Given Directions.

Consider decomposing a vector $\mathbf{r}$ into two components, one parallel to $\mathbf{a}$ and the other parallel to $\mathbf{b}$, where $\mathbf{a} \nparallel \mathbf{b}$. This means determining the coefficients α and β in the decomposition $\mathbf{r} = \alpha\mathbf{a} + \beta\mathbf{b}$. The coefficient α may be obtained by forming the exterior product $\mathbf{r} \wedge \mathbf{b} = (\alpha\mathbf{a} + \beta\mathbf{b}) \wedge \mathbf{b}$ and using $\mathbf{b} \wedge \mathbf{b} = 0$; this results in $\mathbf{r} \wedge \mathbf{b} = \alpha\mathbf{a} \wedge \mathbf{b}$. In the last equation both sides are multiples of $\mathbf{e}_{12}$ and we may write [2]

$$\alpha = \frac{\mathbf{r} \wedge \mathbf{b}}{\mathbf{a} \wedge \mathbf{b}}.$$

Similarly, by evaluating $\mathbf{a} \wedge \mathbf{r}$ we can eliminate α and find

$$\beta = \frac{\mathbf{a} \wedge \mathbf{r}}{\mathbf{a} \wedge \mathbf{b}}.$$

[2] Because of non-commutativity of multiplication in $C\ell_2$ it would be more appropriate to write $\alpha = (\mathbf{r} \wedge \mathbf{b})(\mathbf{a} \wedge \mathbf{b})^{-1}$ and $\beta = (\mathbf{a} \wedge \mathbf{r})(\mathbf{a} \wedge \mathbf{b})^{-1}$.

The coefficients α and β could be obtained visually by comparing the oriented areas (instead of lengths) in the following figure:

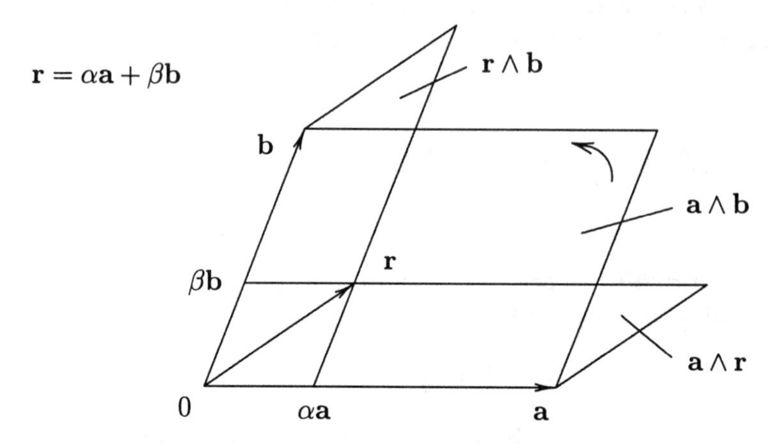

2.2.4 Perpendicular Projections and Reflections.

Let us calculate the component of $\mathbf{a}$ in the direction of $\mathbf{b}$ when the two vectors diverge by an angle φ, $0 < \varphi < 180°$. The parallel component $\mathbf{a}_\parallel$ is a scalar multiple of the unit vector $\mathbf{b}/|\mathbf{b}|$:

$$\mathbf{a}_\parallel = |\mathbf{a}| \cos\varphi \frac{\mathbf{b}}{|\mathbf{b}|} = |\mathbf{a}||\mathbf{b}| \cos\varphi \frac{\mathbf{b}}{|\mathbf{b}|^2}.$$

In other words, the parallel component $\mathbf{a}_\parallel$ is the scalar product $\mathbf{a} \cdot \mathbf{b} = |\mathbf{a}||\mathbf{b}| \cos\varphi$ multiplied by the vector $\mathbf{b}^{-1} = \mathbf{b}/|\mathbf{b}|^2$, called the inverse of the vector $\mathbf{b}$. Thus,

$$\begin{aligned} \mathbf{a}_\parallel &= (\mathbf{a} \cdot \mathbf{b}) \frac{\mathbf{b}}{|\mathbf{b}|^2} \\ &= (\mathbf{a} \cdot \mathbf{b}) \mathbf{b}^{-1} \end{aligned}$$

$\mathbf{a}$ $\mathbf{a}_\perp$ φ $\mathbf{b}$ $\mathbf{a}_\parallel$

The last formula tells us that the length of $\mathbf{b}$ is irrelevant when projecting into the direction of $\mathbf{b}$.

The perpendicular component $\mathbf{a}_\perp$ is given by the difference

$$\begin{aligned} \mathbf{a}_\perp &= \mathbf{a} - \mathbf{a}_\parallel = \mathbf{a} - (\mathbf{a} \cdot \mathbf{b}) \mathbf{b}^{-1} \\ &= (\mathbf{a}\mathbf{b} - \mathbf{a} \cdot \mathbf{b}) \mathbf{b}^{-1} = (\mathbf{a} \wedge \mathbf{b}) \mathbf{b}^{-1}. \end{aligned}$$

Note that the bivector $\mathbf{e}_{12}$ anticommutes with all the vectors in the $\mathbf{e}_1\mathbf{e}_2$-plane, therefore

$$(\mathbf{a} \wedge \mathbf{b}) \mathbf{b}^{-1} = -\mathbf{b}^{-1} (\mathbf{a} \wedge \mathbf{b}) = \mathbf{b}^{-1} (\mathbf{b} \wedge \mathbf{a}) = -(\mathbf{b} \wedge \mathbf{a}) \mathbf{b}^{-1}.$$

The area of the parallelogram with sides $\mathbf{a}, \mathbf{b}$ is seen to be

$$|\mathbf{a}_\perp \mathbf{b}| = |\mathbf{a} \wedge \mathbf{b}| = |\mathbf{a}||\mathbf{b}| \sin\varphi$$

where $0 < \varphi < 180°$.

The reflection of $\mathbf{r}$ across the line $\mathbf{a}$ is obtained by sending $\mathbf{r} = \mathbf{r}_\| + \mathbf{r}_\perp$ to $\mathbf{r}' = \mathbf{r}_\| - \mathbf{r}_\perp$, where $\mathbf{r}_\| = (\mathbf{r}\cdot\mathbf{a})\mathbf{a}^{-1}$. The mirror image $\mathbf{r}'$ of $\mathbf{r}$ with respect to $\mathbf{a}$ is then [3]

$$\begin{aligned}\mathbf{r}' &= (\mathbf{r}\cdot\mathbf{a})\mathbf{a}^{-1} - (\mathbf{r}\wedge\mathbf{a})\mathbf{a}^{-1}\\ &= (\mathbf{r}\cdot\mathbf{a} - \mathbf{r}\wedge\mathbf{a})\mathbf{a}^{-1}\\ &= (\mathbf{a}\cdot\mathbf{r} + \mathbf{a}\wedge\mathbf{r})\mathbf{a}^{-1}\\ &= \mathbf{a}\mathbf{r}\mathbf{a}^{-1}\end{aligned}$$

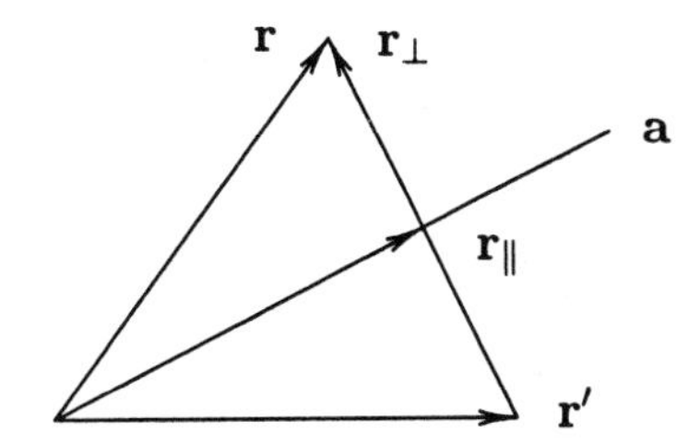

and further

$$\begin{aligned}\mathbf{r}' &= (2\mathbf{a}\cdot\mathbf{r} - \mathbf{r}\mathbf{a})\mathbf{a}^{-1}\\ &= 2\frac{\mathbf{a}\cdot\mathbf{r}}{\mathbf{a}^2}\mathbf{a} - \mathbf{r}\end{aligned}$$

The composition of two reflections, first across $\mathbf{a}$ and then across $\mathbf{b}$ is given by

$$\mathbf{r} \to \mathbf{r}' = \mathbf{a}\mathbf{r}\mathbf{a}^{-1} \to \mathbf{r}'' = \mathbf{b}\mathbf{r}'\mathbf{b}^{-1} = \mathbf{b}(\mathbf{a}\mathbf{r}\mathbf{a}^{-1})\mathbf{b}^{-1} = (\mathbf{b}\mathbf{a})\mathbf{r}(\mathbf{b}\mathbf{a})^{-1}$$

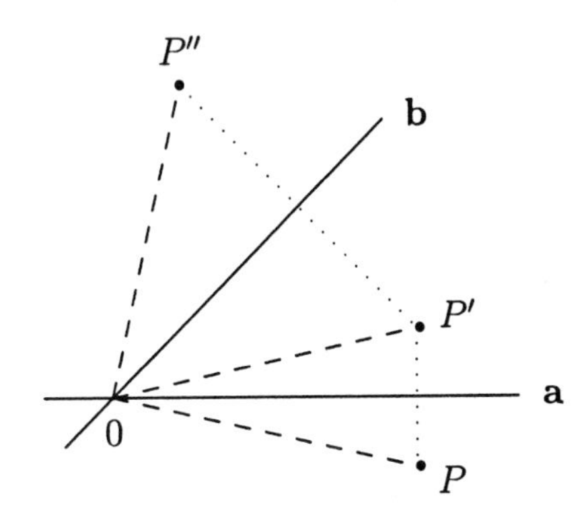

The composition of these two reflections is a rotation by twice the angle between $\mathbf{a}$ and $\mathbf{b}$.

2.2.5 Matrix Representation of $C\ell_2$.

In this section we have introduced the Clifford algebra $C\ell_2$ of the Euclidean plane $\mathbb{R}^2$. The Clifford algebra $C\ell_2$ is a 4-dimensional algebra over the reals $\mathbb{R}$. It is isomorphic, as an associative algebra, to the matrix algebra of real 2×2-matrices

[3] The formula $\mathbf{r}' = \mathbf{a}\mathbf{r}\mathbf{a}^{-1}$ can also be obtained using the Clifford product as follows: $\mathbf{r} = \mathbf{r}_\| + \mathbf{r}_\perp$ where $\mathbf{r}_\| = \alpha\mathbf{a}$ and $\mathbf{a} = \mathbf{a}\mathbf{a}\mathbf{a}^{-1}$ whilst $\mathbf{a}\mathbf{r}_\perp\mathbf{a}^{-1} = -\mathbf{r}_\perp\mathbf{a}\mathbf{a}^{-1} = -\mathbf{r}_\perp$.

$\mathbb{R}(2)$, as can be seen by the correspondences

$$1 \simeq \begin{pmatrix} 1 & 0 \\ 0 & 1 \end{pmatrix}$$

$$\mathbf{e}_1 \simeq \begin{pmatrix} 1 & 0 \\ 0 & -1 \end{pmatrix}, \qquad \mathbf{e}_2 \simeq \begin{pmatrix} 0 & 1 \\ 1 & 0 \end{pmatrix}$$

$$\mathbf{e}_{12} \simeq \begin{pmatrix} 0 & 1 \\ -1 & 0 \end{pmatrix}.$$

However, in the Clifford algebra $C\ell_2$ there is more structure than in the matrix algebra $\mathbb{R}(2)$. In the Clifford algebra $C\ell_2$ we have singled out, by definition a privileged subspace, namely the subspace of vectors or 1-vectors $\mathbb{R}^2 \subset C\ell_2$. No similar privileged subspace is incorporated in the definition of the matrix algebra $\mathbb{R}(2)$.

For arbitrary elements the above correspondences mean that

$$u_0 + u_1\mathbf{e}_1 + u_2\mathbf{e}_2 + u_{12}\mathbf{e}_{12} \simeq \begin{pmatrix} u_0 + u_1 & u_2 + u_{12} \\ u_2 - u_{12} & u_0 - u_1 \end{pmatrix}$$

and

$$\frac{1}{2}[(a+d) + (a-d)\mathbf{e}_1 + (b+c)\mathbf{e}_2 + (b-c)\mathbf{e}_{12}] \simeq \begin{pmatrix} a & b \\ c & d \end{pmatrix}.$$

In this representation the transpose of a matrix

$$\begin{pmatrix} a & b \\ c & d \end{pmatrix}^{\top} = \begin{pmatrix} a & c \\ b & d \end{pmatrix}$$

corresponds to the *reversion*

$$\tilde{u} = u_0 + u_1\mathbf{e}_1 + u_2\mathbf{e}_2 - u_{12}\mathbf{e}_{12}$$

of $u = u_0 + u_1\mathbf{e}_1 + u_2\mathbf{e}_2 + u_{12}\mathbf{e}_{12}$ in $C\ell_2$. The matrix adjoint

$$\operatorname{adj}\begin{pmatrix} a & b \\ c & d \end{pmatrix} = \begin{pmatrix} d & -b \\ -c & a \end{pmatrix}$$

corresponds to the *Clifford-conjugate* [4]

$$\bar{u} = u_0 - u_1\mathbf{e}_1 - u_2\mathbf{e}_2 - u_{12}\mathbf{e}_{12}.$$

2.2.6 Exercises

For those interested, I will list my first hand-outs of exercises in my undergraduate courses on Clifford algebras.

1. Let $a = \mathbf{e}_2 - \mathbf{e}_{12}$, $b = \mathbf{e}_1 + \mathbf{e}_2$, $c = 1 + \mathbf{e}_2$. Compute ab, ac. What is the instruction

[4] In some countries a vector $\mathbf{u} = u_1\mathbf{e}_1 + u_2\mathbf{e}_2 \in \mathbb{R}^2$ is denoted by $\bar{u}$ in hand-writing, but this practice clashes with our notation for the Clifford-conjugation.

of this task?
2. Let $a = \mathbf{e}_2 + \mathbf{e}_{12}$, $b = \frac{1}{2}(1 + \mathbf{e}_1)$. Compute ab, ba. Instruction?
3. Let $a = 1 + \mathbf{e}_1$, $b = -1 + \mathbf{e}_1$, $c = \mathbf{e}_1 + \mathbf{e}_2$. Compute ab, ba, ac, ca, bc and cb. Instruction?
4. Let $a = \frac{1}{2}(1 + \mathbf{e}_1)$, $b = \mathbf{e}_1 + \mathbf{e}_{12}$. Compute a^2, b^2.
5. Let $\mathbf{a} = \mathbf{e}_1 - 2\mathbf{e}_2$, $\mathbf{b} = \mathbf{e}_1 + \mathbf{e}_2$, $\mathbf{r} = 5\mathbf{e}_1 - \mathbf{e}_2$. Compute α, β in the decomposition $\mathbf{r} = \alpha\mathbf{a} + \beta\mathbf{b}$.
6. Let $\mathbf{a} = 8\mathbf{e}_1 - \mathbf{e}_2$, $\mathbf{b} = 2\mathbf{e}_1 + \mathbf{e}_2$. Compute $\mathbf{a}_{\|}$, $\mathbf{a}_{\perp}$.
7. Let $\mathbf{r} = 4\mathbf{e}_1 - 3\mathbf{e}_2$, $\mathbf{a} = 3\mathbf{e}_1 - \mathbf{e}_2$, $\mathbf{b} = 2\mathbf{e}_1 + \mathbf{e}_2$. Reflect $\mathbf{r}$ first across $\mathbf{a}$ and then the result across $\mathbf{b}$.
8. Show that for any $u \in C\ell_2$, $u\bar{u} = \bar{u}u \in \mathbb{R}$, and that u is invertible, if $u\bar{u} \neq 0$, with inverse

$$u^{-1} = \frac{\bar{u}}{u\bar{u}}.$$

9. Let $u = 1 + \mathbf{e}_1 + \mathbf{e}_{12}$. Compute u^{-1}. Show that $u^{-1} = \hat{u}(u\hat{u})^{-1} \neq (u\hat{u})^{-1}\hat{u}$, $u^{-1} = (\hat{u}u)^{-1}\hat{u} \neq \hat{u}(\hat{u}u)^{-1}$ and $u^{-1} = \tilde{u}(u\tilde{u})^{-1} \neq (u\tilde{u})^{-1}\tilde{u}$, $u^{-1} = (\tilde{u}u)^{-1}\tilde{u} \neq \tilde{u}(\tilde{u}u)^{-1}$.

Remark. In completing the exercises, please, note that an arbitrary element of $C\ell_2$ is most easily perceived when written in the order of increasing indices as $u_0 + u_1\mathbf{e}_1 + u_2\mathbf{e}_2 + u_{12}\mathbf{e}_{12}$. ∎

2.2.7 Answers

1. $ab = ac = 1 - \mathbf{e}_1 + \mathbf{e}_2 - \mathbf{e}_{12}$, instruction $ab = ac \not\Rightarrow b = c$.
2. $ab = 0$, $ba = \mathbf{e}_2 + \mathbf{e}_{12}$, instruction $ab = 0 \not\Rightarrow ba = 0$ (and also $ba = a \not\Rightarrow b = 1$).
3. $ab = ba = 0$, $ac = 1 + \mathbf{e}_1 + \mathbf{e}_2 + \mathbf{e}_{12}$, $ca = 1 + \mathbf{e}_1 + \mathbf{e}_2 - \mathbf{e}_{12}$, $bc = 1 - \mathbf{e}_1 - \mathbf{e}_2 + \mathbf{e}_{12}$, $cb = 1 - \mathbf{e}_1 - \mathbf{e}_2 - \mathbf{e}_{12}$, instruction $ab = ba = 0 \not\Rightarrow ac = 0$ or $ca = 0$.
4. $a^2 = a$, $b^2 = 0$.
5. $\mathbf{r} = 2\mathbf{a} + 3\mathbf{b}$.
6. $\mathbf{a}_{\|} = 6\mathbf{e}_1 + 3\mathbf{e}_2$, $\mathbf{a}_{\perp} = 2\mathbf{e}_1 - 4\mathbf{e}_2$.
7. $\mathbf{r}' = \mathbf{a}\mathbf{r}\mathbf{a}^{-1} = 5\mathbf{e}_1$, $\mathbf{r}'' = \mathbf{b}\mathbf{r}'\mathbf{b}^{-1} = 3\mathbf{e}_1 + 4\mathbf{e}_2$.
8. $u\bar{u} = \bar{u}u = u_0^2 - u_1^2 - u_2^2 + u_{12}^2 \in \mathbb{R}$.
9. $u^{-1} = 1 - \mathbf{e}_1 - \mathbf{e}_{12}$ and $(u\hat{u})^{-1}\hat{u} = \tilde{u}(\tilde{u}u)^{-1} = 1 + 3\mathbf{e}_1 - 4\mathbf{e}_2 - 5\mathbf{e}_{12}$ and $\hat{u}(\hat{u}u)^{-1} = (u\tilde{u})^{-1}\tilde{u} = 1 + 3\mathbf{e}_1 + 4\mathbf{e}_2 - 5\mathbf{e}_{12}$.

2.3 Operator Approach to Weyl, Majorana and Dirac Spinors, and their Reconstruction

In this section we will first discuss the role of Clifford algebra in the description of the Dirac equation and focus on various alternatives of spinors: real and complex spinors, spinors in columns and in minimal left ideals, and spinors as operators. The role of complex numbers in quantum mechanics is also scrutinized, and spinors are reconstructed from their bilinear covariants, up to a phase.

2.3.1 Relation to the Convention of Bjorken-Drell.

We start from the standard Bjorken-Drell formulation of the Dirac equation

$$\gamma^\mu(i\partial_\mu - eA_\mu)\psi = m\psi, \qquad \psi \in \mathbb{C}^4$$

where the matrices γ^μ are

$$\gamma_0 = \gamma^0 = \begin{pmatrix} I & 0 \\ 0 & -I \end{pmatrix} \qquad \gamma_k = -\gamma^k = \begin{pmatrix} 0 & -\sigma_k \\ \sigma_k & 0 \end{pmatrix}$$

with

$$\sigma_1 = \begin{pmatrix} 0 & 1 \\ 1 & 0 \end{pmatrix} \quad \sigma_2 = \begin{pmatrix} 0 & -i \\ i & 0 \end{pmatrix} \quad \sigma_3 = \begin{pmatrix} 1 & 0 \\ 0 & -1 \end{pmatrix}.$$

Usually $\psi \in \mathbb{C}^4$ is a column spinor, but we shall also regard it as a 4×4-matrix with only the first column being non-zero; that is, $\psi \in \mathbb{C}(4)f$ where f is the primitive idempotent

$$f = \frac{1}{2}(1+\gamma_0)\frac{1}{2}(1+i\gamma_1\gamma_2).$$

More explicitly, a Dirac spinor might appear as a **column spinor** or as a **square matrix spinor**

$$\psi = \begin{pmatrix} \psi_1 \\ \psi_2 \\ \psi_3 \\ \psi_4 \end{pmatrix} \in \mathbb{C}^4 \quad \text{or} \quad \psi = \begin{pmatrix} \psi_1 & 0 & 0 & 0 \\ \psi_2 & 0 & 0 & 0 \\ \psi_3 & 0 & 0 & 0 \\ \psi_4 & 0 & 0 & 0 \end{pmatrix} \in \mathbb{C}(4)f$$

where $\psi_i \in \mathbb{C}$. In the latter case $\psi = \psi_1 f_1 + \psi_2 f_2 + \psi_3 f_3 + \psi_4 f_4$ where the basis of the complex linear spinor space appears as

$$\begin{aligned}
f_1 &= \tfrac{1}{4}(1+\gamma_0+i\gamma_{12}+i\gamma_{012}) &&= f \\
f_2 &= \tfrac{1}{4}(-\gamma_{13}+i\gamma_{23}-\gamma_{013}+i\gamma_{023}) &&= -\gamma_{13}f \\
f_3 &= \tfrac{1}{4}(\gamma_3-\gamma_{03}+i\gamma_{123}-i\gamma_{0123}) &&= -\gamma_{03}f \\
f_4 &= \tfrac{1}{4}(\gamma_1-i\gamma_2-\gamma_{01}+i\gamma_{02}) &&= -\gamma_{01}f.
\end{aligned}$$

We denote $\gamma_{12} = \gamma_1\gamma_2$ ($\neq i\gamma_1\gamma_2$) and $\gamma_{0123} = \gamma_0\gamma_1\gamma_2\gamma_3$.

Even though as real algebras $\mathbb{C}\otimes C\ell_{1,3} \simeq \mathbb{C}\otimes \mathbb{R}(4)$, the complex conjugations are not the same in $\mathbb{C}\otimes C\ell_{1,3}$ and $\mathbb{C}\otimes\mathbb{R}(4)\simeq\mathbb{C}(4)$. In the matrix algebra $\mathbb{C}(4)$ we take complex conjugates of matrix entries $u^* = (u_{jk})^* = (u^*_{jk})$ whereas in the complexified Clifford algebra $\mathbb{C}\otimes C\ell_{1,3}$ complex conjugation has no effect on the real part $C\ell_{1,3}$ and we have $u^* = (a+ib)^* = a - ib$ for $a, b \in C\ell_{1,3}$. To make this point clear we present the following table of correspondences

	$\mathbb{C}\otimes C\ell_{1,3}$	$\mathbb{C}(4)$	
complex conjugate	u^*	$\gamma_{013}u^*\gamma_{013}^{-1}$	
	$\gamma_{013}u^*\gamma_{013}^{-1}$	u^*	complex conjugate
grade involute	$\hat{u}$	$\gamma_{0123}u\gamma_{0123}^{-1}$	
reverse	$\tilde{u}$	$\gamma_{13}u^\top\gamma_{13}^{-1}$	
conjugate	$\bar{u}$	$\gamma_{02}u^\top\gamma_{02}^{-1}$	
	$\gamma_{13}\tilde{u}\gamma_{13}^{-1}$	$u^\top$	transpose
	$\gamma_0\tilde{u}^*\gamma_0^{-1}$	$u^\dagger = u^{*\top}$	Hermitian conjugate
	$\tilde{u}^*$	$\gamma_0 u^\dagger\gamma_0^{-1}$	Dirac adjoint

For an arbitrary element $u = \langle u\rangle_0 + \langle u\rangle_1 + \langle u\rangle_2 + \langle u\rangle_3 + \langle u\rangle_4 \in C\ell_{1,3}$ decomposed in dimension degrees $\langle u\rangle_k \in \bigwedge^k \mathbb{R}^{1,3}$ we introduce

$$\hat{u} = \langle u\rangle_0 - \langle u\rangle_1 + \langle u\rangle_2 - \langle u\rangle_3 + \langle u\rangle_4 \qquad \text{grade involution}$$
$$\tilde{u} = \langle u\rangle_0 + \langle u\rangle_1 - \langle u\rangle_2 - \langle u\rangle_3 + \langle u\rangle_4 \qquad \text{reversion}$$
$$\bar{u} = \langle u\rangle_0 - \langle u\rangle_1 - \langle u\rangle_2 + \langle u\rangle_3 + \langle u\rangle_4 \qquad \text{Clifford conjugation}$$

where the reversion and conjugation are anti-automorphisms, that is, satisfying $\widetilde{uv} = \tilde{v}\tilde{u}$, $\overline{uv} = \bar{v}\bar{u}$, whereas the grade involution is an automorphism $\widehat{uv} = \hat{u}\hat{v}$. All these involutions are extended to $\mathbb{C}\otimes C\ell_{1,3}$ as complex linear maps, that is, for $\lambda \in \mathbb{C}$ and $u \in C\ell_{1,3}$ we have $(\lambda u)^\wedge = \lambda\hat{u}$, $(\lambda u)^\sim = \lambda\tilde{u}$, $(\lambda u)^- = \lambda\bar{u}$, whereas the complex conjugation is of course anti-linear $(\lambda u)^* = \lambda^* u^*$.

For a column spinor $\psi \in \mathbb{C}^4$ the Dirac adjoint is a row matrix

$$\psi^\dagger\gamma_0 = (\,\psi_1^* \quad \psi_2^* \quad -\psi_3^* \quad -\psi_4^*\,)$$

and for a square matrix spinor $\psi \in \mathbb{C}(4)f$ the Dirac adjoint is a square matrix $\psi^\dagger\gamma_0 = \gamma_0\psi^\dagger\gamma_0^{-1}$ with only the first row being non-zero (we reserve the bar-notation $\bar{\psi}$ for the Clifford conjugation = composition of reversion and grade involution). Note that the real part and the complex conjugate of a Dirac spinor depend on the decomposition (in the real structure singling out the real part)

For $\psi \in \mathbb{C}(4)f$:

$$\mathrm{Re}(\psi) = \begin{pmatrix} \mathrm{Re}(\psi_1) & 0 & 0 & 0 \\ \mathrm{Re}(\psi_2) & 0 & 0 & 0 \\ \mathrm{Re}(\psi_3) & 0 & 0 & 0 \\ \mathrm{Re}(\psi_4) & 0 & 0 & 0 \end{pmatrix} \qquad \psi^* = \begin{pmatrix} \psi_1^* & 0 & 0 & 0 \\ \psi_2^* & 0 & 0 & 0 \\ \psi_3^* & 0 & 0 & 0 \\ \psi_4^* & 0 & 0 & 0 \end{pmatrix}$$

For $\psi \in (\mathbb{C}\otimes C\ell_{1,3})f$ (viewed as a matrix):

$$\mathrm{Re}(\psi) = \frac{1}{2}\begin{pmatrix} \psi_1 & -\psi_2^* & 0 & 0 \\ \psi_2 & \psi_1^* & 0 & 0 \\ \psi_3 & \psi_4^* & 0 & 0 \\ \psi_4 & -\psi_3^* & 0 & 0 \end{pmatrix} \qquad \psi^* = \begin{pmatrix} 0 & -\psi_2^* & 0 & 0 \\ 0 & \psi_1^* & 0 & 0 \\ 0 & \psi_4^* & 0 & 0 \\ 0 & -\psi_3^* & 0 & 0 \end{pmatrix}$$

In other words the Dirac spinor ψ might appear as a column spinor $\psi \in \mathbb{C}^4$ or as a square matrix spinor $\psi \in \mathbb{C}(4)f$ or then as a (Clifford) **algebraic spinor** $\psi \in (\mathbb{C} \otimes C\ell_{1,3})f$ where the last two differ in their real structures.

Important Note. To indicate in what real structure the real part or the complex conjugate is considered we write

$$\mathrm{Re}(\psi) \text{ in } \mathbb{C}(4)f \qquad \text{and} \qquad \psi^* \text{ in } \mathbb{C}(4)f$$

or

$$\mathrm{Re}(\psi) \text{ in } \mathbb{C} \otimes C\ell_{1,3} \qquad \text{and} \qquad \psi^* \text{ in } \mathbb{C} \otimes C\ell_{1,3}.$$

Other contextual indicators are the Hermitian conjugation ($\psi^\dagger \gamma_0$ is either a row spinor or is in $\mathbb{C}(4)$) and for instance the reversion (the composition of the reversion and complex conjugation $\tilde{\psi}^*$ in $\mathbb{C} \otimes C\ell_{1,3}$ corresponds to $\psi^\dagger \gamma_0$ in $\mathbb{C}(4)$).

The reader should also observe that the real part $\mathrm{Re}(\psi)$ in $\mathbb{C} \otimes C\ell_{1,3}$ carries the same information as the original Dirac spinor $\psi \in \mathbb{C}^4$ (in contrast to $\mathrm{Re}(\psi)$ in $\mathbb{C}(4)f$).

Exercises. 1. Show that for $u \in \mathbb{C} \otimes C\ell_{1,3}$ the real part $\mathrm{Re}(u)$ corresponds to $\frac{1}{2}(u + \gamma_{013} u^* \gamma_{013}^{-1}) \in \mathbb{C}(4)$.

2. Show that if $u \in \mathbb{C} \otimes C\ell_{1,3}$ satisfies the condition $u\frac{1}{2}(1 + i\gamma_{12}) = u$ then $u = \mathrm{Re}(u)(1 + i\gamma_{12})$ and $iu = u\gamma_2\gamma_1$.

3. Show that the charge conjugate $\psi_C = -i\gamma_2\psi^*$ of the Dirac spinor $\psi \in \mathbb{C}(4)f$ corresponds to $\psi_C = \hat{\psi}^*\gamma_1 \in (\mathbb{C} \otimes C\ell_{1,3})f$.

4. Show that even though for $\psi \in \mathbb{C}(4)f$, $\mathrm{Re}(\psi) \in \mathbb{C}(4)f$, we have for a non-zero $\psi \in (\mathbb{C} \otimes C\ell_{1,3})f$, $\mathrm{Re}(\psi) \notin (\mathbb{C} \otimes C\ell_{1,3})f$. ∎

Historical Comment. Juvet 1930 and Sauter 1930 replaced column spinors by square matrices in which only the first column was non-zero. Thus spinor spaces became minimal left ideals in a **matrix algebra**. Riesz 1947 used primitive idempotents of Clifford algebras to construct spinor spaces as minimal left ideals in **Clifford algebras**. Kustaanheimo 1964/65 regularized the Kepler motion with spinors, and considered in this connection spinors as operators. Hestenes 1966/67 formulated the Dirac theory with the real Clifford algebra $C\ell_{1,3} \simeq \mathbb{H}(2)$ of the Minkowski time-space $\mathbb{R}^{1,3}$ and replaced spinor spaces by the even subalgebra $C\ell_{1,3}^+$. ∎

2.3.2 Bilinear Covariants.

A column spinor $\psi \in \mathbb{C}^4$ has a **probability density** $\psi^\dagger\psi$ (> 0 for $\psi \neq 0$) and a **current density** $J_k = \psi^\dagger\gamma_0\gamma_k\psi$ ($k = 1,2,3$) which we can combine to a future-oriented vector $\mathbf{J} = J_\mu\gamma^\mu$ with components

$$J_\mu = \psi^\dagger\gamma_0\gamma_\mu\psi \qquad (J_0 = \psi^\dagger\psi).$$

In various formalisms we have

$$J_\mu = \psi^\dagger\gamma_0\gamma_\mu\psi \qquad \psi \in \mathbb{C}^4$$

$$\begin{aligned}
&= \operatorname{trace}(\psi^\dagger\gamma_0\gamma_\mu\psi) && \psi \in \mathbb{C}(4)f \\
&= \operatorname{trace}(\gamma_\mu\psi\psi^\dagger\gamma_0) = 4\langle\gamma_\mu\psi\psi^\dagger\gamma_0\rangle_0 \\
&= 4\langle\gamma_\mu\psi\tilde{\psi}^*\rangle_0 && \psi \in (\mathbb{C}\otimes C\ell_{1,3})f
\end{aligned}$$

where the factor 4 appeared because

$$f = \frac{1}{4}(1+\gamma_0+i\gamma_{12}+i\gamma_{012}) = \begin{pmatrix} 1 & 0 & 0 & 0 \\ 0 & 0 & 0 & 0 \\ 0 & 0 & 0 & 0 \\ 0 & 0 & 0 & 0 \end{pmatrix}$$

has scalar part $\frac{1}{4}$, that is, $\langle f\rangle_0 = \frac{1}{4}$, while $\operatorname{trace}(f) = 1$. The current vector is the resultant

$$\begin{aligned}
\mathbf{J} = \gamma^\mu J_\mu &= \gamma^\mu 4\langle\gamma_\mu\psi\psi^\dagger\gamma_0\rangle_0 && \psi \in \mathbb{C}(4)f \\
&= \gamma^\mu\langle\gamma_\mu\cdot(4\psi\psi^\dagger\gamma_0)\rangle_0 = \gamma^\mu{<}\gamma_\mu, 4\psi\psi^\dagger\gamma_0{>} \\
&= \langle 4\psi\psi^\dagger\gamma_0\rangle_1 && J_\mu = \gamma_\mu\cdot\mathbf{J} \\
&= \langle 4\psi\tilde{\psi}^*\rangle_1 && \psi \in (\mathbb{C}\otimes C\ell_{1,3})f
\end{aligned}$$

Similarly $\psi \in \mathbb{C}^4$ carries a real bivector $\mathbf{S}$ with components

$$S_{\mu\nu} = \psi^\dagger\gamma_0 i\gamma_{\mu\nu}\psi \qquad (\gamma_{\mu\nu} = \gamma_\mu\gamma_\nu \neq i\gamma_\mu\gamma_\nu)$$

for which $S_{\mu\nu} = -\gamma_{\mu\nu}\cdot\mathbf{S}$ and $\mathbf{S} = \frac{1}{2}S_{\mu\nu}\gamma^{\mu\nu}$. In various formalisms

$$\begin{aligned}
S_{\mu\nu} &= \psi^\dagger\gamma_0 i\gamma_{\mu\nu}\psi && \psi \in \mathbb{C}^4 \\
&= \operatorname{trace}(\psi^\dagger\gamma_0 i\gamma_{\mu\nu}\psi) && \psi \in \mathbb{C}(4)f \\
&= \operatorname{trace}(i\gamma_{\mu\nu}\psi\psi^\dagger\gamma_0) = 4\langle i\gamma_{\mu\nu}\psi\psi^\dagger\gamma_0\rangle_0 \\
&= 4\langle i\gamma_{\mu\nu}\psi\tilde{\psi}^*\rangle_0 && \psi \in (\mathbb{C}\otimes C\ell_{1,3})f
\end{aligned}$$

$$\begin{aligned}
\mathbf{S} = \tfrac{1}{2}\gamma^{\mu\nu}S_{\mu\nu} &= \tfrac{1}{2}\gamma^{\mu\nu}4\langle i\gamma_{\mu\nu}\psi\psi^\dagger\gamma_0\rangle_0 && \psi \in \mathbb{C}(4)f \\
&= \tfrac{1}{2}\gamma^{\mu\nu}\langle i\gamma_{\mu\nu}\cdot(4\psi\psi^\dagger\gamma_0)\rangle_0 && S_{\mu\nu} = -\gamma_{\mu\nu}\cdot\mathbf{S} \\
&= \tfrac{1}{2}\gamma^{\mu\nu}{<}-i\gamma_{\mu\nu}, 4\psi\psi^\dagger\gamma_0{>} && {<}u,v{>} = \langle\tilde{u}\lrcorner v\rangle_0 \\
&= \langle -i4\psi\psi^\dagger\gamma_0\rangle_2 = -i\langle 4\psi\psi^\dagger\gamma_0\rangle_2 \\
&= -i\langle 4\psi\tilde{\psi}^*\rangle_2 && \psi \in (\mathbb{C}\otimes C\ell_{1,3})f
\end{aligned}$$

Recall that the Dirac adjoint $\psi^\dagger\gamma_0$ of a column spinor $\psi \in \mathbb{C}^4$ corresponds to $\tilde{\psi}^*$ of an algebraic spinor $\psi \in (\mathbb{C}\otimes C\ell_{1,3})f$ (another notation for the Dirac adjoint is $\bar{\psi}$ but we reserve this bar-notation for conjugation in Clifford algebra). The current vector $\mathbf{J}$ and the bivector $\mathbf{S}$ are examples of **bilinear covariants** listed below for a column spinor $\psi \in \mathbb{C}^4$ and for an algebraic spinor $\psi \in (\mathbb{C}\otimes C\ell_{1,3})f$

$$\begin{aligned}
\sigma &= \psi^\dagger\gamma_0\psi = 4\langle\tilde{\psi}^*\psi\rangle_0 \\
J_\mu &= \psi^\dagger\gamma_0\gamma_\mu\psi = 4\langle\tilde{\psi}^*\gamma_\mu\psi\rangle_0 \\
S_{\mu\nu} &= \psi^\dagger\gamma_0 i\gamma_{\mu\nu}\psi = 4\langle\tilde{\psi}^* i\gamma_{\mu\nu}\psi\rangle_0 && (\gamma_{\mu\nu} = \gamma_\mu\gamma_\nu) \\
K_\mu &= \psi^\dagger\gamma_0 i\gamma_{0123}\gamma_\mu\psi = 4\langle\tilde{\psi}^* i\gamma_{0123}\gamma_\mu\psi\rangle_0 && (\mathbf{K} = K_\mu\gamma^\mu) \\
\omega &= -\psi^\dagger\gamma_0\gamma_{0123}\psi = -4\langle\tilde{\psi}^*\gamma_{0123}\psi\rangle_0 && (\gamma_{0123} = \gamma_0\gamma_1\gamma_2\gamma_3).
\end{aligned}$$

Note that $K^\mu = \psi^\dagger\gamma^0\gamma^\mu i\gamma^{0123}\psi$ and $\omega = \psi^\dagger\gamma^0\gamma^{0123}\psi$. All the bilinear covariants are **real**.

2.3.3 Fierz Identities (Discovered by Pauli and Kofink).

The bilinear covariants satisfy certain quadratic equations called **Fierz identities** (see Holland 1986 p. 276 (2.8))

$$\mathbf{J}^2 = \sigma^2 + \omega^2 \qquad \mathbf{K}^2 = -\mathbf{J}^2$$

$$\mathbf{J}\cdot\mathbf{K} = 0 \qquad \mathbf{J}\wedge\mathbf{K} = -(\omega + \gamma_{0123}\sigma)\mathbf{S}$$

In coordinate form the Fierz identities look as follows (see Crawford 1985 p. 1439 (1.2))

$$J_\mu J^\mu = \sigma^2 + \omega^2 \qquad J_\mu J^\mu = -K_\mu K^\mu$$

$$J_\mu K^\mu = 0 \qquad J_\mu K_\nu - K_\mu J_\nu = -\omega S_{\mu\nu} + \sigma(*S)_{\mu\nu}$$

where $(*S)_{\mu\nu} = -\frac{1}{2}\varepsilon_{\mu\nu\alpha\beta}S^{\alpha\beta}$ (with $\varepsilon_{0123} = 1$) or $*\mathbf{S} = \tilde{\mathbf{S}}\gamma_{0123}$ (in general $*v = \tilde{v}\gamma_{0123}$ given by $u\wedge *v = \langle u, v\rangle\gamma_0\wedge\gamma_1\wedge\gamma_2\wedge\gamma_3$).

In the case when not both σ, $\omega = 0$ the Fierz identities result in (Crawford 1985 p. 1439 (1.3) and also 1986 p. 356 (2.14))

$$\mathbf{S}\cdot\mathbf{J} = \omega\mathbf{K} \qquad \mathbf{S}\cdot\mathbf{K} = \omega\mathbf{J}$$

$$(\gamma_{0123}\mathbf{S})\cdot\mathbf{J} = \sigma\mathbf{K} \qquad (\gamma_{0123}\mathbf{S})\cdot\mathbf{K} = \sigma\mathbf{J}$$

$$\mathbf{S}\cdot\mathbf{S} = \omega^2 - \sigma^2 \qquad (\gamma_{0123}\mathbf{S})\cdot\mathbf{S} = 2\sigma\omega$$

and

$$\mathbf{J}\mathbf{S} = -(\omega + \gamma_{0123}\sigma)\mathbf{K} \qquad \mathbf{K}\mathbf{S} = -(\omega + \gamma_{0123}\sigma)\mathbf{J}$$

$$\mathbf{S}\mathbf{J} = (\omega - \gamma_{0123}\sigma)\mathbf{K} \qquad \mathbf{S}\mathbf{K} = (\omega - \gamma_{0123}\sigma)\mathbf{J}$$

$$\mathbf{S}^2 = (\omega - \gamma_{0123}\sigma)^2 = \omega^2 - \sigma^2 - 2\sigma\omega\gamma_{0123}$$

$$\mathbf{S}^{-1} = -\mathbf{S}(\sigma - \gamma_{0123}\omega)^2/(\sigma^2 + \omega^2)^2 = \mathbf{KSK}/(\sigma^2 + \omega^2)^2$$

In the index-notation some of these identities look like

$$J_\mu S^{\mu\nu} = -\omega K^\nu \qquad J_\mu(*S)^{\mu\nu} = \sigma K^\nu$$

$$\mathbf{S}\cdot\mathbf{S} = -\tfrac{1}{2}S_{\mu\nu}S^{\mu\nu} = \omega^2 - \sigma^2$$

$$(*\mathbf{S})\cdot\mathbf{S} = -\tfrac{1}{2}(*S)_{\mu\nu}S^{\mu\nu} = \tfrac{1}{4}\varepsilon_{\mu\nu\alpha\beta}S^{\mu\nu}S^{\alpha\beta} = -2\sigma\omega$$

Note also that in general $\mathbf{S}\cdot\mathbf{K} = -\mathbf{K}\cdot\mathbf{S}$, $\gamma_{0123}(\mathbf{S}\wedge\mathbf{K}) = (\gamma_{0123}\mathbf{S})\cdot\mathbf{K} = -(*\mathbf{S})\cdot\mathbf{K} = \mathbf{K}\cdot(*\mathbf{S})$, $\gamma_{0123}(\mathbf{S}\wedge\mathbf{S}) = (\gamma_{0123}\mathbf{S})\cdot\mathbf{S} = -(*\mathbf{S})\cdot\mathbf{S}$ and that $(\mathbf{J}\cdot\mathbf{S})\wedge\mathbf{S} = \frac{1}{2}\mathbf{J}\cdot(\mathbf{S}\wedge\mathbf{S})$.

2.3.4 Recovering the Spinor from its Bilinear Covariants.

Let the Dirac spinor ψ have bilinear covariants σ, $\mathbf{J}$, $\mathbf{S}$, $\mathbf{K}$, ω. Take an arbitrary spinor η such that $\tilde{\eta}^*\psi \neq 0$ in $\mathbb{C} \otimes C\ell_{1,3}$ or equivalently $\eta^\dagger\gamma_0\psi \neq 0$ in $\mathbb{C}(4)$. Then the spinor ψ is proportional to

$$\psi \simeq Z\eta \quad \text{where} \quad Z = \sigma + \mathbf{J} + i\mathbf{S} - i\gamma_{0123}\mathbf{K} + \gamma_{0123}\omega$$

that is, ψ and $Z\eta$ differ only by a complex factor. The original Dirac spinor ψ can be recovered by the algorithm (see Crawford 1985)

$$N = \sqrt{\langle\tilde{\eta}^* Z\eta\rangle_0} = \frac{1}{2}\sqrt{\eta^\dagger\gamma_0 Z\eta}$$
$$e^{-i\alpha} = \frac{4}{N}\langle\tilde{\eta}^*\psi\rangle_0 = \frac{1}{N}\eta^\dagger\gamma_0\psi$$
$$\psi = \frac{1}{4N}e^{-i\alpha}Z\eta.$$

(For the choice $\eta = f$ we get simply

$$N = \sqrt{\langle Zf\rangle_0} = \frac{1}{2}\sqrt{\sigma + \mathbf{J}\cdot\gamma_0 - \mathbf{S}\cdot\gamma_{12} - \mathbf{K}\cdot\gamma_3}$$
$$e^{-i\alpha} = \frac{\psi_1}{|\psi_1|}$$

which are not the same N, $e^{-i\alpha}$ as those for an arbitrary η.) Once the spinor ψ has been recovered, we may also write

$$N = 4|\langle\tilde{\eta}^*\psi\rangle_0| = |\eta^\dagger\gamma_0\psi|$$
$$e^{-i\alpha} = \frac{\langle\tilde{\eta}^*\psi\rangle_0}{|\langle\tilde{\eta}^*\psi\rangle_0|} = \frac{\eta^\dagger\gamma_0\psi}{|\eta^\dagger\gamma_0\psi|}.$$

In particular,

a spinor ψ is determined by its bilinear covariants σ, $\mathbf{J}$, $\mathbf{S}$, $\mathbf{K}$, ω

up to a phase factor $e^{-i\alpha}$, and Z projects/extracts out of η the relevant part $\simeq \psi$.

For arbitrary σ, $\mathbf{J}$, $\mathbf{S}$, $\mathbf{K}$, ω (but not both σ, $\omega = 0$) satisfying the Fierz identities Crawford 1985 p. 1439 (2.2) observed the factorization

$$Z = (\sigma + \mathbf{J} + \gamma_{0123}\omega)(1 - i(\sigma + \gamma_{0123}\omega)^{-1}\gamma_{0123}\mathbf{K})$$

of $Z = \sigma + \mathbf{J} + i\mathbf{S} - i\gamma_{0123}\mathbf{K} + \gamma_{0123}\omega$. Using this factorization Crawford proved that if arbitrary σ, $\mathbf{J}$, $\mathbf{S}$, $\mathbf{K}$, ω satisfy the Fierz identities (and $J^0 > 0$ with $4\langle\tilde{\eta}^* Z\eta\rangle_0 = \eta^\dagger\gamma_0 Z\eta > 0$), then σ, $\mathbf{J}$, $\mathbf{S}$, $\mathbf{K}$, ω are bilinear covariants for some spinor ψ, for instance

$$\psi = \frac{1}{4N}Z\eta \qquad (N = \sqrt{\langle\tilde{\eta}^* Z\eta\rangle_0} = \frac{1}{2}\sqrt{\eta^\dagger\gamma_0 Z\eta})$$

(and two such spinors ψ with distinct choices of η differ only by their phases). Hamilton 1984 p. 1827 (4.2) mentioned that $Z = 4\psi\psi^\dagger\gamma_0$, see also Holland 1986 p. 276 (2.9), Hestenes 1986 p. 334 (2.28) and Keller and Rodriguez-Romo 1990 p. 2502 (2.3b).

2.3.5 Boomerangs and the Reconstruction of Spinors.

Definition. If σ, $\mathbf{J}$, $\mathbf{S}$, $\mathbf{K}$, ω satisfy the Fierz identities, then the multivector $Z = \sigma + \mathbf{J} + i\mathbf{S} - i\gamma_{0123}\mathbf{K} + \gamma_{0123}\omega$ is called **Fierz.** ▮

Definition. An element $Z = \sigma + \mathbf{J} + i\mathbf{S} - i\gamma_{0123}\mathbf{K} + \gamma_{0123}\omega$ is called a **boomerang**, if σ, $\mathbf{J}$, $\mathbf{S}$, $\mathbf{K}$, ω are bilinear covariants for some spinor ψ. ▮

Both in the non-null case (not both $\sigma,\ \omega = 0$) and in the null case (both $\sigma,\ \omega = 0$) a spinor ψ is determined up to a phase factor by its bilinear covariant $Z = \sigma + \mathbf{J} + i\mathbf{S} - i\gamma_{0123}\mathbf{K} + \gamma_{0123}\omega$ (as $\psi = \frac{1}{4N}e^{-i\alpha}Z\eta$), which in turn is determined by its spinor ψ as follows $Z = 4\psi\tilde{\psi}^* = 4\psi\psi^\dagger\gamma_0$ (thus we have a boomerang which returns back).

If Z is a boomerang so that $Z = 4\psi\psi^\dagger\gamma_0$ then $Z^2 = 4\sigma Z$ where $\sigma = \langle Z\rangle_0$, because

$$\begin{aligned} Z^2 &= 4\psi\psi^\dagger\gamma_0\, 4\psi\psi^\dagger\gamma_0 = 16\psi(\psi^\dagger\gamma_0\psi)\psi^\dagger\gamma_0 \\ &= 16\,\mathrm{trace}(\psi^\dagger\gamma_0\psi)\psi\psi^\dagger\gamma_0 \qquad [\text{since } \psi^\dagger\gamma_0\psi = \mathrm{trace}(\psi^\dagger\gamma_0\psi)f] \\ &= 16\,\mathrm{trace}(\psi\psi^\dagger\gamma_0)\psi\psi^\dagger\gamma_0 = \mathrm{trace}(4\psi\psi^\dagger\gamma_0)\, 4\psi\psi^\dagger\gamma_0 \end{aligned}$$

Conversely if $\sigma \neq 0$ then $Z^2 = 4\sigma Z$ ensures a boomeranging Z. If Z is Fierz and not both $\sigma,\ \omega = 0$, then it boomerangs back to Z. Crawford's results say that in the case not both $\sigma,\ \omega = 0$ we have a boomeranging Z if and only if Z is Fierz. However, in the null case, $\sigma,\ \omega = 0$, there are such Z which are Fierz but still not boomerang (for instance $Z = \mathbf{J},\ \mathbf{J}^2 = 0,\ \mathbf{J} \neq 0$).

If both $\sigma,\ \omega = 0$ and $\mathbf{J}$, $\mathbf{S}$, $\mathbf{K}$ satisfy the Fierz identities, then for a spinor constructed as follows

$$\psi = \frac{1}{4N}Z\eta \quad \text{where} \quad Z = \mathbf{J} + i\mathbf{S} - i\gamma_{0123}\mathbf{K}$$

we have in general $Z \neq 4\psi\tilde{\psi}^*$ (the Fierz identities are reduced to $\mathbf{J}^2 = \mathbf{K}^2 = 0$, $\mathbf{J}\cdot\mathbf{K} = \mathbf{J}\wedge\mathbf{K} = 0$ which impose no restriction on $\mathbf{S}$). Even if the Fierz identities were supplemented by all the conditions presented in Section entitled 'Fierz identities' (these conditions are, in the non-null case, consequences of the Fierz identities), these extended identities would not result in a boomeranging Z. To handle also the null case $\sigma,\ \omega = 0$ we could replace the Fierz identities by the more restrictive conditions

$$\begin{gathered} Z^2 = 4\sigma Z \qquad Z\gamma_\mu Z = 4J_\mu Z \qquad Zi\gamma_{\mu\nu}Z = 4S_{\mu\nu}Z \\ Zi\gamma_{0123}\gamma_\mu Z = 4K_\mu Z \qquad Z\gamma_{0123}Z = -4\omega Z \end{gathered}$$

(see Crawford 1986 p. 357 (2.16)) but this would result in a tedious checking process. If $Z = \mathbf{J} + i\mathbf{S} - i\gamma_{0123}\mathbf{K}$ is a boomerang, then $Z^2 = 0$ where (arranged by dimension degrees)

$$\begin{aligned} Z^2 = {} & \mathbf{J}^2 - \mathbf{S}\cdot\mathbf{S} - \mathbf{K}^2 \\ & +2\gamma_{0123}(\mathbf{S}\wedge\mathbf{K}) \qquad\qquad \mathbf{K} \text{ in the plane of } \mathbf{S} \end{aligned}$$

$+i2\gamma_{0123}(\mathbf{J} \wedge \mathbf{K})$	$\mathbf{J}$ and $\mathbf{K}$ are parallel
$+i2\mathbf{J} \wedge \mathbf{S}$	$\mathbf{J}$ in the plane of $\mathbf{S}$
$-\mathbf{S} \wedge \mathbf{S}$	$\mathbf{S}$ is simple

The bivector part implies that $\mathbf{J}$ and $\mathbf{K}$ are parallel, the 4-vector part implies that $\mathbf{S}$ is simple, and the vector and 3-vector parts imply that $\mathbf{J}$ and $\mathbf{K}$ are in the plane of $\mathbf{S}$. Altogether we must have

$$Z = \mathbf{J}(1 + i\mathbf{s} + i\gamma_{0123}h)$$

where h is a real number and $\mathbf{s}$ is a space-like vector orthogonal to $\mathbf{J}$, $\mathbf{J} \cdot \mathbf{s} = 0$. We compute again $Z^2 = \mathbf{J}^2(1 + (\mathbf{s} + \gamma_{0123}h)^2) = 0$ and conclude that either

1. $\mathbf{J}^2 = 0$ or else
2. $(\mathbf{s} + \gamma_{0123}h)^2 = -1$.

Neither condition alone is sufficient to enforce Z to become a boomerang (Z is not even Fierz if $\mathbf{J}^2 \neq 0$). However, such a Z is a boomerang if both the conditions are satisfied simultaneously.

Counter-examples. 1. In case $\sigma = 0$, $Z = \mathbf{J} - \gamma_{0123}\omega$, $\mathbf{J}^2 = \omega^2 > 0$, is such that $Z^2 = 0$, but Z is not Fierz.

2. $Z = \mathbf{J} + i\mathbf{S}$ with $\mathbf{J}^2 > 0$, $\mathbf{S} = \gamma_{0123}\mathbf{J}\mathbf{s}$, $\mathbf{J} \cdot \mathbf{s} = 0$, $\mathbf{s}^2 = -1$ is not Fierz, and $Z^2 \neq 0$, but we have $Z\gamma_{0123}Z = 0$.

3. $Z = \mathbf{J} + i\mathbf{S} - i\gamma_{0123}\mathbf{K}$ where $\mathbf{J}^2 = \mathbf{K}^2 = 0$, $\mathbf{J} \cdot \mathbf{K} = 0$, $\mathbf{J} \wedge \mathbf{K} = 0$, $\mathbf{S} \wedge \mathbf{S} \neq 0$, is Fierz but does not satisfy $Z^2 = 0$, $Z\gamma_{0123}Z = 0$.

4. $Z = \mathbf{J}(1 + i\mathbf{s} + i\gamma_{0123}h)$ with $\mathbf{J}^2 = 0$, $\mathbf{J} \cdot \mathbf{s} = 0$, $(\mathbf{s} + \gamma_{0123}h)^2 \neq -1$ is Fierz and satisfies $Z^2 = 0$ and $Z\gamma_{0123}Z = 0$, but still we do not have a boomeranging Z. ∎

Exercise. Do the conditions $Z^2 = 0$ and $Z\gamma_{0123}Z = 0$ imply that Z is Fierz? ∎

Throughout this paper we assume that σ, $\mathbf{J}$, $\mathbf{S}$, $\mathbf{K}$, ω are real multivectors or equivalently that $Z = \sigma + \mathbf{J} + i\mathbf{S} - i\gamma_{0123}\mathbf{K} + \gamma_{0123}\omega$ is Dirac self-adjoint ($\tilde{Z}^* = Z$ or in matrix notation $\gamma_0 Z^\dagger \gamma_0 = Z$). This implies that $\eta^\dagger\gamma_0 Z\eta$ $(= 4\langle\tilde{\eta}^* Z\eta\rangle_0)$ is a real number for all spinors η.

For a boomerang Z we have $\eta^\dagger\gamma_0 Z\eta \geq 0$ for all spinors η and also $J^0 > 0$ (the grade involute $\hat{Z}$ of Z is such that $\langle\hat{Z}\rangle_0 \cdot \gamma_0 < 0$ and $4\langle\tilde{\eta}^*\hat{Z}\eta\rangle_0 = \eta^\dagger\gamma_0\hat{Z}\eta \leq 0$).

Theorem. Let Z be such that $\eta^\dagger\gamma_0 Z\eta \geq 0$ for all spinors η and also $J^0 > 0$. Then the following statements hold

1. Z is a boomerang if and only if $Z\gamma^0\tilde{Z}^* = 4J^0 Z$ or equivalently $ZZ^\dagger\gamma^0 = 4J^0 Z$.
2. In the non-null case (not both σ, $\omega = 0$) Z is a boomerang if and only if it is Fierz.
3. In the null case (both σ, $\omega = 0$) Z is a boomerang if and only if $Z = \mathbf{J}(1 + i\mathbf{s} + i\gamma_{0123}h)$ where $\mathbf{J}$ is a null-vector, $\mathbf{J}^2 = 0$, $\mathbf{s}$ is a space-like vector, $\mathbf{s}^2 < 0$ or $\mathbf{s} = 0$, orthogonal to $\mathbf{J}$, $\mathbf{J} \cdot \mathbf{s} = 0$, and h is a real number so that $h = \pm\sqrt{1 + \mathbf{s}^2}$, $|h| \leq 1$. ∎

The condition $Z\gamma^0\tilde{Z}^* = 4J^0 Z$ could also be written with an arbitrary time-like vector $\mathbf{v}$ as follows $Z\mathbf{v}\tilde{Z}^* = 4(\mathbf{v}\cdot\mathbf{J})Z$.

2.3.6 The Mother of All Real Spinors $\Phi \in C\ell_{1,3}\,\frac{1}{2}(1+\gamma_0)$.

Take two arbitrary elements in the real Clifford algebra $a, b \in C\ell_{1,3}$ in such a way that $\psi = (a+ib)f$, $f = \frac{1}{2}(1+\gamma_0)\frac{1}{2}(1+\gamma_{12})$. Then $\psi\bar{\psi}^* = 0$ and

$$\begin{aligned}\psi\tilde{\psi}^* &= (a+ib)f(\tilde{a}-i\tilde{b}) = af\tilde{a} + bf\tilde{b} + i(bf\tilde{a} - af\tilde{b})\\ &= \tfrac{1}{2}(ag\tilde{a} + bg\tilde{b} - bg\gamma_{12}\tilde{a} + ag\gamma_{12}\tilde{b} + i(ag\gamma_{12}\tilde{a} + bg\gamma_{12}\tilde{b} + bg\tilde{a} - ag\tilde{b}))\end{aligned}$$

where we have denoted $g = \frac{1}{2}(1+\gamma_0)$. Next, we introduce a real spinor, called the **mother spinor**

$$\Phi = (a - b\gamma_{12})(1+\gamma_0) \in C\ell_{1,3}\,\frac{1}{2}(1+\gamma_0).$$

Compute $\Phi\bar{\Phi} = 0$ and

$$\Phi\tilde{\Phi} = 4(a - b\gamma_{12})g(\tilde{a} + \gamma_{12}\tilde{b}) = 4(ag\tilde{a} + ag\gamma_{12}\tilde{b} - bg\gamma_{12}\tilde{a} + bg\tilde{b})$$

to find

$$\frac{1}{2}\Phi\tilde{\Phi} = 4\,\mathrm{Re}(\psi\tilde{\psi}^*), \quad \text{and similarly} \quad \frac{1}{2}\Phi\gamma_{12}\tilde{\Phi} = 4\,\mathrm{Im}(\psi\tilde{\psi}^*).$$

Recall that $Z = 4\psi\tilde{\psi}^*$ is sufficient to reconstruct the original Dirac spinor ψ and conclude that the real mother spinor $\Phi \in C\ell_{1,3}\,\frac{1}{2}(1+\gamma_0)$ carries all the physically relevant information of the Dirac spinor ψ. In fact,

$$\psi = \Phi\,\frac{1}{4}(1 + i\gamma_{12}) \quad \text{and} \quad \Phi = 4\,\mathrm{Re}(\psi)$$

where the real part is taken in the decomposition $\mathbb{C}\otimes C\ell_{1,3}$ (and **not** in the decomposition $\mathbb{C}\otimes \mathbb{R}(4)$).

Denote $Z = P + iQ$ where $P = \sigma + \mathbf{J} + \gamma_{0123}\omega$ and $Q = \mathbf{S} + \mathbf{K}\gamma_{0123}$. We will still show how to recover the real mother spinor Φ from its bilinear covariants $(\tilde{g} = g = \frac{1}{2}(1+\gamma_0))$

$$N = \sqrt{\tfrac{1}{2}\langle \tilde{g}(P - Q\gamma_{12})g\rangle_0} \qquad e^{\gamma_{12}\alpha} = \frac{1}{N}((\tilde{g}\phi)\wedge\gamma_{03})\gamma_{03}^{-1}$$

$$\Phi = \frac{1}{2N}(P - Q\gamma_{12})e^{\gamma_{12}\alpha}g$$

(the same N as for the choice $f \in \mathbb{C}\otimes C\ell_{1,3}$) or for an arbitrary spinor $\eta \in C\ell_{1,3}g$, $\tilde{\eta}\Phi \neq 0$,

$$N = \sqrt{\tfrac{1}{2}\langle \tilde{\eta}(P\eta - Q\eta\gamma_{12})\rangle_0} \qquad e^{\gamma_{12}\alpha} = \frac{1}{N}((\tilde{\eta}\Phi)\wedge\gamma_{03})\gamma_{03}^{-1}$$

$$\Phi = \frac{1}{2N}(P\eta - Q\eta\gamma_{12})e^{\gamma_{12}\alpha}g$$

[$\eta' \in (\mathbb{C} \otimes C\ell_{1,3})f$ and $\eta = 2\,\mathrm{Re}(\eta')$ result in the same numerical value for N]. Note that the role of $i = \sqrt{-1}$ is played by multiplication by $\gamma_2\gamma_1$ from the **right** hand side, that is, $\Phi\gamma_2\gamma_1 = 4\,\mathrm{Re}(i\psi)$.

Exercises. Show that 1. $\mathrm{Im}(\psi) = \mathrm{Re}(\psi)\gamma_{12}$ in $\mathbb{C} \otimes C\ell_{1,3}$
2. $\frac{1}{2}\Phi\gamma_1\bar{\Phi} = 4\,\mathrm{Re}(\psi\gamma_1\bar{\psi}) = -4\,\mathrm{Im}(\psi\gamma_2\bar{\psi})$ (no complex conjugation)
$\frac{1}{2}\Phi\gamma_2\bar{\Phi} = 4\,\mathrm{Re}(\psi\gamma_2\bar{\psi}) = 4\,\mathrm{Im}(\psi\gamma_1\bar{\psi})$
3. $Q_k^2 = -2\sigma P$, $Q_iQ_j = 2\sigma Q_k$ (ijk cycl.) for $Q_k = \frac{1}{2}(\Phi\gamma_k\bar{\Phi})\gamma_{0123}$
($P, Q_1, Q_2, Q_3 = Q$ span a quaternion algebra when $\sigma \neq 0$). ∎

2.3.7 Ideal Spinors $\phi \in C\ell_{1,3}\,\frac{1}{2}(1-\gamma_{03})$.

Hestenes does not use the real mother spinor $\Phi \in C\ell_{1,3}\,\frac{1}{2}(1+\gamma_0)$ but instead the real spinor

$$\phi = \Phi\,\frac{1}{2}(1-\gamma_{03}) \in C\ell_{1,3}\,\frac{1}{2}(1-\gamma_{03})$$

for which $\Phi = \phi(1+\gamma_0)$. This real **ideal spinor** satisfies [Hestenes 1986 p. 334 gives P in (2.26) and $-Q$ in (2.27)]

$$\phi\gamma_0\tilde{\phi} = \phi\gamma_3\tilde{\phi} = P$$
$$\phi\gamma_0\bar{\phi} = \phi\gamma_3\bar{\phi} = -Q\gamma_{0123} \qquad Q = \phi\gamma_{123}\tilde{\phi}$$

and so it carries the same information as the original Dirac spinor $\psi = \phi\frac{1}{2}(1+\gamma_0)\frac{1}{2}(1+i\gamma_{12})$. Hestenes finds the bilinear covariants in the form

$$\sigma = \langle\tilde{\phi}\phi\gamma_0\rangle_0 = (\tilde{\phi}\phi)\cdot\gamma_3$$
$$J_\mu = \langle\tilde{\phi}\gamma_\mu\phi\gamma_0\rangle_0 = (\tilde{\phi}\gamma_\mu\phi)\cdot\gamma_3$$
$$S_{\mu\nu} = -\langle\tilde{\phi}\gamma_{\mu\nu}\phi\gamma_{123}\rangle_0 = -\langle\tilde{\phi}\gamma_{\mu\nu}\phi\rangle_3\cdot\gamma_{123}$$
$$K_\mu = -\langle\tilde{\phi}\gamma_{0123}\gamma_\mu\phi\gamma_{123}\rangle_0$$
$$\omega = -\langle\tilde{\phi}\gamma_{0123}\phi\gamma_0\rangle_0.$$

Exercises. Show that 1. $\phi = \Phi\,\frac{1}{2}(1-\gamma_3)$,
2. $\phi\tilde{\phi}$, $\phi\bar{\phi}$, $\phi\gamma_1\tilde{\phi}$, $\phi\gamma_1\bar{\phi}$, $\phi\gamma_2\tilde{\phi}$, $\phi\gamma_2\bar{\phi}$ all vanish,
3. for a slightly different choice of sign in $\phi = \Phi\,\frac{1}{2}(1+\gamma_{03})$ we have $\phi\gamma_0\tilde{\phi} = P = -\phi\gamma_3\tilde{\phi}$ and $\phi\gamma_3\bar{\phi} = -Q\gamma_{0123} = -\phi\gamma_0\bar{\phi}$.

The ideal spinor $\phi = \Phi\,\frac{1}{2}(1-\gamma_{03}) = 2\,\mathrm{Re}(\psi)(1-\gamma_{03})$ satisfies $\phi\gamma_2\gamma_1 = 2\,\mathrm{Re}(i\psi)(1-\gamma_{03})$ and the Dirac equation has the form

$$\partial\phi = (e\mathbf{A}+m)\phi\gamma_{12}, \qquad \phi \in C\ell_{1,3}\,\frac{1}{2}(1-\gamma_{03}).$$

In contrast to the mother spinor Φ, the ideal spinor ϕ satisfies $\phi\gamma_{0123} = \phi\gamma_2\gamma_1$ and so we could rewrite the Dirac equation in the same way as Hestenes: $\partial\phi\gamma_{0123} = (e\mathbf{A}+m)\phi$.

Decompose the ideal spinor $\phi = \phi_0+\phi_1$ into its even and odd parts and separate the parts

$$\partial\phi_0\gamma_{0123} = e\mathbf{A}\phi_0 + m\phi_1, \qquad \phi_0 = \text{even}(\phi) \in C\ell_{1,3}^+ \frac{1}{2}(1-\gamma_{03})$$

$$\partial\phi_1\gamma_{0123} = e\mathbf{A}\phi_1 + m\phi_0, \qquad \phi_1 = \text{odd}(\phi) \in C\ell_{1,3}^- \frac{1}{2}(1-\gamma_{03})$$

which can be put into the matrix form

$$\begin{pmatrix} \gamma_{0123} & 0 \\ 0 & -\gamma_{0123} \end{pmatrix} \begin{pmatrix} 0 & \partial \\ -\partial & 0 \end{pmatrix} \begin{pmatrix} \phi_0 & 0 \\ \phi_1 & 0 \end{pmatrix}$$

$$= e \begin{pmatrix} 0 & \mathbf{A} \\ -\mathbf{A} & 0 \end{pmatrix} \begin{pmatrix} \phi_0 & 0 \\ \phi_1 & 0 \end{pmatrix} + m \begin{pmatrix} 1 & 0 \\ 0 & -1 \end{pmatrix} \begin{pmatrix} \phi_0 & 0 \\ \phi_1 & 0 \end{pmatrix}$$

where we have used the fact that the matrix

$$\begin{pmatrix} \gamma_{0123} & 0 \\ 0 & -\gamma_{0123} \end{pmatrix} \text{ commutes with } \begin{pmatrix} 0 & \partial \\ -\partial & 0 \end{pmatrix} \text{ and } \begin{pmatrix} \phi_0 & 0 \\ \phi_1 & 0 \end{pmatrix}$$

and thereby takes the role of an overall commuting imaginary unit $\sqrt{-1}$.

2.3.8 Spinor Operators $\Psi \in C\ell_{1,3}^+$.

Decompose the mother spinor into even and odd parts $\Phi = \Phi_0 + \Phi_1 = (\Phi_0 + \Phi_1)\frac{1}{2}(1 + \gamma_0) = \frac{1}{2}(\Phi_0 + \Phi_1\gamma_0) + \frac{1}{2}(\Phi_1 + \Phi_0\gamma_0)$. It follows that, $\Phi_0 = \Phi_1\gamma_0$ and $\Phi_1 = \Phi_0\gamma_0$. The Dirac equation

$$\partial\Phi = (e\mathbf{A} + m)\Phi\gamma_{12}$$

decomposes into parts $(\Phi_0 = \text{even}(\Phi),\ \Phi_1 = \text{odd}(\Phi))$

$$\partial\Phi_0 = e\mathbf{A}\Phi_0\gamma_{12} + m\Phi_1\gamma_{12} \qquad (\Phi_1 = \Phi_0\gamma_0)$$

$$\partial\Phi_1 = e\mathbf{A}\Phi_1\gamma_{12} + m\Phi_0\gamma_{12} \qquad (\Phi_0 = \Phi_1\gamma_0).$$

Thereby, the even part of the mother spinor satisfies the multivector **Dirac equation** in operator form

$$\partial\Psi = (e\mathbf{A}\Psi + m\Psi\gamma_0)\gamma_{12} \qquad \Psi = \text{even}(\Phi) \in C\ell_{1,3}^+$$

(this equation was scrutinized by Hestenes 1967/75 who derived it in 1966 p. 42 (13.13) from a result of Gürsey 1956/58). We may reobtain the mother spinor $\Phi = \Psi(1 + \gamma_0)$ and the ideal spinor $\phi = \Psi\frac{1}{2}(1 + \gamma_0)(1 - \gamma_{03})$ as well as the original Dirac spinor $\psi = \Psi\frac{1}{2}(1 + \gamma_0)\frac{1}{2}(1 + i\gamma_{12})$.

Because of the identities

$$\Psi\tilde{\Psi} = \sigma + \gamma_{0123}\omega$$

$$\Psi\gamma_0\tilde{\Psi} = \mathbf{J}$$

$$\Psi\gamma_{12}\tilde{\Psi} = \mathbf{S} \qquad \Psi\gamma_{03}\tilde{\Psi} = -\mathbf{S}\gamma_{0123}$$

$$\Psi\gamma_3\tilde{\Psi} = \mathbf{K} \qquad \Psi\gamma_{012}\tilde{\Psi} = \mathbf{K}\gamma_{0123}$$

we call Ψ a **spinor operator**. In the non-singular case $\Psi\tilde{\Psi} \neq 0$ the spinor Ψ operates like a Lorentz transformation composed with a dilation (and a duality transformation). We have the following identities

$$\Psi(1+\gamma_0)\tilde{\Psi} = P$$
$$\Psi(1+\gamma_0)\gamma_{12}\tilde{\Psi} = Q \qquad \Psi(1+\gamma_0)\gamma_{ij}\tilde{\Psi} = Q_k \quad (ijk \text{ cycl., } Q_3 = Q)$$
$$\Psi(1+\gamma_0)(1+i\gamma_{12})\tilde{\Psi} = Z \quad (= P + iQ)$$

[Hestenes 1986 p. 334 gives P in (2.26) and $-Q$ in (2.27) and Z in (2.28)] and in coordinate form

$$\sigma + \gamma_{0123}\omega = \Psi\tilde{\Psi} = \tilde{\Psi}\Psi$$
$$J_\mu = \langle\tilde{\Psi}\gamma_\mu\Psi\gamma_0\rangle_0 = (\tilde{\Psi}\gamma_\mu\Psi)\cdot\gamma_0$$
$$S_{\mu\nu} = -\langle\tilde{\Psi}\gamma_{\mu\nu}\Psi\gamma_{12}\rangle_0 = (\tilde{\Psi}\gamma_{\mu\nu}\Psi)\cdot\gamma_{12}$$
$$K_\mu = \langle\tilde{\Psi}\gamma_\mu\Psi\gamma_3\rangle_0 = (\tilde{\Psi}\gamma_{0123}\gamma_\mu\Psi)\cdot\gamma_{012}\,.$$

Exercise. Show that for an algebraic spinor (which is viewed through the matrix window)

$$\psi = \begin{pmatrix} \psi_1 & 0 & 0 & 0 \\ \psi_2 & 0 & 0 & 0 \\ \psi_3 & 0 & 0 & 0 \\ \psi_4 & 0 & 0 & 0 \end{pmatrix} \quad \text{in} \quad (\mathbb{C}\otimes C\ell_{1,3})f$$

the mother spinor and the spinor operator are

$$\Phi = 4\,\mathrm{Re}(\psi) = 2\begin{pmatrix} \psi_1 & -\psi_2^* & 0 & 0 \\ \psi_2 & \psi_1^* & 0 & 0 \\ \psi_3 & \psi_4^* & 0 & 0 \\ \psi_4 & -\psi_3^* & 0 & 0 \end{pmatrix}$$

$$\Psi = \mathrm{even}(\Phi) = \begin{pmatrix} \psi_1 & -\psi_2^* & \psi_3 & \psi_4^* \\ \psi_2 & \psi_1^* & \psi_4 & -\psi_3^* \\ \psi_3 & \psi_4^* & \psi_1 & -\psi_2^* \\ \psi_4 & -\psi_3^* & \psi_2 & \psi_1^* \end{pmatrix}$$ ∎

The theory of spinor operators shows that the Dirac algebra

$$\mathbb{C}(4) \simeq C\ell_{1,3} + iC\ell_{1,3}$$

contains a superfluous component, namely the imaginary part $iC\ell_{1,3}$. In particular, complex numbers are not mandatory in the treatment of Fierz identities [compare this to Keller and Viniegra 1992 p. 439].

Multiplication by the imaginary unit $i = \sqrt{-1}$. We have found that $i\psi = \psi\gamma_2\gamma_1$ ($\neq \psi\gamma_{0123}$) corresponds to $\Phi\gamma_2\gamma_1 = 4\,\mathrm{Re}(i\psi)$ and further to $\phi\gamma_2\gamma_1 = \Phi\gamma_2\gamma_1\frac{1}{2}(1-\gamma_{03})$. When not both σ, $\omega = 0$, denote

$$s = \gamma_{0123}\mathbf{J}\mathbf{K}^{-1} = \mathbf{J}(\mathbf{K}\gamma_{0123})^{-1} = -(\sigma + \omega\gamma_{0123})^{-1}\mathbf{S} = (\sigma + \omega\gamma_{0123})\mathbf{S}^{-1}$$
$$k = -(\omega + \sigma\gamma_{0123})^{-1}\mathbf{K} = (\sigma - \omega\gamma_{0123})^{-1}\gamma_{0123}\mathbf{K} = \mathbf{J}\mathbf{S}^{-1} = -\mathbf{S}\mathbf{J}^{-1}$$

(Using $\rho e_\mu = \Psi\gamma_\mu\tilde{\Psi}$, $\rho^2 = \sigma^2 + \omega^2$ Hestenes 1986 p. 333 (2.23) gives $s = \gamma_{0123}e_3e_0$, Boudet 1985 p. 719 (2.6) gives $-s = e_1e_2$.) Note that

$$s = \Psi(-\gamma_{12})\Psi^{-1} = \Psi\gamma_2\gamma_1\Psi^{-1} \in \textstyle\bigwedge^2 \mathbb{R}^{1,3} \qquad \text{(simple bivector)}$$
$$k = \Psi(-\gamma_{012})\Psi^{-1} = \Psi\gamma_{0123}\gamma_3\Psi^{-1} \in \mathbb{R}^{1,3} + \textstyle\bigwedge^3 \mathbb{R}^{1,3}$$

and $s^2 = -1$, $k^2 = -1$ and $sk = ks$. Both s and k play the role of the imaginary unit (multiplication on the left side)

$i\psi = s\psi = k\psi = \psi\gamma_2\gamma_1 \neq \psi\gamma_{0123}$	Dirac spinor
$s\Phi = k\Phi = \Phi\gamma_2\gamma_1$	mother spinor
$s\phi = k\phi = \phi\gamma_2\gamma_1 = \phi\gamma_{0123} \neq \phi(-\gamma_{012})$	ideal spinor
$s\Psi = k\Psi\gamma_0 = \Psi\gamma_2\gamma_1$	spinor operator

(Hestenes 1986 p. 334 (2.24) reports $s\Psi = \Psi\gamma_2\gamma_1$ and also $s\phi = \phi\gamma_2\gamma_1 = \phi\gamma_{0123}$) and $P = sQ = kQ$. ∎

Let us digress on a metaphor: *Are there superfluous complex numbers in the present formulation of quantum mechanics?* In other words: Is it possible to get rid of *some* complex numbers in QM? Certain physicists believe that complex numbers $x + iy$ are necessary even though they themselves use only the line $y = x$, which could/should be projected – for simplicity and without loss of information – to the real axis $x = \text{Re}(x + ix)$:

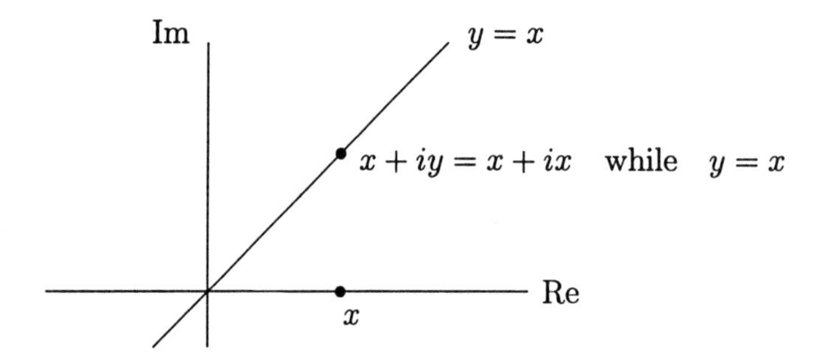

The situation within the present formulation of quantum mechanics is quite analogous to the above picture.

Since the above picture could be criticized by arguing that the product w of two complex numbers of the form $x+ix$ is not of the same type, that is, $\text{Re}(w) \neq \text{Im}(w)$, let me give another analogy: The sums and products of matrices of the following type

$$X = \frac{1}{2}\begin{pmatrix} x & x \\ x & x \end{pmatrix}$$

are of the same type, and give, with respect to addition and multiplication, an isomorphic image of the reals $x \in \mathbb{R}$. It is of course more economical to use the real numbers $x \in \mathbb{R}$ instead of the real 2×2-matrices $X \in \mathbb{R}(2)$.

The idea of redundancy of complex numbers in quantum mechanics could more closely be illustrated as follows: If we have a complex matrix

$$S = \frac{1}{2}\begin{pmatrix} x+iy & -y+ix \\ y-ix & x+iy \end{pmatrix}$$

then the real part, multiplied by two, $Z = 2\,\mathrm{Re}(S)$, that is,

$$Z = \begin{pmatrix} x & -y \\ y & x \end{pmatrix}$$

obeys the same addition and multiplication rules as S and carries the same information as S [contained in the pair (x, y)]. Note that for a complex number $z = x + iy$ we have $S = \frac{z}{2}(I - \sigma_2)$, where the matrix $f = \frac{1}{2}(I - \sigma_2)$ is an idempotent satisfying $f^2 = f$.

Exercises. 1. Under a Lorentz transformation $\mathbf{x} \to R\mathbf{x}\hat{R}^{-1}$, $R \in \mathbf{Pin}(1,3)$, $\mathbf{x} \in \mathbb{R}^{1,3}$, a Dirac spinor $\psi \in \mathbb{C}^4$ transforms as follows $\psi \to R\psi$. Show that a spinor operator $\Psi \in C\ell_{1,3}^+$ transforms like this:

$$\begin{aligned} &\Psi \to R\Psi && \text{when } R \in \mathbf{Spin}(1,3), \\ &\Psi \to R\Psi\gamma_0 && \text{when } R \in \mathbf{Pin}(1,3)\backslash\mathbf{Spin}(1,3). \end{aligned}$$

[Note that the Wigner time-reversal is not represented by any element $R \in \mathbf{Pin}(1,3)\backslash\mathbf{Spin}(1,3)$. Hint: Take $\Psi = \mathrm{even}(4\,\mathrm{Re}(\psi))$ and use $\psi = \psi\frac{1}{2}(1+\gamma_0)\frac{1}{2}(1+i\gamma_{12})$.]

2. By direct computation we can see that

$$\begin{aligned} \mathbf{J}^2 = (\Psi\gamma_0\tilde{\Psi})(\Psi\gamma_0\tilde{\Psi}) &= \Psi\gamma_0\Psi\tilde{\Psi}\gamma_0\tilde{\Psi} = \Psi\gamma_0(\sigma + \gamma_{0123}\omega)\gamma_0\tilde{\Psi} \\ &= \Psi(\sigma - \gamma_{0123}\omega)\gamma_0\gamma_0\tilde{\Psi} = (\sigma - \gamma_{0123}\omega)\Psi\tilde{\Psi} \\ &= (\sigma - \gamma_{0123}\omega)(\sigma + \gamma_{0123}\omega) = \sigma^2 + \omega^2 \end{aligned}$$

which gives one of the Fierz identities. Derive in a similar manner the identity $\mathbf{JK} = -(\omega + \gamma_{0123}\sigma)\mathbf{S}$. ∎

2.3.9 Decomposition and Factorization of Boomerangs.

Denote

$$\begin{aligned} &P = \sigma + \mathbf{J} + \gamma_{0123}\omega \qquad Q = \mathbf{S} + \mathbf{K}\gamma_{0123} = \mathbf{S} - \gamma_{0123}\mathbf{K} \\ &\Sigma = 1 - i\gamma_{0123}\mathbf{J}\mathbf{K}^{-1} = 1 - i\gamma_{0123}\mathbf{K}\mathbf{J}^{-1} \quad (\text{not both } \sigma, \omega = 0) \end{aligned}$$

so that $Z = P + iQ = P\Sigma$. Then $P\Sigma = \Sigma P$ and we have found a solution to the open problem posed by Crawford 1985 p. 1441 ref. (10). [Crawford's second factor in (4.1)

$$\begin{aligned} 1 - i(\sigma + \gamma_{0123}\omega)^{-1}\gamma_{0123}\mathbf{K} &= 1 - i(\omega - \gamma_{0123}\sigma)^{-1}\mathbf{K} \\ &= 1 - i\mathbf{S}^{-1}\mathbf{J} = 1 + i\mathbf{J}^{-1}\mathbf{S} \end{aligned}$$

did not commute with P unless $\omega = 0$.] In the case $\sigma \neq 0$ there is another factorization

$$Z = P(1 + i\frac{1}{2\sigma}Q) = P\frac{1}{2}(1 + i\frac{1}{2\sigma}Q)^2$$

where the factors commute and are Dirac self-adjoint, but in this factorization the second factor is not an idempotent (even though it behaves like one when multiplied by P).

Exercises. Show that for $Z = 4\psi\tilde{\psi}^*$, where $\psi \in (\mathbb{C} \otimes C\ell_{1,3})f$
1. $Z\gamma^0 Z = 4J^0 Z$ $(\neq 0$ for $\psi \neq 0)$ 2. $P\gamma^0 P = 2J^0 P = -Q\gamma^0 Q$
3. $P\gamma_0 Q = Q\gamma_0 P$ 4. $\text{even}(\tilde{Z})Z = 0$, $\text{odd}(\tilde{Z})Z = 0$
5. $\tilde{Z}Z = 0 \Rightarrow P^2 = -Q^2$, $PQ = QP$ (no complex conjugation)
6. $\tilde{Z}Z = 0$, $Z^2 = 4\sigma Z \Rightarrow P^2 = 2\sigma P$, $PQ = 2\sigma Q$
7. $\bar{Z}Z = 0 \Rightarrow \bar{P}P = \bar{Q}Q$, $\bar{P}Q = -\bar{Q}P$
8. $\bar{Z}Z = 0$, $\bar{Z}^*Z = 4\omega\gamma_{0123}Z \Rightarrow \bar{P}P = 2\omega\gamma_{0123}P$, $\bar{P}Q = ?$ ∎

Denote

$$K = \sigma - i\gamma_{0123}\mathbf{K} + \gamma_{0123}\omega \qquad S = \sigma + i\mathbf{S} + \gamma_{0123}\omega$$

$$\Pi = P(\sigma + \gamma_{0123}\omega)^{-1} \qquad \Gamma = K(\sigma + \gamma_{0123}\omega)^{-1}$$

Exercises. Show that 1. $\Sigma = S(\sigma + \gamma_{0123}\omega)^{-1} = 1 - is$
2. $\Gamma = 1 - ik$, $\tilde{\Gamma}^* = 1 + i\tilde{k} = 1 - i(\sigma + \gamma_{0123}\omega)^{-1}\gamma_{0123}\mathbf{K}$
3. $Z = K\Sigma = \Sigma K = \Gamma S = S\tilde{\Gamma}^*$
4. $Z = \Pi S = S\tilde{\Pi} = \Pi K = K\tilde{\Pi}$ (no complex conjugation needed)
$= \Gamma P = P\tilde{\Gamma}^*$ (Crawford's factorization)
$(= P\Gamma$ only if $\omega = 0$ since then $\Gamma = \tilde{\Gamma}^*)$ ∎

2.4 Flags, Poles and Dipoles

Majorana spinors are eigenspinors of the charge conjugation operator, with eigenvalues ± 1. Weyl spinors are eigenspinors of the helicity projection operators $\frac{1}{2}(1 \pm \gamma_{0123})$. In this section we present new spinors, called flag-dipole spinors, residing in between the Weyl, Majorana and Dirac spinors. Unlike Weyl and Majorana spinors, the flag-dipole spinors do not form a real linear subspace, because they are characterized by a quadratic constraint; as a consequence the flag-dipole spinors cannot describe fermions, since the superposition principle is violated.

During the Banff Summer School, Greg Trayling (Windsor) proposed that the flag-dipole spinors might be related to the quark confinement.

2.4.1 A Geometric Classification of Spinors by their Bilinear Covariants.

In the following we shall present a classification of spinors ψ based on properties of their bilinear covariants σ, $\mathbf{J}$, $\mathbf{S}$, $\mathbf{K}$, ω; collected as

$$Z = \sigma + \mathbf{J} + i\mathbf{S} - i\gamma_{0123}\mathbf{K} + \gamma_{0123}\omega.$$

In other words, we classify the boomerangs $Z = 4\psi\tilde{\psi}^*$. Previous attempts to classify spinors have been non-geometric, based on representation theory, irreducible representations of the Lorentz group $SO_+(1,3)$ [Weyl and Majorana spinors are usually introduced by properties of matrices, see Benn and Tucker 1987 and Crumeyrolle 1990]. In contrast, our classification is geometric, based on multivector structure of the Clifford algebra $C\ell_{1,3}$, and it reveals a new class of type spinors, the class 4 of flag-dipole spinors.

Recall that $Z = P + iQ$, $P = \sigma + \mathbf{J} + \omega\gamma_{0123}$, $Q = \mathbf{S} + \mathbf{K}\gamma_{0123}$.

Dirac spinors:

1. $\underline{\sigma \neq 0,\ \omega \neq 0}$: Using $P^2 = 2\sigma P = -Q^2$ we find the relationship $P = \pm(-\frac{1}{2}Q^2)/\sqrt{\langle -\frac{1}{2}Q^2\rangle_0}$ where the sign is given by $J^0 > 0$ (and coincides with the sign of σ). $P = kQ$, where $k = -(\omega + \sigma\gamma_{0123})^{-1}\mathbf{K}$, $i\psi = k\psi$.

2. $\underline{\sigma \neq 0,\ \omega = 0}$: P is a multiple of $\frac{1}{2\sigma}(\sigma + \mathbf{J})$ which looks like a proper **energy projection operator** and which commutes with the **spin projection operator** $\frac{1}{2}(1 - i\gamma_{0123}\mathbf{K}/\sigma)$. $Z = \sigma + \mathbf{J} + i\mathbf{S} - i\gamma_{0123}\mathbf{K} = (\sigma + \mathbf{J})(1 - i\gamma_{0123}\mathbf{K}/\sigma)$, $\mathbf{S} = \gamma_{0123}\mathbf{J}\mathbf{K}/\sigma$. $P = \gamma_{0123}\frac{1}{\sigma}\mathbf{K}Q$, $k = \gamma_{0123}\mathbf{K}/\sigma$. In this class the Yvon-Takabayasi angle β gets only two values, 0 and π; and the charge superselection rule applies.

3. $\underline{\sigma = 0,\ \omega \neq 0}$: Using $P^2 = 2\sigma P$ we find that P is nilpotent $P^2 = 0$. $Z = \mathbf{J} + i\mathbf{S} - i\gamma_{0123}\mathbf{K} + \gamma_{0123}\omega$, $\mathbf{S} = -\mathbf{J}\mathbf{K}/\omega$. $P = -\frac{1}{\omega}\mathbf{K}Q = \pm\mathbf{K}Q/\sqrt{-\mathbf{K}^2}$ (opposite to the sign of ω), $k = -\mathbf{K}/\omega$.

Singular spinors with a light-like pole/current:

4. $\underline{\sigma = \omega = 0,\ \mathbf{K} \neq 0,\ \mathbf{S} \neq 0}$: **Flag-dipole spinors.** $Z = \mathbf{J} + i\mathbf{J}\mathbf{s} - i\gamma_{0123}h\mathbf{J}$, $\mathbf{J}^2 = 0$, $\mathbf{s}$ is a space-like vector, $\mathbf{s}^2 < 0$, orthogonal to $\mathbf{J}$, $\mathbf{J}\cdot\mathbf{s} = 0$, $\mathbf{S} = \mathbf{J}\mathbf{s}$ $(= \mathbf{J}\wedge\mathbf{s})$, $\mathbf{K} = h\mathbf{J}$, $h^2 = 1 + \mathbf{s}^2 < 1$ (h real, $h \neq 0$). $P = \mathbf{J}$, $Q = \mathbf{J}(\mathbf{s} + \gamma_{0123}h)$, $(1 + i\mathbf{s} + i\gamma_{0123}h)Z = 0$. Note that $\frac{1}{2}(1 - i\mathbf{s} - i\gamma_{0123}h)\psi = \psi$ and $(1 + i\mathbf{s} + i\gamma_{0123}h)\psi = 0$. $\tilde{Z}^* = Z$ and $Z^2 = 0$ imply $Z = \mathbf{J}(1 + i\mathbf{s} + i\gamma_{0123}h)$ etc. Let $\psi = \frac{1}{4N}Z\eta$, then $Z = 4\psi\tilde{\psi}^*$ implies $(\mathbf{s} + \gamma_{0123}h)^2 = -1$. $P = (\mathbf{s} + \gamma_{0123}h)Q$, $i\psi = (\mathbf{s} + \gamma_{0123}h)\psi$.

5. $\underline{\sigma = \omega = \mathbf{K} = 0,\ \mathbf{S} \neq 0}$: **Flag-pole spinors** for which $Z = \mathbf{J} + i\mathbf{J}\mathbf{s}$ is a pole $\mathbf{J}$ plus a flag $\mathbf{S} = \mathbf{J}\mathbf{s}$ $(= \mathbf{J}\wedge\mathbf{s})$, $\mathbf{J}\cdot\mathbf{s} = 0$, $\mathbf{s}^2 = -1$. $P = \mathbf{s}Q$, $i\psi = \mathbf{s}\psi$.
— Denote $\mathbf{K}_k = \Psi\gamma_k\tilde{\Psi}$, $\mathbf{S}_k = \Psi\gamma_{ij}\tilde{\Psi}$ (ijk cycl.) with $\mathbf{K}_3 = \mathbf{K}$, $\mathbf{S}_3 = \mathbf{S}$. Then $\mathbf{K}_1 = \mathbf{J}$, $\mathbf{K}_2 = \mathbf{K}_3 = 0$ and $\mathbf{S}_1 = 0$, $\mathbf{S}_2 = \mathbf{J}\mathbf{s}_2$ $(= \mathbf{J}\wedge\mathbf{s}_2)$, $\mathbf{S}_3 = \mathbf{J}\mathbf{s}_3$ where $\mathbf{s}_3 = \mathbf{s}$, $\mathbf{s}_2^2 = -1$, $\mathbf{s}_2\cdot\mathbf{s}_3 = 0$.
— Given an arbitrary Dirac spinor ψ with covariants $\mathbf{J}$, $\mathbf{K}$ (and with $\mathbf{K}_1$, $\mathbf{S}_2$) we may construct, as special cases of flag-pole spinors, two **Majorana spinors**

$\psi_\pm = \frac{1}{2}(\psi \pm \psi_C)$, which are seen to be eigenspinors of the charge conjugation $C(\psi_\pm) = \pm\psi_\pm$, and whose bilinear covariants $\mathbf{J}_\pm$, $\mathbf{S}_\pm$ satisfy $\mathbf{K}\cdot\mathbf{J}_\pm = 0$, $\mathbf{K}\cdot\mathbf{S}_\pm = -\omega\mathbf{J}_\pm$ and $\mathbf{J} = \mathbf{J}_+ + \mathbf{J}_-$, $\mathbf{S} = \mathbf{S}_+ + \mathbf{S}_-$. [Note that $\mathbf{J}_\pm = \frac{1}{2}(\mathbf{J} \pm \mathbf{K}_1)$ and $\mathbf{S}_\pm = \frac{1}{2}(\mathbf{S} \mp \mathbf{S}_2\gamma_{0123})$.] Note that $C(i\psi_+) = -i\psi_+ \neq \pm\psi_-$ and $C(i\psi_-) = i\psi_- \neq \pm\psi_+$.

6. $\sigma = \omega = \mathbf{S} = 0,\ \mathbf{K} \neq 0$: **Weyl spinors** (of massless neutrinos) are eigenspinors of the chirality operator $\gamma_{0123}\psi_\pm = \pm i\psi_\pm$. $Z = \mathbf{J} \mp i\gamma_{0123}\mathbf{J}$, $\mathbf{J} = \pm\mathbf{K}$, $h = \pm 1$, $\psi_\pm = \frac{1}{2}(1 \mp i\gamma_{0123})\psi_\pm$. Note that $\text{even}(\psi_\pm) = \psi_\pm\frac{1}{2}(1 \mp \gamma_{03})$, $\text{odd}(\psi_\pm) = \psi_\pm\frac{1}{2}(1 \pm \gamma_{03})$. $P = \pm\gamma_{0123}Q$.
— Denote $\mathbf{K}_k = \Psi\gamma_k\tilde{\Psi}$, $\mathbf{S}_k = \Psi\gamma_{ij}\tilde{\Psi}$ as before. Then $\mathbf{K}_1 = \mathbf{K}_2 = 0$, $\mathbf{S}_1 = \mathbf{J}\mathbf{s}_1$ $(= \mathbf{J}\wedge\mathbf{s}_1)$, $\mathbf{S}_2 = \mathbf{J}\mathbf{s}_2$ $(= \mathbf{J}\wedge\mathbf{s}_2)$ where $\mathbf{s}_1^2 = \mathbf{s}_2^2 = -1$, $\mathbf{s}_1\cdot\mathbf{s}_2 = 0$.
— Given an arbitrary Dirac spinor ψ with covariants $\mathbf{J}$, $\mathbf{K}$ we may construct two Weyl spinors $\psi_\pm = \frac{1}{2}(1 \mp i\gamma_{0123})\psi$ with covariants $\mathbf{J}_\pm = \frac{1}{2}(\mathbf{J} \pm \mathbf{K})$, $\mathbf{K}_\pm = \frac{1}{2}(\mathbf{K} \pm \mathbf{J})$. Weyl spinors are **pure** $\tilde{\psi}_\pm\gamma_{0123}\gamma_\mu\psi_\pm = 0$ [no complex conjugation; for arbitrary Dirac spinors $\tilde{\psi}\psi = 0$, $\tilde{\psi}\gamma_\mu\psi = 0$, $\tilde{\psi}\gamma_{0123}\psi = 0$ though $\tilde{\psi}\gamma_{\mu\nu}\psi \neq 0$, $\tilde{\psi}\gamma_{0123}\gamma_\mu\psi \neq 0$ (and also $\bar{\psi}\psi = 0$, $\bar{\psi}\gamma_{0123}\gamma_\mu\psi = 0$, $\bar{\psi}\gamma_{0123}\psi = 0$, though $\bar{\psi}\gamma_\mu\psi \neq 0$, $\bar{\psi}\gamma_{\mu\nu}\psi \neq 0$)]. $C(\psi_\pm)$ is of helicity $h = \mp 1$ with covariants $\mathbf{J}_\pm$, $-\mathbf{K}_\pm$.

Comments:

For the classes 1, 2 the element $\frac{1}{4\sigma}Z$ is a primitive idempotent in $\mathbb{C}\otimes C\ell_{1,3}$.

Classes 1, 2, 3 are **Dirac spinors for the electron**. A spinor operator Ψ factors uniquely up to a sign $\Psi = (\sigma + \omega\gamma_{0123})^{\frac{1}{2}}R$, $R \in \mathbf{Spin}_+(1,3)$. In particular, denoting $\mathbf{K}_k = \Psi\gamma_k\tilde{\Psi}$ we have an orthogonal basis $\mathbf{J}$, $\mathbf{K}_1$, $\mathbf{K}_2$, $\mathbf{K}_3$ $(= \mathbf{K})$ of $\mathbb{R}^{1,3}$.

The class 4 consists of **flag-dipole spinors** with a flag $\mathbf{S}$ on a dipole of $\mathbf{J}$ and $\mathbf{K}$. The class 5 consists of **flag-pole spinors** with a flag $\mathbf{S}$ on a pole $\mathbf{J}$. The class 6 consists of **dipole spinors** with a dipole of $\mathbf{J}$ and $\mathbf{K}$.

In the classes 4, 5, 6 the vectors $\mathbf{J}$, $\mathbf{K}_1$, $\mathbf{K}_2$, $\mathbf{K}_3$ no longer form a basis but collapse into a null-line $\mathbf{J}$ (also $\mathbf{S}_1$, $\mathbf{S}_2$, $\mathbf{S}_3$ intersect along $\mathbf{J}$). The even elements $\mathbf{J}(\gamma_0 - s\gamma_{12} - h\gamma_3)$ and Ψ differ only up to a complex factor $x - y\gamma_{12}$ (on the right).

In addition to the electron (classes 1, 2, 3) also the massless neutrino (class 6) has been discussed by Hestenes 1967 p. 808 (8.13) and 1986 p. 343 who quite correctly observed that $\mathbf{J}_\pm = \frac{1}{2}\Psi(\gamma_0 \pm \gamma_3)\tilde{\Psi}$; note also that $\tilde{\Psi}_\pm\gamma_\mu\Psi_\pm = (\gamma_\mu\cdot\mathbf{J}_\pm)(\gamma_0 \mp \gamma_3)$. Hestenes has not discussed the classes 4 and 5. Holland in Found. Phys. 1986, pp. 708-709, does not discuss the types 3, 4, 5, 6 with a nilpotent Z, $Z^2 = 0$, but focuses on a nilpotent ψ, $\psi^2 = 0$.

Majorana spinors $\Psi \in C\ell_{1,3}^+\frac{1}{2}(1 \mp \gamma_{01})$ are not stable under the $U(1)$-gauge transformation $\Psi \to \Psi e^{\alpha\gamma_{12}} \notin C\ell_{1,3}^+\frac{1}{2}(1 \mp \gamma_{01})$. Given a Weyl spinor ψ with bilinear covariants $\mathbf{J}$, $\mathbf{K}$ we can associate to it two Majorana spinors $\psi_\pm = \frac{1}{2}(\psi \pm \psi_C)$ with Penrose flags $Z_\pm = \frac{1}{2}(\mathbf{J} \mp i\mathbf{S}_2\gamma_{0123})$.

The number of parameters in the sets of bilinear covariants (or spinors without $U(1)$-gauge) is seen to be

class n:o	1	2	3	4	5	6
parameters	7	6	6	5	4	3

If the $U(1)$-gauge is taken into consideration, then the number of parameters will be raised by one unit in all classes except in the class number 5 of Majorana spinors [Weyl spinors with $U(1)$-gauge and Majorana spinors both have 4 parameters and can be mapped bijectively onto each other – which enables Penrose flags also to be attached to Weyl spinors].

The Weyl and Majorana spinors can be written with spinor operators in the form

$$\Psi\frac{1}{2}(1+\gamma_0\mathbf{u})$$

where $\mathbf{u} = \pm\gamma_3$ for Weyl spinors and $\mathbf{u} = \pm\gamma_1$ for Majorana spinors [and Ψ is an element in $C\ell_{1,3}^+$]. The flag-pole spinors can be written in a similar form with $\mathbf{u} = \gamma_1\cos\phi + \gamma_2\sin\phi$. It is easy to see that all elements of the form $\Psi\frac{1}{2}(1+\gamma_0\mathbf{u})$, $\Psi \in C\ell_{1,3}^+$ are flag-dipole spinors, when $\mathbf{u}$ is a spatial unit vector, $\mathbf{u}\cdot\gamma_0 = 0$, $\mathbf{u}^2 = -1$, which is not in the γ_3-axis or in the $\gamma_1\gamma_2$-plane. About the converse a participant in the Banff Summer School presented the following:

Conjecture (C. Doran, 1995): All the flag-dipole spinors can be written in the form $\Psi\frac{1}{2}(1+\gamma_0\mathbf{u})$. ∎

When $\mathbf{u}$ varies in the unit sphere S^2 in $\mathbb{R}^3$ (orthogonal to γ_0), the flag-dipole spinor sweeps around the 'paraboloid' $\Psi\tilde{\Psi} = 0$. If the conjecture is true, it would be nice to know the relation between $\mathbf{s}, h$ and $\mathbf{u}$. [Clearly, $h = \mathbf{u}\cdot\gamma_3$.]

2.4.2 Projection Operators in $\mathrm{End}(C\ell_{1,3})$.

Denote as before $P = \sigma + \mathbf{J} + \omega\gamma_{0123}$, $Q = \mathbf{S} + \mathbf{K}\gamma_{0123}$, $Z = P + iQ = P\Sigma = \Sigma P$, $\Sigma = 1 - i\gamma_{0123}\mathbf{J}\mathbf{K}^{-1}$. Then

$$\frac{1}{4\sigma}Z\psi = \psi, \qquad \frac{1}{2\sigma}P\psi = \psi, \qquad \text{when } \sigma \neq 0$$
$$\frac{1}{2}\Sigma\psi = \psi, \qquad \text{when not both } \sigma,\ \omega \neq 0.$$

Define for $u \in C\ell_{1,3}$ (or $u \in \mathbb{C}\otimes C\ell_{1,3}$)

$$P_\pm(u) = \frac{1}{2\sigma}(\sigma \pm \mathbf{J} \pm \omega\gamma_{0123})u \qquad \sigma \neq 0$$
$$\Sigma_\pm(u) = \frac{1}{2}(u \pm \gamma_{0123}\mathbf{J}\mathbf{K}^{-1}u\gamma_{12}) \qquad \Sigma_\pm \in \mathrm{End}(C\ell_{1,3}).$$

Then

$$\begin{array}{llll} P_+(\psi) = \psi & P_+(\Phi) = \Phi & P_+(\phi) = \phi & \\ P_-(\psi) = 0 & P_-(\Phi) = 0 & P_-(\phi) = 0 & \\ \Sigma_+(\psi) = \psi & \Sigma_+(\Phi) = \Phi & \Sigma_+(\phi) = \phi & \Sigma_+(\Psi) = \Psi \\ \Sigma_-(\psi) = 0 & \Sigma_-(\Phi) = 0 & \Sigma_-(\phi) = 0 & \Sigma_-(\Psi) = 0. \end{array}$$

In general, for $u \in C\ell_{1,3}$, $P_\pm^2(u) = P_\pm(u)$, $\Sigma_\pm^2(u) = \Sigma_\pm(u)$ and $P_\pm(\Sigma_\pm(u)) = \Sigma_\pm(P_\pm(u))$, that is, $P_\pm$ and $\Sigma_\pm$ are commuting projection operators. For an arbitrary η in $C\ell_{1,3}\,\frac{1}{2}(1+\gamma_0)$ [or in $C\ell_{1,3}\,\frac{1}{2}(1-\gamma_{03})$] the spinor $P_+(\Sigma_+(\eta))$ is parallel to Φ [or to ϕ], that is, the bilinear covariants of $P_+(\Sigma_+(\eta))$ are proportional to P, Q. However, for an arbitrary $u \in C\ell_{1,3}$, $P_+(\Sigma_+(u)) \notin C\ell_{1,3}\,\frac{1}{2}(1+\gamma_0)$ [or $P_+(\Sigma_+(u)) \notin C\ell_{1,3}\,\frac{1}{2}(1-\gamma_{03})$].

Define

$$\Sigma_\pm^I(u) = \frac{1}{2}(u \mp \gamma_{0123}\mathbf{J}\mathbf{K}^{-1}u\gamma_{0123})$$

where I stands for ideal spinor. Then for an arbitrary $u \in C\ell_{1,3}$ we have $\Sigma_+(\Sigma_+^I(u)) \in C\ell_{1,3}\,\frac{1}{2}(1-\gamma_{03})$, and $P_+(\Sigma_+(\Sigma_+^I(u)))$ is an ideal spinor parallel to ϕ (with bilinear covariants proportional to P, Q). Furthermore, $\Sigma_+^I(\phi) = \phi$, $\Sigma_-^I(\phi) = 0$, and $\Sigma_\pm^I$ are projection operators commuting with $P_\pm$, $\Sigma_\pm$.

Define (O stands for spinor operator)

$$P_\pm^O(u) = \frac{1}{2\sigma}((\sigma \pm \omega\gamma_{0123})u \pm \mathbf{J}u\gamma_0), \qquad u \in C\ell_{1,3}^+$$

which are projection operators commuting with $\Sigma_\pm$.

Exercise. (Inspired by Crawford 1985.) Define

$$\Gamma_\pm(u) = \frac{1}{2}(u \mp (\omega + \sigma\gamma_{0123})^{-1}\mathbf{K}u\gamma_{12})$$

and show that $\Gamma_\pm$ are projection operators commuting with $\Sigma_\pm$ [but not with $P_\pm$ unless $\omega = 0$; recall here the factorization of Crawford]. Show that $P_+(\Gamma_+(\Phi)) = \Phi$, $P_+(\Gamma_+(\phi)) = \phi$. How would you define $\Gamma_\pm$ for a spinor operator Ψ?
[Answer: $\Gamma_\pm^O(u) = \frac{1}{2}(u \mp (\omega + \sigma\gamma_{0123})^{-1}\mathbf{K}u\gamma_{012})$ for $u \in C\ell_{1,3}^+$.] ∎

Remark. Define $\Gamma_\pm$ for an ideal spinor ϕ (I stands for ideal)

$$\Gamma_\pm^I(u) = \frac{1}{2}(u \pm (\omega + \sigma\gamma_{0123})^{-1}\mathbf{K}u\gamma_{0123}), \qquad u \in C\ell_{1,3}\,\frac{1}{2}(1-\gamma_{03}).$$

In the special case $\omega = 0$ of type 2 these become of the form

$$\Gamma_\pm^I(u) = \frac{1}{2}(u \mp \gamma_{0123}\frac{1}{\sigma}\mathbf{K}u\gamma_{0123}), \qquad u \in C\ell_{1,3}\,\frac{1}{2}(1-\gamma_{03})$$

and commute with $P_\pm$ [this special case was also observed by Hestenes 1986 p. 336 (2.32).] ∎

2.4.3 Projection Operators for Majorana and Weyl Spinors.

Treat first the general case (type 4) $\sigma = 0 = \omega$, $\mathbf{K} \neq 0 \neq \mathbf{S}$. Recall that $(1 + i\mathbf{s} + i\gamma_{0123}h)\psi = 0$ or $i\psi = (\mathbf{s} + \gamma_{0123}h)\psi$. Define

$$\Sigma_\pm^G(u) = \frac{1}{2}(u \pm (\mathbf{s} + \gamma_{0123}h)u\gamma_{12}).$$

Then $\Sigma_+^G(\Phi) = \Phi$, $\Sigma_+^G(\phi) = \phi$. Majorana and Weyl spinors are now the limiting cases

$$\Sigma_\pm^M(u) = \frac{1}{2}(u \pm \mathbf{s}u\gamma_{12}) \qquad \Sigma_\pm^W(u) = \frac{1}{2}(u \pm \gamma_{0123}u\gamma_{12}).$$

Exercises. 1. Recall that $\Psi\gamma_2\gamma_1 = \mathbf{s}\Psi\gamma_0 + h\gamma_{0123}\Psi$. How would you define $\Sigma_\pm^G$ for a spinor operator Ψ?
[Answer: $\Sigma_\pm^{GO}(u) = \frac{1}{2}(u \pm \mathbf{s}u\gamma_{012} \pm h\gamma_{0123}u\gamma_{12})$ for $u \in C\ell_{1,3}^+$.]

2. Recall that $\phi\gamma_{0123} = \phi\gamma_2\gamma_1$. How would you define another pair $\Sigma_\pm^G$ for an ideal spinor ϕ?
[Answer: $\Sigma_\pm^{GI}(u) = \frac{1}{2}(u \mp (\mathbf{s} + \gamma_{0123}h)u\gamma_{0123})$, $u \in C\ell_{1,3}\,\frac{1}{2}(1-\gamma_{03})$.]

3. Show that up to a unit complex factor $e^{\gamma_{12}\alpha}$: $\Psi \simeq \frac{1}{4N}(\sigma + \mathbf{J}\gamma_0 - \mathbf{S}\gamma_{12} - \mathbf{K}\gamma_3 + \omega\gamma_{0123})$, when $N = \sqrt{\langle Zf\rangle_0} \neq 0$.

4. Show that the operator form of a Weyl spinor is $\Psi\frac{1}{2}(1 \mp \gamma_{03})$.
[Hint: compute the even part of $4\,\mathrm{Re}(\frac{1}{2}(1 \mp i\gamma_{0123})\psi)$ in the decomposition $\mathbb{C} \otimes C\ell_{1,3}$.]

5. Show that Weyl spinors $\frac{1}{2}(1 \mp i\gamma_{0123})\psi$ correspond to even and odd parts of the ideal spinor $\phi = \phi_0 + \phi_1$. ∎

Bibliography

M.F. Atiyah, R. Bott, A. Shapiro: Clifford modules. *Topology* **3**, suppl. 1 (1964), 3-38. Reprinted in R. Bott: *Lectures on $K(X)$*. Benjamin, New York, 1969, pp. 143-178. Reprinted in *Michael Atiyah: Collected Works*, Vol. 2. Clarendon Press, Oxford, 1988, pp. 301-336.

W. Baylis: *Theoretical Methods in the Physical Science: an introduction to problem solving using MAPLE V*. Birkhäuser, Boston, 1994.

I. M. Benn, R. W. Tucker: *An Introduction to Spinors and Geometry with Applications in Physics*. Adam Hilger, Bristol, 1987.

J.D. Bjorken, S.D. Drell: *Relativistic Quantum Mechanics*. McGraw-Hill, New York, 1964.

R. Boudet: Les algèbres de Clifford et les transformations des multivecteurs, pp. 343-352 in A. Micali et al. (eds.): *Proceedings of the Second Workshop on 'Clifford Algebras and their Applications in Mathematical Physics' (Montpellier, 1989)*. Kluwer, Dordrecht, 1992.

R. Brauer, H. Weyl: Spinors in n dimensions. *Amer. J. Math.* **57** (1935), 425-449. Reprinted in *Selecta Hermann Weyl*. Birkhäuser, Basel, 1956, pp. 431-454.

E. Cartan (exposé d'après l'article allemand de E. Study): Nombres complexes, pp. 329-468 in J. Molk (red.): *Encyclopédie des sciences mathématiques*, Tome **I**, vol. **1**, Fasc. **4**, art. **I5**, 1908.

A. Charlier, A. Bérard, M.-F. Charlier, D. Fristot: *Tensors and the Clifford Algebra, Applications to the Physics of Bosons and Fermions*. Marcel Dekker, New York, 1992.

C. Chevalley: *Theory of Lie Groups*. Princeton University Press, Princeton, 1946.

C. Chevalley: *The Algebraic Theory of Spinors*. Columbia University Press, New York, 1954.

W.K. Clifford: Applications of Grassmann's extensive algebra. *Amer. J. Math.* **1** (1878), 350-358.

W.K. Clifford: On the classification of geometric algebras, pp. 397-401 in R. Tucker (ed.): *Mathematical Papers by William Kingdon Clifford*. Macmillan, London, 1882. (Reprinted by Chelsea, New York, 1968.) Title of talk announced already on p. 135 in *Proc. London Math. Soc.* **7** (1876).

J.P. Crawford: On the algebra of Dirac bispinor densities: Factorization and inversion theorems. *J. Math. Phys.* **26** (1985), 1439-1441.

J.P. Crawford: Dirac equation for bispinor densities, pp. 353-361 in J.S.R. Chisholm, A.K. Common (eds.): *Proceedings of the Workshop on 'Clifford Algebras and their Applications in Mathematical Physics' (Canterbury 1985)*. Reidel, Dordrecht, 1986.

A. Crumeyrolle: *Orthogonal and Symplectic Clifford Algebras, Spinor Structures*. Kluwer, Dordrecht, 1990.

A. Gsponer, J.-P. Hurni: Lanczos' Equation to Replace Dirac's Equation? pp. 509-512 in J.D. Brown (ed.): *Proceedings of the Cornelius Lanczos Centenary Conference (Raleigh, NC, 1993)*.

F. Gürsey: Correspondence between quaternions and four-spinors. *Rev. Fac. Sci. Univ. Istanbul* **A21** (1956), 33-54.

F. Gürsey: Relation of charge independence and baryon conservation to Pauli's transformation. *Nuovo Cimento* **7** (1958), 411-415.

J.D. Hamilton: The Dirac equation and Hestenes' geometric algebra. *J. Math. Phys.* **25** (1984), 1823-1832.

F.R. Harvey: *Spinors and Calibrations*. Academic Press, San Diego, 1990.

D. Hestenes: *Space-Time Algebra*. Gordon and Breach, New York, 1966, 1987, 1992.

D. Hestenes: Real spinor fields. *J. Math. Phys.* **8** (1967), 798-808.

D. Hestenes: Local Observables in the Dirac theory. *J. Math. Phys.* **14** (1973), 893-905.

D. Hestenes: Proper dynamics of a rigid point particle. *J. Math. Phys.* **15** (1974), 1778-1786.

D. Hestenes: Clifford algebra and the interpretation of quantum mechanics, pp. 321-346 in J.S.R. Chisholm, A.K. Common (eds.): *Proceedings of the Workshop on 'Clifford Algebras and their Applications in Mathematical Physics' (Canterbury 1985)*. Reidel, Dordrecht, 1986.

D. Hestenes, G. Sobczyk: *Clifford Algebra to Geometric Calculus*. Reidel, Dordrecht, 1984, 1987.

P.R. Holland: Minimal ideals and Clifford algebras in the phase space representation of spin-$\frac{1}{2}$ fields, pp. 273-283 in J.S.R. Chisholm, A.K. Common (eds.): *Proceedings of the Workshop on 'Clifford Algebras and their Applications in Mathematical Physics' (Canterbury 1985)*. Reidel, Dordrecht, 1986.

P.R. Holland: Relativistic algebraic spinors and quantum motions in phase space. *Found. Phys.* **16** (1986), 708-709.

B. Jancewicz: *Multivectors and Clifford Algebra in Electrodynamics*. World Scientific Publ., Singapore, 1988.

G. Juvet: Opérateurs de Dirac et équations de Maxwell. *Comment. Math. Helv.* **2** (1930), 225-235.

G. Juvet, A. Schidlof: Sur les nombres hypercomplexes de Clifford et leurs applications à l'analyse vectorielle ordinaire, à l'électromagnetisme de Minkowski et à la théorie de Dirac. *Bull. Soc. Neuchat. Sci. Nat.*, **57** (1932), 127-147.

J. Keller, S. Rodriguez-Romo: A multivectorial Dirac equation. *J. Math. Phys.* **31** (1990), 2501-2510.

J. Keller, F. Viniegra: The multivector structure of the matter and interaction field theories, pp. 437-445 in A. Micali et al. (eds.): *Proceedings of the Second Workshop on 'Clifford Algebras and their Applications in Mathematical Physics' (Montpellier 1989).* Kluwer, Dordrecht, 1992.

P. Kustaanheimo, E. Stiefel: Perturbation theory of Kepler motion based on spinor regularization. *J. Reine Angew. Math.* **218** (1965), 204-219.

R. Lipschitz: Principes d'un calcul algébrique qui contient comme espèces particulières le calcul des quantités imaginaires et des quaternions. *C.R. Acad. Sci. Paris* **91** (1880), 619-621, 660-664. Reprinted in *Bull. Soc. Math.* (2) **11** (1887), 115-120.

R. Lipschitz: *Untersuchungen über die Summen von Quadraten.* Max Cohen und Sohn, Bonn, 1886, pp. 1-147. The first chapter of pp. 5-57 translated into French by J. Molk: Recherches sur la transformation, par des substitutions réelles, d'une somme de deux ou troix carrés en elle-même. *J. Math. Pures Appl.* (4) **2** (1886), 373-439. French résumé of all three chapters in *Bull. Sci. Math.* (2) **10** (1886), 163-183.

P. Lounesto: Scalar products of spinors and an extension of Brauer-Wall groups. *Found. Phys.* **11** (1981), 721-740.

R. Penrose, W. Rindler: *Spinors and Space-Time*, Vol. 1. Cambridge University Press, Cambridge, 1984.

I.R. Porteous: *Topological Geometry.* Van Nostrand Reinhold, London, 1969. Cambridge University Press, Cambridge, 1981.

M. Riesz: Sur certain notions fondamentales en théorie quantique relativiste. *C. R.* 10e *Congrès Math. Scandinaves, (Copenhagen, 1946).* Jul. Gjellerups Forlag, Copenhagen, 1947, pp. 123-148. Reprinted in L. Gårding, L. Hörmander (eds.): *Marcel Riesz, Collected Papers.* Springer, Berlin, 1988, pp. 545-570.

M. Riesz: *Clifford Numbers and Spinors.* University of Maryland, 1958. Reprinted as facsimile by Kluwer, Dordrecht, 1993.

F. Sauter: Lösung der Diracschen Gleichungen ohne Spezialisierung der Diracschen Operatoren. *Z. Phys.* **63** (1930), 803-814.

Y. Takahashi: A passage between spinors and tensors. *J. Math. Phys.* **24** (1983), 1783-1790.

K. Th. Vahlen: Über höhere komplexe Zahlen. *Schriften der phys.-ökon. Gesellschaft zu Königsberg* **38** (1897), 72-78.

K. Th. Vahlen: Über Bewegungen und complexe Zahlen. *Math. Ann.* **55** (1902), 585-593.

E. Witt: Theorie der quadratischen Formen in beliebigen Körpern. *J. Reine Angew. Math.* **176** (1937), 31-44.

Chapter 3

Introduction to Geometric Algebras

Garret Sobczyk[1]
Departamento de Fisico-Matematicas,
Universidad de las Americas,
72820 Cholula, Pue., México

Everybody knows that the real number field $\mathbb{R}$ can be extended to the complex number field $\mathbb{C} \equiv \mathbb{R}[i]$ by introducing an imaginary unit i with the property that $i^2 = -1$. But few mathematicians consider the extension of the real number system by a *unipotent* u defined by $u \neq \pm 1$ and $u^2 = 1$, perhaps because the resulting *hyperbolic number system* $\boldsymbol{U}$ is no longer a field but a commutative ring with unity.

The hyperbolic numbers are one of the simplest examples of Clifford's *geometric algebras*, and it is well worth the student's time to become familiar with its properties before going on to the study of more general geometric algebras. The intent of this lecture is to introduce the *unipodal number system*, and then the more general noncommutative geometric algebras of higher dimensions.

3.1 The Unipodal Number System

In the *standard basis* $\{1, u\}$ a *unipodal number* $w \in \boldsymbol{U}$ has the form $w = w_0 + uw_1$ where $u^2 = 1$ and $w_0, w_1 \in \mathbb{C}$. The *idempotent basis* $\{u_+, u_-\}$ is defined by

$$u_+ = \frac{1}{2}(1+u) \quad \text{and} \quad u_- = \frac{1}{2}(1-u),$$

and satisfies $u_+ + u_- = 1$ and $u_+ - u_- = u$. We say that u_+ and u_- are *idempotents* because $u_+^2 = u_+$ and $u_-^2 = u_-$, and they are *mutually annihilating* because $u_+u_- = 0$. Using the projective properties of the idempotent basis we can write

$$w = w(u_+ + u_-) = w_+u_+ + w_-u_-, \tag{3.1}$$

[1] I gratefully acknowledge the support given by INIP of the Universidad de las Americas, and CONACYT (The Mexican National Council for Science and Technology) grant 3803-E.

where $w_+ = w_0 + w_1$ and $w_- = w_0 - w_1$. Conversely, given $w = w_+u_+ + w_-u_-$ for $w_+, w_- \in \mathbb{C}$, we can recover the coordinates with respect to the standard basis by

$$w_0 = \frac{1}{2}(w_+ + w_-) \quad \text{and} \quad w_1 = \frac{1}{2}(w_+ - w_-). \tag{3.2}$$

The special properties of the idempotent basis make it particularly suitable for calculations. For example, the binomial theorem takes the very simple form

$$(w_+u_+ + w_-u_-)^k = (w_+)^k u_+^k + (w_-)^k u_-^k = (w_+)^k u_+ + (w_-)^k u_- \tag{3.3}$$

This formula is valid for *all* real numbers $k \in \mathbb{R}$, and not just the positive integers. For example, for $k = -1$ we find that

$$1/w = w^{-1} = (1/w_+)u_+ + (1/w_-)u_-,$$

a valid formula for the inverse of $w \in \mathbb{U}$, provided that $w_+w_- \neq 0$.

Indeed, the validity of (3.3) allows us to extend the definitions of *all* of the elementary functions to the elementary functions in the complex unipodal plane. If $f(w)$ is such a function for $w = w_+u_+ + w_-u_-$, we define

$$f(w) \equiv f(w_+)u_+ + f(w_-)u_- \tag{3.4}$$

provided that $f(w_+)$ and $f(w_-)$ are defined.

The hyperbolic numbers have been used to find the solution of the cubic equation and to derive simple properties of the Lorentzian geometry of the plane and special relativity [9].[2]

There is an algebra isomorphism between complex 2×2 diagonal matrices and the unipodal numbers. We have the correspondence

$$w = w_+u_+ + w_-u_- \leftrightarrow \begin{pmatrix} w_+ & 0 \\ 0 & w_- \end{pmatrix}$$

As we shall see, the relationship between finite dimensional Clifford algebra and matrix algebra is that of an *algebra isomorphism:* Every finite dimensional Clifford algebra is isomorphic to a matrix algebra, and conversely, every matrix algebra is isomorphic to either a Clifford algebra or a subalgebra of a Clifford algebra, [7], [10].

3.2 Clifford Algebra Matrix Algebra Connection

We discovered at the end of the previous section that there is an algebra isomorphism between the unipodal numbers and diagonal 2×2 matrices. This observation leads to the question of whether it is possible to further extend the unipodal numbers in such a way that the extended number system is isomorphic to the full 2×2 matrix algebra. Of course, if we are to succeed we must introduce *noncommutative* elements.

[2]Editor's note: The references for this part are collected at the end of c. 5.

Let us *split* or *factor* the unipotent element u into the product of two anticommuting units e and f satisfying the properties

$$u = ef = -fe \quad \text{and} \quad e^2 = 1 = -f^2; \tag{3.5}$$

we give $u = ef$ the interpretation of a *bivector* because it is the product of the two anticommuting *vectors* e and f. We say that that the geometric algebra $Cl_{1,1}$ is generated by the anticommuting vectors e and f, and as a linear space has the basis $\{1, e, f, ef\}$. A general element $w \in Cl_{1,1}$ can be written in the form $w = a_0 + a_1 u + e(b_0 + b_1 u)$, and also in the form

$$w = w(u_+ + u_-) = (a_0 + a_1)u_+ + (b_0 - b_1)eu_- + (b_0 + b_1)eu_+ + (a_0 - a_1)u_-.$$

It may be shown that the correspondence

$$w \leftrightarrow \begin{pmatrix} a_0 + a_1 & b_0 - b_1 \\ b_0 + b_1 & a_0 - a_1 \end{pmatrix}$$

is the desired isomorphism.

These ideas have a straight forward generalization: Let us extend the real number system by assuming the existence of n commuting unipotent elements $u_1, u_2, \dots, u_n$, obtaining the unipodal algebra $\mathbb{R}[u_1, u_2, \dots, u_n]$. To exhibit the algebra isomorphism between this unipodal algebra and the $n \times n$ diagonal matrix algebra $M(2^n)$, we construct a set of 2^n mutually annihilating *primitive idempotents* u_{sgn} by taking the products of the 2^n sign combinations of the idempotents $u_{j\pm} \equiv \frac{1}{2}(1 \pm u_j)$. Thus, for example,

$$u_{++\dots+} \equiv u_{1+}u_{2+} \dots u_{n+},$$

and

$$u_{+-+\dots+} \equiv u_{1+}u_{2-}u_{3+} \dots u_{n+}.$$

A general element $w \in \mathbb{R}[u_1, \dots, u_n]$ can now be expressed in the basis of primitive idempotents by

$$w = \sum_{sgn} w_{sgn} u_{sgn}$$

where the 2^n coefficients w_{sgn} are real scalars. The algebra isomorphism between $\mathbb{R}[u_1, \dots, u_n]$ and diagonal matrices in $M(2^n)$ is given by

$$w \leftrightarrow \begin{pmatrix} w_{++\dots+} & 0 & 0 & \dots & 0 \\ 0 & w_{+-++\dots+} & 0 & \dots & 0 \\ \dots & & & & \\ \dots & 0 & \dots & 0 & w_{--\dots-} \end{pmatrix}$$

We can obtain the *neutral* Clifford algebra $Cl_{n,n}$ by splitting or factoring each of the unipotent elements u_j into a product of anticommuting unit vectors

$$u_j = e_j f_j = -f_j e_j, \quad \text{and} \quad e_j^2 = 1 = -f_j^2,$$

for $j = 1, 2, \ldots, n$. The following table gives the algebra isomorphism between the real Clifford algebra $Cl_{n,n}$ and the real matrix algebra $M(2^n)$:

$$\begin{pmatrix} u_{+\ldots+} & e_1 u_{-+\ldots+} & \cdots & e_\lambda^{-1} u_{sgns} & \cdots & e_{1\ldots n}^{-1} u_{-\ldots-} \\ e_1 u_{+\ldots+} & u_{-+\ldots+} & \cdots & e_1 e_\lambda^{-1} u_{sgns} & \cdots & e_1 e_{1\ldots n}^{-1} u_{-\ldots-} \\ \vdots & \vdots & \ddots & \vdots & \ddots & \vdots \\ e_\lambda u_{+\ldots+} & e_\lambda e_1 u_{-+\ldots+} & \cdots & u_{sgns} & \cdots & e_\lambda e_{1\ldots n}^{-1} u_{-\ldots-} \\ \vdots & \vdots & \ddots & \vdots & \ddots & \vdots \\ e_{1\ldots k} u_{+\ldots+} & e_{1\ldots n} e_1 u_{-+\ldots+} & \cdots & e_{1\ldots k} e_\lambda^{-1} u_{sgns} & \cdots & u_{-\ldots-} \end{pmatrix} \tag{3.6}$$

The Clifford numbers in the matrix make up a real 2^{2n}-dimensional basis of the Clifford algebra $Cl_{n,n}$; the position of each element corresponds to a matrix with a 1 in the same position and zeros everywhere else. The elements u_{sgn} are 2^n mutually annihiliating primitive idempotents, and the elements $e_1, \ldots, e_n$ are orthonormal basis vectors with signature $+1$ generating $Cl_{n,0}$. The e_λ represents a product of the generating e_i's corresponding to the *ordered* index set λ, for example for $\lambda = \{1, 2, n\}$, $e_\lambda \equiv e_1 e_2 e_n$. The elements e_λ *conjugate commute* with the idempotent elements u_{sgn}, for example $e_{12n}\, u_{+++\ldots++} = u_{--+\ldots+-}\, e_{12n}$. A more detailed discussion of this isomorphism can be found in [11].

3.3 Geometric algebra

Let V be a not necessarily finite dimensional real vector space endowed with an inner product $v_1 \cdot v_2$ for vectors $v_1, v_2 \in V$. The associative *geometric algebra* $G(V)$ consists of all sums of products $v_1 v_2 \ldots v_k$ of vectors in V subject to the following two principles of *geometric multiplication*:

- $v^2 = v \cdot v$ for all $v \in V$.
- If the vectors v_i are pairwise orthogonal, i.e., $v_i \cdot v_j = 0$ for $i \neq j$, then the geometric product $v_1 v_2 \cdots v_k$ is *skewsymmetric* over the interchange of any pair of vectors, and has the geometric interpretation of a *simple k-vector*.

More generally, a *k-vector* $A_k = B_k + C_k + \ldots$ is the sum of simple k-vectors B_k, C_k, ... We are implicitly assuming that the addition and multiplication of elements in $G(V)$ satisfy all of the usual properties of the real numbers, except the commutative law of multiplication.[3]

[3] If V is endowed with a more general *nondegenerate metric* instead of an inner product, we refer to $Cl(V)$ as the *Clifford algebra* on V with respect to this metric.

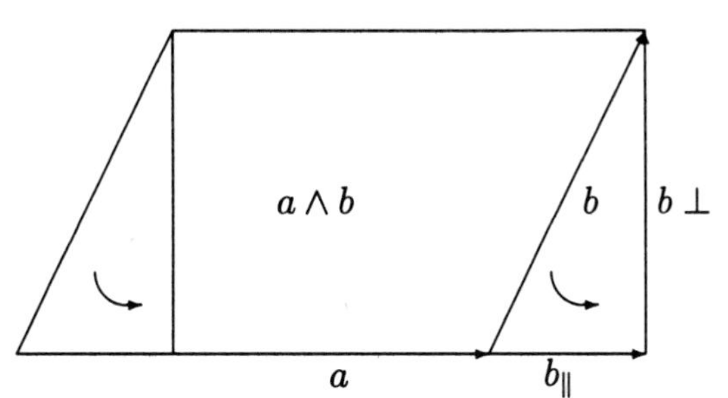

Fig. 3.1. The outer product $a \wedge b = ab_{\perp}$ and is the *directed plane segment* represented either by the parallelogram with sides a and b, or by the rectangle with sides a and $b_{\perp}$.

The geometric algebra $\mathbb{R}_n \equiv G(\mathbb{R}^n)$ is constructed by taking all sums of geometric products of vectors in the n-dimensional euclidean vector space $\mathbb{R}^n$. The 2^n-dimensional geometric algebra $\mathbb{R}_n$ is a *graded* linear space in that

$$\mathbb{R}_n = <\mathbb{R}_n>_0 + <\mathbb{R}_n>_1 + \ldots + <\mathbb{R}_n>_n, \tag{3.7}$$

where $\mathbb{R}_n^k \equiv <\mathbb{R}_n>_k$ denotes the $\binom{n}{k}$-dimensional subspace of all k-vectors of $\mathbb{R}_n$. The geometric algebra $\mathbb{R}_n$ is an *extension* of the real number system $\mathbb{R}$ in-so-far as that $\mathbb{R} \equiv \mathbb{R}_n^0 \subset \mathbb{R}_n$. Note also that the vectors in $\mathbb{R}^n$ are identical to the subspace of 1-vectors, *i.e.*, $\mathbb{R}^n \equiv \mathbb{R}_n^1$.

To find the interpretation of the geometric product ab for arbitrary vectors $a, b \in \mathbb{R}^n$, we first write $b = b_{\|} + b_{\perp}$ where $b_{\|}$ and $b_{\perp}$ are the vector components of b parallel and perpendicular to a, respectively. Using the distributive property and employing the two principles of geometric multiplication, we calculate

$$ab = a(b_{\|} + b_{\perp}) = ab_{\|} + ab_{\perp}, \tag{3.8}$$

where $ab_{\|} \equiv <ab>_0$ is the *scalar* or 0-vector part and $ab_{\perp} \equiv <ab>_2$ is the *bivector* or 2-vector part of the product ab, respectively, *Figure 1.*

There is an even more basic form of this last identity:

$$ab = a \cdot b + a \wedge b, \tag{3.9}$$

where $a \cdot b \equiv \frac{1}{2}(ab + ba) = ab_{\|}$ is the commutative inner product and $a \wedge b \equiv \frac{1}{2}(ab - ba) = ab_{\perp}$ is the anticommutative *outer product* of the vectors a and b.

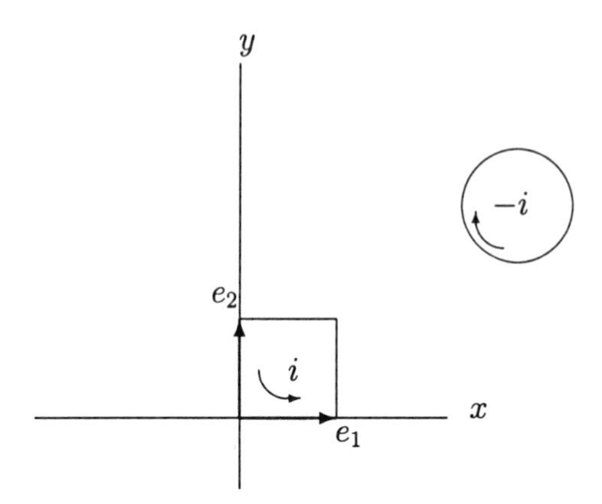

Fig. 3.2. The unit bivector $i = e_1e_2 = -e_2e_1$ in the xy-plane is a *directed plane segment* with *orientation* indicated by the circular arrow. The disk with unit area represents the same bivector, but has *opposite* orientation.

Letting i denote an oriented unit bivector in the plane of a and b so that $i^2 = -1$, we can express the last identity in the form

$$ab = a \cdot b + a \wedge b = |a||b|(\cos\theta + i\sin\theta) = |a||b|\exp i\theta, \tag{3.10}$$

called the *Euler formula* for the geometric multiplication of vectors, *Figure 2.* The Euler formula unites the inner and outer products of vectors into a single quantity, and shows that neither is more nor less fundamental than the other.

All other products are defined in terms of the basic geometric product. If $A_r \in \mathbb{R}_n^r$ and $B_s \in \mathbb{R}_n^s$ for $r, s \geq 1$, then the *inner product* of A_r and B_s is defined by

$$A_r \cdot B_s \equiv < A_rB_s >_{|r-s|}, \tag{3.11}$$

and the *outer product* by

$$A_r \wedge B_s \equiv < A_rB_s >_{r+s} . \tag{3.12}$$

The inner and outer products are linearly extended to arbitrary multivectors A and B in $\mathbb{R}_n$ with vanishing scalar parts.

Probably one of the biggest obstacle to the adoption of Clifford algebra methods in the general scientific community is the seemingly endless number of nontrivial algebraic identities that must be mastered. Many of these identities can be derived as simple consequences of the following basic identity for $a \in \mathbb{R}_n^1$ and $B_r \in \mathbb{R}_n^r$,

$$aB_r = a \cdot B_r + a \wedge B_r \tag{3.13}$$

for $r \geq 1$ and where

$$a \cdot B_r = \frac{1}{2}(aB_r + (-1)^{r+1}B_r a) = (-1)^{r+1}B_r \cdot a$$

and

$$a \wedge B_r = \frac{1}{2}(aB_r + (-1)^r B_r a) = (-1)^r B_r \wedge a.$$

We prove (3.13) in much the same way that we proved (3.9). Without loss of generality we can assume that B_r is a *simple* r-vector. Now write $a = a_\| + a_\perp$ where $a_\|$ is in the subspace of the simple r-vector B_r and $a_\perp$ is in the complementary subspace orthogonal to B_r. It then follows that

$$aB_r = (a_\| B_r + a_\perp B_r) = (-1)^{r+1}B_r a_\| + (-1)^r B_r a_\perp, \tag{3.14}$$

and

$$B_r a = B_r(a_\| + a_\perp) = B_r a_\| + B_r a_\perp. \tag{3.15}$$

Taking $(r-1)$ and $(r+1)$-vector parts of these identities gives

$$a \cdot B_r = a_\| B_r = (-1)^{r+1}B_r a_\|,$$

$$a \wedge B_r = a_\perp B_r = (-1)^r B_r a_\perp,$$

and also

$$B_r a = B_r \cdot a_\| + B_r \wedge a_\perp.$$

The proof is completed by taking the necessary sign combinations of these identities required to show (3.13).

As an application of these fundamental identities, we derive the important identity

$$a \cdot (b \wedge B_r) = (a \cdot b)B_r - b \wedge (a \cdot B_r). \tag{3.16}$$

We have

$$a \cdot (b \wedge B_r) = < a(b \wedge B_r) >_r = \frac{1}{2} < abB_r + (-1)^r aB_r b >_r$$

$$= (a \cdot b)B_r + \frac{1}{2} < -baB_r + (-1)^r aB_r b >_r$$

$$= (a \cdot b)B_r + \frac{1}{2} < -b(a \cdot B_r + a \wedge B_r) + (-1)^r (a \cdot B_r + a \wedge B_r)b >_r$$

$$= (a \cdot b)B_r - b \wedge (a \cdot B_r).$$

A complete discussion of these basic identities and others can be found in [6].

Chapter 4

Linear Transformations

Garret Sobczyk
Departamento de Fisico-Matematicas,
Universidad de las Americas,
72820 Cholula, Pue., México

Let V be a possibly infinite dimensional linear vector space over either the field of real or complex numbers. We shall first be interested in studying general properties of the endomorphism algebra $End_F(V)$ on V, where F is the field of real or complex numbers. The endomorphism algebra $End_F(V)$ exists independently of any further structure which may be assumed on V, such as a nondegenerate inner product induced by a quadratic form of arbitrary signature. Later we shall study the interelationship between the endomorphism algebra $End_F(V)$ and a metric structure $a \cdot b$ on V for vectors $a, b \in V$.

4.1 Structure of a Linear Operator

Let $End(\mathbb{C}^n)$ denote the endomorphism algebra of all complex linear operators on an n-dimensional complex vector space $\mathbb{C}^n$. We denote the identity operator in $End(\mathbb{C}^n)$ simply by 1. For a linear operator $a \in End(\mathbb{C}^n)$, let

$$\psi(\lambda) \equiv \prod_{i=1}^{r}(\lambda - \alpha_i)^{m_i}$$

be its *minimal polynomial*, having the distinct complex roots $\alpha_1, \ldots, \alpha_r \in \mathbb{C}$ with the respective *algebraic multiplicities* m_i. We order these roots so as to satisfy the inequalities $1 \leq m_1 \leq m_2 \leq \ldots \leq m_r$, where possibly the first $h \geq 0$ of them are equal to 1.

By the *structure equation* or *eigenprojector form* of the linear operator a with the minimal polynomial $\psi(\alpha)$, we mean

$$a = \sum_{i=1}^{r}(\alpha_i + q_i)p_i, \tag{4.1}$$

where the commutative operators p_i and q_j satisfy the properties

$$\{p_1 + \ldots + p_r = 1,\ p_i p_j = \delta_{ij} p_i,\ q_k^{m_k - 1} \neq 0 \text{ but } q_k^{m_k} = 0,\ p_k q_k = q_k\}. \tag{4.2}$$

The operators p_i make up a *partition of unity* and are *mutually annihilating idempotents.* The elements q_j are *nilpotents* with the respective *indexes* m_j. The nilpotents q_j are *projectively* related to the corresponding idempotents p_j and hence satisfy $p_j q_j = q_j$. We adopt the convention that $q_j \equiv 0$, for $j = 1, \ldots, h$. The commutative algebra generated by the p_i's and q_j's is just the *factor algebra* $\boldsymbol{C}\{m_1, m_2, \ldots, m_r\}$ (up to an isomorphism) of the principal ideal generated by the minimal polynomial $\psi(\lambda)$, [12].[1]

The *Vandermonde determinant* is efficiently defined in terms of the column vector function $f(\lambda) = (1, \lambda, \lambda^2, \ldots, \lambda^{s-1})^T$ evaluated at the points $\lambda = \lambda_1, \lambda_2, \ldots, \lambda_s \in \boldsymbol{C}$,

$$V \equiv \det\{f(\lambda_1), f(\lambda_2), \ldots, f(\lambda_s)\} = \prod_{1 \leq i < j \leq s} (\lambda_j - \lambda_i). \tag{4.3}$$

The *normalized derivatives* $f^{(k)}(\lambda)$ of the column vector function $f(\lambda)$ are defined by

$$f_\lambda^{(k)} \equiv f^{(k)}(\lambda) = \frac{1}{k!} D_\lambda^k f(\lambda).$$

For example,

$$f_\lambda^{(1)} = (0, 1, 2\lambda, 3\lambda^2, \ldots, (s-1)\lambda^{(s-2)})^T,$$

and

$$f_\lambda^{(2)} = (0, 0, 1, 3\lambda, \ldots, \frac{(s-1)(s-2)}{2}\lambda^{(s-3)})^T.$$

With the above notation in hand, the determinant of the *generalized* Vandermonde matrix[2]

$$W \equiv \{\underbrace{f_1 f_1^{(1)} \ldots f_1^{(m_1 - 1)}}_{m_1\ columns}\ \underbrace{f_2 f_2^{(1)} \ldots f_2^{(m_2 - 1)}}_{m_2\ columns}\ \ldots\ \underbrace{f_r f_r^{(1)} \ldots f_r^{(m_r - 1)}}_{m_r\ columns}\}, \tag{4.4}$$

where $s = m_1 + m_2 + \ldots + m_r$, becomes an elementary excercise of taking and evaluating the appropriate normalized partial derivatives of the equation (4.3), getting

$$\det W = \prod_{1 \leq i < j \leq r} (\lambda_j - \lambda_i)^{m_i m_j}. \tag{4.5}$$

We can now solve the structure equation (4.1) by a simple application of Cramer's rule to a system of operator equations [13]. Classically this structure theorem is known as the *generalized spectral decomposition* of a linear operator.

---

[1] Editor's note: The references for this part are collected at the end of c. 5.

[2] Editor's note: Here, f_j is shorthand for f_{λ_j} and λ_j is identified with the eigenvalue α_j.

Theorem 1 *If $\psi(\alpha)$ is the minimal polynomial of the operator a, then a can be expressed in the eigenprojector form*

$$a = \sum_{j=1}^{r} (\alpha_j + q_j) p_j,$$

where the idempotent and nilpotent operators p_i and q_i, satisfying (4.2), are the unique polynomials in a of degree $< s = m_1 + \ldots m_r$, specified by

$$p_i = \frac{\det\{\overbrace{f_1 \ldots f_1^{(m_1-1)}}^{m_1} \ldots \overbrace{f_a f_i^{(1)} f_i^{(2)} \ldots f_i^{(m_i-1)}}^{m_i} \ldots \overbrace{f_r \ldots f_r^{(m_r-1)}}^{m_r}\}}{\det W},$$

and

$$q_i = \frac{\det\{\overbrace{f_1 \ldots f_1^{(m_1-1)}}^{m_1} \ldots \overbrace{f_i f_a f_i^{(2)} \ldots f_i^{(m_i-1)}}^{m_i} \ldots \overbrace{f_r \ldots f_r^{(m_r-1)}}^{m_r}\}}{\det W},$$

where the $\det W$ *is given in (4.5).*

Proof. We begin by writing

$$a^0 \equiv 1 = \quad p_1 \quad + \quad p_2 \quad + \quad \ldots \quad + \quad p_r,$$

and

$$a = (\alpha_1 + q_1)p_1 \; + \; (\alpha_2 + q_2)p_2 \; + \; \ldots + \; (\alpha_r + q_r)p_r.$$

Using the assumed properties (4.2), we take successive powers of a to complete s powers of a:

$$a^2 = (\alpha_1 + q_1)^2 p_1 + (\alpha_2 + q_2)^2 p_2 + \ldots + (\alpha_r + q_r)^2 p_r$$

$$\vdots \qquad \vdots \qquad \vdots$$

$$a^{s-1} = (\alpha_1 + q_1)^{s-1} p_1 + (\alpha_2 + q_2)^{s-1} p_2 + \ldots + (\alpha_r + q_r)^{s-1} p_r$$

The powers $(\alpha_j + q_j)^k$ can be easily computed, getting

$$(\alpha_j + q_j)^k = \alpha_j^k + \binom{k}{1} \alpha_j^{k-1} q_j + \ldots + \binom{k}{m_j - 1} \alpha_j^{k-m_j+1} q_j^{m_j-1},$$

where $\binom{k}{j}$ are the usual binomial coefficients for $j \leq k$, and $\binom{k}{j} \equiv 0$ for $j > k$. When each of these expansions is substituted back into the successive powers of a, we are led to a system of operator equations which are linear in

$$\{\overbrace{p_1, q_1, q_1^2, \ldots q_1^{m_1-1}}^{m_1}, \ldots, \overbrace{p_r, q_r, q_r^2, \ldots, q_r^{m_r-1}}^{m_r}\}.$$

Just as for a system of linear equations, the system of *operator equations* is consistent and has a unique solution if the determinant $det(W)$ of the coefficient matrix W of the unknown operators is nonvanishing. But W is just the Vandermonde matrix (4.4) given by

$$\begin{pmatrix} 1 & 0 & \ldots & 1 & \ldots & 1 & 0 & \ldots \\ \alpha_1 & 1 & \ldots & \alpha_j & \ldots & \alpha_r & 1 & \ldots \\ \vdots & & & & \vdots & & & \\ \alpha_1^{s-1} & (s-1)\alpha_1^{s-2} & \ldots & \alpha_j^{s-1} & \ldots & \alpha_r^{s-1} & (s-1)\alpha_r^{s-2} & \ldots \end{pmatrix}$$

The determinant (4.5) of this matrix,

$$det(W) = \prod_{i<j} (\alpha_j - \alpha_i)^{m_i m_j} \neq 0,$$

since the roots α_i for $i = 1, \ldots, r$ are distinct. Applying Cramer's rule to this system of linear operator equations, gives the operator unknowns p_i, q_j in terms of polynomials in a as specified in the statement of the theorem.

Q.E.D.

The novel proof of this basic structure theorem given above is apparently missing in the literature. It has many consequences at a basic level, making it necessary to re-examine the proofs of many theorems in linear and multilinear algebra [5], [13], and [14].

4.2 Isometries

The power of geometric algebra as a tool in linear algebra is amply illustrated by the study of skewsymmtric transformations and their corresponding isometries. We shall lay down the basic ideas of this theory, following closely the seminal work of Marcel Riesz [8].

Let V be an n-dimensional vector space provided with a nondegenerate metric $x \cdot y$. We assume we have constructed a pair of *reciprocal bases* $\{e_i\}$ and $\{e^j\}$ satisfying the defining properties that $e_i \cdot e^j = \delta_i^j$. Let $f \in End(V)$ be a linear endomorphism on V.

Definition 1 *The endomorphism f is said to be an isometry of the metric $x \cdot y$ if $f(x) \cdot f(y) = x \cdot y$ for all $x, y \in V$.*

Closely connected to the idea of an isometry is the idea of a *skewsymmetric* transformation:

Definition 2 *An operator $h(x)$ is skewsymmetric on V relative to the metric $x \cdot y$ if*

$$h(x) \cdot y = x \cdot h^{\dagger}(y) = -x \cdot h(y)$$

for all $x, y \in V$.

It immediately follows from Definition 2 that $h(x)$ is skewsymmetric if and only if its *adjoint transformation* $h^{\dagger}(x) = -h(x)$ for all $x \in V$. A skewsymmetric transformation h has the important property that $h^k(x) \cdot y = x \cdot (-h)^k(y)$, from which it follows by a simple argument that $\exp(h)x \cdot y = x \cdot \exp(-h)y$. If we now replace y by $\exp(h)y$, we easily find that

$$\exp(h)x \cdot \exp(h)y = x \cdot y,$$

which establishes that for any skewsymmetric h, $\exp(h)$ is an isometry. If t is a continuous real (or complex) parameter, then we say that $g_t = \exp(th)$ is a one-parameter Abelian group of isometries continuously connected to the identity 1 ($t = 0$). Thus, to every skewsymmetric transformation there belongs a one-parameter group of isometries, and conversely.

Bivectors in Clifford algebra arise in the study of isometries of a given metric because of the one-to-one relationship between bivectors and skewsymmetric transformations: Each skewsymmetric transformation $h(x)$ can be written in the form $h(x) = x \cdot F$, where the bivector F is defined in terms of the vector derivative by $F \equiv \frac{1}{2}\partial_v \wedge h(v)$.[3]

We now give the theorem that specifies the exact relationship between isometries, skewsymmetric transformations and bivectors [8].

Theorem 2 *Let $g_t x = \exp(th)x$ be the one parameter group generated by the skewsymmetric transformation $h(x) = x \cdot F$, then $g_t x = \exp(th)x = \exp(-\frac{1}{2}tF)x\exp(\frac{1}{2}tF)$.*

Proof: Take and evaluate successive derivatives with respect to t at $t = 0$ of both sides of

$$\exp(th)x = \exp(-\frac{1}{2}tF)x\exp(\frac{1}{2}tF),$$

and check the equality of the resulting expressions.

Q.E.D.

[3] The vector derivative ∂_x is defined by

$$\partial_x \equiv \sum_{i=1}^{n} e^i \frac{\partial}{\partial x^i},$$

where for $x = \sum x^i e_i$, $\partial x / \partial x^i = e_i$.

4.3 Minimal Polynomials

The minimal polynomial of a skewsymmetric operator is an important tool for studying its properties.We consider here the minimal polynomials of skewsymmetric operators in the Clifford algebras $Cl_{1,3}$ and $Cl_{2,2}$, although the ideas can be generalized to skewsymmetric operators in Clifford algebras of arbitrary signatures.

Consider first the skewsymmetric operator $f(x) = x \cdot (e_0 \wedge e_1 + e_2 \wedge e_3)$. With respect to the orthonormal basis e_0, e_1, e_2, e_3, we calculate $f(e_0) = e_1$, $f(e_1) = e_0$, $f(e_2) = -e_3$, and $f(e_3) = e_2$, from which it follows that the matrix of f with respect to this basis is

$$\begin{pmatrix} 0 & 1 & 0 & 0 \\ 1 & 0 & 0 & 0 \\ 0 & 0 & 0 & 1 \\ 0 & 0 & -1 & 0 \end{pmatrix}.$$

The minimal polynomial of the subspace determined by the bivector $e_0 \wedge e_1$ is clearly $(\lambda - 1)(\lambda + 1)$, and that of the subspace of the bivector $e_2 \wedge e_3$ is $\lambda^2 + 1$. It follows that the minimal polynomial of f is $(\lambda^2 - 1)(\lambda^2 + 1)$. The bivector $F = e_0 \wedge e_1 + e_2 \wedge e_3$ is the sum of two commuting orthogonal blades $e_0 \wedge e_1$ and $e_1 \wedge e_2$; the blade $e_0 \wedge e_1$ determines the plane of an *hyperbolic rotation*, and the blade $e_2 \wedge e_3$ determines the plane of an *elliptic rotation*. Conversely, if the minimal polynomial of a skewsymmetric transformation in $Cl_{1,3}$ is of the form

$$\psi(\lambda) = (\lambda^2 + \alpha^2)(\lambda^2 - \beta^2)$$

where $\alpha > 0$ and $\beta > 0$, it follows as a consequence of Theorem 1 that F is the sum of two commuting orthogonal blades which determine hyperbolic and elliptic rotations.

Consider now the skewsymmetric operator $f(x) = x \cdot (n \wedge e_2)$ determined by the *parabolic* null bivector $n \wedge e_2$, where $n \equiv e_0 + e_3$ and $\overline{n} \equiv \frac{1}{2}(e_0 - e_3)$ are null vectors. Calculating $f(e_1) = 0 = f(n)$, $f(e_2) = n$, and $f(\overline{n}) = e_2$, we find that the matrix of f with respect to the basis $e_1, n, e_2, \overline{n}$ is

$$\begin{pmatrix} 0 & 0 & 0 & 0 \\ 0 & 0 & 1 & 0 \\ 0 & 0 & 0 & 1 \\ 0 & 0 & 0 & 0 \end{pmatrix},$$

so the minimal polynomial of f is λ^3. It follows that a parabolic bivector takes up 3 dimensions, so the corresponding skewsymmetric transformation can have but one invariant plane.

The difference in the Euclidean and Lorentz cases is due to the possibility of a factor λ^3 in the minimal polynomial. A factor of λ^3 in the minimal polynomial reveals the presence of a *parabolic rotation* in the plane of a *null bivector*, but the decomposition of an arbitrary bivector F into a sum of commuting orthogonal simple bivectors still remains valid for the reasons given above.

The decomposition of an arbitrary bivector F into a sum of commuting orthogonal simple bivectors fails in the case of the *ultrahyperbolic* metric with the signature $+ + - -$, as the following example shows. Take

$$n_1 = \frac{e_1 + f_1}{\sqrt{2}}, \; n_2 = \frac{e_2 + f_2}{\sqrt{2}}, \; n_3 = \frac{e_2 - f_2}{\sqrt{2}}, \; n_4 = \frac{e_1 - f_1}{\sqrt{2}}.$$

Then for $x = \sum x^j n_j$ and $y = \sum y^j n_j$, we have $x \cdot y = x^1y^4 + x^2y^3 + x^3y^2 + x^4y^1$.

Define the skewsymmetric transformation $h(x) = x \cdot F$ for

$$F \equiv n_4 \wedge (n_1 + n_2) + n_3 \wedge n_2 = e_1 f_1 + e_2 f_2 + n_4 n_2,$$

so that $h(n_1) = n_1 + n_2$, $h(n_2) = n_2$, $h(n_3) = -n_3 - n_4$, and $h(n_4) = -n_4$. The vectors n_2 and n_4 are eigenvectors with the respective minimal polynomials $(\lambda - 1)$ and $(\lambda + 1)$. The respective minimal polynomials of n_1 and n_3 are $(\lambda - 1)^2$ and $(\lambda + 1)^2$. It follows that the minimal polynomial of h is $\psi(\lambda) = (\lambda - 1)^2(\lambda + 1)^2$. The bivectors n_1n_2, n_3n_4, and n_2n_4 are eigenbivectors with the corresponding minimal polynomials $(\lambda - 1)^2$, $(\lambda + 1)^2$, and $(\lambda^2 - 1)$. Any other bivectors have minimal polynomials of degree ≥ 3 and hence are not invariants of h. But no two of these bivectors are orthonormal, so a orthogonal decomposition of h is impossible, [8, p. 171].

4.4 Lie Algebras of Bivectors

We will study here skewsymmetric mappings of the neutral Clifford algebra $Cl_{n,n}$. Following the notation in [3], we choose as our basis $e_i, \bar{e}_j$, where $e_i \cdot e_j = \delta_{ij} = -\bar{e}_i \cdot \bar{e}_j$, and $e_i \cdot \bar{e}_j = 0$ for $i, j = 1, 2, \ldots n$. We will be particularly interested in the subalgebras $N = gen\{n_1, n_2, \ldots, n_n\}$ and $\overline{N} = gen\{\overline{n}_1, \ldots, \overline{n}_n\}$ generated by the null vectors $n_i = \frac{1}{\sqrt{2}}(e_i + \overline{e_i})$ and $\overline{n}_i = \frac{1}{\sqrt{2}}(e_i - \bar{e}_i)$.

It is well known that the bivectors in $Cl^2_{n,n}$ form the Lie algebra $so_{n,n}$ of the Lie group $SO_{n,n}$ with the Lie (bracket) product

$$b_1 \otimes b_2 \equiv \frac{1}{2}[b_1, b_2] = \frac{1}{2}(b_1 b_2 - b_2 b_1)$$

for $b_1, b_2 \in Cl^2_{n,n}$. Consider now the Lie subalgebra gl_n of $Cl^2_{n,n}$ generated by the n^2 bivectors

$$gl_n = \; gen\{b_{ij} = n_i \wedge \overline{n_j}\} \tag{4.6}$$

for $i, j = 1, 2, \ldots n$. The structure constants of this Lie algebra are easily checked to be

$$b_{ij} \otimes b_{kl} = \delta_{il} b_{kj} - \delta_{jk} b_{il}.$$

We shall show that gl_n is the Lie algebra of bivectors for the general linear group GL_n on N.

Let us write a general element $B \in gl_n$ in the form $B = \sum_{ij} \beta_{ij} b_{ij}$. The skew-symmetric mapping $B : N \longrightarrow N$ of this bivector, expressed in terms of the basis $\{n\} \equiv \{n_1, n_2, \dots n_n\}$, is

$$B(x) = B \cdot x = B \cdot \{n\} x_{\{n\}} = \{B \cdot n\} x_{\{n\}} = \{n\} B_{\{n\}} x_{\{n\}} \tag{4.7}$$

where the vector $x = \{n\} x_{\{n\}} \in N$ has the *column* scalar components $x_{\{n\}} = (x_1, \quad x_2, \quad \dots \quad , x_n)^\dagger$, and $B_{\{n\}} \equiv (\beta_{ij})$ is the $n \times n$ *matrix* of the transformation $B \cdot x$ with respect to the basis $\{n\}$.

For $x \in N$ and $A, B \in gl_n$ the *Jacobi identity* takes the form

$$(A \otimes B) \otimes x = (A \otimes x) \otimes B + A \otimes (B \otimes x).$$

If $A_{\{n\}}$ and $B_{\{n\}}$ are the matrices of the skewsymmetric mappings $A(x) = A \cdot x$ and $B(x) = B \cdot x$, *i.e.*,

$$A \cdot x = \{n\} A_{\{n\}} x_{\{n\}} \quad \text{and} \quad B \cdot x = \{n\} B_{\{n\}} x_{\{n\}},$$

then we calculate

$$B \cdot (A \cdot x) = B \cdot (\{n\} A_{\{n\}} x_{\{n\}}) = \{n\} B_{\{n\}} A_{\{n\}} x_{\{n\}}.$$

Applying the Jacobi identity, we can then easily show that

$$\{n\}[A_{\{n\}}, B_{\{n\}}] x_{\{n\}} = (A \cdot x) \cdot B - (B \cdot x) \cdot A = \{(A \otimes B) \cdot x\}. \tag{4.8}$$

But the $n \times n$ matrices under the Lie bracket product make up precisely the classical Lie algebra of the general linear group, [4].

One final observation: We can *project* the Lie algebra of bivectors gl_n onto the Lie subalgebra $so_{p,q}$, for $p + q = n$, of the *special orthogonal group* $SO_{p,q}$ by using the projection operator $P_{p,q} : N \longrightarrow Cl_{p,q}$ defined by

$$P_{p,q} X = E^{-1}(E \cdot X) \tag{4.9}$$

where $E \equiv e_1 \wedge \cdots \wedge e_p \wedge \overline{e}_{p+1} \wedge \ldots \wedge \overline{e}_n$,
for $X \in Cl_{p,q}$.

Chapter 5

Directed Integration

Garret Sobczyk
Departamento de Fisico-Matematicas,
Universidad de las Americas,
72820 Cholula, Pue., México

The purpose of this chapter is to introduce the reader to the basic ideas of *simplicial calculus*. For reasons of clarity, the proofs offered here are for the lower dimensions, their generalization to higher dimensions being self-evident. A discussion of some of the basic ideas of Riemannian geometry is included. The reader may wish to refer to [15] and [6] for more details and more general proofs.

5.1 Simplices and Chains

Let $\{a_0, a_1, \ldots, a_k\}$ be an ordered set of points in $\mathbb{R}^n$.

An *oriented k-simplex* is determined by an ordered set of $k+1$ vertices

$$\{a\}_{(k)} \equiv \{a_0, a_1, \ldots, a_k\} = \{a|\ a = \sum_{\mu=0}^{k} t_\mu a_\mu\}, \tag{5.1}$$

where $\sum_{\mu=0}^{n} t_\mu = 1$, and $0 \le t_\mu \le 1$, are the *barycentric coordinates* of a. Interchanging two vertices of $\{a\}_{(k)}$ changes its orientation:

$$\{a_0, \ldots, a_i, \ldots, a_j, \ldots a_k\} = -\{a_0, \ldots, a_j, \ldots, a_i, \ldots a_k\}. \tag{5.2}$$

The *Boundary* of $\{a\}_{(k)}$ is the *chain* of $(k-1)$-simplices

$$\partial\{a\}_{(k)} = \sum_{i=0}^{k} (-1)^i \{\check{a}_i\}_{(k-1)} \tag{5.3}$$

where $\{\check{a}_i\}_{(k-1)}$ is the $(k-1)$-simplex

$$\{\check{a}_i\}_{(k-1)} = \{a_0, a_1, \ldots \check{a}_i \ldots, a_k\},$$

formed by deleting the marked vertex $\breve{a}_i$. For the k-simplices, for $k = 0, 1, 2, 3$, we have

$$\partial\{a\}_{(0)} = \{\},$$

$$\partial\{a\}_{(1)} = \{a_1\} - \{a_0\},$$

$$\partial\{a\}_{(2)} = \{a_1, a_2\} - \{a_0, a_2\} + \{a_0, a_1\},$$

and

$$\partial\{a\}_{(3)} = \{a_1, a_2, a_3\} - \{a_0, a_2, a_3\} + \{a_0, a_1, a_3\} - \{a_0, a_1, a_2\}.$$

A basic result of *homology theory* is

$$\partial^2\{a\}_{(k)} = \{\}, \tag{5.4}$$

i.e, the boundary of the boundary of a k-simplex is the empty set. For example

$$\partial^2\{a_0\} = \partial\{\} = \{\},$$

$$\partial^2\{a_0, a_1\} = \partial\{a_1\} - \partial\{a_0\} = \{\} - \{\} = \{\},$$

$$\partial^2\{a_0, a_1, a_3\} = \partial\{a_1, a_2\} - \partial\{a_0, a_2\} + \partial\{a_0, a_1\}$$

$$= \{a_2\} - \{a_1\} - \{a_2\} + \{a_0\} + \{a_1\} - \{a_0\} = \{\}.$$

The *directed content* of the k-simplex $\{a\}_{(k)}$, is the k-vector

$$a_{(k)} \equiv \frac{1}{k!}(a_1 - a_0) \wedge (a_2 - a_1) \wedge \ldots \wedge (a_k - a_{k-1}), \tag{5.5}$$

for $k \geq 1$. We also write $Dir\{a\}_{(k)} = a_{(k)}$. For $k = 0$ and $k = 1$, we define

$$Dir\{a_0\} \equiv 1, \quad Dir\{\} \equiv 0.$$

The directed content of $\{a\}_{(k)}$ can also be expressed by

$$a_{(k)} = \frac{1}{k!}(a_1 - a_0) \wedge (a_2 - a_0) \wedge \ldots \wedge (a_k - a_0). \tag{5.6}$$

For example,

$$a_{(3)} = \frac{1}{3!}(a_1 - a_0) \wedge (a_2 - a_1) \wedge (a_3 - a_2)$$

$$= \frac{1}{3!}(a_1 - a_0) \wedge (a_2 - a_0 + a_0 - a_1) \wedge (a_3 - a_2)$$

$$= \frac{1}{3!}(a_1 - a_0) \wedge (a_2 - a_0) \wedge (a_3 - a_0 + a_0 - a_2)$$

$$= \frac{1}{3!}(a_1 - a_0) \wedge (a_2 - a_0) \wedge (a_3 - a_0).$$

The *complementary k-simplex* $\{a\}^{(k)}$ of the simplex $\{a\}_{(k)}$ at the point a_0 is defined by

$$\{a\}^{(k)} = \{a_0, a^1, \ldots, a^k\} \tag{5.7}$$

where the vertices $a^1, a^2, \ldots, a^k$ of the k-simplex $\{a\}^{(k)}$ are chosen to satisfy the conditions that $(a^i - a_0) \wedge a_{(k)} = 0$ and $(a_i - a_0) \cdot (a^j - a_0) = \delta_{ij}$ for $i, j = 1, 2, \ldots, k$. In other words, the sets of edge vectors $\{a_i - a_0\}$ and $\{a^j - a_0\}$ make up a pair of *reciprocal frames* with respect to the k-vector $a_{(k)}$. It is natural to define

$$a^{(k)} \equiv (k!)^2 Dir(\{a\}^{(k)}) = k!(a^k - a_0) \wedge \ldots \wedge (a^1 - a_0)$$

so that $a^{(k)} a_{(k)} = 1$. Thus, the k-vector $a^{(k)}$ equals $(k!)^2$ times the directed content of the complementary simplex $\{a\}^{(k)}$.

Theorem 3 *The directed content of the boundary of a k-simplex is zero, i.e.,*

$$Dir\{\partial\{a\}_{(k)}\} = 0. \tag{5.8}$$

Proof: We give the proof here for $k = 0, 1, 2, 3$.
For $k = 0$, we have $Dir(\partial\{a\}) = Dir\{\} = 0$.
For $k = 1$, we have

$$Dir[\partial\{a_0, a_1\}] = Dir\{a_1\} - Dir\{a_0\} = 0.$$

For $k = 2$, we have

$$Dir[\partial\{a_0, a_1, a_2\}] = Dir\{a_1, a_2\} - Dir\{a_0, a_2\} + Dir\{a_0, a_1\}$$

$$= a_2 - a_1 - a_2 + a_0 + a_1 - a_0 = 0.$$

For $k = 3$, we have

$$\partial\{a\}_{(3)} = \{a_1, a_2, a_3\} - \{a_0, a_2, a_3\} + \{a_0, a_1, a_3\} - \{a_0, a_1, a_2\},$$

so that

$$Dir[\partial\{a\}_{(3)}] = \frac{1}{2}\{(a_2 - a_1) \wedge (a_3 - a_1) - (a_2 - a_0) \wedge (a_3 - a_0)$$

$$+(a_1 - a_0) \wedge (a_3 - a_0) - (a_1 - a_0) \wedge (a_2 - a_0)\}$$

$$= \frac{1}{2}\{(a_3 - a_0 - a_1 + a_0) \wedge (a_2 - a_0) + (a_2 - a_1) \wedge (a_3 - a_1) + (a_1 - a_0) \wedge (a_3 - a_0)\}$$

$$= \frac{1}{2}\{(a_3 - a_1) \wedge (a_2 - a_0 - a_2 + a_1) + (a_1 - a_0) \wedge (a_3 - a_0)\}$$

$$= \frac{1}{2}\{(a_3 - a_1) \wedge (a_1 - a_0) + (a_1 - a_0) \wedge (a_3 - a_0)\} = 0.$$

Q.E.D.

The *directed integral* of a geometric-valued function $F(x)$ on an oriented k-surface S is defined by

$$\int_S d^k x F = \lim_{N\to\infty} \sum_{j=1}^{N} \Delta^k x_j F(x_j), \tag{5.9}$$

where $\Delta^k x_j$ is the directed k-vector contents of the k-simplex $\{\Delta^k x_j\}$ of S at x_j, and as $N \to \infty$ each of the k-simplices $\{\Delta^k x_j\}$ shrink to a single point on S in an appropiate manner.[1] Each partition divides up the surface S into a set of approximating k-simplices,

$$S \simeq \sum_{j=1}^{N} \{\Delta^k x_j\}$$

Example 1 *Let $S = \sum_{i=1}^{N} \{a_i\}_{(0)}$. Then, since $Dir(\{a_i\}) = 1$ for each i, it follows that*

$$\int_S d^0 x = \sum_i Dir(\{a_i\}) = N.$$

The following example shows that the the concept of a directed integral is compatible with the concept of the directed content of a simplex.

Example 2 $\int_{\{a\}_{(k)}} d^k x = Dir(\{a\}_{(k)}) = a_{(k)} =$

$$= \frac{1}{k!}(a_1 - a_0) \wedge (a_2 - a_1) \wedge \ldots \wedge (a_k - a_{k-1}).$$

[1] More specifically, we require that the ratios of the radii of the circumscribing and inscribing k-spheres containing each of the k-simplices $\{\Delta^k x_j\}$ are bounded as $N \to \infty$.

Example 3 $\int_{\{a\}_{(1)}} dx\ x \cdot b = \frac{1}{2}(a_1 - a_0)(a_0 + a_1) \cdot b$.

Proof. Let the line $x(t)$ joining the points a_0 and a_1 be parameterized by

$$x(t) = a_0 + t(a_1 - a_0), \quad \text{so that} \quad dx = dt(a_1 - a_0),$$

where $0 \le t \le 1$. We find

$$\int_{\{a\}_{(1)}} dx\ x \cdot b = \int_0^1 dt(a_1 - a_0)[a_0 + t(a_1 - a_0)] \cdot b$$

$$= (a_1 - a_0) \int_0^1 [a_0 \cdot b + t(a_1 - a_0) \cdot b] dt$$

$$= (a_1 - a_0) a_0 \cdot b + 1/2 (a_1 - a_0)(a_1 - a_0) \cdot b$$

$$= \frac{1}{2}(a_1 - a_0)(a_0 + a_1) \cdot b.$$

Q.E.D.

Example 4 $\int_{\partial\{a\}_{(1)}} d^0x\ x \cdot b = (a_1 - a_0) \cdot b$.

Proof. Recalling that $Dir\{a\} = 1$, we find that

$$\int_{\partial\{a_0, a_1\}} d^0x\ \ x \cdot b = \int_{\{a_1\} - \{a_0\}} d^0x\ x \cdot b$$

$$= \int_{\{a_1\}} d^0x\ x \cdot b - \int_{\{a_0\}} d^0x\ x \cdot b = (a_1 - a_0) \cdot b.$$

Q.E.D.

Generalizing example 3 leads to

$$\int_{\{a\}_{(k)}} d^kx\ x \cdot b = \frac{1}{k+1} a_{(k)}(a_0 + \ldots + a_k). \tag{5.10}$$

Similarly, the generalization of example 4 is

$$\int_{\partial\{a\}_{(k)}} d^{k-1}x\ x \cdot b = (-1)^{k+1} a_{(k)} \cdot b. \tag{5.11}$$

5.2 Integral Definition of $\partial_x F(x)$ on S

Let S be a k-surface, x_0 an interior point of S, and $\{x\}_{(k)}$ a k-simplex in S at x_0. Even though $\{x\}_{(k)}$ may be small and a good approximation to S at the point x_0, we are only guaranteed that the vertices of $\{x\}_{(k)}$ will be in S (since S may have curvature at x_0). This poses problems in regard to multivector-valued functions $F(x)$ defined on S; we would like such functions to be defined locally on simplices which approximate S at the point $x_0 \in S$. This leads us to define the *affine approximation* to $F(x)$ on the simplex $\{x\}_{(k)}$ in S at x_0.

Definition 3 *By the affine approximation to $F(x)$ on the simplex $\{x\}_{(k)}$ in S at x_0, we mean the function*

$$f(x) = F(x_0) + \sum_{i=1}^{k} (x - x_0) \cdot (x^i - x_0)(F(x_i) - F(x_0)),$$

where $x^i - x_0$ are edge vectors of the complementary simplex $\{x\}^{(k)}$ defined earlier, and $x \in \{x\}_{(k)}$.

From definition 3, it is seen that $f(x)$ agrees with $F(x)$ on the vertices of $\{x\}_{(k)}$, and is the affine approximation to $F(x)$ for points x in the interior of $\{x\}_{(k)}$.

Evidently, there is a close connection between the affine approximation to $F(x)$ at x_0 and *Taylor's theorem* at x_0. We give Taylor's theorem stated as a Lemma below for a function $F(x)$ differentiable in an open set U containing the point x_0 in the *flat* euclidean space $\mathbb{R}^k$.

Lemma 1 *If $F(x)$ is a continuous and differentiable geometric-valued function at the point $x = x_0 \in \mathbb{R}^k$, then*

$$F(x) = F(x_0) + (x - x_0) \cdot \nabla_{x_0} F(x_0) + |x - x_0| E(x, x - x_0),$$

where $E(x, y)$ is continuous at $x = x_0$ and linear in y, $(x - x_0) \cdot \nabla_{x_0}$ is the ordinary directional derivative in the direction of the vector $x - x_0$, and ∇_{x_0} is the gradient of $\mathbb{R}^k$.

The advantage of definition 1 over lemma 1 is that it is valid for any k-surface embedded in a higher dimensional linear space, whereas lemma 1 makes sense only in the euclidean space $\mathbb{R}^k$. Using definition 1, we now define the *vector derivative* $\partial_x F(x)$ at the point $x = x_0 \in S$.

Definition 4 *For $k \geq 1$,*

$$\partial_{x_0} F(x_0) = (-1)^{k+1} \lim_{\Delta^k x_0 \to 0} \frac{1}{\Delta^k x_0} \int_{\partial\{\Delta^k x_0\}} d^{k-1}x \; f(x),$$

where $f(x)$ is the affine approximation to $F(x)$ at $x = x_0$.

Definition 4 is the natural generalization of the ordinary one dimensional derivative at a point on the x-axis, to the vector derivative at a point on a k-surface. The following theorem shows that the nabla operator ∇_x is equivalent to the vector derivative ∂_x in $\mathbb{R}^k$.

Theorem 4 *If $F(x)$ is differentiable at $x = x_0 \in \mathbb{R}^k$, then $\nabla_x F(x) = \partial_x F(x)$.*

Proof: We will only outline the proof, freely using the fact that in $f(x) \simeq F(x)$ in small neighborhoods of the point x_0. By Taylor's Theorem in $\mathbb{R}^k$,

$$F(x) = F(x_0) + (x - x_0) \cdot \nabla_{x_0} F(x_0) + |x - x_0| E(x, x - x_0). \tag{5.12}$$

Using this, with the help of (5.11), we calculate

$$(-1)^{k+1} \partial_{x_0} F(x_0) = \lim_{\Delta^k x_0 \to 0} \frac{1}{\Delta^k x_0} \int_{\partial \Delta^k x_0} d^{k-1} x f(x)$$

$$= \lim_{\Delta^k x_0 \to 0} \frac{1}{\Delta^k x_0} \int_{\partial \Delta^k x_0} d^{k-1} x [F(x_0) + (x - x_0) \cdot \nabla_{x_0} F(x_0) + |x - x_0| E(x, x - x_0)]$$

$$= (-1)^{k+1} \lim_{\Delta^k x_0 \to 0} \frac{1}{\Delta^k x_0} \Delta^k x_0 \nabla_{x_0} F(x_0) = (-1)^{k+1} \nabla_{x_0} F(x_0),$$

from which the theorem follows.

Q.E.D.

Theorem 5 *BOUNDARY THEOREM*

$$\int_S G(x) d^k x \partial F(x) = (-1)^{k+1} \int_{\partial S} G(x) d^{k-1} x F(x)$$

PROOF:

$$\int_S G d^k x \partial F = \lim_{N \to \infty} \sum_{i=1}^{N} G(x_i) \Delta^k x_i \partial_{x_i} F(x_i)$$

$$= \lim_{N \to \infty} \sum_{i=1}^{N} \int_{\partial \Delta^k x_i} g(x) \Delta^k x_i \frac{(-1)^{k+1}}{\Delta^k x_i} f(x)$$

$$= (-1)^{k+1} \int_{\partial S} G(x) d^{k-1} x F(x).$$

Q.E.D.

5.3 Classical Integration Theorems

The *boundary theorem* for $k = 1, 2, 3$.

For $k = 1$,

$$\int_a^b dx\partial F(x) = \int_{\{b\}-\{a\}} d^0 x f(x) = F(b) - F(a).$$

The left hand side gives

$$\int_a^b dx\partial F(x) = \int_a^b dx \cdot \partial F(x) = \int_a^b dF(x).$$

This is the Fundamental Theorem of Calculus when a and b are points on the x-axis.

For $k = 2$,

$$\int_{S_2} d^2x\partial F(x) = -\int_{\partial S_2} dx F(x).$$

If $F(x)$ is a vector field, we get *Stokes' Theorem*,

$$\int_{S_2} d^2x\partial \wedge F(x) = -\int_{\partial S_2} dx \cdot F(x),$$

or for $i = e_1e_2e_3$,

$$\int_{S_2} d^2xi\partial \times F(x) = -\int_{\partial S_2} dx \cdot F(x),$$

which reduces to

$$\int_{S_2} |d^2x| n \cdot (\nabla \times F(x)) = \int_{\partial S_2} dx \cdot F(x),$$

for the right hand normal vector n.

For $S_3 \subset I\!R^3$, we have $\partial_x = \nabla_x$, and

$$\int_{S_3} d^3x\nabla F(x) = -\int_{\partial S_3} d^2x F(x). \tag{5.13}$$

Multiplying both sides by $-i = -e_1e_2e_3$ gives

$$\int_{S_3} |d^3x|\nabla F(x) = -\int_{\partial S_3} -id^2xF(x) = \int_{\partial S_3} |d^2x| n F(x).$$

Let $F(x)$ be a vector field. Taking scalar parts of (5.13) gives the *Divergence Theorem:*

$$\int_{S_3} |d^3x|\nabla \cdot F(x) = \int_{\partial S_3} |d^2x| n \cdot F(x).$$

5.4 Residue Theorem

Let S be a smooth compact *flat* m-surface lying in E^m. Let

$$g(x,x') = \frac{1}{\Omega}\frac{x-x'}{|x-x'|^m}$$

be the *Green's function* for E^m, where $\Omega = \frac{2\pi^{m/2}}{\Gamma(m/2)}$. The Green's function satisfies

- $\nabla g = -g\nabla'^{\dagger} = \delta(x-x')$, where δ is the famous *delta function distribution.*
- $\int_S |dx'^m|\delta(x-x')F(x') = F(x)$, for a continuous function $F(x)$ at $x \in S$.

Applying the boundary theorem, we get

$$F(x) = -\frac{(-1)^m}{I(x)}\left[\int_S g(x,x')d^m x'\nabla' F(x')\right.$$

$$\left.-\int_{\partial S} g(x,x')d^{m-1}x'F(x')\right]$$

for $x \in S$, and where $I(x)$ is the unit pseudoscalar element of S at x.

Generalized Cauchy's Theorem: If $\nabla F(x) = 0$ in S, then

$$F(x) = \frac{(-1)^m}{I(x)}\int_{\partial S} g(x,x')d^{m-1}x'F(x'),$$

showing that an *analytic* function is determined by its values on the boundary ∂S of S.

Clifford analysis has been highly developed along the lines of functional analysis in [2].

5.5 Riemannian Geometry

Let S be an *abstract Riemannian k-manifold,* a point $x \in S$ coordinatized by $x = x(x^1, x^2, \ldots, x^k)$. Since there are problems with the vector manifold approach as developed in [6] regarding embedding theorems, we wish to give here definitions which do not require an embedding but in which the power of geometric algebra can still be effectively utilized in the study of manifolds.

We wish the gradient operator ∇_x at the point $x \in \mathbb{R}^k$, to become the *intrinsic vector derivative* when generalized to the abstract k-manifold S; for (embedded) vector manifolds we use the vector derivative ∂_x discussed earlier. To accomplish this, we assume that a Riemannian k-manifold comes equipped with a *Riemannian vector derivative* or *Riemannian vector connection* ∇_x characterized by the following three properties:

- $\nabla_x x^i = e^i(x)$ for $i = 1, 2, \ldots, k$ where the *vectors* $e^i(x)$ generate the *tangent geometric algebras* G_x at each point $x \in S$.

  The reciprocal basis $\{e_i\}$ to $\{e^i\}$ define the *tangent coordinate vectors* to the respective coordinate curves at the point $x = x(x^1, x^2, \ldots, x^k)$.

- If $a \in G_x$ is a vector, then $a \cdot \nabla_x\, x = a$.

  By a *constant vector* $a = \sum a^i e_i$ we mean a vector whose *components* $a^1, a^2, \ldots, a^k \in I\!R$ are not a function of $x \in S$, *i.e.*, for each i, $\nabla_x a_i = 0$.

- For constant vectors a and b,

$$a \cdot \nabla_x b = \Gamma_x(a, b),$$

  where $\Gamma_x(a, b)$ is a bilinear vector-valued function in G_x, called the *Christoffel function* at x.

By the *tangent bundle* to the manifold S, we mean

$$G = \cup_{x \in S}\{G_x\}, \tag{5.14}$$

and we call $\nabla = \cup_{x \in S}\{\nabla_x\}$ the *intrinsic vector connection* on the tangent bundle G. The *projection* P_x is then naturally defined by $P_x G = G_x$ and $P_x \nabla = \nabla_x$. The *unit pseudoscalar field* $I(x)$ to S at x is defined by

$$I(x) = e_1 \wedge e_2 \wedge \ldots \wedge e_m / |e_1 \wedge e_2 \wedge \ldots \wedge e_m|.$$

In the special case when $S \subset I\!R^n$, the tangent bundle G becomes a subset of the geometric algebra $I\!R_n \equiv G(I\!R^n)$ of $I\!R^n$, and the projection $P_x(A) = I^{-1}(I \cdot A)$ is the projection onto the sub-geometric algebra G_x at $x \in S$. The vector derivative ∂_x of S can then be written $\partial_x = P_x(\nabla_x)$ where ∇_x is the gradient of $I\!R^n$. Of course, there are many details to consider to guarantee that our theory is compatible with the standard approaches to manifold theory [1].

The main advantage of working in the embedded *vector manifold* is that it is possible to define the bivector-valued *shape operator*

$$S(a) = I^{-1}(a \cdot \partial_x I) = I^{-1} P_a(I)$$

using the vector derivative ∂_x. Slightly more general, for a multivector $A \in G_x$ we can define

$$S(A) \equiv \partial_x P_x(A).$$

We also have the basic relationship

$$\partial_x A = \nabla_x A + S(A)$$

for any multivector field $A(x) \in G$.

Riemannian Curvature is defined by

$$R(a \wedge b) \cdot c \equiv [a \cdot \nabla, b \cdot \nabla]c - [a, b] \cdot \nabla c = (a \wedge b) \cdot (\nabla \wedge \nabla)c.$$

In terms of the shape operator,

$$R(a \wedge b) = P[S_a \times S_b] = P_a(S_b),$$

so curvature is completely determined by $S_a = S(a)$.

The curvature operator is *symmetric*

$$R(A) \cdot B = A \cdot R(B),$$

and satisfies the *Ricci identity*

$$a \cdot R(b \wedge c) + b \cdot R(c \wedge a) + c \cdot R(a \wedge b) = 0.$$

These two conditions are equivalent to

$$\partial_a \wedge R(a \wedge b) = 0.$$

The famous *Bianchi identity* takes the form

$$\nabla_x \wedge R_x(a \wedge b) = 0.$$

In 4-dimensional spacetime the curvature operator is a symmetric endomorphism in the 6-dimensional space of bivectors with the ultra hyperbolic signature $\{+ + + - - -\}$. It would be interesting to carry out a complete classification of this operator via it's minimal polynomial. Such a classification has been carried out for the *Projective Weyl Tensor*

$$W(a \wedge b) \cdot c = R(a \wedge b) \cdot c - \frac{1}{m-1}(a \wedge b) \cdot (\partial_a \cdot R(a \wedge c)),$$

called the *Petrov Classification.*

Let $A_r(x)$ be an r-vector field. The *r-form* of $A_r(x)$ is

$$\alpha_r(dX_r) \equiv dX_r \cdot A_r,$$

where $dX_r = dx_1 \wedge dx_2 \wedge \ldots \wedge dx_r$. The Cartan *exterior derivative* $d\alpha_r$ is defined by

$$d\alpha_r = dX_{r+1} \cdot (\nabla \wedge A_r(x)).$$

The famous formula $d^2\alpha_r = 0$ is a simple consequence of the integrability condition $\nabla \wedge \nabla x^i = \nabla \wedge e^i = 0$:

$$\begin{aligned} d^2\alpha_r &= dX_{r+2} \cdot [\nabla \wedge (\nabla \wedge A_r)] \\ &= dX_{r+2} \cdot [(\nabla \wedge \nabla) \wedge A_r] = 0. \end{aligned}$$

Acknowledgement

I want to thank William Baylis for organizing the most enjoyable Banff Clifford Algebra Summer School and making possible the development of these lectures.

Bibliography

[1] R. L. Bishop, R. J. Crittenden, *Geometry of Manifolds*, Academic Press, New York (1964).

[2] R. Delanghe, F. Sommen and V. Soucek, *Clifford Algebra and Spinor-valued functions: a function theory for the Dirac-operator*, Mathematics and Its Applications 53, Kluwer, Dordrecht (1992).

[3] C. Doran, D. Hestenes, F. Sommen and N. Van Acker, *Lie groups as spin groups*, J. Math. Phys. **34** (8), p. 3642-3669, August 1993.

[4] W. Fulton and Joe Harris, *Representation Theory*, Springer-Verlag 1991.

[5] F.R. Gantmacher, *Matrix Theory*, Vol. 1, Chelsea Publishing Company, New York, 1960.

[6] D. Hestenes and G. Sobczyk, *Clifford Algebra to Geometric Calculus: A Unified Language for Mathematics and Physics*, 2nd edition, Kluwer 1992.

[7] I. R. Porteous, *Topological Geometry, 2nd edition*, Cambridge University Press, 1981.

[8] M. Riesz, *Clifford Numbers and Spinors*, edited by E. Bolinder and P. Lounesto, Kluwer 1993.

[9] G. Sobczyk, Hyperbolic Number Plane, *The College Mathematics Journal*, (to appear) September 1995.

[10] G. Sobczyk, "Clifford Algebra Techniques in Linear Algebra", in CLIFFORD ALGEBRAS AND SPINOR STRUCTURES, editors: Pertti Lounesto and Rafal Ablamowicz, A Volume dedicated to the memory of Albert Crumeyrolle, Mathematics and Its Applications Vol. 321, Kluwer, p.101-110 (1995).

[11] G. Sobczyk, Unipotents, Idempotents, and a Spinor Basis for Matrices, *Advances in Applied Clifford Algebras* [**2**], No. 1 (1992) 53-62.

[12] G. E. Sobczyk, Jordan Form in Clifford Algebras, *Clifford Algebras and their Applications in Mathematical Physics*, Proceedings of the Third International Clifford Algebras Workshop, Edited by Fred Brackx, Richard Delanghe, and Herman Serras, Kluwer, Dordrecht, 1993.

[13] G. E. Sobczyk, Structure of Factor Algebras and Clifford Algebra, *LINEAR ALGEBRA AND ITS APPLICATIONS*, (to appear) 1995.

[14] G. E. Sobczyk, Structure Equation of a Linear Operator, *The College Mathematics Journal*, (to appear).

[15] G. E. Sobczyk, Simplicial Calculus, *Clifford Algebras and their Applications in Mathematical Physics*, Proceedings of the Second International Clifford Algebras Workshop, Kluwer, Dordrecht, 1988.

Chapter 6

Linear Algebra

Chris Doran, Anthony Lasenby, and Stephen Gull
MRAO, Cavendish Laboratory, Madingley Road,
Cambridge, CB3 0HE, UK

We begin by summarising the notations and conventions which we will employ throughout our series of lectures. Summation convention and natural units ($\hbar = c = \epsilon_0 = G = 1$) are employed throughout, except where explicitly stated.

6.1 Geometric Algebra

A geometric algebra is a graded linear space, the elements of which are called multivectors. The names 'scalar', 'vector', 'bivector', 'trivector', 'four-vector' ... are given to the grade-0, grade-1 ... multivectors respectively. The highest-grade element is called the pseudoscalar. Multivectors containing elements of a single grade are termed 'homogeneous'. Homogeneous multivectors form a subspace which is closed under addition and scalar multiplication. Vectors are usually written in lower-case Roman, or lower case Greek for frames in space and spacetime. General multivectors are written in upper-case Roman, or Greek for quantum spinors.

Multivectors are equipped with a product that is associative and distributive over addition, though non-commutative (except for multiplication by a scalar). The geometric product of two multivectors A and B is written as AB. The final axiom (which distinguishes a geometric algebra from other graded algebras) is that the square of any vector in the algebra is a scalar. A simple re-arrangement of the expansion

$$(a+b)^2 = (a+b)(a+b) = a^2 + (ab+ba) + b^2 \tag{6.1}$$

yields

$$ab + ba = (a+b)^2 - a^2 - b^2, \tag{6.2}$$

from which it follows that the symmetrised product of any two vectors is also a scalar. By similar means the properties of the geometric product of arbitrary multivectors can be established inductively from the basic axioms.

The geometric product of a grade-r multivector A_r with a grade-s multivector B_s decomposes into

$$A_r B_s = \langle AB\rangle_{r+s} + \langle AB\rangle_{r+s-2} \cdots + \langle AB\rangle_{|r-s|}. \tag{6.3}$$

The symbol $\langle M\rangle_r$ denotes the projection onto the grade-r component of M. The projection onto the grade-0 (scalar) component of M is written as $\langle M\rangle$. The scalar part of a product of multivectors satisfies the cyclic reordering property

$$\langle A \ldots BC\rangle = \langle CA \ldots B\rangle. \tag{6.4}$$

The '$\cdot$' and '$\wedge$' symbols are used for the lowest-grade and highest-grade terms of the series (6.3), so that

$$\begin{aligned} A_r \cdot B_s &\equiv \langle AB\rangle_{|r-s|} \\ A_r \wedge B_s &\equiv \langle AB\rangle_{s+r}, \end{aligned} \tag{6.5}$$

which are called the inner and outer (or exterior) products respectively. For vectors a and b we have

$$ab = a\cdot b + a\wedge b. \tag{6.6}$$

where

$$a\cdot b \equiv \tfrac{1}{2}(ab+ba), \qquad a\wedge b \equiv \tfrac{1}{2}(ab-ba). \tag{6.7}$$

Note that parallel vectors commute and orthogonal vectors anticommute.

We also employ the scalar product, defined by

$$A{*}B \equiv \langle AB\rangle, \tag{6.8}$$

and the commutator product, defined by

$$A\times B \equiv \tfrac{1}{2}(AB - BA). \tag{6.9}$$

The associativity of the geometric product ensures that the commutator product satisfies the Jacobi identity

$$A\times(B\times C) + B\times(C\times A) + C\times(A\times B) = 0. \tag{6.10}$$

When manipulating chains of products we employ the operator ordering convention that, in the absence of brackets, *inner, outer and scalar products take precedence over geometric products.*

6.2 Spacetime Algebra

The above discussion applies to vector spaces of any dimension, but for many physical applications we are interested in the specific properties of Minkowski spacetime. Accordingly, the geometric (Clifford) algebra generated by Minkowski spacetime is given the special name of *'Spacetime algebra'* (STA). A basis for the STA can be defined from four basis vectors $\{\gamma_\mu\}$, $\mu = 0 \ldots 3$, satisfying

$$\gamma_\mu \cdot \gamma_\nu = \eta_{\mu\nu} = \mathrm{diag}(+\,-\,-\,-). \tag{6.11}$$

The full STA is then spanned by the basis

$$1, \qquad \{\gamma_\mu\}, \qquad \{\sigma_k,\, i\sigma_k\}, \qquad \{i\gamma_\mu\}, \qquad i, \tag{6.12}$$

where

$$\sigma_k \equiv \gamma_k\gamma_0, \qquad k = 1, 2, 3. \tag{6.13}$$

and

$$i \equiv \gamma_0\gamma_1\gamma_2\gamma_3. \tag{6.14}$$

The symbol i is employed for the spacetime pseudoscalar because the square of i is -1. It must not be confused with the unit scalar imaginary employed in quantum mechanics. The pseudoscalar i is a geometrically-significant entity which *anti*commutes with odd-grade elements (vectors and trivectors), and commutes with even-grade elements.

The Spacetime Split The three bivectors $\{\sigma_k\}$, where $\sigma_k \equiv \gamma_k\gamma_0$ (6.13) satisfy

$$\tfrac{1}{2}(\sigma_j\sigma_k + \sigma_k\sigma_j) = -\tfrac{1}{2}(\gamma_j\gamma_k + \gamma_k\gamma_j) = \delta_{jk} \tag{6.15}$$

and therefore generate the geometric algebra of three-dimensional Euclidean space. This is identified as the algebra for the rest-space relative to the timelike vector γ_0. The full algebra for this space is spanned by the set

$$1, \qquad \{\sigma_k\}, \qquad \{i\sigma_k\}, \qquad i, \tag{6.16}$$

which is identifiable as the even subalgebra of the full STA (6.12). The identification of the algebra of relative space with the even subalgebra of the STA simplifies the transition from relativistic quantities to observables in a given frame. Note that the pseudoscalar employed in (6.16) is the same as that employed in spacetime, since

$$\sigma_1\sigma_2\sigma_3 = \gamma_1\gamma_0\gamma_2\gamma_0\gamma_3\gamma_0 = \gamma_0\gamma_1\gamma_2\gamma_3 = i. \tag{6.17}$$

The split of the six spacetime bivectors into relative vectors $\{\sigma_k\}$ and relative bivectors $\{i\sigma_k\}$ is a frame-dependent operation. For example, the spacetime split of the Faraday bivector F in the γ_0-system yields

$$F = \boldsymbol{E} + i\boldsymbol{B} \tag{6.18}$$

where

$$\boldsymbol{E} = \tfrac{1}{2}(F - \gamma_0 F \gamma_0), \qquad i\boldsymbol{B} = \tfrac{1}{2}(F + \gamma_0 F \gamma_0). \tag{6.19}$$

Both $\boldsymbol{E}$ and $\boldsymbol{B}$ are spatial vectors in the γ_0-frame, and $i\boldsymbol{B}$ is a spatial bivector.

Where required, relative (or spatial) vectors in the γ_0-system are written in bold type to record the fact that in the STA they are actually bivectors. This distinguishes them from spacetime vectors, which are left in normal type. No problems arise for the $\{\sigma_k\}$, which are unambiguously spacetime bivectors, and so are left in normal type.

Spatial vectors (such as $\boldsymbol{E}$) and spatial bivectors (such as $i\boldsymbol{B}$) act differently under the operation of spatial reversion. Since this operation coincides with Hermitian conjugation for matrices, we denote with a dagger:

$$M^{\dagger} \equiv \gamma_0 \tilde{M} \gamma_0, \tag{6.20}$$

so that, for example,

$$F^{\dagger} = \boldsymbol{E} - i\boldsymbol{B}. \tag{6.21}$$

The explicit appearance of γ_0 in the definition (6.20) shows that spatial reversion is not a Lorentz-covariant operation.[1]

When working with purely spatial quantities, we often require that the dot and wedge operations drop down to their three-dimensional definitions. For example, given two spatial vectors $\boldsymbol{a}$ and $\boldsymbol{b}$, we would like $\boldsymbol{a} \wedge \boldsymbol{b}$ to denote the spatial bivector swept out by $\boldsymbol{a}$ and $\boldsymbol{b}$. Accordingly we adopt the convention that, *in expressions where both vectors are in bold type, the dot and wedge operations take their three-dimensional meaning.*

Spacetime vectors can also be decomposed by a spacetime split, this time resulting in a scalar and a relative vector. The spacetime split of the vector a is achieved via

$$a\gamma_0 = a \cdot \gamma_0 + a \wedge \gamma_0 \equiv a_0 + \boldsymbol{a}, \tag{6.22}$$

so that a_0 is a scalar (the γ_0-time component of a) and $\boldsymbol{a}$ is the relative spatial vector. For example, the 4-momentum p splits into

$$p\gamma_0 = E + \boldsymbol{p} \tag{6.23}$$

where E is the energy in the γ_0 frame, and $\boldsymbol{p}$ is the 3-momentum. The definition of the relative vector (6.22) ensures that

$$a \cdot b = \langle a\gamma_0\gamma_0 b \rangle = a_0 b_0 - \boldsymbol{a} \cdot \boldsymbol{b}. \tag{6.24}$$

The vector Derivative The fundamental differential operator on spacetime is the derivative with respect to the position vector x. This is known as the *vector derivative* and is given the symbol ∇. It can be defined as

$$\nabla = \gamma^{\mu} \frac{\partial}{\partial x^{\mu}} \tag{6.25}$$

where the $\{x^{\mu}\}$ are a set of scalar Cartesian components. The vector derivative has all the algebraic properties of a grade-1 multivector. For example, Maxwell's equations

$$\nabla \cdot F = J, \quad \text{and} \quad \nabla \wedge F = 0, \tag{6.26}$$

[1] Note added by editor: $\tilde{M}$ is the *reversion* of M, given by

$$\tilde{M} = \langle M \rangle + \langle M \rangle_1 - \langle M \rangle_2 - \langle M \rangle_3 + \langle M \rangle_4 .$$

combine into the single equation

$$\nabla F = J. \tag{6.27}$$

The spacetime split of the vector derivative requires some care. We wish to retain the symbol $\boldsymbol{\nabla}$ for the spatial vector derivative, so that

$$\boldsymbol{\nabla} = \sigma_k \partial_k, \qquad k = 1 \ldots 3. \tag{6.28}$$

This definition of $\boldsymbol{\nabla}$ is inconsistent with the definition (6.22), so for the vector derivative we have to remember that

$$\nabla \gamma_0 = \partial_t - \boldsymbol{\nabla}. \tag{6.29}$$

6.3 Geometric V's Tensor Algebra

Much of modern physics is formulated in the language of tensors. Forces, stresses and strains, polarisation fields, *etc.* are all represented by tensors, as are the physical relationships between these objects. But just how efficient a language is tensor algebra? Can we do better, and develop an index-free notation? The answer is yes, but before seeing why, let's review some key features of tensor analysis.

Tensors are often motivated by first pointing out that mathematical objects exist which look like vectors, but do not behave as vectors. Boas[2], for example, discusses the example of a directed line defined by the axis of a rotation, with length equal to the magnitude of the rotation and orientation defined by the right-hand rule. She argues that such an object is not a vector because when combined with a second rotation the resultant vector is not defined by the vector addition rule. But surely this is nonsense! Vectors can be combined with other vectors by means other than just addition. In this case the vector $\boldsymbol{a}$ is combined with the vector $\boldsymbol{b}$ by combining the rotations they define, so the resultant vector $\boldsymbol{c}$ is given by

$$\boldsymbol{c} = -i \ln(e^{i\boldsymbol{a}}\, e^{i\boldsymbol{b}}). \tag{6.30}$$

Clearly in this expression we have no difficulty in saying that each of $\boldsymbol{a}$, $\boldsymbol{b}$ and $\boldsymbol{c}$ are vectors.

So what then is the motivation for introducing tensors? The reasons can be traced back to the idea that a vector is equally well described as a directed line, or as a set set of three components. Once the components in one frame are known, then the components in any other frame can be easily calculated. For the case of orthonormal frames (Cartesian tensors) the components are related by

$$x'_i = \Lambda_{ij} x_j \tag{6.31}$$

where Λ_{ij} represents an orthogonal transformation,

$$\Lambda_{ij} \Lambda_{ik} = \delta_{jk}. \tag{6.32}$$

[2] *Mathematical Methods in the Physical Sciences*, John Wiley & Sons (1983).

In tensor analysis, a vector is then defined as a set of scalars which transform according to (6.31) under a change of basis frame. This definition then extends to the definition of a rank r tensor as an object carrying r indices and transforming under the action of r copies of Λ_{ij}. For example, the components of a rank-four tensor transform as

$$T'_{\alpha\beta\gamma\delta} = \Lambda_{\alpha i}\Lambda_{\beta j}\Lambda_{\gamma k}\Lambda_{\delta l}T_{ijkl}. \tag{6.33}$$

So what is wrong with this approach? Of course there is nothing wrong mathematically; the issue is how fruitful the approach is. One problem is that, in defining a vector as a set of three scalars with given transformation properties, we have lost contact with the concept of direction. The three components (x_1, x_2, x_3) on their own are not a directed line segment, but the quantity

$$\boldsymbol{x} = x_1\boldsymbol{e}_1 + x_2\boldsymbol{e}_2 + x_3\boldsymbol{e}_3 \tag{6.34}$$

is. If the same vector is decomposed in a second frame, then the components will be related by (6.31), but the vector $\boldsymbol{x}$ is unchanged. The transformation law (6.31) represents a fact that is, basically, inconsequential. A further problem with tensors is that the notation frequently fails to distinguish physically-distinct objects. For example, the bivector $i\boldsymbol{B}$, the linear function $f(\boldsymbol{a})$ and the scalar function of two vectors $\phi(\boldsymbol{a}, \boldsymbol{b})$ are all represented as rank-two tensors. It is only when the concept of direction is included that the differences between them start to emerge. What is required, therefore, is an extension of the geometric language of vectors, bivectors *etc.*, to include linear operators on these objects. This is the subject of this lecture.

6.4 Index-Free Linear Algebra

Many of you probably did some index-free linear algebra when first examining the properties of matrices. I suspect that, like me, many of you were disappointed to learn that the index-free approach was of limited applicability and that tensor algebra was required for handling more advanced physical systems. But in fact it is quite straightforward to to extend the index-free approach and avoid the need for tensor algebra. Suppose that we are interested in a quantity which is a linear map from vectors to vectors in the same space. (Maps between different spaces are just as easy to handle, but will not be needed here.) We write such an object simply as $f(a)$. This satisfies the usual property of linearity

$$f(\lambda a + \mu b) = \lambda f(a) + \mu f(b). \tag{6.35}$$

The components of a rank-two tensor are easily recovered from f by introducing a frame $\{e_i\}$ and defining

$$f_{ij} \equiv e_i \cdot f(e_j). \tag{6.36}$$

It is then simple to verify that the f_{ij} components transform as a rank-two tensor, but this is now a derived consequence of the definition (6.36) and the linearity of f, it is not part of the definition of f. Similarly, a vector valued function of two vectors can be written as $f(a, b)$, or a scalar function of three vectors as $\phi(a, b, c)$, *etc.* In this manner we easily achieve an index-free notation; the question is whether this approach is better.

The Outermorphism For most of this lecture we will concentrate on the properties of linear maps from vectors to vectors. It is useful to write such functions with an underbar, $\underline{f}$. The key to analysing these in geometric algebra is the outermorphism, which extends the action of the function on vectors to the entire geometric algebra. The outermorphism of $\underline{f}$ is defined through its action on blades,

$$\underline{f}(a \wedge b \wedge \cdots \wedge c) \equiv \underline{f}(a) \wedge \underline{f}(b) \wedge \cdots \wedge \underline{f}(c), \tag{6.37}$$

and extension by linearity then defines the action of $\underline{f}$ on arbitrary multivectors. The outermorphism function is grade preserving,

$$\underline{f}(A_r) = \langle \underline{f}(A_r) \rangle_r \tag{6.38}$$

and 'multilinear',

$$\underline{f}(\lambda A + \mu B) = \lambda \underline{f}(A) + \mu \underline{f}(B). \tag{6.39}$$

Example 1 As a simple illustration, consider the rotation

$$\underline{R}(a) = Ra\tilde{R}. \tag{6.40}$$

Acting on the bivector $a \wedge b$ we find that

$$\begin{aligned} \underline{R}(a \wedge b) &= (Ra\tilde{R}) \wedge (Rb\tilde{R}) \\ &= \langle Ra\tilde{R}Rb\tilde{R} \rangle_2 \\ &= Ra \wedge b\tilde{R}. \end{aligned} \tag{6.41}$$

It follows that acting on an arbitrary multivector A we have

$$\underline{R}(A) = RA\tilde{R}. \tag{6.42}$$

An important result for the outermorphism concerns the product of two functions. Suppose that $\underline{h}(a) = \underline{f}[\underline{g}(a)]$. It follows that

$$\begin{aligned} \underline{h}(a \wedge b \wedge \cdots \wedge c) &= \underline{f}[\underline{g}(a)] \wedge \underline{f}[\underline{g}(b)] \wedge \cdots \wedge \underline{f}[\underline{g}(c)] \\ &= \underline{f}[\underline{g}(a) \wedge \underline{g}(b) \wedge \cdots \wedge \underline{g}(c)] \\ &= \underline{f}[\underline{g}(a \wedge b \wedge \cdots \wedge c)], \end{aligned} \tag{6.43}$$

hence the outermorphism of the product of two linear functions is the product of their outermorphisms. In dealing with combinations of linear functions we can therefore write

$$\underline{h}(A) = \underline{f}\,\underline{g}(A), \tag{6.44}$$

as the meaning of the right-hand side is unambiguous.

The Determinant Given the definition of the outermorphism, we can immediately proceed to the definition of the determinant. The pseudoscalar I for any space is unique up to scaling and the outermorphism is grade-preserving, so we define

$$\underline{f}(I) = \det(\underline{f})\, I. \tag{6.45}$$

This should be compared with the tensor algebra definition

$$f_{\alpha i} f_{\beta j} \cdots f_{\gamma k} \epsilon_{\alpha\beta\cdots\gamma} = \det(f)\, \epsilon_{ij\cdots k}. \tag{6.46}$$

The definitions are equivalent, but the geometric algebra form is considerably more compact and intuitive. From equation (6.44) it immediately follows that

$$\det(\underline{f}\,\underline{g})I = \underline{f}\,\underline{g}(I) = \det(\underline{g})\,\underline{f}(I) = \det(\underline{f})\det(\underline{g})\,I, \tag{6.47}$$

which quickly establishes one of the key properties of the determinant.

Example 2 Consider the linear function

$$\underline{f}(a) = a + \alpha a{\cdot}e_1 e_2 \tag{6.48}$$

where α is a scalar and e_1 and e_2 are a pair of arbitrary vectors. Construct the outermorphism of $\underline{f}$ and find its determinant.

We start by forming

$$\begin{aligned} \underline{f}(a{\wedge}b) &= (a + \alpha a{\cdot}e_1 e_2){\wedge}(b + \alpha b{\cdot}e_1 e_2) \\ &= a{\wedge}b + \alpha(b{\cdot}e_1 a - a{\cdot}e_1 b){\wedge}e_2 \\ &= a{\wedge}b + \alpha[(a{\wedge}b){\cdot}e_1]{\wedge}e_2. \end{aligned} \tag{6.49}$$

It follows that

$$\underline{f}(A) = A + \alpha(A{\cdot}e_1){\wedge}e_2. \tag{6.50}$$

The determinant now follows from

$$\begin{aligned} \underline{f}(I) &= I + \alpha(I{\cdot}e_i){\wedge}e_2 \\ &= I + \alpha e_1{\cdot}e_2 I \end{aligned} \tag{6.51}$$

hence $\det(\underline{f}) = 1 + \alpha e_1{\cdot}e_2$.

6.5 Multivector Calculus

Before extending our analysis of linear functions in geometric algebra, we need to make a brief diversion into the subject of differentiation with respect to a multivector. This subject is not essential for linear algebra, but does simplify many formulae and is important for material in later lectures. We first need to establish some simple results for basis vectors. Suppose that the set $\{e_k\}$ form a vector frame. The reciprocal frame is determined by

$$e^j = (-1)^{j-1} e_1 \wedge e_2 \wedge \cdots \wedge \check{e}_j \wedge \cdots \wedge e_n \, e^{-1} \tag{6.52}$$

where

$$e \equiv e_1 \wedge e_2 \wedge \cdots \wedge e_n \tag{6.53}$$

and the check on $\check{e}_j$ denotes that this term is missing from the expression. The $\{e_k\}$ and $\{e^j\}$ frames are related by

$$e_j \cdot e^k = \delta_j^k. \tag{6.54}$$

An arbitrary multivector can be decomposed in terms of this frame into

$$B_{i \cdots j} = B_r \cdot (e_j \wedge \cdots \wedge e_i), \qquad B_r = \sum_{i < \cdots < j} B_{i \cdots j} \, e^i \wedge \cdots \wedge e^j. \tag{6.55}$$

Suppose now that the multivector F is an arbitrary function of some multivector argument X, $F = F(X)$. The derivative of F with respect to X in the A direction is defined by

$$A * \partial_X F(X) \equiv \lim_{\tau \mapsto 0} \frac{F(X + \tau A) - F(X)}{\tau} \tag{6.56}$$

where $A * B = \langle AB \rangle$. The multivector derivative ∂_X is defined in terms of its directional derivatives by

$$\partial_X \equiv \sum_{i < \cdots < j} e^i \wedge \cdots \wedge e^j (e_j \wedge \cdots \wedge e_i) * \partial_X. \tag{6.57}$$

This definition shows how the multivector derivative ∂_X inherits the multivector properties of its argument X, as well as a calculus from equation (6.56).

Most of the properties of the multivector derivative follow from the result that

$$\partial_X \langle XA \rangle = P_X(A), \tag{6.58}$$

where $P_X(A)$ is the projection of A onto the grades contained in X. Leibniz' rule is then used to build up results for more complicated functions. The multivector derivative acts on objects to its immediate right unless brackets are present;[3] for example, in the expression $\partial_X AB$ the ∂_X acts only on A, but in the expression $\partial_X (AB)$ the ∂_X acts on both A and B. If the ∂_X is intended to only act on B then this is written as $\dot{\partial}_X A \dot{B}$, where the overdot denotes the multivector on which the derivative acts. For example, Leibniz' rule can be written as

$$\partial_X (AB) = \dot{\partial}_X \dot{A} B + \dot{\partial}_X A \dot{B}. \tag{6.59}$$

For the rest of this lecture we will mainly employ the derivative with respect to a vector, ∂_a. Combinations of a and ∂_a are used to perform contractions and

[3]Editor's note: This convention is relaxed in the *vector* derivatives below. There, the derivative generally operates on more than the term to the immediate right.

protractions without having to introduce a basis frame. For these the following results are useful:

$$\begin{aligned}\partial_a a\cdot A_r &= rA_r &(6.60)\\ \partial_a a\wedge A_r &= (n-r)A_r &(6.61)\\ \partial_a A_r a &= (-1)^r(n-2r)A_r. &(6.62)\end{aligned}$$

6.6 Adjoints and Inverses

The adjoint to the linear function $\underline{f}$ if written as $\overline{f}$ and defined by

$$a\cdot\underline{f}(b) = \overline{f}(a)\cdot b, \qquad \overline{f}(a) = \partial_b a\cdot\underline{f}(b). \tag{6.63}$$

From

$$\begin{aligned}\overline{f}(a\wedge b) &= \partial_c\wedge\partial_d\, a\cdot\underline{f}(c)\, b\cdot\underline{f}(d)\\ &= \tfrac{1}{2}\partial_c\wedge\partial_d\,(a\wedge b)\cdot\underline{f}(d\wedge c)\end{aligned} \tag{6.64}$$

we see that the outermorphism of the adjoint is the adjoint of the outermorphism. Symmetric functions have $\underline{f} = \overline{f}$, and antisymmetric ones have $\underline{f} = -\overline{f}$. Antisymmetric functions can always be written in the form

$$\underline{f}(a) = a\cdot F, \quad \text{where} \quad F = \tfrac{1}{2}\partial_a\wedge\underline{f}(a). \tag{6.65}$$

Hestenes & Sobczyk call F the characteristic bivector.

Now consider the expression

$$\begin{aligned}\underline{f}(a\wedge b)\cdot c &= \underline{f}(a)\underline{f}(b)\cdot c - \underline{f}(b)\underline{f}(a)\cdot c\\ &= \underline{f}[a\, b\cdot\overline{f}(c) - b\, a\cdot\overline{f}(c)]\\ &= \underline{f}[(a\wedge b)\cdot\overline{f}(c)].\end{aligned} \tag{6.66}$$

This is a special case of the important results:

$$\begin{aligned}A_r\cdot\overline{f}(B_s) &= \overline{f}[\underline{f}(A_r)\cdot B_s] & r\leq s\\ \underline{f}(A_r)\cdot B_s &= \underline{f}[A_r\cdot\overline{f}(B_s)] & r\geq s.\end{aligned} \tag{6.67}$$

The Inverse With the results just established we can now compute the form of the inverse function. Consider the result

$$\det(\underline{f})IB = \underline{f}(I)B = \underline{f}[I\overline{f}(B)]. \tag{6.68}$$

Replacing IB by A we find that

$$\det(\underline{f})A = \underline{f}[I\overline{f}(I^{-1}A)] \tag{6.69}$$

with a similar result holding for the adjoint. It follows that

$$\begin{aligned} \underline{f}^{-1}(A) &= \det(f)^{-1} I \overline{f}(I^{-1}A) \\ \overline{f}^{-1}(A) &= \det(f)^{-1} I \underline{f}(I^{-1}A). \end{aligned} \tag{6.70}$$

This derivation is far quicker than anything available in tensor analysis.

Example 3 Find the inverse of the function defined in Example 6.4.
With

$$\underline{f}(A) = A + \alpha(A \cdot e_1) \wedge e_2 \tag{6.71}$$

we have

$$\langle A_r \underline{f}(B_r)\rangle = \langle A_r B_r\rangle + \alpha\langle A_r (B_r \cdot e_1)\wedge e_2\rangle = \langle A_r B_r\rangle + \alpha\langle e_2 \cdot A_r B_r e_1\rangle \tag{6.72}$$

hence

$$\overline{f}(A) = A + \alpha e_1 \wedge (e_2 \cdot A). \tag{6.73}$$

It follows that

$$\begin{aligned} \underline{f}^{-1}(A) &= (1 + \alpha e_1 \cdot e_2)^{-1} I^{-1}[IA + \alpha e_1 \wedge (e_2 \cdot (IA))] \\ &= (1 + \alpha e_1 \cdot e_2)^{-1}(A + \alpha e_1 \cdot (e_2 \wedge A) \\ &= A - \frac{\alpha}{1 + \alpha e_1 \cdot e_2} e_2 \wedge (e_1 \cdot A). \end{aligned} \tag{6.74}$$

6.7 Eigenvectors and Eigenbivectors

The familiar results for eigenvalues and their associated eigenvectors go through largely unchanged, but many simplifications and extensions are available in geometric algebra. One example is the elimination of complex eigenvalues in favour of eigenbivectors, whose geometric significance is clearer. Suppose that the function $\underline{f}$ has

$$\underline{f}(e_1) = \lambda e_2, \qquad \underline{f}(e_2) = -\lambda e_1. \tag{6.75}$$

Traditionally, one might write that $e_1 \pm j e_2$ are eigenvectors with eigenvalues $\mp j\lambda$, where j is a unit scalar imaginary. But it is more informative to write

$$\underline{f}(e_1 \wedge e_2) = \lambda^2 e_1 \wedge e_2 \tag{6.76}$$

which identifies the plane $e_1 \wedge e_2$ as an eigenbivector of $\underline{f}$.

The Cayley-Hamilton Theorem One of the most impressive demonstrations of the power of geometric algebra is provided by the proof of the Cayley-Hamilton theorem. This theorem states that any linear function satisfies its own characteristic equation. To prove this it is helpful to introduce some more compact notation. We let $\{a_1, a_2 \ldots a_n\}$ be a set of n independent vectors. The derivative with respect to each of these is abbreviated as

$$\partial_{a_k} = \partial_k. \tag{6.77}$$

We also define the 'simplicial' variable

$$a_{(r)} \equiv a_1 \wedge a_2 \wedge \cdots \wedge a_r \tag{6.78}$$

with an associated simplicial derivative

$$\partial_{(r)} \equiv \frac{1}{r!} \partial_r \wedge \partial_{r-1} \wedge \cdots \wedge \partial_1. \tag{6.79}$$

Since

$$\langle A_r \wedge \partial_a a \wedge B_r \rangle = (n-r) \langle A_r B_r \rangle. \tag{6.80}$$

it follows that

$$\partial_{(r)} a_{(r)} = \frac{n!}{(n-r)!\, r!} = \binom{n}{r}. \tag{6.81}$$

We make the further abbreviations

$$\underline{f}(a_j) = \underline{f}_j, \qquad \underline{f}(a_{(r)}) = \underline{f}_{(r)}. \tag{6.82}$$

With these abbreviations we can write

$$\partial_{(1)} \cdot \underline{f}(a_{(1)}) = \partial_{a_1} \cdot \underline{f}(a_1) = \mathrm{Tr}(\underline{f}) \tag{6.83}$$

and

$$\partial_{(n)} \underline{f}_{(n)} = \partial_{(n)} a_{(n)} \det(\underline{f}) = \det(\underline{f}). \tag{6.84}$$

These two invariants are special cases of the range of invariants $\partial_{(r)} \cdot \underline{f}_{(r)}$.

The characteristic polynomial for $\underline{f}$ is found by forming the determinant of the function $\underline{F}(a) = \underline{f}(a) - \lambda a$:

$$\begin{aligned} \det(\underline{F}) &= \partial_{(n)} \underline{F}_{(n)} \\ &= \partial_{(n)} [\underline{f}(a_1) - \lambda a_1] \wedge [\underline{f}(a_2) - \lambda a_2] \wedge \cdots \wedge [\underline{f}(a_n) - \lambda a_n] \\ &= \partial_{(n)} [\underline{f}_{(n)} - n\lambda \underline{f}_{(n-1)} \wedge a_n + \cdots + (-\lambda)^n a_{(n)}. \end{aligned} \tag{6.85}$$

A general term in this expression goes as

$$(-\lambda)^s \binom{n}{s} \partial_{(n)} \cdot [\underline{f}_{(n-s)} \wedge a_{n-s+1} \wedge \cdots \wedge a_n] = (-\lambda)^s \partial_{(n-s)} \cdot \underline{f}_{(n-s)}. \tag{6.86}$$

It follows that the characteristic polynomial is simply

$$C(\lambda) = \sum_{s=0}^{n} (-\lambda)^{n-s} \partial_{(s)} \cdot \underline{f}_{(s)}. \tag{6.87}$$

This clearly illustrates the significance of the invariant quantities $\partial_{(r)} \cdot \underline{f}_{(r)}$.

The Cayley-Hamilton theorem now states that

$$\sum_{s=0}^{n} (-1)^{n-s} \partial_{(s)} \cdot \underline{f}_{(s)} \underline{f}^{n-s}(a) = 0 \tag{6.88}$$

where $\underline{f}^r(a)$ denotes the r-fold application of $\underline{f}$ on a. To prove this we first establish the lemma:

$$\begin{aligned}
(\partial_j \wedge \partial_{(j-1)}) \cdot [\underline{f}_{(j-1)} \wedge \underline{f}(a)]\, a_j \\
&= \langle \partial_{(j-1)} [\underline{f}_{(j-1)} \wedge \underline{f}(a)] \cdot \partial_j \rangle a_j \\
&= \partial_{(j-1)} \cdot \underline{f}_{(j-1)}\, \underline{f}(a) - (\partial_{j-1} \wedge \partial_{(j-2)}) \cdot [\underline{f}_{(j-2)} \wedge \underline{f}(a)]\, \underline{f}(a_{j-1}). \quad (6.89)
\end{aligned}$$

We now proceed by decomposing the term $a\partial_{(n)}\underline{f}_{(n)}$ as follows:

$$\begin{aligned}
a\partial_{(n)}\underline{f}_{(n)} &= \partial_{(n-1)}\underline{f}_{(n-1)} \wedge \underline{f}(a) \\
&= (\partial_n \wedge \partial_{(n-1)}) \cdot [\underline{f}_{(n-1)} \wedge \underline{f}(a)]\, a_n \\
&= \partial_{(n-1)} \cdot \underline{f}_{(n-1)}\, \underline{f}(a) - (\partial_{n-1} \wedge \partial_{(n-2)}) \cdot [\underline{f}_{(n-2)} \wedge \underline{f}(a)]\, \underline{f}(a_{n-1}) \\
&= \partial_{(n-1)} \cdot \underline{f}_{(n-1)}\, \underline{f}(a) - \partial_{(n-2)} \cdot \underline{f}_{(n-2)}\, \underline{f}^2(a) \\
&\quad + (\partial_{n-2} \wedge \partial_{(n-3)}) \cdot [\underline{f}_{(n-3)} \wedge \underline{f}(a)]\, \underline{f}^2(a_{n-2}) \\
&= \partial_{(n-1)} \cdot \underline{f}_{(n-1)}\, \underline{f}(a) - \partial_{(n-2)} \cdot \underline{f}_{(n-2)}\, \underline{f}^2(a) + \cdots + (-1)^{n-1} \underline{f}^n(a). \quad (6.90)
\end{aligned}$$

Taking the final expression over to the left-hand side then establishes the result.

An immediate consequence is that if e is an eigenvector of $\underline{f}$,

$$\underline{f}(e) = \lambda e, \tag{6.91}$$

then λ automatically satisfies the characteristic equation. There is no need to invoke arguments about determinants. The wonderful aspect of this proof is its total generality — it applies for any linear function, in any linear space of any dimension or signature.

Singular Value Decomposition This subject is slightly outside the main topic of this lecture, but is important and worth knowing. Given an arbitrary, non-singular linear function $\underline{f}$ in Euclidean space, we want to find a general canonical form which encodes its properties. This can be viewed as a problem in either tensor or matrix algebra. We start by forming the symmetric function $\overline{f}\,\underline{f}$. This function has n orthogonal eigenvectors with real, positive eigenvalues. The fact that the eigenvalues are positive follows from

$$\overline{f}\,\underline{f}(e) = \lambda e \quad \Rightarrow \quad \underline{f}(e)^2 = \lambda e^2. \tag{6.92}$$

Now suppose that $\underline{R}$ is the rotation that diagonalises $\overline{f}\,\underline{f}$:

$$\underline{R}\,\overline{f}\,\underline{f}\,\overline{R}(a) = \Lambda(a) \tag{6.93}$$

where Λ is a diagonal function. Since Λ can be represented as a diagonal matrix with positive eigenvalues, it has a well-defined square-root, $\Lambda^{1/2}$. Now define

$$\underline{S} = \underline{f}\,\overline{R}\,\Lambda^{-1/2}. \tag{6.94}$$

This satisfies

$$\overline{S}\underline{S} = \Lambda^{-1/2}\underline{R}\overline{f}\,\underline{f}\,\overline{R}\Lambda^{-1/2} = \Lambda^{-1/2}\Lambda\,\Lambda^{-1/2} = \underline{1} \tag{6.95}$$

where $\underline{1}$ is the identity function. It follows that $\underline{S}$ is an orthogonal function, and represents a rotation if $\det(\underline{f}) > 0$. The arbitrary function $\underline{f}$ therefore decomposes as

$$\underline{f} = \underline{S}\,\Lambda^{1/2}\underline{R} \tag{6.96}$$

which represents a dilation sandwiched between two rotations.

6.8 Invariants

A topic of relevance in engineering is the subject of invariants. For example, in computer vision one is interested in projective invariants, since these represent intrinsic features of the object being observed. Mathematically, the problem can be formulated as one of extracting invariant scalar quantities from functions of the form $\phi(a, b, \ldots c)$, where ϕ is a scalar-valued symmetric function of its vector arguments. Two examples of constructing such invariants should demonstrate how easily the general problem is tackled in geometric algebra.

Example 4 $\phi(a, b)$ is a symmetric scalar-valued function of two variables in two dimensions.

We form the linear function

$$\underline{\phi}(a) = \partial_b \phi(a, b). \tag{6.97}$$

Since $\phi(a, b) = \phi(b, a)$ it follows that $\underline{\phi}$ is a symmetric function. Its invariants are therefore its two eigenvalues, or alternatively its trace

$$\partial_a \cdot \underline{\phi}(a) = \partial_a \cdot \partial_b \phi(a, b), \tag{6.98}$$

and its determinant

$$\underline{\phi}(a \wedge b) = \det(\underline{\phi}) a \wedge b = \partial_c \wedge \partial_d \phi(a, c)\phi(b, d). \tag{6.99}$$

Example 5 $\phi(a, b, c)$ is a symmetric scalar-valued function of three variables in three dimensions.

First form the vector-valued function $\Phi(a, b) = \partial_c \phi(a, b, c)$. From this construct the symmetric scalar-valued function

$$f(a, b) = \Phi(a, c) \cdot \Phi(b, \partial_c). \tag{6.100}$$

$f(a, b)$ now defines the symmetric function $\underline{f}(a) = \partial_b f(a, b)$, and the eigenvalues of this function define three invariants. Higher-order invariants also exist and are not hard to find.

6.9 Linear Functions in Spacetime

Much of the preceding analysis goes through unchanged in non-Euclidean spaces, but there are some surprises. The first is that the singular-value decomposition fails, because we can no longer assume that the eigenvalues of $\overline{f}\,\underline{f}$ are positive. The link between symmetric functions and orthogonal eigenvectors is also lost, because of the possibility of the existence of null vectors. Problem 1, for example, contains a symmetric non-singular function which only possesses three distinct eigenvectors.

Further examples of the unusual properties of functions in non-Euclidean space are provided in $(1,1)$ space by the two symmetric functions

$$\underline{g}_1(a) = -12a\cdot\gamma_0\,\gamma_0 + 2a\cdot\gamma_0\,\gamma_1 + 2a\cdot\gamma_1\,\gamma_0 + a\cdot\gamma_1\,\gamma_1 \tag{6.101}$$

and

$$\underline{g}_2(a) = 8a\cdot\gamma_0\,\gamma_0 + a\cdot\gamma_0\,\gamma_1 + a\cdot\gamma_1\,\gamma_0 - a\cdot\gamma_1\,\gamma_1, \tag{6.102}$$

where $\gamma_0^2 = -\gamma_1^2 = 1$ and $\gamma_0\cdot\gamma_1 = 0$. The first of these, $\underline{g}_1$, has no symmetric square root. But the second has two! These are

$$\underline{h}(a) = \frac{1}{\sqrt{3}}[5a\cdot\gamma_0\,\gamma_0 + a\cdot\gamma_0\,\gamma_1 + a\cdot\gamma_1\,\gamma_0 + 2a\cdot\gamma_1\,\gamma_1] \tag{6.103}$$

and

$$\underline{h}'(a) = \frac{1}{\sqrt{15}}[11a\cdot\gamma_0\,\gamma_0 + a\cdot\gamma_0\,\gamma_1 + a\cdot\gamma_1\,\gamma_0 - 4a\cdot\gamma_1\,\gamma_1] \tag{6.104}$$

6.10 Functional Differentiation

A final, more advanced application is to develop a calculus for differentiation with respect to a linear function. This calculus is built up in the same manner as the multivector derivative of Section 3. We start by introducing a fixed frame $\{e_i\}$, and define the scalar coefficients

$$f_{ij} \equiv e_i\cdot\underline{f}(e_j). \tag{6.105}$$

Now consider the derivative with respect to f_{ij} of the scalar $\underline{f}(b)\cdot c$. This is

$$\begin{aligned}\partial_{f_{ij}}\underline{f}(b)\cdot c &= \partial_{f_{ij}}(f_{lk}b^k c^l)\\ &= c^i b^j.\end{aligned} \tag{6.106}$$

Multiplying both sides of this equation by $a\cdot e_j e_i$ we obtain

$$a\cdot e_j e_i \partial_{f_{ij}}\underline{f}(b)\cdot c = a\cdot b\, c, \tag{6.107}$$

which assembles a frame-independent vector on the right-hand side. It follows that the operator $a\cdot e_j\, e_i\partial_{f_{ij}}$ must also be frame-independent. We therefore define the vector functional derivative $\partial_{\underline{f}(a)}$ by

$$\partial_{\underline{f}(a)} \equiv a\cdot e_j\, e_i \partial_{f_{ij}}. \tag{6.108}$$

The essential property of $\partial_{\underline{f}(a)}$ is, from (6.107),

$$\partial_{\underline{f}(a)}\underline{f}(b)\cdot c = a\cdot b\, c. \tag{6.109}$$

This result, together with Leibniz' rule, is sufficient to derive all the required properties of the $\partial_{\underline{f}(a)}$ operator. For example, for a bivector B,

$$\begin{aligned} \partial_{\underline{f}(a)}\langle \underline{f}(b\wedge c)B\rangle &= \dot{\partial}_{\underline{f}(a)}\langle \dot{\underline{f}}(b)\underline{f}(c)B\rangle - \dot{\partial}_{\underline{f}(a)}\langle \dot{\underline{f}}(c)\underline{f}(b)B\rangle \\ &= a\cdot b\underline{f}(c)\cdot B - a\cdot c\underline{f}(b)\cdot B \\ &= \underline{f}[a\cdot(b\wedge c)]\cdot B \end{aligned} \tag{6.110}$$

which extends by linearity to give

$$\partial_{\underline{f}(a)}\langle \underline{f}(A)B\rangle = \underline{f}(a\cdot A)\cdot B, \tag{6.111}$$

where A and B are both bivectors. Proceeding in this manner, we obtain the general formula

$$\partial_{\underline{f}(a)}\langle \underline{f}(A)B\rangle = \sum_r \langle \underline{f}(a\cdot A_r)B_r\rangle_1. \tag{6.112}$$

For a fixed grade-r multivector A_r, we can now write

$$\begin{aligned} \partial_{\underline{f}(a)}\underline{f}(A_r) &= \partial_{\underline{f}(a)}\langle \underline{f}(A_r)X_r\rangle\partial_{X_r} \\ &= \underline{f}(a\cdot A_r)\cdot X_r\,\partial_{X_r} \\ &= (n-r+1)\underline{f}(a\cdot A_r), \end{aligned} \tag{6.113}$$

where we have employed a result from page 58 of "Clifford Algebra to Geometric Calculus".

Equation (6.112) can be used to derive formulae for the functional derivative of the adjoint. The general result is

$$\begin{aligned} \partial_{\underline{f}(a)}\overline{f}(A_r) &= \partial_{\underline{f}(a)}\langle \underline{f}(X_r)A_r\rangle\partial_{X_r} \\ &= \underline{f}(a\cdot\dot{X}_r)\cdot A_r\,\dot{\partial}_{X_r}. \end{aligned} \tag{6.114}$$

When A is a vector, this admits the simpler form

$$\partial_{\underline{f}(a)}\overline{f}(b) = ba. \tag{6.115}$$

If $\underline{f}$ is a symmetric function then $\underline{f} = \overline{f}$. But this fact cannot be exploited when differentiating with respect to $\underline{f}$ since f_{ij} and f_{ji} must be treated as independent variables for the purposes of calculus.

Example 6 Find the derivative with respect to $\underline{f}$ of the determinant $\det(\underline{f})$.
From the definition of the determinant, we find that

$$\begin{aligned} \partial_{\underline{f}(a)}\det(\underline{f}) &= \partial_{\underline{f}(a)}\underline{f}(I)I^{-1} \\ &= \underline{f}(a\cdot I)I^{-1} \\ &= \det(\underline{f})\overline{f}^{-1}(a). \end{aligned} \tag{6.116}$$

which agrees with standard formulae. This derivation is considerably more compact than any available to conventional matrix/tensor methods.

Notes

The most comprehensive treatment of linear functions within geometric algebra is contained in *Clifford Algebra to Geometric Calculus* by D. Hestenes and G. Sobczyk (Reidel, Dordrecht 1984). Further material can be found in D. Hestenes, 'The design of linear algebra and geometry', *Acta Appl. Math.* 23, p. 65 (1991) and in C.J.L. Doran, D. Hestenes, F. Sommen and N. van Acker, 'Lie groups as spin groups', *J. Math. Phys.* 34, p. 3642 (1993).

Exercises

1. The STA (spacetime algebra) is generated by the four orthogonal vectors $\{\gamma_\mu\}$, where $\gamma_0^2 = -\gamma_i^2 = 1$. In this algebra the function $\underline{f}$ is defined by

$$\underline{f}(a) = a + \alpha a\cdot\gamma_+ \gamma_+$$

where γ_+ is the null vector $\gamma_0 + \gamma_3$.

Find the charateristic equation satisfied by $\underline{f}$. What are the roots of the characteristic polynomial? How many independent eigenvectors can you find?

2. By expanding out the left-hand side verify, to third order, that

$$\det[\exp\{\underline{f}\}] = \exp\{\partial_a\cdot\underline{f}(a)\}$$

where the exponential function is defined by the power series

$$\exp\{\underline{f}\}(a) = \sum_{n=0}^{\infty} \frac{1}{n!}\underline{f}^n(a)$$

and $\underline{f}^0(a) = a$. The following result may be helpful:

$$\begin{aligned}\partial_{(s)}\cdot\underline{f}_{(s)} &= \frac{1}{s}[\partial_a\cdot\underline{f}(a)\,\partial_{(s-1)}\cdot\underline{f}_{(s-1)} - \partial_a\cdot\underline{f}^2(a)\,\partial_{(s-2)}\cdot\underline{f}_{(s-2)} + \cdots \\ &\quad +(-1)^{s+1}\partial_a\cdot\underline{f}^s(a)].\end{aligned}$$

Can you construct a general proof along these lines?

3. Prove the following results for the functional derivative:

$$\begin{aligned}\partial_{\underline{f}(a)}\partial_b\cdot\underline{f}^r(b) &= r\underline{f}^{r-1}(a) \qquad r\geq 1 \\ \partial_{\underline{f}(a)}\langle\overline{f}^{-1}(A_r)B_r\rangle &= -\langle\overline{f}^{-1}(a)\cdot B_r\,\overline{f}^{-1}(A_r)\rangle_1.\end{aligned}$$

4. Given a non-singular function $\underline{f}$ in Euclidean space, the function $\underline{e}$ is defined by

$$\underline{e} = \tfrac{1}{2}\ln[\overline{f}\,\underline{f}].$$

The logarithm can be defined either by a power series, or by diagonalising $\overline{f}\,\underline{f}$ and taking the logarithm of the eigenvalues.

Prove that

$$\partial_{\underline{f}(a)}\partial_b\cdot\underline{e}^n(b) = n\overline{f}^{-1}\underline{e}^{n-1}(a).$$

Chapter 7

Dynamics

Stephen Gull, Chris Doran, and Anthony Lasenby
MRAO, Cavendish Laboratory, Madingley Road, Cambridge, CB3 0HE, UK

This chapter is not intended to provide a balanced view of the STA approach to dynamics. We have instead chosen to concentrate on two topics normally covered in the Cambridge second-year undergraduate physics course: rigid body dynamics and elasticity. The level of presentation is variable, ranging from some elementary ideas through to more challenging worked examples.

The section on rigid body dynamics is intended to advertise the utility of the rotor equation. Further reading can be found in *'New Foundations for Classical Mechanics'* by Hestenes (though be warned that his rotor convention differs from the one adopted here). The work on elasticity contains a fresh approach to the subject, one that attempts to extend the validity of the theory to finite elastic displacements. It is only fair to warn that this approach has, so far, not been subjected to any critical scrutiny. The two pieces of work are combined in a final novel application of the rotor equation to the bending and twisting of elastic filaments.

7.1 Rigid Body Dynamics

7.1.1 The Rotor Equation

In three dimensions a rotor R is an even element (a Pauli spinor or quaternion) with unit modulus, $R\tilde{R} = 1$. A rotor generates rotations and can be used to relate an orthonormal frame $\{e_k\}$ to a fixed reference frame $\{\sigma_k\}$ by

$$e_k = R\sigma_k\tilde{R}. \tag{7.1}$$

(Since all applications discussed in this Chapter are to three dimensional problems, the symbols employed are unambiguous and we do not need to write vectors in a bold font.) For rigid body dynamics, the frame $\{e_k\}$ is fixed in the body and is referred to as the 'body' axes. It is related to the fixed reference frame ('space' axes) $\{\sigma_k\}$

by the 'attitude spinor' R. The rotor R may then be used as a dynamical variable to describe the attitude of the rigid body in space.

Consider a body in three dimensions, rotating with instantaneous (vector) angular velocity ω_S. When using the STA, we prefer to say that the body rotates *in a plane*, rather than *about an axis*, and so we will henceforth consider the angular velocity to be a *bivector* $\Omega_S \equiv i\omega_S$. The rate of change of the frame $\{e_k\}$ is given by the vector equation

$$\dot{e}_k = \omega_S \times e_k = -\Omega_S \cdot e_k = -\tfrac{1}{2}(\Omega_S e_k - e_k \Omega_S). \tag{7.2}$$

where $a \times b = -ia \wedge b$ is the vector cross product.

The content of this equation can be expressed in an equivalent and simpler form as the *rotor equation*

$$\dot{R} = -\tfrac{1}{2}\Omega_S R = -\tfrac{1}{2}R\Omega_B. \tag{7.3}$$

The 'space' and 'body' angular velocities (Ω_S and Ω_B respectively) are related by

$$\Omega_S = R\Omega_B\tilde{R}. \tag{7.4}$$

The normalisation $R\tilde{R} = 1$ ensures that the angular velocities are bivectors:

$$0 = \partial_t(R\tilde{R}) = \dot{R}\tilde{R} + R\dot{\tilde{R}} = -\tfrac{1}{2}(\Omega_S R\tilde{R} + R\tilde{R}\tilde{\Omega}_S), \tag{7.5}$$

so that $\Omega_S = -\tilde{\Omega}_S$.

The rotor equation (7.3) can be expressed in terms of either the space or body angular momenta Ω_S and Ω_B, and convenience will usually dictate which form to use. In particular, if either Ω_S or Ω_B is constant, the rotor equation can be integrated immediately. For example, if the body is rotating on a fixed axle, the space angular velocity is constant, so that

$$R(t) = e^{-\Omega_S t/2} R(0). \tag{7.6}$$

7.1.2 Kinetic Energy and the Inertia Tensor

To go further we must specify the nature of the body. We do this by first considering a fixed 'reference copy' of the body. The true spatial position of a point in the body is then obtained by rotating the fixed copy about the origin, and applying a fixed translation. So, with the fixed (body) position described by the vector x_α, The 'space' position x'_α is given by

$$x'_\alpha(t) = R(t)x_\alpha\tilde{R}(t) + x_0(t). \tag{7.7}$$

At the point x_α we assume that the body has mass m_α. It is invariably convenient to choose the origin to coincide with the centre of mass of the fixed body, as this ensures that

$$\sum_\alpha m_\alpha x_\alpha = 0. \tag{7.8}$$

The velocity of mass point α is given by

$$\begin{aligned} v_\alpha &= \dot{R}x_\alpha\tilde{R} + Rx_\alpha\dot{\tilde{R}} + \dot{x}_0 \\ &= -\Omega_S\cdot(Rx_\alpha\tilde{R}) + v_0 \\ &= -R\Omega_B\cdot x_\alpha\tilde{R} + v_0, \end{aligned} \tag{7.9}$$

where $v_0 = \dot{x}_0$ is the velocity of the centre of mass. We next find the kinetic energy $T \equiv \sum_\alpha \frac{1}{2}m_\alpha v_\alpha^2$:

$$\begin{aligned} T &= \tfrac{1}{2}\sum_\alpha m_\alpha \left(v_0^2 - 2v_0\cdot(R\Omega_B\cdot x_\alpha\tilde{R}) + (\Omega_B\cdot x_\alpha)^2\right) \\ &= \tfrac{1}{2}mv_0^2 - \tfrac{1}{2}\sum_\alpha m_\alpha\Omega_B\cdot(x_\alpha x_\alpha\cdot\Omega_B), \end{aligned} \tag{7.10}$$

where $m \equiv \sum_\alpha m_\alpha$ is the total mass of the body. We now define the *inertia tensor* $\mathcal{I}(B)$ by

$$\mathcal{I}(B) \equiv \sum_\alpha m_\alpha x_\alpha\, x_\alpha\cdot B, \tag{7.11}$$

so that the kinetic energy can be written as

$$T = \tfrac{1}{2}mv_0^2 - \tfrac{1}{2}\Omega_B\cdot\mathcal{I}(\Omega_B). \tag{7.12}$$

The inertia tensor relates the angular velocity to the angular momentum, and is the property of a body which determines its dynamics. For a rigid body it is a constant tensor. The inertia tensor is *symmetric*, so that for any two bivectors A and B,

$$A\cdot\mathcal{I}(B) = \mathcal{I}(A)\cdot B. \tag{7.13}$$

If the argument and result of $\mathcal{I}(B)$ are both dualised, one obtains a symmetric linear function mapping vectors to vectors. The eigenvectors of this form an orthonormal frame, and it is convenient to align the initial $\{\sigma_k\}$ frame with these directions. In this case the body frame $\{e_k\}$ coincides with the principal axes.

The angular momentum bivector of the body about its centre of mass is defined by

$$\begin{aligned} L &= \sum_\alpha m_\alpha(x'_\alpha - x_0)\wedge v_\alpha \\ &= -\sum_\alpha m_\alpha Rx_\alpha\wedge(\Omega_B\cdot x_\alpha)\tilde{R} \\ &= R\mathcal{I}(\Omega_B)\tilde{R}. \end{aligned} \tag{7.14}$$

Using this result we can express the kinetic energy as $\frac{1}{2}mv_0^2 - \frac{1}{2}\Omega_S\cdot L$. This can be written in the familiar form $\frac{1}{2}mv_0^2 + \frac{1}{2}\omega_S\cdot l$, where $l = -iL$ is the angular momentum vector.

7.1.3 The Equations of Motion of a Rigid Body

To find the equations of motion of a freely-rotating rigid body, we use the rotational energy defined above as the Lagrangian,

$$\mathcal{L} = -\tfrac{1}{2}\Omega_B \cdot \mathcal{I}(\Omega_B), \tag{7.15}$$

and vary $\int dt\, \mathcal{L}$ with respect to the dynamical variables, in this case the rotor R. This rotor is an element of the even subalgebra of the Pauli algebra so has four components. However, the constraint $R\tilde{R} = 1$ means that there are only three dynamical degrees of freedom. The standard way to approach this variational problem is to parameterise the rotor in terms of three variables (usually the Euler angles) and treat these as the dynamical variables. This approach has the disadvantage of introducing a fixed coordinate system, making it difficult to assemble the final equations into a coordinate-free form. Here we adopt an alternative approach. We replace the rotor R by a spinor ψ, and enforce the constraint $\psi\tilde{\psi} = 1$ through the inclusion of a Lagrange multiplier. This method allows us to use the coordinate-free apparatus of multivector calculus in the variational principle and leads quickly to the full set of Euler equations. (An alternative way of dealing with the $\psi\tilde{\psi} = 1$ constraint is described in an exercise.)

Our modified Lagrangian is therefore

$$\mathcal{L}' = -\tfrac{1}{2}\Omega_B \cdot \mathcal{I}(\Omega_B) - \lambda(\psi\tilde{\psi} - 1), \tag{7.16}$$

where the dynamical variable is now the arbitrary even element (spinor) ψ, and λ is a Lagrange multiplier. The bivector Ω_B is now given in terms of ψ by

$$\Omega_B = -\dot{\tilde{\psi}}\psi + \tilde{\psi}\dot{\psi}, \tag{7.17}$$

which is still a bivector.

In terms of the multivector derivative, the Euler-Lagrange equations reduce to the single multivector equation

$$\partial_\psi \mathcal{L} - \partial_t(\partial_{\dot{\psi}}\mathcal{L}) = 0. \tag{7.18}$$

The symmetry of the inertia tensor simplifies the derivatives, so that[1]

$$\partial_\psi[-\tfrac{1}{2}\Omega_B \cdot \mathcal{I}(\Omega_B)] = -\overset{*}{\partial}_\psi \langle \overset{*}{\Omega}_B\, \mathcal{I}(\Omega_B)\rangle = -2\mathcal{I}(\Omega_B)\dot{\tilde{\psi}}, \tag{7.19}$$

and

$$\partial_{\dot{\psi}}[-\tfrac{1}{2}\Omega_B \cdot \mathcal{I}(\Omega_B)] = -\overset{*}{\partial}_{\dot{\psi}} \langle \overset{*}{\Omega}_B\, \mathcal{I}(\Omega_B)\rangle = 2\mathcal{I}(\Omega_B)\tilde{\psi}. \tag{7.20}$$

Employing these results, the equation of motion from varying with respect to ψ becomes (after reversing)

$$\partial_t[\psi\mathcal{I}(\Omega_B)] + \dot{\psi}\mathcal{I}(\Omega_B) = \lambda\psi, \tag{7.21}$$

[1] Editor's note: Since the dot is used here to indicate differentiation with respect to time, asterisks are employed in place of dots to limit the scope of differentiation. See section 6.5.

and variation with respect to λ sets $\psi\tilde{\psi} = 1$. We can now replace ψ by the rotor R, and arrive at the equation

$$\mathcal{I}(\dot{\Omega}_B) - \Omega_B \mathcal{I}(\Omega_B) = \lambda. \tag{7.22}$$

The scalar part of this equation determines λ and shows that, in the absence of any applied couple, the rotational energy $-\frac{1}{2}\Omega_B \cdot \mathcal{I}(\Omega_B)$ is a constant of the motion.

The equation of motion (equivalent to Euler's equations) is the bivector part of equation (7.22):

$$\mathcal{I}(\dot{\Omega}_B) - \Omega_B \times \mathcal{I}(\Omega_B) = 0. \tag{7.23}$$

Another form should be familiar:

$$\dot{L} = 0. \tag{7.24}$$

We can generalise this second form to $\dot{L} = C$, where C is the external couple on the body.

7.1.4 Free Precession of a Symmetric Top

Suppose that the body frame $\{e_k\}$ coincides with the principal axes, along which the principal moments of inertia are denoted $\{I_k\}$. The angular velocity and angular momentum can now be written in terms of the components $\{\omega_k\}$ as

$$\Omega_B = \sum_k \omega_k i\sigma_k, \qquad \Omega_S = \sum_k \omega_k i e_k, \qquad L = \sum_k I_k \omega_k i e_k. \tag{7.25}$$

In terms of these variables the rotational energy is given by the familiar expression

$$-\tfrac{1}{2}\Omega_S \cdot L = \tfrac{1}{2}\sum_k I_k \omega_k^2. \tag{7.26}$$

A 'symmetric top' is a body with two equal principal moments of inertia, say $I_1 = I_2 \neq I_3$. For this case, the third of the three Euler equations,

$$I_3\dot{\omega}_3 = (I_1 - I_2)\omega_1\omega_2, \tag{7.27}$$

shows that ω_3 is a constant. By manipulating the above expressions we obtain the following simple expression for the angular velocity of a symmetric top:

$$\Omega_S = \left(L + (I_1 - I_3)\omega_3 i e_3\right)/I_1. \tag{7.28}$$

The rotor equation can now be written in the form

$$\dot{R} = -\tfrac{1}{2}\Omega_S R = -\tfrac{1}{2}(\Omega_L R + R\Omega_R), \tag{7.29}$$

where the 'space' and 'body' precession rates Ω_L and Ω_R are defined by

$$\Omega_L \equiv \frac{1}{I_1}L, \qquad \Omega_R \equiv \omega_3 \frac{I_1 - I_3}{I_1} i\sigma_3. \tag{7.30}$$

The key feature of the rotor equation (7.29) is that both Ω_L and Ω_R are constant. Equation (7.29) therefore has the simple solution

$$R(t) = \exp(-\tfrac{1}{2}\Omega_L t)R(0)\exp(-\tfrac{1}{2}\Omega_R t), \tag{7.31}$$

which fully describes the motion of free symmetric top.

7.2 Dynamics of Elastic Media

In this section we construct a coordinate-free version of the elasticity equations. The main idea is the use of a 'displacement field' to describe the elastic distortion. As with the rigid body case, this field can be thought of as acting on a fixed (undistorted) copy of the body, and transforming it to its distorted form. In this picture, all integrals can be taken over the space of the 'undistorted' body. We will also attempt to go further than conventional approaches and treat arbitrary distortions. This is in keeping with David Hestenes' view that physicists often linearise too early. After all, even small local distortions can build up into large displacements, and the elastic body might also be rotating. It is better to write down an explicit transformation that expresses this fact, instead of using the usual approach in which this transformation is only implicit.

Consider, then, a stress-free elastic body having density $\rho(x)$ at 'undistorted' position vector x. This undistorted position is not a function of time t. The actual position $x'(t)$ of the material labelled by x can be written as a function of x,

$$x' = f(x,t), \tag{7.32}$$

and the 'displacement field' $f(x,t)$ will be our main dynamic variable. The directional derivatives of $f(x,t)$ contain information about the local distortion of the material. This information is summarised in the function

$$\underline{f}(a) = \underline{f}(a,x,t) \equiv a\cdot\nabla f(x,t) \tag{7.33}$$

which is a linear function of a and an arbitrary function of x and t. One way to think of the function $\underline{f}(a)$ is as follows: suppose that the material is filled with a series of curves (these could be realised physically using dyes in the formation process, as in a multicoloured eraser). If the tangent vector to one of these curves in the undistorted medium is given by the vector a then, after the distortion, this vector transforms to $\underline{f}(a)$.

We do not expect that a rotational component to $\underline{f}(a)$ could introduce any strains in the material, so the strain can only be a function of $\overline{f}\underline{f}(a)$. This function can be viewed as rotating the principal axes in the material, dilating through different amounts along the resulting directions, and then rotating back again. Since $\overline{f}\underline{f}(a)$ is a positive-definite function, the dilation factors (the eigenvalues) are all positive. We can therefore define further symmetric functions, such as the square root $\overline{f}\underline{f}^{1/2}(a)$ and the natural logarithm $\log\overline{f}\underline{f}(a)$.

We next need a definition of the *strain tensor*, which should conform with the conventional definition and be capable of extension to arbitrary distortions. Since the strain in the undistorted medium must be zero, a sensible definition is

$$\mathcal{E}(a) \equiv \tfrac{1}{2}\log\overline{f}\underline{f}(a). \tag{7.34}$$

This reduces to the usual definition when $\underline{f}$ is close to the identity,

$$\underline{f}(a) = a + \underline{\xi}(a), \quad \Rightarrow \quad \mathcal{E}(a) = \tfrac{1}{2}(\overline{\xi} + \underline{\xi})(a) \tag{7.35}$$

and has the additional advantage that

$$\text{tr}(\mathcal{E}) = \log \det(\underline{f}). \tag{7.36}$$

An *elastic* medium is one which has a well-defined potential energy for any given strain. An *isotropic* elastic medium is one for which the potential energy is dependent only on the eigenvalues of $\mathcal{E}(a)$, and not on the directions of the principal axes, so that

$$\text{Elastic potential energy density} = U(\{\varepsilon_k\}), \tag{7.37}$$

where $\{\varepsilon_k\}$ are the eigenvalues of the strain tensor. We can parameterise these eigenvalues by any set of three scalar invariants, but it will be convenient to use the following:

$$\begin{aligned} \text{tr}(\mathcal{E}) &= \varepsilon_1 + \varepsilon_2 + \varepsilon_3 \\ \text{tr}(\mathcal{E}^2) &= \varepsilon_1^2 + \varepsilon_2^2 + \varepsilon_3^2 \\ \text{tr}(\mathcal{E}^3) &= \varepsilon_1^3 + \varepsilon_2^3 + \varepsilon_3^3. \end{aligned} \tag{7.38}$$

For the important special case in which the elastic energy is quadratic in the strain, the third invariant is not used and we have the explicit form

$$U(\mathcal{E}) = \tfrac{1}{2}\left[2G\,\text{tr}(\mathcal{E}^2) + (B - \tfrac{2}{3}G)\,\text{tr}(\mathcal{E})^2\right], \tag{7.39}$$

where B and G are respectively the bulk and the shear moduli, and may also be functions of x.

The dynamical equations can be written down using a Lagrangian defined as an integral over the space of the 'undistorted' body,

$$\mathcal{L} = \int d^3x \left[\tfrac{1}{2}\rho(x)\dot{f}^2 - U(\mathcal{E})\right]. \tag{7.40}$$

All that remains is to derive the Euler-Lagrange equations for the dynamical variable f. Since the Lagrangian depends on f through only its time and space derivatives, the Euler-Lagrange equations are simply

$$\partial_t(\partial_{\dot{f}}\mathcal{L}) + \partial_a\!\cdot\!\nabla(\partial_{\underline{f}(a)}\mathcal{L}) = 0. \tag{7.41}$$

To evaluate the second term, it is useful to recall the following result from the chapter on linear algebra:

$$\partial_{\underline{f}(a)}\text{tr}(\mathcal{E}^n) = n\overline{f}^{-1}\mathcal{E}^{n-1}(a). \tag{7.42}$$

We now define the *stress tensor* $\mathcal{T}(a)$ via

$$\overline{f}^{-1}\mathcal{T}(a) \equiv \partial_{\underline{f}(a)}U(\mathcal{E}). \tag{7.43}$$

For the quadratic case of equation (7.39) this definition gives

$$\mathcal{T}(a) = 2G\mathcal{E}(a) + (B - \tfrac{2}{3}G)\text{tr}(\mathcal{E})a, \tag{7.44}$$

which agrees with the conventional result. We also see that, for the quadratic case, $U(\varepsilon) = \frac{1}{2}\partial_a \cdot [\mathcal{E}\mathcal{T}(a)]$. More generally, if we treat $U(\mathcal{E})$ as a function of the independent variables $\mathrm{tr}(\mathcal{E})$, $\mathrm{tr}(\mathcal{E}^2)$ and $\mathrm{tr}(\mathcal{E}^3)$, we can write

$$\mathcal{T}(a) = \frac{\partial U}{\partial \mathrm{tr}(\mathcal{E})} a + 2\frac{\partial U}{\partial \mathrm{tr}(\mathcal{E}^2)}\mathcal{E}(a) + 3\frac{\partial U}{\partial \mathrm{tr}(\mathcal{E}^3)}\mathcal{E}^2(a). \tag{7.45}$$

The Euler-Lagrange equations now assemble to give

$$\rho(x)\ddot{f} = \overline{f}^{-1}\mathcal{T}(\overleftarrow{\nabla}), \tag{7.46}$$

where $\overleftarrow{\nabla}$ acts on everything to its left.

7.2.1 Energy Flow

The total energy (kinetic + elastic) of a medium is given by

$$E = \int d^3x\, [\tfrac{1}{2}\rho(x)\dot{f}^2 + U(\mathcal{E})]. \tag{7.47}$$

The rate of change of this energy is

$$\begin{aligned} \partial_t E &= \int d^3x\, [\rho(x)\dot{f}\cdot\ddot{f} + \partial_t U(\mathcal{E})] \\ &= \int d^3x\, \left(\dot{f}\cdot[\overline{f}^{-1}\mathcal{T}(\overleftarrow{\nabla})] + \partial_t U(\mathcal{E})\right), \end{aligned} \tag{7.48}$$

where we have used the equation of motion. Now $\mathcal{E}(a)$ only depends on the displacement field through its derivatives $\underline{f}(a)$, so we can write

$$\begin{aligned} \partial_t U(\mathcal{E}) &= \partial_t \underline{f}(\partial_a)\cdot\partial_{\underline{f}(a)} U(\mathcal{E}) \\ &= \overset{*}{\dot{f}}\cdot\underline{f}^{-1}\mathcal{T}(\overset{*}{\nabla}). \end{aligned} \tag{7.49}$$

We now see that the rate of change of energy is determined by a total derivative, so we can transform to a surface integral and find that

$$\partial_t E = \oint d^2S\, \dot{f}\cdot\overline{f}^{-1}\mathcal{T}(n). \tag{7.50}$$

We accordingly identify $\dot{f}\cdot\overline{f}^{-1}\mathcal{T}(n)$ as the energy flow per unit area perpendicular to n.

7.2.2 Pre-Stressed Media

Not all solids are stress-free when in equilibrium; many are manufactured with stresses already present. The above analysis can easily be extended to describe such pre-stressed media. One first introduces a new linear function $\underline{h}(a) = \underline{h}(a, x)$

to represent the 'equilibrium strains' of the medium. This function need not be derived from a distortion field, because of all the interesting things that can be done to the body during its manufacture. Whether or not $\underline{h}(a)$ can be derived from a distortion is determined by the quantity $\nabla\wedge\overline{h}(a)$. This is zero if and only if $\underline{h}(a)$ is the derivative of a distortion field, since $\nabla\wedge\overline{f}(a) = 0$ for any differential $\underline{f}(a) = a\cdot\nabla f(x)$.

The pre-stresses are incorporated by generalising $\underline{f}(a)$ to $\underline{f}\,\underline{h}(a)$. The strain tensor is now

$$\mathcal{E}(a) = \tfrac{1}{2}\log[\overline{h}\,\overline{f}\,\underline{f}\,\underline{h}](a) \tag{7.51}$$

and the key result for the derivatives of this is

$$\partial_{\underline{f}(a)}\partial_b\cdot\mathcal{E}^n(b) = n\overline{f}^{-1}\overline{h}^{-1}\mathcal{E}^{n-1}\overline{h}(a). \tag{7.52}$$

All the other results, and the dynamical equations, go through unchanged.

If the material is in equilibrium (no externally applied stresses) then $f(x) = x$ and $\underline{f}$ is the identity. In this case the stress tensor is controlled solely by $\underline{h}(a)$, and the field equations reduce to

$$\overset{*}{\mathcal{T}}(\overset{*}{\nabla}) = 0. \tag{7.53}$$

As a final comment, note that for isotropic media there is a rotational 'gauge' freedom in the definition of $\underline{h}$.

7.2.3 Linearised elasticity

The conventional form of the dynamical equations is easily recovered by writing

$$x' = x + \xi(x), \tag{7.54}$$

where the distortion $\xi(x)$ is small. If we work to first order in ξ the strain tensor is

$$\mathcal{E}(a) = \tfrac{1}{2}(\underline{\xi}(a) + \overline{\xi}(a)), \tag{7.55}$$

and the dynamical equations reduce to

$$\rho(x)\ddot{\xi} = \overset{*}{\mathcal{T}}(\overset{*}{\nabla}). \tag{7.56}$$

In evaluating $\overset{*}{\mathcal{T}}(\overset{*}{\nabla})$ the results

$$\overset{*}{\underline{\xi}}(\overset{*}{\nabla}) = \nabla\cdot\nabla\xi = \nabla^2\xi \tag{7.57}$$

and

$$\overset{*}{\overline{\xi}}(\overset{*}{\nabla}) = \partial_a\langle\nabla a\cdot\nabla\xi\rangle = \nabla\nabla\cdot\xi. \tag{7.58}$$

are useful.

In many applications one assumes a harmonic time variation as $\cos(\omega t)$. Substituting this into (7.56) it is then a small step to the *vector Helmholtz equation*, here expressed in terms of transverse and longitudinal components:

$$v_l^2\nabla\nabla\cdot\xi + v_t^2\nabla\cdot(\nabla\wedge\xi) + \omega^2\xi = 0, \tag{7.59}$$

where v_l, v_t are the longitudinal and transverse sound speeds. This equation is used to study many phenomena, ranging from oscillations of an elastic sphere to the propagation of waves created by an earthquake.

7.2.4 The Elastic Filament

As a further, novel application of the rotor equation, we treat the bending and twisting of an elastic filament under static loads. Suppose that the filament is described by the curve $x(\lambda)$. We will choose λ to be the affine parameter along the curve, so that $x' = \partial_\lambda x$ is a unit vector. This vector can be identified with the third vector of an orthonormal frame,

$$x' = e_3 = R(\lambda)\sigma_3\tilde{R}(\lambda). \tag{7.60}$$

The remaining two vectors then determine two directions perpendicular to the filament, and can be used to describe any internal twisting in the filament. With this approach, both the bending and twisting of the filament are described in the single equation for the rotor R.

A thin beam or filament has stiffness to bending. When it is bent, a bending moment (couple) is set up which is linearly related to the curvature. In terms of the two principal directions in the filament, the appropriate formula for the bending moment is

$$M = \frac{YI}{R}, \tag{7.61}$$

where Y is Young's modulus, I is the relevant principal moment of area, and R is the radius of curvature in the plane of the bending. The radius of curvature is determined by the magnitude of the projection of the vector $\dot{e}_3$ into the relevant plane. So the radius of curvature in the e_1e_3 plane, for example, is given by

$$\frac{1}{R_1} = |\dot{e}_3 \cdot (e_1e_3)e_3e_1| = |e_1 \cdot \dot{e}_3|. \tag{7.62}$$

Since we know that $\dot{e}_3 = -\Omega_S \cdot e_3$, where $\Omega_S = -2\dot{R}\tilde{R}$, we see that the radius of curvature just picks out one coefficient of Ω_S.

Equation (7.61) can correspondingly be used to find the curvature induced by an applied couple C. With C and Ω_S given in terms of components by

$$C = \sum_k c_k i e_k, \qquad \Omega_S = \sum_k \omega_k i e_k, \tag{7.63}$$

we find that the curvature and the couple are related by

$$c_1 = YI_1\omega_1, \qquad c_2 = YI_2\omega_2, \tag{7.64}$$

where I_1 is the moment of area measured perpendicular to the e_1 direction.

In addition to its stiffness to bending, the filament has a stiffness to torsion. For the case of elastic behaviour, the twist in the e_1e_2 plane is proportional to the applied torque, and we have

$$c_3 = GI_3\dot{e}_1 \cdot e_2 = GI_3\omega_3. \tag{7.65}$$

The applied couple C and the 'curvature bivectors' Ω_S and Ω_B are therefore related by

$$C = Y(I_1\omega_1\, ie_1 + I_2\omega_2\, ie_2) + GI_3\omega_3\, ie_3 = R\mathcal{I}(\Omega_B)\tilde{R}. \tag{7.66}$$

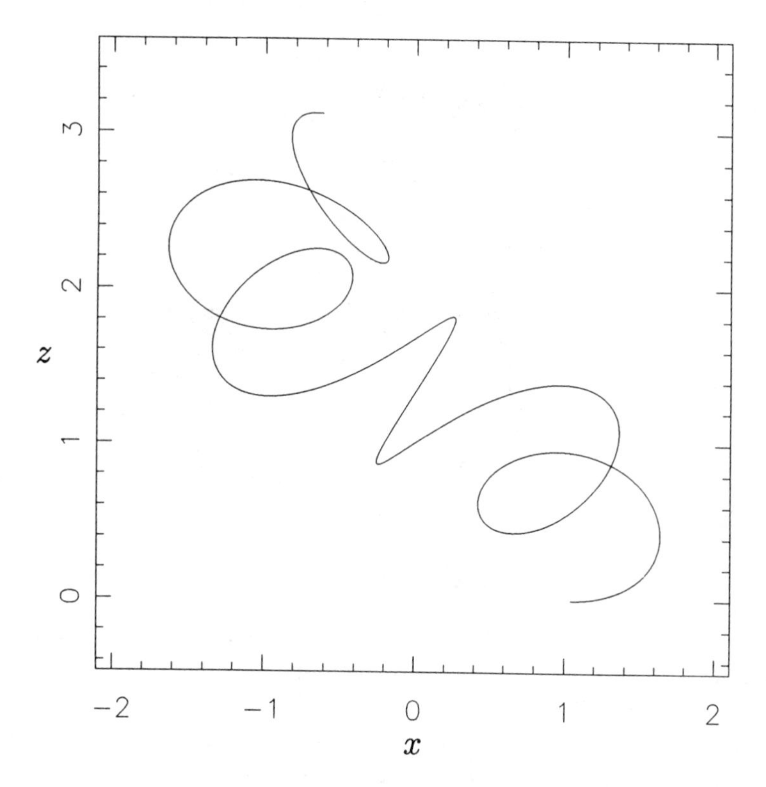

Fig. 7.1. *An example of the shapes that can arise when a light filament is loaded at its two ends. Two directions are shown, though there is also considerable structure in the third. The material has* $I_1 = I_2$ *and a zero Poisson ratio.*

We can invert this relation to give

$$\Omega_B = \mathcal{I}^{-1}(\tilde{R}CR), \tag{7.67}$$

which expresses the curvature bivector Ω_B in terms of the applied couple C and the elastic constants. The full set of equations are now (7.60) and (7.67), together with the rotor equation

$$\frac{dR}{d\lambda} = -\tfrac{1}{2}R(\Omega_B + \Omega_0), \tag{7.68}$$

where the bivector Ω_0 expresses the natural shape of the filament.

An advantage of this set of equations is that locally small distortions of the filament can be allowed to build up into large, global deviations. An interesting simple case is that of a *wrench*, where

$$C(x) = C_0 + F\wedge x, \tag{7.69}$$

where C_0 and F are respectively the couple and force applied at the ends. A wrench such as this describes the general case of a light filament loaded at its ends. Figure 7.1 shows the type of distortion that can result.

Exercises

1. Prove that the inertia tensor is symmetric, $A \cdot \mathcal{I}(B) = \mathcal{I}(A) \cdot B$.

2. For a Lagrangian of the form $\mathcal{L} = \mathcal{L}(\psi, \dot{\psi}, t)$, where ψ is a spinor field, the action is defined by

$$S \equiv \int dt\, \mathcal{L}(\psi, \dot{\psi}, t).$$

Prove that the spinor $\psi(t)$ which extremises this action is found by solving the Euler-Lagrange equation

$$\partial_\psi \mathcal{L} - \partial_t(\partial_{\dot{\psi}} \mathcal{L}) = 0.$$

3. An alternative approach to the problem of incorporating the constraint $\psi\tilde{\psi} = 1$ into the variational procedure is to define Ω_B by

$$\Omega_B = -\psi^{-1}\dot{\psi} + \dot{\tilde{\psi}}\tilde{\psi}^{-1}.$$

Show that the simple Lagrangian $\mathcal{L} = -\frac{1}{2}\Omega_B \cdot \mathcal{I}(\Omega_B)$ now leads directly to the required equation $\dot{L} = 0$. Why does this method work?

4. For a symmetric top show that, provided there is no couple in the $e_1 e_2$ plane, the equations of motion can always be reduced to the form

$$\dot{R}_1 = -\frac{L}{I_1} R_1, \qquad \dot{L} = C.$$

Use this to extend the treatment of the freely precessing symmetric top to the case of a gyroscope in a gravitational field. (A gyroscope is a symmetric top with one end fixed at a point in such a way that it is free to rotate around the point.)

5. Consider an infinite isotropic medium with a spherical hole into which air is pumped. Show that, in the linearised theory, the radial stress τ_r is related to the radius of the hole r as $\tau_r \propto r^{-3}$. Discuss how the full non-linear theory might modify this result.

6. Show that, in the linearised theory, the longitudinal and transverse sound speeds are given by

$$v_l^2 = \frac{1}{\rho}\left(B + \tfrac{4}{3}G\right), \qquad v_t^2 = \frac{G}{\rho}.$$

Chapter 8

Electromagnetism

Stephen Gull, Chris Doran, and Anthony Lasenby
MRAO, Cavendish Laboratory, Madingley Road,
Cambridge, CB3 0HE, UK

In this lecture we discuss applications of STA to problems in electromagnetism. In the STA, we write the electromagnetic field in terms of the 4-potential A as

$$F = \nabla\wedge A = \nabla A - \nabla\cdot A. \tag{8.1}$$

The divergence term $\nabla\cdot A$ is zero in the Lorenz gauge. The field bivector F is expressed in terms of the more familiar electric and magnetic fields by making a space-time split in the γ_0 frame:

$$F = \boldsymbol{E} + i\boldsymbol{B}, \tag{8.2}$$

where

$$\boldsymbol{E} = \tfrac{1}{2}\left(F - \gamma_0 F \gamma_0\right), \qquad i\boldsymbol{B} = \tfrac{1}{2}\left(F + \gamma_0 F \gamma_0\right). \tag{8.3}$$

It is particularly striking that Maxwell's equations can be written in the simple form

$$\nabla F = J, \tag{8.4}$$

where J is the 4-current. Equation (8.4) contains *all* of Maxwell's equations because the ∇ operator is a vector and F is a bivector, so their geometric product has both vector and trivector components. The trivector part is identically zero in the absence of magnetic charges. It is worth emphasising that this compact formula (8.4) is not just a trick of notation. The ∇ operator is invertible and we can solve for F directly:

$$F = \nabla^{-1} J. \tag{8.5}$$

The inverse operator is known to physicists in the guise of the propagators of the Dirac electron theory. We later demonstrate this inversion explicitly for diffraction theory.

As practise in manipulating quantities in the STA, we demonstrate explicitly that the STA equation (8.4) leads to Maxwell's equations in the form familiar to all physics students. Take a space-time split of the equations by pre-multiplying by γ_0:

$$(\partial_t + \boldsymbol{\nabla})\,(\boldsymbol{E} + i\boldsymbol{B}) = \rho - \boldsymbol{J}, \tag{8.6}$$

where ρ is the charge density and $\boldsymbol{J}$ the current density. We expand the LHS using the result that, for the relative vector $\boldsymbol{E}$,

$$\boldsymbol{\nabla}\boldsymbol{E} = \boldsymbol{\nabla}\cdot\boldsymbol{E} + i\boldsymbol{\nabla}\times\boldsymbol{E}, \tag{8.7}$$

where $\times$ denotes the conventional 3d vector product. The resulting 3d multivector equation contains terms of all 4 grades (0 to 3):

$$\begin{array}{rcll} \boldsymbol{\nabla}\cdot\boldsymbol{E} &=& \rho & \text{(scalar)} \\ \partial_t\boldsymbol{E} - \boldsymbol{\nabla}\times\boldsymbol{B} &=& -\boldsymbol{J} & \text{(vector)} \\ i(\partial_t\boldsymbol{B} + \boldsymbol{\nabla}\times\boldsymbol{E}) &=& 0 & \text{(bivector)} \\ i\boldsymbol{\nabla}\cdot\boldsymbol{B} &=& 0 & \text{(trivector).} \end{array} \tag{8.8}$$

We recognise these as the conventional form of Maxwell's equations in free space.

Dielectric and Magnetic Media

The STA form of Maxwell's equation (8.4) is generalised to the case of media with linear permittivity ϵ_r and permeability μ_r by introducing the quantity

$$G \equiv \boldsymbol{D} + i\boldsymbol{H}, \tag{8.9}$$

where

$$\boldsymbol{D} \equiv \epsilon_r\boldsymbol{E}, \qquad \boldsymbol{B} \equiv \mu_r\boldsymbol{H}. \tag{8.10}$$

Maxwell's equations are now given by the pair of equations

$$\begin{array}{rcl} \nabla\wedge F &=& 0 \\ \nabla\cdot G &=& J, \end{array} \tag{8.11}$$

together with the (constitutive) relation between F and G.

The Electromagnetic Stress-Energy Tensor

The stress-energy tensor $T(a)$ is an important concept in field theory. It is a vector-valued linear function of its vector argument a (a 2-component tensor in the standard formalism). For classical fields, $T(a)$ is symmetric and returns the 4-momentum flux transported through a surface perpendicular to a. For the case of the electromagnetic field, $T(a)$ is given by

$$T(a) = -\tfrac{1}{2}FaF. \tag{8.12}$$

As a check, this definition gives the energy density in the γ_0 frame by the familiar formula

$$\gamma_0\cdot T(\gamma_0) = \tfrac{1}{2}(\boldsymbol{E}^2 + \boldsymbol{B}^2). \tag{8.13}$$

In free space the stress-energy tensor gives rise to a set of conserved currents

$$\nabla \cdot T(a) = 0. \tag{8.14}$$

This equation can be written equivalently as

$$\dot{T}(\dot{\nabla}) = 0, \tag{8.15}$$

where the ∇ acts on T.

8.1 Electromagnetic Waves

8.1.1 Stokes Parameters

Consider a plane wave travelling in the $+z$ direction:

$$F = (1+\sigma_3)[(A\sigma_1 + C\sigma_2)\cos(kz - \omega t) + (B\sigma_1 + D\sigma_2)\sin(kz - \omega t)], \tag{8.16}$$

where the constants A, B, C and D are all real. Another way of expressing the polarisation is in terms of the complex Jones vector

$$\boldsymbol{E} = \Re\left((\sigma_1 C_1 + \sigma_2 C_2)e^{j(kz - \omega t)}\right), \tag{8.17}$$

so that $C_1 = A - jB$ and $C_2 = C - jD$. Note that the complexification performed here is purely formal, and there is no geometric significance to the unit imaginary j. As we will soon see, the situation is different for circularly-polarised light.

Following the IAU convention, the Stokes parameters are defined by

$$\begin{aligned} I &= A^2 + B^2 + C^2 + D^2 = |C_1|^2 + |C_2|^2 \\ Q &= A^2 + B^2 - C^2 - D^2 = |C_1|^2 - |C_2|^2 \\ U &= 2(AC + BD) = 2\Re(C_1 C_2^*) \\ V &= 2(BC - AD) = -2\Im(C_1 C_2^*). \end{aligned} \tag{8.18}$$

For partially polarised radiation, we simply make the replacements $AC \longrightarrow \langle AC \rangle$, *etc.* This (plane polarised) representation picks out the direction σ_1 as the polarisation labelled $+Q$. Since the basis set $\{\sigma_1, \sigma_2, \sigma_3\}$ form a right-handed frame, the equivalent for spherical geometry is $\{\sigma_\theta, \sigma_\phi, \sigma_r\}$. So, for a wave travelling radially outwards, the σ_θ direction picks out $+Q$, σ_ϕ represents $-Q$ and $+U$ is half-way between. Plane polarised waves are slightly clumsy in the STA, because of the artificial introduction of the complex structure. The solutions will be needed, however, when we discuss problems involving dielectrics and conductors.

Now consider right-hand circularly polarised waves. These are labelled $+V$, so that one form is given by $A = -D$ and $B = C = 0$. For this we can write

$$\begin{aligned} F &= A(1+\sigma_3)\sigma_1\, e^{-i\sigma_3(kz - \omega t)} \\ &= A(1+\sigma_3)\sigma_1\, e^{i(kz - \omega t)}. \end{aligned} \tag{8.19}$$

In this case the complex structure has arisen geometrically and is provided by the pseudoscalar. This wave has its electric vector rotating clockwise at a fixed plane, when viewed from the direction of the source. (Of course, this means that the end of the $\boldsymbol{E}$ vector describes a left-handed helix in space at any give time!) More generally, for the circularly polarised STA modes we can write

$$F = (1+\sigma_3)\sigma_1 \left(Re^{i(kz - \omega t)} + Le^{-i(kz - \omega t)}\right), \tag{8.20}$$

where the coefficients are now (scalar + pseudoscalar) complex numbers. The correspondence with the plane polarisation notation is

$$\begin{aligned} A &= R_R + L_R \qquad & B &= -R_I + L_I \\ C &= -R_I - L_I \qquad & D &= -R_R + L_R, \end{aligned} \tag{8.21}$$

so that

$$\begin{aligned} I &= 2(|R|^2 + |L|^2) \qquad & V &= 2(|R|^2 - |L|^2) \\ Q &= 4\Re(LR) \qquad & U &= -4\Im(LR). \end{aligned} \tag{8.22}$$

This representation favours V, rather than Q. Note also the factors of 2.

8.1.2 Reflection by a Conducting Plane

A conducting surface imposes the boundary conditions

$$\boldsymbol{E}_{\parallel} = 0, \qquad \boldsymbol{B}_{\perp} = 0. \tag{8.23}$$

Symmetry leads us to consider two special case which have, respectively, $\boldsymbol{E}$ and $\boldsymbol{H}$ in the reflecting plane. The first case we call 'transverse electric' (TE) or, sometimes, 'magnetic' — this latter name arising from the fact that the surface must generate a longitudinal magnetic field to satisfy the boundary conditions. For this case we write the incident and reflected waves as

$$\begin{aligned} &\text{incident:} \qquad && (1 + \cos\!\phi\, \sigma_1 + \sin\!\phi\, \sigma_2)\sigma_3 \cos(k_I \!\cdot\! x) \\ &\text{reflected:} \qquad && R_M(1 - \cos\!\phi\, \sigma_1 + \sin\!\phi\, \sigma_2)\sigma_3 \cos(k_R \!\cdot\! x), \end{aligned} \tag{8.24}$$

where

$$k_I = \omega(\gamma_0 + \cos\!\phi\, \gamma_1 + \sin\!\phi\, \gamma_2) \tag{8.25}$$

and

$$k_R = \omega(\gamma_0 - \cos\!\phi\, \gamma_1 + \sin\!\phi\, \gamma_2). \tag{8.26}$$

At $x = 0$ the σ_3 and $i\sigma_1$ components must vanish for all y and t, so the reflection coefficient R_M must be -1.

For the other case the $\boldsymbol{H}$ field is parallel to σ_3, so the fields are now

$$\begin{aligned} &\text{incident:} \qquad && (1 + \cos\!\phi\, \sigma_1 + \sin\!\phi\, \sigma_2)i\sigma_3 \cos(k_I \!\cdot\! x) \\ &\text{reflected:} \qquad && R_E(1 - \cos\!\phi\, \sigma_1 + \sin\!\phi\, \sigma_2)i\sigma_3 \cos(k_R \!\cdot\! x). \end{aligned} \tag{8.27}$$

This time the $\sigma_1 i\sigma_3 = -i\sigma_2$ component must vanish, implying $R_E = +1$.

For this particularly simple example, the wave is totally reflected, but the reflection coefficients have opposite sign for the TE and TM modes. For a circularly polarised incident mode, the effect is that the reflected wave has the opposite handedness.

8.1.3 Waves in layered media

Now consider the propagation of electromagnetic waves through a layered medium, consisting of $N+1$ parallel layers each normal to the x-axis. The wave has frequency ω and is incident in the x-y plane. The angle of incidence is denoted by ϕ_i, and the transmitted and reflected wave amplitudes are denoted T_i and R_i respectively. The ith layer of the material has permittivity ϵ_i, permeability μ_i, wave impedance $Z_i = \sqrt{(\mu_i/\epsilon_i)}$, refractive index $n_i = \sqrt{(\epsilon_i\mu_i)}$ and wave number $k_i = \omega n_i/c$.

We first solve the matching problem at the 1-2 boundary. Again there are two cases to consider, depending on whether $\boldsymbol{H}$ or $\boldsymbol{E}$ lies in plane of the boundary. We first take the case where $\boldsymbol{H}$ is in the boundary plane:

$$\begin{aligned} T_1 + R_1 &= T_2 + R_2 && \text{(continuity of } H_\parallel) \\ Z_1 \cos\phi_1 (T_1 - R_1) &= Z_2 \cos\phi_2 (T_2 - R_2) && \text{(continuity of } E_\parallel) \\ \epsilon_1 Z_1 \sin\phi_1 (T_1 + R_1) &= \epsilon_2 Z_2 \sin\phi_2 (T_2 + R_2) && \text{(continuity of } D_\perp), \end{aligned} \tag{8.28}$$

where the transmission and reflection amplitudes correspond to the magnitudes of $\boldsymbol{H}$. The continuity of $D_\perp$ implies Snell's Law: $n_i \sin\phi_i =$ constant. These equations can easily be manipulated into the form

$$\begin{pmatrix} T_2 \\ R_2 \end{pmatrix} = \frac{1}{2} \begin{pmatrix} 1+\alpha\beta & 1-\alpha\beta \\ 1-\alpha\beta & 1+\alpha\beta \end{pmatrix} \begin{pmatrix} T_1 \\ R_1 \end{pmatrix}, \tag{8.29}$$

where

$$\alpha \equiv Z_1/Z_2 \qquad \beta \equiv \cos\phi_1/\cos\phi_2. \tag{8.30}$$

The matrix which gives T_2 and R_2 from T_1 and R_1 is the transmission matrix. Note that if $\alpha\beta = 1$ there is no reflection (this is the Brewster condition).

Now consider the other polarisation, where $\boldsymbol{E}$ is in the plane of the boundary:

$$\begin{aligned} \cos\phi_1 (T_1 + R_1) &= \cos\phi_2 (T_2 + R_2) && \text{(continuity of } H_\parallel) \\ Z_1 (T_1 - R_1) &= Z_2 (T_2 - R_2) && \text{(continuity of } E_\parallel) \\ \mu_1 \sin\phi_1 (T_1 - R_1) &= \mu_2 \sin\phi_2 (T_2 - R_2) && \text{(continuity of } B_\perp). \end{aligned} \tag{8.31}$$

The transmission matrix is now

$$\begin{pmatrix} T_2 \\ R_2 \end{pmatrix} = \frac{1}{2} \begin{pmatrix} \alpha+\beta & \beta-\alpha \\ \beta-\alpha & \alpha+\beta \end{pmatrix} \begin{pmatrix} T_1 \\ R_1 \end{pmatrix}. \tag{8.32}$$

The next part of the problem is the propagation across a layer, which is simply solved by a matrix consisting of phase factors

$$\begin{pmatrix} T_2 \\ R_2 \end{pmatrix} = \begin{pmatrix} e^{i\delta} & 0 \\ 0 & e^{-i\delta} \end{pmatrix} \begin{pmatrix} T_1 \\ R_1 \end{pmatrix}, \tag{8.33}$$

where δ is the phase difference across the layer. All that remains to do is to assemble the full boundary/propagation matrix by multiplication of the individual transmission and propagation matrixes for each layer, and then solve for the boundary condition $R_N = 0$.

8.2 Diffraction Theory

We turn now to the diffraction of electromagnetic waves. Suppose that a source produces monochromatic waves incident on an aperture in a surface S. What is the field in the interior volume V? In order to determine this we need the equivalent of Kirchoff's diffraction theory. This general theory is not well known, so we outline it here. The advantages that first-order equations have over their second-order counterparts are clear in this work.

8.2.1 The Boundary-Value Problem in Electrodynamics

Consider the propagation of electromagnetic waves from a surface into free space. The time-dependence of the waves is expressed as

$$F(x) = F(\boldsymbol{x})e^{-i\omega t}, \tag{8.34}$$

so the Maxwell equation $\nabla F = 0$ becomes

$$\boldsymbol{\nabla} F - i\omega F = 0. \tag{8.35}$$

This is our first-order equivalent of the Helmholtz equation.

The key quantity we need is the Green's function $G(\boldsymbol{x}, \boldsymbol{x}')$ which satisfies

$$\boldsymbol{\nabla}_{\boldsymbol{x}'} G(\boldsymbol{x}, \boldsymbol{x}') - i\omega G(\boldsymbol{x}, \boldsymbol{x}') = \delta(\boldsymbol{x} - \boldsymbol{x}'). \tag{8.36}$$

With this Green's function we can calculate F in the interior of some closed region of space with the formula

$$F(\boldsymbol{x}) = \oint |dS(\boldsymbol{x}')| \, \langle \tilde{G}(\boldsymbol{x}, \boldsymbol{x}') \boldsymbol{n}(\boldsymbol{x}') F(\boldsymbol{x}') \rangle_2, \tag{8.37}$$

where $\boldsymbol{n}$ is the normal to the surface pointing *into* the volume V. That equation (8.37) is correct follows from a simple aplication of Green's theorem:

$$\begin{aligned} &\oint |dS(\boldsymbol{x}')| \, \langle \tilde{G}(\boldsymbol{x}, \boldsymbol{x}') \boldsymbol{n}(\boldsymbol{x}') F(\boldsymbol{x}') \rangle_2 && (8.38) \\ &\quad = -\int dV(\boldsymbol{x}') \, \langle \tilde{G}(\boldsymbol{x}, \boldsymbol{x}') \overleftrightarrow{\boldsymbol{\nabla}}_{\boldsymbol{x}'} F(\boldsymbol{x}') \rangle_2 \\ &\quad = \int dV(\boldsymbol{x}') \, \langle [\boldsymbol{\nabla}_{\boldsymbol{x}'} G(\boldsymbol{x}, \boldsymbol{x}') - i\omega G(\boldsymbol{x}, \boldsymbol{x}')]^{\sim} F(\boldsymbol{x}') \rangle_2 \\ &\quad = F(\boldsymbol{x}). && (8.39) \end{aligned}$$

The Greens' function is easily found by standard methods: let

$$G \equiv \boldsymbol{\nabla}\phi + i\omega\phi, \tag{8.40}$$

so that

$$\boldsymbol{\nabla}^2\phi + \omega^2\phi = \delta. \tag{8.41}$$

This is a scalar operator equation, and the Green's function representing outgoing waves is given by

$$\phi = -\frac{1}{4\pi} e^{i\omega r}, \tag{8.42}$$

where

$$r \equiv |\boldsymbol{r}| \qquad \boldsymbol{r} \equiv \boldsymbol{x} - \boldsymbol{x}'. \tag{8.43}$$

We therefore find that

$$G(\boldsymbol{x}, \boldsymbol{x}') = -\frac{1}{4\pi} \left(i\omega e^{i\omega r}/r - \sigma_r \partial_r (e^{i\omega r}/r) \right), \tag{8.44}$$

where σ_r is the unit vector in the direction of $\boldsymbol{r}$.

Substituting this result into equation (8.37) we now find that

$$F(\boldsymbol{x}) = -\frac{1}{4\pi} \oint |dS(\boldsymbol{x}')| \, \langle [i\omega e^{i\omega r}/r - \sigma_r \partial_r (e^{i\omega r}/r)] \boldsymbol{n}(\boldsymbol{x}') F(\boldsymbol{x}') \rangle_2. \tag{8.45}$$

This result contains all the necessary polarisation and obliquity factors, and is equivalent to results derived at considerable length in standard optics texts. A significant advantage of the STA approach is that first-order equations satisfy Huygens' principle. This is evident from (8.37) — F is propagated into the interior simply by multiplying it by a Green's function. This accords with Huygen's original idea of re-radiation of wavelets from any given wavefront.

The traditional approach to the propagation of electromagnetic waves is to construct the wave equation $\nabla^2 A = 0$ for the electromagnetic potential in the Lorenz gauge. The components of A then satisfy the scalar Helmholtz equation

$$\boldsymbol{\nabla}^2 \phi + \omega^2 \phi = 0, \tag{8.46}$$

and the solution of this is

$$\phi(\boldsymbol{x}) = \oint |dS(\boldsymbol{x}')| \, (\phi \boldsymbol{n} \cdot \boldsymbol{\nabla} G - G \boldsymbol{n} \cdot \boldsymbol{\nabla} \phi), \tag{8.47}$$

where G is now the Green's function for the wave equation. But now we are in a quandary: do we use the Dirichlet Green's function ($G = 0$ on S) and take the first term? Or do we set $\boldsymbol{n} \cdot \boldsymbol{\nabla} G = 0$ (Neumann conditions) and use the second term? The two approaches yield different results: the first term has a $\cos\theta$ obliquity factor; the second has none. Actually (as the standard texts state) one should take equal amounts of each term, so as to obtain the correct obliquity factor $\frac{1}{2}(1+\cos\theta)$. There is no such ambiguity for the first-order equation, because the the Green's function *automatically* picks out the appropriate components propagating *into* V rather than away from it.

A subtlety connected with the propagation of electromagnetic waves is that we do not have complete freedom to specify F over the surface of the volume. In particular, the equations

$$\boldsymbol{\nabla} \wedge \boldsymbol{E} = i\omega \boldsymbol{E}, \qquad \boldsymbol{\nabla} \wedge \boldsymbol{B} = i\omega \boldsymbol{B} \tag{8.48}$$

mean that the components of $\boldsymbol{E}$ and $\boldsymbol{B}$ perpendicular to the boundary surface are determined by the derivatives of the components in the surface. This reduces the number of degrees of freedom in the problem from six to four, as is required for electromagnetism.

8.3 The Electromagnetic Field of a Point Charge

As a final example of the power of the STA in relativistic physics, we give a compact formula for the fields of a radiating charge. Suppose that a charge q moves along a world-line $x_0(\tau)$, where τ is the proper time. An observer at spacetime position x receives an electromagnetic influence from the point where the charge's worldline intersects the observer's past light-cone. The vector

$$X \equiv x - x_0(\tau) \tag{8.49}$$

is the separation vector down the light-cone, joining the observer to this intersection point. Since this vector must be null, we can view the equation

$$X^2 = 0 \tag{8.50}$$

as defining a map from spacetime position x to a value of the particle's proper time τ. That is, for every spacetime position x there is a unique value of the proper time along the charge's world-line for which the vector connecting x to the world-line is null. In this sense, we can write $\tau = \tau(x)$, and treat τ as a scalar field.

If the charge is at rest in the observer's frame we have

$$x_0(\tau) = \tau\gamma_0 = (t - r)\gamma_0, \tag{8.51}$$

where r is the 3-space distance from the observer to the charge. The null vector X is therefore

$$X = r(\gamma_0 + e_r). \tag{8.52}$$

For this simple case the 4-potential A is a pure $1/r$ electrostatic field, which we can write as

$$A = \frac{q}{4\pi\epsilon_0} \frac{\gamma_0}{X \cdot \gamma_0}. \tag{8.53}$$

This formula is actually a special case of the general expression

$$A = \frac{q}{4\pi\epsilon_0} \frac{v}{X \cdot v}, \tag{8.54}$$

where $v = \dot{x}_0$ is the velocity of the charge. Equation (8.54) now describes the potential for a charge moving with an arbitrary velocity v. This formula is a particularly compact and clear expression of the Liénard-Wiechert potential.

We now wish to differentiate the potential to find the Faraday bivector. This will involve some general results derived in earlier lectures. The first result follows immediately from the chain rule:

$$\nabla x_0 = \nabla\tau \, v. \tag{8.55}$$

Next, we can differentiate the equation $X^2 = 0$ to obtain

$$\nabla X X + \overset{*}{\nabla} X \overset{*}{X} = 0 = (4 - \nabla\tau \, v)X + (-2X - \nabla\tau \, X v), \tag{8.56}$$

from which it follows that

$$\nabla\tau = \frac{X}{X\cdot v}. \tag{8.57}$$

This explicit expression for $\nabla\tau$ confirms that the particle's proper time can be treated as a scalar field — which is, perhaps, a surprising result.

To differentiate A, we need $\nabla(X\cdot v)$. Using the results already established we have

$$\nabla(vX) = \nabla\tau\,\dot{v}X - 2v - \nabla\tau\,v^2 = \frac{X\dot{v}X - 2X - vXv}{X\cdot v} \tag{8.58}$$

and

$$\nabla(Xv) = \nabla\tau\,X\dot{v} + 4v - \nabla\tau\,v^2 = \frac{2vXv + X}{X\cdot v}, \tag{8.59}$$

which combine to give

$$\nabla(X\cdot v) = \frac{X\dot{v}X - X + vXv}{2(X\cdot v)}. \tag{8.60}$$

Combining these various results we now find that

$$\begin{aligned}\nabla A &= \frac{q}{4\pi\epsilon_0}\left(\frac{\nabla v}{X\cdot v} - \frac{1}{(X\cdot v)^2}\nabla(X\cdot v)v\right)\\ &= \frac{q}{8\pi\epsilon_0(X\cdot v)^3}\left(X\dot{v}vX + Xv - vX\right).\end{aligned} \tag{8.61}$$

The bracketed term is a pure bivector, so $\nabla\cdot A = 0$ and the A field defined by (8.54) is in the Lorenz gauge. The Faraday bivector can now be written in the form

$$F = \frac{q}{4\pi\epsilon_0}\frac{X\wedge v + \frac{1}{2}X\Omega_v X}{(X\cdot v)^3}, \tag{8.62}$$

where Ω_v is the acceleration bivector of the particle,

$$\Omega_v \equiv \dot{v}v = \dot{v}\wedge v. \tag{8.63}$$

The form of the Faraday bivector given by equation (8.62) is very revealing. It displays a clean split into a velocity term proportional to $1/(\text{distance})^2$ and a long-range radiation term proportional to $1/(\text{distance})$. The first term is exactly the Coulomb field in the rest frame of the charge, and the radiation term,

$$F_{rad} = \frac{q}{4\pi\epsilon_0}\frac{\frac{1}{2}X\Omega_v X}{(X\cdot v)^3}, \tag{8.64}$$

is proportional to the rest-frame acceleration projected down the null-vector X.

As a final comment, we note the following curious formulae:

$$A = \frac{-q}{8\pi\epsilon_0}\nabla^2 X, \tag{8.65}$$

and

$$F = \frac{-q}{8\pi\epsilon_0}\nabla^3 X. \tag{8.66}$$

In this expression for F we have expressed a physical field solely in terms of a derivative of the 'information carrying' (adjunct) field X. Expressions such as (8.65) and (8.66) may be of further interest in the elaboration of Wheeler-Feynman type 'action at a distance' ideas.

8.4 Applications

We now consider some applications of the preceding formulae for the fields due to a point charge.

8.4.1 Uniformly Moving Charge

Suppose that a charge moves with uniform velocity v_0. We can assume, without loss of generality, that the charge goes through the origin, so its world-line is given by

$$x_0(\tau) = v_0 \tau. \tag{8.67}$$

Setting $X^2 = 0$, we obtain a quadratic in τ,

$$\tau^2 - 2\tau(x \cdot v_0) + x^2 = 0. \tag{8.68}$$

From this we must select the earliest root to find the value of τ on the past light-cone,

$$\tau = x \cdot v_0 - [(x \cdot v_0)^2 - x^2]^{1/2}, \tag{8.69}$$

from which we obtain

$$X \cdot v = x \cdot v_0 - \tau = [(x \cdot v_0)^2 - x^2]^{\frac{1}{2}} \tag{8.70}$$

We can define an 'effective distance' $d \equiv X \cdot v$, which in this case is given by

$$d = [(x \cdot v_0)^2 - x^2]^{1/2} = |x \wedge v_0|. \tag{8.71}$$

Since $\dot{v} = 0$ we now have

$$F = \frac{q}{4\pi\epsilon_0} \frac{X \wedge v}{(X \cdot v)^3} = \frac{q}{4\pi\epsilon_0} \frac{x \wedge v_0}{|x \wedge v_0|^3}. \tag{8.72}$$

This solution could, of course, be obtained from the Coulomb field simply by replacing γ_0 by v_0. This is achieved by acting on the Coulomb field solution with the boost

$$F \mapsto F' = RF(\tilde{R}xR)\tilde{R} \tag{8.73}$$

where $v_0 = R\gamma_0\tilde{R}$. Poincaré invariance of the Maxwell equations ensures that this procedure yields a new solution.

For many applications we are interested in the space-time split of the solution (8.72) in the γ_0-system, since this tells us the fields seen when the charge and observer are in relative motion. To make this split we write

$$v_0\gamma_0 = \gamma(1 + \boldsymbol{v}_0), \tag{8.74}$$

where

$$\gamma \equiv (1 - \boldsymbol{v}_0{}^2)^{-1/2}. \tag{8.75}$$

Applying the space-time split to F we find, in agreement with Jackson (p. 381),

$$\boldsymbol{E} = \frac{\gamma q}{4\pi\epsilon_0 d^3}(\boldsymbol{x} - \boldsymbol{v}_0 t) \tag{8.76}$$

$$\boldsymbol{B} = \frac{\gamma q}{4\pi\epsilon_0 d^3}\boldsymbol{v}_0 \times \boldsymbol{x} \tag{8.77}$$

where

$$d^2 = b^2 + \gamma^2(\boldsymbol{v}_0 \cdot \boldsymbol{x}/|\boldsymbol{v}_0| - |\boldsymbol{v}_0|t)^2, \tag{8.78}$$

and b is the Euclidean distance of shortest approach. Note that the $\boldsymbol{E}$-field points in the direction of the actual position of the charge at time t, and not its position at the retarded time τ.

8.4.2 Accelerated Charge

Suppose that an accelerated charged particle follows the trajectory

$$x_0(\tau) = a[\sinh(g\tau)\gamma_0 + \cosh(g\tau)\gamma_3], \tag{8.79}$$

where $a = g^{-1}$. The velocity and acceleration are given by

$$v(\tau) = \cosh(g\tau)\gamma_0 + \sinh(g\tau)\gamma_3 = e^{g\tau\sigma_3}\gamma_0 \tag{8.80}$$

and

$$\dot{v}v = g\sigma_3, \tag{8.81}$$

so the charge has a constant acceleration. We again seek the retarded solution of $X^2 = 0$, and in this case we find that

$$e^{g\tau} = \frac{1}{2a(z-t)}\left(a^2 + r^2 - t^2 - [(a^2 + r^2 - t^2)^2 - 4a^2(z^2 - t^2)]^{1/2}\right) \tag{8.82}$$

$$e^{-g\tau} = \frac{1}{2a(z+t)}\left(a^2 + r^2 - t^2 + [(a^2 + r^2 - t^2)^2 - 4a^2(z^2 - t^2)]^{1/2}\right), \tag{8.83}$$

where $\{x, y, z, t\}$ are the usual Cartesian coordinates and $r = |\boldsymbol{x}|$. The condition for the existence of a retarded solution is $z + t > 0$.

We can now calculate the radiation from the charge. First we need the effective distance

$$X \cdot v = \frac{[(a^2 + r^2 - t^2)^2 - 4a^2(z^2 - t^2)]^{1/2}}{2a}, \tag{8.84}$$

which vanishes on the path of the particle ($x = y = 0$ and $z^2 - t^2 = a^2$), as required. The other factor is

$$\begin{aligned} X \wedge v + \tfrac{1}{2}X\dot{v}vX &= \frac{1}{2a}x\sigma_3 x - \frac{a}{2}\sigma_3 \\ &= \frac{1}{2a}(z^2 - \rho^2 - t^2 - a^2)\sigma_3 + \frac{z\rho}{a}\sigma_\rho + \frac{t\rho}{a}i\sigma_\phi, \end{aligned} \tag{8.85}$$

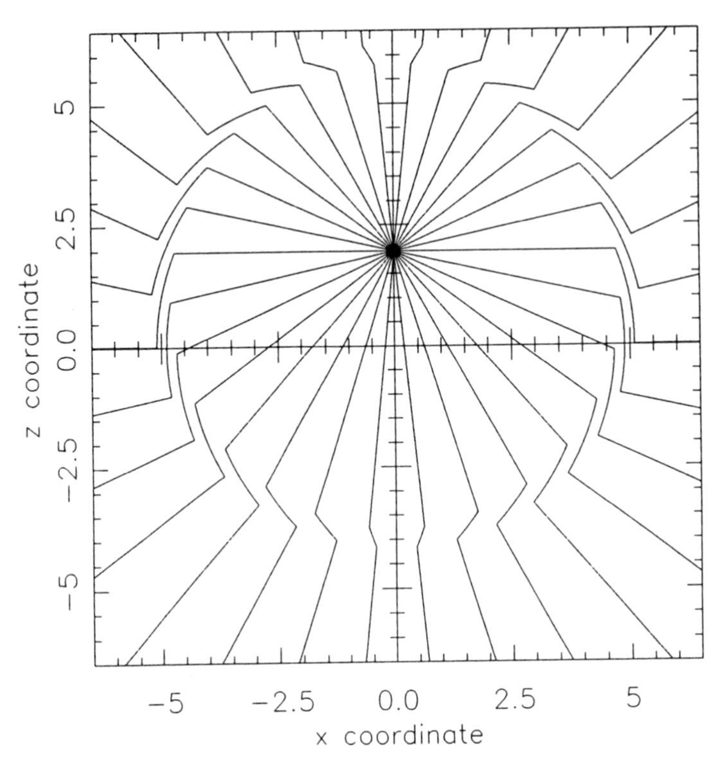

Fig. 8.1. *Electric field lines of a charge at $t = 5a$. The charge accelerated for $-0.2a < t < 0.2a$, leaving an outgoing pulse of transverse radiation field.*

where $\rho \equiv \surd(x^2+y^2)$, and σ_ρ and σ_ϕ are the unit spatial axial and azimuthal vectors respectively. A useful way to display the information contained in the expression for F is to plot the field lines of $\boldsymbol{E}$ at a fixed time. Since $\boldsymbol{\nabla}\cdot\boldsymbol{E} = \rho$, these only terminate on point charges. For the purposes of illustration, we suppose that the charge starts accelerating at $t = t_1$, and stops again at $t = t_2$. There are then discontinuities in the electric field line direction on the two appropriate light spheres. In Figure 8.1 the acceleration is relatively brief, so that a pulse of radiation is sent outwards. In Figure 8.2 the charge began accelerating at $t = -10a$, so that the pattern is well developed, and shows clearly the refocusing of the field lines onto the 'image charge'. The image position corresponds to the place the charge would have reached had it not started accelerating. Of course, the image charge is not actually present, and the field lines diverge after they cross the light sphere corresponding to the start of the acceleration.

A useful approximation is to study the fields a long way from the source. In this region the fields can usually be approximated by simple dipole or higher-order multipole fields. Suppose that the charge accelerates for a short period and emmits a pulse of radiation. In the limit $r \gg a$ the pulse will arrive at some time which, to a good approximation, is centred around the time that minimizes $X\cdot v$. This time is

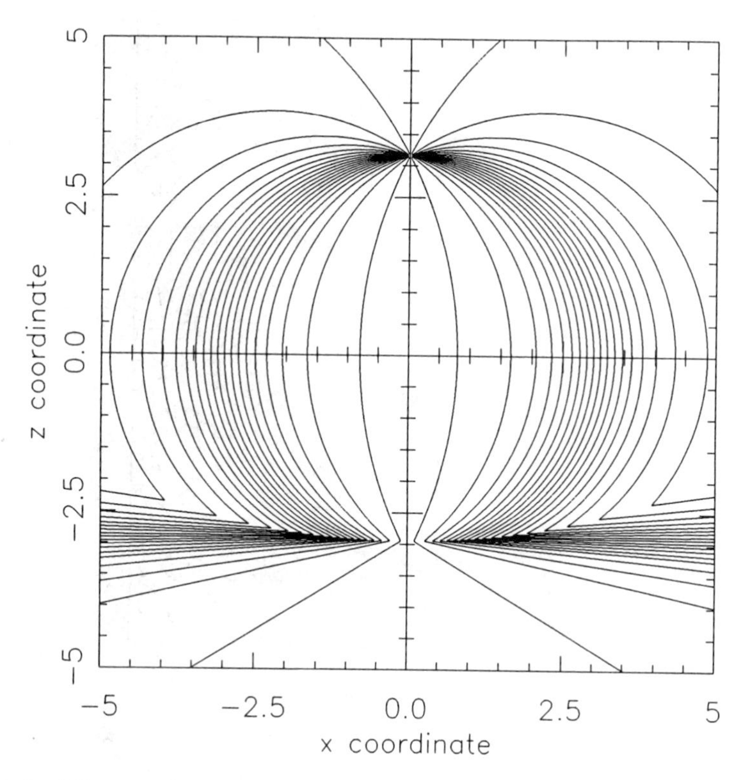

Fig. 8.2. *Electric field lines of an accelerated charge at $t = 3a$. The charge began its acceleration at $t_1 = -10a$ and has thereafter accelerated uniformly.*

given by

$$t_0 \equiv \sqrt{r^2 - a^2}, \tag{8.86}$$

at which time $X \cdot v = \rho$. For this approximation to be sensible we also require that ρ is large, so that the proper distance from the source is large. (For small ρ and $z > a$ a different procedure can be used.) We can now obtain an approximate formula for the radiation field at a fixed location $\boldsymbol{x}$ $(r, \rho \gg a)$ around $t = t_0$. For this we define

$$\delta_t \equiv t - t_0 \tag{8.87}$$

so that

$$X \cdot v \approx \left(\rho^2 + r^2\delta_t^2/a^2\right)^{1/2}, \tag{8.88}$$

and

$$X \wedge v + \tfrac{1}{2}X\dot{v}vX \approx \frac{r\rho}{a}\left(\sigma_\theta + i\sigma_\phi\right), \tag{8.89}$$

which is a pure, outgoing radiation field. The final formula is

$$F \approx \frac{q}{4\pi\epsilon_0}\frac{r\rho}{a}\left(\rho^2 + \frac{r^2\delta_t^2}{a^2}\right)^{-3/2}\left(\sigma_\theta + i\sigma_\phi\right). \tag{8.90}$$

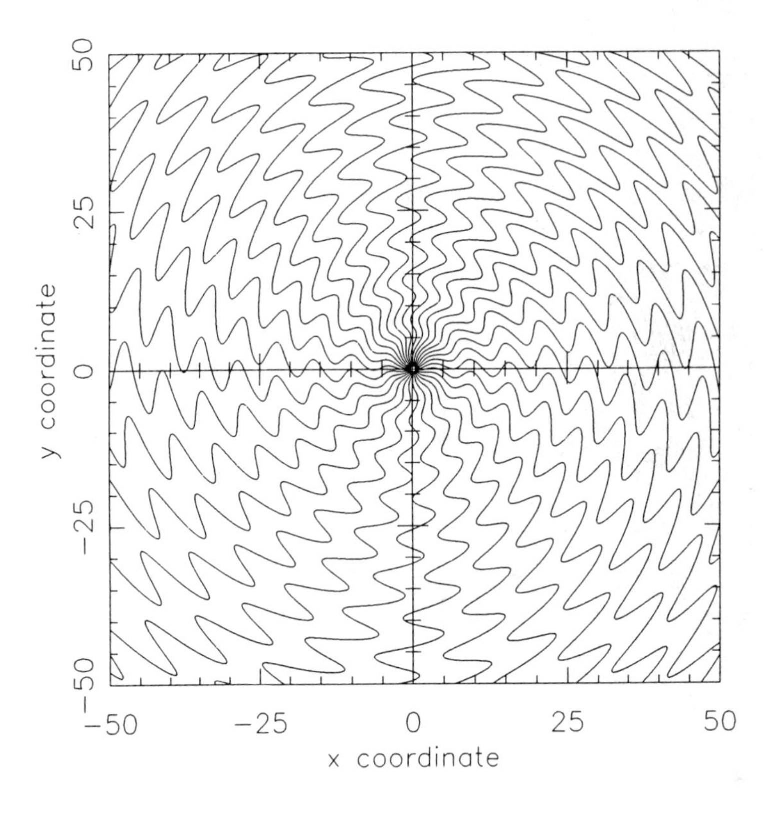

Fig. 8.3. *Field lines of a rotating charge, with rapidity* $u = 0.1$.

8.4.3 Circular Orbits

As a final application, consider a charge in a circular orbit, with a world-line

$$x_0 = \tau \cosh u \, \gamma_0 + a(\cos(\omega\tau)\gamma_1 + \sin(\omega\tau)\gamma_2), \tag{8.91}$$

where $a = \omega^{-1}\sinh u$. The particle velocity is

$$v = \cosh u \, \gamma_0 + \sinh u(-\sin(\omega\tau)\gamma_1 + \cos(\omega\tau)\gamma_2) = R\gamma_0\tilde{R}, \tag{8.92}$$

where

$$R = e^{-\omega\tau i\sigma_3/2} e^{u\sigma_2/2}. \tag{8.93}$$

Our first task is to locate the null vector X. The equation $X^2 = 0$ yields

$$t = \tau \cosh u + \sqrt{r^2 + a^2 - 2a(x\cos\omega\tau + y\sin\omega\tau)} \tag{8.94}$$

which can be viewed as an implicit equation for $\tau(x)$. No simple analytic solution exists, but a numerical solution is easy to achieve. Note that for fixed x, y, z, the mapping bewteen t and τ is monotonic and τ is bounded by the conditions

$$t - \sqrt{r^2 + 2a\rho + a^2} < \tau \cosh u < t - \sqrt{r^2 - 2a\rho + a^2}. \tag{8.95}$$

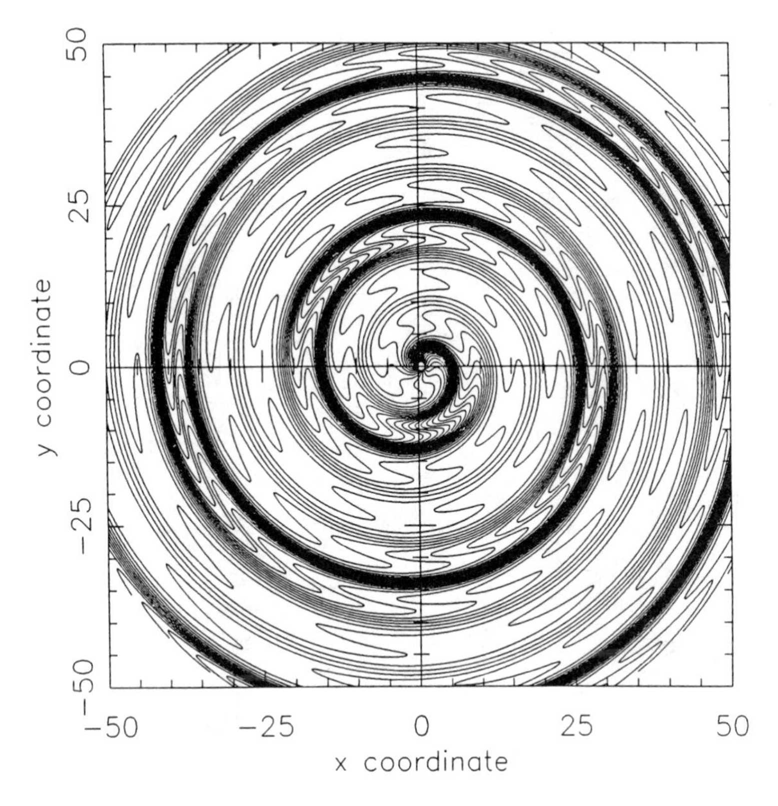

Fig. 8.4. *Field lines of a rotating charge, with rapidity* $u = 0.4$.

Once we have a satisfactory procedure for locating τ on the retarded light-cone, we can straightforwardly employ the formula for F in numerical simulations. The first term required is the effective distance $X \cdot v$, which is given by

$$X \cdot v = \cosh u [r^2 + a^2 - 2a\rho\cos(\omega\tau - \phi)]^{1/2} + \rho \sinh u \sin(\omega\tau - \phi). \tag{8.96}$$

The other term, $X \wedge v + \frac{1}{2}X\dot{v}vX$, is more complicated. The plots show a number of simulations obtained using the general formula numerically. They show the field lines in the equatorial plane of a rotating charge with $\omega = 1$. For 'low' speeds we get a gentle, wavy pattern of field lines (Figure 8.3). The case displayed in Figure 8.4 is for an intermediate velocity ($u = 0.4$), and displays many interesting features. By $u = 1$ (Figure 8.5) the field lines have concentrated into synchrotron pulses, a pattern which continues therafter.

Synchrotron radiation

Synchrotron radiation is important in many areas of physics, from particle physics through to radioastronomy. Synchrotron radiation from a radiogalaxy has, for example, $a \approx 10^8$m and $r \approx 10^{25}$m. A power-series expansion in a/r is therefore quite safe. Typical values of $\cosh u$ are 10^4 for electrons producing radio emission. In the

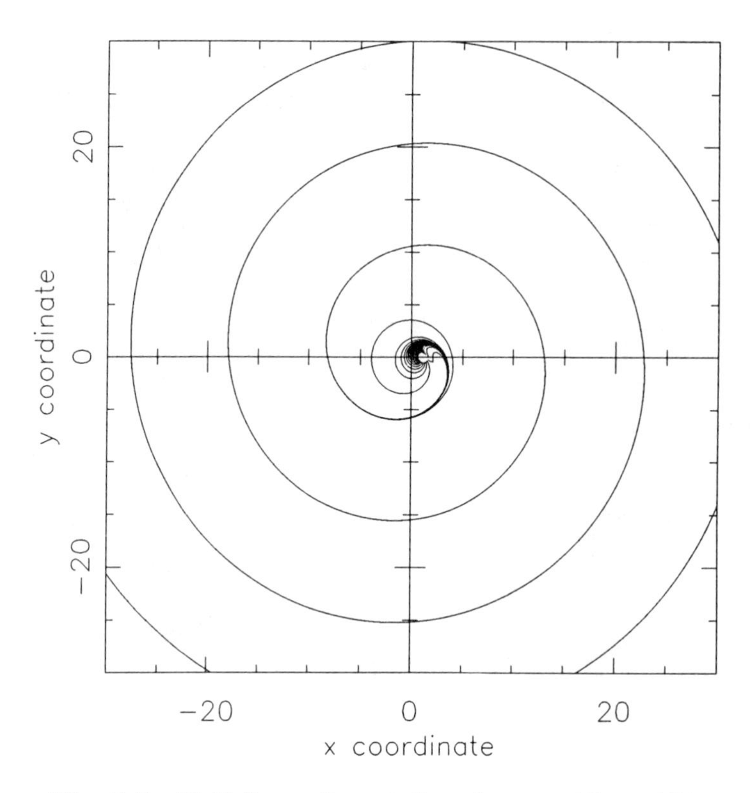

Fig. 8.5. *Field lines of a rotating charge, with rapidity* $u = 1$.

limit $r \gg a$, the relation between t and τ simplifies to

$$t - r \approx \tau \cosh u - a \sin\theta \cos(\omega\tau - \phi), \tag{8.97}$$

where r, θ, ϕ are the standard polar coordinates. This equation is the usual starting point for synchrotron theory, though we are still faced with the problem that τ is only determined implicitly.

The real simplifications are in the effective distance, which reduces to

$$X \cdot v \approx r \cosh u [1 + \tanh u \sin\theta \sin(\omega\tau - \phi)], \tag{8.98}$$

and the null vector $X \approx r(\gamma_0 + e_r)$.. In the expression for F we can clearly ignore the $X \wedge v$ (Coulomb) term relative to the long-range radiation term. For these we need the acceleration bivector

$$\dot{v}v = -\omega \sinh u \cosh u (\cos(\omega\tau)\sigma_1 + \sin(\omega\tau)\sigma_2) + \omega \sinh^2 u \, i\sigma_3. \tag{8.99}$$

The crucial term now simplifies to

$$\begin{aligned} \tfrac{1}{2} X \dot{v} v X \;\approx\; & \omega r^2 \cosh u \sinh u [\cos\theta \cos(\omega\tau - \phi)\sigma_\theta + \sin(\omega\tau - \phi)\sigma_\phi](1 - \sigma_r) \\ & + \omega r^2 \sinh^2 u \, \sin\theta \, \sigma_\phi (1 - \sigma_r). \end{aligned} \tag{8.100}$$

These formulae contain all kinds of wonderful features, but that is another story. Needless to say, the STA derivation beats the traditional 'bits and pieces' approach!

Chapter 9

Electron Physics I

Stephen Gull, Chris Doran, and Anthony Lasenby
MRAO, Cavendish Laboratory, Madingley Road,
Cambridge, CB3 0HE, UK

In this and the following lecture we aim to show that spacetime algebra simplifies the study of the Dirac theory, and that the Dirac theory, once formulated in the spacetime algebra, is a powerful and flexible tool for the analysis of all aspects of electron physics — not just relativistic theory.

Pauli and Dirac column spinors, and the operators that act on them, are formulated and manipulated in the STA. Once the STA formulation is achieved, matrices are eliminated from Dirac theory, and the Dirac equation can be studied and solved entirely within the real STA. A significant result of this work is that the unit imaginary of quantum mechanics is eliminated and replaced by a bivector.

Before proceeding, it is necessary to explain what we mean by a *spinor.* For our purposes, we define a spinor to be an element of a linear space which is closed under left-sided multiplication by a rotor. Thus spinors are acted on by rotor representations of the rotation group. With this in mind, we can proceed directly to study the spinors of relevance to physics.

9.1 Pauli Spinors

The algebra of the Pauli matrices is precisely that of a set of three orthonormal vectors in space under the geometric product. So the Pauli matrices are simply a matrix representation of the geometric algebra of space. But what of the operator action of the Pauli matrices on spinors? This too needs to be represented with the geometric algebra of space. To achieve this aim, we recall the standard representation for the Pauli matrices

$$\hat{\sigma}_1 = \begin{pmatrix} 0 & 1 \\ 1 & 0 \end{pmatrix}, \quad \hat{\sigma}_2 = \begin{pmatrix} 0 & -j \\ j & 0 \end{pmatrix}, \quad \hat{\sigma}_3 = \begin{pmatrix} 1 & 0 \\ 0 & -1 \end{pmatrix}. \tag{9.1}$$

The overhats distinguish these matrix operators from the $\{\sigma_k\}$ vectors whose algebra they represent. The symbol i is reserved for the pseudoscalar, so the symbol j is used for the scalar unit imaginary employed in quantum theory. The $\{\hat{\sigma}_k\}$ operators act on 2-component complex spinors

$$|\psi\rangle = \begin{pmatrix} \psi_1 \\ \psi_2 \end{pmatrix}, \tag{9.2}$$

where ψ_1 and ψ_2 are complex numbers. Quantum states are written with bras and kets to distinguish them from STA multivectors. The set of $|\psi\rangle$'s form a two-dimensional complex vector space. To represent these states as multivectors in the STA we therefore need to find a four-dimensional (real) space on which the action of the $\{\hat{\sigma}_k\}$ operators can be replaced by operations involving the $\{\sigma_k\}$ vectors. There are many ways to achieve this goal, but the simplest is to represent a spinor $|\psi\rangle$ by an element of the even subalgebra of the STA. This space is spanned by the set $\{1, i\sigma_k\}$ and the column spinor $|\psi\rangle$ is placed in one-to-one correspondence with the (Pauli)-even multivector $\psi = \gamma_0\psi\gamma_0$ through the identification

$$|\psi\rangle = \begin{pmatrix} a^0 + ja^3 \\ -a^2 + ja^1 \end{pmatrix} \quad \leftrightarrow \quad \psi = a^0 + a^k i\sigma_k. \tag{9.3}$$

In particular, the spin-up and spin-down basis states become

$$\begin{pmatrix} 1 \\ 0 \end{pmatrix} \quad \leftrightarrow \quad 1 \tag{9.4}$$

and

$$\begin{pmatrix} 0 \\ 1 \end{pmatrix} \quad \leftrightarrow \quad -i\sigma_2. \tag{9.5}$$

The action of the quantum operators $\{\hat{\sigma}_k\}$ and j is now replaced by the operations

$$\hat{\sigma}_k|\psi\rangle \leftrightarrow \sigma_k\psi\sigma_3 \qquad (k = 1, 2, 3) \tag{9.6}$$

$$\text{and} \quad j|\psi\rangle \leftrightarrow \psi i\sigma_3. \tag{9.7}$$

Verifying these relations is a matter of routine computation; for example

$$\hat{\sigma}_1|\psi\rangle = \begin{pmatrix} -a^2 + ja^1 \\ a^0 + ja^3 \end{pmatrix} \quad \leftrightarrow \quad -a^2 + a^1 i\sigma_3 - a^0 i\sigma_2 + a^3 i\sigma_1 = \sigma_1\psi\sigma_3. \tag{9.8}$$

9.1.1 Pauli Observables

We now turn to a discussion of the observables associated with Pauli spinors. These show how the STA formulation requires a shift in our understanding of what constitutes scalar and vector observables at the quantum level. We first need to construct the STA form of the spinor inner product $\langle\psi|\phi\rangle$. It is sufficient just to consider the real part of the inner product, which is given by

$$\Re\langle\psi|\phi\rangle \quad \leftrightarrow \quad \langle\psi^\dagger\phi\rangle, \tag{9.9}$$

so that, for example,

$$\begin{aligned} \langle\psi|\psi\rangle \quad \leftrightarrow \quad \langle\psi^\dagger\psi\rangle &= \langle(a^0 - ia^j\sigma_j)(a^0 + ia^k\sigma_k)\rangle \\ &= (a^0)^2 + a^k a^k. \end{aligned} \tag{9.10}$$

(Note that no spatial integral is implied in our use of the bra-ket notation.) Since

$$\langle\psi|\phi\rangle = \Re\langle\psi|\phi\rangle - j\Re\langle\psi|j\phi\rangle, \tag{9.11}$$

the full inner product becomes

$$\langle\psi|\phi\rangle \quad \leftrightarrow \quad (\psi,\phi)_S \equiv \langle\psi^\dagger\phi\rangle - \langle\psi^\dagger\phi i\sigma_3\rangle i\sigma_3. \tag{9.12}$$

The right hand side projects out the $\{1, i\sigma_3\}$ components from the geometric product $\psi^\dagger\phi$. The result of this projection on a multivector A is written $\langle A\rangle_S$. For Pauli-even multivectors this projection has the simple form

$$\langle A\rangle_S = \tfrac{1}{2}(A - i\sigma_3 A i\sigma_3). \tag{9.13}$$

As an application of (9.12), consider the expectation value of the spin in the k-direction,

$$\langle\psi|\hat{\sigma}_k|\psi\rangle \quad \leftrightarrow \quad \langle\psi^\dagger\sigma_k\psi\sigma_3\rangle - \langle\psi^\dagger\sigma_k\psi i\rangle i\sigma_3. \tag{9.14}$$

Since $\psi^\dagger i\sigma_k\psi$ reverses to give minus itself, it has zero scalar part. The right-hand side of (9.14) therefore reduces to

$$\langle\sigma_k\psi\sigma_3\psi^\dagger\rangle = \sigma_k\cdot\langle\psi\sigma_3\psi^\dagger\rangle_v, \tag{9.15}$$

where $\langle\ldots\rangle_v$ denotes the relative vector component of the term in brackets. (This notation is required because $\langle\ldots\rangle_1$ would denote the spacetime vector part of the term in brackets.) The expression (9.15) has a rather different interpretation in the STA to standard quantum mechanics — it is the σ_k-component of the vector part of $\psi\sigma_3\psi^\dagger$. As $\psi\sigma_3\psi^\dagger$ is both Pauli-odd and Hermitian-symmetric it can contain only a relative vector part, so we define the spin-vector $\boldsymbol{s}$ by

$$\boldsymbol{s} \equiv \psi\sigma_3\psi^\dagger. \tag{9.16}$$

The STA approach thus enables us to work with a single vector $\boldsymbol{s}$, whereas the operator/matrix theory treats only its individual components. We can apply a similar analysis to the momentum operator. The momentum density in the k-direction is given by

$$\langle\psi| - j\partial_k|\psi\rangle \quad \leftrightarrow \quad -\langle\psi^\dagger\sigma_k\cdot\boldsymbol{\nabla}\psi i\sigma_3\rangle - \langle\psi^\dagger\sigma_k\cdot\boldsymbol{\nabla}\psi\rangle i\sigma_3, \tag{9.17}$$

in which the final term is a total divergence and so is ignored. Recombining with the $\{\sigma_k\}$ vectors, we find that the momentum vector field is given by

$$\boldsymbol{p} = -\dot{\boldsymbol{\nabla}}\langle\dot{\psi} i\sigma_3\psi^\dagger\rangle. \tag{9.18}$$

9.1.2 Spinors and Rotations

Further insights into the role of spinors in the Pauli theory are obtained by defining a scalar

$$\rho \equiv \psi\psi^\dagger, \tag{9.19}$$

so that the spinor ψ can be decomposed into

$$\psi = \rho^{1/2} R. \tag{9.20}$$

Here R is defined as

$$R = \rho^{-1/2}\psi \tag{9.21}$$

and satisfies

$$RR^\dagger = 1. \tag{9.22}$$

In three dimensions, all even quantities satisfying (9.22) are rotors. It follows from (9.20) that the spin-vector $\boldsymbol{s}$ can now be written as

$$\boldsymbol{s} = \rho R \sigma_3 R^\dagger, \tag{9.23}$$

which demonstrates that the double-sided construction of the expectation value contains an instruction to rotate the fixed σ_3 axis into the spin direction and dilate it. As an example of the insights afforded by this decomposition, one can now 'explain' why spinors transform single-sidedly under rotations. If the vector $\boldsymbol{s}$ is to be rotated to a new vector $R_0 \boldsymbol{s} R_0^\dagger$ then, according to the rotor group combination law, R must transform to $R_0 R$. This induces the spinor transformation law

$$\psi \mapsto R_0 \psi \tag{9.24}$$

which is the STA equivalent of the quantum transformation law

$$|\psi\rangle \mapsto \exp\{\frac{j}{2}\theta n_k \hat{\sigma}_k\}|\psi\rangle \tag{9.25}$$

where $\{n_k\}$ are the components of a unit vector.

We can also now see why the presence of the σ_3 vector on the right-hand side of the spinor ψ does not break rotational invariance. All rotations are performed by left-multiplication by a rotor, so the spinor ψ effectively shields the σ_3 on the right from the transformation. There is a strong analogy with rigid-body mechanics, which was presented in c. 7. The main results of this section are summarised in Table 9.1.

9.2 Dirac Spinors

The procedures developed for Pauli spinors extend simply to Dirac spinors. Again, we seek to represent complex column spinors, and the matrix operators acting on them, by multivectors and functions in the STA. Dirac spinors are four-component complex entities, so must be represented by objects containing 8 real degrees of

Pauli Matrices	$\hat{\sigma}_1 = \begin{pmatrix} 0 & 1 \\ 1 & 0 \end{pmatrix} \quad \hat{\sigma}_2 = \begin{pmatrix} 0 & -j \\ j & 0 \end{pmatrix} \quad \hat{\sigma}_3 = \begin{pmatrix} 1 & 0 \\ 0 & -1 \end{pmatrix}$
Spinor Equivalence	$\lvert\psi\rangle = \begin{pmatrix} a^0 + ja^3 \\ -a^2 + ja^1 \end{pmatrix} \quad \leftrightarrow \quad \psi = a^0 + a^k i\sigma_k$
Operator Equivalences	$\hat{\sigma}_k\lvert\psi\rangle \leftrightarrow \sigma_k\psi\sigma_3$ $j\lvert\psi\rangle \leftrightarrow \psi i\sigma_3$ $\langle\psi\lvert\psi'\rangle \leftrightarrow \langle\psi^\dagger\psi'\rangle_S$
Observables	$\rho = \psi\psi^\dagger$ $s = \psi\sigma_3\psi^\dagger$

Table 9.1. *Summary of the main results for the STA representation of Pauli spinors*

freedom. The representation that turns out to be most convenient for applications is via the 8-dimensional even subalgebra of the STA. If one recalls that the even subalgebra of the STA is isomorphic to the Pauli algebra, we see that what is required is a map between column spinors and elements of the Pauli algebra. To construct such a map we begin with the γ-matrices in the standard Dirac-Pauli representation,

$$\hat{\gamma}_0 = \begin{pmatrix} I & 0 \\ 0 & -I \end{pmatrix}, \quad \hat{\gamma}_k = \begin{pmatrix} 0 & -\hat{\sigma}_k \\ \hat{\sigma}_k & 0 \end{pmatrix} \quad \text{and} \quad \hat{\gamma}_5 = \begin{pmatrix} 0 & I \\ I & 0 \end{pmatrix}, \tag{9.26}$$

where $\hat{\gamma}_5 = \hat{\gamma}^5 \equiv -j\hat{\gamma}_0\hat{\gamma}_1\hat{\gamma}_2\hat{\gamma}_3$ and I is the 2×2 identity matrix. A Dirac column spinor $\lvert\psi\rangle$ is placed in one-to-one correspondence with an 8-component even element of the STA via

$$\lvert\psi\rangle = \begin{pmatrix} a^0 + ja^3 \\ -a^2 + ja^1 \\ b^0 + jb^3 \\ -b^2 + jb^1 \end{pmatrix} \quad \leftrightarrow \quad \psi = a^0 + a^k i\sigma_k + (b^0 + b^k i\sigma_k)\sigma_3. \tag{9.27}$$

With the spinor $\lvert\psi\rangle$ now replaced by an even multivector, the action of the operators $\{\hat{\gamma}_\mu, \hat{\gamma}_5, j\}$ becomes

$$\begin{aligned} \hat{\gamma}_\mu\lvert\psi\rangle &\leftrightarrow \gamma_\mu\psi\gamma_0 \quad (\mu = 0,\dots,3) \\ j\lvert\psi\rangle &\leftrightarrow \psi i\sigma_3 \\ \hat{\gamma}_5\lvert\psi\rangle &\leftrightarrow \psi\sigma_3. \end{aligned} \tag{9.28}$$

To verify these relations, we note that the map (9.27) can be written more concisely as

$$|\psi\rangle = \begin{pmatrix} |\phi\rangle \\ |\eta\rangle \end{pmatrix} \quad \leftrightarrow \quad \psi = \phi + \eta\sigma_3, \tag{9.29}$$

where $|\phi\rangle$ and $|\eta\rangle$ are two-component spinors, and ϕ and η are their Pauli-even equivalents, as defined by the map (9.3). We can now see, for example, that

$$\hat{\gamma}_k|\psi\rangle = \begin{pmatrix} -\hat{\sigma}_k|\eta\rangle \\ \hat{\sigma}_k|\phi\rangle \end{pmatrix} \quad \leftrightarrow \quad -\sigma_k\eta\sigma_3 + \sigma_k\phi = \gamma_k(\phi + \eta\sigma_3)\gamma_0, \tag{9.30}$$

as required. The map (9.29) shows that the split between the 'large' and 'small' components of the column spinor $|\psi\rangle$ is equivalent to splitting ψ into Pauli-even and Pauli-odd terms in the STA.

9.2.1 Alternative Representations

All algebraic manipulations can be performed in the STA without ever introducing a matrix representation, so equations (9.27) and (9.28) achieve a map to a representation-free language. However, the explicit map (9.27) between the components of a Dirac spinor and the multivector ψ is only relevant to the Dirac-Pauli matrix representation. A different matrix representation requires a different map so that that the effect of the matrix operators is still given by (9.28). The relevant map is easy to construct given the unitary matrix $\hat{S}$ which transforms between the matrix representations via

$$\hat{\gamma}'_\mu = \hat{S}\hat{\gamma}_\mu\hat{S}^{-1}. \tag{9.31}$$

The corresponding spinor transformation is $|\psi\rangle \mapsto \hat{S}|\psi\rangle$, and the map is constructed by transforming the column spinor $|\psi\rangle'$ in the new representation back to a Dirac-Pauli spinor $\hat{S}^\dagger|\psi\rangle'$. The spinor $\hat{S}^\dagger|\psi\rangle'$ is then mapped into the STA in the usual way (9.27). As an example, the Weyl representation is defined by the matrices

$$\hat{\gamma}'_0 = \begin{pmatrix} 0 & -I \\ -I & 0 \end{pmatrix} \quad \text{and} \quad \hat{\gamma}'_k = \begin{pmatrix} 0 & -\hat{\sigma}_k \\ \hat{\sigma}_k & 0 \end{pmatrix}. \tag{9.32}$$

While our translations ensure that the action of the $\{\hat{\gamma}_\mu, \hat{\gamma}_5\}$ matrix operators is always given by (9.28), the same is not true of the operation of complex conjugation. Complex conjugation is a representation-dependent operation, so the STA versions can be different for different representations. For example, complex conjugation in the Dirac-Pauli and Weyl representations is given by

$$|\psi\rangle^* \quad \leftrightarrow \quad -\gamma_2\psi\gamma_2, \tag{9.33}$$

whereas a similar translation of the Majorana representation complex conjugation leads to the STA operation

$$|\psi\rangle^*_{\text{Maj}} \quad \leftrightarrow \quad \psi\sigma_2. \tag{9.34}$$

Rather than think of (9.33) and (9.34) as different representations of the same operation, however, it is simpler to view them as distinct STA operations that can be performed on the multivector ψ.

9.3 The Dirac Equation and Observables

As a simple application of (9.27) and (9.28), consider the Dirac equation

$$\hat{\gamma}^\mu(j\partial_\mu - eA_\mu)|\psi\rangle = m|\psi\rangle. \tag{9.35}$$

The STA version of this equation is, after postmultiplication by γ_0,

$$\nabla\psi i\sigma_3 - eA\psi = m\psi\gamma_0, \tag{9.36}$$

where $\nabla = \gamma^\mu\partial_\mu$ is the spacetime vector derivative. Stripped of the dependence on a matrix representation, equation (9.36) expresses the intrinsic geometric content of the Dirac equation.

In order to discuss the observables of the Dirac theory, we must first consider the spinor inner product. It is necessary at this point to distinguish between the Hermitian and Dirac adjoint. These are written as

$$\begin{array}{lcl} \langle\bar{\psi}| & - & \text{Dirac adjoint} \\ \langle\psi| & - & \text{Hermitian adjoint,} \end{array} \tag{9.37}$$

which are represented in the STA as follows,

$$\begin{array}{lcl} \langle\bar{\psi}| & \leftrightarrow & \tilde{\psi} \\ \langle\psi| & \leftrightarrow & \psi^\dagger = \gamma_0\tilde{\psi}\gamma_0. \end{array} \tag{9.38}$$

One can see clearly from these definitions that the Dirac adjoint is Lorentz-invariant, whereas the Hermitian adjoint requires singling out a preferred timelike vector.

The inner product is handled as in equation (9.12), so that

$$\langle\bar{\psi}|\phi\rangle \quad \leftrightarrow \quad \langle\tilde{\psi}\phi\rangle - \langle\tilde{\psi}\phi i\sigma_3\rangle i\sigma_3 = \langle\tilde{\psi}\phi\rangle_S, \tag{9.39}$$

which is also easily verified by direct calculation. By utilising (9.39) the STA forms of the Dirac spinor bilinear covariants are readily found. For example,

$$\langle\bar{\psi}|\hat{\gamma}_\mu|\psi\rangle \quad \leftrightarrow \quad \langle\tilde{\psi}\gamma_\mu\psi\gamma_0\rangle - \langle\tilde{\psi}\gamma_\mu\psi i\gamma_3\rangle i\sigma_3 = \gamma_\mu\cdot\langle\psi\gamma_0\tilde{\psi}\rangle_1 \tag{9.40}$$

identifies the 'observable' as the γ_μ-component of the vector $\langle\psi\gamma_0\tilde{\psi}\rangle_1$. Since the quantity $\psi\gamma_0\tilde{\psi}$ is odd and reverse-symmetric it can only contain a vector part, so we can define the frame-free vector J by

$$J \equiv \psi\gamma_0\tilde{\psi}. \tag{9.41}$$

The spinor ψ has a Lorentz-invariant decomposition which generalises the decomposition of Pauli spinors into a rotation and a density factor (9.20). Since $\psi\tilde{\psi}$ is even and reverses to give itself, it contains only scalar and pseudoscalar terms. We can therefore define

$$\rho e^{i\beta} \equiv \psi\tilde{\psi}, \tag{9.42}$$

where both ρ and β are scalars. Assuming that $\rho \neq 0$, ψ can now be written as

$$\psi = \rho^{1/2}e^{i\beta/2}R \tag{9.43}$$

Bilinear Covariant	Standard Form	STA Equivalent	Frame-Free Form
Scalar	$\langle\bar{\psi}\vert\psi\rangle$	$\langle\psi\tilde{\psi}\rangle$	$\rho\cos\beta$
Vector	$\langle\bar{\psi}\vert\hat{\gamma}_\mu\vert\psi\rangle$	$\gamma_\mu\cdot(\psi\gamma_0\tilde{\psi})$	$\psi\gamma_0\tilde{\psi}=J$
Bivector	$\langle\bar{\psi}\vert j\hat{\gamma}_{\mu\nu}\vert\psi\rangle$	$(\gamma_\mu\wedge\gamma_\nu)\cdot(\psi i\sigma_3\tilde{\psi})$	$\psi i\sigma_3\tilde{\psi}=S$
Pseudovector	$\langle\bar{\psi}\vert\hat{\gamma}_\mu\hat{\gamma}_5\vert\psi\rangle$	$\gamma_\mu\cdot(\psi\gamma_3\tilde{\psi})$	$\psi\gamma_3\tilde{\psi}=s$
Pseudoscalar	$\langle\bar{\psi}\vert j\hat{\gamma}_5\vert\psi\rangle$	$\langle\psi\tilde{\psi}i\rangle$	$-\rho\sin\beta$

Table 9.2. *Bilinear covariants in the Dirac theory.*

where

$$R=(\rho e^{i\beta})^{-1/2}\psi. \tag{9.44}$$

The even multivector R satisfies $R\tilde{R}=1$ and therefore defines a spacetime rotor. The current J (9.41) can now be written as

$$J=\rho v \tag{9.45}$$

where

$$v\equiv R\gamma_0\tilde{R}. \tag{9.46}$$

The remaining bilinear covariants can be analysed likewise, and the results are summarised in Table 9.2. The final column of this Table employs the quantities

$$s\equiv\psi\gamma_3\tilde{\psi}, \quad \text{and} \quad S\equiv\psi i\sigma_3\tilde{\psi}. \tag{9.47}$$

Double-sided application of R on a vector a produces a Lorentz transformation. The full Dirac spinor ψ therefore contains an instruction to rotate the fixed $\{\gamma_\mu\}$ frame into the frame of observables. The analogy with rigid-body dynamics first encountered in Section 9.1 with Pauli spinors therefore extends to the relativistic theory. In particular, the unit vector v (9.46) is both future-pointing and timelike and has been interpreted as defining an electron velocity. The 'β-factor' appearing in the decomposition of ψ (9.42) has also been the subject of much discussion since, for free-particle states, β determines the ratio of particle to anti-particle solutions.

9.3.1 Plane-Wave States

In most applications of the Dirac theory, the external fields applied to the electron define a rest-frame, which is taken to be the γ_0-frame. The rotor R then decomposes relative to the γ_0 vector into a boost L and a rotation Φ,

$$R=L\Phi, \tag{9.48}$$

where

$$L^\dagger = L \tag{9.49}$$

$$\Phi^\dagger = \tilde{\Phi} \tag{9.50}$$

and $L\tilde{L} = \Phi\tilde{\Phi} = 1$. A positive-energy plane-wave state is defined by

$$\psi = \psi_0 e^{-i\sigma_3 p\cdot x} \tag{9.51}$$

where ψ_0 is a constant spinor. From the Dirac equation (9.36) with $A = 0$, it follows that ψ_0 satisfies

$$p\psi_0 = m\psi_0\gamma_0. \tag{9.52}$$

Postmultiplying by $\tilde{\psi}_0$ we see that

$$p\psi\tilde{\psi} = mJ \tag{9.53}$$

from which it follows that $\exp(i\beta) = \pm 1$. Since p has positive energy we must take the positive solution ($\beta = 0$). It follows that ψ_0 is just a rotor with a normalisation constant. The boost L determines the momentum by

$$p = mL\gamma_0\tilde{L} = mL^2\gamma_0, \tag{9.54}$$

which is solved by

$$L = \sqrt{p\gamma_0/m} = \frac{E + m + \boldsymbol{p}}{\sqrt{2m(E+m)}}, \tag{9.55}$$

where

$$p\gamma_0 = E + \boldsymbol{p}. \tag{9.56}$$

The Pauli rotor Φ determines the 'comoving' spin bivector $\Phi i\sigma_3\tilde{\Phi}$. This is boosted by L to give the spin S as seen in the laboratory frame. A $\Phi\sigma_3\tilde{\Phi}$ gives the relative spin in the rest-frame of the particle, we refer to this as the 'rest-spin'. The rest-spin is equivalent to the 'polarisation' vector defined in the traditional matrix formulation.

Negative energy solutions are constructed in a similar manner, but with an additional factor of i or σ_3 on the right (the choice of which to use is simply a choice of phase). The usual positive- and negative-energy basis states employed in scattering theory are (following the conventions of Itzykson & Zuber [Section 2-2])

$$\text{positive energy} \quad \psi^{(+)}(x) = u_r(p)e^{-i\sigma_3 p\cdot x} \tag{9.57}$$

$$\text{negative energy} \quad \psi^{(-)}(x) = v_r(p)e^{i\sigma_3 p\cdot x} \tag{9.58}$$

with

$$u_r(p) = L(p)\chi_r \tag{9.59}$$

$$v_r(p) = L(p)\chi_r\sigma_3. \tag{9.60}$$

Here $L(p)$ is given by equation (9.55) and $\chi_r = \{1, -i\sigma_2\}$ are spin basis states. The decomposition into a boost and a rotor turns out to be very useful in scattering theory.

The main results for Dirac operators and spinors are summarised in Table 9.3.

Dirac Matrices	$\hat{\gamma}_0 = \begin{pmatrix} I & 0 \\ 0 & -I \end{pmatrix} \quad \hat{\gamma}_k = \begin{pmatrix} 0 & -\hat{\sigma}_k \\ \hat{\sigma}_k & 0 \end{pmatrix} \quad \hat{\gamma}_5 = \begin{pmatrix} 0 & I \\ I & 0 \end{pmatrix}$
Spinor Equivalence	$\vert\psi\rangle = \begin{pmatrix} a^0 + ja^3 \\ -a^2 + ja^1 \\ b^0 + jb^3 \\ -b^2 + jb^1 \end{pmatrix} \quad \leftrightarrow \quad \psi = \begin{matrix} a^0 + a^k i\sigma_k + \\ (b^0 + b^k i\sigma_k)\sigma_3 \end{matrix}$
Operator Equivalences	$\hat{\gamma}_\mu\vert\psi\rangle \leftrightarrow \gamma_\mu\psi\gamma_0$ $j\vert\psi\rangle \leftrightarrow \psi i\sigma_3$ $\hat{\gamma}_5\vert\psi\rangle \leftrightarrow \psi\sigma_3$ $\langle\bar{\psi}\vert\psi'\rangle \leftrightarrow \langle\tilde{\psi}\psi'\rangle_S$
Dirac Equation	$\nabla\psi i\sigma_3 - eA\psi = m\psi\gamma_0$
Observables	$\rho e^{i\beta} = \psi\tilde{\psi} \quad J = \psi\gamma_0\tilde{\psi}$ $S = \psi i\sigma_3\tilde{\psi} \quad s = \psi\gamma_3\tilde{\psi}$
Plane-Wave States	$\psi^{(+)}(x) = L(p)\Phi e^{-i\sigma_3 p\cdot x}$ $\psi^{(-)}(x) = L(p)\Phi\sigma_3 e^{i\sigma_3 p\cdot x}$ $L(p) = (p\gamma_0 + m)/\sqrt{2m(E+m)}$

Table 9.3. *Summary of the main results for the STA representation of Dirac spinors. The matrices and spinor equivalence are for the Dirac-Pauli representation. The spinor equivalences for other representations are constructed via the method outlined in the text.*

9.4 Hamiltonian Form

The problem of how to best formulate operator techniques within the STA is really little more than a question of finding a good notation. We saw earlier that the STA operators often act double-sidedly on the spinor ψ. This is not a problem, as the only permitted right-sided operations are multiplication by γ_0 or $i\sigma_3$, and these operations commute. Our notation can therefore safely suppress these right-sided multiplications and lump all operations on the left. The overhat notation is useful to achieve this and we define

$$\hat{\gamma}_\mu\psi \equiv \gamma_\mu\psi\gamma_0. \tag{9.61}$$

It should be borne in mind that all operations are now defined in the STA, so the $\hat{\gamma}_\mu$ are not intended to be matrix operators, as they were in Section 9.2.

It is also useful to have a symbol for the operation of right-sided multiplication by $i\sigma_3$. The symbol j carries the correct connotations of an operator that commutes with all others and squares to -1, and we define

$$j\psi \equiv \psi i\sigma_3. \tag{9.62}$$

The Dirac equation (9.36) can now be written in the 'operator' form

$$j\hat{\nabla}\psi - e\hat{A}\psi = m\psi. \tag{9.63}$$

where

$$\hat{\nabla}\psi \equiv \nabla\psi\gamma_0, \quad \text{and} \quad \hat{A}\psi \equiv A\psi\gamma_0. \tag{9.64}$$

In many applications we require a Hamiltonian form of the Dirac equation. To express the Dirac equation (9.36) in Hamiltonian form we simply multiply from the left by γ_0. The resulting equation, with the dimensional constants temporarily put back in, is

$$j\hbar\partial_t\psi = c\hat{\boldsymbol{p}}\psi + eV\psi - ce\boldsymbol{A}\psi + mc^2\bar{\psi} \tag{9.65}$$

where

$$\begin{aligned}
\hat{\boldsymbol{p}}\psi &\equiv -j\hbar\boldsymbol{\nabla}\psi && (9.66)\\
\bar{\psi} &\equiv \gamma_0\psi\gamma_0 && (9.67)\\
\text{and} \quad \gamma_0 A &= V - c\boldsymbol{A}. && (9.68)
\end{aligned}$$

Choosing a Hamiltonian is a non-covariant operation, since it picks out a preferred timelike direction. The Hamiltonian relative to the γ_0 direction is the operator on the right-hand side of equation(9.65). We write this operator with the symbol $\mathcal{H}$.

9.5 The Non-Relativistic Reduction

In most modern texts, the non-relativistic approximation is carried out via the Foldy-Wouthuysen transformation. A simpler approach, dating back to Feynman, is to separate out the fast-oscillating component of the waves and then split into separate equations for the Pauli-even and Pauli-odd components of ψ. Thus we write (with $\hbar = 1$ and the factors of c kept in)

$$\psi = (\phi + \eta)e^{-i\sigma_3 mc^2 t} \tag{9.69}$$

where $\bar{\phi} = \phi$ and $\bar{\eta} = -\eta$. The Dirac equation (9.65) now splits into the two equations

$$\begin{aligned}
\mathcal{E}\phi - c\mathcal{O}\eta &= 0 && (9.70)\\
(\mathcal{E} + 2mc^2)\eta - c\mathcal{O}\phi &= 0, && (9.71)
\end{aligned}$$

where

$$\mathcal{E}\phi \equiv (j\partial_t - eV)\phi \tag{9.72}$$
$$\mathcal{O}\phi \equiv (\hat{\boldsymbol{p}} - e\boldsymbol{A})\phi. \tag{9.73}$$

The formal solution to the second equation (9.71) is

$$\eta = \frac{1}{2mc}\left(1 + \frac{\mathcal{E}}{2mc^2}\right)^{-1} \mathcal{O}\phi, \tag{9.74}$$

where the inverse on the right-hand side is understood to denote a power series. The power series is well-defined in the non-relativistic limit as the $\mathcal{E}$ operator is of the order of the non-relativistic energy. The remaining equation for ϕ is

$$\mathcal{E}\phi - \frac{\mathcal{O}}{2m}\left(1 - \frac{\mathcal{E}}{2mc^2} + \cdots\right)\mathcal{O}\phi = 0, \tag{9.75}$$

which can be expanded out to the desired order of magnitude. There is little point in going beyond the first relativistic correction, so we approximate (9.75) by

$$\mathcal{E}\phi + \frac{\mathcal{O}\mathcal{E}\mathcal{O}}{4m^2c^2}\phi = \frac{\mathcal{O}^2}{2m}\phi. \tag{9.76}$$

We seek an equation of the form $\mathcal{E}\phi = \mathcal{H}\phi$, where $\mathcal{H}$ is the non-relativistic Hamiltonian. We therefore need to replace the $\mathcal{O}\mathcal{E}\mathcal{O}$ term in equation (9.76) by a term that does not involve $\mathcal{E}$. To do so we would like to utilise the approximate result that

$$\mathcal{E}\phi \approx \frac{\mathcal{O}^2}{2m}\phi, \tag{9.77}$$

but we cannot use this result directly in the $\mathcal{O}\mathcal{E}\mathcal{O}$ term since the $\mathcal{E}$ does not operate directly on ϕ. Instead we employ the operator rearrangement

$$2\mathcal{O}\mathcal{E}\mathcal{O} = [\mathcal{O}, [\mathcal{E}, \mathcal{O}]] + \mathcal{E}\mathcal{O}^2 + \mathcal{O}^2\mathcal{E} \tag{9.78}$$

to write equation (9.76) in the form

$$\mathcal{E}\phi = \frac{\mathcal{O}^2}{2m}\phi - \frac{\mathcal{E}\mathcal{O}^2 + \mathcal{O}^2\mathcal{E}}{8m^2c^2}\phi - \frac{1}{8m^2c^2}[\mathcal{O}, [\mathcal{E}, \mathcal{O}]]\phi. \tag{9.79}$$

We can now make use of (9.77) to write

$$\mathcal{E}\mathcal{O}^2\phi \approx \mathcal{O}^2\mathcal{E}\phi \approx \frac{\mathcal{O}^4}{2m} + \mathrm{O}(c^{-2}) \tag{9.80}$$

and so approximate (9.76) by

$$\mathcal{E}\phi = \frac{\mathcal{O}^2}{2m}\phi - \frac{1}{8m^2c^2}[\mathcal{O}, [\mathcal{E}, \mathcal{O}]]\phi - \frac{\mathcal{O}^4}{8m^3c^2}\phi, \tag{9.81}$$

which is valid to order c^{-2}. The commutators are easily evaluated, for example

$$[\mathcal{E}, \mathcal{O}] = -je(\partial_t \boldsymbol{A} + \boldsymbol{\nabla} V) = je\boldsymbol{E}. \tag{9.82}$$

There are no time derivatives left in this commutator, so we do achieve a sensible non-relativistic Hamiltonian. The full commutator required in equation (9.81) is

$$\begin{aligned} [\mathcal{O}, [\mathcal{E}, \mathcal{O}]] &= [-j\boldsymbol{\nabla} - e\boldsymbol{A}, je\boldsymbol{E}] \\ &= (e\boldsymbol{\nabla}\boldsymbol{E}) - 2e\boldsymbol{E}\wedge\boldsymbol{\nabla} - 2je^2\boldsymbol{A}\wedge\boldsymbol{E} \end{aligned} \tag{9.83}$$

in which the STA formulation ensures that we are manipulating spatial vectors, rather than performing abstract matrix manipulations.

The various operators (9.72), (9.73) and (9.83) can now be fed into equation (9.81) to yield the STA form of the Pauli equation

$$\begin{aligned} \partial_t \phi i\sigma_3 &= \frac{1}{2m}(\hat{\boldsymbol{p}} - e\boldsymbol{A})^2\phi + eV\phi - \frac{\hat{\boldsymbol{p}}^4}{8m^3c^2}\phi \\ &\quad -\frac{1}{8m^2c^2}[e(\boldsymbol{\nabla}\boldsymbol{E} - 2\boldsymbol{E}\wedge\boldsymbol{\nabla})\phi - 2e^2\boldsymbol{A}\wedge\boldsymbol{E}\phi i\sigma_3], \end{aligned} \tag{9.84}$$

which is valid to $O(c^{-2})$. (We have assumed that $|\boldsymbol{A}| \sim c^{-1}$ to replace the $\mathcal{O}^4$ term by $\hat{\boldsymbol{p}}^4$.) Using the translation scheme of Table 9.1 it is straightforward to check that equation (9.84) is the same as that found in standard texts. In the standard approach, the geometric product in the $\boldsymbol{\nabla}\boldsymbol{E}$ term (9.84) is split into a 'spin-orbit' term $\boldsymbol{\nabla}\wedge\boldsymbol{E}$ and the 'Darwin' term $\boldsymbol{\nabla}\cdot\boldsymbol{E}$. The STA approach reveals that these terms arise from a single source.

A similar approximation scheme can be adopted for the observables of the Dirac theory. For example the current, $\psi\gamma_0\tilde{\psi}$, has a three-vector part

$$\boldsymbol{J} = (\psi\gamma_0\tilde{\psi})\wedge\gamma_0 = \phi\eta^\dagger + \eta\phi^\dagger, \tag{9.85}$$

which is approximated to first-order by

$$\boldsymbol{J} \approx -\frac{1}{m}(\langle\boldsymbol{\nabla}\phi i\sigma_3\phi^\dagger\rangle_v - \boldsymbol{A}\phi\phi^\dagger). \tag{9.86}$$

Not all expositions of the Pauli theory in the literature correctly identify (9.86) as the conserved current in the Pauli theory!

9.6 Angular Eigenstates and Monogenic Functions

Returning to the Hamiltonian of equation (9.65), let us now consider the problem of a central potential $V = V(r)$, $\boldsymbol{A} = 0$, where $r = |\boldsymbol{x}|$. We seek a set of angular-momentum operators which commute with this Hamiltonian. Starting with the scalar operator $B\cdot(\boldsymbol{x}\wedge\boldsymbol{\nabla})$, where B is a spatial bivector, we find that

$$\begin{aligned} [B\cdot(\boldsymbol{x}\wedge\boldsymbol{\nabla}), \mathcal{H}] &= [B\cdot(\boldsymbol{x}\wedge\boldsymbol{\nabla}), -j\boldsymbol{\nabla}] \\ &= j\dot{\boldsymbol{\nabla}}B\cdot(\dot{\boldsymbol{x}}\wedge\boldsymbol{\nabla}) \\ &= -jB\cdot\boldsymbol{\nabla}. \end{aligned} \tag{9.87}$$

But, since $B\cdot\boldsymbol{\nabla} = [B, \boldsymbol{\nabla}]/2$ and B commutes with the rest of $\mathcal{H}$, we can rearrange the commutator into

$$[B\cdot(\boldsymbol{x}\wedge\boldsymbol{\nabla}) - \tfrac{1}{2}B, \mathcal{H}] = 0, \tag{9.88}$$

which gives us the required operator. Since $B\cdot(\boldsymbol{x}\wedge\boldsymbol{\nabla}) - B/2$ is an anti-Hermitian operator, we define a set of Hermitian operators as

$$J_B \equiv j(B\cdot(\boldsymbol{x}\wedge\boldsymbol{\nabla}) - \tfrac{1}{2}B). \tag{9.89}$$

The extra term of $\frac{1}{2}B$ is the term that is conventionally viewed as defining 'spin-1/2'. However, the geometric algebra derivation shows that the result rests solely on the commutation properties of the $B\cdot(\boldsymbol{x}\wedge\boldsymbol{\nabla})$ and $\boldsymbol{\nabla}$ operators. Furthermore, the factor of one-half required in the J_B operators would be present in a space of any dimension.

9.6.1 The Spherical Monogenics

The key ingredients in the solution of the Dirac equation for problems with radial symmetry are the spherical monogenics. These are Pauli spinors (even elements of the Pauli algebra) which satisfy the eigenvalue equation

$$-\boldsymbol{x}\wedge\boldsymbol{\nabla}\psi = l\psi. \tag{9.90}$$

Such functions are called spherical monogenics because they are obtained from the 'monogenic equation'

$$\boldsymbol{\nabla}\Psi = 0 \tag{9.91}$$

by separating Ψ into $r^l\psi(\theta,\phi)$. Equation (9.91) generalises the concept of an analytic function to higher dimensions.

To analyse the properties of equation (9.90) we first note that

$$[J_B, \boldsymbol{x}\wedge\boldsymbol{\nabla}] = 0. \tag{9.92}$$

It follows that ψ can simultaneously be an eigenstate of the $\boldsymbol{x}\wedge\boldsymbol{\nabla}$ operator and one of the J_B operators. To simplify the notation we now define

$$J_k\psi \equiv J_{i\sigma_k}\psi = (i\sigma_k\cdot(\boldsymbol{x}\wedge\boldsymbol{\nabla}) - \tfrac{1}{2}i\sigma_k)\psi i\sigma_3. \tag{9.93}$$

We choose ψ to be an eigenstate of J_3, and provisionally write

$$-\boldsymbol{x}\wedge\boldsymbol{\nabla}\psi = l\psi, \qquad J_3\psi = \mu\psi. \tag{9.94}$$

Before proceeding, we introduce the spherical-polar coordinate system r, θ, ϕ, and corresponding unit vectors $\sigma_r, \sigma_\theta, \sigma_\phi$.

The vector σ_r satisfies

$$\boldsymbol{x}\wedge\boldsymbol{\nabla}\sigma_r = 2\sigma_r. \tag{9.95}$$

It follows that

$$-\boldsymbol{x}\wedge\boldsymbol{\nabla}(\sigma_r\psi\sigma_3) = -(l+2)\sigma_r\psi\sigma_3 \tag{9.96}$$

so, without loss of generality, we can choose l to be positive and recover the negative-l states through multiplying by σ_r. In addition, since

$$\boldsymbol{x}\wedge\boldsymbol{\nabla}(\boldsymbol{x}\wedge\boldsymbol{\nabla}\psi) = l^2\psi \tag{9.97}$$

we find that

$$\frac{1}{\sin\theta}\frac{\partial}{\partial\theta}\left(\sin\theta\frac{\partial\psi}{\partial\theta}\right) + \frac{1}{\sin^2\theta}\frac{\partial^2\psi}{\partial\phi^2} = -l(l+1)\psi. \tag{9.98}$$

Hence, with respect to a constant basis for the STA, the components of ψ are spherical harmonics and l must be an integer for any physical solution.

The next step is to introduce ladder operators to move between different J_3 eigenstates. The conclusions are that, for each value of l, the allowed values of the eigenvalues of J_3 range from $(l+1/2)$ to $-(l+1/2)$. The total degeneracy is therefore $2(l+1)$. The states can therefore be labeled by two *integers* l and m such that

$$-\boldsymbol{x}\wedge\boldsymbol{\nabla}\psi_l^m = l\psi_l^m \qquad l \geq 0 \tag{9.99}$$

$$J_3\psi_l^m = (m+\tfrac{1}{2})\psi_l^m \qquad -1-l \leq m \leq l. \tag{9.100}$$

Labelling the states in this manner is unconventional, but provides for many simplifications in describing the properties of the ψ_l^m.

To find an explicit expression for the ψ_l^m we start from the highest-m eigenstate, which is given by

$$\psi_l^l = \sin^l\theta\, e^{l\phi i\sigma_3}, \tag{9.101}$$

and act on this with the lowering operator J_-. The result is the following, remarkably compact formula:

$$\psi_l^m = [(l+m+1)P_l^m(\cos\theta) - P_l^{m+1}(\cos\theta)i\sigma_\phi]e^{m\phi i\sigma_3}, \tag{9.102}$$

where the associated Legendre polynomials follow the conventions of Gradshteyn & Ryzhik. The expression (9.102) offers a considerable improvement over formulae found elsewhere in terms of both compactness and ease of use. The formula (9.102) is valid for non-negative l and both signs of m. The positive and negative m-states are related by

$$\psi_l^m(-i\sigma_2) = (-1)^m\frac{(l+m+1)!}{(l-m)!}\psi_l^{-(m+1)}. \tag{9.103}$$

The negative-l states are constructed using (9.96) and the J_3 eigenvalues are unchanged by this construction. The possible eigenvalues and degeneracies are summarised in Table 9.4. One curious feature of this table is that we appear to be missing a line for the eigenvalue $l = -1$. In fact solutions for this case do exist, but they contain singularities which render them unnormalisable. For example, the functions

$$\frac{i\sigma_\phi}{\sin\theta}, \quad \text{and} \quad \frac{e^{-i\sigma_3\phi}}{\sin\theta} \tag{9.104}$$

have $l = -1$ and J_3 eigenvalues $+1/2$ and $-1/2$ respectively. Both solutions are singular along the z-axis, however, so are of limited physical interest.

l	Eigenvalues of J_3	Degeneracy
$\vdots$	$\vdots$	$\vdots$
2	$5/2 \cdots -5/2$	6
1	$3/2 \cdots -3/2$	4
0	$1/2 \cdots -1/2$	2
(-1)	?	?
-2	$1/2 \cdots -1/2$	2
$\vdots$	$\vdots$	$\vdots$

Table 9.4. *Eigenvalues and degeneracies for the* ψ_l^m *monogenics.*

9.7 Application — the Coulomb Problem

Having established the properties of the spherical monogenics, we can proceed quickly to the solution of standard case of the hydrogen atom. The Hamiltonian for this problem is

$$\mathcal{H}\psi = \hat{\boldsymbol{p}}\psi - \frac{Z\alpha}{r}\psi + m\bar{\psi}, \tag{9.105}$$

where $\alpha = e^2/4\pi$ is the fine-structure constant and Z is the atomic charge. Since the J_B operators commute with $\mathcal{H}$, ψ can be placed in an eigenstate of J_3. The operator J_iJ_i must also commute with $\mathcal{H}$, but $\boldsymbol{x}\wedge\boldsymbol{\nabla}$ does not, so both the ψ_l^m and $\sigma_r\psi_l^m\sigma_3$ monogenics are needed in the solution.

Though $\boldsymbol{x}\wedge\boldsymbol{\nabla}$ does not commute with $\mathcal{H}$, the operator

$$K = \hat{\gamma}_0(1 - \boldsymbol{x}\wedge\boldsymbol{\nabla}) \tag{9.106}$$

does, as follows from

$$\begin{aligned} [\hat{\gamma}_0(1 - \boldsymbol{x}\wedge\boldsymbol{\nabla}), \boldsymbol{\nabla}] &= 2\hat{\gamma}_0\boldsymbol{\nabla} - \hat{\gamma}_0\dot{\boldsymbol{\nabla}}\dot{\boldsymbol{x}}\wedge\boldsymbol{\nabla} \\ &= 0. \end{aligned} \tag{9.107}$$

We can therefore work with eigenstates of the K operator, which means that the spatial part of ψ goes either as

$$\psi(\boldsymbol{x}, l+1) = \psi_l^m u(r) + \sigma_r\psi_l^m v(r) i\sigma_3 \tag{9.108}$$

or as

$$\psi(\boldsymbol{x}, -(l+1)) = \sigma_r\psi_l^m\sigma_3 u(r) + \psi_l^m iv(r). \tag{9.109}$$

In both cases the second label in $\psi(\boldsymbol{x}, l+1)$ specifies the eigenvalue of K. The functions $u(r)$ and $v(r)$ are initially 'complex' superpositions of a scalar and an $i\sigma_3$ term. It turns out, however, that the scalar and $i\sigma_3$ equations decouple, and it is sufficient to treat $u(r)$ and $v(r)$ as scalars.

We now substitute the trial functions (9.108) and (9.109) into the Hamiltonian (9.105). Using the results that

$$-\boldsymbol{\nabla}\psi_l^m = l/r\,\sigma_r\psi_l^m, \quad -\boldsymbol{\nabla}\sigma_r\psi_l^m = -(l+2)/r\,\psi_l^m, \tag{9.110}$$

and looking for stationary-state solutions of energy E, the radial equations reduce to

$$\begin{pmatrix} u' \\ v' \end{pmatrix} = \begin{pmatrix} (\kappa-1)/r & -(E+Z\alpha/r+m) \\ E+Z\alpha/r-m & (-\kappa-1)/r \end{pmatrix} \begin{pmatrix} u \\ v \end{pmatrix}, \tag{9.111}$$

where κ is the eigenvalue of K. (κ is a non-zero positive or negative integer.) The solution of these radial equations can be found in many textbooks. The solutions can be given in terms of confluent hypergeometric functions, and the energy spectrum is obtained from the equation

$$E^2 = m^2 \left[1 - \frac{(Z\alpha)^2}{n^2 + 2n\nu + (l+1)^2}\right], \tag{9.112}$$

where n is a positive integer and

$$\nu = [(l+1)^2 + (Z\alpha)^2]^{1/2}. \tag{9.113}$$

Whilst this analysis does not offer any new results, it should demonstrate how easy it is to manipulate expressions involving the spherical monogenics.

Chapter 10

Electron Physics II

Stephen Gull, Chris Doran, and Anthony Lasenby
MRAO, Cavendish Laboratory, Madingley Road,
Cambridge, CB3 0HE, UK

10.1 Propagation and Characteristic Surfaces

One of the simplest demonstrations of the insights provided by the STA formulation of both Maxwell and Dirac theories is in the treatment of characteristic surfaces. Suppose that we have a generic equation of the type

$$\nabla\psi = f(\psi, x), \tag{10.1}$$

where $\psi(x)$ is any multivector field (not necessarily a spinor field) and $f(\psi, x)$ is some arbitrary, known function. If we are given initial data over some 3-D surface, are there any obstructions to us propagating this information off the surface? If so, the surface is a characteristic surface. We start at a point on the surface and pick three independent vectors $\{a, b, c\}$ tangent to the surface at the chosen point. Knowledge of ψ on the surface enables us to calculate

$$a\cdot\nabla\psi, \quad b\cdot\nabla\psi \quad \text{and} \quad c\cdot\nabla\psi. \tag{10.2}$$

We next form the trivector $a\wedge b\wedge c$ and dualise to define

$$n \equiv ia\wedge b\wedge c. \tag{10.3}$$

We can now multiply equation (10.1) by n and use

$$\begin{aligned} n\nabla\psi &= n\cdot\nabla\psi + n\wedge\nabla\psi \\ &= n\cdot\nabla\psi + i(a\wedge b\wedge c)\cdot\nabla\psi & (10.4) \\ &= n\cdot\nabla\psi + i(a\wedge b\, c\cdot\nabla\psi - a\wedge c\, b\cdot\nabla\psi + b\wedge c\, a\cdot\nabla\psi), & (10.5) \end{aligned}$$

to obtain

$$n\cdot\nabla\psi = nf(\psi, x) - i(a\wedge b\, c\cdot\nabla\psi - a\wedge c\, b\cdot\nabla\psi + b\wedge c\, a\cdot\nabla\psi). \tag{10.6}$$

All of the terms on the right-hand side of equation (10.6) are known, so we can find $n\cdot\nabla\psi$ and use this to propagate ψ in the n direction (*i.e.* off the surface). The only situation in which we fail to propagate, therefore, is when n remains in the surface. This occurs when

$$\begin{aligned} n\wedge(a\wedge b\wedge c) &= 0 \\ \Rightarrow \quad n\wedge(ni) &= 0 \\ \Rightarrow \quad n\cdot n &= 0. \end{aligned} \tag{10.7}$$

Hence we only fail to propagate when $n^2 = 0$, and it follows immediately that the characteristic surfaces of equation (10.1) are *null* surfaces. This result applies to any first-order equation based on the vector derivative ∇, including the Maxwell and Dirac equations. The fundamental significance of null directions in these theories is transparent in their STA form.

10.2 Spinor Potentials and Propagators

A simple method to generate propagators for the Dirac theory is to introduce a spinor potential satisfying a scalar second-order equation. Suppose that ψ satisfies the Dirac equation

$$\nabla\psi i\sigma_3 - m\psi\gamma_0 = 0. \tag{10.8}$$

ψ can be generated from the (odd multivector) potential ϕ via

$$\psi = \nabla\phi i\sigma_3 + m\phi\gamma_0 \tag{10.9}$$

provided that

$$(\nabla^2 + m^2)\phi = 0. \tag{10.10}$$

The standard second-order theory can then be applied to ϕ, and then used to recover ψ. For constant-energy waves

$$\psi = \psi(\boldsymbol{x})e^{-i\sigma_3 Et}. \tag{10.11}$$

The Dirac equation then becomes

$$\boldsymbol{\nabla}\psi i\sigma_3 + E\psi - m\bar{\psi} = 0 \tag{10.12}$$

which is solved by

$$\psi = -\boldsymbol{\nabla}\phi i\sigma_3 + E\phi + m\bar{\phi} \tag{10.13}$$

where

$$\phi(\boldsymbol{x}) = -\frac{1}{4\pi}\oint |dS'|\, \boldsymbol{n}'\psi(\boldsymbol{x}')\frac{e^{i\sigma_3 pr}}{r}. \tag{10.14}$$

In this integral the initial data $\psi(\boldsymbol{x}')$ is given over some closed spatial surface with normal $\boldsymbol{n}' = \boldsymbol{n}(\boldsymbol{x}')$, and p and r are defined by

$$p \equiv \sqrt{E^2 + m^2} \quad \text{and} \quad r \equiv |\boldsymbol{x} - \boldsymbol{x}'|. \tag{10.15}$$

10.3 Scattering Theory

We now take a brief look at how the matrix approach to scattering theory is handled in the STA. We continue to employ the symbol j for $i\sigma_3$ in places where it simplifies the notation. In particular, we employ the j symbol in the exponential terms introduced by Fourier transforming to momentum space. Where the j's play a more significant geometric role they are left in the $i\sigma_3$ form.

We start by rewriting the Dirac equation as an integral equation:

$$\psi(x) = \psi_i(x) + e\int d^4x'\, S_F(x-x')A(x')\psi(x') \tag{10.16}$$

where ψ_i is the asymptotic in-state which solves the free-particle equation, and $S_F(x-x')$ is the STA form of the Feynman propagator. Substituting (10.16) into the Dirac equation, we find that $S_F(x-x')$ must satisfy

$$\nabla_x S_F(x-x')M(x')i\sigma_3 - mS_F(x-x')M(x')\gamma_0 = \delta(x-x')M(x') \tag{10.17}$$

for an arbitrary multivector $M(x')$. The solution to this equation is

$$S_F(x-x')M(x') = \int \frac{d^4p}{(2\pi)^4}\frac{pM(x')+mM(x')\gamma_0}{p^2-m^2}e^{-jp\cdot(x-x')} \tag{10.18}$$

where, for causal propagation, the dE integral must arrange that positive-frequency waves propagate into the future ($t > t'$) and negative-frequency waves propagate into the past ($t' > t$). The result of performing the dE integral is

$$\begin{aligned} S_F(x-x')M &= -\theta(t-t')\int \frac{d^3\boldsymbol{p}}{(2\pi)^3}\frac{1}{2E}(pM+mM\gamma_0)i\sigma_3 e^{-jp\cdot(x-x')} \\ &\quad +\theta(t'-t)\int \frac{d^3\boldsymbol{p}}{(2\pi)^3}\frac{1}{2E}(pM-mM\gamma_0)i\sigma_3 e^{jp\cdot(x-x')} \end{aligned} \tag{10.19}$$

where $E = +\sqrt{\boldsymbol{p}^2+m^2}$ and $M = M(x')$.

With $\psi_{\text{diff}}(x)$ defined by

$$\psi_{\text{diff}}(x) \equiv \psi(x) - \psi_i(x) \tag{10.20}$$

we find that, as t tends to $+\infty$, $\psi_{\text{diff}}(x)$ is given by

$$\psi_{\text{diff}}(x) = -e\int d^4x' \int \frac{d^3\boldsymbol{p}}{(2\pi)^3}\frac{1}{2E}[pA(x')\psi(x') + mA(x')\psi(x')\gamma_0]i\sigma_3 e^{-jp\cdot(x-x')}. \tag{10.21}$$

We therefore define a set of final states $\psi_f(x)$ by

$$\psi_f(x) \equiv -e\int \frac{d^4x'}{2E_f}[p_fA(x')\psi(x') + mA(x')\psi(x')\gamma_0]i\sigma_3 e^{-jp_f\cdot(x-x')}, \tag{10.22}$$

which are plane-wave solutions to the free-field equations with momentum p_f. The spinor $\psi_{\text{diff}}(x)$ can now be expressed as a superposition of these plane-wave states,

$$\psi_{\text{diff}}(x) = \int \frac{d^3\boldsymbol{p}_f}{(2\pi)^3}\psi_f(x). \tag{10.23}$$

10.3.1 The Born Approximation and Coulomb Scattering

In order to find $\psi_f(x)$ we must evaluate the integral (10.22). In the Born approximation, we simplify the problem by approximating $\psi(x')$ by $\psi_i(x')$. In this case, since

$$\psi_i(x') = \psi_i e^{-jp_i \cdot x'}, \quad \text{and} \quad m\psi_i \gamma_0 = p_i \psi_i, \tag{10.24}$$

we can write

$$\psi_f(x) = -e \int \frac{d^4x'}{2E_f} [p_f A(x') + A(x') p_i] \psi_i i\sigma_3 e^{jq\cdot x'} e^{-jp_f \cdot x}, \tag{10.25}$$

where

$$q \equiv p_f - p_i. \tag{10.26}$$

The integral in (10.25) can now be evaluated for any given A-field.

As a simple application consider Coulomb scattering, for which $A(x')$ is given by

$$A(x') = \frac{Ze}{4\pi|\boldsymbol{x}'|}\gamma_0. \tag{10.27}$$

Inserting this in (10.25) and carrying out the integrals, we obtain

$$\psi_f(x) = -S_{fi}\psi_i i\sigma_3 \frac{(2\pi)^2}{E_f}\delta(E_f - E_i) \tag{10.28}$$

where

$$\begin{aligned} S_{fi} &\equiv \frac{Z\alpha}{4\pi}[p_f\gamma_0 + \gamma_0 p_i] \int d^3x \, \frac{e^{-j\boldsymbol{q}\cdot\boldsymbol{r}}}{r} \\ &= \frac{Z\alpha}{\boldsymbol{q}^2}(2E + \boldsymbol{q}). \end{aligned} \tag{10.29}$$

Here $E = E_f = E_i$ and $\alpha = e^2/(4\pi)$ is the fine-structure constant. The quantity S_{fi} contains all the information about the scattering process. Its magnitude determines the cross-section via

$$\frac{d\sigma}{d\Omega_f} = S_{fi}\tilde{S}_{fi} \tag{10.30}$$

and the remainder of S_{fi} determines the change of momentum and spin vectors. This is clear from (10.28), which shows that S_{fi} must contain the rotor $R_f\tilde{R}_i$, where R_i and R_f are the rotors for the initial and final plane-wave states.

Substituting (10.29) into (10.30) we immediately recover the Mott scattering cross-section

$$\frac{d\sigma}{d\Omega_f} = \frac{Z^2\alpha^2}{\boldsymbol{q}^4}(4E^2 - \boldsymbol{q}^2) = \frac{Z^2\alpha^2}{4\boldsymbol{p}^2\beta^2\sin^4(\theta/2)}\left(1 - \beta^2\sin^2(\theta/2)\right), \tag{10.31}$$

where

$$\boldsymbol{q}^2 = (\boldsymbol{p}_f - \boldsymbol{p}_i)^2 = 2\boldsymbol{p}^2(1 - \cos\theta) \quad \text{and} \quad \beta = |\boldsymbol{p}|/E. \tag{10.32}$$

The notable feature of this derivation is that no spin sums are required. Instead, all the spin dependence is contained in the directional information in S_{fi}. As well as being computationally more efficient, the STA method for organising cross-section calculations offers deeper insights into the structure of the theory. For example, for Mott-scattering the directional information is contained entirely in the quantity

$$S'_{fi} = \frac{1}{m}[p_f\gamma_0 + \gamma_0 p_i] = L_f^2 + \tilde{L}_i^2 \tag{10.33}$$

where L_f and L_i are the boosts contained in R_f and R_i respectively. The algebraic structure

$$S_{fi} = p_f M + M p_i, \tag{10.34}$$

where M is some odd multivector, is common to many scattering problems.

Since S_{fi} contains the quantity $R_f \tilde{R}_i$, we obtain a spatial rotor by removing the two boost terms. We therefore define the (unnormalised) rotor

$$U'_{fi} \equiv \tilde{L}_f(L_f^2 + \tilde{L}_i^2)L_i = L_f L_i + \tilde{L}_f \tilde{L}_i, \tag{10.35}$$

so that $U_{fi} = U'_{fi}/|U_{fi}|$ determines the rotation from initial to final rest-spins. A simple calculation gives

$$U'_{fi} = 2[(E+m)^2 + \boldsymbol{p}_f \boldsymbol{p}_i], \tag{10.36}$$

hence the rest-spin vector precesses in the $\boldsymbol{p}_f \wedge \boldsymbol{p}_i$ plane through an angle δ, where

$$\tan(\delta/2) = \frac{\sin\theta}{(E+m)/(E-m) + \cos\theta}. \tag{10.37}$$

10.4 Plane Waves at Potential Steps

We now turn to a discussion of the matching of Dirac plane waves at a potential step. The case of perpendicular incidence is a standard problem and is treated in most texts. In order to demonstrate the power of the STA approach we treat the more general case of oblique incidence,. A number of applications are given as illustrations, including the tunnelling of monochromatic waves and spin precession on total reflection at a barrier. We conclude the section with a discussion of the Klein paradox.

The problem of interest is that of plane waves incident on a succession of potential steps. The steps are taken as lying along the x direction, with infinite extent in the y and z directions. Since the spatial components of the incoming and outgoing wavevectors lie in a single plane, the matching problem can be reduced to one in two dimensions. The analysis is simplified further if the wavevectors are taken to lie in the $i\sigma_3$ plane. (Other configurations can always be obtained by applying a rotation.) The arrangement is illustrated in Figure 10.1. The waves all oscillate at a single frequency E, and the Dirac equation in the ith region is

$$(E - eV_i)\psi = -\boldsymbol{\nabla}\psi i\sigma_3 + m\gamma_0\psi\gamma_0. \tag{10.38}$$

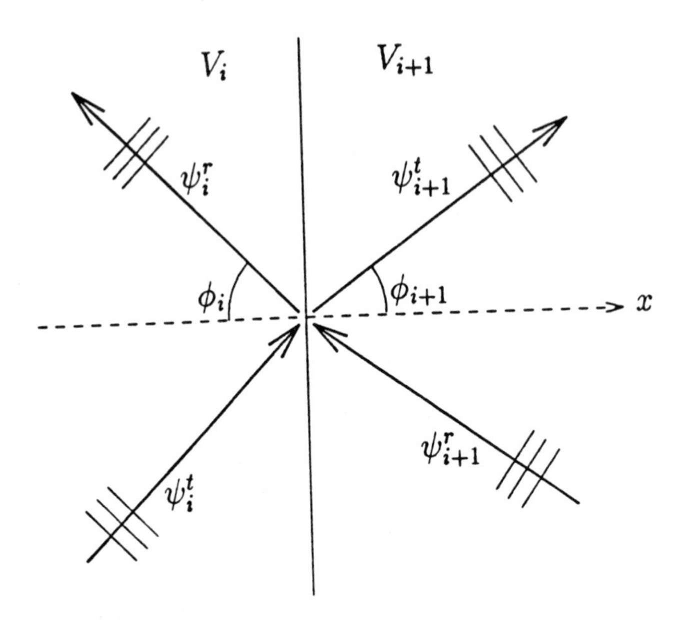

Fig. 10.1. *Plane waves at a potential step. The spatial component of the wavevector lies in the x–y plane and the step lies in the y–z plane.*

By continuity of ψ at each boundary, the y component of the wavevector, p_y, must be the same in all regions. For the ith region we define

$$E_i' \equiv E - eV_i \tag{10.39}$$

and, depending on the magnitude of V_i, the waves in the this region will be either travelling or evanescent. For travelling waves we define (dropping the subscripts)

$$p_x^2 \equiv E'^2 - p_y^2 - m^2 \qquad |E - eV| > \sqrt{p_y^2 + m^2}. \tag{10.40}$$

In terms of the angle of incidence ϕ we also have

$$p_x = p\cos\phi, \qquad p_y = p\sin\phi, \qquad E'^2 = p^2 + m^2. \tag{10.41}$$

For evanescent waves we write

$$\kappa^2 \equiv -E'^2 + p_y^2 + m^2 \qquad |E - eV| < \sqrt{p_y^2 + m^2}. \tag{10.42}$$

In all the cases that we study, the incoming waves are assumed to be positive-energy travelling waves in a region where $eV < E - \sqrt{p_y^2 + m^2}$. Recalling the plane-wave solutions found previously, the travelling waves are given by

$$\begin{aligned} \psi^t &= [\cosh(u/2) + \sinh(u/2)(\cos\phi\,\sigma_1 + \sin\phi\,\sigma_2)]\Phi e^{-i\sigma_3(Et - p_x x - p_y y)}T \\ \psi^r &= [\cosh(u/2) + \sinh(u/2)(-\cos\phi\,\sigma_1 + \sin\phi\,\sigma_2)]\Phi e^{-i\sigma_3(Et + p_x x - p_y y)}R, \end{aligned} \tag{10.43}$$

where

$$\tanh(u/2) = p/(E' + m) \tag{10.44}$$

so that

$$\sinh u = p/m, \qquad \cosh u = E'/m. \tag{10.45}$$

The transmission and reflection coefficients T and R are scalar $+i\sigma_3$ combinations, always appearing on the right-hand side of the spinor. The fact that p_y is the same in all regions gives the electron equivalent of Snell's law,

$$\sinh u \, \sin\phi = \text{constant}. \tag{10.46}$$

The Pauli spinor Φ describes the rest-spin of the particle, with $\Phi = 1$ giving spin-up and $\Phi = -i\sigma_2$ spin down. Other situations are, of course, built from superpositions of these basis states. For these two spin basis states the spin vector is $\pm\sigma_3$, which lies in the plane of the barrier and is perpendicular to the plane of motion. Choosing the states so that the spin is aligned in this manner simplifies the analysis, as the two spin states completely decouple.

There are three matching situations to consider, depending on whether the transmitted waves are travelling, evanescent or in the Klein region ($eV > E + \sqrt{p_y^2 + m^2}$). We consider each of these in turn.

10.4.1 Matching Conditions for Travelling Waves

The situation of interest here is when there are waves of type (10.43) in both regions. The matching condition in all of these problems is simply that ψ is continuous at the boundary. The work involved is therefore, in principle, less than for the equivalent non-relativistic problem. The matching is slightly different for the two spins, so we consider each in turn.

Spin-Up ($\Phi = 1$) We simplify the problem initially by taking the boundary at $x = 0$. Steps at other values of x are then dealt with by inserting suitable phase factors. The matching condition at $x = 0$ reduces to

$$\begin{aligned}&[\cosh(u_i/2) + \sinh(u_i/2)\sigma_1 e^{\phi_i i\sigma_3}]T_i^\uparrow \\ &+ [\cosh(u_i/2) - \sinh(u_i/2)\sigma_1 e^{-\phi_i i\sigma_3}]R_i^\uparrow \\ &\qquad = [\cosh(u_{i+1}/2) + \sinh(u_{i+1}/2)\sigma_1 e^{\phi_{i+1} i\sigma_3}]T_{i+1}^\uparrow \\ &\qquad\quad + [\cosh(u_{i+1}/2) - \sinh(u_{i+1}/2)\sigma_1 e^{-\phi_{i+1} i\sigma_3}]R_{i+1}^\uparrow.\end{aligned} \tag{10.47}$$

Since the equations for the reflection and transmission coefficients involve only scalar and $i\sigma_3$ terms, it is again convenient to replace the $i\sigma_3$ bivector with the symbol j. If we now define the 2×2 matrix

$$\boldsymbol{A}_i \equiv \begin{pmatrix} \cosh(u_i/2) & \cosh(u_i/2) \\ \sinh(u_i/2)e^{j\phi_i} & -\sinh(u_i/2)e^{-j\phi_i} \end{pmatrix} \tag{10.48}$$

we find that equation (10.47) can be written concisely as

$$\boldsymbol{A}_i \begin{pmatrix} T_i^\uparrow \\ R_i^\uparrow \end{pmatrix} = \boldsymbol{A}_{i+1} \begin{pmatrix} T_{i+1}^\uparrow \\ R_{i+1}^\uparrow \end{pmatrix}. \tag{10.49}$$

The $\boldsymbol{A}_i$ matrix has a straightforward inverse, so equation (10.49) can be easily manipulated to describe various physical situations. For example, consider plane waves incident on a single step. The equation describing this configuration is simply

$$\boldsymbol{A}_1 \begin{pmatrix} T_1^{\uparrow} \\ R_1^{\uparrow} \end{pmatrix} = \boldsymbol{A}_2 \begin{pmatrix} T_2^{\uparrow} \\ 0 \end{pmatrix} \tag{10.50}$$

so that

$$\begin{pmatrix} T_1^{\uparrow} \\ R_1^{\uparrow} \end{pmatrix} = \frac{T_2^{\uparrow}}{\sinh u_1 \cos\phi_1} \begin{pmatrix} \sinh(\frac{1}{2}u_1)\cosh(\frac{1}{2}u_2)e^{-j\phi_1} + \cosh(\frac{1}{2}u_1)\sinh(\frac{1}{2}u_2)e^{j\phi_2} \\ \sinh(\frac{1}{2}u_1)\cosh(\frac{1}{2}u_2)e^{j\phi_1} - \cosh(\frac{1}{2}u_1)\sinh(\frac{1}{2}u_2)e^{j\phi_2} \end{pmatrix} \tag{10.51}$$

from which the reflection and transmission coefficients can be read off. The case of perpendicular incidence is particularly simple as equation (10.49) can be replaced by

$$\begin{pmatrix} T_{i+1} \\ R_{i+1} \end{pmatrix} = \frac{1}{\sinh u_{i+1}} \begin{pmatrix} \sinh\frac{1}{2}(u_{i+1}+u_i) & \sinh\frac{1}{2}(u_{i+1}-u_i) \\ \sinh\frac{1}{2}(u_{i+1}-u_i) & \sinh\frac{1}{2}(u_{i+1}+u_i) \end{pmatrix} \begin{pmatrix} T_i \\ R_i \end{pmatrix}, \tag{10.52}$$

which is valid for all spin orientations. So, for perpendicular incidence, the reflection coefficient $r = R_1/T_1$ and transmission coefficient $t = T_2/T_1$ at a single step are

$$r = \frac{\sinh\frac{1}{2}(u_1-u_2)}{\sinh\frac{1}{2}(u_1+u_2)}, \qquad t = \frac{\sinh u_1}{\sinh\frac{1}{2}(u_1+u_2)}, \tag{10.53}$$

which agree with the results given in standard texts.

Spin-Down ($\Phi = -i\sigma_2$) The matching equations for the case of opposite spin are

$$\begin{aligned} &[\cosh(u_i/2) + \sinh(u_i/2)\sigma_1 e^{\phi_i i\sigma_3}](-i\sigma_2)T_i^{\downarrow} \\ &+ [\cosh(u_i/2) - \sinh(u_i/2)\sigma_1 e^{-\phi_i i\sigma_3}](-i\sigma_2)R_i^{\downarrow} \\ &\qquad = [\cosh(u_{i+1}/2) + \sinh(u_{i+1}/2)\sigma_1 e^{\phi_{i+1} i\sigma_3}](-i\sigma_2)T_{i+1}^{\downarrow} \\ &\qquad\quad + [\cosh(u_{i+1}/2) - \sinh(u_{i+1}/2)\sigma_1 e^{-\phi_{i+1} i\sigma_3}](-i\sigma_2)R_{i+1}^{\downarrow}. \end{aligned} \tag{10.54}$$

Pulling the $i\sigma_2$ out on the right-hand side just has the effect of complex-conjugating the reflection and transmission coefficients, so the matrix equation (10.47) is unchanged except that it now relates the complex conjugates of the reflection and transmission coefficients. The analog of equation (10.49) is therefore

$$\boldsymbol{A}_i^* \begin{pmatrix} T_i^{\downarrow} \\ R_i^{\downarrow} \end{pmatrix} = \boldsymbol{A}_{i+1}^* \begin{pmatrix} T_{i+1}^{\downarrow} \\ R_{i+1}^{\downarrow} \end{pmatrix}. \tag{10.55}$$

As mentioned earlier, the choice of alignment of spin basis states ensures that there is no coupling between them.

One can string together series of barriers by including suitable 'propagation' matrices. For example, consider the set-up described in Figure 10.2. The matching equations for spin-up are, at the first barrier,

$$\boldsymbol{A} \begin{pmatrix} T^{\uparrow} \\ R^{\uparrow} \end{pmatrix} = \boldsymbol{A}_1 \begin{pmatrix} T_1^{\uparrow} \\ R_1^{\uparrow} \end{pmatrix} \tag{10.56}$$

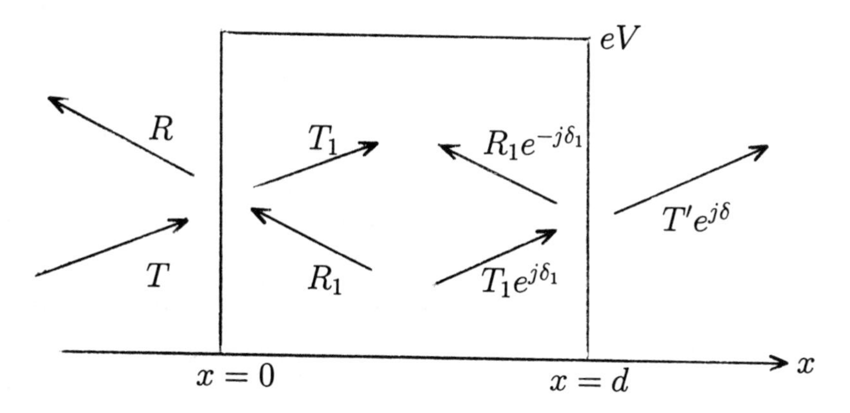

Fig. 10.2. *Plane waves scattering from a barrier. The barrier has height eV and width d. Quantities inside the barrier are labeled with a subscript 1, and the free quantities have no subscripts. The phases are given by $\delta_1 = md\,\mathrm{sinh}u_1\,\mathrm{cos}\phi_1$ and $\delta = p_x d$.*

and at the second barrier,

$$\boldsymbol{A}_1 \begin{pmatrix} e^{j\delta_1} & 0 \\ 0 & e^{-j\delta_1} \end{pmatrix} \begin{pmatrix} T_1^\uparrow \\ R_1^\uparrow \end{pmatrix} = \boldsymbol{A} \begin{pmatrix} e^{j\delta} & 0 \\ 0 & e^{-j\delta} \end{pmatrix} \begin{pmatrix} T^{\uparrow\prime} \\ 0 \end{pmatrix}. \tag{10.57}$$

Equation (10.57) demonstrates neatly how matrices of the type

$$\mathbf{P} = \begin{pmatrix} e^{j\delta} & 0 \\ 0 & e^{-j\delta} \end{pmatrix}, \tag{10.58}$$

where $\delta = p_x d$ and d is the distance between steps, can be used to propagate from one step to the next. In this case the problem is reduced to the equation

$$\begin{pmatrix} T^\uparrow \\ R^\uparrow \end{pmatrix} = \left[\mathrm{cos}\delta_1\,\boldsymbol{I} - j\,\mathrm{sin}\delta_1\,\boldsymbol{A}^{-1}\boldsymbol{A}_1 \begin{pmatrix} 1 & 0 \\ 0 & -1 \end{pmatrix} \boldsymbol{A}_1^{-1}\boldsymbol{A}\right] \begin{pmatrix} e^{j\delta}T^{\uparrow\prime} \\ 0 \end{pmatrix}, \tag{10.59}$$

which quickly yields the reflection and transmission coefficients.

10.4.2 Matching onto Evanescent Waves

Before studying matching onto evanescent waves, we must first solve the Dirac equation in the evanescent region. Again, the two spin orientations behave differently and are treated separately. Taking spin-up first and looking at the transmitted (decaying) wave in the evanescent region, the solution takes the form

$$\psi^t = [\cosh(u/2) + \sinh(u/2)\sigma_2]e^{-\kappa x}e^{-i\sigma_3(Et - p_y y)}T. \tag{10.60}$$

Substituting this into the Dirac equation yields

$$(E'\gamma_0 + p_y\gamma_2)e^{u\sigma_2/2} = e^{u\sigma_2/2}(m\gamma_0 - \kappa\gamma_2) \tag{10.61}$$

which is consistent with the definition of κ (10.42). From equation (10.61) we find that

$$\tanh(u/2) = \frac{E' - m}{p_y - \kappa} = \frac{p_y + \kappa}{E' + m}, \tag{10.62}$$

which completes the solution. For the incoming (growing) wave we flip the sign of κ. We therefore define $u^{\pm}$ via

$$\tanh(u^{\pm}/2) = \frac{p_y \pm \kappa}{E' + m} \tag{10.63}$$

and write the outgoing and incoming spin-up waves in the evanescent region as

$$\begin{aligned} \psi^t &= [\cosh(u^+/2) + \sinh(u^+/2)\sigma_2] e^{-\kappa x} e^{-i\sigma_3(Et - p_y y)} T^{\uparrow} \\ \psi^r &= [\cosh(u^-/2) + \sinh(u^-/2)\sigma_2] e^{\kappa x} e^{-i\sigma_3(Et - p_y y)} R^{\uparrow}. \end{aligned} \tag{10.64}$$

If we now consider matching at $x = 0$, the continuity equation becomes

$$\begin{aligned} &[\cosh(u_i/2) + \sinh(u_i/2)\sigma_1 e^{\phi_i i\sigma_3}] T_i^{\uparrow} \\ &+ [\cosh(u_i/2) - \sinh(u_i/2)\sigma_1 e^{-\phi_i i\sigma_3}] R_i^{\uparrow} \\ &\qquad = [\cosh(u_{i+1}^+/2) + \sinh(u_{i+1}^+/2)\sigma_2] T_{i+1}^{\uparrow} \\ &\qquad\quad + [\cosh(u_{i+1}^-/2) + \sinh(u_{i+1}^-/2)\sigma_2)] R_{i+1}^{\uparrow}. \end{aligned} \tag{10.65}$$

On defining the matrix

$$\boldsymbol{B}_i^+ \equiv \begin{pmatrix} \cosh(u_i^+/2) & \cosh(u_i^-/2) \\ j\sinh(u_i^+/2) & +j\sinh(u_i^-/2) \end{pmatrix}, \tag{10.66}$$

we can write equation (10.65) compactly as

$$\boldsymbol{A}_i \begin{pmatrix} T_i^{\uparrow} \\ R_i^{\uparrow} \end{pmatrix} = \boldsymbol{B}_{i+1}^+ \begin{pmatrix} T_{i+1}^{\uparrow} \\ R_{i+1}^{\uparrow} \end{pmatrix}. \tag{10.67}$$

Again, either of the matrices can be inverted to analyse various physical situations. For example, the case of total reflection by a step is handled by

$$\boldsymbol{A}_1 \begin{pmatrix} T_1^{\uparrow} \\ R_1^{\uparrow} \end{pmatrix} = \boldsymbol{B}_2^+ \begin{pmatrix} T_2^{\uparrow} \\ 0 \end{pmatrix}, \tag{10.68}$$

from which one finds the reflection coefficient

$$r^{\uparrow} = -\frac{\tanh(u^+/2) + \tanh(u/2) j e^{j\phi}}{\tanh(u^+/2) - \tanh(u/2) j e^{-j\phi}} \tag{10.69}$$

which has $|r^{\uparrow}| = 1$, as expected. The subscripts on u_1, $u_2^{\pm}$ and ϕ_1 are all obvious, and have been dropped.

The case of spin-down requires some sign changes. The spinors in the evanescent region are now given by

$$\begin{aligned} \psi^t &= [\cosh(u^-/2) + \sinh(u^-/2)\sigma_2](-i\sigma_2) e^{-\kappa x} e^{-i\sigma_3(Et - p_y y)} T^{\downarrow} \\ \psi^r &= [\cosh(u^+/2) + \sinh(u^+/2)\sigma_2](-i\sigma_2) e^{\kappa x} e^{-i\sigma_3(Et - p_y y)} R^{\downarrow} \end{aligned} \tag{10.70}$$

and, on defining

$$B_i^- \equiv \begin{pmatrix} \cosh(u_i^-/2) & \cosh(u_i^+/2) \\ j\sinh(u_i^-/2) & -j\sinh(u_i^+/2) \end{pmatrix}, \tag{10.71}$$

the analog of equation (10.67) is

$$A_i^* \begin{pmatrix} T_i^\downarrow \\ R_i^\downarrow \end{pmatrix} = B_{i+1}^{-*} \begin{pmatrix} T_{i+1}^\downarrow \\ R_{i+1}^\downarrow \end{pmatrix}. \tag{10.72}$$

These formulae are now applied to two situations of physical interest.

10.5 Spin Precession at a Barrier

When a monochromatic wave is incident on a single step of sufficient height that the wave cannot propagate there is total reflection. In the preceding section we found that the reflection coefficient for spin-up is given by equation (10.69), and the analogous calculation for spin-down yields

$$r^\downarrow = -\frac{\tanh(u^-/2) - \tanh(u/2)je^{-j\phi}}{\tanh(u^-/2) + \tanh(u/2)je^{j\phi}}. \tag{10.73}$$

Both $r^\uparrow$ and $r^\downarrow$ are pure phases, but there is an overall phase difference between the two. If the rest-spin vector $\boldsymbol{s} = \Phi\sigma_3\tilde{\Phi}$ is not perpendicular to the plane of incidence, then this phase difference produces a precession of the spin vector. To see how, suppose that the incident wave contains an arbitrary superposition of spin-up and spin-down states,

$$\Phi = \cos(\theta/2)e^{i\sigma_3\phi_1} - \sin(\theta/2)i\sigma_2 e^{i\sigma_3\phi_2} = e^{i\sigma_3\phi/2}e^{-i\sigma_2\theta/2}e^{-i\sigma_3\epsilon/2}, \tag{10.74}$$

where

$$\phi = \phi_1 - \phi_2, \qquad \epsilon = \phi_1 + \phi_2, \tag{10.75}$$

and the final pure-phase term is irrelevant. After reflection, suppose that the separate up and down states receive phase shifts of $\delta^\uparrow$ and $\delta^\downarrow$ respectively. The Pauli spinor in the reflected wave is therefore

$$\begin{aligned} \Phi^r &= \cos(\theta/2)e^{i\sigma_3(\phi_1+\delta^\uparrow)} - \sin(\theta/2)i\sigma_2 e^{i\sigma_3(\phi_2+\delta^\downarrow)} \\ &= e^{i\sigma_3\delta/2}e^{i\sigma_3\phi/2}e^{-i\sigma_2\theta/2}e^{-i\sigma_3\epsilon'/2}, \end{aligned} \tag{10.76}$$

where

$$\delta = \delta^\uparrow - \delta^\downarrow \tag{10.77}$$

and again there is an irrelevant overall phase. The rest-spin vector for the reflected wave is therefore

$$\boldsymbol{s}^r = \Phi^r\sigma_3\tilde{\Phi}^r = e^{i\sigma_3\delta/2}\boldsymbol{s}e^{-i\sigma_3\delta/2}, \tag{10.78}$$

so the spin-vector precesses in the plane of incidence through an angle $\delta^\uparrow - \delta^\downarrow$. If $\delta^\uparrow$ and $\delta^\downarrow$ are defined for the asymptotic (free) states then this result for the spin precession is general.

To find the precession angle for the case of a single step we return to the formulae (10.69) and (10.73) and write

$$\begin{aligned} e^{j\delta} &= r^{\uparrow} r^{\downarrow *} \\ &= \frac{(\tanh(u^+/2) + \tanh(u/2) j e^{j\phi})(\tanh(u^-/2) + \tanh(u/2) j e^{j\phi})}{(\tanh(u^+/2) - \tanh(u/2) j e^{-j\phi})(\tanh(u^-/2) - \tanh(u/2) j e^{-j\phi})} \end{aligned} \tag{10.79}$$

If we now recall that

$$\tanh(u/2) = p/(E+m), \qquad \tanh(\frac{u^{\pm}}{2}) = \frac{p_y \pm \kappa}{E - eV + m}, \tag{10.80}$$

we find that

$$e^{j\delta} = e^{2j\phi} \frac{m\cos\phi - jE\sin\phi}{m\cos\phi + jE\sin\phi}. \tag{10.81}$$

The remarkable feature of this result is that all dependence on the height of the barrier has vanished, so that the precession angle is determined solely by the incident energy and direction. To proceed we write

$$m\cos\phi - jE\sin\phi = \rho e^{j\alpha} \tag{10.82}$$

so that

$$\tan\alpha = -\cosh u \tan\phi. \tag{10.83}$$

Equation (10.81) now yields

$$\tan(\delta/2 - \phi) = -\cosh u \tan\phi, \tag{10.84}$$

from which we obtain the final result that

$$\tan(\delta/2) = -\frac{(\cosh u - 1)\tan\phi}{1 + \cosh u \tan^2\phi}. \tag{10.85}$$

The formula (10.85) agrees with equation (10.37) from Section 10.3, since the angle θ employed there is related to the angle of incidence ϕ by

$$\theta = \pi - 2\phi. \tag{10.86}$$

Since the decomposition of the plane-wave spinor into a boost term and a Pauli spinor term is unique to the STA, it is not at all clear how the conventional approach can formulate the idea of the rest-spin. In fact, the rest-spin vector is contained in the standard approach in the form of the 'polarisation operator' which, in the STA, is given by

$$\hat{O}(\boldsymbol{n}) \equiv -\frac{j}{\hat{\boldsymbol{p}}^2}[i\hat{\boldsymbol{p}}\hat{\boldsymbol{p}}\cdot\boldsymbol{n} + \hat{\gamma}_0 i(\boldsymbol{n}\wedge\hat{\boldsymbol{p}})\cdot\hat{\boldsymbol{p}}] \tag{10.87}$$

where $\boldsymbol{n}$ is a unit spatial vector. This operator is Hermitian, squares to 1 and commutes with the free-field Hamiltonian. If we consider a free-particle plane-wave state, then the expectation value of the $\hat{O}(\sigma_i)$ operator is

$$\begin{aligned} \frac{\langle \psi^\dagger \hat{O}(\sigma_i)\psi\rangle}{\langle \psi^\dagger\psi\rangle} &= \frac{m}{E\boldsymbol{p}^2}\langle \tilde{\Phi} L(\sigma_i\cdot\boldsymbol{p}\boldsymbol{p}L\Phi\sigma_3 + \sigma_i\wedge\boldsymbol{p}\boldsymbol{p}\tilde{L}\Phi\sigma_3)\rangle \\ &= \frac{m}{E\boldsymbol{p}^2}[L(\sigma_i\cdot\boldsymbol{p}\boldsymbol{p}L + \sigma_i\wedge\boldsymbol{p}\boldsymbol{p}\tilde{L})]\cdot\boldsymbol{s} \end{aligned} \tag{10.88}$$

where L is the boost

$$L(\boldsymbol{p}) = \frac{E+m+\boldsymbol{p}}{\sqrt{2m(E+m)}} \tag{10.89}$$

and $\hat{\boldsymbol{p}}$ is replaced by its eigenvalue $\boldsymbol{p}$. To manipulate equation (10.88) we use the facts that L commutes with $\boldsymbol{p}$ and satisfies

$$L^2 = (E+\boldsymbol{p})/m \tag{10.90}$$

to construct

$$\begin{aligned}\frac{m}{E\boldsymbol{p}^2}[L(\sigma_i{\cdot}\boldsymbol{p}\boldsymbol{p}L + \sigma_i{\wedge}\boldsymbol{p}\boldsymbol{p}\tilde{L})] &= \frac{m}{E\boldsymbol{p}^2}[\sigma_i{\cdot}\boldsymbol{p}\frac{(E+\boldsymbol{p})}{m}\boldsymbol{p} - \sigma_i{\cdot}\boldsymbol{p}\boldsymbol{p} \\ &\quad + \frac{E-m}{2m}(E+m+\boldsymbol{p})\sigma_i(E+m-\boldsymbol{p})]\end{aligned} \tag{10.91}$$

Since only the relative vector part of this quantity is needed in equation (10.88) we are left with

$$\begin{aligned}\frac{E-m}{2E\boldsymbol{p}^2}(2\sigma_i{\cdot}\boldsymbol{p}\boldsymbol{p} + (E+m)^2\sigma_i - \boldsymbol{p}\sigma_i\boldsymbol{p}) &= \frac{1}{2E(E+m)}(\boldsymbol{p}^2\sigma_i + (E+m)^2\sigma_i) \\ &= \sigma_i.\end{aligned} \tag{10.92}$$

The expectation value of the 'polarisation' operators is therefore simply

$$\frac{\langle\psi^\dagger\hat{O}(\sigma_i)\psi\rangle}{\langle\psi^\dagger\psi\rangle} = \sigma_i{\cdot}\boldsymbol{s}, \tag{10.93}$$

which just picks out the components of the rest-spin vector, as claimed.

For the case of the potential step, $\hat{O}(\boldsymbol{n})$ still commutes with the full Hamiltonian when $\boldsymbol{n}$ is perpendicular to the plane of incidence. In their paper Fradkin & Kashuba decompose the incident and reflected waves into eigenstates of $\hat{O}(\boldsymbol{n})$, which is equivalent to aligning the spin in the manner adopted in this section. As we have stressed, removing the boost and working directly with Φ simplifies many of these manipulations, and removes any need for the polarisation operator.

10.6 Tunnelling of Plane Waves

Suppose now that a continuous beam of plane waves is incident on a potential barrier of finite width. We know that, quantum-mechanically, some fraction of the wave tunnels through to the other side. To answer this we will need to combine plane-wave solutions to construct a wavepacket, so here we give the results for plane waves. The physical set-up is illustrated in Figure 10.3.

The matching equation at the $x=0$ boundary is, for spin-up,

$$\boldsymbol{B}^+\begin{pmatrix} T_1^\uparrow \\ R_1^\uparrow \end{pmatrix} = \boldsymbol{A}\begin{pmatrix} T^{\uparrow\prime} \\ 0 \end{pmatrix}, \tag{10.94}$$

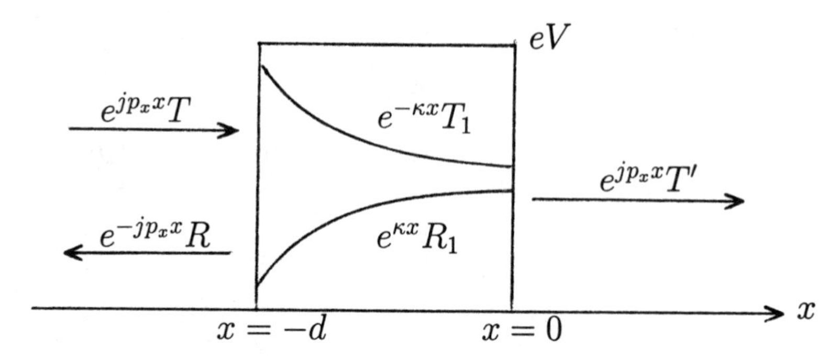

Fig. 10.3. *Schematic representation of plane-wave tunnelling.*

where the $\boldsymbol{A}$ and $\boldsymbol{B}^+$ matrices are as defined in (10.48) and (10.66) respectively. All subscripts can be dropped again as $\boldsymbol{A}$ always refers to free space and $\boldsymbol{B}^+$ to the barrier region. The matching conditions at $x = -d$ require the inclusion of suitable propagators and the resulting equation is

$$\boldsymbol{A}\begin{pmatrix} e^{-jdp_x} & 0 \\ 0 & e^{jdp_x} \end{pmatrix}\begin{pmatrix} T^\uparrow \\ R^\uparrow \end{pmatrix} = \boldsymbol{B}^+\begin{pmatrix} e^{\kappa d} & 0 \\ 0 & e^{-\kappa d} \end{pmatrix}\begin{pmatrix} T_1^\uparrow \\ R_1^\uparrow \end{pmatrix}. \tag{10.95}$$

Equation (10.95) shows that the relevant propagator matrix for evanescent waves is

$$\boldsymbol{P} = \begin{pmatrix} e^{\kappa d} & 0 \\ 0 & e^{-\kappa d} \end{pmatrix}. \tag{10.96}$$

The problem now reduces to the matrix equation

$$\begin{pmatrix} e^{-jdp_x}T^\uparrow \\ e^{jdp_x}R^\uparrow \end{pmatrix} = \left[\cosh(\kappa d) + \sinh(\kappa d)\boldsymbol{A}^{-1}\boldsymbol{B}^+\begin{pmatrix} 1 & 0 \\ 0 & -1 \end{pmatrix}\boldsymbol{B}^{+^{-1}}\boldsymbol{A}\right]\begin{pmatrix} T^{\uparrow\prime} \\ 0 \end{pmatrix}, \tag{10.97}$$

from which the reflection and transmission coefficients are easily obtained.

Applications deal mainly with perpendicular incidence, so we now specialise to this situation. For perpendicular incidence we can set $u^+ = -u^- = u'$, where

$$\tanh u' = \frac{\kappa}{E' + m}. \tag{10.98}$$

It follows that the equations for spin-up and spin-down are the same, and we can remove the up-arrows from the preceding equations. Equation (10.94) now yields

$$\begin{pmatrix} T_1 \\ R_1 \end{pmatrix} = \frac{T'}{\sinh u'}\begin{pmatrix} \sinh(u'/2)\cosh(u/2) - j\cosh(u'/2)\sinh(u/2) \\ \sinh(u'/2)\cosh(u/2) + j\cosh(u'/2)\sinh(u/2) \end{pmatrix} \tag{10.99}$$

and from T_1 and R_1 the current in the evanescent region can be constructed. The ratio J_1/J_0 may be interpreted as defining a 'velocity' inside the barrier.

Multiplying out the matrices in equation (10.97) is straightforward, and yields

$$\begin{pmatrix} e^{-jdp}T \\ e^{jdp}R \end{pmatrix} = T'\begin{pmatrix} \cosh(\kappa d) - j\sinh(\kappa d)(EE' - m^2)/(\kappa p) \\ -j\sinh(\kappa d)eVm/(\kappa p) \end{pmatrix}, \tag{10.100}$$

which solves the problem. The transmission coefficient is

$$t = \frac{\kappa p e^{-jdp}}{\kappa p \cosh(\kappa d) - j(p^2 - eVE)\sinh(\kappa d)} \tag{10.101}$$

which recovers the familiar non-relativistic formula in the limit $E \approx m$.

10.7 The Klein Paradox

In the Klein region, $eV - E > \sqrt{p_y^2 + m^2}$, travelling wave solutions exist again. To find these we observe that plane-wave solutions must now satisfy

$$(p - eV\gamma_0)\psi = m\psi\gamma_0 \tag{10.102}$$

and, as $p - eV\gamma_0$ has a negative time component, $\psi\tilde{\psi}$ must now be -1. We could achieve this flip by inserting a 'β-factor', but this would mix the rest-spin states. It is more convenient to work with solutions given by

$$\psi^t = [\cosh(u/2) + \sinh(u/2)(\cos\phi\sigma_1 - \sin\phi\sigma_2)]\sigma_1\Phi e^{-i\sigma_3(Et + p_x x - p_y y)}T \tag{10.103}$$

$$\psi^r = [\cosh(u/2) + \sinh(u/2)(-\cos\phi\sigma_1 - \sin\phi\sigma_2)]\sigma_1\Phi e^{-i\sigma_3(Et - p_x x - p_y y)}R \tag{10.104}$$

where the choice of σ_1 or σ_2 on the right-hand side of the boost is merely a phase choice. To verify that ψ^t is a solution we write the Dirac equation as

$$[(E - eV)\gamma_0 - p_x\gamma_1 + p_y\gamma_2]e^{(\cos\phi\sigma_1 - \sin\phi\sigma_2)u/2}\sigma_1 = me^{(\cos\phi\sigma_1 - \sin\phi\sigma_2)u/2}\sigma_1\gamma_0 \tag{10.105}$$

which holds provided that

$$\tanh(u/2) = \frac{p}{m + eV - E}. \tag{10.106}$$

It follows that

$$m\cosh u = eV - E, \qquad m\sinh u = p. \tag{10.107}$$

The current obtained from ψ^t is found to be

$$\psi^t\gamma_0\tilde{\psi}^t = (eV - E)\gamma_0 + p_x\gamma_1 - p_y\gamma_2 \tag{10.108}$$

which is future pointing (as it must be) and points in the positive-x direction. It is in order to obtain the correct direction for the current that the sign of p_x is changed in (10.103) and (10.104). Some texts on quantum theory miss this argument and match onto a solution inside the barrier with an incoming group velocity! The result is a reflection coefficient greater than 1. This is interpreted as evidence for pair production, though in fact the effect is due to the choice of boundary conditions.

To find the correct reflection and transmission coefficients for an outgoing current, we return to the matching conditions. This time we define the matrix

$$\boldsymbol{C}_i \equiv \begin{pmatrix} \sinh(u_i/2)e^{j\phi_i} & -\sinh(u_i/2)e^{-j\phi_i} \\ \cosh(u_i/2) & \cosh(u_i/2) \end{pmatrix} \tag{10.109}$$

so that the matching conditions can be written as (exercise)

$$\boldsymbol{A}_i \begin{pmatrix} T_i^{\uparrow} \\ R_i^{\uparrow} \end{pmatrix} = \boldsymbol{C}_{i+1} \begin{pmatrix} T_{i+1}^{\uparrow} \\ R_{i+1}^{\uparrow} \end{pmatrix} \quad \text{and} \quad \boldsymbol{A}_i^* \begin{pmatrix} T_i^{\downarrow} \\ R_i^{\downarrow} \end{pmatrix} = \boldsymbol{C}_{i+1}^* \begin{pmatrix} T_{i+1}^{\downarrow} \\ R_{i+1}^{\downarrow} \end{pmatrix}. \tag{10.110}$$

The Klein 'paradox' occurs at a single step, for which the matching equation is

$$\boldsymbol{A}_1 \begin{pmatrix} T_1^{\uparrow} \\ R_1^{\uparrow} \end{pmatrix} = \boldsymbol{C}_2 \begin{pmatrix} T_2^{\uparrow} \\ 0 \end{pmatrix}. \tag{10.111}$$

Inverting the $\boldsymbol{A}_1$ matrix yields

$$\begin{pmatrix} T_1^{\uparrow} \\ R_1^{\uparrow} \end{pmatrix} = \frac{T_2^{\uparrow}}{\text{sinh}u \cos\phi} \begin{pmatrix} \cosh(u+u')/2 \\ \sinh(u/2)\sinh(u'/2)e^{2j\phi} - \cosh(u/2)\cosh(u'/2) \end{pmatrix} \tag{10.112}$$

from which the reflection and transmission coefficients can be read off. In particular, for perpendicular incidence, we recover

$$r = -\frac{\cosh(u-u')/2}{\cosh(u+u')/2}, \qquad t = \frac{\text{sinh}u}{\cosh(u+u')/2}. \tag{10.113}$$

The reflection coefficient is always ≤ 1, as it must be from current conservation with these boundary conditions. But, although a reflection coefficient ≤ 1 appears to ease the paradox, some difficulties remain. In particular, the momentum vector inside the barrier points in an opposite direction to the current.

A more complete understanding of the Klein barrier requires quantum field theory since, as the barrier height is $> 2m$, we expect pair creation to occur. An indication that this must be the case comes from an analysis of boson modes based on the Klein-Gordon equation. There one finds that superradiance ($r > 1$) does occur, which has to be interpreted in terms of particle production. For the fermion case the resulting picture is that electron-positron pairs are created and split apart, with the electrons travelling back out to the left and the positrons moving into the barrier region. If a single electron is incident on such a step then it is reflected and, according to the Pauli principle, the corresponding pair-production mode is suppressed.

A complete analysis of the Klein barrier has been given by Manogue to which readers are referred for further details. Manogue concludes that the fermion pair-production rate is given by

$$\Gamma = \int \frac{d^2\boldsymbol{k}}{(2\pi)^2} \int \frac{d\omega}{2\pi} \frac{\text{sinh}u'}{\text{sinh}u} \sum_i |T^i|^2 \tag{10.114}$$

where the integrals run over the available modes in the Klein region, and the sum runs over the two spin states. This formula gives a production rate per unit time, per unit area, and applies to any shape of barrier. The integrals in (10.114) are not easy to evaluate, but a useful expression can be obtained by assuming that the barrier height is only slightly greater than $2m$,

$$eV = 2m(1+\epsilon). \tag{10.115}$$

Then, for the case of a single step, we obtain a pair-production rate of

$$\Gamma = \frac{\pi m^3 \epsilon^3}{32}, \tag{10.116}$$

to leading order in ϵ. The dimensional term is m^3 which, for electrons, corresponds to a rate of 10^{48} particles per second, per square meter. Such an enormous rate would clearly be difficult to sustain in any physically-realistic situation!

Chapter 11

STA and the Interpretation of Quantum Mechanics

Anthony Lasenby, Stephen Gull, and Chris Doran
MRAO, Cavendish Laboratory, Madingley Road,
Cambridge, CB3 0HE, UK

The STA is a mathematical system, rather than of itself containing new physics. However, when we employ it in the description of physical phenomena, we usually find that some fresh insight is obtained, often on old questions. The oldest question of 20th century physics is the interpretation of quantum mechanics, and in this lecture we aim to discuss some of the light that an STA approach can throw upon this issue. This will be undertaken in the context of specific examples, and so in the main these lecture notes contain the details of these, rather than the comments that will be made about issues of interpretation. We think these examples stand as interesting and important applications in their own right, irrespective of whether one thinks there is a problem to be solved in the interpretation of quantum theory. This is particularly true for the beginnings of a *multiparticle* STA approach which we describe here, since it seems undeniable that there is a need for a more coherent and justifiable rationale for the set of recipes and operational procedures that currently constitute relativistic multiparticle theory.

11.1 Tunnelling Times

In the lectures on electron physics we studied tunnelling of a continuous plane wave through a potential barrier, and equations were found for the growing and decaying waves in the barrier region for spin-up and spin-down electrons. Restricting to the case of perpendicular incidence we find that, for arbitrary spin, the wavefunction in the barrier region is

$$\begin{aligned}\psi_1 \;=\; & [\cosh(u'/2)\Phi + \sinh(u'/2)\sigma_2\sigma_3\Phi\sigma_3]e^{-\kappa x}e^{-i\sigma_3 Et}\alpha + \\ & [\cosh(u'/2)\Phi - \sinh(u'/2)\sigma_2\sigma_3\Phi\sigma_3]e^{\kappa x}e^{-i\sigma_3 Et}\alpha^* \qquad (11.1)\end{aligned}$$

where

$$\alpha \equiv \frac{T'}{\sinh u'}[\sinh(u'/2)\cosh(u/2) - i\sigma_3\cosh(u'/2)\sinh(u/2)] \tag{11.2}$$

and

$$\tanh(u'/2) = \frac{\kappa}{E'+m} = \frac{\kappa}{E-eV+m}, \qquad \kappa^2 = m^2 - E'^2. \tag{11.3}$$

The current in the barrier region is

$$\begin{aligned}\psi_1\gamma_0\tilde{\psi}_1 &= \frac{|T'|^2}{m\kappa^2}[m^2eV\cosh(2\kappa x) + E'(p^2 - Eev) \\ &\quad + p\kappa^2\sigma_1 - m\kappa eV\sinh(2\kappa x)(i\sigma_1)\cdot \boldsymbol{s}]\gamma_0\end{aligned} \tag{11.4}$$

from which we can define a 'velocity'

$$\frac{dx}{dt} \equiv \frac{J\cdot\gamma^1}{J\cdot\gamma_0} = \frac{p\kappa^2}{m^2eV\cosh(2\kappa x) + E'(p^2 - Eev)}. \tag{11.5}$$

In fact, the velocity (11.5) does not lead to a sensible definition of a tunnelling time for an individual particle. As we shall see shortly, an additional phenomenon underlies wavepacket tunnelling, leading to much shorter times than those predicted from (11.5). To study wavepacket tunnelling it is useful, initially, to simplify to a one-dimensional problem. To achieve this we must eliminate the transverse current in (11.4) by setting $\boldsymbol{s} = \pm\sigma_1$. This is equivalent to aligning the spin vector to point in the direction of motion. (In this case there is no distinction between the laboratory and comoving spin.) With Φ chosen so that $\boldsymbol{s} = \sigma_1$ it is a now a simple matter to superpose solutions at $t = 0$ to construct a wavepacket centred to the left of the barrier and moving towards the barrier. The wavepacket at later times is then reassembled from the plane-wave states, whose time evolution is known. The density $J^0 = \gamma_0\cdot J$ can then be plotted as a function of time and the result of such a simulation is illustrated in Figure 11.1.

The Dirac current $J = \psi\gamma_0\tilde{\psi}$ is conserved even in the presence of an electromagnetic field. It follows that J defines a set of streamlines which never end or cross. Furthermore, the time-component of the current is positive-definite so the tangents to the streamlines are always future-pointing timelike vectors. According to the standard interpretation of quantum mechanics, $J^0(\boldsymbol{x}, t)$ gives the probability density of locating a particle at position $\boldsymbol{x}$ at time t. But, considering a flux tube defined by adjacent streamlines, we find that

$$\rho(t_0, x_0)dx_0 = \rho(t_1, x_1)dx_1 \tag{11.6}$$

where (t_0, x_0) and (t_1, x_1) are connected by a streamline. It follows that the density J^0 flows along the streamlines without 'leaking' between them. So, in order to study the tunnelling process, we should follow the streamlines from the initial wavepacket through spacetime. A sample set of these streamlines is shown in Figure 11.2. A significant feature of this plot is that a continuously-distributed set of initial input conditions has given rise to a disjoint set of outcomes (whether or not a streamline passes through the barrier). Hence the deterministic evolution of the wavepacket

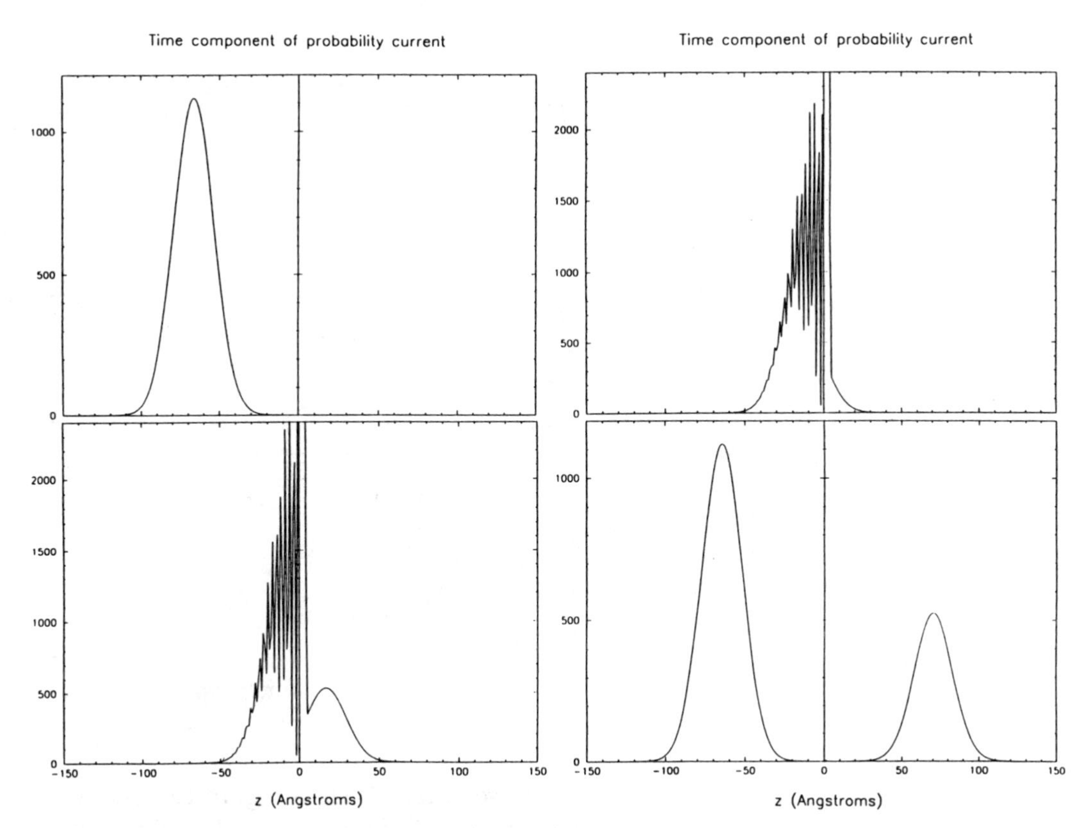

Fig. 11.1. *Evolution of the density J^0 as a function of time. The initial packet is a Gaussian of width $\Delta k = 0.04\text{Å}^{-1}$ and energy 5eV. The barrier starts at the origin and has width 5Å and height 10eV. The top line shows the density profile at times -0.5×10^{-14}s and -0.1×10^{-14}s, and the bottom line shows times 0.1×10^{-14}s and 0.5×10^{-14}s. In all plots the vertical scale to the right of the barrier is multiplied by 10^4 to enhance the features of the small, transmitted packet.*

alone is able to explain the discrete results expected in a quantum measurement, and all notions of wavefunction collapse are avoided. This is of fundamental significance to the interpretation of quantum mechanics, although the results presented here are, of course, independent of any particular interpretation. For exmaple, we certainly do not need some sort of relativistic version of the apparatus of the Bohm/de Broglie theory in order to accept the validity of predictions obtained from the current streamlines.

The second key feature of the streamline plot in Figure 11.2 is that it is only the streamlines starting near the front of the initial wavepacket that pass through the barrier. Relative to the centre of the packet, they therefore have a 'head start' in their arrival time at some chosen point on the far side of the barrier. Over the front part of the barrier, however, the streamlines slow down considerably, as can be seen by the change in their slope. These two effects, of picking out the front end

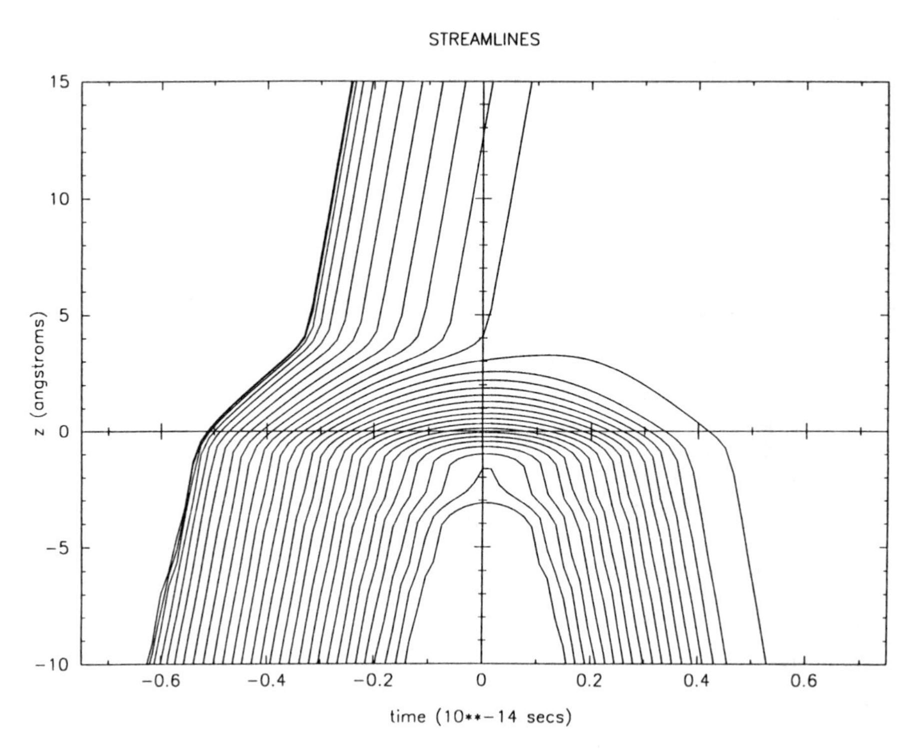

Fig. 11.2. *Particle streamlines for the packet evolution shown in Figure 11.1. Only the streamlines from the very front of the packet cross the barrier, with the individual streamlines slowing down as they pass through.*

of the packet and then slowing it down, compete against each other and it is not immediately obvious which dominates. To establish this, we return to Figure 11.1 and look at the positions of the wavepacket peaks. At $t = 0.5 \times 10^{-14}$s, the peak of the transmitted packet lies at $x = 70$Å, whereas the peak of the initial packet would have been at $x = 66$Å had the barrier not been present. In this case, therefore, the peak of the transmitted packet is slightly advanced, a phenomenon often interpreted as showing that tunnelling particles speed up, sometimes to velocities greater than c. The plots presented here show that such an interpretation is completely mistaken. There is no speeding up, as all that happens is that it is only the streamlines from the front of the wavepacket that cross the barrier (slowing down in the process) and these reassemble to form a localised packet on the far side. The reason that tunnelling particles may be transmitted faster than free particles is due entirely to the spread of the initial wavepacket.

There is considerable interest in the theoretical description of tunnelling processes because it is now possible to obtain measurements of the times involved. The clearest experiments conducted to date have concerned photon tunnelling, where an ingenious 2-photon interference technique is used to compare photons that pass through a barrier with photons that follow an unobstructed path. The discussion of the results of photon-tunnelling experiments usually emphasise packet reshaping, but miss the arguments about the streamlines. Thus many articles concentrate on a comparison of the peaks of the incident and transmitted wavepackets and discuss whether the experiments show particles travelling at speeds $> c$. As we have seen, a full relativistic study of the streamlines followed by the electron probability density show clearly that no superluminal velocities are present.

Ever since the possibility of tunnelling was revealed by quantum theory, people have attempted to define how long the process takes. Most approaches attempt to define a single tunnelling time for the process, rather than a distribution of possible outcomes as is the case here. Quite why one should believe that it is possible to define a single time in a probabilistic process such as tunnelling is unclear, but the view is still regularly expressed in the modern literature. A further flaw in many other approaches is that they attempt to define how long the particle spent in the barrier region, with answers ranging from the implausible (zero time) to the utterly bizarre (imaginary time). From the streamline plot presented here, it is clearly possible to obtain a distribution of the times spent in the barrier for the tunnelling particles, and the answers will be relatively long as the particles slow down in the barrier. But such a distribution neglects the fact that the front of the packet is preferentially selected, and anyway does not appear to be accessible to direct experimental measurement. As the recent experiments show, it is the arrival time at a point on the far side of the barrier that is measurable, and not the time spent in the barrier.

11.2 Spin Measurements

We now turn to a second application of the local observables approach to quantum theory, namely to determine what happens to a wavepacket when a spin measurement is made. The first attempts to answer this question were made by Dewdney *et al.*, who used the Pauli equation for a particle with zero charge and an anomalous magnetic moment to provide a model for a spin-1/2 particle in a Stern-Gerlach apparatus. Written in the STA, the relevant equation is

$$\partial_t \Phi i\sigma_3 = -\frac{1}{2m}\boldsymbol{\nabla}^2\Phi - \mu \boldsymbol{B}\Phi\sigma_3 \tag{11.7}$$

and the current employed by Dewdney *et al.* is

$$\boldsymbol{J} = -\frac{1}{m}\dot{\boldsymbol{\nabla}}\langle\dot{\Phi} i\sigma_3 \Phi^\dagger\rangle. \tag{11.8}$$

Dewdney *et al.* parameterise the Pauli spinor Φ in terms of a density and three 'Euler angles'. In the STA, this parameterisation takes the transparent form

$$\Phi = \rho^{1/2} e^{i\sigma_3\phi/2} e^{i\sigma_1\theta/2} e^{i\sigma_3\psi/2}, \tag{11.9}$$

where the rotor term is precisely that needed to parameterise a rotation in terms of the Euler angles. With this parameterisation, it is a simple matter to show that the current becomes

$$J = \frac{\rho}{2m}(\nabla\psi + \cos\theta\nabla\phi). \tag{11.10}$$

But, as was noted in the first lecture on electron physics, the current defined by equation (11.8) is not consistent with that obtained from the Dirac theory through a non-relativistic reduction. In fact, the two currents differ by a term in the curl of the spin vector.

To obtain a fuller understanding of the spin measurement process, an analysis based on the Dirac theory is required. Such an analysis is presented here. As well as dealing with a well-defined current, basing the analysis in the Dirac theory is important if one intends to proceed to study correlated spin measurements performed over spacelike intervals (*i.e.* to model an EPR-type experiment). To study such systems it is surely essential that one employs relativistic equations so that causality and the structure of spacetime are correctly built in.

11.2.1 A Relativistic Model of a Spin Measurement

The modified Dirac equation for a neutral particle with an anomalous magnetic moment μ is

$$\nabla\psi i\sigma_3 - i\mu F\psi\gamma_3 = m\psi\gamma_0. \tag{11.11}$$

This is the equation we use to study the effects of a spin measurement, and it is not hard to show that equation (11.11) reduces to (11.7) in the non-relativistic limit. Following Dewdney *et al.* we model the effect of a spin measurement by applying an impulsive magnetic field gradient,

$$F = Bz\delta(t)i\sigma_3. \tag{11.12}$$

The other components of $\boldsymbol{B}$ are ignored, as we are only modelling the behaviour of the packet in the z-direction. Around $t = 0$ equation (11.11) is approximated by

$$\partial_t\psi i\sigma_3 = \Delta p\, z\delta(t)\gamma_3\psi\gamma_3, \tag{11.13}$$

where

$$\Delta p \equiv \mu B. \tag{11.14}$$

To solve (11.13) we decompose the initial spinor ψ_0 into

$$\psi^{\uparrow} \equiv \tfrac{1}{2}(\psi_0 - \gamma_3\psi_0\gamma_3), \qquad \psi^{\downarrow} \equiv \tfrac{1}{2}(\psi_0 + \gamma_3\psi_0\gamma_3). \tag{11.15}$$

Equation (11.13) now becomes, for $\psi^{\uparrow}$

$$\partial_t\psi^{\uparrow} = \Delta p\, z\delta(t)\psi^{\uparrow}i\sigma_3 \tag{11.16}$$

with the opposite sign for $\psi^{\downarrow}$. The solution is now straightforward, as the impulse just serves to insert a phase factor into each of $\psi^{\uparrow}$ and $\psi^{\downarrow}$:

$$\psi^{\uparrow} \to \psi^{\uparrow}e^{i\sigma_3\Delta p\, z}, \qquad \psi^{\downarrow} \to \psi^{\downarrow}e^{-i\sigma_3\Delta p\, z}. \tag{11.17}$$

If we now suppose that the initial ψ consists of a positive-energy plane-wave

$$\psi_0 = L(\boldsymbol{p})\Phi e^{i\sigma_3(\boldsymbol{p}\cdot\boldsymbol{x} - Et)} \tag{11.18}$$

then, immediately after the shock, ψ is given by

$$\psi = \psi^{\uparrow} e^{i\sigma_3(\boldsymbol{p}\cdot\boldsymbol{x} + \Delta p\, z)} + \psi^{\downarrow} e^{i\sigma_3(\boldsymbol{p}\cdot\boldsymbol{x} - \Delta p\, z)}. \tag{11.19}$$

The spatial dependence of ψ is now appropriate to two different values of the 3-momentum, $\boldsymbol{p}^{\uparrow}$ and $\boldsymbol{p}^{\downarrow}$, where

$$\boldsymbol{p}^{\uparrow} \equiv \boldsymbol{p} + \Delta p\, \sigma_3, \qquad \boldsymbol{p}^{\downarrow} \equiv \boldsymbol{p} - \Delta p\, \sigma_3. \tag{11.20}$$

The boost term $L(\boldsymbol{p})$ corresponds to a different momentum, however, so both positive and negative frequency waves are required for the future evolution. After the shock, the wavefunction therefore propagates as

$$\psi = \psi_+^{\uparrow} e^{-i\sigma_3 p^{\uparrow}\cdot x} + \psi_-^{\uparrow} e^{i\sigma_3 \bar{p}^{\uparrow}\cdot x} + \psi_+^{\downarrow} e^{-i\sigma_3 p^{\downarrow}\cdot x} + \psi_-^{\downarrow} e^{i\sigma_3 \bar{p}^{\downarrow}\cdot x} \tag{11.21}$$

where

$$p^{\uparrow}\gamma_0 = E^{\uparrow} + \boldsymbol{p}^{\uparrow}, \tag{11.22}$$

$$\bar{p}^{\uparrow}\gamma_0 = E^{\uparrow} - \boldsymbol{p}^{\uparrow}, \tag{11.23}$$

and

$$E^{\uparrow} = (m^2 + \boldsymbol{p}^{\uparrow 2})^{1/2}. \tag{11.24}$$

Both $p^{\downarrow}$ and $E^{\downarrow}$ are defined similarly.

Each term in (11.21) must separately satisfy the free-particle Dirac equation, so it follows that

$$p^{\uparrow}\psi_+^{\uparrow} = m\psi_+^{\uparrow}\gamma_0, \tag{11.25}$$

$$-\bar{p}^{\uparrow}\psi_-^{\uparrow} = m\psi_-^{\uparrow}\gamma_0, \tag{11.26}$$

which are satisfied together with

$$\psi^{\uparrow} = \psi_+^{\uparrow} + \psi_-^{\uparrow}. \tag{11.27}$$

The same set of equations hold for $\psi^{\downarrow}$. Dropping the arrows, we find that

$$\psi_+ = \frac{1}{2E}(p\gamma_0\psi + m\bar{\psi}) \tag{11.28}$$

$$\psi_- = \frac{1}{2E}(\bar{p}\gamma_0\psi - m\bar{\psi}) \tag{11.29}$$

which hold for both $\psi^{\uparrow}$ and $\psi^{\downarrow}$.

The effect of the magnetic shock on a monochromatic wave is to split the wave into four components, each with a distinct momentum. The positive frequency waves are transmitted by the device and split into two waves, whereas the negative frequency states are reflected. The appearance of the antiparticle states must ultimately be attributed to pair production, and only becomes significant for large $\boldsymbol{B}$-fields. We examine this effect after looking at more physical situations.

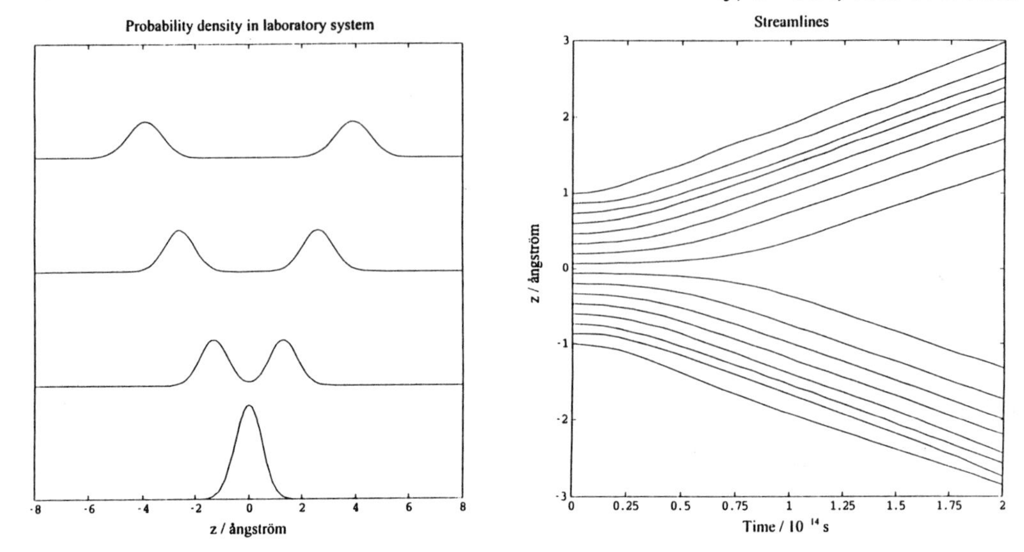

Fig. 11.3. *Splitting of a wavepacket caused by an impulsive $\boldsymbol{B}$-field. The initial packet has a width of $1 \times 10^{-24} kg\, m\, s^{-1}$ in momentum space, and receives an impulse of $\Delta p = 1 \times 10^{-23} kg\, m\, s^{-1}$. The figure on the left shows the probability density J_0 at $t = 0, 1.3, 2.6, 3.9 \times 10^{-14} s$, with t increasing up the figure. The figure on the right shows streamlines in the (t, z) plane.*

11.2.2 Wavepacket Simulations

For computational simplicity we take the incident particle to be localised along the field direction only, with no momentum components transverse to the field. This reduces the dimensionality of the problem to one spatial coordinate and the time coordinate. This was the set-up considered by Dewdney *et al.* and is sufficient to demonstrate the salient features of the measurement process. The most obvious difference between this model and a real experiment where the electron is moving is that, in our model, all four packets have group velocities along the field direction.

The initial packet is built up from plane-wave solutions of the form

$$\psi = e^{u\sigma_3/2} \Phi e^{i\sigma_3(pz - Et)} \tag{11.30}$$

which are superposed numerically to form a Gaussian packet. After the impulse, the future evolution is found from equation (11.21) and the behaviour of the spin vector and the streamlines can be found for various initial values of Φ. The results of these simulations are plotted on the next few pages. These plots were prepared by Anthony Challinor.

In Figures 11.3 and 11.4 we plot the evolution of a packet whose initial spin vector points in the σ_1 direction ($\Phi = \exp\{-i\sigma_2\pi/4\}$). After the shock, the density splits neatly into two equal-sized packets, and the streamlines bifurcate at the origin. As with the tunnelling simulations, we see that disjoint quantum outcomes are entirely consistent with the causal wavepacket evolution defined by the Dirac equation. The plot of the spin vector $s \wedge \gamma_0$ shows that immediately after the shock

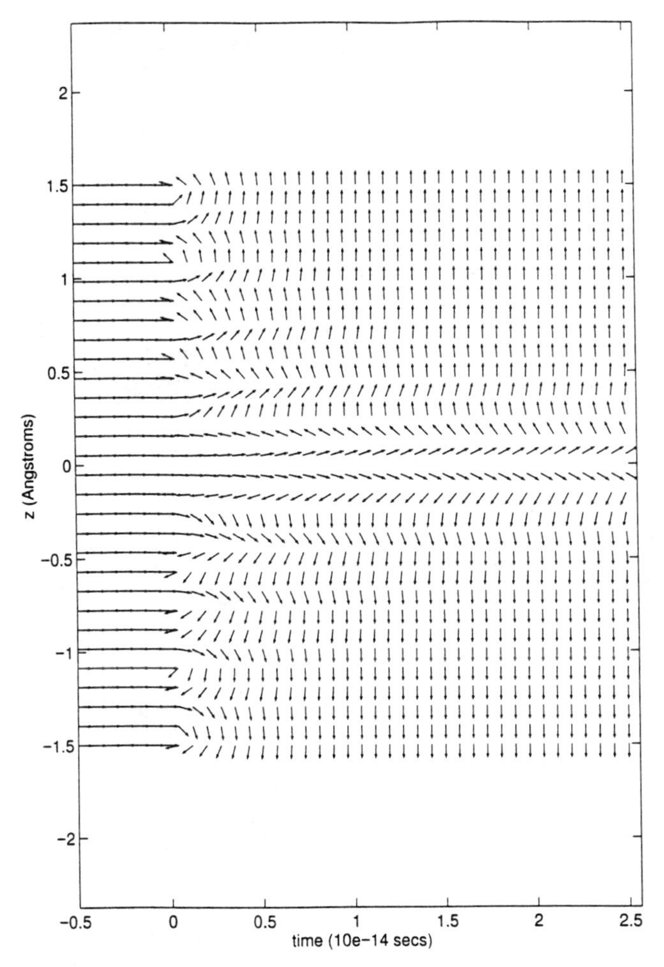

Fig. 11.4. *Evolution of the relative spin-vector* $s\wedge\gamma_0$*, in projection in the* (x, z) *plane. Immediately after the shock the spin-vectors point in all directions, but after about* $2\times 10^{-14}s$ *they sort themselves into the two packets, pointing in the* $+z$ *and* $-z$ *directions.*

the spins are disordered, but that after a little time they sort themselves into one of the two packets, with the spin vector pointing in the direction of motion of the deflected packet. These plots are in good qualitative agreement with those obtained by Dewdney *et al.*, who also found that the choice of which packet a streamline enters is determined by its starting position in the incident wavepacket.

Figure 11.5 shows the results of a similar simulation, but with the initial spinor now containing unequal amounts of spin-up and spin-down components. This time we observe an asymmetry in the wavepacket split, with more of the density travelling in the spin-up packet. It is a simple matter to compute the ratio of the sizes of the two packets, and to verify that the ratio agrees with the prediction of standard quantum theory.

As a final, novel, illustration of our approach, we consider a strong shock applied

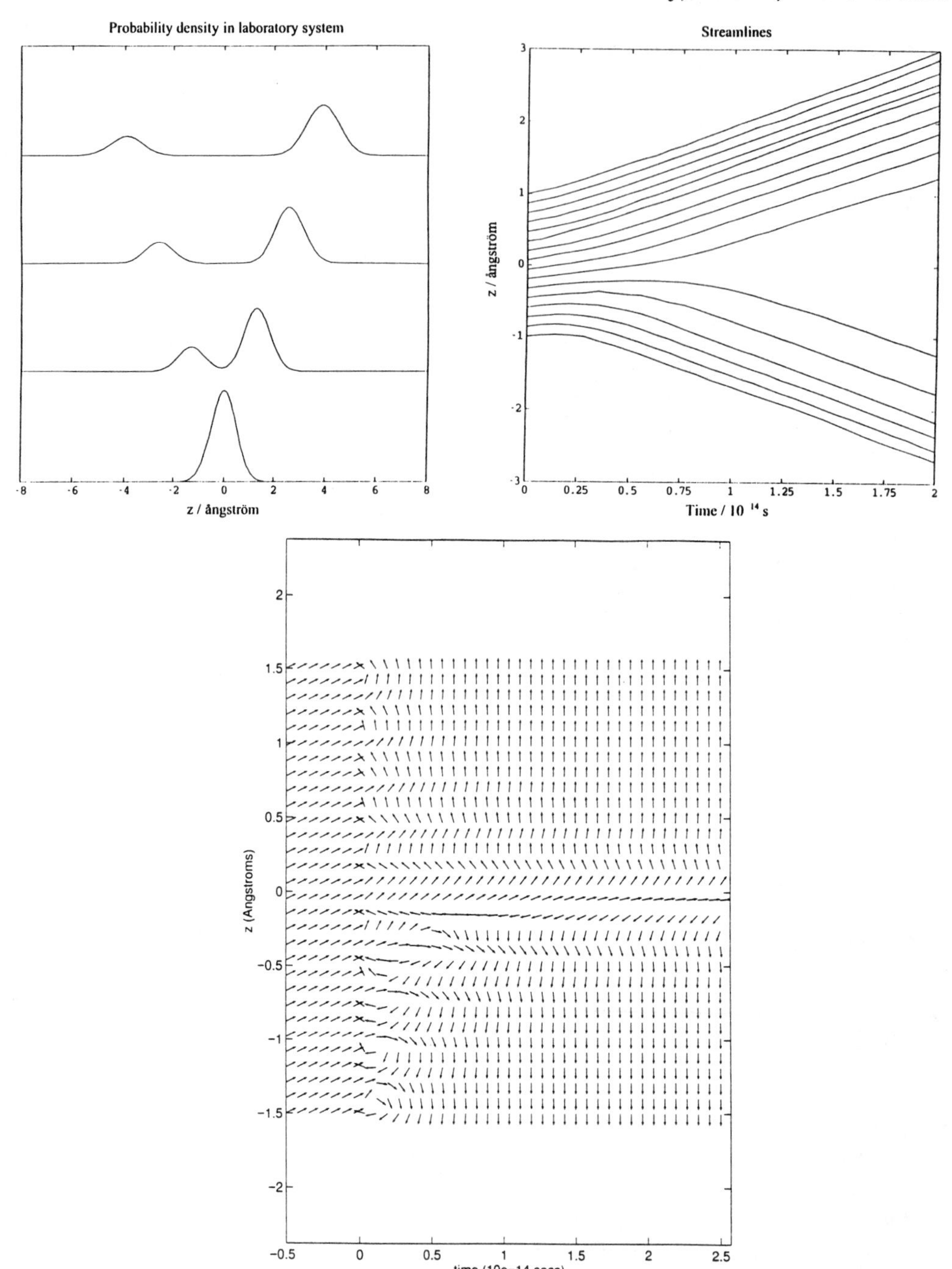

Fig. 11.5. *Splitting of a wavepacket with unequal mixtures of spin-up and spin-down components. The initial packet has $\Phi = 1.618 - i\sigma_2$, so more of the streamlines are deflected upwards, and the bifurcation point lies below the $z = 0$ plane. The evolution of the spin vector is shown in the bottom plot.*

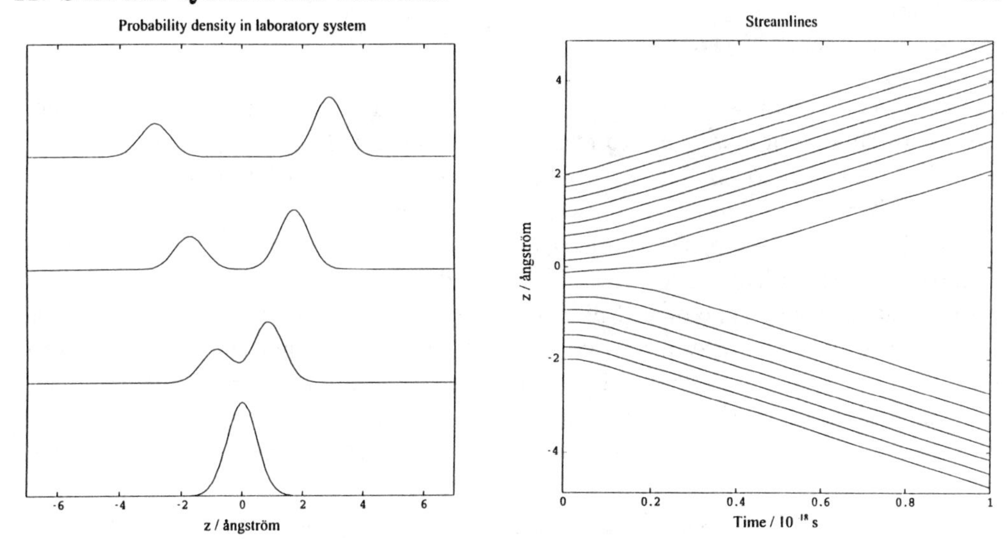

Fig. 11.6. *Creation of antiparticle states by a strong magnetic shock. The impulse used is $\Delta p = 1 \times 10^{-18} kg\, m\, s^{-1}$ and the initial packet is entirely spin-up ($\Phi = 1$). The packet travelling to the left consists of negative energy (antiparticle) states.*

to a packet which is already aligned in the spin-up direction. For a weak shock the entire packet is deflected but, if the shock is sufficiently strong that the antiparticle states have significant amplitude, we find that a second packet is created. The significant feature of Figure 11.6 is that the antiparticle states are deflected in the opposite direction, despite the fact that their spin is still oriented in the $+z$ direction. The antiparticle states thus behave as if they have a magnetic moment to mass ratio of opposite sign. A more complete understanding of this phenomenon requires a field-theoretic treatment. The appearance of antiparticle states would then be attributed to pair production, with the antiparticle states having the same magnetic moment, but the opposite spin. (One of the crucial effects of the field quantisation of fermionic systems is to flip the signs of the charges and spins of antiparticle states.)

The conclusions reached in this section are in broad agreement with those of Dewdney *et al.*. From the viewpoint of the local observables of the Dirac wavefunction (the current and spin densities), a Stern-Gerlach apparatus does not fulfil the role of a classical measuring device. Instead, it behaves much more like a polariser, where the ratio of particles polarised up and down is dependent on the initial wavefunction. The $\boldsymbol{B}$-field dramatically alters the wavefunction and its observables, though in a causal manner that is entirely consistent with the predictions of standard quantum theory. The implications of these observations for the interpretation of quantum mechanics are profound, though they are only slowly being absorbed by the wider physics community.

11.3 The Multiparticle STA

So far we have dealt with the application of the STA to single-particle quantum theory. In this section we turn to multiparticle theory. The aim here is to develop the STA approach so that it is capable of encoding multiparticle wavefunctions, and describing the correlations between them. Given the advances in clarity and insight that the STA brings to single-particle quantum mechanics, we expect similar advances in the multiparticle case. This is indeed what we have found, although the field is relatively unexplored as yet. Here we highlight some areas where the multiparticle STA promises a new conceptual approach, rather than attempting to reproduce the calculational techniques employed in standard approaches to many-body or many-electron theory. In particular, we concentrate on the unique geometric insights that the multiparticle STA provides — insights that are lost in the matrix theory.

The n-particle STA is created simply by taking n sets of basis vectors $\{\gamma_\mu^i\}$, where the superscript labels the particle space, and imposing the geometric algebra relations

$$\begin{aligned} \gamma_\mu^i\gamma_\nu^j + \gamma_\nu^i\gamma_\mu^j &= 0, \qquad & i \neq j \\ \gamma_\mu^i\gamma_\nu^j + \gamma_\nu^i\gamma_\mu^j &= 2\eta_{\mu\nu} \qquad & i = j. \end{aligned} \tag{11.31}$$

These relations are summarised in the single formula

$$\gamma_\mu^i\cdot\gamma_\nu^j = \delta^{ij}\eta_{\mu\nu}. \tag{11.32}$$

The fact that the basis vectors from distinct particle spaces anticommute means that we have constructed a basis for the geometric algebra of a $4n$-dimensional configuration space. There is nothing uniquely quantum-mechanical in this idea — a system of three classical particles could be described by a set of three trajectories in a single space, or one path in a nine-dimensional space. The extra dimensions serve simply to label the properties of each individual particle, and should not be thought of as existing in anything other than a mathematical sense. This construction enables us, for example, to define a rotor which rotates one particle whilst leaving all the others fixed. The unique feature of the multiparticle STA is that it implies a separate copy of the time dimension for each particle, as well as the three spatial dimensions. To our knowledge, this is the first attempt to construct a solid conceptual framework for a multi-time approach to quantum theory. Clearly, if successful, such an approach will shed light on issues of locality and causality in quantum theory.

The $\{\gamma_\mu^i\}$ serve to generate a geometric algebra of enormously rich structure. Here we illustrate just a few of the more immediate features of this algebra. It is our belief that the multiparticle STA will prove rich enough to encode all aspects of multiparticle quantum field theory, including the algebra of the fermionic creation/annihilation operators.

Throughout, Roman superscripts are employed to label the particle space in which the object appears. So, for example, ψ^1 and ψ^2 refer to two copies of the same 1-particle object ψ, and not to separate, independent objects. Separate objects are given distinct symbols, or subscripts if they represent a quantity such as the current or spin-vector, which are vectors in configuration space with different projections into the separate copies of the STA. The absence of superscripts denotes that all objects have been collapsed into a single copy of the STA. As always, Roman and Greek subscripts are also used as frame indices, though this does not interfere with the occasional use of subscripts to determine separate projections.

11.3.1 2-Particle Pauli States and the Quantum Correlator

As an introduction to the properties of the multiparticle STA, we first consider the 2-particle Pauli algebra and the spin states of pairs of spin-1/2 particles. As in the single-particle case, the 2-particle Pauli algebra is just a subset of the full 2-particle STA. A set of basis vectors is defined by

$$\sigma_i^1 = \gamma_i^1\gamma_0^1 \tag{11.33}$$
$$\sigma_i^2 = \gamma_i^2\gamma_0^2 \tag{11.34}$$

which satisfy

$$\sigma_i^1\sigma_j^2 = \gamma_i^1\gamma_0^1\gamma_j^2\gamma_0^2 = \gamma_i^1\gamma_j^2\gamma_0^2\gamma_0^1 = \gamma_j^2\gamma_0^2\gamma_i^1\gamma_0^1 = \sigma_j^2\sigma_i^1. \tag{11.35}$$

So, in constructing multiparticle Pauli states, the basis vectors from different particle spaces commute rather than anticommute. Using the elements $\{1, i\sigma_k^1, i\sigma_k^2, i\sigma_j^1 i\sigma_k^2\}$ as a basis, we can construct 2-particle states. Here we have introduced the abbreviation

$$i\sigma_i^1 \equiv i^1\sigma_i^1 \tag{11.36}$$

since, in most expressions, it is obvious which particle label should be attached to the i. In cases where there is potential for confusion, the particle label is put back on the i. The basis set $\{1, i\sigma_k^1, i\sigma_k^2, i\sigma_j^1 i\sigma_k^2\}$ spans a 16-dimensional space, which is twice the dimension of the direct product space of two 2-component complex spinors. For example, the outer-product space of two spin-1/2 states can be built from complex superpositions of the set

$$\begin{pmatrix}1\\0\end{pmatrix}\otimes\begin{pmatrix}1\\0\end{pmatrix}, \quad \begin{pmatrix}0\\1\end{pmatrix}\otimes\begin{pmatrix}1\\0\end{pmatrix}, \quad \begin{pmatrix}1\\0\end{pmatrix}\otimes\begin{pmatrix}0\\1\end{pmatrix}, \quad \begin{pmatrix}0\\1\end{pmatrix}\otimes\begin{pmatrix}0\\1\end{pmatrix}, \tag{11.37}$$

which forms a 4-dimensional complex space (8 real dimensions). The dimensionality has doubled because we have not yet taken the complex structure of the spinors into account. While the role of j is played in the two single-particle spaces by right multiplication by $i\sigma_3^1$ and $i\sigma_3^2$ respectively, standard quantum mechanics does not distinguish between these operations. A projection operator must therefore be included to ensure that right multiplication by $i\sigma_3^1$ or $i\sigma_3^2$ reduces to the same operation. If a 2-particle spin state is represented by the multivector ψ, then ψ must satisfy

$$\psi i\sigma_3^1 = \psi i\sigma_3^2 \tag{11.38}$$

from which we find that

$$\begin{aligned} \psi &= -\psi i\sigma_3^1 i\sigma_3^2 \\ \Rightarrow \quad \psi &= \psi\tfrac{1}{2}(1 - i\sigma_3^1 i\sigma_3^2). \end{aligned} \tag{11.39}$$

On defining

$$E = \tfrac{1}{2}(1 - i\sigma_3^1 i\sigma_3^2), \tag{11.40}$$

we find that

$$E^2 = E \tag{11.41}$$

so right multiplication by E is a projection operation. (The relation $E^2 = E$ means that E is technically referred to as an 'idempotent' element.) It follows that the 2-particle state ψ must contain a factor of E on its right-hand side. We can further define

$$J = Ei\sigma_3^1 = Ei\sigma_3^2 = \tfrac{1}{2}(i\sigma_3^1 + i\sigma_3^2) \tag{11.42}$$

so that

$$J^2 = -E. \tag{11.43}$$

Right-sided multiplication by J takes on the role of j for multiparticle states.

The STA representation of a direct-product 2-particle Pauli spinor is now given by $\psi^1\phi^2 E$, where ψ^1 and ϕ^2 are spinors (even multivectors) in their own spaces. A complete basis for 2-particle spin states is provided by

$$\begin{aligned}
\begin{pmatrix}1\\0\end{pmatrix} \otimes \begin{pmatrix}1\\0\end{pmatrix} &\quad\leftrightarrow\quad E \\
\begin{pmatrix}0\\1\end{pmatrix} \otimes \begin{pmatrix}1\\0\end{pmatrix} &\quad\leftrightarrow\quad -i\sigma_2^1 E \\
\begin{pmatrix}1\\0\end{pmatrix} \otimes \begin{pmatrix}0\\1\end{pmatrix} &\quad\leftrightarrow\quad -i\sigma_2^2 E \\
\begin{pmatrix}0\\1\end{pmatrix} \otimes \begin{pmatrix}0\\1\end{pmatrix} &\quad\leftrightarrow\quad i\sigma_2^1 i\sigma_2^2 E.
\end{aligned} \tag{11.44}$$

This procedure extends simply to higher multiplicities. All that is required is to find the 'quantum correlator' E_n satisfying

$$E_n i\sigma_3^j = E_n i\sigma_3^k = J_n \qquad \text{for all } j,\ k. \tag{11.45}$$

E_n can be constructed by picking out the $j = 1$ space, say, and correlating all the other spaces to this, so that

$$E_n = \prod_{j=2}^{n} \tfrac{1}{2}(1 - i\sigma_3^1 i\sigma_3^j). \tag{11.46}$$

The value of E_n is independent of which of the n spaces is singled out and correlated to. The complex structure is defined by

$$J_n = E_n i\sigma_3^j, \tag{11.47}$$

where $i\sigma_3^j$ can be chosen from any of the n spaces. To illustrate this consider the case of $n = 3$, where

$$E_3 = \tfrac{1}{4}(1 - i\sigma_3^1 i\sigma_3^2)(1 - i\sigma_3^1 i\sigma_3^3) \tag{11.48}$$

$$= \tfrac{1}{4}(1 - i\sigma_3^1 i\sigma_3^2 - i\sigma_3^1 i\sigma_3^3 - i\sigma_3^2 i\sigma_3^3) \tag{11.49}$$

and

$$J_3 = \tfrac{1}{4}(i\sigma_3^1 + i\sigma_3^2 + i\sigma_3^3 - i\sigma_3^1 i\sigma_3^2 i\sigma_3^3). \tag{11.50}$$

Both E_3 and J_3 are symmetric under permutations of their indices.

A significant feature of this approach is that all the operations defined for the single-particle STA extend naturally to the multiparticle algebra. The reversion operation, for example, still has precisely the same definition — it simply reverses the order of vectors in any given multivector. The spinor inner product (equations 9.12 and 9.39) also generalises immediately, to

$$(\psi, \phi)_S = \langle E_n \rangle^{-1} [\langle \psi^\dagger \phi \rangle - \langle \psi^\dagger \phi J_n \rangle i\sigma_3], \tag{11.51}$$

where the right-hand side is projected onto a single copy of the STA. The factor of $\langle E_n \rangle^{-1}$ is included so that the state '1' always has unit norm, which matches with the inner product used in the matrix formulation.

11.3.2 Multiparticle Wave Equations

In order to extend the local-observables approach to quantum theory to the multiparticle domain, we need to construct a relativistic wave equation satisfied by an n-particle wavefunction. This is a subject that is given little attention in the literature, with most textbooks dealing solely with the field-quantised description of an n-particle system. An n-particle wave equation is essential, however, if one aims to give a relativistic description of a bound system (where field quantisation and perturbation theory on their own are insufficient).

The most popular previous equation for a relativistic 2-particle system has been the Bethe-Salpeter equation. Written in the STA, this equation becomes

$$(j\nabla^1 - m_1)(j\nabla^2 - m_2)\psi(r, s) = I(r, s)\psi(r, s) \tag{11.52}$$

where j represents right-sided multiplication by J, $I(r, s)$ is an integral operator representing the inter-particle interaction, and

$$\nabla^1 \equiv \gamma_\mu^1 \frac{\partial}{\partial r^\mu}, \qquad \nabla^2 \equiv \gamma_\mu^2 \frac{\partial}{\partial s^\mu}, \tag{11.53}$$

with r and s the 4-D positions of the two particles. Strictly, we should have written ∇_r^1 and ∇_s^2 instead of simply ∇^1 and ∇^2. In this case, however, the subscripts can safely be ignored.

The problem with equation (11.52) is that it is not first-order in the 8-dimensional vector derivative $\nabla = \nabla^1 + \nabla^2$. We are therefore unable to generalise many of the simple first-order propagation techniques we have discussed elsewhere. Clearly, we would like to find an alternative to (11.52) which retains the first-order nature of the single-particle Dirac equation. Here we will simply assert what we believe to be a good candidate for such an equation, and then work out its consequences. The equation we shall study, for two free spin-1/2 particles of masses m_1 and m_2 respectively, is

$$\left(\frac{\nabla^1}{m_1} + \frac{\nabla^2}{m_2}\right) \psi(x) \left(i\gamma_3^1 + i\gamma_3^2\right) = 2\psi(x). \tag{11.54}$$

We can assume, *a priori*, that ψ is not in the correlated subspace of the the direct-product space. But, since E commutes with $i\gamma_3^1 + i\gamma_3^2$, any solution to (11.54) can be reduced to a solution in the correlated space simply by right-multiplying by E. Written out explicitly, the vector x in equation (11.54) is

$$x = r^1 + s^2 = \gamma_\mu^1 r^\mu + \gamma_\mu^2 s^\mu \tag{11.55}$$

where $\{r^\mu, s^\mu\}$ are a set of 8 independent components for ψ. Of course, all particle motions ultimately occur in a single space, in which the vectors r and s label two independent position vectors. We stress that in this approach there are two time-like coordinates, r^0 and s^0, which is necessary if our 2-particle equation is to be Lorentz covariant. The derivatives ∇^1 and ∇^2 are as defined by equation (11.53), and the 8-dimensional vector derivative $\nabla = \nabla_x$ is given by

$$\nabla = \nabla^1 + \nabla^2. \tag{11.56}$$

Equation (11.54) can be derived from a Lorentz-invariant action integral in 8-dimensional configuration space in which the $1/m_1$ and $1/m_2$ factors enter via a linear distortion of the vector derivative ∇. We write this as

$$\left(\frac{\nabla^1}{m_1} + \frac{\nabla^2}{m_2}\right) = \overline{h}\,(\nabla), \tag{11.57}$$

where $\overline{h}$ is the linear mapping of vectors to vectors defined by

$$\overline{h}(a) = \frac{1}{m_1}\left(a\cdot\gamma^{\mu\,1}\right)\gamma_\mu^1 + \frac{1}{m_2}\left(a\cdot\gamma^{\mu\,2}\right)\gamma_\mu^2. \tag{11.58}$$

This distortion is of the type used in the gauge theory approach to gravity developed by us, and it is extremely suggestive that mass enters equation (11.54) via this route.

Any candidate 2-particle wave equation must be satisfied by factored states of the form

$$\psi = \phi^1(r^1)\chi^2(s^2)E, \tag{11.59}$$

where ϕ^1 and χ^2 are solutions of the separate single-particle Dirac equations,

$$\nabla\phi = -m_1\phi i\gamma_3, \qquad \nabla\chi = -m_2\chi i\gamma_3. \tag{11.60}$$

To verify that our equation (11.54) meets this requirement, we substitute in the direct-product state (11.59) and use (11.60) to obtain

$$\left(\frac{\nabla^1}{m_1} + \frac{\nabla^2}{m_2}\right)\phi^1\chi^2 E\left(i\gamma_3^1 + i\gamma_3^2\right) = -\phi^1\chi^2 E\left(i\gamma_3^1 + i\gamma_3^2\right)\left(i\gamma_3^1 + i\gamma_3^2\right), \tag{11.61}$$

where we have used the result that ∇^2 commutes with ϕ^1. Now, since $i\gamma_3^1$ and $i\gamma_3^2$ anticommute, we have

$$\left(i\gamma_3^1 + i\gamma_3^2\right)\left(i\gamma_3^1 + i\gamma_3^2\right) = -2 \tag{11.62}$$

so that

$$\left(\frac{\nabla^1}{m_1} + \frac{\nabla^2}{m_2}\right)\phi^1\chi^2 E\left(i\gamma_3^1 + i\gamma_3^2\right) = 2\phi^1\chi^2 E, \tag{11.63}$$

and (11.54) is satisfied. Equation (11.54) is only satisfied by direct-product states as a result of the fact that vectors from separate particle spaces anticommute. Hence equation (11.54) does not have an equivalent expression in terms of the direct-product matrix formulation, which can only form *commuting* operators from different spaces.

11.3.3 The Pauli Principle

In quantum theory, indistinguishable particles must obey either Fermi-Dirac or Bose-Einstein statistics. For fermions this requirement results in the Pauli exclusion principle that no two particles can occupy a state in which their properties are identical. At the relativistic multiparticle level, the Pauli principle is usually encoded in the anticommutation of the creation and annihilation operators of fermionic field theory. Here we show that the principle can be successfully encoded in a simple geometrical manner at the level of the relativistic wavefunction, without requiring the apparatus of quantum field theory.

We start by introducing the grade-4 multivector

$$I \equiv \Gamma_0\Gamma_1\Gamma_2\Gamma_3, \tag{11.64}$$

where

$$\Gamma_\mu \equiv \frac{1}{\sqrt{2}}\left(\gamma_\mu^1 + \gamma_\mu^2\right). \tag{11.65}$$

It is a simple matter to verify that I has the properties

$$I^2 = -1, \tag{11.66}$$

and

$$I\gamma_\mu^1 I = \gamma_\mu^2, \qquad I\gamma_\mu^2 I = \gamma_\mu^1. \tag{11.67}$$

It follows that I functions as a geometrical version of the particle exchange operator. In particular, acting on the 8-dimensional position vector $x = r^1 + s^2$ we find that

$$IxI = r^2 + s^1 \tag{11.68}$$

where

$$r^2 = \gamma_\mu^2 r^\mu, \qquad s^1 = \gamma_\mu^1 s^\mu. \tag{11.69}$$

So I can certainly be used to interchange the coordinates of particles 1 and 2. But, if I is to play a fundamental role in our version of the Pauli principle, we must first confirm that it is independent of our choice of initial frame. To see that it is, suppose that we start with a rotated frame $\{R\gamma_\mu\tilde{R}\}$ and define

$$\Gamma'_\mu = \frac{1}{\sqrt{2}}\left(R^1\gamma_\mu^1\tilde{R}^1 + R^2\gamma_\mu^2\tilde{R}^2\right) = R^1R^2\Gamma_\mu\tilde{R}^2\tilde{R}^1. \tag{11.70}$$

The new Γ'_μ give rise to the rotated 4-vector

$$I' = R^1R^2I\tilde{R}^2\tilde{R}^1. \tag{11.71}$$

But, acting on a bivector in particle space 1, we find that

$$Ia^1 \wedge b^1 I = -(Ia^1 I) \wedge (Ib^1 I) = -a^2 \wedge b^2, \tag{11.72}$$

and the same is true of an arbitrary even element in either space. More generally, $I \ldots I$ applied to an even element in one particle space flips it to the other particle space and changes sign, while applied to an odd element it just flips the particle space. It follows that

$$I\tilde{R}^2 \tilde{R}^1 = \tilde{R}^1 I \tilde{R}^1 = \tilde{R}^1 \tilde{R}^2 I, \tag{11.73}$$

and substituting this into (11.71) we find that $I' = I$, so I is indeed independent of the chosen orthonormal frame.

We can now use the 4-vector I to encode the Pauli exchange principle geometrically. Let $\psi(x)$ be a wavefunction for two electrons. Our suggested relativistic generalization of the Pauli principle is that $\psi(x)$ should be invariant under the operation

$$\psi(x) \mapsto I\psi(IxI)I. \tag{11.74}$$

For n-particle systems the extension is straightforward: the wavefunction must be invariant under the interchange enforced by the I's constructed from each pair of particles.

We must first check that (11.74) is an allowed symmetry of the 2-particle Dirac equation. With x' defined as IxI it is simple to verify that

$$\nabla_{x'} = \nabla_r^2 + \nabla_s^1 = I\nabla I, \tag{11.75}$$

and hence that

$$\nabla = I\nabla_{x'} I. \tag{11.76}$$

So, assuming that $\psi(x)$ satisfies the 2-particle equation (11.54) with equal masses m, we find that

$$\begin{aligned} \nabla[I\psi(IxI)I]\left(i\gamma_3^1 + i\gamma_3^2\right) &= -I\nabla_{x'}\psi(x')I\left(i\gamma_3^1 + i\gamma_3^2\right) \\ &= mI\psi(x')\left(i\gamma_3^1 + i\gamma_3^2\right)I\left(i\gamma_3^1 + i\gamma_3^2\right). \end{aligned} \tag{11.77}$$

But $i\gamma_3^1 + i\gamma_3^2$ is odd and symmetric under interchange of its particle labels. It follows that

$$I\left(i\gamma_3^1 + i\gamma_3^2\right)I = \left(i\gamma_3^1 + i\gamma_3^2\right) \tag{11.78}$$

and hence that

$$\nabla[I\psi(IxI)I]\left(i\gamma_3^1 + i\gamma_3^2\right) = 2mI\psi(IxI)I. \tag{11.79}$$

So, if $\psi(x)$ is a solution of the 2-particle equal-mass Dirac equation, then so to is $I\psi(IxI)I$.

Next we must check that the proposed relativistic Pauli principle deals correctly with well-known elementary cases. Suppose that two electrons are in the same spatial state. Then we should expect our principle to enforce the condition that they are in an antisymmetric spin state. For example, consider $i\sigma_2^1 - i\sigma_2^2$, the spin singlet state. We find that

$$I(i\sigma_2^1 - i\sigma_2^2)I = -i\sigma_2^2 + i\sigma_2^1, \tag{11.80}$$

recovering the original state, which is therefore compatible with our principle. On the other hand

$$I(i\sigma_2^1 + i\sigma_2^2)I = -(i\sigma_2^1 + i\sigma_2^2), \tag{11.81}$$

so no part of this state can be added in to the wavefunction, which again is correct. In conclusion, given some 2-particle solution $\psi(x)$, the corresponding state

$$\psi_I \equiv \psi(x) + I\psi(IxI)I \tag{11.82}$$

still satisfies the Dirac equation and is invariant under $\psi(x) \mapsto I\psi(IxI)I$. We therefore claim that the state ψ_I is the correct relativistic generalisation of a state satisfying the Pauli principle. In deference to standard quantum theory, we refer to equation (11.82) as an antisymmetrisation procedure.

The final issue to address is the Lorentz covariance of the antisymmetrisation procedure (11.82). Suppose that we start with an arbitrary wavefunction $\psi(x)$ satisfying the 2-particle equal-mass equation (11.54). If we boost this state via

$$\psi(x) \mapsto \psi'(x) \equiv R^1R^2\psi(\tilde{R}^2\tilde{R}^1xR^1R^2) \tag{11.83}$$

then $\psi'(x)$ also satisfies the same equation (11.54). The boosted wavefunction $\psi'(x)$ can be thought of as corresponding to a different observer in relative motion. The boosted state $\psi'(x)$ can also be antisymmetrised to yield a solution satisfying our relativistic Pauli principle. But, for this procedure to be covariant, the same state must be obtained if we first antisymmetrise the original $\psi(x)$, and then boost the result. Thus we require that

$$S\psi(\tilde{S}xS) + IS\psi(I\tilde{S}xSI)I = S[\psi(\tilde{S}xS) + I\psi(\tilde{S}IxIS)I] \tag{11.84}$$

where $S \equiv R^1R^2$. Equation (11.84) reduces to the requirement that

$$IS\psi(I\tilde{S}xSI) = SI\psi(\tilde{S}IxIS) \tag{11.85}$$

which is satisfied provided that

$$IS = SI \tag{11.86}$$

or

$$R^1R^2I = IR^1R^2. \tag{11.87}$$

But we proved precisely this equation in demonstrating the frame-invariance of I, so our relativistic version of the Pauli principle is Lorentz invariant. This is important as, rather like the inclusion of the quantum correlator, the Pauli procedure discussed here looks highly non-local in character.

11.3.4 8-Dimensional Streamlines and Pauli Exclusion

For a single Dirac particle, a characteristic feature of the STA approach is that the probability current is a rotated/dilated version of the γ_0 vector, $J = \psi\gamma_0\tilde{\psi}$. This current has zero divergence and can therefore be used to define streamlines, as discussed in Section 11.1. Here we demonstrate how the same idea extends to the 2-particle case. We find that the conserved current is now formed from ψ acting on the $\gamma_0^1 + \gamma_0^2$ vector, and therefore exists in 8-dimensional configuration space. This current can be used to derive streamlines for two particles in correlated motion. This approach should ultimately enable us to gain a better insight into what happens in experiments of the Bell type, where spin measurements on pairs of particles are performed over spacelike separations. We saw in Section 11.2 how the local observables viewpoint leads to a radical re-interpretation of what happens in a single spin-measurement, and we can expect an equally radical shift to occur in the analysis of spin measurements of correlated particles. As a preliminary step in this direction, here we construct the current for two free particles approaching each other head-on. The streamlines for this current are evaluated and used to study both the effects of the Pauli antisymmetrisation and the spin-dependence of the trajectories. This work generalises that of Dewdney *et al.* to the relativistic domain.

We start with the 2-particle Dirac equation (11.54), and multiply on the right by E to ensure the total wavefunction is in the correlated subspace. Also, since we want to work with the indistinguishable case, we assume that both masses are m. In this case our basic equation is

$$\nabla\psi E\left(i\gamma_3^1 + i\gamma_3^2\right) = 2m\psi E \tag{11.88}$$

and, since

$$E\left(i\gamma_3^1 + i\gamma_3^2\right) = J(\gamma_0^1 + \gamma_0^2), \tag{11.89}$$

equation (11.88) can be written in the equivalent form

$$\nabla\psi E\left(\gamma_0^1 + \gamma_0^2\right) = -2m\psi J. \tag{11.90}$$

Now, assuming that ψ satisfies $\psi = \psi E$, we obtain

$$\nabla\psi(\gamma_0^1 + \gamma_0^2)\tilde{\psi} = -2m\psi J\tilde{\psi}, \tag{11.91}$$

and adding this equation to its reverse yields

$$\nabla\psi(\gamma_0^1 + \gamma_0^2)\tilde{\psi} + \psi(\gamma_0^1 + \gamma_0^2)\dot{\tilde{\psi}}\dot{\nabla} = 0. \tag{11.92}$$

The scalar part of this equation gives

$$\nabla\cdot\langle\psi(\gamma_0^1 + \gamma_0^2)\tilde{\psi}\rangle_1 = 0, \tag{11.93}$$

which shows that the current we seek is

$$\mathcal{J} = \langle\psi(\gamma_0^1 + \gamma_0^2)\tilde{\psi}\rangle_1. \tag{11.94}$$

The vector $\mathcal{J}$ has components in both particle-1 and particle-2 spaces, which we write as

$$\mathcal{J} = \mathcal{J}_1^1 + \mathcal{J}_2^2. \tag{11.95}$$

The current $\mathcal{J}$ is conserved in *eight*-dimensional space, so its streamlines never cross there. The streamlines of the individual particles, however, are obtained by integrating $\mathcal{J}_1$ and $\mathcal{J}_2$ in ordinary 4-d space, and these can of course cross. An example of

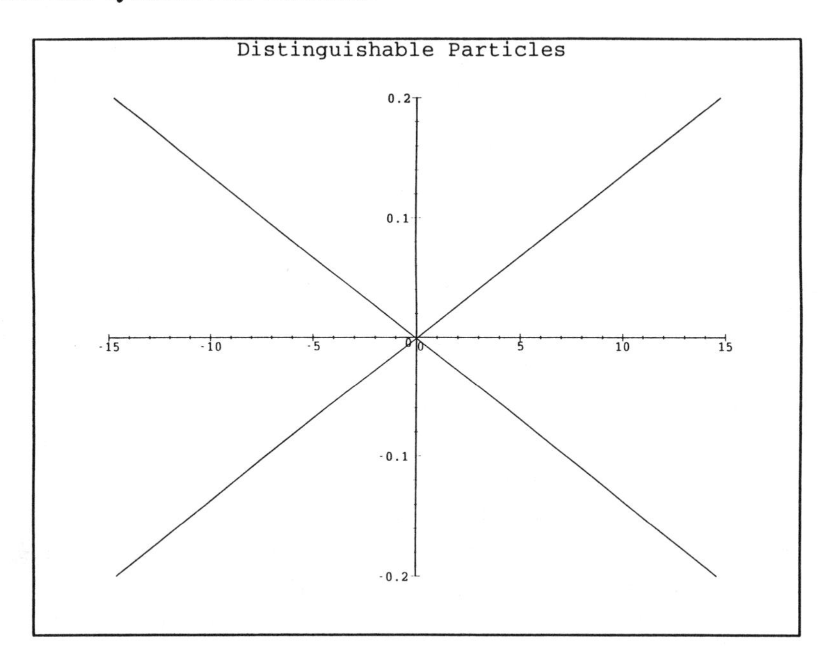

Fig. 11.7. *Streamlines generated by the unsymmetrised 2-particle wavefunction* $\psi = \phi^1(r^1)\chi^2(s^2)E$. *Time is shown on the vertical axis.* ϕ *and* χ *are Gaussian wavepackets moving in opposite directions and the 'collision' is arranged to take place at* $t = 0$. *The lack of any antisymmetrisation applied to the wavefunction means that the streamlines pass straight through each other.*

this is illustrated in Figure 11.7, which shows the streamlines corresponding to distinguishable particles in two Gaussian wavepackets approaching each other head-on. The wavefunction used to produce this figure is just

$$\psi = \phi^1(r^1)\chi^2(s^2)E, \tag{11.96}$$

with ϕ and χ being Gaussian wavepackets, moving in opposite directions. Since the distinguishable case is assumed, no Pauli antisymmetrisation is used. The individual currents for each particle are given by

$$\mathcal{J}_1(r,s) = \phi(r)\gamma_0\tilde{\phi}(r)\,\langle\chi(s)\tilde{\chi}(s)\rangle, \qquad \mathcal{J}_2(r,s) = \chi(s)\gamma_0\tilde{\chi}(s)\,\langle\phi(r)\tilde{\phi}(r)\rangle \tag{11.97}$$

and, as can be seen, the streamlines (and the wavepackets) simply pass straight through each other.

An interesting feature emerges in the individual currents in (11.97). One of the main problems with single-particle Dirac theory is that the current is always positive-definite so, if we wish to interpret it as a charge current, it fails to represent antiparticles correctly. The switch of sign of the current necessary to represent positrons is put into conventional theory essentially 'by hand', via the anticommutation and normal ordering rules of fermionic field theory. In equation (11.97), however, the norm $\langle\chi\tilde{\chi}\rangle$ of the second state multiplies the current for the first, and vice versa. Since $\langle\chi\tilde{\chi}\rangle$ can be negative, it is possible to obtain currents which flow backwards in time. This suggests that the required switch of signs can be accomplished whilst remaining wholly within a wavefunction-based approach. An apparent problem is that, if only one particle has a negative norm state — say for example χ has $\langle\chi\tilde{\chi}\rangle < 0$ — then it is the ϕ current which is reversed, and not the χ current. However, it is easy to see that this objection is not relevant to indistinguishable particles, and it is to these we now turn.

We now apply the Pauli symmetrization procedure of the previous subsection to the wavefunction of equation (11.96), so as to obtain a wavefunction applicable to indistinguishable particles. This yields

$$\psi = \left(\phi^1(r^1)\chi^2(s^2) - \chi^1(r^2)\phi^2(s^1)\right) E, \tag{11.98}$$

from which we form $\mathcal{J}_1$ and $\mathcal{J}_2$, as before. We must next decide which spin states to use for the two particles. We first take both particles to have their spin vectors pointing in the positive z-direction, with all motion in the $\pm z$-direction. The resulting streamlines are shown in Figure 11.8(a). The streamlines now 'repel' one another, rather than being able to pass straight through. The corrugated appearance of the lines near the origin is the result of the streamlines having to pass through a region of highly-oscillatory destructive interference, since the probability of both particles occupying the same position (the origin) with the same spin state is zero. If instead the particles are put in *different* spin states then the streamlines shown in Figure 11.8(b) result. In this case there is no destructive interference near the origin, and the streamlines are smooth there. However, they still repel! The explanation for this lies in the symmetry properties of the 2-particle current. Given that the wavefunction ψ has been antisymmetrised according to our version of the Pauli principle, then it is straightforward to show that

$$I\mathcal{J}(IxI)I = \mathcal{J}(x). \tag{11.99}$$

It follows that at the same spacetime position, encoded by $IxI = x$ in the 2-particle algebra, the two currents $\mathcal{J}_1$ and $\mathcal{J}_2$ are equal. Hence, if two streamlines ever met, they could never separate again. For the simulations presented here, it follows from the symmetry of the set-up that the spatial currents at the origin are both zero and so, as the particles approach the origin, they are forced to slow up. The delay means that they are then swept back in the direction they have just come from by the wavepacket travelling through from the other side. We therefore see that 'repulsion' as measured by streamlines has its origin in indistinguishability, and that the spin of the states exerts only a marginal effect.

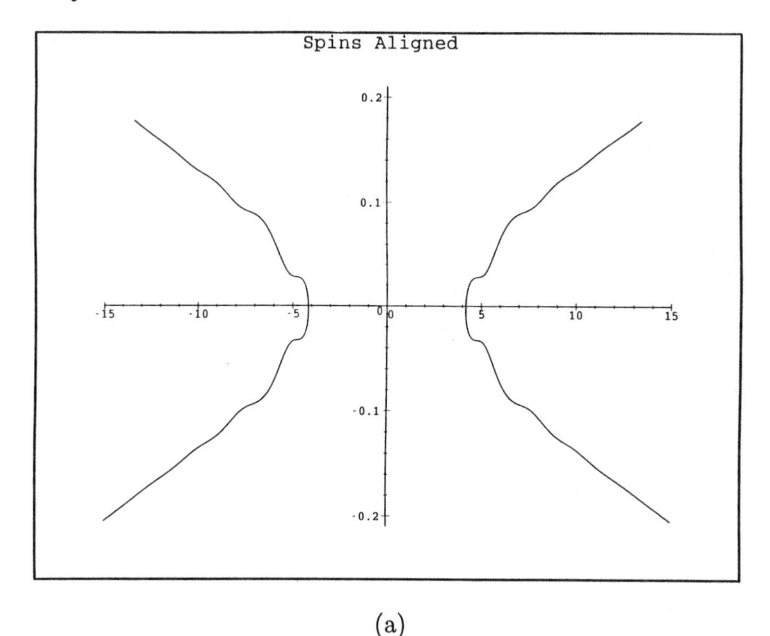

(a)

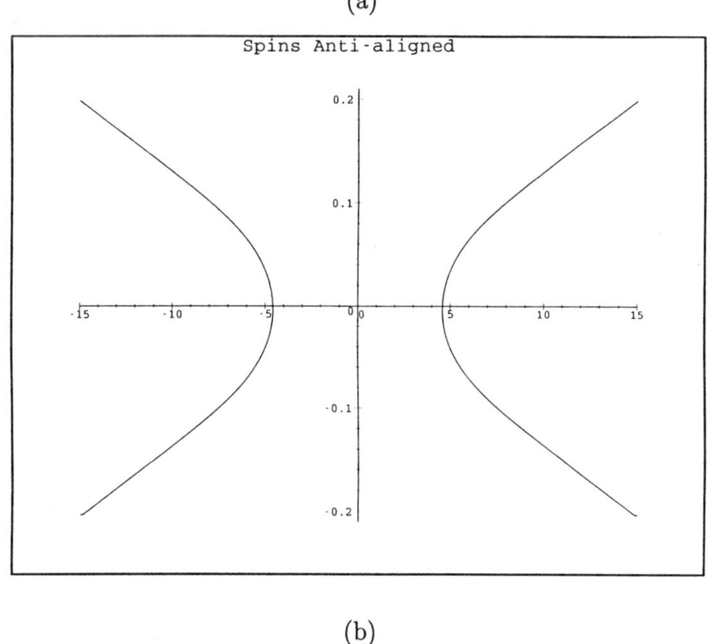

(b)

Fig. 11.8. *Streamlines generated by the antisymmetrised 2-particle wavefunction* $\psi = [\phi^1(r^1)\chi^2(s^2) - \chi^1(r^2)\phi^2(s^1)]E$. *The individual wavepackets pass through each other, but the streamlines from separate particles do not cross. Figure (a) has both particles with spins aligned in the* $+z$*-direction, and Figure (b) shows particles with opposite spins, with* ϕ *in the* $+z$ *direction, and* χ *in the* $-z$ *direction. Both wavepackets have energy 527KeV and a spatial spread of* $\sim$ *20pm. The spatial units are* $10^{-12}m$ *and the units of time are* $10^{-18}s$. *The effects of the antisymmetrisation are only important where there is significant wavepacket overlap.*

Chapter 12

Gravity I — Introduction

Anthony Lasenby, Chris Doran, and Stephen Gull
MRAO, Cavendish Laboratory, Madingley Road,
Cambridge, CB3 0HE, UK

This series of five lectures introduces an approach to gravity which we have recently developed in Cambridge[1]. For many years physicists and mathematicians have attempted to express general relativity (GR) as a *gauge theory*,indexgravity!gauge theory of with the aim of placing GR on a similar conceptual footing to the modern theories of strong and electroweak interactions. These attempts have met with mixed success and GR continues to be taught, and understood, in terms of the classical differential geometry of manifolds. Our approach aims to bridge this gap. We will show that it is both conceptually and mathematically simpler to adopt a gauge theory approach at the outset. This will enable us to reach advanced applications without ever mentioning topics such as manifolds or tangent spaces. Indeed, we will never think of spacetime itself as doing anything peculiar, such as expanding, or deforming like a rubber sheet.

The full theory we have developed derives the gravitational gauge fields from the Dirac Lagrangian, so deals immediately with the coupling of gravitational fields to spin-1/2 particles. In this sense our theory is already more general than classical GR. Though the full formal development is quite technical, we hope to cover it in sufficient detail that it is comprehensible. However, our main aim here is reach interesting applications in a minimum amount of time. In particular, although no knowledge of GR will be assumed, we will address many standard applications of GR (usually taught in courses lasting many times longer) as well as some research-level topics, such as cosmic strings.

In doing this you need not fear that you will be learning a new theory unchecked by experiment. For many of the applications presented here, the predictions of our approach are the same as those of GR, and therefore have already been tested in

[1]Further material, including references, can be found in 'Astrophysical and Comological Consequences of a Gauge Theory of Gravity', in N. Sanchez and A. Zichichi eds. *Advances in Astrofundamental Physics, Erice 1994* (Kluwer Academic, Dordrecht, 1995), p. 359, and in 'Gravity, Gauge Theories and Geometric Algebra', submitted to Phys. Rev. D. (1995).

a variety of experimental and observational contexts. For these cases you can take what is presented here as being a clearer and quicker analysis of the physics than is afforded by GR. For example, in GR some confusion persists as to what constitutes a real physical prediction, whereas our approach concisely eliminates such ambiguities. The applications leading to predictions that are different from, or inaccessible to, GR will be emphasized when we reach them. Whilst it is currently not possible to test any of these differences experimentally, this is an exciting prospect for the future.

12.1 Gauge Theories

The simplest example of a gauge theory is electromagnetism, so we start by analysing this in its STA form. Consider the free-particle Dirac equation:

$$\nabla \psi i\sigma_3 = m\psi\gamma_0. \tag{12.1}$$

Since $i\sigma_3$ commutes with γ_0, a *global symmetry* of this equation is the transformation

$$\psi \mapsto \psi' = \psi e^{i\sigma_3\theta}, \tag{12.2}$$

where θ is a constant. This is a symmetry because if equation (12.1) holds for ψ, it also holds for ψ'. The symmetry is 'global' because θ has the same value everywhere in space and time. The quantity $\exp\{i\sigma_3\theta\}$ is conventionally called a phase factor and thought of as belonging to the 'internal symmetry group' $U(1)$. Here we can see that it is also a spatial rotor, corresponding to rotations in the $\gamma_2\gamma_1$ plane through angle 2θ. We write this rotor as R.

Now what if θ is not a constant, but depends on spacetime position x, $\theta = \theta(x)$? In this case ψ' will no longer be a solution of the equation if ψ is, since

$$\nabla\psi' = (\nabla\psi)R + (\nabla\theta)\psi R i\sigma_3 \tag{12.3}$$

and so $\nabla\psi' \neq m\psi'\gamma_0$. Hence (12.2) is not a *local* symmetry of equation (12.1) as θ cannot be varied arbitrarily from point to point.

At this stage of the development it is argued that we do want equation (12.2) to work as a local symmetry, and therefore the Dirac equation (12.1) must be modified in some way. The reason why equation (12.2) is expected to define a local symmetry comes from a consideration of the **observables** of the Dirac theory. There are, in essence, two types of physical statement that can be extracted from the Dirac theory. The first lies in the values of observables, which are formed from inner products between spinors. In the STA these are all of the form

$$(\psi, \phi)_S = \langle\tilde{\psi}\phi\rangle - \langle\tilde{\psi}\phi i\sigma_3\rangle i\sigma_3. \tag{12.4}$$

The second is statements about the equality of two spinor expressions. For example the expression

$$\psi = \psi_1 + \psi_2, \tag{12.5}$$

may decompose ψ into two orthogonal eigenstates of some operator. In both cases, if all spinors pick up the same locally-varying phase factor (rotor) then the physical predictions are unchanged.

So, having understood the motivation, how do we change equation (12.1) to make (12.2) a local symmetry? We clearly need to change the ∇ operator in some way, so as to cancel out the term due to the derivative of R in equation (12.3). Let us first rewrite ∇ as

$$\nabla = \partial_a a\cdot\nabla, \tag{12.6}$$

so that we separate out its vector and derivative characteristics clearly. Then we define a new, 'covariant' derivative operator D by including an extra piece in ∇:

$$D\psi = \partial_a \left(a\cdot\nabla\psi + \tfrac{1}{2}\psi\Omega(a)\right) \tag{12.7}$$

(the factor 1/2 is inserted for later convenience). Here $\Omega(a)$ is a multivector field of some kind, the nature and transformation properties of which we are going to discover from the requirement that the new derivative D does what we want. It is included to the right of ψ since, if we rewrite (12.3) as

$$\nabla\psi' = \partial_a \left(a\cdot\nabla\psi R + \psi a\cdot\nabla R\right), \tag{12.8}$$

we can see that the $a\cdot\nabla R$ term we wish to remove lies to the right of ψ. Our modified Dirac equation is therefore

$$D\psi i\sigma_3 = m\psi\gamma_0. \tag{12.9}$$

The behaviour we require is that under a local rotation D should transform in such a way that ψR is still a solution of this equation. Hence, with D transforming to D', we require that

$$D'(\psi R) \equiv (D\psi)R, \tag{12.10}$$

for any R. We expect that D' should have the same functional form as D (this is one way of expressing the 'minimal coupling' procedure) so we must have

$$D'\phi = \partial_a[a\cdot\nabla\phi + \tfrac{1}{2}\phi\Omega'(a)], \tag{12.11}$$

for general ϕ. So, from (12.10),

$$D'(\psi R) = \partial_a \left(a\cdot\nabla\psi R + \psi a\cdot\nabla R + \tfrac{1}{2}\psi R\Omega'(a)\right) = \partial_a \left(a\cdot\nabla\psi + \tfrac{1}{2}\psi\Omega(a)\right) R, \tag{12.12}$$

and this we can read off that

$$\Omega'(a) = \tilde{R}\Omega(a)R - 2\tilde{R}a\cdot\nabla R. \tag{12.13}$$

Now R is a rotor, so satisfies $\tilde{R}R = 1$. Hence

$$a\cdot\nabla\tilde{R}R + \tilde{R}a\cdot\nabla R = 0, \tag{12.14}$$

and we see that $\tilde{R}a\cdot\nabla R$ is equal to minus its own reverse, so is therefore a bivector. The transformation $\tilde{R}\ldots R$ always preserves grade, and so from its transformation law (12.13) we see that $\Omega(a)$ must be a *bivector* field. This is therefore what we have to introduce in order to make rotation by R a local symmetry of the Dirac equation.

Note that we have only used the form of the $-2\tilde{R}a\cdot\nabla R$ term in (12.13) to say what type of object $\Omega(a)$ is — we are *not* asserting that $\Omega(a)$ is equal to $-2\tilde{R}a\cdot\nabla R$. On the contrary, as will become apparent soon, if $\Omega(a)$ was given by the gradient of a rotor like this it would give rise to a vanishing field strength and therefore be of no physical interest. This step, of taking a term arising from a derivative (like $-2\tilde{R}a\cdot\nabla R$ here), and generalizing it to a field *not* in general derivable from a derivative, is the essence of the gauging process.

Returning to electromagnetism, we are concerned with the restricted class of rotations which take place wholly in the $\gamma_2\gamma_1$ plane. In this case, using the parametrization in terms of θ, we have

$$\begin{aligned} -2\tilde{R}a\cdot\nabla R &= -2e^{-i\sigma_3\theta}a\cdot(\nabla\theta)e^{i\sigma_3\theta}i\sigma_3 \\ &= -2a\cdot(\nabla\theta)i\sigma_3. \end{aligned} \tag{12.15}$$

So, in generalizing to Ω in the same fashion as above, we can deduce that

$$\Omega(a) = -2a\cdot Ai\sigma_3 \quad \text{say}, \tag{12.16}$$

where A is a general 4-d vector. If A was in fact the gradient of a scalar, then we would expect the field strength to vanish. Having reached this point we are back on familiar ground of course, since this is just the statement that $\nabla\wedge A$ vanishes if $A = \nabla\phi$.

In a gauge theory, how do we define the 'field strength' generally? In electromagnetism we know $F = \nabla\wedge A$, but where does this come from if we are using a gauge theory route? The answer is as a commutator of covariant derivatives. Writing $D\psi = \partial_a D_a\psi$, where

$$D_a\psi = a\cdot\nabla\psi + \tfrac{1}{2}\psi\Omega(a), \tag{12.17}$$

we define a 'field strength tensor' $F(a\wedge b)$ via

$$\psi F(a\wedge b) = [D_a, D_b]\psi. \tag{12.18}$$

This is possible since all the derivatives of ψ cancel when the commutator is taken (this is left as an exercise) and we are left with just

$$F(a\wedge b) = \tfrac{1}{2}\left(b\cdot\nabla\Omega(a) - a\cdot\nabla\Omega(b) + \Omega(a)\times\Omega(b)\right). \tag{12.19}$$

Returning to electromagnetism, $\Omega(a)$ is of the form $-2a\cdot Ai\sigma_3$, so $F(a\wedge b)$ becomes

$$F(a\wedge b) = (a\cdot\nabla b\cdot A - b\cdot\nabla a\cdot A)\, i\sigma_3 = -(a\wedge b)\cdot(\nabla\wedge A)i\sigma_3. \tag{12.20}$$

The fact that the field strength tensor is in fact a linear mapping of bivectors to bivectors of course gets lost in standard treatments of electromagnetism, where (12.20) just gets abbreviated to

$$F_{\mu\nu} = \partial_\mu A_\nu - \partial_\nu A_\mu, \tag{12.21}$$

but the full bivector characteristics of this mapping are significant for gravity.

As a final point, we should establish the transformation properties of $F(a\wedge b)$ under local rotations $\psi \mapsto \psi R$. To find these we substitute the new Ω' from (12.13) into (12.19), yielding (exercise),

$$F'(a\wedge b) = \tilde{R}F(a\wedge b)R. \qquad (12.22)$$

In electromagnetism this rotation is hidden because R commutes with $F(a\wedge b)$. This is not the case for general non-abelian gauge theories (such as gravity or electroweak).

12.2 Gauge Principles and Gravitation

Having now successfully derived electromagnetism, we turn attention to gravity. Our aim is to derive and describe gravitational forces within the context of the Spacetime Algebra, which deals so successfully with electromagnetism and Dirac theory. We wish to put forward a new theory of gravity and, as with any physical theory, our starting point is a set of mathematical and physical assumptions. The origin of these assumptions lies outside the theory, of course, and they must be based on criteria of reasonableness and plausibility and, where possible, experimental fact.

Our first two basic assumptions are as follows:

(i) All physical quantities and processes can be modelled by fields and their relations defined in the STA. (For the applications of interest here the fields are all simple multivectors.)

(ii) Since all physical quantities can be represented by fields, the mathematical structure of the STA itself cannot correspond to anything measurable. Hence it is meaningless to talk about the absolute position of a field, or its absolute orientation, in the STA.

Before proceeding, let us look at the traditional view regarding Point (ii). This is usually taken as showing that the equations of physics should be invariant under Poincaré transformations which, for a multivector field $M(x)$, are performed by

$$M(x) \mapsto M' = RM(x')\tilde{R} \qquad (12.23)$$

where

$$x' = \tilde{R}xR + a \qquad (12.24)$$

and a and R constant. Hence we demand that if a set of multivector fields $\{M_i\}$ satisfy some physical relationship, then the fields $\{M_i'\}$, transformed according to (12.23), should satisfy the same relationship. Enforcing this invariance property then restricts the type of allowable physical model.

But does Poincaré invariance achieve the aim of (ii)? The answer to this must be no! Consider a relation of the form $a(x) = b(x)$, where $a(x)$ and $b(x)$ are two spacetime vector fields. An equation of equality such as this corresponds to a clear physical statement. But, considered as a relation between fields, the physical relationship expressed by this statement is completely independent of where we choose to think of x as lying in spacetime. In particular, we can associate each position x with some new position $x' = f(x)$ and rewrite the relation as $a(x') = b(x')$, and the equation still has precisely the same content. Poincaré invariance does not embody this principle, since it only deals with rigid translations — once the value of a field is fixed at one point, it is fixed at all others. Thus Poincaré invariance still affords physical significance to aspects of the absolute space of the STA.

A similar argument applies to rotations. As with the discussion of electromagnetism above, the *intrinsic* content of a relation such as $a(x) = b(x)$ at a given point x_0 is unchanged if we rotate each of a and b by the same amount. That is, the equation $Ra(x_0)\tilde{R} = Rb(x_0)\tilde{R}$ has the same physical content as the equation $a(x_0) = b(x_0)$. For example, scalar product relations, from which we can derive angles, are completely unaffected by this change. These arguments apply to any physical relation between any type of multivector field. But again, Poincaré transformations alone fail to encode this invariance. Not only do they only deal with rigid rotations, the Poincaré transformations described by (12.23) also mix field rotations and displacements. Any attempt at gauging this group fails to decouple the two, so cannot be responsible for enforcing the local symmetry we seek. Clearly, the Poincaré group does not provide a suitable starting point for our theory, and to achieve Point (ii) some further assumptions are needed. These are

(iii) The physical content of a field equation in the STA must be invariant under arbitrary local displacements of the fields. (This is called position-gauge invariance)

(iv) The physical content of a field equation in the STA must be invariant under arbitrary local rotations of the fields. (This is called rotation-gauge invariance)

This completes our list of assumptions, but we would obviously do well to dwell on the final two a while longer! By now all sorts of objections will probably be forming. One might argue, for example, that of course we *can* measure the angle between a vector at one position, and a vector at another. But this requires an operational procedure for transporting the vectors to a common point for comparison. Such a procedure might involve gyroscopes, or light paths, for example. When these physical processes are described in our theory we then find that the predictions do agree with what is measured. It is only the quantities that cannot be defined operationally that are not predicted uniquely, since these depend on a choice of 'gauge'.

Similarly, we may believe intuitively that the relative positions of points in a spacetime grid is something measurable — we just put our ruler down and measure it. Even in GR the situation is much the same — instead of a flat spacetime, GR substitutes spacetime represented by a curved manifold, but the structure of this is rigid (for a given matter distribution), and we have a picture of measuring these distances with a succession of small rulers (small so that they can lie tangent to the manifold, and not stick out of it). However, this is again mixing up two entities — the physical measurements we can carry out in the world, and our mathematical description of it. In our theory predictions for physically measurable distances and angles (scalar products between vectors at a point) must be derived from gauge-invariant relations between the field quantities themselves, and not from the properties of the STA itself.

A final introductory point: we have said that the mapping of fields onto spacetime positions is arbitrary, but it must also be the case that the fields be well-defined over the whole of spacetime. Every point in the STA must correspond to a position in spacetime and, except for some special cases, we cannot allow the fields to be singular anywhere. Here we see the starting point for a range of potential disagreements with GR: we insist that the whole of flat spacetime have fields allocated to it, including gravitational fields, in all cases. Because GR works with curved manifolds, which can be topologically distinct from Minkowski space, there is no equivalent assumption in GR. In our view, one of the weaknesses of GR is that it has no control over the topology of the manifolds it uses (the field equations do not determine this topology at all), whereas we have. This means that our theory is much more restrictive (and therefore more predictive) than GR.

12.3 The Gravitational Gauge Fields

We now examine the consequences of the local gauging just discussed. In exactly the same way as with electromagnetism, what we have to study is the effects on *derivatives*, since all non-derivative relations are already locally invariant.

We start by considering a scalar field $\phi(x)$ and form its vector derivative $\nabla\phi(x)$. Suppose now that $\phi(x)$ is replaced by the new field $\phi'(x)$,

$$\phi'(x) \equiv \phi(x') \tag{12.25}$$

where

$$x' = f(x) \tag{12.26}$$

and $f(x)$ is an arbitrary (differentiable) map between spacetime position vectors. The map $f(x)$ should not be thought of as a map between manifolds, or as moving points around. The function $f(x)$ is merely a rule for relating one position vector to another within a single vector space. If we now consider ∇ acting on the new scalar field ϕ' we form the quantity $\nabla\phi[f(x)]$. To evaluate this we return to the definition of the vector derivative and construct

$$\begin{aligned} a\cdot\nabla\phi[f(x)] &= \lim_{\epsilon\to 0}\frac{1}{\epsilon}\left(\phi f(x+\epsilon a) - \phi f(x)\right) \\ &= \lim_{\epsilon\to 0}\frac{1}{\epsilon}\left(\phi[f(x)+\epsilon\underline{f}(a)] - \phi f(x)\right) \\ &= \underline{f}(a)\cdot\nabla_{x'}\phi(x'), \end{aligned} \tag{12.27}$$

where $\underline{f}(a) = \underline{f}_x(a) = a\cdot\nabla f(x)$ and the subscript on $\nabla_{x'}$ records that the derivative is now with respect to the new vector position variable x'. It follows that

$$\nabla_x = \overline{f}(\nabla_{x'}) \tag{12.28}$$

and hence that $\nabla\phi'(x) = \overline{f}[\nabla_{x'}\phi(x')]$.

In physics we frequently need to relate the gradient of some scalar to a vector field. In three dimensions, for example, a static electric field can be written as the gradient of the scalar potential ϕ. We have said that we wish to move to a situation where all relations are unaffected by arbitrary displacements. To achieve this we must introduce a gauge field which assembles with the vector derivative to form an object which, under a local position transformation, simply re-evaluates to the new position. Specifically, let $A(x)$ be the modified derivative, then we want the same behaviour for A as for the scalar field above, i.e.

$$A'(x) = A(x'). \tag{12.29}$$

We construct such an object by replacing ∇ with a new derivative $\overline{h}(\nabla)$, where $\overline{h}(a)$ has an arbitrary position dependence and is a linear function of a. If we wish to make the position dependence explicit we will write $\overline{h}_x(a)$ or $\overline{h}(a,x)$ (recalling that $\overline{h}(a,x)$ is linear on a and non-linear on x). Under local displacements the gauge field $\overline{h}(a)$ is defined to transform to the new field $\overline{h}'(a)$, where

$$\overline{h}'(a,x) \equiv \overline{h}(\overline{f}^{-1}(a), f(x)) = \overline{h}_{x'}\overline{f}^{-1}(a) \tag{12.30}$$

so that

$$\overline{h}_x(\nabla_x) \mapsto \overline{h}_{x'}\overline{f}^{-1}(\nabla_x) = \overline{h}_{x'}(\nabla_{x'}). \tag{12.31}$$

This transformation law ensures that the vector $\overline{h}[\nabla\phi(x)]$ transforms as

$$\overline{h}[\nabla\phi(x)] \mapsto \overline{h}_{x'}[\nabla_{x'}\phi(x')]. \tag{12.32}$$

Hence the resultant vector $\overline{h}[\nabla\phi(x)]$ just changes its vector position under *arbitrary* local displacements. This is the type of behaviour we are after, since ultimately all that should matter are the relations between fields at a point, and the actual position vector of that point (and its relation to nearby points) should be irrelevant. The same intrinsic relations should therefore hold if the fields are moved arbitrarily from one vector position to another. The introduction of the $\overline{h}$-field ensures that derivatives can also be moved around arbitrarily. The $\overline{h}$-field is not a connection in the conventional Yang-Mills sense. The coupling to derivatives is different, as is the transformation law (12.30). This is unsurprising, since the group of arbitrary displacements is infinite-dimensional (if we were considering maps between manifolds then this would form the group of diffeomorphisms). There is no doubt, however, that the $\overline{h}$-field embodies the idea of replacing a global symmetry by a local one, so clearly deserves to be called a gauge field.

Given that $\overline{h}(a)$ can be viewed as a gauge field, it is interesting to ask under what conditions it can be transformed to the identity. If such a transformation existed, it would give

$$\overline{h}_{x'}(a)\overline{f}^{-1}(a) = a \tag{12.33}$$

$$\Rightarrow \overline{h}_{x'}(a) = \overline{f}(a). \tag{12.34}$$

But from the definition of $\underline{f}(a)$ it follows that (for constant a)

$$\begin{aligned}\overline{f}(a) &= \partial_b\langle a\underline{f}(b)\rangle \\ &= \partial_b\langle ab\cdot\nabla f(x)\rangle \\ &= \nabla\langle f(x)a\rangle, \end{aligned} \tag{12.35}$$

and hence that

$$\nabla\wedge\overline{f}(a) = 0. \tag{12.36}$$

So, if the $\overline{h}(a)$ field can be transformed to the identity, it must satisfy

$$\nabla_x\wedge\overline{h}_{x'}(a) = 0. \tag{12.37}$$

This condition can be simplified by using equation (12.34) to write ∇_x as $\overline{h}_{x'}(\nabla_{x'})$ and obtain

$$\overline{h}_{x'}(\nabla_{x'})\wedge\overline{h}_{x'}(a) = 0. \tag{12.38}$$

This must hold for all x', so we can equally well replace x' with x and write

$$\overline{h}(\nabla)\wedge\overline{h}(a) = 0 \tag{12.39}$$

$$\Rightarrow \quad \overline{h}(\dot{\nabla})\wedge\dot{\overline{h}}\,\overline{h}^{-1}(a) = 0 \tag{12.40}$$

$$\Rightarrow \quad -\overline{h}[\nabla\wedge\overline{h}^{-1}(a)] = 0. \tag{12.41}$$

Thus we finally obtain the 'pure gauge' condition in the simple form

$$\nabla\wedge\overline{h}^{-1}(a) = 0. \tag{12.42}$$

An arbitrary $\overline{h}$-field will not satisfy this equation, so in general there is no way to assign position vectors so that the effects of the $\overline{h}$-field vanish. In the light of equations (12.36) and (12.42) it might look more natural to introduce the gauge field as $\overline{h}^{-1}(\nabla)$, instead of $\overline{h}(\nabla)$. There is little to choose between these conventions, though there are some points in favour of our choice. These will emerge when we consider the gravitational field equations.

12.3.1 The Rotation-Gauge Field

We now consider how the derivative must be modified to allow the rotational freedom from point to point, as discussed in Point (iv) above. We will give two derivations of the required fields, one based on the properties of classical fields, and one on the properties of quantum spinor fields. A study of the observables formed from spinors then shows how the two approaches are unified.

We have already seen that the gradient of a scalar field is modified to $\overline{h}(\nabla\phi)$ to achieve covariance under displacements. But objects such as temperature gradients are certainly physical, and can be equated with other physical quantities. Consequently vectors such as $\overline{h}(\nabla\phi)$ must transform under rotations in the same manner as all other physical fields. It follows that, under local spacetime rotations, the $\overline{h}$-field must transform as

$$\overline{h}(a) \mapsto R\overline{h}(a)\tilde{R}. \tag{12.43}$$

Now consider an equation such as Maxwell's equation $\nabla F = J$. Once the position-gauge field is introduced, this equation becomes

$$\overline{h}(\nabla)\mathcal{F} = \mathcal{J}, \tag{12.44}$$

where

$$\mathcal{F} \equiv \overline{h}(F) \quad \text{and} \quad \mathcal{J} \equiv \det(\underline{h})\, \underline{h}^{-1}(J). \tag{12.45}$$

(The reasons behind these definitions will be explained in Lecture III of this series. The use of a calligraphic letter for certain covariant fields is a convention we have found very useful.) The definitions of $\mathcal{F}$ and $\mathcal{J}$ ensure that under local rotations they transform as

$$\mathcal{F} \mapsto R\mathcal{F}\tilde{R} \quad \text{and} \quad \mathcal{J} \mapsto R\mathcal{J}\tilde{R}. \tag{12.46}$$

Any (multi)vector that transforms in this manner under rotations and is covariant under displacements is referred to as a *covariant* (multi)vector.

Equation (12.44) is covariant under arbitrary displacements, and we now need to make it covariant under local rotations as well. To achieve this we replace $\overline{h}(\nabla)$ by $\overline{h}(\partial_a)a\cdot\nabla$ and focus attention on the term $a\cdot\nabla\mathcal{F}$. Under a position-dependent rotation we find that

$$a\cdot\nabla(R\mathcal{F}\tilde{R}) = Ra\cdot\nabla\mathcal{F}\tilde{R} + a\cdot\nabla R\,\mathcal{F}\tilde{R} + R\mathcal{F}a\cdot\nabla\tilde{R}. \tag{12.47}$$

As in electromagnetism, since $R\tilde{R} = 1$ for a rotation we have

$$a\cdot\nabla R\tilde{R} = -Ra\cdot\nabla\tilde{R} \tag{12.48}$$

hence $a\cdot\nabla R\tilde{R}$ is equal to minus its reverse and so must be a bivector. We can therefore write

$$a\cdot\nabla(R\mathcal{F}\tilde{R}) = Ra\cdot\nabla\mathcal{F}\tilde{R} + 2(a\cdot\nabla R\tilde{R})\times(R\mathcal{F}\tilde{R}). \tag{12.49}$$

To construct a covariant derivative we must therefore add a 'connection' term to $a\cdot\nabla$ to construct the operator

$$\mathcal{D}_a = a\cdot\nabla + \Omega(a)\times . \tag{12.50}$$

Here $\Omega(a) = \Omega(a,x)$ is a bivector-valued linear function of a with an arbitrary x-dependence. The operation of commuting a multivector with a bivector is grade-preserving, so, even though it is not a scalar operator, $\mathcal{D}_a$ preserves the grade of the multivector on which it acts.

Under local rotations we demand that $\Omega(a)$ transforms as

$$\Omega(a) \mapsto \Omega'(a) = R\Omega(a)\tilde{R} - 2a\cdot\nabla R\tilde{R}. \tag{12.51}$$

Since $\Omega(a)$ is now an arbitrary function of position, however, it cannot in general be transformed away by the application of a rotor. The equations (12.50) and (12.51) ensure that under under a local rotation

$$\mathcal{D}_a{}'(R\mathcal{F}\tilde{R}) = R\mathcal{D}_a\mathcal{F}\tilde{R}. \tag{12.52}$$

We finally reassemble with the $\overline{h}(\partial_a)$ term to form the final equation

$$\overline{h}(\partial_a)\mathcal{D}_a\mathcal{F} = \mathcal{J}, \tag{12.53}$$

and the transformation properties of the $\overline{h}(a)$, $\mathcal{F}$, $\mathcal{J}$ and $\Omega(a)$ fields ensure that this equation is now covariant under rotations as well as displacements.

For completeness, we note that under local displacements $\Omega(a)$ must transform in the same way as $a\cdot\nabla R\tilde{R}$, hence

$$\Omega_x(a) \mapsto \Omega_{x'}\underline{f}(a) = \Omega(\underline{f}(a), f(x)), \tag{12.54}$$

where the subscript is again used to label position dependence. It follows that

$$\begin{aligned}\overline{h}(\partial_a)\Omega_x(a)\times\mathcal{F}(x) &\mapsto \overline{h}_{x'}\overline{f}^{-1}(\partial_a)\Omega_{x'}\underline{f}(a)\times\mathcal{F}(x') \\ &= \overline{h}_{x'}(\partial_a)\Omega_{x'}(a)\times\mathcal{F}(x')\end{aligned} \tag{12.55}$$

as required for covariance under local displacements.

General considerations have led us to the introduction of two new gauge fields, the $\overline{h}(a,x)$ linear function and the $\Omega(a,x)$ bivector-valued linear function, both of which are arbitrary functions of position vector x. This gives a total of $4\times4+4\times6=40$ scalar degrees of freedom. The $\overline{h}(a)$ and $\Omega(a)$ fields are incorporated into the vector derivative to form the operator $\mathcal{D}=\overline{h}(\partial_a)\mathcal{D}_a$, which acts covariantly on multivector fields. Thus we can begin to construct equations whose intrinsic content is free from the manner in which we choose to represent spacetime positions with vectors. We next see how these fields arise in the setting of the Dirac theory. This enables us to derive the properties of the $\mathcal{D}$ operator from more primitive considerations of the properties of spinors and the means by which observables are constructed from them.

Spinor Fields and the Dirac Equation

Let us now return to the STA form of the free-particle Dirac equation (12.1),

$$\nabla\psi i\sigma_3 = m\psi\gamma_0. \tag{12.56}$$

Suppose now that ψ is replaced by the locally displaced field $\psi'(x)$, defined by

$$\psi'(x) = \psi(x'), \tag{12.57}$$

where $x'=f(x)$. Following the previous discussion, to make equation (12.56) covariant under displacements we must introduce the $\overline{h}$-field and replace $\nabla\psi$ by $\overline{h}(\nabla)\psi$.

Under spacetime rotations, the spinor ψ transforms as

$$\psi \mapsto R\psi, \tag{12.58}$$

which is the spacetime analogue of the rotations discussed in Section 12.1. Since the $\overline{h}(a)$ function transforms to $R\overline{h}(a)\tilde{R}$, the modified equation

$$\overline{h}(\nabla)\psi i\sigma_3 = m\psi\gamma_0 \tag{12.59}$$

is clearly covariant under constant rotations. To turn this into a local symmetry we must add in a connection term in the same manner as discussed in Section 12.1. The only difference is that now the connection term must multiply ψ from the left, so the directional derivatives $a\cdot\nabla\psi$ must be replaced by

$$D_a\psi = a\cdot\nabla\psi + \tfrac{1}{2}\Omega(a)\psi. \tag{12.60}$$

Following the same route as Section 12.1 shows that $\Omega(a)$ must have precisely the transformation property as already found in (12.51) above. We now discuss this link between the derivatives of classical and quantum fields in more detail.

12.4 Observables and Covariant Derivatives

As well as keeping everything within the real STA, representing Dirac spinors with elements of the even subalgebra offers many advantages when forming observables. Observables are formed by the double-sided application of a Dirac spinor ψ to some combination of the fixed $\{\gamma^\mu\}$ frame vectors. So, for example, the charge current is given by $\mathcal{J} = \psi\gamma_0\tilde{\psi}$ and the spin current by $s = \psi\gamma_3\tilde{\psi}$. In general, an observable is of the form

$$M \equiv \psi\Gamma\tilde{\psi}, \tag{12.61}$$

where Γ is a constant multivector formed from the $\{\gamma^\mu\}$. All observables considered are invariant under phase rotations, so Γ must be invariant under rotations in the $i\sigma_3$ plane. Hence Γ can only consist of combinations of γ_0, γ_3, $i\sigma_3$ and their duals (*i.e.* multiplied by i). There is a strong analogy with rigid-body mechanics here, with the $\{\gamma^\mu\}$ representing a set of fixed reference axes and ψ operating on these to form the spatial observables. An important point is that, in forming the observable M, the Γ multivector is completely shielded from any covariant multivector. Thus we can replace Γ by $R\Gamma\tilde{R}$ and ψ by $\psi\tilde{R}$, and the observable M will be unchanged. This is why the appearance of γ_0 and $i\sigma_3$ on the right-hand side of the spinor ψ in the Dirac equation does not compromise Lorentz invariance, and does not pick out a preferred direction in space. Of course, when we come to solve the Dirac equation, we invariably line up the $\{\gamma^\mu\}$ frame with a set of vectors in space, thus leading to a particular form of ψ. But it is only the observables that have any real significance, and these are independent of how we chose to align the 'internal' frame.

Under displacements and rotations the observables formed in the above manner (12.61) inherit the transformation properties of the spinor ψ. So under displacements the observable $M = \psi\Gamma\tilde{\psi}$ transforms from $M(x)$ to $M(x')$, and under rotations M transforms to $R\psi\Gamma\tilde{\psi}\tilde{R} = RM\tilde{R}$. The observable M is therefore covariant, and has the same transformation properties as, for example, $\bar{h}(\nabla)\phi$. We can therefore start to build intrinsic relations between Dirac observables and other covariant quantities. One might wonder why it is that the observables are *in*variant under phase rotations, but only *co*variant under spatial rotations. In fact, the $\bar{h}$ field enables us to form quantities like $\underline{h}(A)$, which is now invariant under spatial rotations. This gives an alternative insight into the role of the $\bar{h}$-field. We will find that both covariant observables (M) and their rotationally-invariant forms ($\underline{h}(M)$ and $\overline{h}^{-1}(M)$) have important roles to play in the theory constructed here.

If we next consider directional derivatives of M, we find that these can be written as

$$a\cdot\nabla M = (a\cdot\nabla\psi)\Gamma\tilde{\psi} + \psi\Gamma(a\cdot\nabla\psi)\tilde{}. \tag{12.62}$$

This immediately tells us how to turn the directional derivative $a\nabla M$ into a covariant derivative: simply replace the spinor directional derivatives by covariant derivatives. We thus form

$$\begin{aligned}(D_a\psi)\Gamma\tilde{\psi} &+ \psi\Gamma(D_a\psi)\tilde{} \\ &= (a\cdot\nabla\psi)\Gamma\tilde{\psi} + \psi\Gamma(a\cdot\nabla\psi)\tilde{} + \tfrac{1}{2}\Omega(a)\psi\Gamma\tilde{\psi} - \tfrac{1}{2}\psi\Gamma\tilde{\psi}\Omega(a) \\ &= a\cdot\nabla(\psi\Gamma\tilde{\psi}) + \Omega(a)\times(\psi\Gamma\tilde{\psi}).\end{aligned} \tag{12.63}$$

We therefore define the covariant derivative for observables by

$$\mathcal{D}_a M \equiv a\cdot\nabla M + \Omega(a)\times M. \tag{12.64}$$

We have thus recovered the covariant derivative found in Section 12.2, though now it is understood as arising from the more basic transformation properties of the spinor ψ. The operator (12.64) acts on any covariant multivector and has the important property of satisfying Leibniz' rule,

$$\mathcal{D}_a(AB) = (\mathcal{D}_a A)B + A(\mathcal{D}_a B). \tag{12.65}$$

Hence $\mathcal{D}_a$ is a *derivation*, as follows from the identity

$$\Omega(a)\times(AB) = (\Omega(a)\times A)B + A(\Omega(a)\times B). \tag{12.66}$$

We have now introduced a number of different derivatives,indexderivatives!summary of in the STA so we finish this section by summarising these and introducing some useful notation. For spinors we have the covariant directional derivative

$$D_a\psi \equiv a\cdot\nabla\psi + \tfrac{1}{2}\Omega(a)\psi, \tag{12.67}$$

from which we assemble the spinor covariant form of the vector derivative

$$D\psi \equiv \overline{h}(\partial_a)D_a\psi = \overline{h}(\nabla)\psi + \tfrac{1}{2}\overline{h}(\partial_a)\Omega(a)\psi. \tag{12.68}$$

For covariant multivectors, the covariant directional derivative is given by

$$\mathcal{D}_a A \equiv a\cdot\nabla A + \Omega(a)\times A \tag{12.69}$$

from which we assemble the covariant vector derivative

$$\mathcal{D}A \equiv \overline{h}(\partial_a)\mathcal{D}_a A. \tag{12.70}$$

The covariant vector derivative contains a grade-raising and a grade-lowering component, so we write

$$\mathcal{D}A = \mathcal{D}\cdot A + \mathcal{D}\wedge A, \tag{12.71}$$

where

$$\begin{aligned} \mathcal{D}\cdot A &\equiv \overline{h}(\partial_a)\cdot(\mathcal{D}_a A) & (12.72)\\ \mathcal{D}\wedge A &\equiv \overline{h}(\partial_a)\wedge(\mathcal{D}_a A). & (12.73)\end{aligned}$$

As with the vector derivative, $\mathcal{D}$ inherits the algebraic properties of a vector.

It is frequently useful to include the $\overline{h}$-field in the covariant directional derivatives. This maintains covariance under local displacements as well as rotations. To this end, we define

$$\omega(a) = \Omega\underline{h}(a), \tag{12.74}$$

and write

$$a\cdot\mathcal{D}A \equiv a\cdot\overline{h}(\partial_b)\mathcal{D}_b A = a\cdot\overline{h}(\nabla)A + \omega(a)\times A, \tag{12.75}$$

so that $\underline{h}^{-1}(a)\cdot\mathcal{D} = \mathcal{D}_a$. With this operator we can also write

$$\mathcal{D}A = \partial_a a\cdot\mathcal{D}A, \tag{12.76}$$

which separates the algebraic contraction from the covariant derivative.

Using the above notation, and all that we learnt above about making the derivatives in the free-particle Dirac equation fully covariant, our final reassembled form of the Dirac equation is

$$D\psi i\sigma_3 - e\mathcal{A}\psi = m\psi\gamma_0, \tag{12.77}$$

where $\mathcal{A} = \overline{h}(A)$. We shall call this the 'minimally-coupled' Dirac equation (minimally coupled to both electromagnetism and gravity). In the next lecture we shall rederive the Dirac equation from an action principle, and requiring that it have the form found here will be shown to lead to a surprising restriction on the form of the gravitational action.

Chapter 13

Gravity II — Field Equations

Chris Doran, Anthony Lasenby, and Stephen Gull
MRAO, Cavendish Laboratory, Madingley Road,
Cambridge, CB3 0HE, UK

In Lecture I we introduced the $\underline{h}$ and $\Omega(a)$ gauge fields. Their transformation properties are summarised in Table 13.1. From these, we need to construct 'covariant' quantities which transform in the same way that physical fields do. We start by defining the field strength $R(a\wedge b)$ by

$$\tfrac{1}{2}R(a\wedge b)\psi \equiv [D_a, D_b]\psi, \tag{13.1}$$

$$\Rightarrow R(a\wedge b) = a\cdot\nabla\Omega(b) - b\cdot\nabla\Omega(a) + \Omega(a)\times\Omega(b). \tag{13.2}$$

$R(a\wedge b)$ is a bivector-valued linear function of its bivector argument $a\wedge b$. This extends by linearity to the function $R(B)$, where B is an arbitrary bivector. Where necessary, the position dependence of $R(B)$ is made explicit by writing $R(B,x)$ or $R_x(B)$.

Under local rotations $R(B)$ transforms as

$$R(B) \mapsto R'(B) = R\,R(B)\,\tilde{R}. \tag{13.3}$$

Under local displacements we find that

$$\begin{aligned} R'(a\wedge b) &= a\cdot\nabla\Omega_{x'}\underline{f}(b) - b\cdot\nabla\Omega_{x'}\underline{f}(a) + \Omega_{x'}\underline{f}(a)\times\Omega_{x'}\underline{f}(b) \\ &= R_{x'}\underline{f}(a\wedge b). \end{aligned} \tag{13.4}$$

Local Symmetry	Transformed Fields	
	$\overline{h}'(a,x)$	$\Omega'(a,x)$
Displacements	$\overline{h}_{x'}\overline{f}^{-1}(a)$	$\Omega_{x'}\underline{f}(a)$
Rotations	$R\overline{h}(a)\tilde{R}$	$R\Omega(a)\tilde{R} - 2a\cdot\nabla R\tilde{R}$

Table 13.1. Transformation properties of the gravitational gauge fields.

This result rests on the fact that

$$\dot{\underline{f}}[B\cdot\dot{\nabla}] = 0. \tag{13.5}$$

A covariant quantity is therefore constructed by defining

$$\mathcal{R}(B) \equiv R\underline{h}(B). \tag{13.6}$$

Under local displacements and rotations $\mathcal{R}(B)$ has the following transformation laws:

$$\begin{aligned} \text{Displacements:} \quad \mathcal{R}'(B,x) &= \mathcal{R}(B,x') \\ \text{Rotations:} \quad \mathcal{R}'(B,x) &= R\mathcal{R}(\tilde{R}BR,x)\,\tilde{R}. \end{aligned} \tag{13.7}$$

We refer to any linear function with transformation laws of this type as a covariant tensor. ($\mathcal{R}(B)$ is the gauge theory equivalent of the Riemann tensor.) As usual, covariant quantities such as $\mathcal{R}(B)$ and $\mathcal{D}$ are written with calligraphic ('curly') symbols.

From $\mathcal{R}(B)$ we define the following contractions:

$$\begin{aligned} \text{Ricci Tensor:} \quad \mathcal{R}(b) &= \partial_a\cdot\mathcal{R}(a\wedge b) \\ \text{Ricci Scalar:} \quad \mathcal{R} &= \partial_a\cdot\mathcal{R}(a) \\ \text{Einstein Tensor:} \quad \mathcal{G}(a) &= \mathcal{R}(a) - \tfrac{1}{2}a\mathcal{R}. \end{aligned} \tag{13.8}$$

The argument of $\mathcal{R}$ denotes whether it represents the Riemann or Ricci tensors, or the Ricci scalar. Both $\mathcal{R}(a)$ and $\mathcal{G}(a)$ are also covariant tensors, since they inherit the transformation properties of $\mathcal{R}(B)$. All of these objects represent quantities whose position dependence can be changed by arbitrary displacements.

13.1 The Gravitational Field Equations

The Ricci scalar is invariant under rotations, so is our first candidate for a Lagrangian for the gravitational gauge fields. We therefore assume that the overall action integral is of the form

$$S = \int |d^4x| \, \det(\underline{h})^{-1}(\tfrac{1}{2}\mathcal{R} - \kappa\mathcal{L}_m), \tag{13.9}$$

where $\mathcal{L}_m$ describes the matter content and $\kappa = 8\pi G$. The independent dynamical variables are $\overline{h}(a)$ and $\Omega(a)$, and in terms of these

$$\mathcal{R} = \langle \overline{h}(\partial_b\wedge\partial_a)[a\cdot\nabla\Omega(b) - b\cdot\nabla\Omega(a) + \Omega(a)\times\Omega(b)]\rangle. \tag{13.10}$$

We also assume that $\mathcal{L}_m$ contains no second-order derivatives, so that $\underline{h}$ and $\Omega(a)$ appear undifferentiated in the matter Lagrangian.

The $\overline{h}$-Equation The $\overline{h}$-field is undifferentiated in the entire action, so its Euler-Lagrange equation is simply

$$\partial_{\overline{h}(a)}[\det(\underline{h})^{-1}(\mathcal{R}/2 - \kappa\mathcal{L}_m)] = 0. \tag{13.11}$$

Now,

$$\partial_{\overline{h}(a)} \det(\underline{h})^{-1} = -\det(\underline{h})^{-1}\underline{h}^{-1}(a) \tag{13.12}$$

and

$$\partial_{\overline{h}(a)}\mathcal{R} = 2\overline{h}(\partial_b)\cdot R(a\wedge b) = 2\mathcal{R}\underline{h}^{-1}(a). \tag{13.13}$$

So, defining the covariant matter stress-energy tensor $\mathcal{T}(a)$ by

$$\det(\underline{h})\partial_{\overline{h}(a)}(\mathcal{L}_m \det(\underline{h})^{-1}) = \mathcal{T}\underline{h}^{-1}(a), \tag{13.14}$$

we arrive at the equation

$$\mathcal{G}(a) = \kappa\mathcal{T}(a). \tag{13.15}$$

This is the gauge theory statement of Einstein's equation. As yet nothing is implied about the symmetries of $\mathcal{G}(a)$ or $\mathcal{T}(a)$, and the properties of spacetime are untouched in this derivation.

The $\Omega(a)$-Equation The Euler-Lagrange field equation from $\Omega(a)$ is, after multiplying through by $\det(\underline{h})$,

$$\partial_{\Omega(a)}\mathcal{R} - \det(\underline{h})\partial_b\cdot\nabla[\partial_{\Omega(a),b}\mathcal{R}\det(\underline{h})^{-1}] = 2\kappa\partial_{\Omega(a)}\mathcal{L}_m, \tag{13.16}$$

where $\partial_{\Omega(a)}$ and $\partial_{\Omega(a),b}$ are defined by obvious extension of the work covered in the earlier lecture on linear algebra. The only properties needed here are the following:

$$\begin{aligned}\partial_{\Omega(a)}\langle\Omega(b)M\rangle &= a\cdot b\langle M\rangle_2 &(13.17)\\ \partial_{\Omega(b),a}\langle c\cdot\nabla\Omega(d)M\rangle &= a\cdot c\, b\cdot d\langle M\rangle_2. &(13.18)\end{aligned}$$

From these we derive

$$\begin{aligned}\partial_{\Omega(a)}\langle\overline{h}(\partial_d\wedge\partial_c)\Omega(c)\times\Omega(d)\rangle &= \Omega(d)\times\overline{h}(\partial_d\wedge a) + \overline{h}(a\wedge\partial_c)\times\Omega(c)\\ &= 2\Omega(b)\times\overline{h}(\partial_b\wedge a)\\ &= 2[\omega(b)\cdot\partial_b)]\wedge\overline{h}(a) + 2\partial_b\wedge[\omega(b)\cdot\overline{h}(a)] \quad (13.19)\end{aligned}$$

and

$$\partial_{\Omega(a),b}\langle\overline{h}(\partial_d\wedge\partial_c)[c\cdot\nabla\Omega(d) - d\cdot\nabla\Omega(c)]\rangle = \overline{h}(a\wedge b) - \overline{h}(b\wedge a) = 2\overline{h}(a\wedge b). \tag{13.20}$$

The right-hand side of (13.16) defines the 'spin' of the matter,indexspin!of matter in gravitational theory

$$S(a) \equiv \partial_{\Omega(a)}\mathcal{L}_m, \tag{13.21}$$

where $S(a)$ is a bivector-valued linear function of a. Combining (13.16), (13.19) and (13.20) yields

$$\mathcal{D}\wedge\overline{h}(a) + \det(\underline{h})\mathcal{D}_b[\overline{h}(\partial_b)\det(\underline{h})^{-1}]\wedge\overline{h}(a) = \kappa S(a). \tag{13.22}$$

To make further progress we contract this equation with $\underline{h}^{-1}(\partial_a)$. To achieve this requires the results that

$$\begin{aligned}\underline{h}^{-1}(\partial_a)\cdot(\mathcal{D}\wedge\overline{h}(a)) &= \mathcal{D}_a\overline{h}(\partial_a) - \overline{h}(\partial_b)\underline{h}^{-1}(\partial_a)\cdot(\mathcal{D}_b\cdot\overline{h}(a))\\ &= \mathcal{D}_a\overline{h}(\partial_a) - \overline{h}(\partial_b)\langle\underline{h}^{-1}(\partial_a)b\cdot\nabla\overline{h}(a)\rangle \qquad (13.23)\end{aligned}$$

and

$$\begin{aligned}\langle b\cdot\nabla\overline{h}(a)\underline{h}^{-1}(\partial_a)\rangle &= \det(\underline{h})^{-1}\langle b\cdot\nabla\overline{h}(\partial_a)\,\partial_{\overline{h}(a)}\det(\underline{h})\rangle \\ &= \det(\underline{h})^{-1}b\cdot\nabla\det(\underline{h}).\end{aligned} \tag{13.24}$$

It follows that

$$\det(\underline{h})\mathcal{D}_b[\overline{h}(\partial_b)\det(\underline{h})^{-1}] = -\tfrac{1}{2}\kappa\partial_a\cdot\mathcal{S}(a) \tag{13.25}$$

where $\mathcal{S}(a) \equiv S\overline{h}^{-1}(a)$ is the covariant spin tensor.

Now consider the covariantly-coupled Dirac action

$$S = \int |d^4x| \det(\underline{h})^{-1}\langle\overline{h}(\partial_a)(a\cdot\nabla + \tfrac{1}{2}\Omega(a))\psi i\gamma_3\tilde{\psi} - \overline{h}(A)\psi\gamma_0\tilde{\psi} - m\psi\tilde{\psi}\rangle. \tag{13.26}$$

The spin bivector obtained from this is

$$\mathcal{S}(a) = \mathcal{S}\cdot a, \tag{13.27}$$

where $\mathcal{S}$ is the spin trivector, $\mathcal{S} = \psi i\gamma_3\tilde{\psi}/2$. It follows that, for spin generated by the Dirac action,

$$\partial_a\cdot\mathcal{S}(a) = (\partial_a\wedge a)\cdot\mathcal{S} = 0 \tag{13.28}$$

and hence

$$\mathcal{D}_a[\overline{h}(\partial_a)\det(\underline{h})^{-1}] = 0. \tag{13.29}$$

There is a remarkable consistency loop at work here, because the equation obtained from varying (13.26) with respect to ψ is

$$D\psi i\sigma_3 - e\mathcal{A}\psi = m\psi\gamma_0 - \tfrac{1}{2}\det(\underline{h})\mathcal{D}_a[\overline{h}(\partial_a)\det(\underline{h})^{-1}]\psi i\gamma_3. \tag{13.30}$$

This only reduces to the minimally-coupled equation if equation (13.29) is satisfied. Hence the Dirac action is of just the right type to ensure that the minimally-coupled action produces the minimally coupled equation. *But this is only true if the gravitational action is given by the Ricci scalar*! No higher-order gravitational action is consistent in this way. The only freedom in the action for the gravitational fields is the possible inclusion of a cosmological constant. Hence we have successfully constrained both the gravitational action and the torsion type.

Given that the spin is entirely of Dirac type, equation (13.22) now takes the form

$$\mathcal{D}\wedge\overline{h}(a) = \kappa\mathcal{S}\cdot\overline{h}(a), \tag{13.31}$$

valid for all constant vectors a. If a is a function of position then we write

$$\mathcal{D}\wedge\overline{h}(a) = \overline{h}(\nabla\wedge a) + \kappa\mathcal{S}\cdot\overline{h}(a). \tag{13.32}$$

13.2 Covariant Forms of the Field Equations

Henceforth we simplify matters by setting the spin to zero. It is not hard to make the necessary generalisations in the presence of spin. Since $\bar{h}(\nabla)$ and $\omega(a)$ are both covariant under local displacements, these are the quantities we must focus on. To this end we define the operator

$$L_a \equiv a\cdot\bar{h}(\nabla), \tag{13.33}$$

and, for the remainder of this section, the vectors a, b *etc.* are assumed to be arbitrary functions of position. From equation (13.32) (with $\mathcal{S}=0$) we write

$$\begin{aligned} \bar{h}(\dot{\nabla})\wedge\dot{\bar{h}}(c) &= -\partial_d\wedge[\omega(d)\cdot\bar{h}(c)] \\ \Rightarrow \langle b\wedge a\,\bar{h}(\dot{\nabla})\wedge\dot{\bar{h}}(c)\rangle &= -\langle b\wedge a\,\partial_d\wedge[\omega(d)\cdot\bar{h}(c)]\rangle \\ \Rightarrow [\dot{L}_a\dot{\underline{h}}(b) - \dot{L}_b\dot{\underline{h}}(a)]\cdot c &= [a\cdot\omega(b) - b\cdot\omega(a)]\cdot\bar{h}(c) \end{aligned} \tag{13.34}$$

where the overdots determine the scope of a differential operator. It follows that the commutator of L_a and L_b is

$$\begin{aligned} [L_a, L_b] &= [L_a\underline{h}(b) - L_b\underline{h}(a)]\cdot\nabla \\ &= [\dot{L}_a\dot{\underline{h}}(b) - \dot{L}_b\dot{\underline{h}}(a)]\cdot\nabla + (L_a b - L_b a)\cdot\bar{h}(\nabla) \\ &= [a\cdot\omega(b) - b\cdot\omega(a) + L_a b - L_b a]\cdot\bar{h}(\nabla). \end{aligned} \tag{13.35}$$

We can therefore write

$$[L_a, L_b] = L_c \tag{13.36}$$

where

$$c = a\cdot\omega(b) - b\cdot\omega(a) + L_a b - L_b a = a\cdot\mathcal{D}b - b\cdot\mathcal{D}a. \tag{13.37}$$

This 'bracket' structure summarises the intrinsic content of (13.31).

The general technique we use for studying the field equations is to let $\omega(a)$ contain a set of arbitrary functions, and then use (13.37) to find relations between them. Central to this approach is the construction of the Riemann tensor $\mathcal{R}(B)$, which contains a great deal of covariant information. From the definition of the Riemann tensor (13.2) we find that

$$\begin{aligned} \mathcal{R}(a\wedge b) &= \dot{L}_a\dot{\Omega}\underline{h}(a) - \dot{L}_b\dot{\Omega}\underline{h}(a) + \omega(a)\times\omega(b) \\ &= L_a\omega(b) - L_b\omega(a) + \omega(a)\times\omega(b) - \Omega(L_a\underline{h}(b) - L_b\underline{h}(a)), \end{aligned} \tag{13.38}$$

hence

$$\mathcal{R}(a\wedge b) = L_a\omega(b) - L_b\omega(a) + \omega(a)\times\omega(b) - \omega(c), \tag{13.39}$$

where c is given by equation (13.37). Equation (13.39) now enables $\mathcal{R}(B)$ to be calculated in terms of position-gauge covariant variables.

Solution of the 'Wedge' Equation Equation (13.31) (with $\mathcal{S}$=0) can be solved to obtain $\Omega(a)$ as a function of $\bar{h}$ and its derivatives. We define

$$H(a) \equiv \bar{h}(\nabla\wedge\bar{h}^{-1}(a)) = -\bar{h}(\dot{\nabla})\wedge\dot{\bar{h}}\bar{h}^{-1}(a), \tag{13.40}$$

so that equation (13.31) becomes

$$\partial_b\wedge[\omega(b)\cdot a] = H(a) \tag{13.41}$$

This is protracted to give

$$\partial_a\wedge\partial_b\wedge[\omega(b)\cdot a] = 2\partial_b\wedge\omega(b) = \partial_b\wedge H(b) \tag{13.42}$$

and then dotted with a, leaving

$$\omega(a) - \partial_b\wedge(a\cdot\omega(b)) = \tfrac{1}{2}a\cdot[\partial_b\wedge H(b)]. \tag{13.43}$$

Hence, using equation (13.41) again, we find that

$$\omega(a) = -H(a) + \tfrac{1}{2}a\cdot(\partial_b\wedge H(b)). \tag{13.44}$$

In the presence of spin the term $\frac{1}{2}\kappa a\cdot\mathcal{S}$ is added to the right-hand side.

13.3 Symmetries and Invariants of $\mathcal{R}(B)$

In letting $\omega(a)$ be an arbitrary function we lose some of the information contained in the 'wedge' equation (13.31). This information is recovered by employing the symmetry properties of $\mathcal{R}(B)$ and the Bianchi identities. The rotation gauge is also fixed at the level of $\mathcal{R}(B)$, so it is important to understand its general structure. We start with the result that, for an arbitrary multivector $A(x)$,

$$\mathcal{D}\wedge\mathcal{D}\wedge A = \bar{h}(\nabla\wedge\nabla\wedge\bar{h}^{-1}(A)) = 0. \tag{13.45}$$

It then follows from the fact that

$$\begin{aligned}\bar{h}(\partial_a)\wedge\mathcal{D}_a[\bar{h}(\partial_b)\wedge\mathcal{D}_bA] &= \bar{h}(\partial_a)\wedge\bar{h}(\partial_b)\wedge[\mathcal{D}_a\mathcal{D}_bA] \\ &= \tfrac{1}{2}\bar{h}(\partial_a)\wedge\bar{h}(\partial_b)\wedge[R(a\wedge b)\times A],\end{aligned} \tag{13.46}$$

that

$$\partial_a\wedge\partial_b\wedge(\mathcal{R}(a\wedge b)\times A) = 0, \tag{13.47}$$

for any multivector A.

Now, setting A to the vector c and protracting with ∂_c, we find that

$$\partial_c\wedge\partial_a\wedge\partial_b\wedge(\mathcal{R}(a\wedge b)\times c) = -2\partial_a\wedge\partial_b\wedge\mathcal{R}(a\wedge b) = 0. \tag{13.48}$$

Taking the inner product of the term on the right-hand side with c we obtain

$$c\cdot[\partial_a\wedge\partial_b\wedge\mathcal{R}(a\wedge b)] = \partial_b\wedge\mathcal{R}(c\wedge b) - \partial_a\wedge\mathcal{R}(a\wedge c) - \partial_a\wedge\partial_b\wedge[\mathcal{R}(a\wedge b)\times c], \tag{13.49}$$

in which both the left-hand side and the final term on the right-hand side vanish. We are therefore left with the simple expression

$$\partial_a\wedge\mathcal{R}(a\wedge b) = 0, \tag{13.50}$$

which summarises all the symmetries of $\mathcal{R}(B)$. This equation says that the trivector $\partial_a \wedge \mathcal{R}(a \wedge b)$ vanishes for all values of the vector b, so gives a set of $4 \times 4 = 16$ equations. These reduce the number of independent degrees of freedom in $\mathcal{R}(B)$ from 36 to the familiar 20.

The Weyl Tensor In our gauge theory six of the degrees of freedom in $\mathcal{R}(B)$ can be removed by arbitrary rotations. It follows that $\mathcal{R}(B)$ can only contain 14 physical degrees of freedom. To see how these are encoded in $\mathcal{R}(B)$ we decompose it into Weyl and 'matter' terms. By forming the contraction

$$\partial_a\cdot[\mathcal{R}(a)\wedge b + a\wedge\mathcal{R}(b)] = 2\mathcal{R}(b) + b\mathcal{R}, \tag{13.51}$$

we see that

$$\partial_a\cdot[\tfrac{1}{2}\left(\mathcal{R}(a)\wedge b + a\wedge\mathcal{R}(b)\right) - \tfrac{1}{6}a\wedge b\mathcal{R}] = \mathcal{R}(b). \tag{13.52}$$

We can therefore write

$$\mathcal{R}(a\wedge b) = \mathcal{W}(a\wedge b) + \tfrac{1}{2}[\mathcal{R}(a)\wedge b + a\wedge\mathcal{R}(b)] - \tfrac{1}{6}a\wedge b\mathcal{R} \tag{13.53}$$

where $\mathcal{W}(B)$ is the Weyl tensor, and must satisfy

$$\partial_a\cdot\mathcal{W}(a\wedge b) = 0. \tag{13.54}$$

Returning to equation (13.50) and contracting, we obtain

$$\partial_b\cdot[\partial_a\wedge\mathcal{R}(a\wedge b)] = \partial_a\wedge\mathcal{R}(a) = 0 \tag{13.55}$$

which shows that the Ricci tensor $\mathcal{R}(a)$ is symmetric. It follows that

$$\partial_a\wedge[\tfrac{1}{2}\left(\mathcal{R}(a)\wedge b + a\wedge\mathcal{R}(b)\right) - \tfrac{1}{6}a\wedge b\mathcal{R}] = 0 \tag{13.56}$$

and hence that

$$\partial_a\wedge\mathcal{W}(a) = 0. \tag{13.57}$$

Equations (13.54) and (13.57) combine to give the single equation

$$\partial_a\mathcal{W}(a\wedge b) = 0. \tag{13.58}$$

The Weyl tensor is therefore 'tractionless' in the language of Hestenes & Sobczyk. Equation (13.53) decomposes $\mathcal{R}(B)$ into the tractionless term $\mathcal{W}(B)$, and a term specified solely by the matter stress-energy tensor (which determines $\mathcal{R}(a)$ through the Einstein tensor $\mathcal{G}(a)$). There is no generally accepted name for the part of $\mathcal{R}(B)$ that is not given by the Weyl tensor so, as it is entirely determined by the matter stress-energy tensor, we refer to it as the matter term.

Duality To study the consequences of equation (13.58) it is useful to employ the fixed $\{\gamma_\mu\}$ frame, so that equation (13.58) produces the four equations

$$\begin{aligned}
\sigma_1\mathcal{W}(\sigma_1) + \sigma_2\mathcal{W}(\sigma_2) + \sigma_3\mathcal{W}(\sigma_3) &= 0 &\quad (13.59)\\
\sigma_1\mathcal{W}(\sigma_1) - i\sigma_2\mathcal{W}(i\sigma_2) - i\sigma_3\mathcal{W}(i\sigma_3) &= 0 &\quad (13.60)\\
-i\sigma_1\mathcal{W}(i\sigma_1) + \sigma_2\mathcal{W}(\sigma_2) - i\sigma_3\mathcal{W}(i\sigma_3) &= 0 &\quad (13.61)\\
-i\sigma_1\mathcal{W}(i\sigma_1) - i\sigma_2\mathcal{W}(i\sigma_2) + \sigma_3\mathcal{W}(\sigma_3) &= 0. &\quad (13.62)
\end{aligned}$$

Summing the final three equations, and using the first, produces

$$i\sigma_k \mathcal{W}(i\sigma_k) = 0 \tag{13.63}$$

and substituting this into each of the final three equations produces

$$\mathcal{W}(i\sigma_k) = i\mathcal{W}(\sigma_k). \tag{13.64}$$

It follows that the Weyl tensor satisfies

$$\mathcal{W}(iB) = i\mathcal{W}(B) \tag{13.65}$$

and so is 'self-dual'. Equation (13.65) means that $\mathcal{W}(B)$ can be analysed as a function on a three-dimensional complex space, rather than as a linear function on a real six-dimensional space. (This is why complex formalisms, such as the Newman-Penrose formalism, are so successful in analysing the Weyl tensor.) Given this self-duality, the remaining content of equations (13.59)–(13.62) is summarised by the relation

$$\sigma_k \mathcal{W}(\sigma_k) = 0. \tag{13.66}$$

This equation says that, viewed as a three-dimensional complex linear function, $\mathcal{W}(B)$ is symmetric and traceless. This gives $\mathcal{W}(B)$ five complex, or ten real degrees of freedom. (Combinations of the form scalar + pseudoscalar are referred to loosely as complex scalars.) The gauge-invariant ('intrinsic') information in $\mathcal{W}(B)$ is contained in its complex eigenvalues and, since the sum of these is zero, only two are independent. This leaves a set of four intrinsic scalar quantities.

Overall $\mathcal{R}(B)$ has 20 degrees of freedom, 6 of which are then soaked up in the freedom to perform arbitrary local rotations. Of the remaining 14 physical degrees of freedom, four are contained in the two complex eigenvalues of $\mathcal{W}(B)$, and a further four in the real eigenvalues of the matter stress-energy tensor. The six remaining physical degrees of freedom determine the rotation between the frame that diagonalises $\mathcal{G}(a)$ and the frame that diagonalises $\mathcal{W}(B)$. This identification of the physical degrees of freedom contained in $\mathcal{R}(B)$ is novel and potentially very significant.

The Petrov Classification The algebraic properties of the Weyl tensor are traditionally encoded in its Petrov type. This classification is based on the solutions of the 'eigenbivector' equation

$$\mathcal{W}(B) = \alpha B, \tag{13.67}$$

in which B is the eigenbivector and α is a complex scalar. There are five Petrov types: *I*, *II*, *III*, *D* and *N*. Type *I* are the most general, with two independent eigenvalues and three independent orthogonal eigenbivectors. Such tensors have the general form

$$\mathcal{W}(B) = \tfrac{1}{2}\alpha_1(B + 3F_1BF_1) + \tfrac{1}{2}\alpha_2(B + 3F_2BF_2) \tag{13.68}$$

where α_1, α_2 are complex scalars and F_1, F_2 are unit bivectors ($F_1^2 = F_2^2 = 1$).

Type D (degenerate) are a special case of type I tensors where two of the eigenvalues are the same. Physical examples are provided by the Schwarzschild solution, which has

$$\mathcal{R}(B) = \mathcal{W}(B) = -\frac{M}{2r^3}(B + 3\sigma_r B \sigma_r), \tag{13.69}$$

and the Kerr solution, which has

$$\mathcal{R}(B) = \mathcal{W}(B) = -\frac{M}{2(r + iL\cos\theta)^3}(B + 3\sigma_r B \sigma_r). \tag{13.70}$$

Note that $\mathcal{R}(B)$ for the Kerr solution differs from the Schwarzschild solution only in the fact that its eigenvalues contain an imaginary term governed by the angular momentum L. Verifying that (13.69) and (13.70) are tractionless is simple, requiring only the result that

$$\partial_a B a \wedge b = \partial_a B(ab - a \cdot b) = -bB. \tag{13.71}$$

For tensors of Petrov type other than I, null bivectors play a significant role. Type II tensors have eigenvalues α_1, $-\alpha_1$ and 0 and two independent eigenbivectors, one timelike and one null. Type III and type N have all three eigenvalues zero, and satisfy

$$\text{type } III\text{:} \quad \mathcal{W}^3(B) = 0, \quad \text{type } N\text{:} \quad \mathcal{W}^2(B) = 0. \tag{13.72}$$

An example of a type N tensor is provided by gravitational radiation: for plane waves in the γ_3 direction one form of $\mathcal{R}(B)$ is

$$\mathcal{R}(B) = \mathcal{W}(B) = \tfrac{1}{4} f(t - z)\, \gamma_+ (\gamma_1 B \gamma_1 - \gamma_2 B \gamma_2) \gamma_+. \tag{13.73}$$

where $\gamma_+ = \gamma_0 + \gamma_3$.

13.4 The Bianchi Identity

As well as the symmetries of $\mathcal{R}(B)$, information from the wedge equation (13.31) is also contained in the Bianchi identity. One form of this follows from a simple application of the Jacobi identity:

$$[\mathcal{D}_a, [\mathcal{D}_b, \mathcal{D}_c]]A + \text{cyclic permutations} = 0 \tag{13.74}$$

$$\Rightarrow \mathcal{D}_a R(b \wedge c) + \text{cyclic permutations} = 0, \tag{13.75}$$

but more work is required to achieve a fully covariant relation. We start by forming the adjoint relation to (13.75),

$$\partial_a \wedge \partial_b \wedge \partial_c \langle [a \cdot \nabla R(b \wedge c) + \Omega(a) \times R(b \wedge c)] B \rangle = 0 \tag{13.76}$$

where B is a constant bivector. We next use the result

$$B_1 \cdot \mathcal{R}(B_2) = B_2 \cdot \mathcal{R}(B_1), \tag{13.77}$$

which follows from equation (13.50), to obtain

$$\nabla \wedge \bar{h}^{-1}[\mathcal{R}(B)] - \partial_a \wedge \bar{h}^{-1}[\mathcal{R}(\Omega(a) \times B)] = 0. \tag{13.78}$$

Finally, acting on this equation with $\overline{h}$ and using equation (13.31), we establish the covariant result

$$\mathcal{D}\wedge\mathcal{R}(B) - \partial_a\wedge\mathcal{R}(\omega(a)\times B) = 0. \tag{13.79}$$

This result takes a more natural form when B becomes an arbitrary function of position and we write the Bianchi identity as

$$\partial_a\wedge[a\cdot\mathcal{D}\mathcal{R}(B) - \mathcal{R}(a\cdot\mathcal{D}B)] = 0. \tag{13.80}$$

We can extend the overdot notation in the obvious manner to write equation (13.80) as

$$\dot{\mathcal{D}}\wedge\dot{\mathcal{R}}(B) = 0, \tag{13.81}$$

which is very compact, but somewhat symbolic and hard to apply without unwrapping into the form of equation (13.80).

The self-duality of the Weyl tensor implies that

$$\dot{\mathcal{D}}\wedge\dot{\mathcal{W}}(iB) = -i\dot{\mathcal{D}}\cdot\dot{\mathcal{W}}(B). \tag{13.82}$$

So, in situations where the matter vanishes and $\mathcal{W}(B)$ is the only contribution to $\mathcal{R}(B)$, the Bianchi identity reduces to

$$\dot{\mathcal{D}}\dot{\mathcal{W}}(B) = 0. \tag{13.83}$$

The contracted Bianchi identity is

$$\dot{\mathcal{D}}\cdot\dot{\mathcal{G}}(a) = 0 \tag{13.84}$$

or, written out in full,

$$\partial_a\cdot[L_a\mathcal{G}(b) - \mathcal{G}(L_a b) + \omega(a)\times\mathcal{G}(b) - \mathcal{G}(\omega(a)\times b)] = 0. \tag{13.85}$$

13.5 Symmetries and Conservation Laws

Since our gauge theory is founded on an action principle in a 'flat' vector space, it follows that all the familiar equations relating symmetries of the action to conserved quantities hold without modification. Any symmetry transformation of the total action integral parameterised by a continuous scalar will therefore result in a vector which is conserved with respect to the vector derivative ∇. To every such vector their corresponds a covariant equivalent, as is seen from the simple re-arrangement

$$\begin{aligned}\mathcal{D}\cdot\mathcal{J} &= i\mathcal{D}\wedge(i\mathcal{J})\\ &= \det(\underline{h})i\nabla\wedge[\overline{h}^{-1}(i\mathcal{J})]\\ &= \det(\underline{h})\nabla\cdot[\underline{h}(\mathcal{J})\det(\underline{h})^{-1}].\end{aligned} \tag{13.86}$$

So, if J satisfies $\nabla\cdot J = 0$, then the covariant equivalent

$$\mathcal{J} = \underline{h}^{-1}(J)\det(\underline{h}) \tag{13.87}$$

satisfies the covariant equation $\mathcal{D}\cdot\mathcal{J} = 0$.

A second important feature of our gauge theory is that the field equations can be recast in integral form. This form is often non-covariant, so is of limited applicability. But the integral equation form is very well suited to probing the properties of the 'singularities' in the gravitational field (*c.f.* the use of Gauss law in electromagnetism.) This is particularly significant for the Kerr solution, where the integral equations reveal a *disk* of matter generating the fields.

Many other features of GR can be taken across intact into our gauge theory. For example, Killing's equation for a Killing vector K becomes

$$a\cdot(b\cdot\mathcal{D}K) + b\cdot(a\cdot\mathcal{D}K) = 0. \tag{13.88}$$

However, a general feature of this procedure is that many objects which are afforded great significance in GR (such as the metric, or the Lie derivative) are of secondary importance in our theory.

Chapter 14

Gravity III — First Applications

Anthony Lasenby, Chris Doran, and Stephen Gull
MRAO, Cavendish Laboratory, Madingley Road,
Cambridge, CB3 0HE, UK

After the development of the theory, it is now time to look at some applications. Historically the first solution to Einstein's field equations was found by Schwarzschild, in 1916, who considered the fields of a mass-point and a spherically-symmetric star. Here we start with the vacuum part of this problem, and its extension to black holes. We will do this by fairly elementary techniques, keeping the entries in the $\bar{h}$-function to the fore, and thereby foregoing manifest gauge covariance. In the next lecture you will see how the derivations can be improved considerably, by what we call the 'intrinsic method', in which explicit position-gauge covariance is maintained throughout.

14.1 Spherically-Symmetric Static Solutions

In order to find spherically-symmetric static solutions to the field equations we first introduce a set of polar coordinates $\{t, r, \theta, \phi\}$,

$$\begin{aligned} t &\equiv x \cdot \gamma_0 & \cos\theta &\equiv x \cdot \gamma^3 / r \\ r &\equiv \sqrt{(x \wedge \gamma_0)^2} & \tan\phi &\equiv (x \cdot \gamma^2)/(x \cdot \gamma^1). \end{aligned} \tag{14.1}$$

From these we define the coordinate frame

$$\begin{aligned} e_t &= \partial_t x = \gamma_0 \\ e_r &= \partial_r x = \sin\theta \cos\phi \gamma_1 + \sin\theta \sin\phi \gamma_2 + \cos\theta \gamma_3 \\ e_\theta &= \partial_\theta x = r(\cos\theta \cos\phi \gamma_1 + \cos\theta \sin\phi \gamma_2 - \sin\theta \gamma_3) \\ e_\phi &= \partial_\phi x = r \sin\theta (-\sin\phi \gamma_1 + \cos\phi \gamma_2), \end{aligned} \tag{14.2}$$

and the dual-frame vectors are denoted as $\{e^t, e^r, e^\theta, e^\phi\}$. We will also frequently employ the unit vectors $\hat{\theta}$ and $\hat{\phi}$ defined by

$$\hat{\theta} \equiv e_\theta / r, \qquad \hat{\phi} \equiv e_\phi / (r \sin\theta). \tag{14.3}$$

Associated with these unit vectors are the unit timelike bivectors

$$\sigma_r \equiv e_r e_t, \qquad \sigma_\theta \equiv \hat{\theta} e_t, \qquad \sigma_\phi \equiv \hat{\phi} e_t, \tag{14.4}$$

which satisfy $\sigma_r \sigma_\theta \sigma_\phi = e_t e_r \hat{\theta}\hat{\phi} = i$. The dual spatial bivectors are given by

$$i\sigma_r = -\hat{\theta}\hat{\phi}, \qquad i\sigma_\theta = e_r \hat{\phi}, \qquad i\sigma_\phi = -e_r \hat{\theta}. \tag{14.5}$$

For our initial ansätz we choose $\overline{h}(a)$ to be of the form

$$\begin{aligned} \overline{h}(e^t) &= f_1 e^t + f_2 e^r & \overline{h}(e^\theta) &= e^\theta \\ \overline{h}(e^r) &= g_1 e^r + g_2 e^t & \overline{h}(e^\phi) &= e^\phi, \end{aligned} \tag{14.6}$$

where f_i and g_i are functions of r only. We can write $\overline{h}$ in a more compact form as

$$\overline{h}(n) = n + n\cdot e_t[(f_1 - 1)e^t + f_2 e^r] + n\cdot e_r[(g_1 - 1)e^r + g_2 e^t]. \tag{14.7}$$

We also take a trial form for the bivector field Ω; abbreviating $\Omega(e_\mu)$ to Ω_μ, this is

$$\begin{aligned} \Omega_t &= a e_r e_t & \Omega_\theta &= (b_1 e_r + b_2 e_t) e_\theta / r \\ \Omega_r &= 0 & \Omega_\phi &= (b_1 e_r + b_2 e_t) e_\phi / r, \end{aligned} \tag{14.8}$$

where a and b_i are also scalar functions of r only. This is not the most general form of spherically-symmetric Ω, but the general form can be generated from (14.8) by local gauge transformations.

The first of the field equations can be written as

$$\overline{h}(e^\mu)\wedge(\mathcal{D}_\mu \overline{h}(e^\nu)) = 0, \tag{14.9}$$

where $\mathcal{D}_\mu = \partial_\mu + \Omega_\mu \times$. Substituting (14.6) and (14.8) into (14.9) generates the four equations

$$g_2 f_2' - g_1 f_1' - a {f_1}^2 + a {f_2}^2 = 0 \tag{14.10}$$

$$g_1 g_2' - g_1' g_2 + a f_1 g_2 - a f_2 g_1 = 0 \tag{14.11}$$

$$g_1 = b_1 + 1 \tag{14.12}$$

$$g_2 = -b_2, \tag{14.13}$$

where the primes denote differentiation with respect to r. We use (14.12) and (14.13) to eliminate b_1 and b_2. Next, calculating the 6 quantities $R_{\mu\nu} \equiv R(e_\mu \wedge e_\nu)$, yields

$$R_{tr} = -a' e_r e_t \tag{14.14}$$

$$R_{t\theta} = a(g_1 e_t - g_2 e_r) e_\theta / r \tag{14.15}$$

$$R_{r\theta} = (g_1' e_r - g_2' e_t) e_\theta / r \tag{14.16}$$

$$R_{\theta\phi} = ({g_1}^2 - {g_2}^2 - 1) e_\theta e_\phi / r^2, \tag{14.17}$$

with $R_{t\phi}$, $R_{r\phi}$ having the same form as $R_{t\theta}$, $R_{r\theta}$ respectively. By forming the contraction $\overline{h}(e^\mu)\cdot R_{\mu\nu}$ and setting the result equal to zero, we find that

$$2a + a'r = 0 \tag{14.18}$$

$$2g_1' + f_1 a' r = 0 \tag{14.19}$$

$$2g_2' + f_2 a' r = 0 \tag{14.20}$$

$$ar(f_1 g_1 - f_2 g_2) + r(g_1 g_1' - g_2 g_2') + {g_1}^2 - {g_2}^2 - 1 = 0. \tag{14.21}$$

Equation (14.18) yields a immediately, and equations (14.19) and (14.20) define f_1 and f_2 in terms of g_1 and g_2. On combining these with (14.10) we find that

$$\det(\underline{h}) = \overline{h}(i)i^{-1} = f_1 g_1 - f_2 g_2 = \text{constant}, \tag{14.22}$$

and we set this constant equal to 1, since $\overline{h}$ is required to reduce to the identity at large distances. All that remains is a simple equation for ${g_1}^2 - {g_2}^2$, and the full solution to our field equations is

$$\begin{aligned} a &= GM/r^2 \\ {g_1}^2 - {g_2}^2 &= 1 - 2GM/r \\ GMf_1 &= r^2 g_1' \\ GMf_2 &= r^2 g_2', \end{aligned} \tag{14.23}$$

subject to the boundary conditions that

$$\left.\begin{aligned} f_1, g_1 &\to 1 \\ f_2, g_2 &\to 0 \end{aligned}\right\} \quad \text{as} \quad r \to \infty. \tag{14.24}$$

These boundary conditions guarantee that at large distances the effects of the fields fall away to zero. The solution (14.23) contains a single arbitrary function, g_2 say, subject to the condition that

$${g_2}^2(r) \geq 2GM/r - 1, \tag{14.25}$$

together with the boundary conditions. From our initial restricted choice of $\overline{h}$ and Ω we have found a one-parameter family of solutions. This is extended to a four-parameter family by considering spherically-symmetric gauge transformations. The four classes of transformation which preserve radial symmetry are

$$\begin{aligned} \text{Radial boost}: &\quad R = \exp(\alpha(r)\sigma_r/2) \\ \text{Rotation}: &\quad R = \exp(\alpha(r)i\sigma_r/2) \\ \text{Time translation}: &\quad f(x) = x + \alpha(r)e_t \\ \text{Radial dilation}: &\quad f(x) = x{\cdot}e_t e_t + \alpha(r)e_r. \end{aligned} \tag{14.26}$$

These transformations induce more general forms of $\overline{h}$ and Ω, and combinations of the first and third can be used to move within the one-parameter family described by (14.23). All transformations in (14.26) leave ${g_1}^2 - {g_2}^2$ and Ω_t unchanged.

We are now in a position to compare our solutions with the Schwarzschild metric of general relativity. The 'line element' in our flat-space gauge theory is given by:

$$\begin{aligned} ds^2 &= \underline{h}^{-1}(e_\mu){\cdot}\underline{h}^{-1}(e_\nu)dx^\mu\, dx^\nu \\ &= (1 - 2GM/r)\, dt^2 + 2(f_1 g_2 - f_2 g_1)\, dr\, dt - ({f_1}^2 - {f_2}^2)\, dr^2 \\ &\quad - r^2(d\theta^2 + \sin^2\theta\, d\phi^2). \end{aligned} \tag{14.27}$$

We see that the exterior Schwarzschild metric is recovered by setting $g_2 = 0$, which can only be done outside the horizon ($r > 2GM$). If we wish to extend the same line element inside the horizon, we must set $g_1 = 0$ for $r < 2GM$. However this solution is then strongly discontinuous at the horizon: something has gone wrong! To see exactly what, we focus on the cross-term in the line element, since standard treatments of the Schwarzschild solution always assume that this term can be transformed away. The coefficient of this term is $f_1g_2 - f_2g_1$, and since our solutions have $g_1 = \pm g_2$ at the horizon and $f_1g_1 - f_2g_2 = 1$ everywhere, we find that

$$f_1g_2 - f_2g_1 = \pm 1 \quad \text{at } r = 2GM. \tag{14.28}$$

The assumption that $f_1g_2 - f_2g_1$ can be transformed away fails at the horizon, so that the standard form of the Schwarzschild solution is not admissible in our theory. We shall shortly see that this removes much of the pathological behaviour of test particles at the horizon of a Schwarzschild black hole. The reason for our more restrictive class of solutions is that we retain a notion of position in a flat spacetime, and demand that the $\overline{h}$ and Ω functions be well-defined throughout this spacetime (except possibly where a point source is present). General relativity, by contrast, does not place such restrictions on the components of the metric. These components are scalar functions which can be transformed by a coordinate transformation. General relativity admits coordinate transformations which result in patchs of spacetime not being covered; such transformations have no counterpart in our theory.

The shift from the Schwarzschild solution, with $f_1g_2 - f_2g_1 = 0$, to two distinct families of solutions, with $f_1g_2 - f_2g_1 = \pm 1$ at the horizon, is characteristic of the transition from second-order to first-order theories. A similar phenomenon is seen in the theory of propagation of electromagnetic waves, for example, where the first-order formulation correctly fixes the obliquity factors which have to be put in by hand in the second-order theory. Furthermore, the appearance of two disconnected families of solutions is what one expects for a first-order theory, since, as we shall soon see, the disconnected families are related by the discrete symmetry of time-reversal.

14.1.1 Point-Particle Trajectories

The dynamics of a Dirac fermion in a gravitational background are described by the Dirac equation together with the quantum-mechanical rules for constructing observables. For many applications, however, it is useful to work with semi-classical and classical approximations to the full quantum theory. The idea is to specialise to motion along a single streamline defined by the Dirac current $\psi\gamma_0\tilde{\psi}$. Thus the particle is described by a trajectory $x(\tau)$, together with a spinor $\psi(\tau)$ which contains information about the velocity and spin of the particle. The covariant velocity is $\underline{h}^{-1}(\dot{x})$ and this is identified with $\psi\gamma_0\tilde{\psi}$. A Lagrange multiplier p is included in the action integral to enforce this identification. Finally, an einbein e is introduced to guarantee reparameterisation invariance and the resultant action is

$$S = \int d\tau \, \langle \dot{\psi} i\sigma_3 \tilde{\psi} + \tfrac{1}{2}\Omega(\dot{x})\psi i\sigma_3\tilde{\psi} + p(v - e\psi\gamma_0\tilde{\psi}) + me \rangle, \tag{14.29}$$

where

$$v \equiv \underline{h}^{-1}(\dot{x}). \tag{14.30}$$

The equations of motion arising from (14.29) will not be discussed here, but an effect worth noting is that, due to the spin of the particle, the velocity v and momentum p are not collinear.

We can make a full classical approximation by neglecting the spin (dropping all the terms containing ψ) and replacing $\psi\gamma_0\tilde{\psi}$ by p/m. This process leads to the action

$$S = \int d\tau \, \langle p\underline{h}^{-1}(\dot{x}) - \tfrac{1}{2}e(p^2 - m^2)\rangle. \tag{14.31}$$

The equations of motion from (14.31) are

$$\begin{aligned} v &= ep \\ p^2 &= m^2 \\ \partial_\tau \overline{h}^{-1}(p) &= \overset{*}{\nabla}\, p \cdot \overset{*}{\underline{h}}{}^{-1}(\dot{x}), \end{aligned} \tag{14.32}$$

where for this section only we use overstars in place of overdots for the scope of a differential operator. The latter equation yields

$$\begin{aligned} \dot{p} &= \overline{h}(\overset{*}{\nabla})p \cdot \overset{*}{\underline{h}}{}^{-1}(\dot{x}) - \dot{x} \cdot \overset{*}{\nabla}\, \overline{h}\, \overset{*}{\overline{h}}{}^{-1}(p) \\ &= \overline{h}[(\overset{*}{\nabla} \wedge \overset{*}{\overline{h}}{}^{-1}(p)) \cdot \dot{x}] \\ &= \overline{h}[\overline{h}^{-1}(H(v)) \cdot \underline{h}(v)] \\ &= H(v) \cdot v/e, \end{aligned} \tag{14.33}$$

where $H(a)$ was defined in the previous lecture. From the results given there we see that $a\cdot\omega(a) = -a\cdot H(a)$, hence

$$e\partial_\tau(v/e) = -\omega(v)\cdot v. \tag{14.34}$$

This is the gauge theory equivalent of the geodesic equation. It takes its simplest form when τ is the affine parameter for the trajectory, in which case $e = 1/m$ and the equation becomes

$$\dot{v} = -\omega(v)\cdot v, \tag{14.35}$$

or, in manifestly covariant form,

$$v \cdot \mathcal{D} v = 0. \tag{14.36}$$

The geodesic equation is usually obtained from the action

$$S = m \int d\tau \, \sqrt{\underline{h}^{-1}(\dot{x})^2}, \tag{14.37}$$

which is obtained from (14.31) be eliminating p and e with their respective equations of motion. The Hamiltonian form (14.31) is rarely seen in conventional GR, since it requires the introduction of a vierbein. Despite this, the action (14.31) has a number of very useful features, especially when it comes to extracting conservation laws. For example, Noether's theorem appled to the translation $x \mapsto= x + \epsilon a$ quickly yields

$$\partial_\tau[a\cdot\overline{h}^{-1}(p)] = a\cdot\overset{*}{\nabla}\, p\cdot\underline{\overset{*}{h}}^{-1}(\dot{x}), \tag{14.38}$$

where a is a constant vector. It follows that if the $\overline{h}$-field is invariant under translations in the direction a then the quantity $a\cdot\overline{h}^{-1}(p)$ is conserved.

Later, in Lecture 5, we will require a formula for the gravitational redshift induced by the $\overline{h}$-field. Suppose that a source of radiation follows a worldline $x_1(\tau_1)$, with covariant velocity $v_1 = \underline{h}^{-1}(\dot{x}_1)$. The radiation is received by an observer $x_2(\tau_2)$ and follows a null geodesic with a covariant velocity u. The spectral shift z is determined by the ratio of the frequency observed at the source, $u(x_1)\cdot v_1$ and the frequency observed at the receiver, $u(x_2)\cdot v_2$, by

$$1 + z \equiv \frac{u(x_1)\cdot v_1}{u(x_2)\cdot v_2}. \tag{14.39}$$

14.1.2 Particle Motion in a Spherically-Symmetric Background

Returning to the specific case of a static spherically-symmetric source, we now consider the motion of a point particle as defined by equation (14.36). To simplify the problem, suppose that all motion takes place in the azimuthal plane ($\theta = \pi/2$). We can then write

$$\dot{x} = \dot{t}e_t + \dot{r}e_r + \dot{\phi}e_\phi, \tag{14.40}$$

and the equations of motion give

$$r^2\dot{\phi} = L \quad \text{(constant)} \tag{14.41}$$

$$\dot{r}^2 = A^2 - 1 - (1 - 2GM/r)L^2/r^2 + 2GM/r \tag{14.42}$$

$$\dot{t}(1 - 2GM/r) = A - \dot{r}(f_1g_2 - f_2g_1). \tag{14.43}$$

Equations (14.41) and (14.42) agree with those found in general relativity using the Schwarzschild metric, and (14.42) can be differentiated to give

$$\ddot{r} - r\dot{\phi}^2 = -(1 + 3L^2/r^2)GM/r^2. \tag{14.44}$$

This will reproduce the standard results for the shape of the orbit and the precession rate per orbit, which have been checked to high accuracy by measurements on binary pulsar systems. Equation (14.43), which describes the rate of change of coordinate time with respect to proper time, differs from that of the Schwarzschild metric through the inclusion of the $\dot{r}(f_1g_2 - f_2g_1)$ term. We saw in Section 3 that $f_1g_2 - f_2g_1$ must equal ± 1 at the horizon, so, taking the negative sign and radial infall ($\dot{r} < 0$, $A > 0$), we find that

$$A - \dot{r}(f_1g_2 - f_2g_1) = 0 \quad \text{at } r = 2GM. \tag{14.45}$$

This result removes the pole present in the corresponding equation for the Schwarzschild metric, and allows particles to cross the horizon in *finite* external coordinate time.

We can give a specific illustration of this in a gauge that we have found very well adapted to black hole physics, but for which the GR metric version of it does not seem to be known. This gauge will emerge naturally from the 'intrinsic approach' followed in the next lecture, but here we just define it in terms of our existing functions. We call it the 'Newtonian gauge', because in this gauge the equations describing spherically-symmetric systems are almost totally Newtonian in form! It is defined by

$$g_1 = 1, \quad g_2 = -\sqrt{2M/r}, \quad f_1 = 1, \quad \text{and} \quad f_2 = 0. \tag{14.46}$$

In this gauge the $\overline{h}$-function takes the remarkably simple form

$$\overline{h}(a) = a - \sqrt{2M/r}\, a\cdot e_r e_t, \tag{14.47}$$

which only differs from the identity through a single term. The geodesic equation for a radially-infalling particle (with unit mass) reduces to

$$\dot{r}^2 = 2M/r + \sinh^2 u_0 \tag{14.48}$$

$$(1-2M/r)\dot{t} = \cosh u_0 + \dot{r}\sqrt{2M/r}. \tag{14.49}$$

The $\dot{r}$ equation shows immediately that $\ddot{r} = -M/r^2$ and the constant $\sinh^2 u_0$ can be identified with twice the initial kinetic energy of the infalling particle. At the horizon $(r = 2M)$ $\dot{r} = -\cosh u_0$, so there is no pole in $\dot{t}$ (14.49). All particles cross the horizon and reach the singularity in a finite coordinate time. Some possible trajectories are illustrated in Figure 14.1.

In the case where the particle is dropped from rest at $r = \infty$ equations (14.48) and (14.49) reduce to

$$\dot{r} = -\sqrt{2M/r}, \quad \dot{t} = 1, \tag{14.50}$$

and we recover an entirely Newtonian description of the motion. The properties of a black hole are so simple in the gauge defined by (14.47) that it is astonishing that it is almost never seen in the literature. This is presumably because the line element associated with (14.47) does not look as natural as the $\overline{h}$-function itself, and hides the underlying simplicity of the system. In the Newtonian gauge one hardly needs to modify classical reasoning at all to understand the processes involved — all particles just cross the horizon and fall into the singularity in a finite coordinate time. And the horizon is located at $r = 2M$ precisely because we can apply Newtonian arguments! The only departures from Newtonian physics lie in relativistic corrections to the proper-time taken for infall, and in modifications to the equations for angular motion which lead to the familiar results for orbital precession.

Horizons and Birkhoff's Theorem

The bulk of the literature on black holes works with the Schwarzschild solution, which is obtained by setting $g_2 = 0$. In this case

$$g_1 = \sqrt{1-2M/r} \tag{14.51}$$

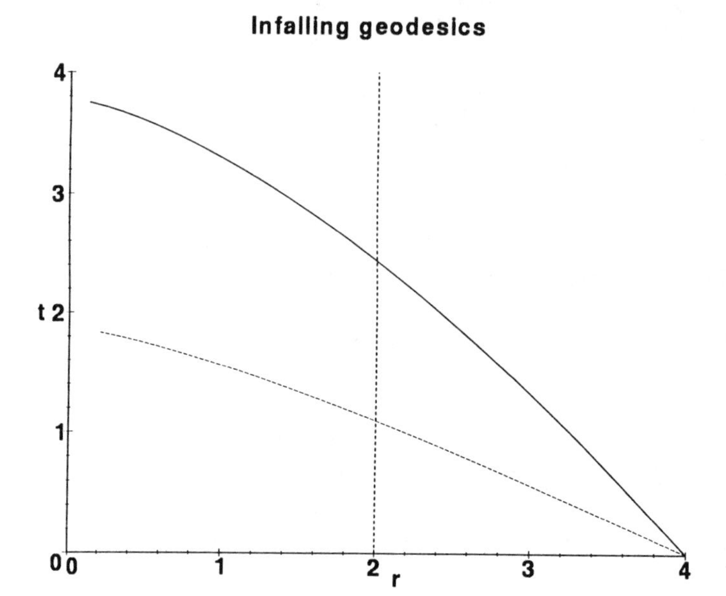

Fig. 14.1. *Possible particle trajectories for radially-infalling particles in the Newtonian gauge. The upper curve is for a particle released with* $\dot{r} = -\sqrt{2M/r}$ *at* $r = 4M$*, while the lower curve is for a photon. The vertical dotted line indicates the horizon.*

which is only defined for $r > 2M$. GR starts to invoke coordinate transformations at this point, usually resulting in the advanced or retarded Eddington-Finkelstein form of the solution. In our gauge theory, however, the picture is different; $g_2 = 0$ is simply *not* an allowed solution since it does not result in a globally-defined $\overline{h}$-function. It is over such questions regarding the global nature of fields that differences between our 'flatspace' gauge theory and GR start to emerge. Here the differences manifest themselves quite clearly. We are forced to a solution in which $g_2 \neq 0$, so $\overline{h}(a)$ is not diagonal and hence not time-reverse symmetric. Time reversal is achieved by combining the displacement

$$f(x) = -e_t x e_t = x' \tag{14.52}$$

with the reflection

$$\overline{h}'(a) = -e_t \overline{h}(a) e_t. \tag{14.53}$$

Hence the time-reversed solution is given by

$$\overline{h}^*(a) = e_t \overline{h}_{x'}(e_t a e_t) e_t. \tag{14.54}$$

The effect of this transformation is to swap the signs of f_2 and g_2 so, applied to the solution (14.47), we obtain

$$\overline{h}^*(a) = a + \sqrt{2M/r}\, a\cdot e_r e_t. \tag{14.55}$$

The result is a solution in which particles inside the horizon are swept out. Once outside, the force is still attractive but particles cannot re-enter back through the horizon. Despite the fact that the Riemann tensor is unchanged by time-reversal, it is impossible to find an $\overline{h}$-function that is also time-reverse symmetric. In our theory the black hole has more memory about its formation than simply its mass M. It also remembers that it was formed in a particular time direction. The appearance of a horizon is thus associated with the onset of time asymmetry, which is very satisfying from a physical viewpoint.

Another form of the Schwarzschild solution that runs into difficulties in our theory is that obtained by using Kruskal coordinates. These coordinates introduce a 'double-cover' of spacetime so that each value of r determines two distinct hypersurfaces. This is clearly not possible in our theory without some radical redefinition of how r is viewed as a function of spacetime position x. In GR the Kruskal solution is the maximal continuation of the Schwarzschild metric and is viewed as giving the complete description of a spherically-symmetric black hole. The fact that it is ruled out of our gauge theory means that our allowed solutions are not 'maximal' and forces us to address the issue of geodesic incompleteness. For the solution (14.47) geodesics exist which cannot be extended into the past for all values of their affine parameter. But, if we adopt the view that the black hole must have formed in the past from some collapse process, then there must have been a time before which the horizon did not exist. Any geodesic must therefore have come from a region in the past where no horizon was present, so there is no question of the geodesics being incomplete. A model of such a collapse process is discussed in the final lecture. We therefore arrive at a consistent picture in which the formation of a horizon retains information about the direction of time in which collapse occurred, and all geodesics from the past emanate from a period before the horizon formed. This picture is very different from GR, which is happy to deal with eternal, time-symmetric black holes. This shift to a picture with a fixed time direction is typical of the transition from a second-order theory to a first-order one.

It is finally worth commenting on how the above affects Birkhoff's theorem. One form of Birkhoff's theorem is that the gravitational fields outside any spherically-symmetric distribution of matter are necessarily static. This is seen immediately from the fact that the Riemann tensor is a function of r only. However, Birkhoff's theorem is frequently used to argue that the line element outside a spherically-symmetric body can always be brought to the form

$$ds^2 = (1 - 2M/r)\, dt^2 - (1 - 2M/r)^{-1}\, dr^2 - r^2(d\theta^2 + \sin^2\theta\, d\phi^2). \tag{14.56}$$

As we have seen, this ceases to be correct in our theory if a horizon is present. In this case $\overline{h}(a)$ is independent of t, but cannot be time-reverse symmetric.

14.2 Electromagnetism in a Gravitational Background

The electromagnetic vector potential A is the generalisation of the gradient of a scalar and transforms under $U(1)$ phase rotations as

$$A \mapsto A - \nabla\phi. \tag{14.57}$$

It follows that under local displacements A transforms as $\nabla\phi$, and so maps to $\overline{f}(A(x'))$. The covariant form of the vector potential is therefore $\mathcal{A} = \overline{h}(A)$, which is the form that appeared in the Dirac equation. To construct an action integral for the electromagnetic field, however, we need a $U(1)$-invariant field. This is the Faraday bivector

$$F \equiv \nabla\wedge A. \tag{14.58}$$

The definition of F implies that under local displacements

$$\begin{aligned} F(x) \mapsto \nabla\wedge\overline{f}A(x') &= \dot{\nabla}\wedge\overline{f}\dot{A}(x') \\ &= \overline{f}(\nabla_{x'}\wedge A(x')) \\ &= \overline{f}F(x'). \end{aligned} \tag{14.59}$$

We therefore form the covariant bivector

$$\mathcal{F} = \overline{h}(F) \tag{14.60}$$

so that $\mathcal{F}$ is covariant under local displacements and rotations, as well as being $U(1)$-invariant.

The minimally-coupled action for a free electromagnetic field is now

$$S = \int |d^4x| \det(\underline{h})^{-1} \tfrac{1}{2}\mathcal{F}\cdot\mathcal{F}. \tag{14.61}$$

We can include a source term by adding in an $\mathcal{A}\cdot\mathcal{J}$ term, where $\mathcal{J}$ is a covariant vector (not a 'tensor density'). Such a term is also found in the Dirac action, where $\mathcal{J}$ was given by the Dirac current $\psi\gamma_0\tilde{\psi}$. The full action integral is therefore

$$S = \int |d^4x| \det(\underline{h})^{-1} (\tfrac{1}{2}\mathcal{F}\cdot\mathcal{F} + \mathcal{A}\cdot\mathcal{J}), \tag{14.62}$$

which is varied with respect to A with $\overline{h}$ and $\mathcal{J}$ treated as external fields. The result of this variation is the equation

$$\nabla\cdot\left(\underline{h}\overline{h}(\nabla\wedge A)\det(\underline{h})^{-1}\right) = J, \tag{14.63}$$

where

$$J \equiv \det(\underline{h})^{-1}\underline{h}(\mathcal{J}). \tag{14.64}$$

Equation (14.63) combines with the identity $\nabla\wedge F = 0$ to form the full set of Maxwell equations in a gravitational background. Some insight into these equations is provided by performing a space-time split and writing

$$\begin{aligned} \boldsymbol{E} + ci\boldsymbol{B} &\equiv F && (14.65)\\ \boldsymbol{D} + i\boldsymbol{H}/c &\equiv \epsilon_0 \underline{h}\overline{h}(F)\det(\underline{h})^{-1}, && (14.66) \end{aligned}$$

where we have temporarily included the factors of c and ϵ_0. In terms of these variables Maxwell's equations can be written in the familiar form

$$\begin{aligned} &\boldsymbol{\nabla}\cdot\boldsymbol{B} = 0 && \boldsymbol{\nabla}\cdot\boldsymbol{D} = \rho \\ &\boldsymbol{\nabla}\wedge\boldsymbol{E} + i\partial_t\boldsymbol{B} = 0 && \partial_t\boldsymbol{D} + \boldsymbol{\nabla}\cdot(i\boldsymbol{H}) = -\boldsymbol{J} \end{aligned} \qquad (14.67)$$

where $J\gamma_0 = \rho + \boldsymbol{J}$. This form of the equations shows how the $\overline{h}$-field defines the dielectric properties of the space through which the electromagnetic fields propagate.

So far, however, we have failed to achieve a covariant form of the Maxwell equations. We have, furthermore, failed to unite the separate equations into a single equation. In the absence of gravitational effects the equations $\nabla\cdot F = J$ and $\nabla\wedge F = 0$ combine into the single equation

$$\nabla F = J. \qquad (14.68)$$

The significance of this equation is that the ∇ operator is invertible, whereas the separate $\nabla\cdot$ and $\nabla\wedge$ operators are not.

To find a covariant equation, we first extend the 'wedge' equation as follows:

$$\begin{aligned} \mathcal{D}\wedge\overline{h}(B) &= (\mathcal{D}\wedge\overline{h}(a))\wedge\overline{h}(b) - \overline{h}(a)\wedge\mathcal{D}\wedge\overline{h}(b) \\ &= \overline{h}(\nabla\wedge a)\wedge\overline{h}(b) - \overline{h}(a)\wedge\overline{h}(\nabla\wedge b) + \kappa(\overline{h}(a)\cdot\mathcal{S})\wedge\overline{h}(b) - \kappa\overline{h}(a)\wedge(\overline{h}(b)\cdot\mathcal{S}) \\ &= \overline{h}(\nabla\wedge B) - \kappa\overline{h}(B)\times\mathcal{S}, \end{aligned} \qquad (14.69)$$

where $B = a\wedge b$ and, for generality, we have included the spin term. We can therefore replace the equation $\nabla\wedge F = 0$ with

$$\mathcal{D}\wedge\mathcal{F} - \kappa\mathcal{S}\times\mathcal{F} = 0. \qquad (14.70)$$

Next, we use a double-duality transformation to write the left-hand side of equation (14.63) as

$$\begin{aligned} \nabla\cdot(\underline{h}(\mathcal{F})\det(\underline{h})^{-1}) &= i\nabla\wedge(i\underline{h}(\mathcal{F})\det(\underline{h})^{-1}) \\ &= i\nabla\wedge(\overline{h}^{-1}(i\mathcal{F})) \\ &= i\overline{h}^{-1}[\mathcal{D}\wedge(i\mathcal{F}) + \kappa(i\mathcal{F})\times\mathcal{S}]. \end{aligned} \qquad (14.71)$$

Equation (14.63) now becomes

$$\mathcal{D}\cdot\mathcal{F} - \kappa\mathcal{S}\cdot\mathcal{F} = i\overline{h}(Ji) = \mathcal{J}, \qquad (14.72)$$

and equations (14.70) and (14.72) combine into the single equation

$$\mathcal{D}\mathcal{F} - \kappa\mathcal{S}\mathcal{F} = \mathcal{J}, \qquad (14.73)$$

which achieves our objective. Equation (14.73) is manifestly covariant; indeed the appearance of the $\mathcal{D}\mathcal{F}$ term is precisely what was found in the first lecture. The appearance of the spin term is a surprise, however. Gauge arguments alone would not have discovered this term and it is only through the construction of a gauge-invariant action integral that the term is found.

To complete the description of electromagnetism in a gravitational background we need a formula for the free-field stress-energy tensor. Applying the definition given in Lecture II, we construct

$$\begin{aligned}\mathcal{T}_{\mathrm{em}}\underline{h}^{-1}(a) &= \tfrac{1}{2}\det(\underline{h})\partial_{\overline{h}(a)}\langle\overline{h}(F)\overline{h}(F)\det(\underline{h})^{-1}\rangle \\ &= \overline{h}(a\cdot F)\cdot\mathcal{F} - \tfrac{1}{2}\underline{h}^{-1}(a)\mathcal{F}\cdot\mathcal{F}.\end{aligned} \tag{14.74}$$

Hence,

$$\begin{aligned}\mathcal{T}_{\mathrm{em}}(a) &= \overline{h}(\underline{h}(a)\cdot F)\cdot\mathcal{F} - \tfrac{1}{2}a\mathcal{F}\cdot\mathcal{F} \\ &= (a\cdot\mathcal{F})\cdot\mathcal{F} - \tfrac{1}{2}a\mathcal{F}\cdot\mathcal{F} \\ &= -\tfrac{1}{2}\mathcal{F}a\mathcal{F},\end{aligned} \tag{14.75}$$

which is the natural covariant extension of the gravitational-free form, $-FaF/2$.

The forms (14.67,14.63) offer some insight into how the gravitational field interacts with the electromagnetic field. The tensor $\det(\underline{h})^{-1}\underline{h}\overline{h}$ is a generalized permittivity/permeability tensor and defines the properties of the space through which the electromagnetic field propagates. For example, the bending of light by the sun can be easily understood in terms of the properties of the dielectric defined by the $\overline{h}$-field exterior to it.

14.2.1 Application to a Black-Hole Background

The problem of interest here is to find the fields around a point source at rest outside the horizon of a spherically-symmetric black hole. The $\overline{h}$-function in this case can be taken as that of equation (14.47). The solution to this problem can be found by adapting the work of Copson and Linet to the present gauge choices. Assuming that the charge is placed at a distance $a > 2M$ along the z-axis, the vector potential can be written in terms of a single scalar potential $V(r,\theta)$ as

$$A = V(r,\theta)(e_t + \frac{\sqrt{2Mr}}{r-2M}e_r), \tag{14.76}$$

so that

$$\mathcal{F} = -\frac{\partial V}{\partial r}e_r e_t - \frac{1}{r-2M}\frac{\partial V}{\partial \theta}\hat{\theta}(e_t + \sqrt{2M/r}e_r) \tag{14.77}$$

and

$$G = \boldsymbol{D} = -\frac{\partial V}{\partial r}e_r e_t - \frac{1}{r-2M}\frac{\partial V}{\partial \theta}\hat{\theta}e_t. \tag{14.78}$$

The Maxwell equations now reduce to the single partial differential equation

$$\frac{1}{r^2}\frac{\partial}{\partial r}\left(r^2\frac{\partial V}{\partial r}\right) + \frac{1}{r(r-2M)}\frac{1}{\sin\theta}\frac{\partial}{\partial\theta}\left(\sin\theta\frac{\partial V}{\partial\theta}\right) = -\rho \tag{14.79}$$

where $\rho = q\delta(\boldsymbol{x} - \boldsymbol{a})$ is a δ-function at $z = a$. This was the problem originally tackled by Copson who obtained a solution that was valid locally in the vicinity of the charge, but contained an additional pole at the origin. Linet modified Copson's solution by removing the singularity at the origin to produce a potential $V(r, \theta)$ whose only pole is on the z-axis at $z = a$. Linet's solution is

$$V(r,\theta) = \frac{q}{ar}\frac{(r-M)(a-M) - M^2\cos^2\theta}{D} + \frac{qM}{ar}, \tag{14.80}$$

where

$$D = [r(r-2M) + (a-M)^2 - 2(r-M)(a-M)\cos\theta + M^2\cos^2\theta]^{1/2}. \tag{14.81}$$

The novel feature we wish to stress here is that once (14.80) is inserted back into (14.77) the resultant $\mathcal{F}$ is both finite and continuous at the horizon. Working in the Newtonian gauge enables us to discuss the *global* properties of the solution, rather than having to resort to the 'membrane paradigm' to exclude the region $r < 2M$. One simple way to illustrate the global properties of $\mathcal{F}$ and G is to plot the streamlines of $\boldsymbol{D}$, which are conserved by equation (14.67). The streamlines should therefore spread out from the charge and cover all space. Furthermore, since the distance scale r was chosen to agree with the gravitationally-defined distance, the streamlines of $\boldsymbol{D}$ convey genuine intrinsic information. Hence the plots are completely unaffected by our choice for the g_1 or g_2 functions, or indeed our choice of t-coordinate. Figure 14.2 shows streamline plots for charges held at different distances above the horizon. Similar plots were first obtained by Hanni & Ruffini, though they were unable to extend their plots through the horizon. The physical interpretation of these plots differs somewhat from that advanced in the 'Membrane Paradigm'.

An interesting feature of the above solution is the existence of a repulsive 'polarisation' force, the effect of which is that a smaller force is needed to keep a charged particle at rest outside a black hole than an uncharged one. In their derivation of this force, Smith & Will employed a complicated energy argument which involved renormalising various divergent integrals. In fact, as we show elsewhere, this repulsion is due entirely to the term that Linet added to Copson's formula. This is the second term in equation (14.80), and produces an outward-directed force on the charge of magnitude q^2M/a^3 — the same magnitude as found by Smith & Will. This is an example of the importance of finding global solutions. The polarisation force is felt outside the horizon, yet the correction term that led to it was motivated by the properties of the field at the origin.

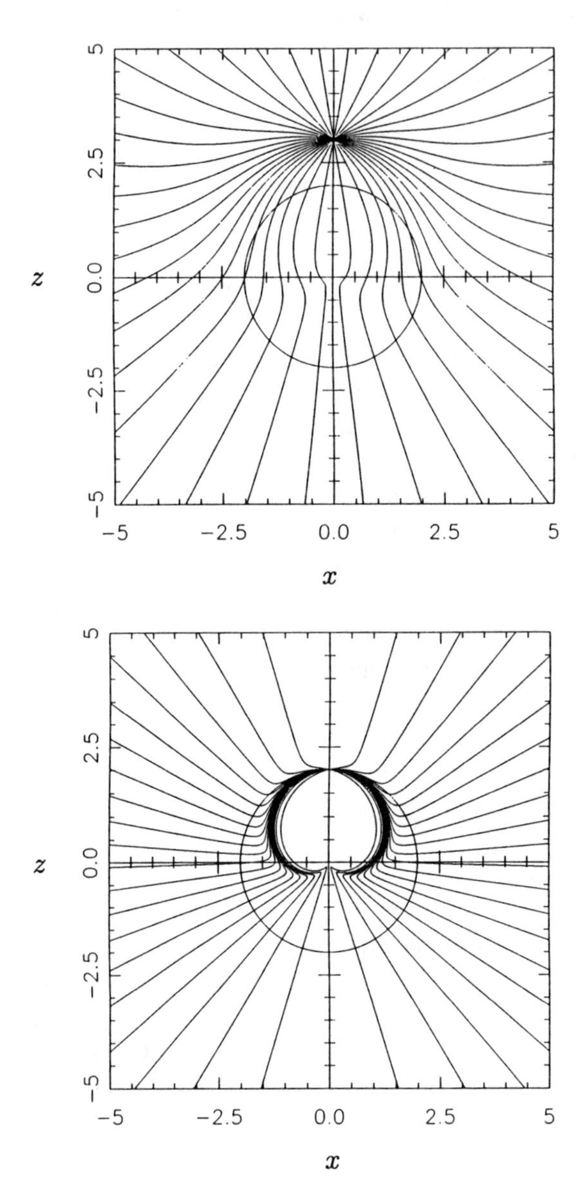

Fig. 14.2. *Streamlines of the* $\boldsymbol{D}$ *field. The horizon is at* $r = 2$ *and the charge is placed on the* z*-axis. The charge is at* $z = 3$ *and* $z = 2.01$ *for the top and bottom diagrams respectively. The streamlines are seeded so as to reflect the magnitude of* $\boldsymbol{D}$*. The streamlines are attracted towards the origin but never actually meet it. Note the appearance of a 'cardiod of avoidance' as the charge gets very close to the horizon.*

Chapter 15

Gravity IV — The 'Intrinsic' Method

Chris Doran, Anthony Lasenby, and Stephen Gull
*MRAO, Cavendish Laboratory, Madingley Road,
Cambridge, CB3 0HE, UK*

In this Lecture we discuss a new approach to studying the gravitational field equations. In GR one works with the metric, usually encoded in the form of the line element,

$$ds^2 = g_{\mu\nu} dx^\mu \, dx^\nu. \tag{15.1}$$

The quantity $g_{\mu\nu}$ is derived from the $\overline{h}$-function via

$$g_{\mu\nu} = \underline{h}^{-1}(e_\mu) \cdot \underline{h}^{-1}(e_\nu) \tag{15.2}$$

where the $\{e_\mu\}$ are a coordinate frame. Hence, in forming $g_{\mu\nu}$, all reference to the rotation gauge is lost — GR deals solely with quantities which transform as scalars under rotation-gauge transformations.

Our approach is precisely the opposite. We keep the rotation-gauge field explicit, and work entirely with quantities that transform covariantly under position-gauge transformations. Such quantities include $\overline{h}(\nabla)$, $\omega(a)$ and $\mathcal{R}(B)$. To achieve such a formalism we work with directional derivatives of the form $L_a = a \cdot \overline{h}(\nabla)$, and treat $\omega(a)$ as an arbitrary field. The relationship between $\overline{h}$ and $\omega(a)$ is then encoded in the commutation relations of the L_a. $\mathcal{R}(B)$ is then constructed from abstract first-order derivatives of the terms in $\omega(a)$ and additional quadratic terms. Once $\mathcal{R}(B)$ is found in this manner, the rotation-gauge freedom is fixed up by specifying the precise form that $\mathcal{R}(B)$ takes. For example, one can arrange that $\mathcal{W}(B)$ is diagonal in some suitable frame. This gauge fixing is crucial to arriving at a set of equations that are not under-constrained.

Once $\mathcal{R}(B)$ is fixed, one arrives at a set of relations between first-order abstract derivatives of the terms in $\omega(a)$, and quadratic and matter terms. The final step is to impose the Bianchi identities, which ensure overall consistency of the equations with the bracket structure. Once all this is achieved, one arrives at a fully 'intrinsic' set of equations. Solving these equations usually involves searching for natural 'integrating factors'. These integrating factors provide 'intrinsic' coordinates, and many of the fields can be expressed as functions of these alone. The final step is to 'coordinatise' with an explicit choice of the $\overline{h}$-function. The natural choice is usually to ensure that the coordinates used in parameterising $\overline{h}$ match the intrinsic coordinates defined by the integrating factors.

Here we outline two applications, time-dependent spherically-symmetric systems, and stationary axially-symmetric systems. In both cases we model the matter as a perfect fluid, with stress-energy tensor

$$\mathcal{T}(a) = (\rho + p)a{\cdot}vv - pa \tag{15.3}$$

where ρ is the energe density, p is the pressure, and v is the covariant fluid velocity ($v^2 = 1$).

15.1 Spherically-Symmetric Systems

We continue to employ the notation for a spherical-polar coordinate system introduced in the preceding lecture.

The $\overline{h}$-Function The first step is to choose a suitably general form for the $\overline{h}$-function which is consistent with spherical symmetry. With B a constant spatial bivector ($e_t{\cdot}B = 0$) and

$$R = e^{B/2}, \qquad x' = \tilde{R}xR \tag{15.4}$$

the gravitational fields will be spherically symmetric if rotating $\overline{h}(a)$ to $R\overline{h}(a)\tilde{R}$ and displacing it to the back-rotated position x' leaves $\overline{h}(a)$ unchanged. Hence rotational symmetry is enforced through the requirement that

$$R\overline{h}_{x'}(\tilde{R}aR)\tilde{R} = \overline{h}(a). \tag{15.5}$$

A general form for $\overline{h}(a)$ consistent with this requirement is given by

$$\begin{aligned} \overline{h}(e^t) &= f_1 e^t + f_2 e^r \\ \overline{h}(e^r) &= g_1 e^r + g_2 e^t \\ \overline{h}(e^\theta) &= \alpha e^\theta \\ \overline{h}(e^\phi) &= \alpha e^\phi, \end{aligned} \tag{15.6}$$

where f_1, f_2, g_1, g_2 and α are all functions of t and r only. Some rotation-gauge freedom has been absorbed in making $\overline{h}(a)$ diagonal on $\hat{\theta}$ and $\hat{\phi}$, but we are still free to carry out a boost in the σ_r direction without altering the functional form of $\overline{h}(a)$.

The $\omega(a)$-Function To find a general form for $\omega(a)$ consistent with (15.6) we substitute (15.6) into the equation for $\omega(a)$ as a function of $\overline{h}(a)$ derived in Lecture II. Where the coefficients in $\omega(a)$ contain derivatives of terms from $\overline{h}(a)$ new symbols are introduced. Undifferentiated terms from $\overline{h}(a)$ in $\omega(a)$ are left in explicitly. These

	σ_r	σ_θ	σ_ϕ
$\mathcal{D}_t$	0	$G\,i\sigma_\phi$	$-Gi\sigma_\theta$
$\mathcal{D}_r$	0	$F\,i\sigma_\phi$	$-Fi\sigma_\theta$
$\mathcal{D}_{\hat{\theta}}$	$T\,\sigma_\theta - S\,i\sigma_\phi$	$-T\,\sigma_r$	$S\,i\sigma_r$
$\mathcal{D}_{\hat{\phi}}$	$T\,\sigma_\phi + S\,i\sigma_\theta$	$-S\,i\sigma_r$	$-T\,\sigma_r$

Table 15.1. *Covariant derivatives of the polar-frame unit timelike bivectors.*

arise from frame derivatives and the algebra is simplified if they are included. This procedure results in the following form for $\omega(a)$:

$$\begin{aligned} \omega(e_t) &= Ge_re_t \\ \omega(e_r) &= Fe_re_t \\ \omega(\hat{\theta}) &= S\hat{\theta}e_t + (T-\alpha/r)e_r\hat{\theta} \\ \omega(\hat{\phi}) &= S\hat{\phi}e_t + (T-\alpha/r)e_r\hat{\phi}, \end{aligned} \tag{15.7}$$

where G, F, S and T are functions of t and r only. Substituting this definition for $\omega(a)$ into the bracket relation derived in Lecture II we obtain:

$$\begin{aligned} [L_t, L_r] &= GL_t - FL_r & \qquad [L_r, L_{\hat{\theta}}] &= -TL_{\hat{\theta}} \\ [L_t, L_{\hat{\theta}}] &= -SL_{\hat{\theta}} & [L_r, L_{\hat{\phi}}] &= -TL_{\hat{\phi}} \\ [L_t, L_{\hat{\phi}}] &= -SL_{\hat{\phi}} & [L_\theta, L_\phi] &= 0, \end{aligned} \tag{15.8}$$

where

$$\begin{aligned} L_t &\equiv e_t\cdot\overline{h}(\nabla) & \qquad L_{\hat{\theta}} &\equiv \hat{\theta}\cdot\overline{h}(\nabla) \\ L_r &\equiv e_r\cdot\overline{h}(\nabla) & L_{\hat{\phi}} &\equiv \hat{\phi}\cdot\overline{h}(\nabla). \end{aligned} \tag{15.9}$$

A set of bracket relations such as (15.8) is precisely what one aims to achieve — all reference to the $\overline{h}$-function is removed, and one deals only with quantities which transform as covariant scalars under local displacements.

The Riemann Tensor Given $\omega(a)$ we can now calculate $\mathcal{R}(B)$. This derivation employs the results listed in Table 15.1. A delicate point is the use of the bracket relations (15.8) to simplify terms involving derivatives of α/r:

$$L_t(\alpha/r) = L_tL_{\hat{\theta}}\theta = [L_t, L_{\hat{\theta}}]\theta = -S\alpha/r \tag{15.10}$$

$$L_r(\alpha/r) = L_rL_{\hat{\theta}}\theta = [L_r, L_{\hat{\theta}}]\theta = -T\alpha/r. \tag{15.11}$$

With these results we find that the Riemann tensor is given by

$$\begin{aligned} \mathcal{R}(\sigma_r) &= (L_rG - L_tF + G^2 - F^2)\sigma_r \\ \mathcal{R}(\sigma_\theta) &= (-L_tS + GT - S^2)\sigma_\theta + (L_tT + ST - SG)i\sigma_\phi \\ \mathcal{R}(\sigma_\phi) &= (-L_tS + GT - S^2)\sigma_\phi - (L_tT + ST - SG)i\sigma_\theta \\ \mathcal{R}(i\sigma_\phi) &= (L_rT + T^2 - FS)i\sigma_\phi - (L_rS + ST - FT)\sigma_\theta \\ \mathcal{R}(i\sigma_\theta) &= (L_rT + T^2 - FS)i\sigma_\theta + (L_rS + ST - FT)\sigma_\phi \\ \mathcal{R}(i\sigma_r) &= (-S^2 + T^2 - (\alpha/r)^2)i\sigma_r. \end{aligned} \tag{15.12}$$

The Matter Field and Gauge Fixing As previously stated, we assume that the matter is modelled by an ideal fluid (15.3). Radial symmetry means that v can only lie in the e_t and e_r directions, so v must take the form

$$v = \cosh u \, e_t + \sinh u \, e_r. \tag{15.13}$$

But the $\overline{h}$-function of equation (15.6) retained the gauge freedom to perform arbitrary radial boosts. This freedom can now be employed to set $v = e_t$, so that the matter stress-energy tensor becomes

$$\mathcal{T}(a) = (\rho + p)a\cdot e_t e_t - pa. \tag{15.14}$$

There is no physical content in this step, since all we have done is chosen a convenient gauge to work in. In fixing $\mathcal{T}(a)$ to the form (15.14) all rotational gauge freedom has now been absorbed. From this form of $\mathcal{T}(a)$, $\mathcal{G}(a)$ and $\mathcal{R}(a)$ are fixed, and $\mathcal{R}(B)$ is restricted to the form

$$\mathcal{R}(B) = \mathcal{W}(B) + 4\pi[(\rho + p)B\cdot e_t e_t - \tfrac{2}{3}\rho B]. \tag{15.15}$$

Comparing this with equation (15.12) we find that $\mathcal{W}(B)$ has the general form

$$\begin{aligned} \mathcal{W}(\sigma_r) &= \alpha_1 \sigma_r & \mathcal{W}(i\sigma_r) &= \alpha_4 i\sigma_r \\ \mathcal{W}(\sigma_\theta) &= \alpha_2 \sigma_\theta + \beta_1 i\sigma_\phi & \mathcal{W}(i\sigma_\theta) &= \alpha_3 i\sigma_\theta + \beta_2 \sigma_\phi \\ \mathcal{W}(\sigma_\phi) &= \alpha_2 \sigma_\phi - \beta_1 i\sigma_\theta & \mathcal{W}(i\sigma_\phi) &= \alpha_3 i\sigma_\phi - \beta_2 \sigma_\theta. \end{aligned} \tag{15.16}$$

But $\mathcal{W}(B)$ must be self-dual, so $\alpha_1 = \alpha_4$, $\alpha_2 = \alpha_3$ and $\beta_1 = -\beta_2$. It must also be symmetric, which forces $\beta_1 = \beta_2$, and it follows that $\beta_1 = \beta_2 = 0$. Finally, $\mathcal{W}(B)$ must be traceless, which requires that $\alpha_1 + 2\alpha_2 = 0$. Taken together, these conditions reduce $\mathcal{W}(B)$ to the form

$$\mathcal{W}(B) = \frac{\alpha_1}{4}(B + 3\sigma_r B \sigma_r) \tag{15.17}$$

which is of Petrov type D. It follows that if we set

$$A \equiv \tfrac{1}{4}(-S^2 + T^2 - (\alpha/r)^2) \tag{15.18}$$

then the full Riemann tensor must take the form

$$\mathcal{R}(B) = (A + \tfrac{2}{3}\pi\rho)(B + 3\sigma_r B\sigma_r) + 4\pi[(\rho + p)B\cdot e_t e_t - \tfrac{2}{3}\rho B]. \tag{15.19}$$

Comparing this with equation (15.12) we now obtain the following set of equations:

$$L_t S = 2A + GT - S^2 - 4\pi p \tag{15.20}$$
$$L_t T = S(G - T) \tag{15.21}$$
$$L_r S = T(F - S) \tag{15.22}$$
$$L_r T = -2A + FS - T^2 - 4\pi\rho \tag{15.23}$$
$$L_r G - L_t F = F^2 - G^2 + 4A + 4\pi(\rho + p). \tag{15.24}$$

The Bianchi Identity Before forming the full Bianchi identity, it is usually easier to start with the contracted identity. For a perfect fluid this yields

$$\begin{aligned} \mathcal{D}\cdot(\rho v) + p\mathcal{D}\cdot v = 0 \\ (\rho + p)(v\cdot\mathcal{D}v)\wedge v - (\mathcal{D}p)\wedge v = 0. \end{aligned} \tag{15.25}$$

Since $(v\cdot\mathcal{D}v)\wedge v$ is the acceleration bivector, the latter of these equations relates the acceleration to the pressure gradient. For the case of radially-symmetric fields equations (15.25) reduce to

$$L_t\rho = -(F + 2S)(\rho + p) \tag{15.26}$$
$$L_r p = -G(\rho + p), \tag{15.27}$$

the latter of which identifies G as the radial acceleration. The full Bianchi identities now turn out to be satisfied as a consequence of the contracted identities and the bracket relation

$$[L_t, L_r] = GL_t - FL_r. \tag{15.28}$$

Equations (15.10), (15.11), (15.20)–(15.24), the contracted identities (15.26) and (15.27) and the bracket condition (15.28) now form the complete set of intrinsic equations. The structure is closed, in that it is easily verified that the bracket relation (15.28) is consistent with the known derivatives. The derivation of such a set of equations is the basic aim of the 'intrinsic method'.

Integrating Factors If we form the derivatives of A we find that

$$L_tA + 3SA = 2\pi Sp, \qquad L_rA + 3TA = -2\pi T\rho. \tag{15.29}$$

These results, and equations (15.21) and (15.22), suggest that we should look for an integrating factor X with the properties:

$$L_tX = SX, \qquad L_rX = TX. \tag{15.30}$$

Such a function can only exist if its derivatives are consistent with the bracket relation (15.28). This is checked as follows:

$$\begin{aligned} [L_t, L_r]X &= L_t(TX) - L_r(SX) \\ &= X(L_tT - L_rS) \\ &= X(SG - FT) \\ &= GL_tX - FL_rX. \end{aligned} \tag{15.31}$$

In fact, equations (15.10) and (15.11) show that r/α already has the properties required of X, so it is r/α that emerges as the intrinsic distance scale. We therefore make the position-gauge choice that r corresponds to the intrinsic distance by setting $\alpha = 1$. This lifts r from the status of an arbitrary coordinate to that of a physically-measurable quantity.

With the position-gauge choice $r = X$, $\alpha = 1$ we can now make some further simplifications. From equation (15.30), and the form of $\overline{h}(a)$ (15.6), we see that

$$g_1 = L_r r = Tr, \qquad g_2 = L_t r = Sr, \tag{15.32}$$

which solves for two of the functions in $\overline{h}(a)$. We also define

$$M \equiv -2r^3 A = \tfrac{1}{2} r ({g_2}^2 - {g_1}^2 + 1), \tag{15.33}$$

which satisfies

$$L_t M = -4\pi r^2 g_2 p \tag{15.34}$$

$$L_r M = 4\pi r^2 g_1 \rho. \tag{15.35}$$

Thus M clearly plays the role of an intrinsic mass.

The 'Newtonian' Gauge So far a natural distance scale has been identified, but no natural time coordinate has yet emerged. We therefore need some additional criteria to motivate this choice. An indication of how this choice should be made is obtained from the derivatives of M, which invert to yield

$$\frac{\partial M}{\partial t} = \frac{-4\pi g_1 g_2 r^2 (\rho + p)}{f_1 g_1 - f_2 g_2} \tag{15.36}$$

$$\frac{\partial M}{\partial r} = \frac{4\pi r^2 (f_1 g_1 \rho + f_2 g_2 p)}{f_1 g_1 - f_2 g_2}. \tag{15.37}$$

The second of these equations reduces to a simple classical relation if we choose $f_2 = 0$, as we then obtain

$$\partial_r M = 4\pi r^2 \rho, \tag{15.38}$$

which says that $M(r,t)$ is determined by the amount of mass-energy in a sphere of radius r. There are other reasons for choosing the time variable such that $f_2 = 0$. For example, we can then use the bracket structure to solve for f_1. With $f_2 = 0$ we have

$$L_t = f_1 \partial_t + g_2 \partial_r \tag{15.39}$$

$$L_r = g_1 \partial_r \tag{15.40}$$

and the bracket relation (15.28) implies that

$$\begin{aligned} L_r f_1 &= -G f_1 \\ \Rightarrow \partial_r f_1 &= -\frac{G}{g_1} f_1 \\ \Rightarrow f_1 &= \epsilon(t) \exp\{-\int^r \frac{G}{g_1} dr\}. \end{aligned} \tag{15.41}$$

The function $\epsilon(t)$ can be absorbed by a further rescaling of t, so with $f_2 = 0$ we can reduce to a system in which

$$f_1 = \exp\{-\int^r \frac{G}{g_1} dr\}. \tag{15.42}$$

Another reason why $f_2 = 0$ is a natural gauge choice is seen when the pressure is zero. In this case equation (15.27) forces G to be zero, and equation (15.42) then sets $f_1 = 1$. A free-falling particle with $v = e_t$ (*i.e.* comoving with the fluid) then has

$$\dot{t}e_t + \dot{r}e_r = e_t + g_2 e_r, \tag{15.43}$$

where the dots denote differentiation with respect to the affine parameter. Since $\dot{t} = 1$ the time coordinate t matches the proper time of all observers comoving with the fluid. So, in the absence of pressure, we are able to recover a global 'Newtonian' time on which all observers can agree. Furthermore, it is also clear from (15.43) that g_2 is the velocity of the particle. Hence equation (15.36), which reduces to

$$\frac{\partial M}{\partial t} = -4\pi g_2 r^2 \rho \tag{15.44}$$

in the absence of pressure, has a simple Newtonian interpretation — it equates the work with the rate of flow of energy density.

For these and other reasons, $f_2 = 0$ is the natural position-gauge choice for the time coordinate. Because of the transparent form of the physical equations in this gauge, we refer to it as the 'Newtonian' gauge. The equations in this gauge are summarised in Table 15.2. One significant feature is that given initial data in the form of the density $\rho(r, t_0)$ and the velocity $g_2(r, t_0)$, together with an equation of state, the future evolution of the system is fully determined. To see this, one just uses the $L_t M$ and $L_t g_2$ equations to update the data from one timeslice to the next.

15.2 Two Applications

Static Matter Distributions For a static configuration, such as a radially-symmetric star, ρ and p are functions of r only. It follows from the derivatives of M that

$$M(r) = \int_0^r 4\pi r'^2 \rho(r')\, dr' \tag{15.45}$$

and

$$L_t M = 4\pi r^2 g_2 \rho = -4\pi r^2 g_2 p. \tag{15.46}$$

But, for any physical matter distribution, ρ and p must both be positive. Equation (15.46) can therefore only be satisfied if $g_2 = 0$, which in turn forces $F = 0$. It follows that g_1 is given simply by

$${g_1}^2 = 1 - 2M(r)/r, \tag{15.47}$$

recovering contact with the line-element employed in GR.

The remaining equation of use is that for $L_t g_2$, which now gives

$$G g_1 = M(r)/r^2 + 4\pi r p. \tag{15.48}$$

The $\overline{h}$-function	$\overline{h}(e^t) = f_1 e^t$ $\overline{h}(e^r) = g_1 e^r + g_2 e^t$ $\overline{h}(e^\theta) = e^\theta$ $\overline{h}(e^\phi) = e^\phi$
The ω-function	$\omega(e_t) = G e_r e_t$ $\omega(e_r) = F e_r e_t$ $\omega(\hat{\theta}) = g_2/r\,\hat{\theta} e_t + (g_1 - 1)/r\, e_r \hat{\theta}$ $\omega(\hat{\phi}) = g_2/r\,\hat{\phi} e_t + (g_1 - 1)/r\, e_r \hat{\phi}$
Directional derivatives	$L_t = f_1 \partial_t + g_2 \partial_r$ $L_r = g_1 \partial_r$
Equations relating the $\overline{h}$- and ω-functions	$L_t g_1 = G g_2$ $L_r g_2 = F g_1$ $f_1 = \exp\{\int^r -G/g_1\, dr\}$
Definition of M	$M \equiv \frac{1}{2} r({g_2}^2 - {g_1}^2 + 1)$
Remaining derivatives	$L_t g_2 = G g_1 - M/r^2 - 4\pi r p$ $L_r g_1 = F g_2 + M/r^2 - 4\pi r \rho$
Matter derivatives	$L_t M = -4\pi r^2 g_2 p$ $\quad L_t \rho = -(2g_2/r + F)(\rho + p)$ $L_r M = 4\pi r^2 g_1 \rho$ $\quad L_r p = -G(\rho + p)$
Riemann tensor	$\mathcal{R}(B) = 4\pi[(\rho + p) B \cdot e_t e_t - 2\rho/3\, B] - \frac{1}{2}(M/r^3 - 4\pi\rho/3)(B + 3\sigma_r B \sigma_r)$
Fluid stress-energy tensor	$\mathcal{T}(a) = (\rho + p) a \cdot e_t e_t - p a$

Table 15.2. *Equations governing a radially-symmetric perfect fluid.*

Equations (15.47) and (15.48) now combine with that for $L_r p$ to yield the famous Oppenheimer-Volkov equation

$$\frac{\partial p}{\partial r} = -\frac{(\rho + p)(M(r) + 4\pi r^3 p)}{r(r - 2M(r))}. \tag{15.49}$$

Black Holes For a point-source solution, or black hole, matter is concentrated at a single point ($r = 0$). Away from this source the matter equations reduce to

$$\left.\begin{array}{l} L_t M = 0 \\ L_r M = 0 \end{array}\right\} \Rightarrow M = \text{constant}. \tag{15.50}$$

Maintaining the symbol M for this constant we now find that the equations reduce to

$$L_t g_1 = G g_2, \qquad L_r g_2 = F g_1 \tag{15.51}$$

and

$$g_1{}^2 - g_2{}^2 = 1 - 2M/r. \tag{15.52}$$

There are no further equations which yield new information, so the vacuum equations are under-determined. This is because the Riemann tensor reduces to

$$\mathcal{R}(B) = -\frac{M}{2r^3}(B + 3\sigma_r B \sigma_r), \tag{15.53}$$

which is now invariant under boosts in the σ_r plane.

Given this new gauge freedom, we look for a choice of g_1 and g_2 which simplifies the physics as far as possible. The 'Schwarzschild' choice ($g_2 = 0$) fails for $r < 2M$, so is no good. It turns out that, for collapsing dust, g_1 controls the energy of infalling matter at $r = \infty$. A sensible gauge choice is therefore to set $g_1 = 1$ so that

$$\begin{array}{ll} f_1 = 1 & G = 0 \\ g_2 = -\sqrt{2M/r} & F = -M/(g_2 r^2) \end{array} \tag{15.54}$$

In this gauge the $\overline{h}$-function takes the remarkably simple form

$$\overline{h}(a) = a - \sqrt{2M/r}\, a \cdot e_r e_t, \tag{15.55}$$

which is the form employed in Lecture III.

15.3 Stationary, Axially-Symmetric Systems

It is an astonishing fact, that in the 70 years since the discovery of GR, no one has found an analytic solution for the gravitational fields inside a rotating star. This subject therefore provides an ideal test for our intrinsic method. This is very much 'research in progress' but we can make some preliminary observations here. Again, we start by postulating a suitable form of $\overline{h}(a)$. This time $\overline{h}(a)$ should be unchanged by rigid rotations in the $i\sigma_3$ plane only, and the coefficients in $\overline{h}(a)$ should be functions of r and θ. In addition, we make the assumption that the solution is symmetric under simultaneous replacement of t by $-t$ and ϕ by $-\phi$. (This rules out the existence of a horizon.) The most general form of $\overline{h}$ consistent with these criteria is

$$\begin{aligned} \overline{h}(e^t) &= f_1 e^t + f_4 e^\phi \\ \overline{h}(e^r) &= g_1 e^r + g_3 e^\theta \\ \overline{h}(e^\theta) &= i_1 e^\theta + i_3 e^r \\ \overline{h}(e^\phi) &= h_1 e^\phi + h_2 e^t, \end{aligned} \tag{15.56}$$

where the f_i, g_i, h_i and i_i are all scalar functions of r and θ only. This position gauge field generates the following $\omega(a)$:

$$\begin{aligned} \omega_t &= -(T + iJ) e_r e_t - (S + iK)\hat{\theta} e_t + h_2 i\sigma_3 \\ \omega_r &= (S' + iK') e_r \hat{\theta} - i_3 e_r \hat{\theta} \\ \omega_{\hat{\theta}} &= (G' + iJ') e_r \hat{\theta} - (i_1/r) e_r \hat{\theta} \\ \omega_{\hat{\phi}} &= (H + iK)\hat{\theta}\hat{\phi} + (G + iJ) e_r \hat{\phi} + h_1/(r \sin\theta)\, i\sigma_3. \end{aligned} \tag{15.57}$$

(We apologise for the lack of consistency in our labelling schemes.)

The bracket relations generated by (15.57) are as follows:

$$\begin{aligned} &[L_t, L_r] = -TL_t - (K+K')L_{\hat{\phi}} && [L_r, L_{\hat{\theta}}] = -S'L_r - G'L_{\hat{\theta}} \\ &[L_t, L_{\hat{\theta}}] = -SL_t + (J-J')L_{\hat{\phi}} && [L_r, L_{\hat{\phi}}] = -(K-K')L_t - GL_{\hat{\phi}} \\ &[L_t, L_{\hat{\phi}}] = 0 && [L_{\hat{\theta}}, L_{\hat{\phi}}] = (J+J')L_t - HL_{\hat{\phi}}. \end{aligned} \tag{15.58}$$

and the Riemann tensor is

$$\begin{aligned} &\mathcal{R}(e_r e_t) = \\ &\quad [-L_r(T+iJ) + (S+iK)(S'+iK') + T(T+iJ) + i(K+K')(H+iK)]e_r e_t \\ &\quad +[-L_r(S+iK) - (S'+iK')(T+iJ) + T(S+iK) - i(K+K')(G+iJ)]\hat{\theta}e_t \\ &\mathcal{R}(\hat{\theta}e_t) = \\ &\quad [-L_{\hat{\theta}}(S+iK) - (G'+iJ')(T+iJ) + S(S+iK) + i(J-J')(G+iJ)]\hat{\theta}e_t \\ &\quad +[-L_{\hat{\theta}}(T+iJ) + (G'+iJ')(S+iK) + S(T+iJ) - i(J-J')(H+iK)]e_r e_t \\ &\mathcal{R}(\hat{\phi}e_t) = -[(G+iJ)(T+iJ) + (S+iK)(H+iK)]\hat{\phi}e_t \\ &\mathcal{R}(e_r\hat{\theta}) = [L_r(G'+iJ') - L_{\hat{\theta}}(S'+iK') + G'(G'+iJ') + S'(S'+iK')]e_r\hat{\theta} \\ &\mathcal{R}(e_r\hat{\phi}) = \\ &\quad [L_r(G+iJ) - (S'+iK')(H+iK) + G(G+iJ) + i(K-K')(S+iK)]e_r\hat{\phi} \\ &\quad +[L_r(H+iK) + (S'+iK')(G+iJ) + G(H+iK) - i(K-K')(T+iJ)]\hat{\theta}\hat{\phi} \\ &\mathcal{R}(\hat{\theta}\hat{\phi}) = \\ &\quad [L_{\hat{\theta}}(H+iK) + (G'+iJ')(G+iJ) + H(H+iK) + i(J+J')(T+iJ)]\hat{\theta}\hat{\phi} \\ &\quad +[L_{\hat{\theta}}(G+iJ) - (G'+iJ')(H+iK) + H(G+iJ) - i(J+J')(S+iK)]e_r\hat{\phi}. \end{aligned} \tag{15.59}$$

Again we assume that the matter stress-energy tensor is that of a perfect fluid, and this time v is restricted to the form

$$v = \cosh\chi\, e_t + \sinh\chi\, \hat{\phi}, \tag{15.60}$$

where χ is a function of r and θ only.

Gauge Fixing From (15.59) it is clear the most general form for $\mathcal{W}(B)$ consistent with its symmetry properties is

$$\begin{aligned} \mathcal{W}(\sigma_r) &= \alpha_1\sigma_r + \beta\sigma_\theta \\ \mathcal{W}(\sigma_\theta) &= \alpha_2\sigma_\theta + \beta\sigma_r \\ \mathcal{W}(\sigma_\phi) &= -(\alpha_1+\alpha_2)\sigma_\phi \end{aligned} \tag{15.61}$$

where α_1, α_2 and β are all complex scalar+pseudoscalar combinations. But, returning to the initial $\overline{h}$-function (15.56), we see that its structure is unchanged by rotations in the $i\sigma_\phi$ plane and boosts in the σ_ϕ direction. We can use this rotation-gauge freedom to eliminate two degrees of freedom from $\mathcal{W}(B)$, and the simplest choice is to remove the β-terms so that $\mathcal{R}(B)$ becomes diagonal. With the rotation gauge chosen such that $\beta = 0$, the Riemann tensor for an axially-symmetric field becomes

$$\begin{aligned}\mathcal{R}(B) &= -\alpha_1\sigma_r\,\sigma_r\times B - \alpha_2\sigma_\theta\,\sigma_\theta\times B + (\alpha_1+\alpha_2)\sigma_\phi\,\sigma_\phi\times B \\ &\quad +4\pi[(\rho+p)B\cdot vv - \frac{2}{3}\rho B],\end{aligned} \tag{15.62}$$

from which all gauge freedom has been eliminated. The velocity defined by tanhχ is therefore now intrinsic, as it measures the fluid velocity relative to the frame in which the Weyl tensor is diagonal. That it is possible to give a local definition of the fluid velocity in this manner is highly significant for any discussion of Mach's principle in our theory.

15.4 The Kerr Solution

Further analysis of the intrinsic equations is complicated and best performed with a symbolic-algebra package (we prefer to use Maple). It is instructive, however, to see how the Kerr solution fits into this general scheme. In looking for vacuum solutions we find that, after the Bianchi identities are used, we have a set of 16 real equations for the derivatives of the 10 terms in $\omega(a)$. We are therefore four relations short of the full set, and the vacuum setup is under-determined. The Kerr solution is reached by setting each of the four primed quantities equal to their unprimed counterparts, resulting in the beautiful set of equations shown in Table 15.3. Whilst there is some mathematical justification for this step, it seems highly unlikely that it is forced. Indeed, the same simplification applied to the full case with matter forces us to a situation in which both $\rho + 3p$ is constant and the star is rigidly rotating. The simplification leading to the Kerr solution is clearly not consistent with the freedom to impose an arbitrary equation of state.

The structure described by Table 15.3 is truly remarkable and involves abstract mathematics about which little is known (though there is a possible link with current work on integrable systems). The derivatives of the six intrinsic variables $\{G, J, T, S, K, H\}$ are all known, and are all consistent with the bracket relation. The structure is therefore 'complete'. To analyse it we first note that

$$L_r\alpha = -3(G+iJ)\alpha, \qquad L_{\hat\theta}\alpha = 3(S+iK)\alpha \tag{15.63}$$

so, as with the radially-symmetric case, the eigenvalue of the Weyl tensor defines an integrating factor, which plays the role of the intrinsic distance scale. In this case the required integrating factor is a complex quantity, Z, defined by

$$Z \equiv (\alpha_0\alpha)^{-1/3}, \tag{15.64}$$

where α_0 is an arbitrary constant. Z satisfies

$$L_r Z = (G+iJ)Z, \qquad L_{\hat\theta} Z = -(S+iK)Z. \tag{15.65}$$

Bracket relation	$[L_r, L_{\hat{\theta}}] = -SL_r - GL_{\hat{\theta}}$
Riemann tensor	$\mathcal{R}(B) = \frac{1}{2}\alpha(B + 3\sigma_r B\sigma_r)$ $\alpha = (G+iJ)(T+iJ) + (S+iK)(H+iK)$

L_r-Derivatives

$$
\begin{aligned}
&L_r(G+iJ) = -(G+iJ)^2 - T(G+iJ)\\
&L_r(T+iJ) = (S+iK)^2 - [2(G+iJ) - T](T+iJ) - 2S(H+iK)\\
&L_r(S+iK) = -iJ(S+iK) - 2iK(G+iJ)\\
&L_r(H+iK) = -(G+iJ)(S+iK) - G(H+iK)
\end{aligned}
$$

$L_{\hat{\theta}}$-Derivatives

$$
\begin{aligned}
&L_{\hat{\theta}}(S+iK) = (S+iK)^2 + H(S+iK)\\
&L_{\hat{\theta}}(H+iK) = -(G+iJ)^2 + [2(S+iK) - H](H+iK) + 2G(T+iJ)\\
&L_{\hat{\theta}}(G+iJ) = iK(G+iJ) + 2iJ(S+iK)\\
&L_{\hat{\theta}}(T+iJ) = (G+iJ)(S+iK) + S(T+iJ)
\end{aligned}
$$

Table 15.3. *Equations governing the Kerr solution.*

It turns out that the field equations can be recast in the form of complex equations for the variable Z, and that a simple analytic structure underlies the solution. Here we just state the results. On defining

$$\rho \equiv (r^2 + L^2\cos^2\theta)^{1/2}, \qquad \Delta \equiv g(r)^2 = r^2 - 2Mr + L^2, \tag{15.66}$$

the terms in $\omega(a)$ are given by

$$
\begin{array}{ll}
G + iJ = \Delta^{1/2}/(\rho Z) & T - G = -(r-M)/(\rho\Delta^{1/2})\\
S + iK = -iL\sin\theta(\rho Z) & H - S = \cos\theta(\rho\sin\theta),
\end{array} \tag{15.67}
$$

where $Z = r - iL\cos\theta$, M is the mass and L is the angular momentum. The Riemann tensor has the simple form

$$\mathcal{R}(B) = -\frac{M}{2(r - iL\cos\theta)^3}(B + 3\sigma_r B\sigma_r). \tag{15.68}$$

The $\overline{h}$-function leading to this solution is

$$
\begin{aligned}
\overline{h}(e^t) &= (r^2 + L^2)/(\rho\Delta^{1/2})\, e^t + Lr\sin^2\theta/\rho\, e^\phi\\
\overline{h}(e^r) &= \Delta^{1/2}/\rho\, e^r\\
\overline{h}(e^\theta) &= r/\rho\, e^\theta\\
\overline{h}(e^\phi) &= r/\rho\, e^\phi + L/(\rho\Delta^{1/2})\, e^t,
\end{aligned} \tag{15.69}
$$

which gives rise to the standard Kerr metric in Boyer-Lindquist form.

Chapter 16

Gravity V — Further Applications

Anthony Lasenby, Chris Doran, and Stephen Gull
MRAO, Cavendish Laboratory, Madingley Road,
Cambridge, CB3 0HE, UK

We end this series of lectures by looking at three further applications of our approach to gravity — collapsing dust, cosmology, and cosmic strings.

16.1 Collapsing Dust and Black Hole Formation

The simplest model for collapsing matter is one in which the pressure is set to zero so that the situation describes collapsing dust. Applying the formulae summarised in Table 2 of the preceding lecture, we see that $p = 0$ implies that $G = 0$ and $f_1 = 1$. It follows that t is the time measured by freely-falling observers (from ∞). The equations derived in the previous lecture reduce to

$$F = \partial_r g_2 \tag{16.1}$$

$$M(r,t) = \int_0^r 4\pi r'^2 \rho(r',t)\, dr', \tag{16.2}$$

which define F and M on a time slice, together with the update equations

$$\partial_t g_2 + g_2 \partial_r g_2 = -M/r^2 \tag{16.3}$$

$$\partial_t M + g_2 \partial_r M = 0. \tag{16.4}$$

Since g_2 is the velocity of a particle comoving with the fluid, equations (16.3) and (16.4) provide a completely Newtonian description of the fluid. Equation (16.3) is the Euler equation with an inverse-square gravitational force, and (16.4) is the equation for conservation of mass. It is also worth noting that the L_t derivative now plays the role of the 'matter' or 'comoving' derivative for the fluid.

Equations (16.3) and (16.4) enable ρ and M to be propagated from one time-slice to the next so, given suitable initial conditions, one can propagate into the future. The system of equations is in fact soluble analytically, but here we just consider a numerical simulation. From an initial density profile $\rho(r, t_0)$ and velocity profile

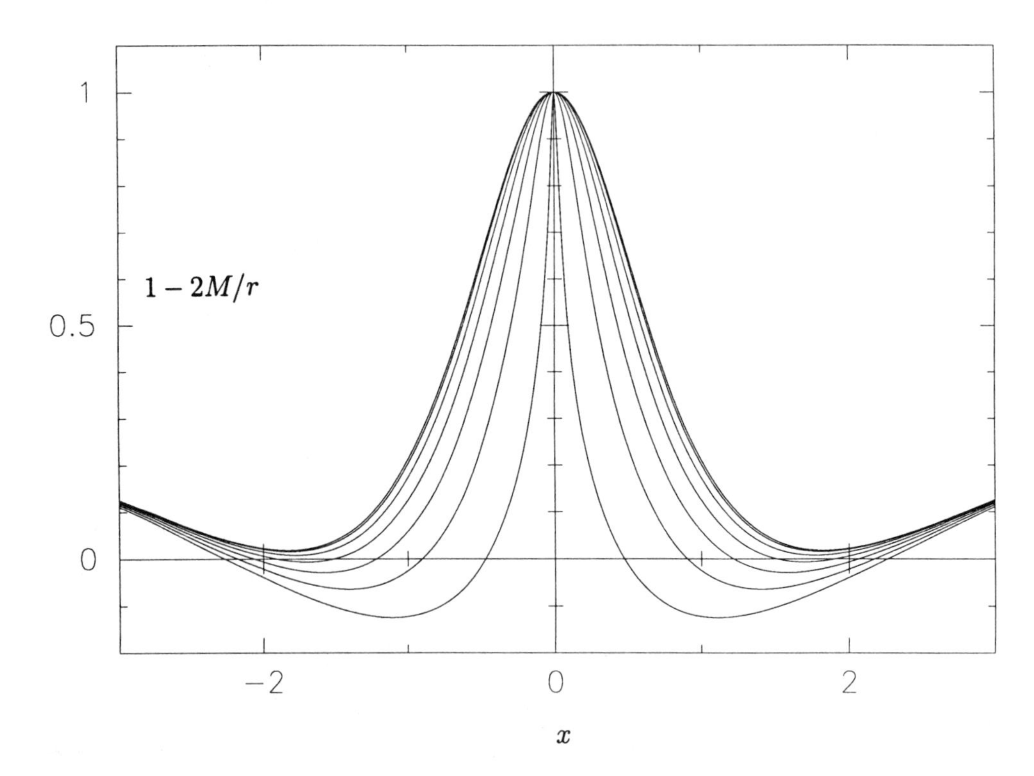

Fig. 16.1. *Simulation of collapsing dust in the Newtonian gauge. Successive time slices for the horizon function* $(1 - 2M(r,t)/r)$ versus x *are shown, with the top curve corresponding to* $t = 0$ *and lower curves to successively later times. The initial velocity is zero, and the initial density profile is of the form* $\rho = \rho_0/(1 + (r/r_0)^2)^2$, *with* $\rho_0 = 0.22$ *and* $r_0 = 1$. *There is no horizon present initially, but a trapped region quickly forms, since in regions where* $1 - 2M/r < 0$ *photons can only move inwards.*

$g_2(r, t_0)$, equation (16.2) is used to obtain $M(r, t_0)$. Equations (16.3) and (16.4) are then used to update M and g_2. On any given time-slice, ρ and F can be recovered using equations (16.1) and (16.2), and g_1 can be recovered from g_2 and M. The results of such a simulation are displayed in Figure 16.1.

Any particle on a radial path has a covariant velocity of the form

$$v = \cosh u \, e_t + \sinh u \, e_r. \tag{16.5}$$

The underlying trajectory has $\dot{x} = \underline{h}(v)$, so the radial motion is determined by

$$\dot{r} = \cosh u \, g_2 + \sinh u \, g_1. \tag{16.6}$$

Since g_2 is negative for collapsing matter, the particle can only achieve an outward velocity if ${g_1}^2 > {g_2}^2$. A horizon therefore forms at the point where

$$2M(r,t)/r = 1. \tag{16.7}$$

The appearance of this horizon is illustrated in Figure 16.1. It is conventional to extend the horizon back in time along the past light-cone to the origin ($r = 0$), since any particle inside this surface could not have reached the point where $2M/r - 1$ first drops to zero, and hence is also trapped.

If pressure is included the pressure gradient causes the internal clock carried by the fluid to run at a different rate from t. The overall picture is similar, however. The horizon forms in a finite external coordinate time and matter has no difficulty crossing the horizon and falling onto the central singularity. The resulting solution is 'maximal' and the end result is a solution of the type discussed in Lecture III. Again, the 'Newtonian' gauge makes the physics so simple that it surely deserves a prominent place in the description of black hole physics.

16.2 Cosmology

The equations derived in the previous lecture are sufficiently general to deal with cosmology as well as astrophysics. In recent years, however, it has once more become fashionable to include a cosmological constant in the field equations. The derivation is largely unaffected by the inclusion of the cosmological term, and only a few modifications are required. The full set of equations with a cosmological constant incorporated are summarised in Table 16.1.

In cosmology we are interested in homogeneous solutions to the equations of Table 16.1. Such solutions are found by setting ρ and p to functions of t only, so it follows immediately from the $L_r p$ equation that

$$G = 0 \quad \Rightarrow f_1 = 1. \tag{16.8}$$

For homogeneous fields the Weyl component of the Riemann tensor must vanish since this contains directional information through the e_r vector. The vanishing of this term requires that

$$M(r,t) = \frac{4}{3}\pi r^3 \rho, \tag{16.9}$$

which is consistent with the $L_r M$ equation. The $L_t M$ and $L_t \rho$ equations now reduce to

$$F = g_2/r \tag{16.10}$$

and

$$\dot{\rho} = -3g_2(\rho + p)/r. \tag{16.11}$$

But we know that $L_r g_2 = F g_1$, which can only be consistent with (16.10) if

$$F = H(t), \qquad g_2(r,t) = rH(t). \tag{16.12}$$

The $L_t g_2$ equation now reduces to a simple equation for $\dot{H}$,

$$\dot{H} + H^2 - \Lambda/3 = -\frac{4\pi}{3}(\rho + 3p). \tag{16.13}$$

The $\overline{h}$-function	The ω-function
$\overline{h}(e^t) = f_1 e^t$	$\omega(e_t) = G e_r e_t$
$\overline{h}(e^r) = g_1 e^r + g_2 e^t$	$\omega(e_r) = F e_r e_t$
$\overline{h}(e^\theta) = e^\theta$	$\omega(\hat{\theta}) = g_2/r\, \hat{\theta} e_t + (g_1 - 1)/r\, e_r \hat{\theta}$
$\overline{h}(e^\phi) = e^\phi$	$\omega(\hat{\phi}) = g_2/r\, \hat{\phi} e_t + (g_1 - 1)/r\, e_r \hat{\phi}$

Directional derivatives	$L_t = f_1 \partial_t + g_2 \partial_r$ $L_r = g_1 \partial_r$
Equations relating the $\overline{h}$- and ω-functions	$L_t g_1 = G g_2$ $L_r g_2 = F g_1$ $f_1 = \exp\{\int^r -G/g_1\, dr\}$
Definition of M	$M \equiv \frac{1}{2} r ({g_2}^2 - {g_1}^2 + 1 - \Lambda r^2/3)$
Remaining derivatives	$L_t g_2 = G g_1 - M/r^2 + r\Lambda/3 - 4\pi r p$ $L_r g_1 = F g_2 + M/r^2 - r\Lambda/3 - 4\pi r \rho$
Matter derivatives	$L_t M = -4\pi g_2 r^2 p \quad L_t \rho = -(2g_2/r + F)(\rho + p)$ $L_r M = 4\pi g_1 r^2 \rho \quad L_r p = -G(\rho + p)$
Riemann tensor	$\mathcal{R}(B) = 4\pi(\rho + p) B \cdot e_t e_t - \frac{1}{3}(8\pi\rho + \Lambda) B$ $-\frac{1}{2}(M/r^3 - 4\pi\rho/3)(B + 3\sigma_r B \sigma_r)$
Fluid stress-energy tensor	$\mathcal{T}(a) = (\rho + p) a \cdot e_t e_t - pa$

Table 16.1. *Equations governing a radially-symmetric perfect fluid — case with a non-zero cosmological constant* Λ. *The shaded equations differ from those of the previous table.*

Finally, we are left with the following pair of equations for g_1:

$$L_t g_1 = 0 \tag{16.14}$$

$$L_r g_1 = ({g_1}^2 - 1)/r. \tag{16.15}$$

The latter equation yields ${g_1}^2 = 1 + r^2 \phi(t)$ and the former reduces to

$$\dot{\phi} = -2H(t)\phi. \tag{16.16}$$

Hence g_1 is given by

$$ {g_1}^2 = 1 - kr^2 \exp\{-2 \int^t H(t')\, dt'\}, \tag{16.17}$$

The $\bar{h}$-function	$\bar{h}(a) = a + a\cdot e_r[(g_1 - 1)e^r + H(t)re^t]$ ${g_1}^2 = 1 - kr^2 \exp\{-2\int^t H(t')\,dt'\}$
The ω-function	$\omega(a) = H(t)a\wedge e_t - (g_1 - 1)/r\, a\wedge(e_r e_t)e_t$
The density	$\frac{8\pi}{3}\rho = H(t)^2 - \Lambda/3 + k\exp\{-2\int^t H(t')\,dt'\}$
Dynamical equations	$\dot{H} + H^2 - \Lambda/3 = -4\pi/3\,(\rho + 3p)$ $\dot{\rho} = -3H(t)(\rho + p)$

Table 16.2. *Equations governing a homogeneous perfect fluid.*

where k is an arbitrary constant of integration. It is straightforward to check that (16.17) is consistent with the equations for $\dot{H}$ and $\dot{\rho}$. The full set of equations describing a homogeneous perfect fluid are summarised in Table 16.2.

At first sight, the equations of Table 16.2 do not resemble the usual Friedmann equations. The Friedmann equations are recovered straight-forwardly, however, by setting

$$H(t) = \frac{\dot{S}(t)}{S(t)}. \tag{16.18}$$

With this substitution we find that

$${g_1}^2 = 1 - kr^2/S^2 \tag{16.19}$$

and the $\dot{H}$ and density equations become

$$\frac{\ddot{S}}{S} - \frac{\Lambda}{3} = -\frac{4\pi}{3}(\rho + 3p) \tag{16.20}$$

and

$$\frac{\dot{S}^2 + k}{S^2} - \frac{\Lambda}{3} = \frac{8\pi}{3}\rho, \tag{16.21}$$

recovering the Friedmann equations in their standard form. The intrinsic treatment has therefore led us to work directly with the 'Hubble velocity' $H(t)$, rather than the 'distance' scale $S(t)$. There is a good reason for this. Once the Weyl tensor is set to zero, the Riemann tensor reduces to

$$\mathcal{R}(B) = 4\pi(\rho + p)B\cdot e_t e_t - \frac{1}{3}(8\pi\rho + \Lambda)B, \tag{16.22}$$

and we have now lost contact with an intrinsically-defined distance scale. We can therefore rescale the radius variable r with an arbitrary function of t (or r) without altering the Riemann tensor. The Hubble velocity, on the other hand, is intrinsic and it is therefore unsurprising that our treatment has led directly to equations for this.

Amongst the class of radial rescalings a particularly useful one is to rescale r to $r' = S(t)r$. This is achieved with the transformation

$$f(x) = x\cdot e_t e_t + Sx\wedge e_t e_t \tag{16.23}$$

so that the transformed $\overline{h}$-function is

$$\overline{h}'(a) = a\cdot e_t e_t + \frac{1}{S}[(1-kr^2)^{1/2} a\cdot e_r e^r + a\wedge(e_r e_t)e_t e_t]. \tag{16.24}$$

The function (16.24) reproduces the standard line element used in cosmology. We can therefore use the transformation (16.23) to move between the 'Newtonian' gauge developed here and the traditional 'static' gauge. The differences between these gauges can be understood by considering geodesic motion. A particle at rest with respect to the cosmological frame (defined by the cosmic microwave background) has $v = e_t$. In the standard 'static' gauge such a particle is not moving in the flatspace background (the distance variable r is equated with the comoving coordinate of GR). For this reason we refer to this gauge as being static, even though the associated line element is usually thought of as defining an expanding spacetime. In the Newtonian gauge, on the other hand, comoving particles are moving outwards radially at a velocity $\dot{r} = H(t)r$, though this expansion centre is not an intrinsic feature. Of course, attempting to distinguish these pictures is a pointless exercise, since all observables must be gauge invariant. All that is of physical relevance is that, if two particles are at rest with respect to the cosmological frame (defined by the cosmic microwave background), then the light-travel time between these particles is an increasing function of time and light is redshifted as it travels between them. The value of this redshift is independent of the gauge in which it is calculated, and attempting to assign the redshift to an expansion of spacetime, or a change in the speed of light, or any other property of the background space is a gauge-dependent description. The fact that alternative gauge choices exist, and can indeed be useful, is ignored in most treatments of cosmology.

Redshifts

As an illustration of the utility of the Newtonian gauge in cosmology, we consider photon paths. We can simplify the problem by specialising to motion in the $\theta = \pi/2$ plane. In this case the photon's momentum can be written as

$$P = \Phi R(\gamma_0 + \gamma_1)\tilde{R}, \tag{16.25}$$

where

$$R = e^{\alpha/2\, i\sigma_3} \tag{16.26}$$

and Φ is the frequency measured by observers comoving with the fluid. We restrict to the pressureless case, so $G = 0$ and $f_1 = 1$, but to add interest we will allow ρ

to be r-dependent and only specialise to the homogeneous case at the end of the calculation.

A simple application of equation (36) from Lecture III produces

$$\partial_\tau \Phi = -\Phi^2 \Big(\frac{g_2}{r} \sin^2 \chi + \partial_r g_2 \cos^2 \chi \Big), \tag{16.27}$$

where $\chi = \phi + \alpha$. But, since $f_1 = 1$, we find that $\partial_\tau t = \Phi$, so

$$\frac{d\Phi}{dt} = -\Phi \Big(\frac{g_2}{r} \sin^2 \chi + \partial_r g_2 \cos^2 \chi \Big), \tag{16.28}$$

which holds in any spherically-symmetric pressureless fluid.

For the case of a cosmological background we have $g_2 = H(t)r$, so the angular terms drop out of equation (16.28), and we are left with the simple equation

$$\frac{d\Phi}{dt} = -H(t)\Phi = \dot{\rho}/3\rho. \tag{16.29}$$

This integrates to give the familiar redshift versus density relation

$$1 + z = (\rho_1/\rho_0)^{1/3}, \tag{16.30}$$

where ρ_0 is the density at the present epoch, and ρ_1 is the density at the epoch of emission.

16.2.1 The Dirac Equation in a Cosmological Background

As another application of the theory outlined here we consider the Dirac equation in a cosmological background. At this point it is slightly easier to work with the $\overline{h}$-function in the more conventional diagonal form of (16.24). But clearly there is a problem here if k is positive, since the square root in (16.24) is ill-defined for $r > k^{-1/2}$. This can be viewed as an artefact of the means by which this solution maps a 3-sphere onto Euclidean space. We can remove the difficulty by using a stereographic projection to achieve a globally-defined $\overline{h}$-function, though this is not necessary for the application discussed here. (The stereographic projection is useful in other applications, however, and is discussed further below.)

The Dirac equation using our intrinsic notation is

$$\partial_a [L_a + \tfrac{1}{2}\omega(a)] \psi i\sigma_3 = m\psi\gamma_0, \tag{16.31}$$

and we seek solutions to this in the background of a dust model for cosmology. With the $\overline{h}$-function in the form of (16.24) the relevant equation is

$$\begin{aligned} &\left(e^t \partial_t + \frac{1}{S}[(1-kr^2)^{1/2} e^r \partial_r + e^\theta \partial_\theta + e^\phi \partial_\phi] \right) \psi i\sigma_3 \\ &\quad + \tfrac{1}{2}(3H(t)e_t - 2[(1-kr^2)^{1/2} - 1]e_r)\psi i\sigma_3 = m\psi\gamma_0. \end{aligned} \tag{16.32}$$

The question we wish to address is this: can we find solutions to (16.32) such that the observables are homogeneous? There is clearly no difficulty if $k = 0$, since equation (16.32) is solved by

$$\psi = \rho^{1/2} e^{-i\sigma_3 mt} \tag{16.33}$$

and the observables are fixed vectors which just scale as $\rho(t)$ in magnitude. But what happens when $k \neq 0$? It turns out that the solution (16.33) must now be modified to

$$\psi = \frac{\rho^{1/2}}{1+\sqrt{1-kr^2}} e^{-i\sigma_3 mt}. \tag{16.34}$$

But this is no longer homogeneous, and the observables obtained form (16.34) all contain additional r-dependence as well as scaling as $\rho(t)$. In principle, therefore, one could determine the origin of this space with measurements of the current density. This clearly violates the principle of homogeneity, though it is not inconsistent with experiment. (One consequence is that it is only possible to construct self-consistent solutions to the coupled Einstein-Dirac equations if $k = 0$.)

So how can it be that classical phenomena do not see this 'preferred' direction in $k \neq 0$ models, but the quantum spinor ψ does? The answer lies in the gauge structure of the theory. The 'minimal-coupling' procedure couples Dirac spinors directly to the ω-function, which in this case is not homogeneous. Dirac spinors can therefore probe the theory directly at the level of the ω-field, whereas classical quantities only interact at the level of the covariant quantities obtained from the gravitational fields (which are homogeneous). Whilst the above result is not conclusive, it does strongly suggest that cosmological models with $k \neq 0$ are inconsistent with the assumption of homogeneity. (This does not rule out spatially-flat universes with a non-zero cosmological constant.)

16.2.2 Point Charge in a $k > 0$ Cosmology

As a final application of the cosmology equations, we consider the fields due to a point charge in a $k > 0$ background. To study these we must first address the problem that the solution (16.24) is only defined $r < k^{-1/2}$. To obtain a global solution we apply the displacement

$$f(x) = x \cdot e_t e_t + \frac{r}{1+kr^2/4} e_r, \tag{16.35}$$

which results in the simple solution

$$\overline{h}'(a) = a \cdot e_t e_t + \frac{1}{S}(1+kr^2/4) a \wedge e_t e_t. \tag{16.36}$$

This solution can be viewed as resulting from a stereographic projection of a 3-sphere. It is well-defined for all x and generates an 'isotropic' line element.

To find the fields due to a point charge in the background defined by (16.36) we start with the natural ansätz

$$A = \frac{1}{S} V(\boldsymbol{x})\, e_t, \tag{16.37}$$

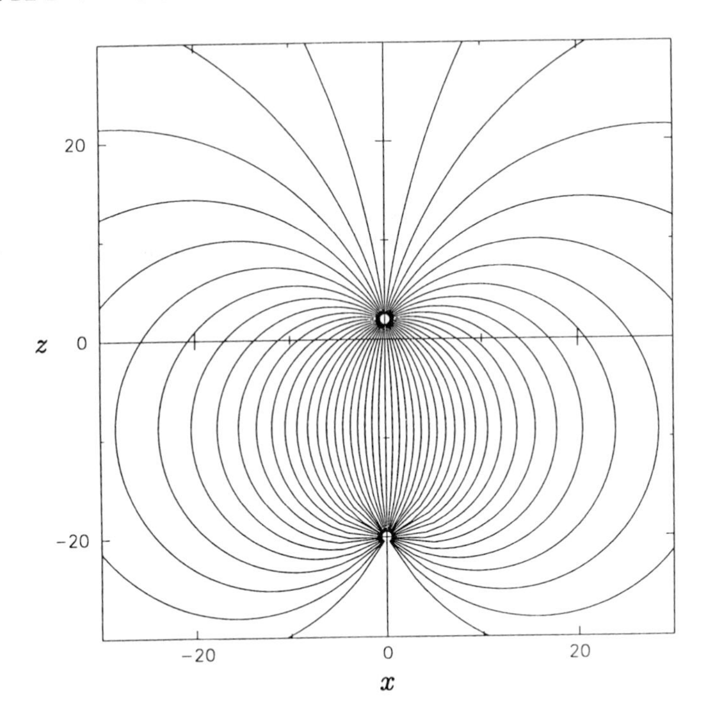

Fig. 16.2. *Fieldlines of the* $\boldsymbol{D}$*-field for a point charge in a* $k > 0$ *universe. The fieldlines follow null geodesics, which are circles in this gauge. The existence of an image charge is clear.*

so that

$$\boldsymbol{E} = -\frac{1}{S}\boldsymbol{\nabla} V \tag{16.38}$$

and

$$\boldsymbol{D} = -[1 + kr^2/4]^{-1}\boldsymbol{\nabla} V. \tag{16.39}$$

It follows that the equation we need to solve is simply

$$-\boldsymbol{\nabla}\cdot\Big([1 + kr^2/4]^{-1}\boldsymbol{\nabla} V\Big) = q\delta(\boldsymbol{x} - \boldsymbol{a}), \tag{16.40}$$

where the charge q is located at $\boldsymbol{x} = \boldsymbol{a}$. The solution to this equation turns out to be

$$V = \frac{1 + ka^2/4 - ks/2}{[s(1 + ka^2/4 - ks/4)]^{1/2}}, \tag{16.41}$$

where

$$s = \frac{(\boldsymbol{x} - \boldsymbol{a})^2}{1 + kr^2/4}, \qquad a = \sqrt{\boldsymbol{a}^2}. \tag{16.42}$$

Fieldline plots of the $\boldsymbol{D}$ field defined from (16.41) are shown in Figure 16.2. The fieldlines follow null geodesics and clearly reveal the existence of an image charge. The reason for this is can be seen in the denominator of V, which is singular at both $s = 0$ and

$$1 + ka^2/4 - ks/4 = 0. \tag{16.43}$$

So, if the charge lies on the z axis, its image charge is located at

$$z = -4/(ka). \tag{16.44}$$

If we try to remove this image charge by adding a point source at its position, we find that the fields vanish everywhere, since the new source has its own image which cancels the original charge. The image charge is therefore an unavoidable feature of $k > 1$ cosmologies. The existence of these image charges clearly raises problems in attempting to take $k > 0$ universes seriously.

16.3 Cosmic Strings

For our final application we consider solutions representing a 'cosmic string'. Cosmic strings are an example of a field configuration known as a 'topological defect', which are currently of great interest because of their potential to seed for structure formation in the early universe. As well as their significance in cosmology, the study of cosmic string solutions also demonstrates how our theory can handle topologically interesting solutions without ever leaving a flat spacetime.

We start with the introduction of a cylindrical-polar coordinate system:

$$\begin{aligned} t &\equiv x\cdot\gamma_0 & \tan\phi &\equiv (x\cdot\gamma^2)/(x\cdot\gamma^1) \\ \rho &\equiv \sqrt{-(x\wedge\sigma_3)^2} & z &\equiv x\cdot\gamma^3. \end{aligned} \tag{16.45}$$

Note that we have now adopted the symbol ρ for the distance from the z-axis, and from now on we will use the symbol ε for the energy density. The associated coordinate frame is

$$\begin{aligned} e_t &= \gamma_0 & e_\phi &\equiv \rho(-\sin\phi\,\gamma_1 + \cos\phi\,\gamma_2) \\ e_\rho &= \cos\phi\,\gamma_1 + \sin\phi\,\gamma_2 & e_z &\equiv \gamma_3, \end{aligned} \tag{16.46}$$

with the dual-frame vectors denoted as $\{e^t, e^\rho, e^\phi, e^z\}$. We also define the unit bivector

$$\sigma_\rho \equiv e_\rho e_t. \tag{16.47}$$

We are interested in stationary systems for which the z-axis drops out of consideration, so that we effectively simulate gravity in 2+1 dimensions. We will also impose the restriction that our system is symmetric under simultaneous reversal of the t and ϕ directions. This requires that there be no coupling between the e_ρ and e_t or e_ϕ directions, and therefore rules out systems in which a horizon is present. The most general form of $\overline{h}$-function meeting these criteria is given by

$$\begin{aligned} \overline{h}(e^t) &= f_1 e^t + \rho f_4 e^\phi & \overline{h}(e^\phi) &= \rho h_1 e^\phi + h_2 e^t \\ \overline{h}(e^\rho) &= g_1 e^\rho & \overline{h}(e^z) &= e^z, \end{aligned} \tag{16.48}$$

where the ρ's have been included on f_4 and h_1 to simplify some later equations. The scalar functions appearing in (16.48) are all functions of ρ only.

Following the route described in Lecture IV, we next introduce the ω-field

$$\begin{aligned} \omega_t &= -T\sigma_\rho + (K+h_2)i\sigma_3 & \omega_{\hat\phi} &= K\sigma_\rho + (h_1 - G)i\sigma_3 \\ \omega_\rho &= K'\sigma_\phi & \omega_z &= 0. \end{aligned} \tag{16.49}$$

Again, the new scalar functions appearing here (T, K, K', G) are functions of ρ alone. Since all expressions involving L_z must vanish, there are only three non-vanishing bracket relations implied by (16.49). These are

$$\begin{aligned} [L_\rho, L_t] &= TL_t + (K+K')L_{\hat\phi} \\ [L_\rho, L_{\hat\phi}] &= -(K-K')L_t - GL_{\hat\phi} \\ [L_t, L_{\hat\phi}] &= 0, \end{aligned} \tag{16.50}$$

and since neither L_t or $L_{\hat\phi}$ contain derivatives with respect to ρ, these bracket relations immediately yield

$$\begin{aligned} L_\rho f_1 &= Tf_1 + (K+K')f_4 & L_\rho f_4 &= -Gf_4 - (K-K')f_1 \\ L_\rho h_1 &= -Gh_1 - (K-K')h_2 & L_\rho h_2 &= Th_2 + (K+K')h_1. \end{aligned} \tag{16.51}$$

The Riemann tensor has the form

$$\begin{aligned} \mathcal{R}(\sigma_\rho) &= \alpha_1\sigma_\rho + \beta i\sigma_3 \\ \mathcal{R}(i\sigma_3) &= \alpha_2 i\sigma_3 - \beta\sigma_\rho \\ \mathcal{R}(\sigma_\phi) &= \alpha_3\sigma_\phi, \end{aligned} \tag{16.52}$$

with all other terms being zero. The scalar functions appearing in $\mathcal{R}(B)$ are defined by

$$\alpha_1 = -L_\rho T + T^2 - K(K+2K') \tag{16.53}$$

$$\alpha_2 = L_\rho G + G^2 - K(K-2K') \tag{16.54}$$

$$\alpha_3 = K^2 - GT \tag{16.55}$$

$$\beta = L_\rho K + G(K+K') - T(K-K'). \tag{16.56}$$

The Bianchi identity reduces to the single scalar equation

$$L_\rho\alpha_3 + T(\alpha_2 - \alpha_3) + G(\alpha_3 - \alpha_1) - 2K\beta = 0 \tag{16.57}$$

which is satisfied automatically by virtue of equations (16.53)—(16.56).

The Einstein tensor is given by a simple rearrangement of the above terms:

$$\begin{aligned} \mathcal{G}(e_t) &= -\alpha_2 e_t - \beta\hat\phi \\ \mathcal{G}(e_\rho) &= -\alpha_3 e_\rho \\ \mathcal{G}(\hat\phi) &= -\alpha_1\hat\phi + \beta e_t \\ \mathcal{G}(e_z) &= -(\alpha_1 + \alpha_2 + \alpha_3)e_z, \end{aligned} \tag{16.58}$$

and this is to be equated with the matter stress-energy tensor. Two features of the Einstein tensor are immediately apparent. The first is that the off-diagonal term in both $\mathcal{R}(B)$ and $\mathcal{G}(a)$ is controlled by β. The same frame diagonalises both the

Riemann tensor and the Einstein tensor, and there is no mutual rotation between them. This contrasts with the axisymmetric case, and is a result of the fact that we are effectively working in a 2+1 dimensional system. The second feature of $\mathcal{G}(a)$ is that the coefficient of $\mathcal{G}(e_z)$ is determined algebraically by the other three coefficients. The same must be true of the matter stress-energy tensor, and the e_z-component of the Einstein equations contains no new information.

The $\overline{h}$-function (16.48) contains a single rotation-gauge freedom to perform a boost in the σ_ϕ plane. We can use this to set $\beta = 0$, so that all rotation-gauge freedom has now been absorbed. Next we must choose a form for the matter stress-energy tensor. We will only consider the simplest system, in which there is no transverse pressure and the string is not rotating. In this case $\mathcal{T}(a)$ can be written neatly as

$$\mathcal{T}(a) = \tfrac{1}{2}\varepsilon(\rho)(a - i\sigma_3\, a\, i\sigma_3), \tag{16.59}$$

and the Riemann tensor therefore has the compact form

$$\mathcal{R}(B) = 8\pi\varepsilon\, B\cdot i\sigma_3\, i\sigma_3. \tag{16.60}$$

We now see that $\alpha_1 = \alpha_3 = \beta = 0$, and equation (16.57) immediately sets $T = 0$. The remaining equations then set K and K' to zero, and all that remains is the single intrinsic equation

$$L_\rho G + G^2 = -8\pi\varepsilon, \tag{16.61}$$

together with equations (16.51). From equation (16.51) we see that both f_1 and h_2 are constant, and a global rotation and time dilatation can be used to set $f_1 = 1$ and $h_2 = 0$.

The remaining equations from (16.51) are

$$L_\rho h_1 = -Gh_1, \qquad L_\rho f_4 = -Gf_4, \tag{16.62}$$

and it follows that $f_4 = \lambda h_1$, where λ is an arbitrary constant. But the requirement that $\overline{h}(a)$ be well-defined on the z-axis implies that $\lambda = 0$, so we are left with $\overline{h}(a)$ in the simple form

$$\overline{h}(a) = a + (g_1 - 1)a\cdot e_\rho e^\rho + (\rho h_1 - 1)a\cdot e_\phi e^\phi. \tag{16.63}$$

The remaining equations are now

$$\begin{aligned} L_\rho h_1 &= -Gh_1 & (16.64)\\ L_\rho G &= -8\pi\varepsilon - G^2, & (16.65) \end{aligned}$$

and to complete the solution we must make a gauge choice for g_1. To see how to make this we note that

$$\partial_\rho(G/h_1) = -8\pi\varepsilon(g_1 h_1)^{-1}, \tag{16.66}$$

hence

$$G/h_1 = 1 - 4\mu(\rho) \tag{16.67}$$

where $\mu(\rho)$ is the covariant mass per unit length,

$$\mu(r) = \int_0^\rho 2\pi s\varepsilon \frac{ds}{sh_1 g_1}. \tag{16.68}$$

The gauge choice usually employed in the literature is $g_1 = 1$, but the analogy with spherically-symmetric systems suggests that a better choice is

$$\rho h_1 = 1, \tag{16.69}$$

since with this choice we have

$$g_1{}^2 = 1 - 8M(\rho), \tag{16.70}$$

where

$$M(\rho) = \int_0^\rho 2\pi s\varepsilon \, ds. \tag{16.71}$$

Our final solution is now

$$\overline{h}(a) = a + [\sqrt{1 - 8M(\rho)} - 1]a\cdot e_\rho e^\rho, \tag{16.72}$$

which works for any density distribution, provided that $M(\rho) < 1/8$. The exterior solution is defined by

$$\overline{h}(a) = a + [\sqrt{1 - 8M} - 1]a\cdot e_\rho e^\rho, \tag{16.73}$$

where M is the value of $M(\rho)$ at the boundary. The Riemann tensor from (16.73) vanishes, so one might expect to find a gauge transformation which takes (16.73) to the identity. It turns out that the required transformation involves combining the rotation defined by

$$R = \exp\{2\mu\phi i\sigma_3\}, \qquad \mu = \tfrac{1}{4}(1 - \sqrt{1 - 8M}) \tag{16.74}$$

with the displacement

$$f(x) = \tilde{R}xR. \tag{16.75}$$

But the displacement (16.75) presents us with a problem, because the new angle ϕ' defined from $x' = f(x)$ is given by

$$\phi' = (1 - 4\mu)\phi. \tag{16.76}$$

Since $0 < \phi < 2\pi$, the new angle ϕ' only runs from zero through to $2\pi(1-4\mu)$, which is less than 2π. The gauge transformation is therefore not well-defined over all space, as it opens up a 'deficit' angle of $8\pi\mu$. This is the gauge theory equivalent of the standard picture of the exterior string metric as representing flat spacetime with a wedge of space removed and the edges identified. From the gauge theory perspective, the solution (16.73) has a vanishing Riemann tensor, but no globally-defined gauge transformation exists which takes (16.73) to the identity. This is precisely analogous to the case of the electromagnetic gauge field A in the region exterior to a solenoid, as used, for example, in an Aharonov–Bohm type experiment. In this case the field strength F vanishes in the external region, but A still has physical effects there since no global gauge choice exists that sets $A = 0$ throughout the exterior while keeping it matched at the boundary of the solenoid.

Chapter 17

The Paravector Model of Spacetime

William E. Baylis
Department of Physics, University of Windsor,
Windsor, Ont., Canada N9B 3P4

The geometric (or Clifford) algebra $C\ell_3$ of three-dimensional Euclidean space is endowed with a natural complex structure on a four-dimensional space. If paravectors, which are formed sums of scalars and vectors, are taken as the "real" elements of the space, then the space can be shown to have a Minkowski spacetime metric, and the paravectors may be identified with spacetime vectors. Physical Lorentz transformations of spacetime vectors are described by spin transformations of the paravectors. The transformation elements are unimodular elements of the algebra, and they form the six-parameter group SL(2,C), the two-fold covering group of restricted Lorentz transformations, $SO_+(1,3)$. Its elements are also reducible spinors, whose elements carry a reducible spin representation of SL(2,C). The spin representation is reduced by splitting elements into complementary minimal left ideals. The complex spin space that results has a symplectic structure and its elements belong to Sp(2).

In this lecture, the mathematics, notations, and interpretation of the paravector model is developed. It will be seen to provide a covariant description of relativistic phenomena and to emphasize the importance of the relation of events relative to an observer. The eigenspinor concept will be introduced in the second lecture and applied there to problems in classical electrodynamics. Extensions to quantum theory follow in the third lecture.

17.1 A Brief Introduction to the Pauli Algebra

The Pauli algebra $C\ell_3$ is Clifford's *geometric algebra* of three-dimensional Euclidean space. Its essential feature is an associative product of vectors, distributive over addition, which allows one to construct inverses, square roots, and other functions of vectors much as you would with real numbers. The main differences between the algebra of vectors and a field of numbers are (1) that the product of noncollinear vectors is not abelian (commutative): $\mathbf{ab} \neq \mathbf{ba}$ and (2) there exist zero-divisors[1], that is, elements which are not zero themselves, but when multiplied together equal zero: $\mathbf{uv} = 0,\ \mathbf{u} \neq 0,\ \mathbf{v} \neq 0$. This feature endows the algebra of vectors with a richer structure than a field.

The principal advantage of geometric algebra is that it provides economy and clarity in the expression of geometrical relationships on which physics in intimately based. Rotations and reflections, for example, are easily expressed and calculated. The algebra also avoids cumbersome component notation with its implicit dependence on an assumed set of coordinates.

17.1.1 Generating $\mathcal{C}\ell_3$

The algebra is generated as shown in the introduction by vectors of real three-dimensional space $\mathbb{R}^3$ together with an associative, bilinear vector product which satisfies the basic axiom that the square of a vector is its length squared:

$$\mathbf{u}^2 = \mathbf{u} \cdot \mathbf{u} \tag{17.1}$$

for any vector $\mathbf{u} \in \mathbb{R}^3$, where the dot product is the usual symmetric scalar product: $\mathbf{u} \cdot \mathbf{v} = \mathbf{v} \cdot \mathbf{u} \in \mathbb{R}$. The axiom (17.1) is sufficient to determine the structure of the algebra. In terms of orthonormal basis vectors $\mathbf{e}_j,\ j = 1, 2, 3$:

$$\mathbf{e}_j\mathbf{e}_k + \mathbf{e}_k\mathbf{e}_j = 2\delta_{jk}\,. \tag{17.2}$$

Exercise 17.1.1 *Derive (17.2) from (17.1).*

Exercise 17.1.2 *Show that any product of three vectors in $\mathbb{R}^3$ is a combination of a vector and a trivector, representing a volume, and that all trivectors are scalar multiples of the basis trivector $\mathbf{e}_1\mathbf{e}_2\mathbf{e}_3$.*

Exercise 17.1.3 *Find the size of the bivector and trivector spaces in an n-dimensional Euclidean space. Verify that when $n = 3$, the sizes are 3 and 1, respectively.*

Exercise 17.1.4 *Consider a two-dimensional pseudo-Euclidean space with a metric tensor g_{jk}, whose basis elements $\mathbf{e}_1, \mathbf{e}_2$ obey*

$$\mathbf{e}_j\mathbf{e}_k + \mathbf{e}_k\mathbf{e}_j = 2g_{jk} \tag{17.3}$$

where $g_{11} = -g_{22} = 1$ and $g_{12} = g_{21} = 0$. Find the dimensionality of the bivector space and find the square $(\mathbf{e}_j\mathbf{e}_k)^2$ of any unit bivector $(j \neq k)$.

All elements in $\mathcal{C}\ell_3$ can be expressed as a linear combination of a scalar, a vector, a bivector, and a trivector, which are sometimes referred to as elements of grade 0, 1, 2, and 3, respectively. A full basis for the real algebra of three-dimensional space thus comprises eight elements:

$$\{1, \mathbf{e}_j, \mathbf{e}_j\mathbf{e}_k, \mathbf{e}_1\mathbf{e}_2\mathbf{e}_3\}\,, \quad 1 \leq j < k \leq 3 \tag{17.4}$$

[1] Except for the three geometric algebras identified with the real and complex fields and the quaternions.

17.1.2 Bivectors as Operators

One of the most important concepts in geometric algebra that is missing from traditional vector treatments is the bivector. As discussed in the introduction, it can be interpreted simply as a patch of rotation plane, but its use as an operator for rotations and reflections may be more important. Thus, the unit vector obtained from $\mathbf{e}_1$ by a rotation by θ in the $\mathbf{e}_1\mathbf{e}_2$ plane is

$$\mathbf{e}_\theta := \mathbf{e}_1 \cos\theta + \mathbf{e}_2 \sin\theta = \mathbf{e}_1 \exp\left(\mathbf{e}_1\mathbf{e}_2\theta\right) = \exp\left(\mathbf{e}_2\mathbf{e}_1\theta\right)\mathbf{e}_1\,, \tag{17.5}$$

and this can be rearranged to give

$$\mathbf{e}_1\mathbf{e}_\theta = \exp\left(\mathbf{e}_1\mathbf{e}_2\theta\right) = \cos\theta + \left(\mathbf{e}_1\mathbf{e}_2\right)\sin\theta\,. \tag{17.6}$$

To rotate a vector $\mathbf{r}$ that does not necessarily lie in the rotation plane, one can use

$$\begin{aligned} \mathbf{r}' &= \exp\left(\mathbf{e}_2\mathbf{e}_1\theta/2\right)\mathbf{r}\exp\left(\mathbf{e}_1\mathbf{e}_2\theta/2\right) \\ &= \left(\mathbf{e}_{\theta/2}\mathbf{e}_1\right)\mathbf{r}\left(\mathbf{e}_1\mathbf{e}_{\theta/2}\right) \\ &= \left(\mathbf{e}_{\theta/2}\mathbf{e}_3\right)\left(\mathbf{e}_3\mathbf{e}_1\right)\mathbf{r}\left(\mathbf{e}_1\mathbf{e}_3\right)\left(\mathbf{e}_3\mathbf{e}_{\theta/2}\right)\,. \end{aligned} \tag{17.7}$$

The last line, obtained by inserting $\left(\mathbf{e}_3\right)^2 = 1$, demonstrates the equivalence of a rotation to a pair of successive reflections in intersecting planes. The angle of rotation is seen to be twice the angular opening between the planes.

Exercise 17.1.5 *Prove that the transformation (17.7) rotates bivectors as well as vectors.*

Exercise 17.1.6 *Show that bivectors are invariant under rotations in the plane of the bivector.*

Exercise 17.1.7 *Verify that the equations for reflections and rotations work for Euclidean spaces of any dimension n. Show that they also work if the metric is negative definite: $\left(\mathbf{e}_j\right)^2 = -1,\ j = 1, 2, 3, \cdots$.*

17.1.3 Complex Structure

The size of the $\mathcal{C}\ell_3$ basis can be cut in half if its natural complex structure is recognized. As shown in the introduction, the *canonical* element $\mathbf{e}_1\mathbf{e}_2\mathbf{e}_3$ squares to -1 and commutes with all other elements. Its function in the algebra is therefore exactly that of the imaginary i (or $-i$). We choose to set

$$\mathbf{e}_1\mathbf{e}_2\mathbf{e}_3 = i \tag{17.8}$$

for a right-handed coordinate system: $\mathbf{e}_1 \times \mathbf{e}_2 = \mathbf{e}_3$. The element i in the algebra represents a *handedness*. For example, a twist from $\mathbf{e}_1$ to $\mathbf{e}_2$ is coupled to a thrust along $\mathbf{e}_3$. Under inversion of the basis vectors $\mathbf{e}_j \to -\mathbf{e}_j$ the coordinate system becomes left-handed and the product $\mathbf{e}_1\mathbf{e}_2\mathbf{e}_3 \to -i$. With the basis trivector associated with i, all trivectors of the algebra become imaginary scalars. The real scalars and the trivectors together form what is called the *center* of the algebra, that is the subalgebra that commutes with all elements; it may be identified with the field of complex numbers, and as a result, the Pauli algebra is said to possess a natural complex structure. Every bivector of $\mathcal{C}\ell_3$ can be written as an imaginary vector, for example

$$\mathbf{e}_1\mathbf{e}_2 = \mathbf{e}_1\mathbf{e}_2\mathbf{e}_3/\mathbf{e}_3 = i/\mathbf{e}_3 = i\mathbf{e}_3 . \tag{17.9}$$

The bivector $\mathbf{e}_1\mathbf{e}_2$ and the vector $\mathbf{e}_3$ are said to be (Hodge) *duals* of each other because their product is the canonical element:

$$(\mathbf{e}_1\mathbf{e}_2)\,\mathbf{e}_3 = i = \mathbf{e}_3\,(\mathbf{e}_1\mathbf{e}_2) . \tag{17.10}$$

More generally, the cross product of vectors is dual to the bivector formed from them:

$$\mathbf{a} \wedge \mathbf{b} = i\mathbf{a} \times \mathbf{b} . \tag{17.11}$$

Having recognized the natural complex structure of $\mathcal{C}\ell_3$, we can now express every element as a complex linear combination of just four basis elements: $1, \mathbf{e}_1, \mathbf{e}_2, \mathbf{e}_3$. The geometric algebra generated by a real three-dimensional Euclidean space thus operates in a complex four-dimensional vector space which contains both the complex field and the original three-dimensional space as subspaces. A general element $p \in \mathcal{C}\ell_3$ is the sum of a scalar p_0 and a vector $\mathbf{p}$, both of which may be complex:

$$p = p_0 + \mathbf{p} . \tag{17.12}$$

It is often called a *paravector.*[2] We should remember, however, the geometrical significance of i in the algebra: an imaginary scalar represents a twist in a plane coupled to a thrust along a perpendicular direction, whereas an imaginary vector represents a twist in the spatial plane perpendicular to the vector.

17.1.4 Involutions of $\mathcal{C}\ell_3$

The transformation of *complex conjugation* is an important concept in complex numbers. Similar conjugations, known as *involutions*, appear in the Pauli algebra. Because the algebra has a richer structure than complex numbers, it has more than one conjugation. An involution of $\mathcal{C}\ell_3$ is an invertible transformation within $\mathcal{C}\ell_3$, that is a one-to-one mapping $p \to \bar{p} \in \mathcal{C}\ell_3$ and whose square is unity: $\bar{p} \to \bar{\bar{p}} = p$. Because each element of $\mathcal{C}\ell_3$ is mapped into a unique element, an equality of elements is preserved by the mapping: $p = \pi \Rightarrow \bar{p} = \bar{\pi}$.

The involution of *spatial reversal,* also called *clifford conjugation,* reverses the sign on the (complex) vector part of the element:

$$p = p_0 + \mathbf{p} \to \bar{p} = p_0 - \mathbf{p} \tag{17.13}$$

[2]The name *paravector* for the sum of a scalar and a vector, is common in all Clifford algebras. It is due to J. G. Maks, Doctoral Dissertation, Technische Universiteit Delft (the Netherlands), 1989, p. 22. A general element of a Clifford algebra is called a *clifford number* or a *cliffor.*

It is readily verified that the spatial reversal of a product reverses the order of multiplication[3]: $\overline{pq} = \bar{q}\bar{p}$. With this involution, any element can be split into scalar and vector parts:

$$\begin{aligned} p &= \tfrac{1}{2}(p + \bar{p}) + \tfrac{1}{2}(p - \bar{p}) \\ &=: \langle p \rangle_S + \langle p \rangle_V \end{aligned} \tag{17.14}$$

where $\langle p \rangle_S$ and $\langle p \rangle_V$ are respectively the (possibly complex) scalar and vector parts of the element p. The scalar and vector parts of a product may be thought of as extensions of the scalar (dot) and exterior (wedge) products of vectors to general cliffors of $\mathcal{C}\ell_3$:

$$\begin{aligned} \langle pq \rangle_S &:= \tfrac{1}{2}(pq + \overline{pq}) \\ \langle pq \rangle_V &:= \tfrac{1}{2}(pq - \overline{pq}) \,. \end{aligned} \tag{17.15}$$

A cliffor p is a scalar if and only if p is its own spatial reversal. In particular, $p\bar{p} = \overline{p\bar{p}}$ is always a scalar. As we will show in the third lecture, spatial reversal is not to be confused with *spatial inversion.*

Hermitean conjugation, $p \to p^\dagger$, is another important involution. It is like complex conjugation except that it reverses the order of products: $(pq)^\dagger \to q^\dagger p^\dagger$. It changes every i to $-i$ and reverses the order of multiplication. If an element $p \in \mathcal{C}\ell_3$ is expanded in the *tetrad basis* $\{\mathbf{e}_0 \equiv 1, \mathbf{e}_1, \mathbf{e}_2, \mathbf{e}_3\}$, then Hermitean conjugation is effected by taking the complex conjugate (indicated here by an asterisk) of each coefficient:

$$p = p^\mu \mathbf{e}_\mu \to p^\dagger = p^{\mu *} \mathbf{e}_\mu \tag{17.16}$$

where repeated indices in any term are summed over their allowed values; in (17.16), the sums are over $\mu = 0, 1, 2, 3$. Hermitean conjugation is used to split elements into *real* and *imaginary* parts:

$$\begin{aligned} p &= \tfrac{1}{2}(p + p^\dagger) + \tfrac{1}{2}(p - p^\dagger) \\ &=: \langle p \rangle_\Re + \langle p \rangle_\Im \,. \end{aligned} \tag{17.17}$$

Hermitean conjugation is also known as *reversal,* since the Hermitean conjugate of every product of real spatial vectors is simply the product in reverse order and every basis element (17.4) is such a product. By using both involutions of spatial reversal and Hermitean conjugation, all four grades of a general element (real scalar, imaginary scalar, real vector, imaginary vector) can be isolated. For example, the real scalar part of p is

$$\langle p \rangle_{\Re S} = \frac{1}{4}\left(p + p^\dagger + \bar{p} + \bar{p}^\dagger\right) \,. \tag{17.18}$$

A cliffor p is real *if*and only if p is its own hermitean conjugate. In particular, the product $pp^\dagger = \left(pp^\dagger\right)^\dagger$ is always real.

The combination of spatial reversal and Hermitean conjugation changes the signs of the real vectors and the imaginary scalars. In terms of the eight-dimensional real

[3]An invertible mapping of elements into elements that preserves the order of multiplication is called an *automorphism.* If the order is reversed, the mapping is said to be an *antiautomorphism.* Some mathematicians define involutions to be antiautomorphic, but the term is also often applied to automorphisms (that are their own inverse).

Pauli algebra, these are the *odd* elements, that is those constructed from products of an odd number of real vectors. The *even* elements are unchanged. Any element $p \in \mathcal{C}\ell_3$ can thus be split into even and odd parts:

$$\begin{aligned} p &= \tfrac{1}{2}\left(p + \bar{p}^{\dagger}\right) + \tfrac{1}{2}\left(p - \bar{p}^{\dagger}\right) \\ &=: \langle p \rangle_{\mathcal{E}} + \langle p \rangle_{\mathcal{O}} \end{aligned} , \tag{17.19}$$

and the transformation $p \to \bar{p}^{\dagger}$ is called the *grade automorphism.*

The same subscript notation can be used to specify part of the algebra. For example, $\langle \mathcal{C}\ell_3 \rangle_{\Im S}$ is the part containing only imaginary scalars. Of the possible pieces of the Pauli algebra, $\langle \mathcal{C}\ell_3 \rangle_{\Re S} \equiv \mathbb{R}$ (the real numbers), $\langle \mathcal{C}\ell_3 \rangle_{S} \equiv \mathbb{C}$ (the complex numbers), and $\langle \mathcal{C}\ell_3 \rangle_{\mathcal{E}} \equiv \mathbb{H}$ (the real *quaternion algebra,* formed from the even elements of $\mathcal{C}\ell_3$) are important subalgebras of $\mathcal{C}\ell_3$: they are themselves algebras.

17.2 Inverses and the metric

An element $p \in \mathcal{C}\ell_3$ has an inverse if and only if there exists another element q of $\mathcal{C}\ell_3$ whose product with p is a nonzero scalar. The inverse is then $p^{-1} = q/(pq)$ which is both a left and a right inverse, since $pq = \langle pq \rangle_S = \langle qp \rangle_S = qp$. If an involution exists which both takes $p \to q$ and $q \to p$, then the scalar pq may be interpreted as a square "length" of p and can be used to define the metric of the space. The procedure should be familiar from the field of complex numbers where to every number c another number c^* exists such that cc^* is a real scalar and the inverse of c is $c^{-1} = c^*/(cc^*)$. If p is a three-vector: $p = \mathbf{p}$, then q can also be taken equal to $\mathbf{p}$ and its inverse is $\mathbf{p}^{-1} = \mathbf{p}/(\mathbf{pp})$.

The product of a general element $p \in \mathcal{C}\ell_3$ with itself is generally not a scalar, but $p\bar{p}$ is, since it is a scalar: $p\bar{p} = \langle p\bar{p} \rangle_S$. As long as $p\bar{p} \neq 0$, the inverse of p exists and is given by

$$p^{-1} = \bar{p}/(p\bar{p}) . \tag{17.20}$$

The metric determined by the "square length" ("square modulus")

$$p\bar{p} = p_0^2 - \mathbf{p}^2 \tag{17.21}$$

is just the *Minkowski spacetime metric.* It is given more explicitly by a product rule analogous to (17.2):

$$\langle \mathbf{e}_\mu \bar{\mathbf{e}}_\nu \rangle_S = \frac{1}{2}\left(\mathbf{e}_\mu \bar{\mathbf{e}}_\nu + \mathbf{e}_\nu \bar{\mathbf{e}}_\mu\right) = \eta_{\mu\nu} , \tag{17.22}$$

where for the orthonormal basis (tetrad) $\{\mathbf{e}_\mu\}$, η is the diagonal 4×4 tensor (the *spacetime metric tensor*)

$$\eta = \begin{pmatrix} 1 & 0 & 0 & 0 \\ 0 & -1 & 0 & 0 \\ 0 & 0 & -1 & 0 \\ 0 & 0 & 0 & -1 \end{pmatrix} \tag{17.23}$$

The metric reflects the fact that spacetime vectors in relativity may be represented by real elements (*real paravectors*) of $C\ell_3$. In three-dimensional Euclidean space, the elements are vectors, the relation corresponding to (17.22) is

$$\langle \mathbf{e}_j \mathbf{e}_k \rangle_S = \delta_{jk} \tag{17.24}$$

[see (17.2)], and the metric tensor can be taken to be the Kronecker delta.

Exercise 17.2.1 *Show that the part of (17.22) restricted to spatial basis vectors* $\mathbf{e}_k$, $k = 1, 2, 3$ *implies (17.24).*

Exercise 17.2.2 *Show that the spatial reversal* $\bar{p}$ *of any element* $p \in C\ell_3$ *is given by*

$$\bar{p} = -\frac{1}{2}\mathbf{e}_\mu p \mathbf{e}^\mu,$$

where the reciprocal basis vectors $\mathbf{e}^\mu$ *are defined by* $\mathbf{e}^\mu = \eta^{\mu\nu}\mathbf{e}_\nu$ *and* $\eta^{\mu\nu}\eta_{\nu\sigma} = \delta^\mu{}_\sigma$. *(Note: since the matrix* $\eta^{-1} = \eta$ *for an orthonormal basis, the elements* $\eta_{\mu\nu}$ *and* $\eta^{\mu\nu}$ *are equal and* $\mathbf{e}^\mu = \bar{\mathbf{e}}_\mu$*.)*

Examples of spacetime vectors (paravectors of $C\ell_3$)[4], to be discussed in more detail below, include

- the proper velocity $u = \gamma + \mathbf{u} = \gamma(1 + \mathbf{v})$, with $u\bar{u} = 1$,
- the momentum $p = E + \mathbf{p}$, with $p\bar{p} = m^2$,
- the vector potential $A = \phi + \mathbf{A}$,
- the current density $j = \rho + \mathbf{j}$.

In these examples, as generally in what follows, we choose units with the speed of light $c = 1$.

Note that the "square length" in spacetime is not necessarily positive. Indeed there are three possibilities:

$$\begin{array}{ll} p\bar{p} > 0 & p \text{ is timelike} \\ p\bar{p} = 0 & p \text{ is lightlike or null} \\ p\bar{p} < 0 & p \text{ is spacelike.} \end{array} \tag{17.25}$$

[4]The term *spacetime vector* refers to the physical nature and transformation properties of the element, whereas *paravector* refers to the mathematical role of the object in the Pauli algebra, where it is the sum of a scalar and a spatial vector.

17.3 The Spacetime Manifold

In special relativity, a point x is often thought of as a vector in spacetime with components x^μ relative to an origin of coordinates. However, spacetime is in general not flat and one cannot join all pairs of points with vectors in the manifold. It is therefore preferable to view x simply as a *point* in spacetime, not as a vector. The basis vectors are given by the partial derivative of the point x with respect to the coordinates:

$$\mathbf{e}_\mu = \frac{\partial x}{\partial x^\mu} =: \partial_\mu x \tag{17.26}$$

which states in mathematical form that the μth basis element is *tangent* to the curve formed in spacetime when all coordinates except x^μ are held fixed. It is a *tangent vector*, which exists not in the spacetime manifold itself, but in the *tangent space* to the manifold at the point x.

The observer never confuses her space and time coordinates. Any particle (persistent point) at rest in her laboratory will change only its time value. More generally, a moving particle follows a *path* $x(\tau)$ in spacetime where the parameter τ may be taken to be the *proper time* of the particle, that is, the time as it would be measured in its own rest frame. The tangent vector to the path is the spacetime velocity

$$u := \frac{dx}{d\tau} = \partial_\mu x \frac{dx^\mu}{d\tau} = \mathbf{e}_\mu \frac{dx^\mu}{d\tau} \tag{17.27}$$

which lies in the tangent space of the spacetime manifold at x. The spacetime velocity of a particle in its rest frame is simply $\mathbf{e}_0 = 1$.

With the help of the partial derivatives ∂_μ, we can introduce an algebraic gradient in $C\ell_3$:

$$\partial = \partial^\mu \mathbf{e}_\mu = \partial_t - \nabla. \tag{17.28}$$

Note the sign on the nabla (∇), which follows from the relation

$$V_\mu = \eta_{\mu\nu} V^\nu \tag{17.29}$$

between spacetime vector components with upper ("contravariant") and lower ("covariant") indices, and which ensures that $\partial \langle k\bar{x}\rangle_S = k$. Here, η is the metric tensor (17.23).

Exercise 17.3.1 *Find the paravector* $\partial(xt + yz)$, *where* $t = x^0$, $x = x^1$, $y = x^2$, *and* $z = x^3$. *[Answer:* $x - t\mathbf{e}_1 - z\mathbf{e}_2 - y\mathbf{e}_3$.*]*

Exercise 17.3.2 *Expand* ∂ *and* $\langle k\bar{x}\rangle_S$ *show that the gradient of* $\partial \langle k\bar{x}\rangle_S$ *is* k.

17.4 Lorentz Transformations I

Linear transformations of elements which preserve the reality of spacetime vectors and leave their spacetime length invariant are called *Lorentz transformations* or *spacetime rotations.* They include spatial rotations, boosts (velocity transformations), and any combination of boosts and rotations.

Lorentz transformations in special relativity relate different inertial frames. They can be *active* and change the frame of the system being observed, or they can be *passive* and correspond to changing the observer of a fixed system. Physically, it is only the *relative velocity and orientation* of the observed system (object frame) relative to the observer (lab frame) which is significant. The Lorentz transformation of a spacetime vector p in the object frame to the spacetime vector p' relative to the lab can be written

$$p \to p' = LpL^{\dagger}. \tag{17.30}$$

Here L gives the motion and orientation of the object frame with respect to the observer. It has the form

$$L = \exp(\mathbf{W}/2), \tag{17.31}$$

with $\mathbf{W} = \mathbf{w} - i\boldsymbol{\theta}$. For an active transformation, if $\mathbf{w} = 0$, then L describes a pure rotation of the observed system in the plane $i\boldsymbol{\theta}$ by the angle θ, whereas if $\theta = 0$, then L is a boost of the object with the *rapidity* (or *boost parameter*) $\mathbf{w}$. In a passive transformation, the transformation L gives the motion and orientation of the initial observer (in the object frame) relative to the final observer (in the lab frame). Every proper Lorentz transformation is taken to be *unimodular*:

$$L\bar{L} = 1 \tag{17.32}$$

and conversely, every unimodular element can be interpreted as a Lorentz transformation. Thus, $\bar{L} = L^{-1}$ and the *inverse transformation* to (17.30) is

$$p = \bar{L}p'\bar{L}^{\dagger}. \tag{17.33}$$

If L is also real ($L = L^{\dagger}$), it represents a boost,

$$B = \exp(\mathbf{w}/2) = \cosh(w/2) + \hat{\mathbf{w}}\sinh(w/2) \tag{17.34}$$

whereas if it is unitary ($L = \bar{L}^{\dagger}$), it represents a rotation:

$$R = \exp(-i\boldsymbol{\theta}/2) = \cos\theta/2 - i\widehat{\boldsymbol{\theta}}\sin\theta/2. \tag{17.35}$$

Every transformation L can be written as the product BR of a boost $B = \left(LL^{\dagger}\right)^{1/2}$ and a rotation $R = \bar{B}L$.

Note that the Lorentz transformation (17.30) is *linear*: for any spacetime vectors p, q and any scalar c,

$$\begin{aligned} L(p+q)L^{\dagger} &= LpL^{\dagger} + LqL^{\dagger} \\ LcpL^{\dagger} &= cLpL^{\dagger} \end{aligned} \tag{17.36}$$

The linearity lets us simplify the calculation of any boost or rotation by breaking the spacetime vector p to be transformed into parts which commute $(p^0 + \mathbf{p}_{\parallel})$and anticommute $(\mathbf{p}_{\perp} \equiv \mathbf{p} - \mathbf{p}_{\parallel})$with the direction $\hat{\mathbf{w}}$ of the boost B or the axis $\widehat{\boldsymbol{\theta}}$ of rotation R. Here $\mathbf{p}_{\parallel}$ is the part of $\mathbf{p}$ parallel to $\hat{\mathbf{w}}$ or $\widehat{\boldsymbol{\theta}}$. Thus,

$$\begin{aligned} BpB &= B^2\left(p^0 + \mathbf{p}_{\parallel}\right) + \mathbf{p}_{\perp} \\ RpR^{\dagger} &= \left(p^0 + \mathbf{p}_{\parallel}\right) + R^2\mathbf{p}_{\perp} \end{aligned} \tag{17.37}$$

and, for example,

$$\begin{aligned} R^2\mathbf{p}_\perp &= e^{-i\boldsymbol{\theta}}\mathbf{p}_\perp = \left(\cos\theta - i\widehat{\boldsymbol{\theta}}\sin\theta\right)\mathbf{p}_\perp \\ &= \mathbf{p}_\perp\cos\theta + \widehat{\boldsymbol{\theta}}\times\mathbf{p}_\perp\sin\theta \end{aligned} \tag{17.38}$$

By applying the involutions to the basic Lorentz transformation of spacetime vectors (17.30) , equivalent transformations can be derived:

$$\begin{aligned} \bar{p} &\rightarrow \bar{L}^\dagger\bar{p}\bar{L} \\ p^\dagger &\rightarrow Lp^\dagger L^\dagger \\ \bar{p}^\dagger &\rightarrow \bar{L}^\dagger\bar{p}^\dagger\bar{L} \end{aligned} \tag{17.39}$$

Transformations of products of elements are particularly simple when elements are multiplied by spatial reversal of elements:

$$p\bar{q} \rightarrow Lp\bar{q}\bar{L} \tag{17.40}$$

Note that the scalar part $\langle p\bar{q}\rangle_S$ of the paravector product commutes with all elements and is therefore invariant under Lorentz transformation. The vector part $\langle p\bar{q}\rangle_V$, called a *parabivector*, has three real plus three imaginary vector components and represents a spacetime plane; it transforms like $p\bar{q}$.

The ability to perform Lorentz transformations easily has great practical advantage. It means we can work out a problem in the simplest frame, often the rest frame, and then transform the result to the desired frame, usually the lab frame. If the information sought is a Lorentz invariant, the calculation is even simpler since no Lorentz transformation is needed. For example, the scalar product $\langle A\bar{u}\rangle_S$ of the vector potential A with the proper velocity is just its value in the rest frame where $u = 1$: $\langle A\bar{u}\rangle_S = \langle A_{rest}\rangle_S = \phi_{rest}$.

17.5 Vector Notation and Rotations

A major advantage of component-free vector notation is that a single symbol can represent an object that has different components on any number of basis systems. We want our choice of an algebra to offer this advantage in four-dimensional spacetime. However, we need a clear understanding of the interplay of vectors and transformations with objects and observers, and to avoid the complicating issues of the indefinite metric and Minkowski spacetime and its representation we first consider spatial vectors in three-dimensional Euclidean space.

Any two different right-handed orthonormal bases $\{\mathbf{e}_k\}$ and $\{\mathbf{u}_k\}$ are related by a rotation

$$\mathbf{u}_k = R\mathbf{e}_k R^\dagger, \tag{17.41}$$

where the $SU(2)$ rotation operator R has the form

$$R(\boldsymbol{\theta}) = \exp(-i\boldsymbol{\theta}/2) \tag{17.42}$$

for a rotation by θ in the plane $i\hat{\boldsymbol{\theta}}$ (or equivalently, about the axis $\hat{\boldsymbol{\theta}}$.) Any number of such orthonormal bases can be defined, each corresponding to a distinct $\boldsymbol{\theta}$. A given vector $\mathbf{v}$ can be expanded in any such basis:

$$\mathbf{v} = v^k \mathbf{e}_k = v'^j \mathbf{u}_j \,. \tag{17.43}$$

This is a *passive* transformation: it relates the components of one vector on two different vector bases. The relation between components is

$$v'^j = v^k \mathbf{e}_k \cdot \mathbf{u}_j = v^k \left\langle \mathbf{e}_k R \mathbf{e}_j R^\dagger \right\rangle_S \,. \tag{17.44}$$

17.5.1 The Merry-Go-Round

Consider next an *active* transformation, involving a merry-go-round (MGR) frame that is rotating at the angular velocity $\boldsymbol{\omega}$ with respect to the inertial (I) frame. At time $t = 0$ the bases used in the two frames coincide and are given by $\{\mathbf{e}_k\}$. A corotating observer in the MGR frame will naturally use a basis that is fixed in her frame, and the inertial observer will choose a basis fixed in the I frame. Since at $t = 0$ each observer uses the coincident basis $\{\mathbf{e}_k\}$, each can rightly claim to continue to use the same basis at all times. Thus, each observer can claim $\{\mathbf{e}_k\}$ as a time-independent basis, although each will see the other's basis as rotating. Let

$$\mathbf{v} = v^k \mathbf{e}_k \tag{17.45}$$

be a stationary vector in the MGR frame. In the I frame, the basis vectors $\mathbf{e}_k$ of the MGR frame are rotating:

$$\mathbf{u}_k = R(\boldsymbol{\omega} t)\, \mathbf{e}_k R^\dagger (\boldsymbol{\omega} t) \tag{17.46}$$

and so is the vector:

$$\mathbf{v}' = R(\boldsymbol{\omega} t)\, \mathbf{v} R^\dagger (\boldsymbol{\omega} t) = v^k R(\boldsymbol{\omega} t)\, \mathbf{e}_k R^\dagger (\boldsymbol{\omega} t) = v^k \mathbf{u}_k \,. \tag{17.47}$$

The rotating vector can be expanded in the $\mathbf{e}_k$:

$$\mathbf{v}' = v^k R \mathbf{e}_k R^\dagger = v'^j \mathbf{e}_j \tag{17.48}$$

with

$$v'^j = v^k \mathbf{u}_k \cdot \mathbf{e}_j \,. \tag{17.49}$$

Note that for the observer in the I frame, the basis $\{\mathbf{e}_k\}$ refers to his fixed frame, even though its elements enter the transformation (17.48) as the basis vectors of the MGR frame. Furthermore, the same physical vector is $\mathbf{v}$ to the observer in the MGR frame and $\mathbf{v}'$ to the observer in frame I. A vector $\mathbf{v}$, given by (17.45), for the observer fixed in I is constant and physically distinct from $\mathbf{v}$ as seen by the observer corotating with MGR. The passive transformation obtained by equating (17.47) and (17.48)

$$\mathbf{v}' = v^k \mathbf{u}_k = v'^j \mathbf{e}_j \tag{17.50}$$

is valid as seen by the observer in frame I, where $\{\mathbf{u}_k\}$ and $\{\mathbf{e}_k\}$ are two different frames as viewed by the one observer. Alternatively, the two bases $\{\mathbf{u}_k\}$ and $\{\mathbf{e}_k\}$ can represent the same MGR frame as seen by different observers (in I and MGR, respectively), but then the equality (17.50) is not meaningful.

17.5.2 Observers, Frames, and Vector Bases

The different possible uses of a vector basis can easily lead to confusion unless we carefully distinguish *observers, frames,* and the *vector bases* with which observers specify the various frames. A single observer may use different frames and describe each frame by a vector basis; she can relate components on the different frames by means of passive transformations such as (17.50). A natural frame for the observer is one with which she is at rest (the observer's rest frame), but she may also consider non-static frames in order to relate observations made from different frames. The transformation (17.47) can represent either an *active* transformation as viewed by a single observer as the vector $\mathbf{v}$ is rotated, or equivalently, it can represent the *passive* transformation of a single vector and relate the vector as seen by an observer fixed in frame I to that seen by another fixed in MGR. In both interpretations, the vectors before ($\mathbf{v}$) and after ($\mathbf{v}'$) transformation are distinct mathematically: they do not represent absolute vectors, but rather vectors *relative to an observer.* Often in applications of passive transformations in the form of (17.50), the observer can be implicit and play only a silent role, but in other cases his/her role cannot be ignored.

Basis vectors in particular are only relative. The same basis vector $\mathbf{e}_k$ used by different observers may be physically distinct. Conversely, a single direction may be labeled with different basis vectors (or different linear combinations thereof) by different observers. The basis vectors for a given frame may be static for one observer and changing in time for another.

17.6 Lorentz Transformations II

Similar considerations hold for Lorentz transformations in spacetime. Two identical particles with different proper velocities will have different momenta (and if they are charged, different electromagnetic fields), even though there will exist inertial frames in which each is instantaneously at rest. Observers in these two inertial frames will both view the momentum of the corresponding particle as $p = m$, even though the momenta as viewed by any single observer are distinct. A Lorentz transformation will relate any pair of orthonormal tetrads and therefore the observations by observers fixed in the different frames.

A natural tetrad basis $\{\mathbf{e}_\mu\}$of paravector space for a given observer is a static one, for which $\mathbf{e}_0 = 1$, but other bases may be used. A different basis $\{\mathbf{u}_\mu\}$, where

$$\mathbf{u}_\mu = L\mathbf{e}_\mu L^\dagger \tag{17.51}$$

represents a frame moving with proper velocity $\mathbf{u}_0 = LL^\dagger$ with respect to the observer. Any spacetime vector p can be expanded in the two bases

$$p = p^\mu \mathbf{e}_\mu = p'^\nu \mathbf{u}_\nu \tag{17.52}$$

Since the Lorentz scalar product of two spacetime vectors p and q has the form $\langle p\bar{q}\rangle_S$, we can define *reciprocal basis vectors* $\mathbf{u}^\nu$ by

$$\langle \mathbf{u}_\mu \bar{\mathbf{u}}^\nu \rangle_S = \delta_\mu^\nu\,. \tag{17.53}$$

All vectors can be expanded in the basis $\{\mathbf{u}_\lambda\}$, we write

$$\mathbf{u}^\nu = c^{\nu\lambda}\mathbf{u}_\lambda\,, \tag{17.54}$$

where the coefficients $c^{\lambda\nu}$ are to be determined. Thus from (17.53),

$$c^{\nu\lambda}\left\langle \mathbf{u}_\mu \bar{\mathbf{u}}_\lambda \right\rangle_S = \delta_\mu^\nu\,, \tag{17.55}$$

and since $\{\mathbf{u}_\mu\}$ is an orthonormal basis, $\left\langle \mathbf{u}_\mu \bar{\mathbf{u}}_\lambda \right\rangle_S = \eta_{\mu\lambda}$ and $c^{\nu\lambda} = \eta^{\nu\lambda}$. With the reciprocal basis vectors $\{\mathbf{u}^\nu\}$ one can relate the components of p :

$$p'^\nu = p^\mu \left\langle \mathbf{e}_\mu \bar{\mathbf{u}}^\nu \right\rangle_S . \tag{17.56}$$

The basis vectors $\mathbf{e}_\mu$ and $\mathbf{u}_\nu$ are not absolute, but represent two frames relative to a single observer. However, the Lorentz transformation elements

$$\mathcal{L}_\mu^{\ \nu} = \left\langle \mathbf{e}_\mu \bar{\mathbf{u}}^\nu \right\rangle_S = \left\langle \mathbf{u}^\nu \bar{\mathbf{e}}_\mu \right\rangle_S = \left\langle L\mathbf{e}^\nu L^\dagger \bar{\mathbf{e}}_\mu \right\rangle_S \tag{17.57}$$

that relate the frames are Lorentz scalars and are therefore the same for every inertial observer. The inverse transformation is

$$p^\mu = p'^\nu \left\langle \mathbf{u}_\nu \bar{\mathbf{e}}^\mu \right\rangle_S = p'^\nu \mathcal{L}^\mu_{\ \nu} \tag{17.58}$$

Note

$$\begin{aligned} \mathbf{u}_\nu &= \mathbf{e}_\mu \mathcal{L}^\mu_{\ \nu} \\ \bar{\mathbf{u}}^\nu &= \bar{\mathbf{e}}^\mu \mathcal{L}_\mu^{\ \nu} \end{aligned} \tag{17.59}$$

from which the identity follows:

$$\begin{aligned} \left\langle \mathbf{u}_\nu \bar{\mathbf{u}}^\lambda \right\rangle_S &= \delta_\nu^\lambda = \left\langle \mathbf{e}_\mu \bar{\mathbf{e}}^\sigma \right\rangle_S \mathcal{L}^\mu_{\ \nu} \mathcal{L}_\sigma^{\ \lambda} \\ &= \mathcal{L}^\mu_{\ \nu} \mathcal{L}_\mu^{\ \lambda} \end{aligned} \tag{17.60}$$

To summarize, a single transformation (rotation, boost) applied to the rest-frame tetrad $\{\mathbf{e}_\mu\}$ of the observer yields the tetrad $\{\mathbf{u}_\mu = L\mathbf{e}_\mu L^\dagger\}$ of the frame transformed with respect to the observer. The transformation may be considered as *active* (the observer frame is unchanged, and the object frame is transformed), passive (the object frame is the same, but the observer has been transformed backwards), or a mixture of the two. Only the *relative* orientation and motion of the observer and object frames is important.

Exercise 17.6.1 *Show that the transformation matrix* $\mathcal{L}_\mu^{\ \nu}$ *can also be written* $\left\langle \bar{L}\mathbf{e}_\mu \bar{L}^\dagger \bar{\mathbf{e}}^\nu \right\rangle_S$. *Interpret physically.*

The matrices $\mathcal{L}_\mu^{\ \nu}$ form the group $SO_+(1,3)$ of spacetime rotations (or restricted Lorentz transformations). They are formed from elements L of the double covering group SL(2,C) of complex, unimodular ("special") elements of $\mathcal{C}\ell_3$.

Note that the basis paravectors represent spacetime vectors corresponding to the separation of a pair of *events* rather than the enduring spatial vectors familiar from Galilean mechanics. The latter sweep out spacetime planes in time and are represented by timelike bivectors in Minkowski spacetime.

17.6.1 Proper Velocity

A simple example of a spacetime-vector transformation is the transformation of the proper velocity from one frame to another. In the rest frame of the particle, its proper velocity is $u_{rest} = 1$. In the lab frame, it is $u = Lu_{rest}L^{\dagger} = LL^{\dagger}$. If L is written as a product of a boost with a rotation: $L = BR$, then u is independent of the rotation:

$$u = LL^{\dagger} = B^2. \tag{17.61}$$

We can identify any boost with the timelike square root of a proper velocity:

$$B = u^{1/2}. \tag{17.62}$$

In an active transformation, B boosts an object frame at rest up to the paravector velocity u (the *paravelocity*), whereas in a passive transformation, the boost changes the observer's frame to one in which the original observer moves with paravelocity u. Since only the relative motion of the observer and the object is important, these transformations are fully equivalent. Note that boosts are more commonly referred to by the paravelocity $u = B^2$ they induce in a rest frame (or by the corresponding spatial velocity $\mathbf{v} = \langle u \rangle_V / \langle u \rangle_S$, see below) than by the rapidity $\mathbf{w}$.

The Lorentz-invariant product $u\bar{u} = u_{rest}\bar{u}_{rest} = 1$, which also follows from (17.61) and the unimodularity (17.32) of proper Lorentz transformations. As a result, u is a unimodular paravector: $u^{-1} = \bar{u}$. It is said to be a *unit spacetime vector.* Since the proper velocity u of an object is the tangent vector $u = dx/d\tau$ of its world line $x(\tau)$, its scalar part γ gives the relative rates of coordinate and rest-frame clocks:

$$\langle u \rangle_S = \gamma = \frac{dt}{d\tau} \tag{17.63}$$

and its vector part is related to the coordinate velocity $\mathbf{v} = d\mathbf{x}/dt$ by

$$\langle u \rangle_V = \mathbf{u} = \frac{d\mathbf{x}}{d\tau} = \gamma\mathbf{v} \tag{17.64}$$

which can be combined with the previous result in the form

$$u = \gamma(1 + \mathbf{v}) \tag{17.65}$$

The unit size of u means that γ and $\mathbf{v}$ are related by

$$u\bar{u} = \gamma^2\left(1 - \mathbf{v}^2\right) = 1 \tag{17.66}$$

and that the invariant interval of proper time can also be expressed

$$(d\tau)^2 = (ud\tau)(\bar{u}d\tau) = dx\,d\bar{x} \tag{17.67}$$

from which its Lorentz invariance is obvious. Alternatively, one can write

$$d\tau = \langle \bar{u}ud\tau \rangle_S = \langle \bar{u}dx \rangle_S \tag{17.68}$$

which also shows that $d\tau$ is identical with the interval of time in the rest frame.

Consider two frames $\mathcal{S}$ and $\mathcal{S}'$ related by a pure boost $L = B$ so that any proper vector p in $\mathcal{S}$ is seen in $\mathcal{S}'$ to be $p' = LpL^\dagger = BpB$. The inverse transformation gives spacetime vectors in $\mathcal{S}$ in terms of those in $\mathcal{S}'$: $p = \bar{L}p'\bar{L}^\dagger = \bar{B}p'\bar{B}$. In particular, the proper velocity of $\mathcal{S}$ in $\mathcal{S}'$ is $u = B^2$ whereas that of $\mathcal{S}'$ in $\mathcal{S}$ is $\bar{B}^2 = \bar{u}$. Thus, a particle of mass m at rest in $\mathcal{S}$ has spacetime momentum

$$p = E + \mathbf{p} = mu \tag{17.69}$$

in $\mathcal{S}'$, where $E = m\gamma$ is the rest energy plus the kinetic energy.

Exercise 17.6.2 *In frame $\mathcal{S}$, a beam of monoenergetic neutrons moves with coordinate velocity $\mathbf{v} = 0.36\mathbf{e}_1 + 0.48\mathbf{e}_2$. Determine the proper velocity u in $\mathcal{S}$. Check your result by verifying the unimodularity of u. By about how much is the mean lifetime of a neutron in the beam going to be increased beyond its value of about 15 minutes when at rest? Now find the proper velocity and the coordinate velocity of the neutrons in a frame $\mathcal{S}'$ in which frame $\mathcal{S}$ moves with velocity $\mathbf{V} = 0.8\mathbf{e}_1$.*

17.6.2 Covariant vs. Invariant

Traditional treatments of relativity often consider spacetime vectors and bivectors to be well-defined quantities independent of any observer, and the equations of physics relating such quantities are said to be *invariant.* The only transformations considered within such a framework are passive.

In our treatment, we wish to consider active as well as passive transformations, and perhaps transformations where both the observer and the observed object are transformed. Our treatment makes immediate the important result that only the relative orientation and motion of the observer and the observed object is significant. Our equations have quantities which are generally different for different pairs of object/observer frames, but their forms are the same for all object/observer pairs: both sides of any fundamental physical relation transform in the same way, that is *covariantly,* so that the equation is the same even if the objects are described differently by the different observers. Every observer will naturally use a tetrad at rest, in which $\mathbf{e}_0 = 1$ and will observe the tetrad of another frame as having a time axis $\mathbf{u}_0$ given by the frame's spacetime velocity.

17.6.3 Spacetime Diagrams

A simple application of transformations is to drawing spacetime diagrams. Consider a tetrad $\{\mathbf{e}_\mu\}$ at rest with respect to observer $\mathcal{S}$, but moving with velocity $\mathbf{v} = v\mathbf{e}_3$ with respect to observer $\mathcal{S}'$. The moving tetrad is seen by B as $\mathbf{u}_\mu = L\mathbf{e}_\mu L^\dagger$. The transformation L has the form $L = B = \exp(\mathbf{e}_3 w/2)$ and commutes with $\mathbf{e}_0, \mathbf{e}_3$ while anticommuting with $\mathbf{e}_1, \mathbf{e}_2$. Thus,

$$\begin{aligned}
\mathbf{u}_0 &= B\mathbf{e}_0 B = \gamma\,(1 + v\mathbf{e}_3)\,\mathbf{e}_0 = \gamma\,(\mathbf{e}_0 + v\mathbf{e}_3) \\
\mathbf{u}_1 &= \mathbf{e}_1
\end{aligned}$$

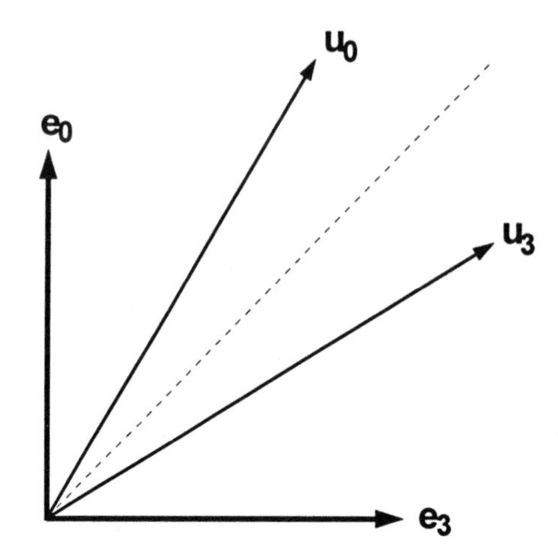

Fig. 17.1. Spacetime diagram for $v = 0.6$. Here, u_0 is the world line of the moving tetrad. The dashed line is the world line of a light signal.

$$
\begin{aligned}
\mathbf{u}_2 &= \mathbf{e}_2 \\
\mathbf{u}_3 &= u\mathbf{e}_3 = \gamma\left(1 + v\mathbf{e}_3\right)\mathbf{e}_3 = \gamma\left(\mathbf{e}_3 + v\mathbf{e}_0\right) .
\end{aligned}
\tag{17.70}
$$

Consider the case $v = 0.6$, $\gamma = (1 - 0.36)^{-1/2} = 1.25$, $\gamma v = 0.75$. The spacetime diagram (Fig. 1.1) shows part of the moving tetrad $\{\mathbf{u}_0, \mathbf{u}_3\}$ in terms of $\mathbf{e}_0$ and $\mathbf{e}_3$. Note that $\mathbf{u}_0, \mathbf{u}_3$ are still unit paravectors, because their square "lengths" are given by $\mathbf{u}_0\bar{\mathbf{u}}_0 = \gamma^2\left(1 - v^2\right) = 1 = -\mathbf{u}_3\bar{\mathbf{u}}_3$.

Exercise 17.6.3 *Draw the spacetime diagram of the tetrads of the two frames as seen by observer $\mathcal{S}$.*

Exercise 17.6.4 *Relate the six parabivectors $\langle \mathbf{e}_\mu \bar{\mathbf{e}}_\nu \rangle_V$ of the observer's rest frame with those $\langle \mathbf{u}_\mu \bar{\mathbf{u}}_\nu \rangle_V$ of the moving frame by taking products of the tetrad elements above. Add the basis vector $\mathbf{e}_1 = \mathbf{u}_1$ to Fig. (17.1) and sketch the spacetime planes $\mathbf{e}_1\bar{\mathbf{e}}_0$ and $\mathbf{e}_1\bar{\mathbf{e}}_3$ together with the transformed $\mathbf{u}_1\bar{\mathbf{u}}_0$ and $\mathbf{u}_1\bar{\mathbf{u}}_3$. Show that the plane $\mathbf{e}_0\bar{\mathbf{e}}_3$ is invariant under the boost along $\mathbf{e}_3$.*

Chapter 18

Eigenspinors in Electrodynamics

William E. Baylis
Department of Physics, University of Windsor,
Windsor, Ont., Canada N9B 3P4

The eigenspinor concept is a powerful tool for finding the motion of charges in electromagnetic fields. This lecture examines the concept and illustrates its use in classical electrodynamics. The concept is extended to quantum fields in the third lecture.

18.1 Basic Electrodynamics

The electromagnetic field is an oriented spacetime plane, that is, a spacetime bivector, which is modeled by a parabivector in $\mathcal{C}\ell_3$. Its real part is the electric field $\mathbf{E}$ and its imaginary part is the magnetic field $i\mathbf{B}$:

$$\mathbf{F} = \mathbf{E} + i\mathbf{B}. \tag{18.1}$$

The electric field $\mathbf{E}$ is a persistent spatial vector which sweeps out a timelike plane in spacetime. (A spacetime bivector $\mathbf{F}$ is said to be timelike if $\langle \mathbf{F}^2 \rangle_{\Re} > 0$.) On the other hand, the magnetic field is more accurately thought of as acting in a spacelike ($\langle \mathbf{F}^2 \rangle_{\Im} < 0$) plane perpendicular to $\mathbf{B}$.

The scalar/vector split of parabivectors like $\mathbf{F}$ is Lorentz invariant, and indeed, $\mathbf{F}$ is always a (spatial) vector with no scalar part. However, the real/imaginary "split" of $\mathbf{F}$ into electric- and magnetic-field parts is different for different observers. In particular, an electromagnetic field $\mathbf{F}$ in the frame $\mathcal{S}$ is transformed to

$$\mathbf{F} \to \mathbf{F}' = L\mathbf{F}\bar{L} = u^{1/2}\mathbf{F}\bar{u}^{1/2} = \mathbf{F}_{\parallel} + u\mathbf{F}_{\perp} \tag{18.2}$$

in the frame $\mathcal{S}'$ in which the frame $\mathcal{S}$ moves with proper velocity u. (In $\mathcal{S}$, frame $\mathcal{S}'$ moves with proper velocity $\bar{u}$.) Here, $\mathbf{F}_{\parallel} = \mathbf{F} \cdot \hat{\mathbf{u}}\hat{\mathbf{u}}$ is the part of the field parallel to the spatial direction of u, and $\mathbf{F}_{\perp} = \mathbf{F} - \mathbf{F}_{\parallel}$ is that part perpendicular. Note that $\mathbf{F}^2 = (\mathbf{E}^2 - \mathbf{B}^2) + 2i\mathbf{E} \cdot \mathbf{B}$ is Lorentz invariant.

The foundations of classical electrodynamics are summarized in just two equations: Maxwell's equation,

$$\bar{\partial}\mathbf{F} = \frac{1}{\varepsilon_0}\bar{j}\,, \tag{18.3}$$

where $\varepsilon_0^{-1} = 4\pi \times 29.9792458 = 376.73\cdots$ Ohm is the inverse of the vacuum permittivity, and the Lorentz-force equation, which the Pauli algebra is able to render in the covariant, component-free form

$$\dot{p} = \langle e\mathbf{F}u\rangle_{\Re}\,, \tag{18.4}$$

where the dot indicates differentiation with respect to the proper time τ of the charge. However, the *spinorial form*, which we introduce next, is even easier to interpret and solve.

18.2 Eigenspinors

The eigenspinor Λ relates the rest-frame of the charge to the lab frame: any spacetime-vector property q_r in the rest frame of the charge is seen to in the lab frame as $q = \Lambda q_r \Lambda^\dagger$. In particular, the time axis $\mathbf{e}_0 = 1$ in the rest frame is transformed to the proper velocity in the lab frame:

$$u = \Lambda\Lambda^\dagger\,. \tag{18.5}$$

The eigenspinor Λ is the SL(2,C) ("spinorial") form of the Lorentz transformation between the frames. As for the Lorentz transformations discussed in the last lecture, it can be written $\Lambda = \exp\left[\mathbf{W}/2\right]$, where $\mathbf{W} = \mathbf{w} - i\boldsymbol{\theta}$ determines the *spacetime rotation* of the charge as observed in the lab. We will see below that $\mathbf{W}$ is a spacetime bivector.

The eigenspinor of a particle is different for different observers. Suppose Λ_{AP} is the eigenspinor of the charge P as seen by observer A, so that A sees the charge moving with proper velocity

$$u_{AP} = \Lambda_{AP}\Lambda_{AP}^\dagger\,. \tag{18.6}$$

Let L_{BA} transform the observer from frame A to frame B, or equivalently an object from frame B to frame A. The proper velocity of the charge as seen by observer B is then found by transforming u_{AP} :

$$u_{BP} = L_{BA}u_{AP}L_{BA}^\dagger = L_{BA}\Lambda_{AP}\Lambda_{AP}^\dagger L_{BA}^\dagger\,. \tag{18.7}$$

The eigenspinor of the charge for observer B can thus be taken to be

$$\Lambda_{BP} = L_{BA}\Lambda_{AP}. \tag{18.8}$$

The transformation of an eigenspinor under a change in observer generally takes the form

$$\Lambda \to L\Lambda\,, \tag{18.9}$$

which is a *spinor*-type transformation. Spinors like Λ, which are themselves Lorentz transformations, are carriers of a representation of the group SL(2,C), the universal covering group of restricted Lorentz transformations. The spinor transformations (18.9), in which L acts from the left, make the eigenspinors elements of the *left regular representation* of the group algebra of SL(2,C).

Since only the *relative* motion and orientation of frames is significant, we can interpret the transformation (18.9) in terms of a single observer. The cliffor L then describes a spacetime rotation of the *moving* charge as seen by the observer. The same physical transformation can be expressed in terms of the equivalent spacetime rotation L_r as seen by an observer commoving with the initial particle rest frame:

$$\Lambda \to \Lambda L_r\,, \tag{18.10}$$

and this makes Λ an element of the *right regular representation* of the algebra of SL(2,C). Here, $L_r L_r^\dagger$ is the proper velocity of the transformed charge as viewed from its former rest frame, and L_r is its eigenspinor after transformation as seen in its former rest frame.

The relation (18.8) can be rearranged to give the Lorentz transformation L_{BA} relating A to B in terms of the eigenspinors of P:

$$L_{BA} = \Lambda_{BP}\bar{\Lambda}_{AP}\,. \tag{18.11}$$

The same relation (18.11) holds for different particles P following other world lines: Λ here could represent the eigenspinor of any particle. The only reason for distinguishing eigenspinors from other Lorentz transformations is to emphasize that the inertial frames they relate are on different footings: one is instantaneously commoving with the observed object while the other is an observer's frame, as shown schematically:

$$\text{lab frame} : \Lambda : \text{particle frame}\,. \tag{18.12}$$

As a result, one usually considers further Lorentz transformations to only one of the two connected frames.

18.3 The Group SL(2,C): Diagrams

The spinor transformations (18.9) and (18.10) describe the *same* physical transformation if and only if

$$L = \Lambda L_r \bar{\Lambda}\,. \tag{18.13}$$

This relates the Lorentz transformation L_r in the rest frame of the charge to the equivalent transformation L in the lab frame and shows how Lorentz transformations normally transform. If L and L_r are written as exponentials $L = \exp\left(\frac{1}{2}\mathbf{W}\right)$ and $L_r = \exp\left(\frac{1}{2}\mathbf{W}_r\right)$, then power-series expansions give

$$\mathbf{W} = \Lambda \mathbf{W}_r \bar{\Lambda}\,, \tag{18.14}$$

so that the parabivector $\mathbf{W}$ transforms as any spacetime bivector.

Since a given Lorentz transformation L has different coefficients in different frames, when we transform between frames, how do we know which frame to use in the expression of L? Consider $L_{BA} = \exp\left[W^k \mathbf{e}_k/2\right]$, which transforms the observer from A to B or, for the observer in B, the object from B to A, in the sense that properties of objects at rest in B are transformed by L_{BA} into ones that are commoving with A at paravelocity $u_{BA} = L_{BA}L_{BA}^{\dagger}$ in B. Let the spacetime bivectors $\mathbf{e}_k = \mathbf{e}_k\bar{\mathbf{e}}_0, i\mathbf{e}_k$ be those at rest in B. Those commoving with A are proportional to $L_{BA}\mathbf{e}_k\bar{L}_{BA} =: \mathbf{u}_k$, and

$$L_{BA} = L_{BA}L_{BA}\bar{L}_{BA} = \exp\left[\frac{1}{2}L_{BA}W^k\mathbf{e}_k\bar{L}_{BA}\right] = \exp\left[\frac{1}{2}W^k\mathbf{u}_k\right] \tag{18.15}$$

has the same coefficients in both frames.

The restricted Lorentz transformations in spinorial form, L, constitute a *group*. The group is unimodular (*special*) and, as noted above, is called SL(2,C). Since both L and $-L$ induce the same Lorentz transformation of a spacetime vector, SL(2,C) is said to be the two-fold covering group of $SO_+(1,3)$, the group of restricted Lorentz transformations of spacetime vectors. Even elements ($L = \bar{L}^{\dagger}$) of SL(2,C) are also unitary and form the subgroup SU(2), which is the two-fold covering group $SO_+(3)$ of rotations in 3-dimensional Euclidean space. The real (hermitean) elements of SL(2,C) do *not* form a group since the product of hermitean elements is not generally hermitean. In other words, the product of boosts is not generally a pure boost but may contain a rotational component. Various abelian (commutative) subgroups of SL(2,C) can be identified, such as the group of "rifle transformations", which rotate about the boost axis.

Every physically accessible inertial frame is uniquely determined relative to an arbitrary origin by a restricted Lorentz transformation, and thus by the six real parameters $\left(w^k, \theta^k\right)$, with $k = 1, 2, 3$, in the domains $w^k \in \mathbb{R}$, $0 \leq \theta^k < 2\pi$. Every point in *frame space* (the 6-dimensional parameter space restricted to these domains) thus represents an inertial frame. Every ordered pair of points corresponds, within a sign (a factor of ± 1), to a member $L \in$ SL(2,C) which relates the two inertial frames. In the abstract depiction of the parameter space in figure 18.1, points A and B represent the inertial fames of observers A and B, and the line connecting them represents (to within a sign) both the transformation L_{AB}, which gives the motion of B as seen by A, and L_{BA}, which gives the motion of A relative to B. The two transformations, which are inverses of one another, are distinguished by placing an arrow on the line: an arrow from A to B denotes L_{AB}, whereas one from B to A represents L_{BA}. The direction of the arrow makes sense from the standpoint of an active transformation on the frame tetrad: L_{AB} applied to a tetrad $\{\mathbf{e}_\mu\}$ initially at rest in the observer's frame A transforms it to one $\{\mathbf{e}'_\mu\}$ commoving with B as observed by A.

Moving from point to point in frame space corresponds to applying successive Lorentz transformations, that is, to multiplying the spinorial elements L together. The multiplication order from left to right is followed as one proceeds in the direction of the arrows.[1] The product of successive transformations corresponding to a closed

[1]From the standpoint of an observer, the actual order of *application* is reversed, that is in the

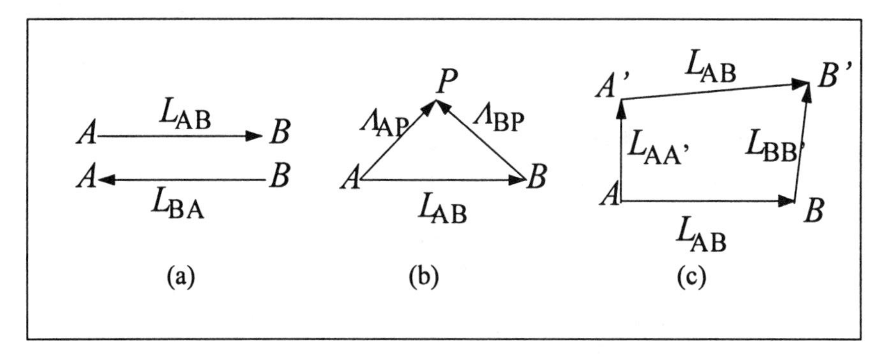

Fig. 18.1. Lorentz transformations in frame space. The product of transformations on a closed loop is ± 1.

loop in frame space must equal ± 1. In the simplest case, we consider only two frames A and B, and $L_{AB}L_{BA} = 1$ [see Fig. 18.1(a)]. Thus, $L_{BA} = \bar{L}_{AB}$. If we consider three inertial frames, say two observers A and B and a particle P [see Fig. 18.1(b)], then $L_{BA}\Lambda_{AP}\bar{\Lambda}_{BP} = 1$. With four inertial frames, A, A', B, B' one finds $L_{AA'}L_{A'B'} = L_{AB}L_{BB'}$, and if B' is defined to be related to A' by the same transformation which relates B to A, that is, if $L_{AB} =: L_{A'B'}$, then $L_{BB'} = L_{BA}L_{AA'}\bar{L}_{BA}$ [see Fig. 18.1(c)], which gives the result of the transformation at A of $L_{AA'}$ as seen by observer B.

18.4 Time Evolution of Eigenspinor

For any given observer, the eigenspinor depends on the proper time τ of the charge, and the motion of the charge is determined by the dependence $\Lambda(\tau)$. Given the eigenspinor at two proper times, the product $L(\tau_2, \tau_1) := \Lambda(\tau_2)\bar{\Lambda}(\tau_1)$ is a Lorentz transformation that can be identified as the "time-evolution" operator relating the eigenspinor at the different proper times:

$$\Lambda(\tau_2) = L(\tau_2, \tau_1)\Lambda(\tau_1)\,. \tag{18.16}$$

The proper time-rate of change of the eigenspinor can always be written

$$\dot{\Lambda} = \frac{1}{2}\Omega\Lambda\,, \tag{18.17}$$

where $\Omega := 2\dot{\Lambda}\bar{\Lambda}$. The spatial reversal of (18.17) is

$$\dot{\bar{\Lambda}} = \frac{d}{d\tau}\bar{\Lambda} = \frac{1}{2}\bar{\Lambda}\bar{\Omega}. \tag{18.18}$$

Note that since the proper time is a scalar, the proper-time derivative of a spatially reversed element is the same as the spatial reversal of the proper-time derivative.

right-to-left order. This is necessary since the observer will apply each transformation from his/her frame.

Since $\Lambda\bar{\Lambda} = 1$,

$$\frac{d}{d\tau}\Lambda\bar{\Lambda} = \dot{\Lambda}\bar{\Lambda} + \Lambda\dot{\bar{\Lambda}} = \frac{1}{2}\left(\Omega + \bar{\Omega}\right) = 0\,, \tag{18.19}$$

and thus the scalar part of Ω vanishes. It is therefore a pure complex vector. Since the proper time τ is a Lorentz scalar, Ω transforms as a covariant spacetime bivector:

$$\begin{aligned} \Omega &= 2\dot{\Lambda}\bar{\Lambda} \\ &\to L\Omega\bar{L} \end{aligned}\,. \tag{18.20}$$

Physically, Ω may be interpreted as the spacetime rotation rate of the particle frame.

If (18.17) can be solved, then the time evolution of any spacetime vector or bivector which is fixed in the particle frame can be found. For example, the momentum p has the fixed value m in the rest frame of the charge. Its proper-time derivative is

$$\begin{aligned} \dot{p} &= \dot{\Lambda}m\Lambda^\dagger + \Lambda m\dot{\Lambda}^\dagger \\ &= \langle\Omega p\rangle_{\Re}\,. \end{aligned} \tag{18.21}$$

Exercise 18.4.1 *Show that in the special case that $\Omega = -i\boldsymbol{\omega}$ is purely imaginary, (18.21) describes a pure rotation at the angular rate $\boldsymbol{\omega}$ and reduces to the familiar form*

$$\dot{\mathbf{p}} = \boldsymbol{\omega} \times \mathbf{p}\,,\, \dot{E} = 0\,. \tag{18.22}$$

18.4.1 Thomas Precession

As an important application, consider a motion that we describe at each point in terms of a boost $B = u^{1/2}$. In this case, (18.17) becomes

$$\dot{B} = \frac{1}{2}\Omega B \tag{18.23}$$

with a spacetime rotation rate $\Omega = 2\dot{B}\bar{B}$. Now since $B\bar{B} = 1$,

$$B(B + \bar{B}) = u + 1 = 2B\langle B\rangle_S\,, \tag{18.24}$$

the scalar part of which is $\gamma + 1 = 2\langle B\rangle_S^2$, the boost may be written

$$B = \frac{u+1}{\sqrt{2\,(\gamma+1)}}\,. \tag{18.25}$$

We find that Ω contains a spatial rotation part, known as the proper Thomas precession rate, that describes the spatial rotation undergone by the sequence of boosted frames:

$$\begin{aligned} \omega_{Th} &= i\langle\Omega\rangle_{\Im} \\ &= i\langle\left[\frac{d}{d\tau}\left(\frac{u+1}{\sqrt{\gamma+1}}\right)\right]\frac{\bar{u}+1}{\sqrt{\gamma+1}}\rangle_{\Im} \\ &= \frac{\dot{\mathbf{u}} \times \mathbf{u}}{\gamma+1}\,. \end{aligned} \tag{18.26}$$

18.5 Spinorial Lorentz-Force Equation

A comparison of (18.21) with the Lorentz-force equation (18.4) shows that the spacetime rotation rate may be identified with the electromagnetic field $\mathbf{F}$ at the position of the charge

$$\Omega = \frac{e}{m}\mathbf{F}, \tag{18.27}$$

and that the Lorentz-force equation itself can be expressed in the form

$$\dot{\Lambda} = \frac{e}{2m}\mathbf{F}\Lambda = \frac{e}{2m}\Lambda\mathbf{F}_r \tag{18.28}$$

since the field $\mathbf{F}$ in the lab is related to the field $\mathbf{F}_r$ in the frame of the charge by $\mathbf{F} = \Lambda\mathbf{F}_r\bar{\Lambda}$.

18.5.1 Solutions

According to (18.20), Ω^2 and hence $\mathbf{F}^2$ are Lorentz invariants. Now

$$\mathbf{F}^2 = \left(\mathbf{E}^2 - \mathbf{B}^2\right) + 2i\mathbf{E}\cdot\mathbf{B}, \tag{18.29}$$

so that the invariance of $\mathbf{F}^2$ implies the invariance independently of both $\mathbf{E}^2 - \mathbf{B}^2$ and $\mathbf{E}\cdot\mathbf{B}$. If $\langle\Omega^2\rangle_{\Re} > 0$, the spacetime rotation is about a timelike plane and is predominantly a boost, whereas if $\langle\Omega^2\rangle_{\Re} < 0$, the spacetime rotation is about a spacelike plane and is predominantly a spatial rotation. If the electromagnetic field $\mathbf{F}$ at the position of the charge is constant, the spinorial Lorentz-force equation (18.28) is easily integrated to give

$$\Lambda(\tau) = \exp\left(\frac{e}{2m}\mathbf{F}\tau\right)\Lambda(0)\ . \tag{18.30}$$

In a pure constant electric field $\mathbf{F} = \mathbf{E}$, a charge released from rest $[\Lambda(0) = 1]$ experiences a constant field in its rest frame given by

$$\mathbf{F}_r = \bar{\Lambda}\mathbf{F}\Lambda = \mathbf{F} \tag{18.31}$$

since $\Lambda = \exp(e\mathbf{F}\tau/2m)$ commutes with $\mathbf{F}$. The proper velocity in the lab frame is

$$u = \Lambda\Lambda^\dagger = \exp(\mathbf{a}\tau)\ , \tag{18.32}$$

where $\mathbf{a} := (e/m)\,\mathbf{F}$, and its integral gives the spacetime position r. If we take $r(0) = \mathbf{a}^{-1}$, then

$$r(\tau) = \mathbf{a}^{-1}\exp(\mathbf{a}\tau)\ , \tag{18.33}$$

with spatial-vector and scalar parts that obey $-r\bar{r} = \mathbf{r}^2 - t^2 = a^{-2}$, where a is the length of $\mathbf{a}$. Since , a spacetime plot of $\mathbf{r}(t)$ gives a hyperbola. For example, with $\mathbf{a} = \mathbf{e}_1$, $[x, t] = [\cosh\tau, \sinh\tau]$, gives the plot in Fig. 18.2.
The particle approaches the origin from positive x, slowing down as it comes. It is instantaneously at rest when the world line crosses the x axis and then accelerates away. The motion is called *hyperbolic.*

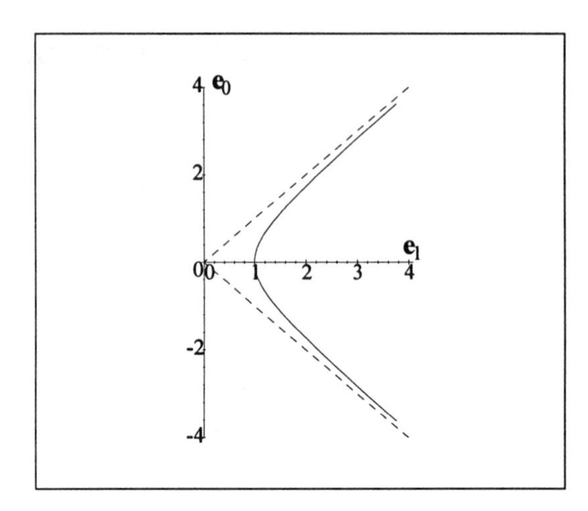

Fig. 18.2. Hyperbolic motion of particle with constant rest-frame acceleration.

On the other hand, in a constant magnetic field $\mathbf{F} = i\mathbf{B}$, the Lorentz transformation Λ is a pure spatial rotation, and the velocity of the charge rotates at proper angular velocity $\boldsymbol{\omega} = -e\mathbf{B}/m$:

$$u(\tau) = e^{-i\boldsymbol{\omega}\tau/2} u(0) e^{i\boldsymbol{\omega}\tau/2} . \tag{18.34}$$

Because the length of a vector is unchanged by a spatial rotation, the speed and hence the energy of the charge is constant in the magnetic field. The particle follows a helix in what is called cyclotron motion, with the component $\mathbf{v}_{\parallel}$ of the coordinate velocity $\mathbf{v}$ parallel to $\mathbf{B}$ remaining fixed and the perpendicular component $\mathbf{v}_{\perp}$ rotating in a circle:

$$\begin{aligned} \mathbf{v}(t) &= e^{-i\boldsymbol{\omega}t/2\gamma} \mathbf{v}(0) e^{i\boldsymbol{\omega}t/2\gamma} = \mathbf{v}_{\parallel}(0) + e^{-i\boldsymbol{\omega}t/\gamma} \mathbf{v}_{\perp}(0) \\ \gamma &= \sqrt{1 + \mathbf{u}^2} = 1/\sqrt{1 - \mathbf{v}^2} . \end{aligned} \tag{18.35}$$

The particle is locked to the magnetic field; its path twists around the field lines. It cannot move far in the plane perpendicular to $\mathbf{B}$ unless other forces (or collisions) are present. Note that the proper angular velocity $\boldsymbol{\omega}$ (and therefore the proper period $2\pi/\omega$) is independent of the speed, and consequently higher energy particles follow larger orbits. The path is found by integrating (18.35):

$$\begin{aligned} \mathbf{r}(t) &= \mathbf{r}_c + \mathbf{v}_{\parallel}(0) t + e^{-i\boldsymbol{\omega}t/\gamma} \mathbf{r}_0 \\ \mathbf{r}_0 &:= -\gamma \boldsymbol{\omega}^{-1} \times \mathbf{v}(0) , \end{aligned} \tag{18.36}$$

where $\mathbf{r}_c + \mathbf{v}_{\parallel}(0) t$ is the centre of the orbit and $\mathbf{r}_0$ is the initial radius vector of the helix.

If both electric and magnetic fields are nonzero constant vectors and if $\mathbf{F}^2 \neq 0$, a "drift frame" can be found in which the electric and magnetic fields are parallel (or in which one of them vanishes if $\mathbf{E}\cdot\mathbf{B}=0$), and the resulting motion is rifle-like: the rotation (if any) is about the direction of the boost (if any).

In the case $\mathbf{F}^2=0$, the drift frame moves at the speed of light and cannot be reached by massive charges or observers. However, the solution (18.30) has a simple expansion in this case:

$$\Lambda(\tau)=\left[1+\frac{e}{2m}\mathbf{F}\tau\right]\Lambda(0), \quad \mathbf{F}^2=0. \tag{18.37}$$

If the field at the position of the charge is not a constant, but its direction is:

$$\mathbf{F}(\tau)=\mathbf{F}_0 f(\tau), \tag{18.38}$$

then the equation of motion (18.28) can still be integrated to give

$$\Lambda(\tau)=\exp(\mathbf{F}_0 s/2)\Lambda(0), \tag{18.39}$$

where $s=(e/m)\int_0^\tau d\tau' f(\tau')$. This solution will prove useful in the following sections where we consider the motion of charges in an electromagnetic plane wave.

18.6 Electromagnetic Waves in Vacuum

Before studying the motion of charges in electromagnetic plane waves, we introduce the paravector description of such waves and relate the formalism to standard (if comparatively awkward) complex notation.

18.6.1 Maxwell's Equation in a Vacuum

Recall Maxwell's equation $\bar{\partial}\mathbf{F}(x)=0$ when the source term vanishes $j=0$. In terms of the paravector potential A, the relation $\mathbf{F}=\partial\bar{A}$ gives the wave equation for A,

$$\partial\bar{\partial}A=\square A=\left(\partial_t^2-\nabla^2\right)A=0, \tag{18.40}$$

where the Lorenz gauge condition $\langle\bar{\partial}A\rangle_S=0$ has been imposed. Since the D'Alembertian $\square$ is a scalar operator, the wave equation holds for each component of A. It also holds for each component of the electromagnetic field $\mathbf{F}$, as is seen by differentiating Maxwell's equation in a source-free region: $\square\mathbf{F}=\partial\bar{\partial}\mathbf{F}=0$.

What implications does the wave equation have for the paravector potential A? Suppose for the moment that A depends on position through the scalar product[2]

$$\langle\bar{k}x\rangle_S=k^\lambda x^\mu\langle\bar{\mathbf{e}}_\lambda\mathbf{e}_\mu\rangle_S=k^\lambda x^\mu\eta_{\lambda\mu}=k_\mu x^\mu=\omega t-\mathbf{k}\cdot\mathbf{x}, \tag{18.41}$$

where $k=\omega+\mathbf{k}$ is a constant spacetime vector. The paravector potential A is thus the same everywhere on the *hypersurface* $\langle\bar{k}x\rangle_S=const$, and at a given time t, it

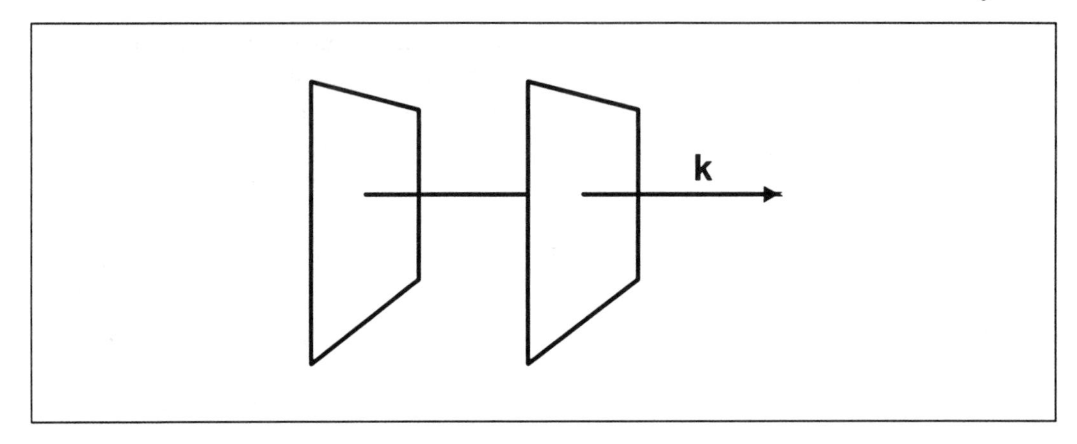

Fig. 18.3. The constant spatial planes of a directed plane wave propagate with velocity $\omega/\vec{k}$.

is the same everywhere on the spatial surface $\mathbf{k}\cdot\mathbf{x} = const$ (see Figure 18.3).

During the interval dt, every constant surface moves a distance $d\mathbf{x}$ perpendicular to the surface and hence parallel to $\mathbf{k}$ by an amount which satisfies $\omega dt = \mathbf{k}\cdot d\mathbf{x}$. Thus, $d\mathbf{x} = \mathbf{k}^{-1}\omega\, dt$. The velocity $d\mathbf{x}/dt = \mathbf{k}^{-1}\omega$ is called the *wave (phase) velocity* of the wave. Then since

$$\partial = \partial^{\mu}\mathbf{e}_{\mu} = \partial_{\mu}\mathbf{e}^{\mu} = \mathbf{e}^{\mu}\frac{\partial}{\partial x^{\mu}}, \tag{18.42}$$

the field is

$$\mathbf{F} = \partial\bar{A}\left(\left\langle\bar{k}x\right\rangle_S\right) = \left(\partial\left\langle\bar{k}x\right\rangle_S\right)\bar{A}'\left(\left\langle\bar{k}x\right\rangle_S\right) = k\bar{A}'\left(\left\langle\bar{k}x\right\rangle_S\right) \tag{18.43}$$

where the prime indicates differentiation with respect to the scalar argument. Similarly,

$$\partial\bar{\partial}A\left(\left\langle\bar{k}x\right\rangle_S\right) = \Box A\left(\left\langle\bar{k}x\right\rangle_S\right) = k\bar{k}A''\left(\left\langle\bar{k}x\right\rangle_S\right). \tag{18.44}$$

The wave equation (18.40) tells us that this vanishes, so that unless $A'' = 0$ at all positions, in which case A is either constant throughout spacetime or grows without limit as $\left\langle\bar{k}x\right\rangle_S$, then k must be null:

$$k\bar{k} = \omega^2 - \mathbf{k}^2 = 0. \tag{18.45}$$

Any function A of $\left\langle\bar{k}x\right\rangle_S$, where k is any constant null spacetime vector, is a solution of the wave equation (18.40). More generally, since (18.40) is linear, any

---

[2] Recall that repeated indices are summed over. Usually one of the indices is an upper ("contravariant") index while the other is a lower ("covariant") one. The result does not depend on which factor has the upper and which the lower index. The metric tensor $\eta_{\mu\nu}$ can be used to raise and lower indices. It follows that

$$\eta^{\mu}_{\nu} = \langle\mathbf{e}^{\mu}\bar{\mathbf{e}}_{\nu}\rangle_S = \langle\bar{\mathbf{e}}^{\mu}\mathbf{e}_{\nu}\rangle_S = \delta^{\mu}_{\nu}$$

is the Kronecker delta.

superposition of solutions with different values of $k = \omega\left(1 + \hat{\mathbf{k}}\right)$is also a solution. The gauge freedom means that many different choices of the vector potential A give rise to exactly the same electromagnetic field $\mathbf{F}$. In particular, $\mathbf{F}$ is invariant under the gauge transformation

$$A \to A + \partial\chi\,, \tag{18.46}$$

where χ is any scalar function. The Lorenz-gauge condition $\langle\bar{\partial}A\rangle_S = 0$ limits the gauge freedom.

18.7 Projectors

Since k is not only null (18.45) but also real, it can be factored into

$$k = \omega\left(1 + \hat{\mathbf{k}}\right) = 2\omega P_{\hat{\mathbf{k}}}\,, \tag{18.47}$$

where $P_{\hat{\mathbf{k}}} = \frac{1}{2}\left(1 + \hat{\mathbf{k}}\right)$ is a *projector,* defined to be a *real primitive idempotent*[3] with the properties

$$\begin{aligned} P_{\hat{\mathbf{k}}}^2 &= P_{\hat{\mathbf{k}}} = P_{\hat{\mathbf{k}}}^\dagger \\ P_{\hat{\mathbf{k}}}\bar{P}_{\hat{\mathbf{k}}} &= 0 = \bar{P}_{\hat{\mathbf{k}}}P_{\hat{\mathbf{k}}} \\ P_{\hat{\mathbf{k}}} + \bar{P}_{\hat{\mathbf{k}}} &= 1\,. \end{aligned} \tag{18.48}$$

Projectors are examples of *zero-divisors,* elements which are not zero but which are nevertheless factors of zero. It is the existence of such elements plus the noncommutivity of nonparallel vectors which keeps the algebra from a being a field. The existence of projectors adds richness to the algebra and allows "magic" manipulations, operations which are not possible in fields.

One useful bit of such magic is the ability of a projector to absorb a unit vector with which it commutes:

$$\begin{aligned} \hat{\mathbf{k}}P_{\hat{\mathbf{k}}} &= P_{\hat{\mathbf{k}}} \\ \hat{\mathbf{k}}\bar{P}_{\hat{\mathbf{k}}} &= -\bar{P}_{\hat{\mathbf{k}}}\,. \end{aligned} \tag{18.49}$$

This bit of magic can be applied over and over again, so that if $f(\mathbf{k})$ is any function of $\mathbf{k} = \omega\hat{\mathbf{k}}$ with a power series expansion, then

$$f(\mathbf{k})\,P_{\hat{\mathbf{k}}} = f(\omega)\,P_{\hat{\mathbf{k}}} \tag{18.50}$$

and

$$f(\mathbf{k})\,\bar{P}_{\hat{\mathbf{k}}} = f(-\omega)\,\bar{P}_{\hat{\mathbf{k}}}\,. \tag{18.51}$$

[3] The first equality of the properties defines an *idempotent,* the second equality defines a *real* element, and an idempotent is *primitive* if it is not zero and if it cannot be written as the sum of two other idempotents. In particular from line 3 of the properties, 1 is an idempotent but not a primitive idempotent.

Equations (18.50) and (18.51) may be considered *eigenvalue equations*, where $f(\pm\omega)$ are the *eigenvalues* and the projectors $P_{\hat{\mathbf{k}}}$ and $\bar{P}_{\hat{\mathbf{k}}}$ are *eigenprojectors* of $f(\mathbf{k})$. Since $P_{\hat{\mathbf{k}}} + \bar{P}_{\hat{\mathbf{k}}} = 1$, the function $f(\mathbf{k})$ can be expanded in eigenprojectors

$$f(\mathbf{k}) = f(\omega) P_{\hat{\mathbf{k}}} + f(-\omega) \bar{P}_{\hat{\mathbf{k}}}. \tag{18.52}$$

Such an expansion is known as the *spectral decomposition* of $f(\mathbf{k})$. Similar decompositions can be written down for analytic functions of any real paravector.

For any projector P, the sandwich PxP of an arbitrary cliffor $x = x^0 + \mathbf{x}$ is just the projection of x along P in the sense

$$PxP = \left(Px + \bar{x}\bar{P}\right) P = 2 \langle Px \rangle_S P, \tag{18.53}$$

where if $P = \frac{1}{2}(1 + \hat{\mathbf{n}})$, then $2\langle Px \rangle_S = x^0 + \mathbf{x} \cdot \hat{\mathbf{n}}$. Similarly,

$$Px\bar{P} = Px(1 - P) = P\mathbf{x}_\perp = \mathbf{x}_\perp \bar{P}, \tag{18.54}$$

where, since $\hat{\mathbf{n}}P = P$,

$$\mathbf{x}_\perp := \mathbf{x} - \mathbf{x} \cdot \mathbf{n}\,\hat{\mathbf{n}}. \tag{18.55}$$

18.8 Directed Plane Waves

The paravector potential discussed above, which depends on x only through the scalar $\langle \bar{k}x \rangle_S$, is a directed plane wave. Consider a particularly simple case in which A is a vector of constant magnitude whose direction at x is rotated by the angle $\langle \bar{k}x \rangle_S$:

$$A(x) = \mathbf{a}e^{\pm i\hat{\mathbf{k}}\langle \bar{k}x \rangle_S}, \tag{18.56}$$

where $A(0) = \mathbf{a}$ is a constant vector perpendicular to $\hat{\mathbf{k}}$.

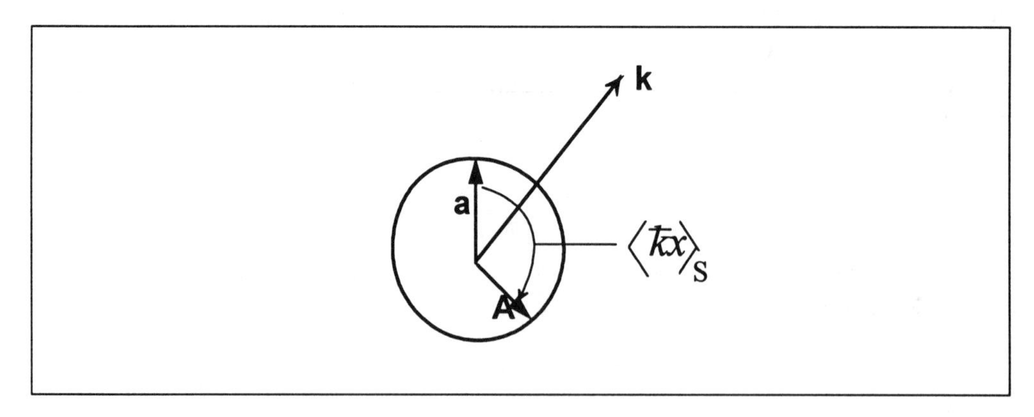

Fig. 18.4. The vector potential of a plane wave is a real rotating vector or a linear combination thereof.

The sign of the rotation angle is the *helicity* of the wave. Thus the wave $\mathbf{a}e^{i\hat{\mathbf{k}}\langle \bar{k}x \rangle_S}$ is said to have positive helicity whereas the wave $\mathbf{a}e^{-i\hat{\mathbf{k}}\langle \bar{k}x \rangle_S}$ has negative helicity.

The waves (18.56) are said to be *monochromatic* because they rotate (or oscillate) with a single angular frequency (color). The corresponding electromagnetic field can be written

$$\mathbf{F}(x) = \left\langle \partial \bar{A} \right\rangle_V = \pm ik\hat{\mathbf{k}}\mathbf{A}(x) , \tag{18.57}$$

which, since $k\hat{\mathbf{k}} = k$, can be simplified to

$$\mathbf{F}(x) = \pm ike^{\mp i\hat{\mathbf{k}}\langle \bar{k}x\rangle_S}\mathbf{a} = \pm ik\mathbf{a}e^{\mp i\langle \bar{k}x\rangle_S} . \tag{18.58}$$

This relates a spatial rotation of $\mathbf{F}(0) = \pm ik\mathbf{a}$ to a rotation in the complex plane. The latter mixes $\mathbf{F}$ with its dual $-i\mathbf{F}$ and is therefore sometimes called a duality rotation.

Linear combinations of the two solutions for various ω but a given direction $\hat{\mathbf{k}}$ can be expressed in the form

$$\mathbf{F}(x) = \mathbf{F}(0) f\left(\langle \bar{k}x\rangle_S\right) , \tag{18.59}$$

where $f(x)$ is a complex scalar function with $f(0) = 1$. This still describes a directed plane wave, but not generally a monochromatic wave. Maxwell's equation for source-free space, namely $\bar{\partial}\mathbf{F}(x) = 0$, gives

$$\bar{k}\mathbf{F} = (\omega - \mathbf{k})\mathbf{F} = 0 = \mathbf{F}k , \tag{18.60}$$

the scalar part of which gives the orthogonality of the fields with $\mathbf{k}$

$$\mathbf{k} \cdot \mathbf{F} = 0 , \tag{18.61}$$

and the vector part of which, $\mathbf{F} = \hat{\mathbf{k}}\mathbf{F} = i\hat{\mathbf{k}} \times \mathbf{F}$, relates real and imaginary parts $-\hat{\mathbf{k}} \times \mathbf{B} = \mathbf{E}$, $i\hat{\mathbf{k}} \times \mathbf{E} = i\mathbf{B}$. Since k is proportional to $P_{\hat{\mathbf{k}}}$ and $P_{\hat{\mathbf{k}}} + \bar{P}_{\hat{\mathbf{k}}} = 1$, (18.60) shows that the electromagnetic field can be written

$$\mathbf{F} = P_{\hat{\mathbf{k}}}\mathbf{F}\bar{P}_{\hat{\mathbf{k}}} . \tag{18.62}$$

It follows from $P\bar{P} = 0$ that

$$\mathbf{F}^2 = 0 , \tag{18.63}$$

which will be important when we solve for particle motion in the electromagnetic wave.

The condition $\mathbf{F}^2 = 0$ expresses conditions which hold for *any* directed electromagnetic plane wave in a vacuum: $\mathbf{E}$ and $\mathbf{B}$ are of equal magnitude and perpendicular to each other. As we saw above, $\mathbf{F}^2 = 0$ follows from the more general condition $\mathbf{F}k = 0$ which also implies that the vectors $\{\mathbf{E}, \mathbf{B}, \mathbf{k}\}$ form a right-handed orthogonal vector basis of three-dimensional space. However, plane waves do not need to have perpendicular electric and magnetic fields if waves of opposite propagation directions $\hat{\mathbf{k}}$ are combined.[4]

[4]Indeed, standing waves can have parallel electric and magnetic fields. See, for example, W. E. Baylis and G. Jones, *J. Phys. A. (Math, Gen)***22**,17-29 (1989).

18.8.1 Polarization

Consider the positive-helicity case: at a fixed position $\mathbf{x}$ in space, $\mathbf{A}$ rotates around $\hat{\mathbf{k}}$ in a right-handed sense, but at any given time, a "snapshot" of the wave would show a left-handed spiral, that is a spiral which rotates to the left as one progresses in the direction of $\hat{\mathbf{k}}$. The wave of positive helicity is therefore sometimes said to be circularly polarized in a "right-handed" sense, and sometimes in a "left-handed" sense. To avoid misunderstanding, we use the helicity (which most authors agree on) to describe the sense of circular polarization.

Linear combinations of two waves of the same k but opposite helicity give waves of elliptical polarization:

$$\boxed{\begin{aligned} A(x) &= \hat{\mathbf{a}}\left(a_+ e^{i\hat{\mathbf{k}}\langle\bar{k}x\rangle_S} + a_- e^{-i\hat{\mathbf{k}}\langle\bar{k}x\rangle_S}\right) \\ &= \mathbf{a}\cos\left\langle\bar{k}x\right\rangle_S + \mathbf{b}\sin\left\langle\bar{k}x\right\rangle_S \end{aligned}} \tag{18.64}$$

with semimajor axis $\mathbf{a} = (a_+ + a_-)\,\hat{\mathbf{a}}$, semiminor axis $\mathbf{b} = (a_+ - a_-)\,\hat{\mathbf{b}}$, and $\hat{\mathbf{b}} = i\hat{\mathbf{a}}\hat{\mathbf{k}} = \hat{\mathbf{k}} \times \hat{\mathbf{a}}$, where $a_\pm$ can be taken real and positive, and the origin $x = 0$ of coordinates has been chosen to lie in one of planes where $\mathbf{A}$ is oriented along the semimajor axis $\mathbf{a}$.

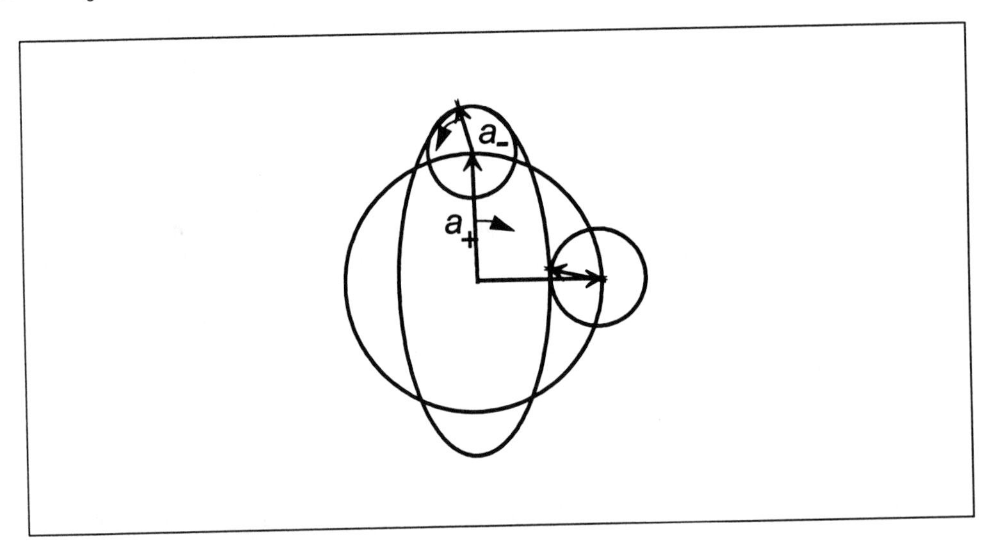

Fig. 18.5. Elliptical polarization is the sum of counter-rotating circular polarizations.

If one of a_+, a_- vanishes, then $\mathbf{a}^2 = \mathbf{b}^2$ and the wave is circularly polarized, whereas if $a_+ = \pm a_-$, then one of $\mathbf{a}$ or $\mathbf{b}$ vanishes and the wave is linearly polarized. The electromagnetic field corresponding to $A(x)$ in (18.64) is

$$\mathbf{F}(x) = ik\hat{\mathbf{a}}\left(a_+ e^{-i\langle\bar{k}x\rangle_S} - a_- e^{i\langle\bar{k}x\rangle_S}\right). \tag{18.65}$$

It is often convenient to refer the major axis direction $\hat{\mathbf{a}}$ to a fixed spatial direction $\widehat{\boldsymbol{\xi}}$ in the plane $i\mathbf{k}$. If $\hat{\mathbf{a}}$ is related to $\widehat{\boldsymbol{\xi}}$ by a rotation about $\hat{\mathbf{k}}$ by the angle δ, then $\hat{\mathbf{a}} = \exp\left(-i\hat{\mathbf{k}}\delta\right)\widehat{\boldsymbol{\xi}}$ and

$$\mathbf{F}(x) = k\widehat{\boldsymbol{\xi}}\alpha_k(x) , \tag{18.66}$$

where $\alpha_k(x)$ is the complex scalar function

$$\alpha_k(x) = ie^{-i\delta}\left(a_+ e^{-i\langle\bar{k}x\rangle_S} - a_- e^{i\langle\bar{k}x\rangle_S}\right) . \tag{18.67}$$

The polarization of a wave is traditionally defined in terms of the electric field $\mathbf{E} = \langle\mathbf{F}\rangle_\Re$ which can be expressed in terms of a real amplitude E_0 and a complex polarization vector $\boldsymbol{\epsilon}$ as

$$\mathbf{E} = \left(\mathbf{F} + \mathbf{F}^\dagger\right)/2 = E_0\left\langle\boldsymbol{\epsilon}e^{-i\langle\bar{k}x\rangle_S}\right\rangle_\Re . \tag{18.68}$$

The polarization vector is normalized so that

$$\boldsymbol{\epsilon}\boldsymbol{\epsilon}^\dagger = 1 . \tag{18.69}$$

A straightforward calculation shows

$$E_0\boldsymbol{\epsilon} = ik\widehat{\boldsymbol{\xi}}a_+e^{-i\delta} + i\widehat{\boldsymbol{\xi}}ka_-e^{i\delta} \tag{18.70}$$

18.9 Motion of Charges in Plane Waves

We consider now the motion of a charge e in a directed plane wave as given above (18.59). Since $\mathbf{F}^2 = 0$, the eigenspinor solution (18.37) found above for motion in crossed electric and magnetic fields of equal magnitude can be used here with $s = (e/m)\int d\tau\, f\left(\langle\bar{k}x\rangle_S\right)$. The resulting proper velocity

$$\begin{aligned} u(\tau) = \Lambda\Lambda^\dagger &= \left[1 + \mathbf{F}(0)\, s/2\right] u(0)\left[1 + \mathbf{F}(0)^\dagger s^*/2\right] \\ &= u(0) + \langle\mathbf{F}(0)\, u(0)\, s\rangle_\Re + \tfrac{1}{4}\mathbf{F}(0)\, u(0)\,\mathbf{F}(0)^\dagger |s|^2 \end{aligned} \tag{18.71}$$

displays an unexpected and curious trait: since $\bar{k}\mathbf{F} = 0 = \mathbf{F}^\dagger\bar{k}$, the Lorentz-invariant

$$\langle\bar{k}u(\tau)\rangle_S = \frac{d}{d\tau}\langle\bar{k}x\rangle_S = \omega_0 = \langle\bar{k}u(0)\rangle_S \tag{18.72}$$

is constant. This states that the charge moves in such a way that the Doppler-shifted frequency ω_0 of the wave, as seen in the rest frame of the particle, is fixed. Integration thus gives the argument $\langle\bar{k}x\rangle_S = \omega_0\tau + \phi$ of f:

$$\dot{s} = \frac{e}{m}f\left(\langle\bar{k}x\rangle_S\right) = \frac{e}{m}f\left(\omega_0\tau + \phi\right) , \tag{18.73}$$

where $\phi = \langle\bar{k}x_0\rangle_S$ and x_0 is the position of the charge at $\tau = 0$. For a given functional dependence of f, $\dot{s}$ can be integrated and $u(\tau)$ found directly.

For example, let the directed plane wave be a monochromatic wave of circular polarization:

$$f\left(\left\langle\bar{k}x\right\rangle_S\right) = \exp\left(-i\left\langle\bar{k}x\right\rangle_S\right) .$$

Integration of $\dot{s}$ (18.73) then gives

$$s = \frac{ie}{m\omega_0}\left[e^{-i\omega_0\tau} - 1\right]e^{-i\phi}$$

and

$$|s|^2 = 2\left(\frac{e}{m\omega_0}\right)^2\left(1 - \cos\omega_0\tau\right) . \tag{18.74}$$

If the initial velocity has no component perpendicular to **k**, then the energy of the particle is

$$m\gamma = m\left\langle u\left(\tau\right)\right\rangle_S = m\gamma_0 + m\frac{\omega}{\omega_0}\left(\frac{e\mathbf{E}\left(0\right)}{m\omega}\right)^2\left(1 - \cos\omega_0\tau\right) . \tag{18.75}$$

The energy gained $m\left(\gamma - \gamma_0\right)$ can be quite large for a charge traveling with the wave, since then, the Doppler-shifted frequency (the wave frequency as seen in the frame of the charge) $\omega_0 \ll \omega$:

$$\frac{\omega}{\omega_0} = \left(\frac{1+v}{1-v}\right)^{1/2} , \tag{18.76}$$

where $v = \mathbf{v}\cdot\hat{\mathbf{k}}$ is the particle speed, and the proper time ω_0^{-1} spent "riding" the wave is relatively long. The effect is known as "surfing."

Chapter 19

Eigenspinors in Quantum Theory

William E. Baylis
Department of Physics, University of Windsor,
Windsor, Ont., Canada N9B 3P4

19.1 Introduction

Eigenspinors carry information giving the velocity and angular orientation of a frame. Compound systems, with distinguishable parts, generally require several eigenspinors to describe their state of motion. A classical elementary particle might be defined as an object that requires only one eigenspinor to describe its motion.

In the last lecture, we saw that the motion of charges in electromagnetic fields is described by linear time-evolution equations reminiscent of quantum formulas. In this lecture, we attempt to make the connection more precise and complete. We start with the relatively simple nonrelativistic case of a distribution of spins. With the strong association this case suggests between the classical eigenspinor and the quantum wave function, we then progress to the fully relativistic approach, the resulting Dirac equation, and its interpretation.

19.2 Spin

Spin is usually described as a pure quantum phenomenon with no classical analog.[1] Because it flows naturally from the Dirac equation, it is often also associated with relativity. Here we argue that there is a simple classical analog, one that yields exact quantum relations for such quantities as the spin density. Quantum behavior, such as the splitting of a beam of silver atoms in the field of a Stern-Gerlach magnet, is manifested by linear superpositions of classical eigenspinors that arise naturally as solutions to linear eigenspinor equations of motion.

[1]See, for exeample, A. Messiah, *Quantum Mechanics* Vol. II., (North Holland, Amsterdam, 1961), p. 540.

The accuracy of the classical model suggests changes to usual interpretations of quantum measurements. A prepared state may have a precise spin direction, not just a known spin component, and exploiting a strong analogy to polarized optics, an experimental procedure involving static and slowly rotating magnetic fields can be devised to measure the spin direction.

19.2.1 Magic of the Pauli Hamiltonian

The nonrelativistic Pauli Hamiltonian, traditionally written

$$H_{\text{Pauli}} = \frac{(\mathbf{p}\cdot\boldsymbol{\sigma})^2}{2m} + V(\mathbf{r}) , \tag{19.1}$$

includes spin interactions. Here $\boldsymbol{\sigma}$ is the "vector spin operator" whose components are the Pauli spin matrices. From the relations $(\mathbf{a}\cdot\boldsymbol{\sigma})\ (\mathbf{b}\cdot\boldsymbol{\sigma}) = \mathbf{a}\cdot\mathbf{b} + i\mathbf{a}\times\mathbf{b}\cdot\boldsymbol{\sigma}$ and (with $\hbar = c = 1$) $\mathbf{p} = -i\nabla - e\mathbf{A}$, H_{Pauli} includes the coupling of the spin to the magnetic field $\mathbf{B} = \boldsymbol{\nabla}\times\mathbf{A}$

$$H_{\text{Pauli}} = \frac{(i\nabla + e\mathbf{A})^2}{2m} - \frac{e}{2m}\boldsymbol{\sigma}\cdot\mathbf{B} + V(\mathbf{r}) \tag{19.2}$$

with the correct magnetic moment $\boldsymbol{\mu} = e\boldsymbol{\sigma}/(2m)$. The interaction of the spin with the magnetic field appears as if by magic, and the origin of the spin is shrouded in mystery.

To anyone working in Clifford algebras,

$$\mathbf{p}\cdot\boldsymbol{\sigma} = p_x\sigma_x + p_y\sigma_y + p_z\sigma_z = \begin{pmatrix} p_z & p_- \\ p_+ & -p_z \end{pmatrix}, \tag{19.3}$$

with $p_\pm = p_x \pm ip_y$ is simply misleading notation for the matrix representation of the *vector* $\mathbf{p}$ in $C\ell_3$. However, is it not curious that a *classical unit vector* should be identified as the quantum-mechanical operator for one *component* of the electron spin?

19.2.2 Classical Spin Distribution

In $C\ell_3$,: $\mathbf{v}' = R\mathbf{v}R^\dagger$, $R = \exp(-i\boldsymbol{\theta}/2)$, rotates vector $\mathbf{v}$ by θ about $\hat{\boldsymbol{\theta}}$. A distribution of "spin" directions can be represented classically by

$$\rho\boldsymbol{\sigma}_{\text{cl}} := \rho R\mathbf{e}_3R^\dagger, \tag{19.4}$$

where $\rho = \rho(\mathbf{r},t)$ is a normalized density and $R = R(\mathbf{r},t)$ gives the rotation of the direction from $\mathbf{e}_3$ to $\boldsymbol{\sigma}_{\text{cl}} = R\mathbf{e}_3R^\dagger$as a function of position and time. It is easier to work with distributions of scalars. The distribution of the component of the spin direction onto the direction $\mathbf{n}$ is

$$\rho\boldsymbol{\sigma}_{\text{cl}}\cdot\mathbf{n} = \rho\left\langle R\mathbf{e}_3R^\dagger\mathbf{n}\right\rangle_S . \tag{19.5}$$

The classical distribution will be shown equivalent to the quantum expression of the spin density measured with a quantization axis $\mathbf{n}$.

19.2.3 Quantum Form

In terms of the projector $P = \frac{1}{2}(1 + \mathbf{e}_3)$,

$$\begin{aligned} R\mathbf{e}_3 R^\dagger &= R\left(P - \bar{P}\right) R^\dagger \\ &= (RP)(RP)^\dagger - \overline{(RP)(RP)^\dagger} . \end{aligned} \tag{19.6}$$

Thus,

$$\begin{aligned} \rho \left\langle R\mathbf{e}_3 R^\dagger \mathbf{n} \right\rangle_S &= 2\rho \left\langle (RP)^\dagger \mathbf{n} (RP) \right\rangle_S \\ &= \operatorname{tr}\left\{ \psi^\dagger \mathbf{n} \psi \right\} , \end{aligned} \tag{19.7}$$

where $\psi = \rho^{1/2} RP$ has the standard matrix representation

$$\psi = e^{-i\gamma/2}\rho^{1/2} \begin{pmatrix} e^{-i\alpha/2}\cos\beta/2 & 0 \\ e^{i\alpha/2}\sin\beta/2 & 0 \end{pmatrix} \tag{19.8}$$

identical with that for a nonrelativistic spinor wave function with an extra (inconsequential) column of zeros. If we omit the second column, we can also drop the tr symbol. If we also denote the matrix representation of $\mathbf{n}$ by $\boldsymbol{\sigma} \cdot \mathbf{n}$, the spin density is

$$\psi^\dagger \boldsymbol{\sigma} \cdot \mathbf{n} \psi , \tag{19.9}$$

identical to the quantum form.

19.2.4 Stern-Gerlach Filter

The Stern-Gerlach magnet acts on the classical eigenspinor field in the same way that a birefringent medium acts on a beam of light. The key is that any rotation can be written

$$\begin{aligned} R &= \exp(-i\boldsymbol{\theta}/2)\exp(-i\mathbf{e}_3\gamma/2) \\ &= \left(\cos\theta/2 - i\hat{\boldsymbol{\theta}}\sin\theta/2\right)\exp(-i\mathbf{e}_3\gamma/2) , \end{aligned} \tag{19.10}$$

where the rotation axis $\hat{\boldsymbol{\theta}}$ lies in the $\mathbf{e}_1\mathbf{e}_2$ plane and the rotation element $\exp(-i\boldsymbol{\theta}/2)$ has been expanded into a linear superposition of a rotation by 0°, namely 1, and a rotation of 180° about $\hat{\boldsymbol{\theta}}$, namely $-i\hat{\boldsymbol{\theta}}$. The expansion is analogous to splitting a spinor wave function into a linear combination of parts representing spin up and spin down. Such an expansion can be made with respect to any direction. Because the equation of motion is real linear, the evolution of each part can be found separately and the results superimposed. In the field of a Stern-Gerlach magnet, the spin-up and spin-down parts of the spinor are separated in space, just as beams of light of differing polarizations are usually split by a birefringent material. By selecting one spatial part, a polarization filter is created.

The analogy between electron spin and light polarization is reinforced by the quantum description of the spin of a particle in a pure state in terms of the spin density matrix

$$D = \psi\psi^\dagger = \rho RPR^\dagger = \frac{1}{2}\rho\left(1 + \boldsymbol{\sigma}_{\text{cl}}\right) , \tag{19.11}$$

whose time-development, given by the Schrödinger equation $i\hbar\dot{D} = [H, D]$, shows that even quantum mechanically, a magnetic-field interaction $\mu_0 \boldsymbol{\sigma}_{\text{cl}} \cdot \mathbf{B}$ induces a precession

$$\dot{\boldsymbol{\sigma}}_{\text{cl}} = \boldsymbol{\omega} \times \boldsymbol{\sigma}_{\text{cl}} \tag{19.12}$$

that can be used to detect unambiguously all components of the spin in a beam of similarly prepared fermions, where $\boldsymbol{\omega} = -2\mu_0 \mathbf{B}/\hbar$ and $\mu_0 = e\hbar/(2m)$ is the Bohr magneton. A consistent interpretation of fermion spin seems to be that it has a definite direction, much like the polarization direction of a classical electromagnetic field. A single measurement can determine the components along a single direction, and a sequence of measurements can unambiguously determine all components of a beam, even a very dilute one, of fermions prepared in a pure spin state.

19.2.5 Linear Combinations of Spatial Rotations

A spatial rotation is an even unimodular cliffor. Since even elements are all equal to their bar-daggers, a rotation is also unitary. Indeed, any even unimodular cliffor can be written in the form $\exp(-i\boldsymbol{\theta}/2)$ and interpreted as a rotation element. Such rotations play dual roles as operators and as spinors that specify the orientation of the system relative to a given laboratory frame. Above, we associated rotational spinors and quantum wave functions. In this section, we prove three useful theorems about linear combinations of rotational spinors.

Theorem 1. Two rotations have orthogonal cliffors if and only if they are related by a 180-degree rotation.

Proof: R_1 and R_2 are orthogonal $\Leftrightarrow \langle R_1 \bar{R}_2 \rangle_S = 0$. Now $R_1 \bar{R}_2$ is the rotation which takes R_2 into $R_2 : R_1 = (R_1 \bar{R}_2) R_2 = \exp(-i\boldsymbol{\theta}_{12}/2) R_2$ and $\langle R_1 \bar{R}_2 \rangle_S = \cos(\boldsymbol{\theta}_{12}/2)$. Evidently R_1 and R_2 can be orthogonal in $\mathcal{C}\ell_3$ if and only if $\theta_{12} = \pi$ and they are related by a rotation of $\theta_{12} = 180°$ (mod 2π).

Theorem 2. Any rotation R can be decomposed into a real linear combination of an arbitrary rotation R_1 and another rotation R_2 orthogonal to it.

Proof: Consider the rotation $R\bar{R}_1$. Let $R\bar{R}_1 = \exp(-i\mathbf{n}\theta/2) = \cos\theta/2 - i\mathbf{n}\sin\theta/2$, where $\mathbf{n}$ is a real unit vector. Any rotation can be expressed in this form. Multiply from the right by R_1 to obtain

$$R = R_1 \cos(\theta/2) + R_2 \sin(\theta/2) , \tag{19.13}$$

where $R_2 = -i\mathbf{n}R_1 = \exp(-i\mathbf{n}\pi/2) R_1$ is related to R_1 by a 180° rotation.

The spinor space of R is reducible, and the reduction amounts to splitting every R into complementary ideals $\mathcal{C}\ell_3 P$ and $\mathcal{C}\ell_3 \bar{P}$:

$$R = RP + R\bar{P} , \tag{19.14}$$

where P is a projector $P = \frac{1}{2}(1 + \mathbf{e})$ with $\mathbf{e}$ a real unit vector. The minimum left ideal $\mathcal{C}\ell_3 P$ is a space of two-component spinors and is spanned by a basis $\{\alpha_0, \alpha_1\}$, where we can choose $\alpha_0 = P$ and $\alpha_1 = \mathbf{n}P$, with $\mathbf{n}$ a real unit vector perpendicular to $\mathbf{e}$.

Theorem 3. The basis spinors α_0 and α_1 are related by a 180° rotation in the **en** plane.

Proof: $\alpha_1 = \mathbf{n}\alpha_0 = \mathbf{ne}\alpha_0 = \exp(\mathbf{ne}\pi/2)\,\alpha_0$, since $(\mathbf{ne})^2 = -1$.

Any rotation element R can be reconstructed from the two-component spinor $\eta = R\alpha_0$:

$$\eta + \bar{\eta}^\dagger = R\alpha_0 + R\bar{\alpha}_0 = R\,. \tag{19.15}$$

19.3 Covariant Eigenspinors

19.3.1 Generalized Unimodularity

Linear transformations that leave the norm invariant are called Lorentz transformations. Lorentz transformations of a spacetime vector $p = p_0 + \mathbf{p}$ can be combinations of the following three forms:

$$\begin{aligned} p &\to LpL^\dagger \\ p &\to -p \\ p &\to \bar{p}\,, \end{aligned} \tag{19.16}$$

where to preserve the norm of p,

$$\left|L\bar{L}\right|^2 = 1\,. \tag{19.17}$$

We replace the usual unimodularity condition for L (namely $L\bar{L} = 1$) by the more general condition

$$L\bar{L} = \bar{L}L = e^{i\beta}\,. \tag{19.18}$$

With $\beta = 0$ one obtains the usual unimodularity condition, and with $\beta = \pi$ an antiunimodularity condition is fulfilled. In classical mechanics, one usually considers only positive-energy particles for which we can restrict the transformations L to be unimodular, i.e., $\beta = 0$. Formally, one can replace L by $L' = Le^{i\beta/2}$ in (19.18); the new L' is then unimodular, the paravector transformation $p \to LpL^\dagger$ is unchanged, and $L' \in$ SL(2,C). However, since we need to include both unimodular and antiunimodular eigenspinors, we retain the generalized condition (19.18).

19.3.2 Eigenspinor of an Elementary Particle

As discussed in the previous lecture, the eigenspinor of an electron (or other classical "elementary particle") Λ is the Lorentz transformation L that relates the *rest frame*[2] of the electron to the lab frame of the observer. Thus, for an observer in the rest frame of the electron, the spacetime momentum is $p_{rest} = m$, and it is transformed to

$$p = \Lambda m \Lambda^\dagger = m\Lambda\Lambda^\dagger = mu \tag{19.19}$$

for an electron moving with respect to the observer. This equation is the basis for the Dirac equation.[3]

[2]The rest frame of an elementary particle is the instantaneously commoving inertial frame. It does not rotate. See below in this section.

[3]W. E. Baylis, *Phys. Rev. A* **45**, 4293-4302 (1992).

The eigenspinor satisfies the generalized unimodularity condition (19.18) and can be written as the product of a phase factor $\exp(i\beta/2)$, a real unimodular boost $B = B^\dagger = \exp(\mathbf{w}/2)$, and a unitary rotation $R = \bar{R}^\dagger = \exp(-i\boldsymbol{\theta}/2)$:

$$\Lambda = e^{i\beta/2} BR. \tag{19.20}$$

One can think of Λ as the transformation that transforms properties of the electron from the values it would have when at rest in the lab frame to the values it has after boosts and rotations bring it to whatever motion and orientation it actually has in the lab frame at proper time τ. In the limiting case that $\beta = 0 \bmod 2\pi$, Λ is unimodular and hence timelike. If $\beta = \pi \bmod 2\pi$, Λ is antiunimodular and hence spacelike.

The transformation (19.19) implies a positive-energy particle iff $m > 0$, because

$$\langle \Lambda\Lambda^\dagger \rangle_S = \sum_{\mu=0}^{3} |\Lambda^\mu|^2 > 0\,. \tag{19.21}$$

One traditionally takes $m > 0$. Negative-energy particles then obey

$$p = -\Lambda m \Lambda^\dagger\,, \tag{19.22}$$

which can be viewed either as an orthochronous transformation from $-m$ (a positron at rest) or a nonorthochronous transformation from m (an electron at rest). In the former case, the transformation can be a gradual transformation of an antiparticle, that is, one built of many incremental transformations; in the latter, it is an abrupt transformation in which the particle is scattered backward in time. Both views are at times useful. The electron-positron particles are "unified" if we view the positron as a transformed electron. The transformation (19.22) is then nonorthochronous and hence abrupt. One can combine relations (19.19) and (19.22) by adding a phase factor $e^{i\beta}$:

$$p = \Lambda m e^{i\beta} \Lambda^\dagger\,, \tag{19.23}$$

where $m > 0$ and for positive-energy particles, $\beta = 0$ (modulus 2π) and for negative-energy ones, $\beta = \pi$ (modulus 2π).

If we apply the generalized unimodularity condition (19.18) to $L = \Lambda$ and assume that the phase angle β is the same in equations (19.23) and (19.18), then (19.23) can be put in the form

$$p\bar{\Lambda}^\dagger = m\Lambda\,. \tag{19.24}$$

This is the classical Dirac equation. It remains true if multiplied on both sides by a real scalar $\rho_r^{1/2}$ and is also invariant under a fixed (global) rotation of the rest frame: $\Lambda \to \Lambda R_0$. We replace Λ by a *spinor field* Ψ by making Λ function of spacetime position and putting

$$\Psi = \rho_r^{1/2}\Lambda = e^{i\beta/2}\rho_r^{1/2} BR\,, \tag{19.25}$$

where for reference we included the explicit factorization (19.20) of Λ. The Dirac equation (19.24) becomes

$$p\bar{\Psi}^\dagger = m\Psi \tag{19.26}$$

with the normalization $\Psi\bar{\Psi} = \bar{\Psi}\Psi = \rho_r e^{i\beta}$.

The particle current density is traditionally (in first-quantized Dirac theory)

$$j = \Psi\Psi^\dagger = \rho_r \Lambda\Lambda^\dagger = \rho_r B^2 \tag{19.27}$$

and is thus a future-directed time-like paravector:

$$\bar{j}j = \left|\bar{\Psi}\Psi\right|^2 = \rho_r^2 \geq 0 \tag{19.28}$$

and

$$\rho := \langle j \rangle_S = \left\langle \Psi\Psi^\dagger \right\rangle_S = \rho_r \left\langle \Lambda\Lambda^\dagger \right\rangle_S \geq 0, \tag{19.29}$$

with $\rho = 0$ if and only if $\rho_r = 0$. In the Feynman-Stückelberg interpretation of antiparticles as "particles moving backward in time," the proper velocity of the particle is

$$u = p/m = \Lambda e^{i\beta}\Lambda^\dagger = e^{i\beta}B^2 . \tag{19.30}$$

Since the proper velocity is the tangent vector to the world line $x(\tau)$ of the particle, a particle has an *Yvon-Takabayasi angle* β of $\beta = 0$ for positive-energy solutions and is moving forward in time: $dt/d\tau = \left\langle \Lambda\Lambda^\dagger \right\rangle > 0$, whereas $\beta = \pi$ for negative-energy ones which therefore move backward: $dt/d\tau = -\left\langle \Lambda\Lambda^\dagger \right\rangle < 0$.

19.4 Differential-Operator Form

The Dirac equation (19.26) takes its quantum form as a differential equation with the identification[4]

$$p\bar{\Psi}^\dagger = i\partial\bar{\Psi}^\dagger \mathbf{e} - eA\bar{\Psi}^\dagger, \tag{19.31}$$

where $\mathbf{e}$ is the unit vector in the direction of the spin in the rest frame and A is the paravector potential. Note that p is the *kinetic* momentum and $p + eA$ is the *conjugate* momentum.[5] The total energy of the system corresponds to the conjugate component $E = p^0 + eA^0 = p^0 + V$. The differential form of the classical Dirac equation (19.26) is thus

$$i\partial\bar{\Psi}^\dagger \mathbf{e} - eA\bar{\Psi}^\dagger = m\Psi . \tag{19.32}$$

There is no complete *derivation* of the identification (19.31) and, indeed, its equivalent standard form for column-spinor wave functions, namely $p = i\partial - eA$, is usually taken as axiomatic in quantum theory. Its ultimate justification must lie in agreement with experiment, but there are a couple of relevant facts:

---

[4]In fact, since this relation should hold for any spinor Ψ and any orientation of the coordinates, we impose the equivalent relation for each component:

$$p^\mu\bar{\Psi}^\dagger = i\partial^\mu\bar{\Psi}^\dagger \mathbf{e} - eA^\mu\bar{\Psi}^\dagger .$$

[5]For an electron, $e < 0$.

- Many experiments confirm the existence of de Broglie waves for particle beams of well-defined momentum. These waves manifest themselves by interference phenomena when coherent beams following different paths overlap. The operator relation for the momentum should reproduce such waves.

- Equations (19.23) to (19.30) are all invariant under an initial rotation. Of course, this reflects their insensitivity to the initial orientation of the electron frame. However, they are *not* invariant under a change in the complex phase of Ψ, since the equations involve both Ψ and $\bar{\Psi}^\dagger$.

We want to relate different spacetime positions of a continuous spinor field. The spinor can describe different orientations and velocities at different spacetime positions, but these should be continuously related. Recall first the classical eigenspinor Λ of a *free,* positive-energy particle in its rest frame, which relates the particle at different points on its world line and is parameterized by the proper time τ of the particle. The variation is constrained by the conditions that the eigenspinor is normalized and, since the particle is free, there is no acceleration:

$$\frac{d}{d\tau}\left(\Lambda\bar{\Lambda}\right) = \frac{d}{d\tau}\left(\Lambda\Lambda^\dagger\right) = 0\,. \tag{19.33}$$

If we *define* $\Omega_r = 2\bar{\Lambda}\dot{\Lambda}$, then it follows that Ω_r is an imaginary vector. We put

$$\Omega_r := -i\mathbf{e}\omega\,. \tag{19.34}$$

Here ω is the rate of spin of the particle in its "rest frame" (*i.e.*, its instantaneously commoving inertial frame) and $-i\mathbf{e}$ gives both the spin plane and the sense of rotation in that plane. Since there is no preferred time origin, ω must be independent of time. The solution to the time-development equation $\dot{\Lambda} = \frac{1}{2}\Lambda\Omega_r$ is

$$\Lambda\left(\tau\right) = \Lambda\left(0\right)\exp\left(-i\mathbf{e}\omega\tau/2\right)\,,\ \ \mathbf{e}\omega = const., \tag{19.35}$$

In fact, the proper spin rate of a free particle, ω, should be Lorentz invariant as well as constant.

The quantum spinor field should be consistent with such time development along all possible paths through spacetime. From the homogeneity of spacetime, we expect there to be a solution for which j (19.27) and the normalization are constant:

$$\Psi\bar{\Psi} = const.,\quad \Psi\Psi^\dagger = const. \tag{19.36}$$

In analogy with the work above on the classical eigenspinor, we put

$$\Omega_\mu = 2\Psi^{-1}\left(\partial_\mu\Psi\right)\,, \tag{19.37}$$

to find

$$\partial_\mu\Psi = \frac{1}{2}\Psi\Omega_\mu,\ \bar{\Omega}_\mu = \Omega_\mu^\dagger = -\Omega_\mu\,. \tag{19.38}$$

Thus, as above for Ω_r, each Ω_μ is an imaginary vector. The proper-time dependence along the world line with tangent vector u is given by

$$\dot{\Psi} = u^\mu \partial_\mu \Psi = \frac{1}{2} u^\mu \Psi \Omega_\mu . \tag{19.39}$$

To be consistent with (19.35) along the world line of a classical particle,

$$u^\mu \Psi \Omega_\mu = \Psi u^\mu \Omega_\mu = -i\omega \Psi \mathbf{e} . \tag{19.40}$$

Since for a free particle, $\omega \mathbf{e}$ is constant and Ψ is invertible,

$$u^\mu Q_\mu = \omega , \tag{19.41}$$

where $Q_\mu := i\Omega_\mu \mathbf{e}$.The only spacetime quantities that contract with the proper velocity u^μ to give a Lorentz invariant are proportional to the spacetime momentum or to the proper velocity, itself. Evidently, Q_μ must be proportional to the μth component of the spacetime momentum. We write

$$Q_\mu = 2kp_\mu , \tag{19.42}$$

where k is a Lorentz-invariant constant. Combining Eqs. (19.38) and (19.42), we find

$$\partial_\mu \Psi \mathbf{e} = \frac{1}{2} \Psi \Omega_\mu \mathbf{e} = ikp_\mu \Psi , \tag{19.43}$$

which can be solved to give the momentum operator:

$$\bar{p} \Psi = ik^{-1} \bar{\partial} \Psi \mathbf{e} . \tag{19.44}$$

We will usually assume that the constant k can be absorbed into the units: $k = 1$. The rotation rate in the rest frame is then the *Zitterbewegung frequency* $2m$.

19.4.1 The Electromagnetic Gauge Field

By combining (19.44) with (19.26), we obtain the differential form of the Dirac equation for a free electron:

$$\bar{p} \Psi = i\bar{\partial} \Psi \mathbf{e} = m \bar{\Psi}^\dagger . \tag{19.45}$$

The equation is invariant under any fixed rotation of the rest frame about $\mathbf{e}$. In order to extend such *global invariance* to the more demanding *local invariance*, we must add a compensating gauge field $A(x)$. Thus consider the same equation (19.45) for a spinor field with an additional rest-frame rotation $\exp(-i\chi \mathbf{e})$, where χ is a scalar function of the spacetime position x. Version (19.45) is not invariant under such a rotation, but

$$i\bar{\partial} \Psi \mathbf{e} - e\bar{A} \Psi = m \bar{\Psi}^\dagger \tag{19.46}$$

is, provided the rotation is accompanied by the *gauge transformation*

$$eA \rightarrow eA + \partial \chi \tag{19.47}$$

that eliminates the extra term. Thus we identify the momentum with the operator

$$\bar{p}\Psi = i\bar{\partial}\Psi\mathbf{e} - e\bar{A}\Psi\,, \qquad (19.48)$$

where the gauge field A is the electromagnetic paravector potential. The momentum operator (19.48) can be combined with (19.26) to yield the differential form of the Dirac equation (19.32).

The addition of the gauge field is equivalent to changing the rest-frame rotation rate from $\omega = 2m$ to $2(m + e\phi_{rest}) = 2\langle (p + eA)\,\bar{u}\rangle_S$ in (19.35). The rotation angle is thus the usual action integral and the standard stationary-phase argument for superpositions can lead to the classical Euler-Lagrange equations.

19.4.2 Linearity and Superposition

The Dirac equation in $C\ell_3$

$$\bar{p}\Psi = i\partial\Psi\mathbf{e} - eA\Psi = m\bar{\Psi}^{\dagger} \qquad (19.49)$$

is *real linear* but not complex linear. Consequently, if Ψ_k are solutions, then so is the linear combination $a^k\Psi_k$ provided the scalar coefficients a^k are real. Linear combinations with *complex* coefficients are not generally solutions. However, if Ψ is a solution, then so is $\Psi e^{-i\mathbf{e}\phi}$, that is, Ψ with an additional rotation of the rest frame about the spin axis. Thus, complex linear combinations of the form $a^k e^{-i\phi_k}\psi_k$ in traditional quantum formalism can be replaced in $C\ell_3$ by real linear combinations of rotated solutions

$$a^k\Psi_k e^{-i\mathbf{e}\phi_k}\,. \qquad (19.50)$$

Linear combinations of positive and negative-energy solutions are also possible, but because of linearity, the two solutions evolve independently and their frequency of interference is usually so high that at lower frequencies than required for pair creation, the interference is averaged away.

Note that many common relations for complex phase factors can be extended to rotations about a fixed axis. For example,

$$\int_{-\infty}^{\infty} dk\, e^{i\mathbf{e}kx} = \int_{-\infty}^{\infty} dk \cos kx = 2\pi\delta(x)\,. \qquad (19.51)$$

19.5 Basic Symmetry Transformations

Spatial reversal is not the same as parity inversion P. Since only odd elements of the algebra should change sign under P, we represent inversion by the grade automorphism $p \to \bar{p}^{\dagger}$. There are also simple forms for space-time inversion (PT) and its product with charge conjugation C, namely CPT:

$$P : \Psi \to \bar{\Psi}^{\dagger} \qquad (19.52)$$

$$PT : \Psi \to \Psi e^{i\pi\mathbf{n}/2} \qquad (19.53)$$

$$CPT : \Psi \to i\Psi \, . \tag{19.54}$$

These reproduce the standard transformation forms in the usual Dirac theory with column bispinors and have simple interpretations: PT rotates the rest frame by 180° degrees about an axis perpendicular to the spin and thus flips the spin direction. CPT changes $\beta \to \beta + \pi$. The above transformations may be combined, for example:

$$T : \Psi \to \bar{\Psi}^{\dagger} e^{i\pi\mathbf{n}/2} \tag{19.55}$$

and

$$C : \Psi \to \Psi\mathbf{n} \, . \tag{19.56}$$

Note that $T^2 = (PT)^2 = (CPT)^2 = -1$ whereas $P^2 = C^2 = 1$. Furthermore, $C(\Psi)$ and $PT(\Psi)$ are seen to be orthogonal to Ψ, in that

$$\langle \bar{\Psi}\Psi\mathbf{n} \rangle_S = 0 = \langle \bar{\Psi}\Psi i\mathbf{n} \rangle_S \, . \tag{19.57}$$

It is easy to verify the effect of such symmetry transformations on the Dirac equation (19.32). Parity inversion (19.52) changes p to $\bar{p}$ and hence

$$P : i\bar{\partial}\Psi\mathbf{e} - e\bar{A}\Psi = m\bar{\Psi}^{\dagger} \to i\partial\Psi\mathbf{e} - eA\Psi = m\bar{\Psi}^{\dagger} \tag{19.58}$$

after an additional bar-dagger has been taken. Note that in the rest frame of a particle, $B = 1$ and $\Psi = \rho_r e^{i\beta/2} R = e^{i\beta}\bar{\Psi}^{\dagger}$. Thus, positive-energy particles, with $\beta = 0$, have $\Psi = \bar{\Psi}^{\dagger}$ and are have *positive intrinsic parity*, whereas negative-energy ones, with $\beta = \pi$, have $\Psi = -\bar{\Psi}^{\dagger}$ and possess *negative intrinsic parity.* Under charge conjugation, after multiplying from the left by $\mathbf{n}$, which is perpendicular to $\mathbf{e}$,

$$C : i\bar{\partial}\Psi\mathbf{e} - e\bar{A}\Psi = m\bar{\Psi}^{\dagger} \to i\bar{\partial}\Psi\mathbf{e} + e\bar{A}\Psi = m\bar{\Psi}^{\dagger} \, . \tag{19.59}$$

This is equivalent to simply changing the sign of the charge. Finally, under CPT (19.54),

$$CPT : i\bar{\partial}\Psi\mathbf{e} - e\bar{A}\Psi = m\bar{\Psi}^{\dagger} \to i\bar{\partial}\Psi\mathbf{e} + e\bar{A}\Psi = -m\bar{\Psi}^{\dagger} \, . \tag{19.60}$$

19.6 Relation to Standard Form

The differential equations for the components of the standard "Dirac bispinor" are found by splitting (19.32) into the two complementary minimal left ideals $\mathcal{C}\ell_3 P$, $\mathcal{C}\ell_3 \bar{P}$ where

$$P = \frac{1}{2}(1 + \mathbf{e}) \, . \tag{19.61}$$

After noting that $\mathbf{e}P = P$, we obtain the two equations by applying P to (19.32) and its bar-dagger from the right:

$$i\bar{\partial}\Psi P - e\bar{A}\Psi P = m\bar{\Psi}^{\dagger} P \tag{19.62}$$

and

$$i\partial\bar{\Psi}^{\dagger} P - eA\bar{\Psi}^{\dagger} P = m\Psi P \, . \tag{19.63}$$

An equivalent set of two equations can be obtained with the complementary projector $\bar{P}$. These may be combined into one equation of the usual quantum form

$$\gamma^{\mu}\left(i\partial_{\mu} - eA_{\mu}\right)\psi = m\psi\,, \tag{19.64}$$

where in the Weyl representation ψ is the four-row bispinor

$$\psi^{(W)} = \frac{1}{\sqrt{2}}\begin{pmatrix}\bar{\Psi}^{\dagger}P\\ \Psi P\end{pmatrix} \tag{19.65}$$

and, when $\mathbf{e}$ is represented by the 2×2 matrix σ_3, the gamma matrices are given as in the usual Weyl representation by

$$\gamma^0 = \gamma_0 = \begin{pmatrix}0 & 1\\ 1 & 0\end{pmatrix},\ \gamma^k = -\gamma_k = \begin{pmatrix}0 & \sigma_k\\ -\sigma_k & 0\end{pmatrix},\ \text{Weyl rep} \tag{19.66}$$

and each entry is itself a 2×2 matrix. Although $\psi^{(W)}$ is a 4×2 matrix, the second column is zero. It may therefore be represented by a single column. The barred bispinor is

$$\bar{\psi}^{(W)} := \psi^{(W)\dagger}\gamma_0 = \frac{P}{\sqrt{2}}\left(\Psi^{\dagger}, \bar{\Psi}\right). \tag{19.67}$$

19.6.1 Weyl Spinors

The two-component *Weyl spinors* can be introduced in $\mathcal{C}\ell_3$ by

$$\eta := \Psi\alpha_0\,,\ \xi := \Psi\alpha_1\,, \tag{19.68}$$

where, as above, $\alpha_0 := P$ and $\alpha_1 := \mathbf{n}P$ constitute a spinor basis $\{\alpha_0, \alpha_1\}$ of the minimal left ideal $\mathcal{C}\ell_3 P$. Like all spinors in $\mathcal{C}\ell_3 P$, η and ξ obey

$$\eta\bar{\eta} = \bar{\eta}\eta = 0\,,\ \eta\bar{\xi} = 0\,,\ \eta = \eta P\,. \tag{19.69}$$

They can also be expanded

$$\eta = \eta^0\alpha_0 + \eta^1\alpha_1 = \left(\eta^0 + \eta^1\mathbf{n}\right)\alpha_0 := \eta^A\alpha_A\,. \tag{19.70}$$

We can reconstitute Ψ from (19.68) if η and ξ are known:

$$\Psi = \eta + \xi\mathbf{n}\,. \tag{19.71}$$

The Weyl bispinor $\psi^{(W)}$ (19.65) is

$$\psi^{(W)} = \frac{1}{\sqrt{2}}\begin{pmatrix}-\bar{\xi}^{\dagger}\mathbf{n}\\ \eta\end{pmatrix}, \tag{19.72}$$

and the Dirac-Pauli bispinor is related by

$$\psi^{(DP)} = \frac{1}{\sqrt{2}}\begin{pmatrix}1 & 1\\ -1 & 1\end{pmatrix}\psi^{(W)} = \frac{1}{2}\begin{pmatrix}\eta - \bar{\xi}^{\dagger}\mathbf{n}\\ \eta + \bar{\xi}^{\dagger}\mathbf{n}\end{pmatrix} = \begin{pmatrix}\Phi^{(+)}P\\ \Phi^{(-)}P\end{pmatrix}, \tag{19.73}$$

where $\Phi^{(\pm)} = \frac{1}{2}\left(\Psi \pm \bar{\Psi}^{\dagger}\right)$ are the even and odd parts of Ψ. In the low-velocity limit, $\Psi \approx e^{i\beta}\bar{\Psi}^{\dagger}$, which means that the lower half of $\psi^{(DP)}$ vanishes in the low velocity limit for positive-energy ($\beta = 0$) particles, whereas as the upper half vanishes in this limit for negative-energy ($\beta = \pi$) ones.

19.6.2 Momentum Eigenstates

Consider solutions of (19.31) in free space ($A = 0$) characterized by a given constant momentum p:

$$\Psi_p(x) = \Psi_p(0) \exp\left(-i \langle p\bar{x}\rangle_S \mathbf{e}\right) . \tag{19.74}$$

Since $\langle p\bar{x}\rangle_S = m\tau$, where we have assumed coincident coordinate origins in the particle and lab frames, (19.74) describes a rapid rotation at the *Zitterbewegung* frequency $\omega = 2m$ about the direction $\mathbf{e}$ in the particle frame. The spinor field is a plane wave of phase velocity $E/\mathbf{p}$, whose magnitude is greater than the speed of light. The factor $\Psi_p(0) = e^{i\beta/2}\rho_r(0) B(0) R(0)$ may describe an additional constant rotation and a boost. Substitution into the Dirac equation (19.32) gives

$$p\bar{\Psi}_p^\dagger(0) = m\Psi_p(0) . \tag{19.75}$$

We multiply from the left by $\Psi_p^\dagger(0)/m$ and use the normalization $\Psi\bar{\Psi} = \rho_r e^{i\beta}$ to obtain the constant current

$$j_p = \Psi_p(0)\Psi_p^\dagger(0) = \rho_r B^2 = \frac{p}{m}\rho_r e^{-i\beta} . \tag{19.76}$$

The solution may be normalized to a rest-frame density of unity: $\rho_r = 1$. For a positive-energy solution, $\beta = 0$ and we can write

$$\Psi_p(0) = (p/m)^{1/2} R(0) = \frac{p+m}{\sqrt{2m(E+m)}} R(0) , \quad E > 0 \tag{19.77}$$

where $R(0)$ is an arbitrary rotation which, together with the boost $(p/m)^{1/2}$, determines the direction of the spin in the lab frame. The standard matrix representation of $\Psi(x)$ in (19.74) and (19.77) is

$$\Psi_p(x) = \begin{pmatrix} m+E+p_z & p_x - ip_y \\ p_x + ip_y & m+E-p_z \end{pmatrix} \frac{R(0)\exp\left(-i\langle p\bar{x}\rangle_S \mathbf{e}\right)}{\sqrt{2m(E+m)}} . \tag{19.78}$$

Negative-energy solutions ($\beta = \pi$) are similar, but with $p = E + \mathbf{p}$ and E replaced by $-p = |E| - \mathbf{p}$ and $-E = |E|$, respectively.

In the limit of low velocities for positive energies,

$$\Psi_{m+\mathbf{p}}(x) = \left(1 + \frac{\mathbf{p}}{2m}\right) R(0) \exp\left(-imt\mathbf{e} + i\mathbf{p}\cdot\mathbf{x}\mathbf{e}\right) . \tag{19.79}$$

19.6.3 Standing Waves

A linear superposition of plane waves of the same spin and energy but opposite spatial momenta gives standing waves

$$\Psi(x) = \left[\Psi_{m+\mathbf{p}}(x)\cos\alpha + \Psi_{m-\mathbf{p}}(x)\sin\alpha\right] \tag{19.80}$$

with a probability current at low velocities given by

$$\Psi\Psi^\dagger = 1 + \frac{\mathbf{p}}{m}\cos 2\alpha + \left(\cos 2\mathbf{p}\cdot\mathbf{x} + \frac{\mathbf{p}\times\mathbf{s}}{m}\sin 2\mathbf{p}\cdot\mathbf{x}\right)\sin 2\alpha . \tag{19.81}$$

Note the sheets of current lying in planes parallel to the plane of reflection and moving in the directions $\pm\mathbf{p}\times\mathbf{s}$. Sheets of opposite current densities are separated by twice the de Broglie wavelength. They are associated with the spin and arise when the spin is not aligned with $\mathbf{p}$ and where the density gradient does not vanish.

19.6.4 Zitterbewegung

Linear superpositions of the positive- and negative-energy solutions (19.79) with the same $\mathbf{p}$ give *Zitterbewegung* at the angular frequency $2m$ in the spin direction $\mathbf{s} = R(0)\,\mathbf{e}\bar{R}(0)$:

$$\Psi(x) = \left[\Psi_{m+\mathbf{p}}(x)\cos\alpha + \Psi_{-m+\mathbf{p}}(x)\sin\alpha\right] \tag{19.82}$$

with a probability current (19.27)

$$j = \Psi\Psi^{\dagger} = 1 + \frac{\mathbf{p}}{m}\cos 2\alpha - \mathbf{s}\sin 2mt\sin 2\alpha\,, \tag{19.83}$$

where terms of order $\mathbf{p}^2/m^2$ have been ignored. Note that the current of a purely positive-energy state ($\alpha = 0$) is $1 + \mathbf{p}/m$ whereas that of a purely negative-energy state ($\alpha = \pi/2$) is $1 - \mathbf{p}/m$. The current of a negative-energy state is thus in the opposite direction to its vector momentum $\mathbf{p}$. The rapid Zitterbewegung oscillations arise solely from the interference of the positive- and negative-energy components.

When momentum eigenspinors of opposite momenta p are superimposed, we find

$$\Psi = \cos\alpha\Psi_{m+\mathbf{p}}(x) + \sin\alpha\Psi_{-m-\mathbf{p}}(x) \tag{19.84}$$

with a current

$$j = \Psi\Psi^{\dagger} = 1 + \frac{\mathbf{p}}{m} - \left(\frac{\mathbf{p}}{m}\cdot\mathbf{s} + \mathbf{s}\right)\sin 2\alpha\sin 2\left\langle p\bar{x}\right\rangle_S\,. \tag{19.85}$$

19.7 Hamiltonians

The Hamiltonian operator in $C\ell_3$ Dirac theory is defined by

$$H\Psi = i\partial_t\Psi\mathbf{e}\,. \tag{19.86}$$

From (19.32),

$$H\Psi = (V + \mathbf{p})\,\Psi + m\bar{\Psi}^{\dagger}. \tag{19.87}$$

19.7.1 Stationary States

We look for energy eigenstates of the form

$$i\partial_t\Psi\mathbf{e} = E\Psi. \tag{19.88}$$

These describe *normal modes* of the system in which all parts rotate at a single frequency $2E$ about the particle-frame direction $\mathbf{e}$:

$$\Psi(x) = \Psi(\mathbf{x})\exp(-i\mathbf{e}Et)\,. \tag{19.89}$$

(Even classical macroscopic systems often have boundary conditions which allow only discrete normal modes of vibration.) Note that a negative-energy state is equivalent to a positive-energy one rotating about the opposite direction in the particle frame.

The time-independent form of the Dirac equation (19.32) in the Weyl eigenspinor representation is thus

$$(E - V - \mathbf{p})\,\Psi = m\bar{\Psi}^{\dagger}\,. \tag{19.90}$$

From the sum and difference of the last equation with its bar-dagger conjugate, one obtains the Dirac-Pauli equations for the normal rotational modes:

$$\mathbf{p}\Phi^{(\mp)} = (E \mp m - V)\,\Phi^{(\pm)}. \tag{19.91}$$

Eliminating the "small" component $\Phi^{(-)}$, we obtain a relation

$$\mathbf{p}\,(E + m - V)^{-1}\,\mathbf{p}\Phi^{(+)} = (E - m - V)\,\Phi^{(+)} \tag{19.92}$$

that in the low-velocity limit, $E + m - V \approx 2m$, becomes the Pauli-Schrödinger equation

$$\left[\frac{\mathbf{p}^2}{2m} + V\right]\Phi^{(+)} = (E - m)\,\Phi^{(+)}\,. \tag{19.93}$$

19.7.2 Landau Levels

Although the nonrelativistic equations can be solved when $B = \mathit{const.} \neq 0$ and the potential $V = 0$, it is just as easy to solve the exact relativistic equations for this case. From (19.92) with $V = 0$,

$$\mathbf{p}^2\phi = \left(E^2 - m^2\right)\phi\,, \tag{19.94}$$

where with the two-component spinor $\phi = \Phi^{+}\,(1 + \mathbf{e})\,/2$,

$$\mathbf{p}^2\phi = \left[-\nabla^2 + \frac{1}{4}B^2(x^2 + y^2) - eB\,(L_3 + \mathbf{e}_3)\right]\phi\,, \tag{19.95}$$

so that the two components of ϕ are decoupled. The solution contains products of circular solutions of the two-dimensional harmonic oscillator (Landau levels) with plane waves along the $\mathbf{e}_3$ direction.

19.8 Fierz Identities of Bilinear Covariants

The bilinear covariants have the traditional form

$$B_{\mu\nu} = \bar{\psi}\Gamma_{\mu\nu}\psi\,, \tag{19.96}$$

where the $\Gamma_{\mu\nu}$ are the sixteen linearly independent products of gamma matrices, which can be represented in direct-product form by $\Gamma_{\mu\nu} = \sigma_\mu \bigotimes \sigma_v$. In the Weyl representation, the bilinear covariant $B_{\mu\nu}$ is the nonvanishing (upper left) element of

$$\frac{1}{2}P\left(\Psi^\dagger,\bar{\Psi}\right)\Gamma_{\mu\nu}\begin{pmatrix}\bar{\Psi}^\dagger\\ \Psi\end{pmatrix}P\,, \tag{19.97}$$

which we can extract as the trace of the full 2×2 matrix. Algebraically, the trace is equivalent to twice the scalar part:

$$B_{\mu\nu} = \left\langle \left(\Psi^\dagger,\bar{\Psi}\right)\Gamma_{\mu\nu}\begin{pmatrix}\bar{\Psi}^\dagger\\ \Psi\end{pmatrix}P\right\rangle_S . \tag{19.98}$$

Physically, the bilinear covariants represent basic properties of the rest frame, such as the spacetime vectors $\mathbf{e}_0 = 1$ and $\mathbf{e}$, and the parabivector formed from them, multiplied by the density and transformed to the lab frame. The spacetime scalar (Lorentz scalar) is

$$\Psi\bar{\Psi} = \bar{\Psi}\Psi = \rho_r e^{i\beta} = B_{00} - B_{30}\,. \tag{19.99}$$

The vector is the positive-definite current density (components $\bar{\psi}\gamma_\mu\psi$ in traditional notation)

$$\Psi\Psi^\dagger = j = B_{10} + iB_{2k}\bar{e}^k\,. \tag{19.100}$$

The spin bivector (usual components $\bar{\psi}\frac{i}{2}\left(\gamma_\mu\gamma_\nu - \gamma_\nu\gamma_\mu\right)\psi = \bar{\psi}\sigma_{\mu\nu}\psi$) is

$$i\Psi\mathbf{e}\bar{\Psi} = i\Psi\mathbf{e}\bar{\mathbf{e}}_0\bar{\Psi} = 2\rho_r e^{i\beta}\mathbf{S} = i\left(B_{0k} - B_{3k}\right)\bar{e}^k\,. \tag{19.101}$$

The spin dual (a pseudovector with components $\bar{\psi}i\gamma_5\gamma_\mu\psi$) is

$$i\Psi\mathbf{e}\Psi^\dagger = 2i\rho_r s = -\left(B_{20} + iB_{1k}\bar{e}^k\right)\,, \tag{19.102}$$

and the pseudoscalar is $i\Psi\bar{\Psi}$ (traditionally $\bar{\psi}i\gamma_5\psi$). The spin density is

$$\rho_r s = k = \frac{1}{2}\Psi\mathbf{e}\Psi^\dagger\,. \tag{19.103}$$

For reference, the relations are summarized by

$$\begin{aligned}\Psi\left(1+\mathbf{e}\right)\Psi^\dagger &= \left(B_{1\nu} + iB_{2\nu}\right)\bar{e}^\nu\\ \Psi\left(1-\mathbf{e}\right)\Psi^\dagger &= \left(B_{1\nu} - iB_{2\nu}\right)e^\nu\\ \Psi\left(1+\mathbf{e}\right)\bar{\Psi} &= \left(B_{0\nu} - B_{3\nu}\right)\bar{e}^\nu\,.\end{aligned} \tag{19.104}$$

The Fierz identities follow trivially. In particular,

$$j\bar{j} = \left|\Psi\bar{\Psi}\right|^2 = \rho_r^2 \tag{19.105}$$

$$\mathbf{S}^2 = s\bar{s} = -1/4 \tag{19.106}$$

$$\mathbf{S}j = i\rho_r e^{i\beta}s = ik/2,\ \mathbf{S}k = ij/2 \tag{19.107}$$

$$j\bar{k} = i\rho_r^2\mathbf{S},\ \langle j\bar{s}\rangle_S = 0\,. \tag{19.108}$$

In this lecture we have shown how the eigenspinor nicely bridges the gap from the classical realm to the quantum one. First-quantized Dirac theory is fully reproduced, with all its impressive results as well as with its problems, namely the definition of the charge current and the Klein paradox of nonconservation of charge in the presence of high electric fields. Many physical properties and geometrical transformations are seen to be more transparent when expressed in the Pauli algebra.

Chapter 20

Eigenspinors in Curved Spacetime

William E. Baylis
Department of Physics, University of Windsor,
Windsor, Ont., Canada N9B 3P4

This chapter presents the final lecture in the series on paravectors and eigenspinors in physics. Its purpose is to establish that the Pauli algebra $\mathcal{C}\ell_3$ and eigenspinors expressed in it provide an efficient description of curved spacetime. Much of the work applying the Pauli algebra to general relativity is due to Dr. George Jones.[1]

We start by observing how standard spinors and their symplectic spaces are encompassed in $\mathcal{C}\ell_3$. We finish by studying the application of eigenspinors in $\mathcal{C}\ell_3$ to curved spacetime and looking for clearer physical insights that may result in general relativity.

20.1 Ideals, Spinors, and Symplectic Spaces

As we saw in the last lecture, two-component spinor spaces appear in the Pauli algebra as minimal ideals. Any real direction $\mathbf{e} \in \mathbb{R}^3$ can be taken to define two complementary spinor spaces in $\mathcal{C}\ell_3$

$$\mathcal{S}_+ = \mathcal{C}\ell_3 P \text{ and } \mathcal{S}_- = \mathcal{C}\ell_3 \bar{P} \tag{20.1}$$

where $P = \frac{1}{2}(1+\mathbf{e})$ and $\bar{P} = \frac{1}{2}(1-\mathbf{e})$ are projectors:

$$P + \bar{P} = 1, \; P\bar{P} = 0, \; P = P^2 = P^\dagger \tag{20.2}$$

The space $\mathcal{S}_+$ is spanned by the basis $\{\alpha_0 = P, \alpha_1 = \mathbf{n}P\}$ over $\mathbb{C}$ where the real unit vector $\mathbf{n}$ is perpendicular to $\mathbf{e}$: $\mathbf{n}P = \bar{P}\mathbf{n}$. This is proved by noting

$$P = \mathbf{e}P \,, \; \mathbf{n}P = \mathbf{ne}P \tag{20.3}$$

[1] George L. Jones, Ph.D. Dissertation, University of Windsor, 1993.

and every $p \in C\ell_3$ can be expanded

$$p = p^0 + p^3\mathbf{e} + p^1\mathbf{n} + ip^2\mathbf{ne} \tag{20.4}$$

where the coefficients p^μ are complex scalars. It follows that

$$pP = \left(p^0 + p^3\right) P + \left(p^1 + ip^2\right) \mathbf{n}P \tag{20.5}$$

and this has the desired form.

A symplectic linear structure can be defined on $\mathcal{S}_+$ by

$$\langle \eta, \xi \rangle = \eta^0\xi^1 - \xi^0\eta^1 = -\langle \xi, \eta \rangle = 2\left\langle \bar{\eta}\xi \left(\bar{\alpha}_0\alpha_1\right)^\dagger \right\rangle_S \tag{20.6}$$

for any $\eta = \eta^A\alpha_A \in \mathcal{S}_+$ and $\xi = \xi^A\alpha_A \in \mathcal{S}_+$. Thus $\mathcal{S}_+$ is a two-dimensional complex symplectic vector space. A similar structure is also analogously constructed for $\mathcal{S}_-$. If $\eta \in \mathcal{S}_+$, then $\eta = \eta P$ and $\eta\mathbf{n} = \eta P\mathbf{n} = \eta\mathbf{n}\bar{P}$, and thus $\eta\mathbf{n} \in \mathcal{S}_-$. Note that if and only if $\langle \eta, \xi \rangle = 0$, then $\eta^0/\eta^1 = \xi^0/\xi^1$ and consequently $\xi = c\eta$ where $c \in \mathbb{C}$.

One can use the basis spinors to define light-like paravectors:

$$\alpha_0\alpha_0^\dagger = P\,,\ \alpha_1\alpha_1^\dagger = \bar{P} \tag{20.7}$$

are real, and

$$\alpha_0\alpha_1^\dagger = P\mathbf{n}\,,\ \alpha_1\alpha_0^\dagger = \mathbf{n}P \tag{20.8}$$

are complex. Linear combinations can be taken to form an orthonormal tetrad $\{\mathbf{e}_0, \mathbf{e}_1, \mathbf{e}_2, \mathbf{e}_3\}$ with

$$\begin{aligned} \mathbf{e}_0 &= 1 = \alpha_0\alpha_0^\dagger + \alpha_1\alpha_1^\dagger \\ \mathbf{e}_3 &= \mathbf{e} = \alpha_0\alpha_0^\dagger - \alpha_1\alpha_1^\dagger \\ \mathbf{e}_1 &= \mathbf{n} = \alpha_0\alpha_1^\dagger + \alpha_1\alpha_0^\dagger \\ i\mathbf{e}_2 &= \mathbf{en} = \alpha_0\alpha_1^\dagger - \alpha_1\alpha_0^\dagger\,. \end{aligned} \tag{20.9}$$

Under restricted Lorentz transformations ("spin" transformations), spinors in $\mathcal{S}_\pm$ transform as

$$\eta \to L\eta\,,\ L \in SL(2, \mathbb{C})\,. \tag{20.10}$$

Spatial reverses transform as

$$\bar{\xi} \to \bar{\xi}\bar{L}\,. \tag{20.11}$$

If η and ξ are any two spinors in the same two-dimensional spinor space $\mathcal{S}_+$, then

$$\eta = \eta P \text{ and } \xi = \xi P \tag{20.12}$$

and the product

$$\eta\bar{\xi} = \eta P\bar{P}\xi = 0\,. \tag{20.13}$$

Now $\langle \eta\bar{\xi} \rangle_S = \langle \bar{\xi}\eta \rangle_S = \left(\bar{\xi}\eta + \bar{\eta}\xi\right)/2$, and as a consequence of (20.13), products of the form

$$\bar{\xi}\eta = -\bar{\eta}\xi = \langle \xi, \eta \rangle\, \bar{\alpha}_0\alpha_1 \tag{20.14}$$

are parabivectors that represent spacetime planes. The same arguments, of course, hold for any two spinors in $\mathcal{S}_-$.

Since L is unimodular, spinor products of the form $\bar{\xi}\eta$ are invariant under such transformations. Thus, $\bar{\xi}\eta = \bar{P}\bar{\xi}\eta P$ is an invariant spacetime plane. Similarly, the skew-scalar product $\langle \eta, \xi \rangle$ is also invariant. The unimodularity is indeed necessary and sufficient to preserve the skew-scalar product. Therefore, $SL(2, \mathbb{C})$ can be identified with the symplectic group of two dimensions:

$$SL(2, \mathbb{C}) \approx Sp(2) \tag{20.15}$$

that is defined to leave invariant the skew product of a complex two-dimensional spinor space.

It is useful to note that any element η of the ideal $\mathcal{S}_+$ can be written as the projection of an even element of $C\ell_3$. To prove, consider

$$\eta = pP \tag{20.16}$$

where p is an arbitrary element of $C\ell_3$. Then we can split p into even and odd parts, and because $P = \mathbf{e}P$,

$$\begin{aligned} \eta &= \left(\langle p \rangle_+ + \langle p \rangle_-\right) P \\ &= \left(\langle p \rangle_+ + \langle p \rangle_- \mathbf{e}\right) P \\ &= \left(\langle p \rangle_+ + \langle p\mathbf{e} \rangle_+\right) P \\ &= \langle p + p\mathbf{e} \rangle_+ P \end{aligned} \tag{20.17}$$

which proves the assertion. Now any even cliffor in $C\ell_3$ can be uniquely expressed as the product of a real, positive scalar $\rho^{1/2}$ and rotation R:

$$\eta = \rho^{1/2} R P \,. \tag{20.18}$$

We emphasize that such a decomposition is possible for any element of $\mathcal{S}_+ = C\ell_3 P$, even one formed by projecting a boost. It is always possible to replace $\rho^{1/2}$ by a boost: If $w = \ln \rho$, then

$$\eta = e^{w/2} R P = e^{\mathbf{w}/2} R P \tag{20.19}$$

where $\mathbf{w} = w\hat{\mathbf{w}}$ and

$$\hat{\mathbf{w}} = R\mathbf{e}R^\dagger \,. \tag{20.20}$$

Furthermore, it can be shown that for any null cliffor there exists a unique projector P that admits such a decomposition.

20.2 Bispinors

A bispinor can be formed from the sum of spinors in the two complementary spinor spaces:

$$\Lambda = \eta + \xi \mathbf{n} \tag{20.21}$$

where $\eta, \xi \in \mathcal{S}_+$ and $\xi\mathbf{n} \in \mathcal{S}_-$. Indeed, any Pauli cliffor can be split into elements of the two minimum ideals $\mathcal{S}_+$ and $\mathcal{S}_-$:

$$\begin{aligned} \Lambda &= \Lambda\left(P+\bar{P}\right) = \Lambda P + \Lambda\mathbf{n}P\mathbf{n} && (20.22)\\ &= \eta + \xi\mathbf{n}\,. && (20.23) \end{aligned}$$

Here we identify

$$\begin{aligned} \eta &= \Lambda P = \Lambda\alpha_0 \\ \xi &= \Lambda\mathbf{n}P = \Lambda\alpha_1\,. \end{aligned} \qquad (20.24)$$

If $\Lambda \in SL(2,C)$, it may be viewed as a spinorial Lorentz transformation, and η and ξ are then the Lorentz-transformed basis vectors of $\mathcal{S}_+$. If Λ is an eigenspinor, one would identify the proper velocity u and the spin paravector s by

$$\begin{aligned} u &= \Lambda\Lambda^\dagger = \eta\eta^\dagger + \xi\xi^\dagger \\ s &= \Lambda\mathbf{e}\Lambda^\dagger = \eta\eta^\dagger - \xi\xi^\dagger\,. \end{aligned} \qquad (20.25)$$

Unimodularity implies

$$\begin{aligned} \Lambda\bar{\Lambda} &= 1 = \xi\mathbf{n}\bar{\eta} - \eta\mathbf{n}\bar{\xi} \\ \bar{\Lambda}\Lambda &= 1 = \bar{\eta}\xi\mathbf{n} - \mathbf{n}\bar{\xi}\eta\,. \end{aligned} \qquad (20.26)$$

Note that s and u are orthogonal:

$$\langle s\bar{u}\rangle_S = \left\langle \Lambda\mathbf{e}\bar{\Lambda}\right\rangle_S = 0\,, \qquad (20.27)$$

and that the spin paravector s is related to the spin parabivector

$$\mathbf{S} = \Lambda\mathbf{e}_1\mathbf{e}_2\bar{\Lambda} = -i\Lambda\mathbf{e}\bar{\Lambda} \qquad (20.28)$$

by

$$s = i\mathbf{S}u\,. \qquad (20.29)$$

20.3 Flagpoles and Flags

Any ideal spinor $\eta = \eta P$ generates a null paravector $\eta\eta^\dagger$ known as its *flagpole*:

$$\eta\eta^\dagger\left(\overline{\eta\eta^\dagger}\right) = \eta\left(\eta^\dagger\bar{\eta}^\dagger\right)\bar{\eta} = \eta\bar{\eta}\left(\eta^\dagger\bar{\eta}^\dagger\right) = 0\,. \qquad (20.30)$$

In terms of (20.24),

$$\eta\eta^\dagger = u + s = \Lambda P\Lambda^\dagger\,. \qquad (20.31)$$

Similarly, the flagpole generated by ξ is

$$\xi\xi^\dagger = u - s = \Lambda\bar{P}\Lambda^\dagger\,. \qquad (20.32)$$

Furthermore,

$$\begin{aligned} \eta\xi^\dagger &= \Lambda P \mathbf{n} \Lambda^\dagger \\ \xi\eta^\dagger &= \Lambda \mathbf{n} P \Lambda^\dagger \end{aligned} \tag{20.33}$$

and consequently,

$$\begin{aligned} \eta\xi^\dagger + \xi\eta^\dagger &= \Lambda \mathbf{n} \Lambda^\dagger =: q \\ \eta\xi^\dagger - \xi\eta^\dagger &= \Lambda \mathbf{e}\mathbf{n} \Lambda^\dagger =: ir\,. \end{aligned} \tag{20.34}$$

which give the transformed transverse directions. The relation between η and ξ ensures that the invariant product is

$$\bar{\eta}\xi = \bar{P}\mathbf{n}P = \mathbf{n}P = \bar{\alpha}_0\alpha_1\,, \tag{20.35}$$

which means that the antisymmetric product is of unit value:

$$\langle \eta, \xi \rangle = 1\,. \tag{20.36}$$

Note that q and r are orthogonal to u and s. The parabivector formed by q and the flagpole $u+s$ is known as the *flag*:

$$q\,(\overline{u+s}) = \Lambda\left(\mathbf{n}\bar{P}\right)\bar{\Lambda}\,. \tag{20.37}$$

The combination of the flagpole and the flag is known as the *null flag*.

20.4 Spinor Pairs

We saw above that any ideal spinor could be written as a dilated rotational spinor (20.18) and (20.19). If we put

$$\eta = e^{w/2}RP\,,\ \xi = e^{w'/2}R'P\,, \tag{20.38}$$

then by (20.35),

$$\bar{\eta}\xi = e^{(w+w')/2}\bar{P}\bar{R}R'P = \mathbf{n}P\,. \tag{20.39}$$

One solution ξ of this relation is

$$w' = -w,\ R' = R\mathbf{n}\mathbf{e} = R\exp\left(\mathbf{n}\mathbf{e}\pi/2\right) \tag{20.40}$$

in other words

$$\xi = e^{-w/2}R\mathbf{n}P\,. \tag{20.41}$$

Any other solution ξ' of (20.35) satisfies

$$\bar{\eta}\left(\xi' - \xi\right) = 0 \tag{20.42}$$

and hence by the above, $\xi' = \xi + c\eta$ where $c \in \mathbb{C}$. An ideal spinor may be said to be *normalized* if the dilation factor $e^{w/2}$ is unity. The flagpoles

$$\begin{aligned} \eta\eta^\dagger &= e^w RPR^\dagger \\ \xi\xi^\dagger &= e^{-w} R\bar{P}R^\dagger \end{aligned} \tag{20.43}$$

are idempotent projectors if and only if the spinors are normalized.

If the spinors are normalized,

$$\eta\eta^\dagger + \xi\xi^\dagger = R\left(P + \bar{P}\right)R^\dagger = 1\,. \tag{20.44}$$

On the other hand, unnormalized spinors give

$$\begin{aligned} \eta\eta^\dagger + \xi\xi^\dagger &= R\left(e^w P + e^{-w}\bar{P}\right)R^\dagger \\ &= \cosh w + R\mathbf{e}R^\dagger \sinh w \\ &= \exp\left(wR\mathbf{e}R^\dagger\right). \end{aligned} \tag{20.45}$$

More generally,

$$\begin{aligned} \xi'\xi'^\dagger &= \xi\xi^\dagger + |c|^2\,\eta\eta^\dagger + 2\left\langle c\eta\xi^\dagger\right\rangle_{\Re} \\ &= e^w RPR^\dagger + |c|^2\,e^w RPR^\dagger + 2\left\langle cRP\mathbf{n}R^\dagger\right\rangle_{\Re} \end{aligned} \tag{20.46}$$

An alternative form of ideal spinors in $\mathcal{S}_+$ is

$$\eta = c\hat{\mathbf{v}}P \tag{20.47}$$

where $\hat{\mathbf{v}}$ is a real unit vector and $c \in \mathbb{C}$. A proof follows from the expansion in the spinor basis $\{P, \mathbf{n}P\}$:

$$\begin{aligned} \eta &= \left(\eta^0 + \eta^1\mathbf{n}\right)P \\ &= \eta^0\left(\mathbf{e} + re^{i\phi}\mathbf{n}\right)P \\ &= \eta^0\left(\mathbf{e} + r\mathbf{m}\right)P \\ &= c\hat{\mathbf{v}}P\,, \end{aligned} \tag{20.48}$$

where $r = \eta^1/\eta^0$, $\mathbf{m} = \mathbf{n}e^{\mathbf{ie}\phi}$ is the unit vector obtained from $\mathbf{n}$ by a rotation of 2ϕ about $\mathbf{e}$, $c = \eta^0\sqrt{1+r^2}$, and $\hat{\mathbf{v}} = (\mathbf{e} + r\mathbf{m})/\sqrt{1+r^2}$. The flagpole is

$$\eta\eta^\dagger = |c|^2\,\hat{\mathbf{v}}P\hat{\mathbf{v}}\,. \tag{20.49}$$

20.5 Time Evolution

The linear time evolution of the eigenspinor

$$\dot{\Lambda} = \frac{1}{2}\Omega\Lambda \tag{20.50}$$

extends immediately to the minimal spinors:

$$\dot{\eta} = \frac{1}{2}\Omega\eta\,,\ \dot{\xi} = \frac{1}{2}\Omega\xi\,. \tag{20.51}$$

These give the same equation of motion for the proper velocity as for the spin paravector (20.25):

$$\begin{aligned} \dot{u} &= \langle \Omega u \rangle_{\Re} \\ \dot{s} &= \langle \Omega s \rangle_{\Re} \end{aligned} \tag{20.52}$$

These equations are uncoupled and imply $g = 2$ for the "elementary" particle. Compound particles may have more than one eigenspinor and hence other g factors, and they may couple spin and velocity through such terms as $\nabla(\mu \cdot \mathbf{B})$.

20.6 Bispinor Basis of $\mathbb{C}^4$

A bispinor basis of $\mathbb{C}^4$ can be defined

$$\left\{\alpha_0 = P, \alpha_1 = \mathbf{n}P, \alpha_2 = \bar{P}, \alpha_3 = \mathbf{n}\bar{P}\right\} \tag{20.53}$$

with the projectors P, $\bar{P}$, and the perpendicular unit vector $\mathbf{n}$. Thus any bispinor can be expanded

$$\Lambda = \Lambda^a \alpha_a\,,\ a = 0,1,2,3. \tag{20.54}$$

To separate the various components, note the Lorentz-invariant expressions

$$\begin{aligned} 2\left\langle \bar{\alpha}_a \alpha_b \left(\bar{\alpha}_0 \alpha_1\right)^{\dagger}\right\rangle_S &= \delta_{a0}\delta_{b1} - \delta_{a1}\delta_{b0} \\ 2\left\langle \bar{\alpha}_a \alpha_b \left(\bar{\alpha}_2 \alpha_3\right)^{\dagger}\right\rangle_S &= \delta_{a2}\delta_{b3} - \delta_{a3}\delta_{b2}\,. \end{aligned} \tag{20.55}$$

Thus,

$$\begin{aligned} \Lambda^0 &= 2\left\langle \bar{\Lambda}\alpha_1 \left(\bar{\alpha}_0 \alpha_1\right)^{\dagger}\right\rangle_S \\ \Lambda^1 &= 2\left\langle \bar{\alpha}_0 \Lambda \left(\bar{\alpha}_0 \alpha_1\right)^{\dagger}\right\rangle_S \\ \Lambda^2 &= 2\left\langle \bar{\Lambda}\alpha_3 \left(\bar{\alpha}_2 \alpha_3\right)^{\dagger}\right\rangle_S \\ \Lambda^3 &= 2\left\langle \bar{\alpha}_2 \Lambda \left(\bar{\alpha}_2 \alpha_3\right)^{\dagger}\right\rangle_S\,. \end{aligned} \tag{20.56}$$

The scalar product of two bispinors

$$\begin{aligned} \left\langle \Psi\bar{\Phi}\right\rangle_S &= \Psi^a \Phi^b \left\langle \alpha_a \bar{\alpha}_b \right\rangle_S \\ &= \frac{1}{2}\Psi^a \Phi^b g_{ab}\,, \end{aligned} \tag{20.57}$$

where

$$g_{ab} = 2\left\langle \alpha_a \bar{\alpha}_b \right\rangle_S = \begin{pmatrix} 0 & 0 & 1 & 0 \\ 0 & 0 & 0 & -1 \\ 1 & 0 & 0 & 0 \\ 0 & -1 & 0 & 0 \end{pmatrix} \tag{20.58}$$

is the metric tensor whose signature is $(2,2)$, *i.e.*, its eigenvalues are $1, 1, -1, -1$:

$$\det \begin{pmatrix} g & 0 & 1 & 0 \\ 0 & g & 0 & -1 \\ 1 & 0 & g & 0 \\ 0 & -1 & 0 & g \end{pmatrix} = g^4 - 2g^2 + 1 = (g-1)^2 (g+1)^2 = 0\,. \tag{20.59}$$

Under restricted Lorentz transformations,

$$\alpha_a \to L\alpha_a, \text{where } L \in SL(2, \mathbb{C})\ , \tag{20.60}$$

the bispinor scalar product is seen to be invariant.

Another spinor mapping to bispinors is given by the Dirac-Pauli representation of Dirac theory. There,

$$\Phi = \begin{pmatrix} \langle \Lambda \rangle_+ \\ \langle \Lambda \rangle_- \end{pmatrix} P\,, \tag{20.61}$$

where

$$\langle \Lambda \rangle_\pm = \left(\Lambda \pm \bar{\Lambda}^\dagger\right)/2 \tag{20.62}$$

are the even and odd parts of Λ. Acting in the space $\mathbb{C}^{2,2}$ of Φ, L is seen to be unitary. This establishes the six-parameter group $SL(2, \mathbb{C})$ as a subgroup of the 15-parameter group $SU(2,2) \approx C(1,3)$.

20.7 Twistors

Flat-space twistors are bispinors in which the 2-component spinor parts may be thought of as representing the momentum and angular momentum of a zero-mass particle:

$$Z = \omega_- + \pi_+\,, \tag{20.63}$$

where

$$\pi_+ = \sqrt{2E}RP \tag{20.64}$$

and R is a rotation operator, gives the momentum

$$p = \pi_+ \pi_+^\dagger = E\left(1 + \hat{\mathbf{p}}\right) \tag{20.65}$$

with a spatial part which is directed along

$$\hat{\mathbf{p}} = R\mathbf{e}R^\dagger\,, \tag{20.66}$$

and where

$$\mathbf{M} = 2i\left\langle \omega_- \bar{\pi}_+ \right\rangle_V \tag{20.67}$$

is the angular momentum. The scalar $Z\bar{Z} = 2\langle \omega_- \bar{\pi}_+ \rangle_S$ gives the helicity.

The comparison with the eigenspinor approach is striking. Consider

$$\Lambda = \eta + \xi \mathbf{n} = \eta P + \xi P \mathbf{n} \tag{20.68}$$

as a restricted Lorentz transformation connecting the rest frame of the particle to the observer's frame with a normalization

$$\Lambda\bar{\Lambda} = \pm m = \xi \mathbf{n} \bar{\eta} - \eta \mathbf{n} \bar{\xi}\,. \tag{20.69}$$

The momentum of the particle is then

$$p = \Lambda\Lambda^\dagger = \eta\eta^\dagger + \xi\xi^\dagger\,, \tag{20.70}$$

whereas its laboratory spin angular momentum is

$$\begin{aligned} \mathbf{M} &= -\frac{i}{2m}\Lambda \mathbf{e} \bar{\Lambda} = -\frac{i}{2m}\Lambda\left(P - \bar{P}\right)\bar{\Lambda} \\ &= \frac{i}{2m}\left(\eta \mathbf{n} \bar{\xi} + \xi \mathbf{n} \bar{\eta}\right) \\ &= \frac{i}{m}\langle \xi \mathbf{n} \bar{\eta} \rangle_V\,. \end{aligned} \tag{20.71}$$

The Pauli-Lubanski spin paravector is

$$\begin{aligned} i\mathbf{M}p &= \frac{1}{2m}\Lambda \mathbf{e} \bar{\Lambda}\Lambda\Lambda^\dagger = \pm\frac{1}{2}\Lambda\left(P - \bar{P}\right)\Lambda^\dagger \\ &= \pm\frac{1}{2}\left(\eta\eta^\dagger - \xi\xi^\dagger\right) \end{aligned} \tag{20.72}$$

and its scalar part should be the helicity times the magnitude of the momentum. If $\xi_- = 0$, the particle has zero mass, its momentum is the null paravector $\eta\eta^\dagger$, and its helicity is $\pm\frac{1}{2}$ for particle and antiparticle. On the other hand, if $\eta_+ = 0$, the particle has zero mass, momentum paravector $\xi_-\xi_-^\dagger$ and helicity $\mp\frac{1}{2}$ for particle and antiparticle.

20.8 Relation to $SO_+(1,3)$

As seen in the earlier lectures, the quadratic form of paravectors:

$$\langle p\bar{p} \rangle_S = p\bar{p} = \left(p^0 + \mathbf{p}\right)\left(p^0 - \mathbf{p}\right) = \left(p^0\right)^2 - \mathbf{p}^2\,, \tag{20.73}$$

gives the Minkowski (flat) spacetime metric. Lorentz transformations leave the quadratic form invariant and can be expressed in terms of components by

$$p = p^\mu \mathbf{e}_\mu \mapsto p^\nu \mathbf{u}_\nu\,,\ \mathbf{u}_\nu = L\mathbf{e}_\nu L^\dagger\,, \tag{20.74}$$

where $\{\mathbf{e}_\mu\} = \{1, \mathbf{e}_k\}$ is the observer's orthonormal tetrad (at rest with observer), and $\{\mathbf{u}_\mu\}$ is the orthonormal tetrad of the moving, rotated frame (as viewed by the observer) reached by the Lorentz (spin) transformation $L = BR$. In particular $\mathbf{u}_0 = L\mathbf{e}_0 L^\dagger = LL^\dagger = B^2$ is the proper velocity of the moving tetrad. The spin transformations $L \in \mathrm{SL}(2,\mathbb{C})$ are related to elements of restricted Lorentz transformations $\mathcal{L} \in \mathrm{SO}_+(1,3)$ with the help of the reciprocal tetrad $\{\mathbf{e}^\nu\}$ where

$$\langle \mathbf{e}_\mu \bar{\mathbf{e}}^\nu \rangle_S = \delta_\mu^\nu \tag{20.75}$$

by

$$p \mapsto p^\nu \mathcal{L}_\nu^{\ \mu} \mathbf{e}_\mu \tag{20.76}$$

with

$$\begin{aligned} \mathcal{L}_\nu^{\ \mu} &= \langle \mathbf{u}_\nu \bar{\mathbf{e}}^\mu \rangle_S = \left\langle \left(L\mathbf{e}_\nu L^\dagger \right) \bar{\mathbf{e}}^\mu \right\rangle_S \\ &= \left\langle \mathbf{u}_\nu \overline{\bar{L}\mathbf{u}^\mu \bar{L}^\dagger} \right\rangle_S = \left\langle \mathbf{e}_\nu \overline{\bar{L}\mathbf{e}^\mu \bar{L}^\dagger} \right\rangle_S . \end{aligned} \tag{20.77}$$

20.9 Spinors in Curved Spacetime

In curved spacetime, an orthonormal tetrad basis can be found to span the tangent space at each point on the manifold. We assume the existence of a continuous global tetrad field comprising such orthonormal bases. Lorentz transformations relate different tetrads generally only at the same spacetime point. Let $\{\mathbf{e}_\nu(x)\}$ be the orthonormal tetrad field at a point x fixed by the coordinates of a given patch, and let the set $\{\mathbf{u}_\mu(s)\}$ be a path-dependent tetrad frame that is *parallel transported* along its path $x(s)$, where s is a scalar parameter. Because both tetrads are orthonormal,

$$\eta_{\mu\nu} = \langle \mathbf{u}_\mu \bar{\mathbf{u}}_\nu \rangle_S = \langle \mathbf{e}_\mu \bar{\mathbf{e}}_\nu \rangle_S . \tag{20.78}$$

The covariant derivative of each $\mathbf{u}_\mu$ along the tangent vector v to the path then vanishes by definition:

$$\nabla_v \mathbf{u}_\mu = 0 , \ v = dx/ds . \tag{20.79}$$

At any point $x(s)$ along the path, the tetrads will be related by a spin transformation $\Lambda(s)$ that clearly depends on the path:

$$\mathbf{u}_\mu [x(s)] = \Lambda(s) \mathbf{e}_\mu(s) \Lambda^\dagger(s) . \tag{20.80}$$

Unimodularity, $\Lambda\bar{\Lambda} = 1$, implies a covariant derivative of the form

$$\nabla_v \bar{\Lambda} = \frac{d}{ds} \bar{\Lambda}(s) = \frac{1}{2} \Gamma_v \bar{\Lambda}(s) \tag{20.81}$$

where Γ_v is a parabivector (the "spin connection"). The term $\bar{\Gamma}_\nu = -\Gamma_v$ is seen to give the spacetime rotation rate of the parallel-transported tetrad $\{\mathbf{u}_\mu\}$ with respect to the tetrad field $\{\mathbf{e}_\mu\}$. Note that if Γ_v is real, motion is rotation–free (*Fermi-Walker transport*) in the rest frame (with $\mathbf{e}_0 = 1$), and if Γ_v is imaginary, motion is *geodesic* (free fall).

The components of Γ_v on unit spacetime planes at x are the usual connection coefficients: $\Gamma_v = \frac{1}{2}\Gamma_v^{\alpha\beta}\mathbf{e}_\alpha\bar{\mathbf{e}}_\beta$ as may be seen by evaluating

$$\nabla_v \mathbf{e}_\mu = \frac{1}{2}\left(\Gamma_v \bar{\Lambda}\mathbf{u}_\mu \bar{\Lambda}^\dagger + \bar{\Lambda}\mathbf{u}_\mu \bar{\Lambda}^\dagger \Gamma_v^\dagger\right) = \langle \Gamma_v \mathbf{e}_\mu \rangle_{\Re} \,. \tag{20.82}$$

By taking a path that makes an infinitesimal loop, we relate the total spacetime rotation of the parallel-transported tetrad to the spacetime curvature, here given by a set of six Riemann parabivectors:

$$\mathcal{R}_{\mu\nu} = \nabla_\mu \Gamma_\nu - \nabla_\nu \Gamma_\mu - \frac{1}{2}\left[\Gamma_\mu \Gamma_\nu - \Gamma_\nu \Gamma_\mu\right] - \Gamma_{[\mu,\nu]} \,, \tag{20.83}$$

which through the four Ricci paravectors:

$$\mathcal{R}_\nu = \langle \mathcal{R}_{\mu\nu} \mathbf{e}^\nu \rangle_{\Re} \tag{20.84}$$

and curvature scalar

$$\mathcal{R} = \mathcal{R}_\mu \bar{\mathbf{e}}^\mu \tag{20.85}$$

give the Einstein equation

$$\mathcal{R}_\nu - \frac{1}{2}\mathcal{R} = 8\pi G T \,. \tag{20.86}$$

We thus see how the machinery of general relativity is efficiently presented in the Pauli algebra $\mathcal{C}\ell_3$.

20.10 Conclusions

This concludes the set of lectures on paravectors and eigenspinors in physics. We have sought to show that eigenspinors are common to classical physics, general relativity, and quantum (Dirac) theory. Quantum interference phenomena arise from the superposition of eigenspinors that have followed different paths as solutions of linear equations of motion. The eigenspinors suggest a common thread for possible formulations of quantum gravity in $\mathcal{C}\ell_3$.

The author thanks the Natural Sciences and Engineering Research Council of Canada for support of the research that formed the basis for these lectures.

Bibliography

On foundations of the Pauli algebra:

- W. E. Baylis and G. Jones, "The Pauli-Algebra Approach to Special Relativity", *J. Phys.* A**22**, 1–16 (1989).

- W. E. Baylis J. Huschilt, and Jiansu Wei, "Why *i*?", *Am. J. Phys.* **60**, 788–797 (1992).

On spinors:

- G. Jones and W. E. Baylis, "Crumeyrolle-Chevalley-Riesz Spinors and Covariance", in *Clifford Algebras and Spinors*, edited by P. Lounesto and R. Abłamowicz, Kluwer Academic (1994).

On the Dirac Equation:

- W. E. Baylis, "Classical Eigenspinors and the Dirac Equation", *Phys. Rev.* A**45**, 4293-4302 (1992).

Chapter 21

Spinors: Lorentz Group

James P. Crawford
Department of Physics, Penn State - Fayette,
Uniontown, PA 15401 USA

21.1 Introduction

We take here as the physically relevant definition that spinor fields are elements of the carrier spaces of the fundamental representations of the Lorentz Group. Hence we begin with an informal (and therefore non-rigorous) discussion of the structure and the various representations of this physically important group. In particular, this lecture includes a discussion of the fundamental and general representations of the Lorentz group as well as the special cases of self-dual and real representations. This will clarify what precisely is meant by the term "spinor," and permit analogies between the notions of self-dual tensors and Weyl spinors as well as real antisymmetric tensors and Majorana spinors.

21.2 Lorentz Group

The Lorentz group L is defined to be the group of all transformations which preserve the Minkowski metric:

$$V'^{\alpha} = \Lambda^{\alpha}{}_{\beta} V^{\beta} \quad , \quad U'_{\alpha} V'^{\alpha} = \eta_{\alpha\beta} \Lambda^{\alpha}{}_{\gamma} \Lambda^{\beta}{}_{\delta} U^{\gamma} V^{\delta} = U_{\alpha} V^{\alpha} \tag{21.1}$$

$$\Rightarrow \quad \eta_{\alpha\beta} \Lambda^{\alpha}{}_{\gamma} \Lambda^{\beta}{}_{\delta} = \eta_{\gamma\delta} \tag{21.2}$$

where the Minkowski metric is given by $\eta_{\alpha\beta} = diag(1, -1, -1, -1)$. The Lorentz group is composed of four disconnected components. From Equation (21.2) it follows that $\det(\Lambda) = \pm 1$, so we have $L = L_{+} \oplus L_{-}$, where L_{+} is the subgroup composed of the elements with $\det(\Lambda) = +1$, and L_{-} is the set of elements with $\det(\Lambda) = -1$. Note that L_{-} contains the elements P (parity transformation) and T (time reversal transformation), whereas L_{+} contains the identity and PT. The elements of L_{+}

are called *proper Lorentz transformations,* whereas the elements of L_- are called *improper Lorentz transformations.* We may also define the *orthochronous subgroup* $L^\uparrow$ to be the group of transformations with $\Lambda^0{}_0 \geq 1$, and the set of *antichronous transformations* $L^\downarrow$ to be the set of all transformations with $\Lambda^0{}_0 \leq -1$. So we have $L = L^\uparrow \oplus L^\downarrow$, where $L^\uparrow$ contains the identity and P, whereas $L^\downarrow$ contains T and PT. The complete separation of the Lorentz group may now be written as:

$$L = L_+^\uparrow \oplus L_-^\uparrow \oplus L_+^\downarrow \oplus L_-^\downarrow = L_+^\uparrow \otimes \{\mathbf{1}, T\} \otimes \{\mathbf{1}, P\}, \tag{21.3}$$

where the second equality is a heuristic reminder only, since neither T nor P commute with all elements of $L_+^\uparrow$. Note that the set of all *proper orthochronous transformations* form the subgroup $L_+^\uparrow$, and this subgroup is continuously connected to the identity. This subgroup is clearly isomorphic to the group $SO(1,3)$, and although this group is therefore technically the *proper orthochronous Lorentz group* we will lapse into the incorrect but convenient terminology that $SO(1,3)$ is the Lorentz group.

Since this group is continuously connected to the identity we may represent an arbitrary element as follows:

$$\Lambda^\gamma{}_\delta = [\exp(\frac{i}{2}\lambda^{\alpha\beta}\sigma_{\alpha\beta})]^\gamma{}_\delta \tag{21.4}$$

where $\lambda^{\alpha\beta}$ are the real group parameters, and the matrices $[\sigma_{\alpha\beta}]^\gamma{}_\delta$ are found to be given by:

$$[\sigma_{\alpha\beta}]^\gamma{}_\delta = i(\delta_\alpha^\gamma \eta_{\beta\delta} - \delta_\beta^\gamma \eta_{\alpha\delta}). \tag{21.5}$$

21.2.1 Lorentz Lie Algebra

The Lie algebra $so(1,3)$ is easily obtained from Equation (21.5) and is found to be:

$$[\sigma_{\alpha\beta}, \sigma_{\gamma\delta}] = -i(\eta_{\alpha\gamma}\sigma_{\beta\delta} - \eta_{\alpha\delta}\sigma_{\beta\gamma} - \eta_{\beta\gamma}\sigma_{\alpha\delta} + \eta_{\beta\delta}\sigma_{\alpha\gamma}) \tag{21.6}$$

The algebra may be decomposed into physically relevant segments with the following definitions:

$$\sigma_{0i} \equiv B_i \quad , \quad \sigma_{ij} \equiv \epsilon_{ijk} R_k \Leftrightarrow R_i = \frac{1}{2}\epsilon_{ijk}\sigma_{jk} \tag{21.7}$$

where B_i and R_i are respectively the generators of boosts and rotations. The subscript 0 denotes the temporal component and the subscripts $i = 1,2,3$ denote the spatial components. In terms of the boost and rotation generators the Lie algebra takes the form:

$$[R_i, R_j] = i\epsilon_{ijk} R_k \ , \ [B_i, B_j] = -i\epsilon_{ijk} R_k \ , \ [R_i, B_j] = i\epsilon_{ijk} B_k \tag{21.8}$$

The mathematical structure of this algebra is easily identified if we make the following "chiral" definitions:

$$J_i \equiv \frac{1}{2}(R_i + iB_i) \quad , \quad K_i \equiv \frac{1}{2}(R_i - iB_i) \tag{21.9}$$

in which case Equations (21.8) may be written as:

$$[J_i, J_j] = i\epsilon_{ijk}J_k \ , \ [K_i, K_j] = i\epsilon_{ijk}K_k \ , \ [J_i, K_j] = 0. \tag{21.10}$$

Therefore we find the following isomorphism (where the algebras are assumed to be defined over the complex field):

$$so(1,3) \sim so(3) \oplus so(3) \sim su(2) \oplus su(2) \tag{21.11}$$

where the second isomorphism $so(3) \sim su(2)$ is well known. This Lie algebra isomorphism will imply a similar Lie group isomorphism, but the isomorphism for the group contains a subtlety due to the presence of i in equation (21.9).

21.2.2 Lorentz Group Representations

We are interested in finite dimensional representations of the Lorentz group, in which case the generators of the group $\sigma_{\alpha\beta}$ form a particular set of matrices. Any such set of matrices will be associated with a specific representation of the group. (For example, the particular set given in Equation (21.5) corresponds to the vector representation.) If this set is not reducible to block diagonal form by a common similarity transformation it is called irreducible.

Associated with any representation of a group is a *carrier space*. The carrier space may be regarded as the space of vectors (possibly complex) upon which the group matrices may act.

Consider an element of the group $SO(1,3)$ given by:

$$S(\lambda) \equiv \exp \frac{i}{2}\lambda^{\alpha\beta}\sigma_{\alpha\beta} \equiv \exp i\lambda. \tag{21.12}$$

(Note that this equation is the same as Equation (21.4), but Equation (21.4) was written for the specific case of vector representations.) Now with the physical definitions (φ_i are the boost parameters and ϑ_i are the rotation parameters):

$$\lambda^{0i} = -\lambda_{0i} \equiv -\varphi_i \quad , \quad \lambda^{ij} = \lambda_{ij} \equiv \epsilon_{ijk}\vartheta_k \tag{21.13}$$

we have:

$$\lambda = (\vartheta_i + i\varphi_i)J_i + (\vartheta_i - i\varphi_i)K_i \equiv \phi_i J_i + \phi_i^* K_i \tag{21.14}$$

where we have defined ϕ_i to be the *complex* group parameters. The group element now becomes:

$$S(\lambda) = \exp i(\phi_i J_i + \phi_i^* K_i) = (\exp i\phi_i J_i)(\exp i\phi_i^* K_i) \equiv S_J(\lambda)S_K(\lambda) \tag{21.15}$$

where we have used the commutativity of J_i and K_i (see Equation (21.10)) in obtaining the last equality. Therefore we obtain the group isomorphism and the covering group isomorphism:

$$SO(1,3) \approx SO(3,C) \ , \ SU(2,C) \approx SL(2,C). \tag{21.16}$$

This is the canonical result. Note that Equation (21.15) may naively be interpreted as yielding the group isomorphism $SO(1,3) \approx SO(3,C) \otimes SO(3,C)$. However, since the same six real group parameters appear in each factor (in the form of ϕ_i in the first factor and ϕ_i^* in the second), these factors constitute transformations of the *same group*, and therefore the isomorphism is that of a single $SO(3,C)$. The second isomorphism essentially follows from the vanishing trace of the Lie algebra generators.

Now the finite dimensional Hermitian representations of the Lie algebra $su(2)$ are well known from the study of angular momentum, and the representations of J_i and K_i may each be chosen independently, so the irreducible representations of the group $SU(2,C)$ are delineated by:

$$(j,k) \qquad j = \frac{n}{2}\,,\ k = \frac{m}{2} \quad n,m = 0,1,2,3... \tag{21.17}$$

The dimension of the carrier space of such a representation is:

$$D(j,k) = (2j+1)(2k+1), \tag{21.18}$$

so that the group elements may be chosen to be $D \times D$ matrices. We now make an important observation: In general, since the group parameters are complex, *these representations are complex* even for the case of integer j and k.

The angular momentum of representation (j,k) may be found by considering the rotation generators. From Equation (21.9) we have:

$$R_i = J_i + K_i \;\Rightarrow\; l = j+k,\ j+k-1, \cdots,\ |j-k| \tag{21.19}$$

and we see that in general an irreducible representation of the Lorentz group contains several angular momentum states. This is as expected since the rotation group $SO(3)$ is a subgroup of the Lorentz group.

Spinor Representations

There are two inequivalent fundamental (spinor) representations of the Lorentz group, and correspondingly there are two distinct spinor carrier spaces, each spanned by two complex numbers. These spinors are commonly denoted as follows:

$$(\tfrac{1}{2},0): \quad \zeta^a = \begin{pmatrix} \zeta^1 \\ \zeta^2 \end{pmatrix} \qquad (0,\tfrac{1}{2}): \quad \zeta^{\dot{a}} = \begin{pmatrix} \zeta^{\dot{1}} \\ \zeta^{\dot{2}} \end{pmatrix} \tag{21.20}$$

The conventional "dotted" and "undotted" spinor indices are used here to distinguish these two spinors. To be brutally explicit (in the hope of avoiding confusion later), a specific transformation law for the undotted spinor may be obtained by choosing the matrix representation for the Lie algebra elements to be:

$$J_i = \frac{1}{2}[\sigma_i]^a{}_b \quad \Rightarrow \quad S(\lambda) = S_J(\lambda)\cdot 1 \equiv S^a{}_b \quad \Rightarrow \quad \zeta'^a = S^a{}_b\zeta^b \tag{21.21}$$

where σ_i are the conventional Pauli matrices:

$$\sigma_1 = \begin{pmatrix} 0 & 1 \\ 1 & 0 \end{pmatrix} \quad , \quad \sigma_2 = \begin{pmatrix} 0 & -i \\ i & 0 \end{pmatrix} \quad , \quad \sigma_3 = \begin{pmatrix} 1 & 0 \\ 0 & -1 \end{pmatrix} . \tag{21.22}$$

Conversely, a convenient specific transformation law for the dotted spinor may be obtained by choosing:

$$K_i = \frac{1}{2}[-\sigma_i^*]^{\dot{a}}{}_{\dot{b}} \quad \Rightarrow \quad S(\lambda) = 1 \cdot S_K(\lambda) \equiv S^{\dot{a}}{}_{\dot{b}} \quad \Rightarrow \quad \zeta'^{\dot{a}} = S^{\dot{a}}{}_{\dot{b}}\zeta^{\dot{b}} \tag{21.23}$$

where again σ_i are the conventional Pauli matrices, but now written with dotted indices. We have also made use of the following observation concerning representations of Lie algebras. For any Lie algebra with real structure factors, that is if $[L_i, L_j] = if_{ij}{}^k L_k$, $f_{ij}^{*\,k} = f_{ij}{}^k$, and if $[L_i]^A{}_B$ forms a matrix representation of this algebra, then the matrices $[M_i]^A{}_B = [-L_i^*]^A{}_B$ also form a matrix representation of this algebra. The proof follows easily by taking the complex conjugate. Finally note that although it is *not necessary* to choose this particular representation for the K_i, it is *very convenient*, for with this definition we have the following result:

$$\begin{aligned} S_J = \exp \tfrac{i}{2}\phi_i\sigma_i \ , \ S_K = \exp -\tfrac{i}{2}\phi_i^*\sigma_i^* \\ \Rightarrow \ S_J^* = S_K \ , \ S_K^* = S_J \ . \end{aligned} \tag{21.24}$$

This particular choice is also consistent with the convention of interchanging dotted and undotted indices under complex conjugation. We will generalize this result in the section concerning real representations.

We may define an invariant inner product on the spinor spaces as follows:

$$h(\zeta, \xi) \equiv h_{ab}\zeta^a\xi^b = h_{ab}\zeta'^a\xi'^b \quad \Rightarrow \quad h_{ab} = h_{cd}S^c{}_a S^d_b \tag{21.25}$$

from which we obtain:

$$[h]_{ab} = \begin{bmatrix} 0 & 1 \\ -1 & 0 \end{bmatrix} , \ [h]^{ab} = \begin{bmatrix} 0 & -1 \\ 1 & 0 \end{bmatrix} \Rightarrow h^{ac}h_{cb} = \delta^a_b \tag{21.26}$$

Note that since this metric is antisymmetric we have:

$$h_{ab}\zeta^a\xi^b \equiv \zeta^a\xi_a = -\xi^b\zeta_b \quad \Rightarrow \quad \zeta^a\zeta_a = 0 \tag{21.27}$$

so that all of these spinors are null. The metric for the dotted spinors is the same: $h_{\dot{a}\dot{c}} = h_{ac}$.

Real Representations

We may find real representations as follows. First we denote a particular Hermitian representation of $su(2)$ as $\rho_i(l)$, where $l = \frac{n}{2}$ and $n = 0, 1, 2, \dots$. Furthermore we may *choose* the matrices J_i and K_i to be given by:

$$J_i = \rho_i(j) \quad , \quad K_i = -\rho_i^*(k). \tag{21.28}$$

Now from Equation (21.15) we have:

$$\begin{aligned} S_{J(j)}(\lambda) = \exp i\phi_i \rho_i(j) \ , \ S_{K(k)}(\lambda) = \exp -i\phi_i^* \rho_i^*(k) \\ \Rightarrow \quad S_{J(j)}^*(\lambda) = S_{K(j)}(\lambda) \ , \ S_{K(k)}^*(\lambda) = S_{J(k)}(\lambda). \end{aligned} \tag{21.29}$$

This equation is the generalization of Equation (21.24). Then for the elements of the carrier spaces we have

$$(j,k)' = S_{J(j)} S_{K(k)} (j,k)$$

$$\Rightarrow (j,k)^{*'} = S_{J(j)}^* S_{K(k)}^* (j,k)^* = S_{K(j)} S_{J(k)} (j,k)^*$$

$$\Rightarrow \quad (j,k)^* = (k,j). \tag{21.30}$$

Therefore, a real representation must be one of the following two forms:

$$type\ (1) = (j,j) \ , \ type\ (2) = (j,k) \oplus (k,j) \tag{21.31}$$

Type (1) is the only possible form for a real irreducible representation of the Lorentz group. We will see that many real tensor representations are type (2), and therefore are reducible in the sense of these general complex representations. However, if we insist on only real representations, as is usually the case for the tensor representations, then type (2) representations should also be considered irreducible.

Tensor Representations

We are now in position to consider the tensor representations of the Lorentz group. The scalar representation is trivially given by:

$$scalar, \ \Phi : (0,0). \tag{21.32}$$

Since this representation is type (1), a scalar field may be either real or complex, as expected.

The vector representation may be intuitively identified by noting that a Lorentz vector contains a total of four components, and the 1+3 dimensional (temporal + spatial) decomposition contains angular momentum zero (spatial scalar) and angular momentum one (spatial vector). Therefore the vector representation is given by:

$$vector, \ A^\alpha : (\tfrac{1}{2}, \tfrac{1}{2}) \quad , \quad D = 4 \quad , \quad l = 0, 1. \tag{21.33}$$

Since this representation is type (1), a vector field may be either real or complex, as expected. If we wish to explicitly display the spinor indices we define an invariant object with one vector and two spinor indices:

$$\Sigma_\alpha'^{\ a\dot{a}} = \Lambda_\alpha^{\ \beta} S^a_{\ c} S^{\dot{a}}_{\ \dot{c}} \Sigma_\beta^{\ c\dot{c}} = \Sigma_\alpha^{\ a\dot{a}}. \tag{21.34}$$

This has the solution

$$\Sigma_0^{\ a\dot{a}} = \delta^{a\dot{a}} \ , \ \Sigma_i^{\ a\dot{a}} = [\sigma_i]^{a\dot{a}} \tag{21.35}$$

where again σ_i are the conventional Pauli matrices, but this time with mixed spinor indices. A vector may now be written explicitly in terms of the spinor indices as follows:

$$A^{a\dot{a}} = A^{\alpha}\Sigma_{\alpha}^{\ \ a\dot{a}} \tag{21.36}$$

and the transformation of $A^{a\dot{a}}$ implies the correct transformation of A^{α}, as expected:

$$A'^{a\dot{a}} = S^{a}_{\ \ c}S^{\dot{a}}_{\ \ \dot{c}}A^{c\dot{c}} \equiv A'^{\alpha}\Sigma_{\alpha}^{\ \ a\dot{a}} \quad \Rightarrow \quad A'^{\alpha} = \Lambda^{\alpha}_{\ \ \beta}A^{\beta}. \tag{21.37}$$

We now recognize the transformation of $A^{a\dot{a}}$ as an $SL(2,C)$ transformation:

$$S^{\dot{b}}_{\ \ \dot{d}} = S^{*b}_{\ \ \ d} = S^{\dagger\, b}_{\ d} \quad \Rightarrow \quad A' = SAS^{\dagger}. \tag{21.38}$$

Clearly, all higher order tensors may also be written in terms of the spinor indices by making use of $\Sigma_{\alpha}^{\ \ a\dot{a}}$, and certain calculations are indeed simpler in spinor form. However, for the remainder of this section we will not do this.

The Lorentz group representation structure of general tensors may be found by considering products of the vector representations. For the case of a two tensor consider

$$(\tfrac{1}{2},\tfrac{1}{2}) \otimes (\tfrac{1}{2},\tfrac{1}{2}) = (0,0) \oplus (0,1) \oplus (1,0) \oplus (1,1) \tag{21.39}$$

where we have used the familiar rules for "addition of angular momentum" for both j and k separately. This may be decomposed as follows

$$T_{\alpha\beta} = \overbrace{T^{\gamma}_{\ \ \gamma}\eta_{\alpha\beta}}^{(0,0)} \oplus \overbrace{\frac{1}{2}(T_{\alpha\beta} + T_{\beta\alpha} - 2T^{\gamma}_{\ \ \gamma}\eta_{\alpha\beta})}^{(1,1)} \oplus \overbrace{\frac{1}{2}(T_{\alpha\beta} - T_{\beta\alpha})}^{((1,0)\oplus(0,1))} . \tag{21.40}$$

We see that a general two tensor comprising sixteen independent components may be decomposed into a trace: $(0,0)$, one component; a traceless symmetric tensor: $(1,1)$, nine components; and an antisymmetric tensor: $(1,0)\oplus(0,1)$, six components. Note that all parts of this tensor may be real as expected. For example, the electromagnetic field strength tensor $F_{\alpha\beta}$ is composed of two real spatial-vector fields, the electric field and the magnetic field. A physical example of a symmetric tensor field is the energy-momentum tensor $\Theta_{\alpha\beta}$, this forming a real reducible representation of the Lorentz group, $(0,0)\oplus(1,1)$. Clearly this procedure may be used to deduce the representation structure of arbitrary tensor fields.

Self-Dual Representations

Self-dual tensor fields occur in a variety of places in theoretical physics, including instantons in Yang-Mills theory, and most recently in the Ashtekar formulation of General Relativity. In the case of instantons, one considers four dimensional Euclidean space where $SO(4) \approx SO(3) \otimes SO(3)$ and a self-dual tensor may be real. However, in Minkowski space where $SO(1,3) \approx SO(3,C)$ a self-dual tensor is necessarily complex.

First we define the dual to an antisymmetric tensor as follows:

$$\tilde{F}_{\alpha\beta} \equiv \frac{i}{2}\epsilon_{\alpha\beta\gamma\delta}F^{\gamma\delta} \tag{21.41}$$

where $\epsilon_{\alpha\beta\gamma\delta}$ is the totally antisymmetric tensor with $\epsilon^{0123} = -\epsilon_{0123} = +1$. Now recall the identity:

$$\epsilon_{\alpha\beta\zeta\eta}\epsilon^{\zeta\eta\gamma\delta} = -2(\delta^\gamma_\alpha\delta^\delta_\beta - \delta^\gamma_\beta\delta^\delta_\alpha) \tag{21.42}$$

so we find:

$$\frac{i}{2}\epsilon_{\alpha\beta\gamma\delta}\tilde{F}^{\gamma\delta} = \frac{i}{2}\epsilon_{\alpha\beta\gamma\delta}\frac{i}{2}\epsilon^{\gamma\delta\zeta\eta}F_{\zeta\eta} = F_{\alpha\beta}. \tag{21.43}$$

So that the dual operation is its own inverse. This is the reason for the factor of i in the definition (Equation (21.41)). In four dimensional Euclidean space there is no minus sign in Equation (21.42) and hence the factor of i is not required.

To describe the self-dual fields it proves convenient to first define the self-dual and antiself-dual projection operators:

$$P_{R\alpha\beta}{}^{\gamma\delta} \equiv \frac{1}{2}\left(\frac{1}{2}(\delta^\gamma_\alpha\delta^\delta_\beta - \delta^\gamma_\beta\delta^\delta_\alpha) + \frac{i}{2}\epsilon_{\alpha\beta}{}^{\gamma\delta}\right) \equiv \frac{1}{2}(\mathbf{1} + i\tilde{\gamma})_{\alpha\beta}{}^{\gamma\delta} \tag{21.44}$$

$$P_{L\alpha\beta}{}^{\gamma\delta} \equiv \frac{1}{2}\left(\frac{1}{2}(\delta^\gamma_\alpha\delta^\delta_\beta - \delta^\gamma_\beta\delta^\delta_\alpha) - \frac{i}{2}\epsilon_{\alpha\beta}{}^{\gamma\delta}\right) \equiv \frac{1}{2}(\mathbf{1} - i\tilde{\gamma})_{\alpha\beta}{}^{\gamma\delta} \tag{21.45}$$

where we have defined $\mathbf{1}_{\alpha\beta}{}^{\gamma\delta}$ to be the identity operator for the space of antisymmetric tensors, and we have chosen a suggestive notational definition for the dual operator:

$$\begin{aligned} \mathbf{1}_{\alpha\beta}{}^{\gamma\delta} &\equiv \tfrac{1}{2}(\delta^\gamma_\alpha\delta^\delta_\beta - \delta^\gamma_\beta\delta^\delta_\alpha)\ ,\ i\tilde{\gamma}_{\alpha\beta}{}^{\gamma\delta} \equiv \tfrac{i}{2}\epsilon_{\alpha\beta}{}^{\gamma\delta} \\ \Rightarrow\quad i\tilde{\gamma}_{\alpha\beta}{}^{\zeta\eta}i\tilde{\gamma}_{\zeta\eta}{}^{\gamma\delta} &= \mathbf{1}_{\alpha\beta}{}^{\gamma\delta}. \end{aligned} \tag{21.46}$$

This last statement is a reiteration of the dual operation being its own inverse. Henceforth we shall only explicitly display the indices when it is necessary for clarity. The self-dual and antiself-dual operators satisfy the following relations:

$$\begin{aligned} P_R P_R = P_R\ ,\ P_L P_L = P_L \\ P_R P_L = 0 = P_L P_R\ ,\ P_R + P_L = \mathbf{1} \end{aligned} \tag{21.47}$$

and therefore are projection operators. We also note the following identities:

$$i\tilde{\gamma}P_R = P_R\ ,\ i\tilde{\gamma}P_L = -P_L. \tag{21.48}$$

Now any antisymmetric tensor field may be split into its self-dual and antiself-dual parts:

$$F = P_R F + P_L F = F_R + F_L \tag{21.49}$$

$$i\tilde{\gamma}F_R = F_R \quad ,\quad i\tilde{\gamma}F_L = -F_L \tag{21.50}$$

To gain insight into what this split means consider the self-dual and antiself-dual parts of the Lorentz generators:

$$\lambda = \frac{1}{2}\lambda^{\alpha\beta}\sigma_{\alpha\beta} = \frac{1}{2}\lambda^{\alpha\beta}\left(P_{R\alpha\beta}{}^{\gamma\delta}\sigma_{\gamma\delta} + P_{L\alpha\beta}{}^{\gamma\delta}\sigma_{\gamma\delta}\right) \tag{21.51}$$

$$\frac{1}{2}\lambda^{\alpha\beta}P_{R\alpha\beta}{}^{\gamma\delta}\sigma_{\gamma\delta} = \phi_i J_i \quad ,\quad \frac{1}{2}\lambda^{\alpha\beta}P_{L\alpha\beta}{}^{\gamma\delta}\sigma_{\gamma\delta} = \phi_i^* K_i\,. \tag{21.52}$$

These last two equations should be compared to Equation (21.14). The self-dual and antiself-dual parts of the Lorentz generators $\sigma_{\alpha\beta}$ are precisely the "chiral" generators J_i and K_i. The complete connection to the chiral projection operators will appear in the discussion of the Clifford algebra.[1]

As an illustration consider the electromagnetic field strength tensor:

$$F_{\alpha\beta} \,:\, F_{0i} = E_i \quad , \quad F_{ij} = -\epsilon_{ijk} B_k \Leftrightarrow B_i = -\frac{1}{2}\epsilon_{ijk} F_{jk} \tag{21.53}$$

where we have made the usual definitions for the electric and magnetic fields. (Note that for this section B_i represents the magnetic field, not the boost generator as in Equation (21.7).) This tensor is real, and its Lorentz representation is $(1,0)\oplus(0,1)$. Now computing the self-dual and antiself-dual parts we find:

$$F_{R\,0i} = \frac{1}{2}\left(E_i + iB_i\right) \quad , \quad F_{R\,ij} = \frac{1}{2}\epsilon_{ijk}\left(iE_i - B_i\right) = i\epsilon_{ijk} F_{R\,0k} \tag{21.54}$$

$$F_{L0i} = \frac{1}{2}\left(E_i - iB_i\right) \quad , \quad F_{L\,ij} = \frac{1}{2}\epsilon_{ijk}\left(-iE_i - B_i\right) = -i\epsilon_{ijk} F_{L\,0k} \tag{21.55}$$

so that both $F_{R\alpha\beta}$ and $F_{L\,\alpha\beta}$ each contain all of the information contained in $F_{\alpha\beta}$ but now encoded in complex form. (Had we considered a *complex* antisymmetric tensor, with twelve independent functions contained in the *complex* fields E_i and B_i, then this split does indeed project the tensor into two independent pieces, each with only six independent components.) In fact, as should be clear from Equations (21.51) and (21.52), one can show that $F_{R\alpha\beta}$ forms the $(1,0)$ representation of the Lorentz group, and $F_{L\,\alpha\beta}$ forms the $(0,1)$ representation. Now since $F_{R\alpha\beta}$ (or $F_{L\,\alpha\beta}$) contains all of the information in $F_{\alpha\beta}$, it should be possible to write Maxwell's equations in self-dual (antiself-dual) form. Of course, this will just be equivalent to writing Maxwell's equations in complex form.

21.3 Summary

The basic physical definition of a spinor is as an element of a carrier space of a fundamental representation of the Lorentz group. There are two inequivalent fundamental representations of the Lorentz group and therefore two distinct Lorentz spinors: $(\frac{1}{2},0)$ and $(0,\frac{1}{2})$. In general, arbitrary representations of the Lorentz group are complex, but real representations also exist under certain circumstances. In particular, it is no surprise to find that all of the standard tensor fields encountered in theoretical physics may be real. However, self-dual tensor fields are necessarily complex.

Bibliography

[1] A.O. Barut, *Electrodynamics and Classical Theory of Fields and Particles*, Dover, 1980.

[1] Editor's note: see c. 22.

[2] J.P. Crawford, "Clifford Algebra: Notes on the spinor metric and Lorentz, Poincaré, and conformal groups" J. Math. Phys., 32, 1991, 576-583.

[3] H. Georgi, *Lie Algebras in Particle Physics*, Benjamin/Cummings, 1982.

[4] R. Gilmore, *Lie Groups, Lie Algebras, and Some of Their Applications*, Wiley, 1974.

[5] C. Itzykson and J.-B. Zuber, *Quantum Field Theory*, McGraw-Hill, 1980.

[6] R. Penrose and W. Rindler, *Spinors and space-time*, Cambridge University Press, 1984.

[7] P. Ramond, *Field Theory: A Modern Primer*, Benjamin/Cummings, 1981.

[8] H. Weyl, *Classical Groups*, 2nd ed., Princeton University Press, 1946.

[9] E.P. Wigner, "On unitary representations of the inhomogeneous Lorentz group" *Ann. Math.*, **40**, 1939, 149-189.

Chapter 22

Spinors: Clifford Algebra

James P. Crawford
Department of Physics, Penn State - Fayette,
Uniontown, PA 15401 USA

22.1 Introduction

This lecture contains an introduction to Clifford algebra and a discussion of its relationship to the Lorentz, Poincaré, and conformal groups. Here the Dirac spinor is defined to be an element of the carrier space of the representation of the Clifford algebra. We will consider the relationship of the various Clifford algebra spinors to the Lorentz spinors including the Dirac (reducible complex), Weyl (irreducible complex = chiral), and Majorana (irreducible real) cases. This will serve to make these relationships clear as well as suggesting some possible generalizations of General Relativity.

22.2 Clifford Algebra

The basic defining relation for the Clifford algebra may be introduced by asking for elements γ_α such that for $V \equiv V^\alpha \gamma_\alpha$ we have $V^2 \equiv VV = V^\alpha V_\alpha \mathbf{1}$; we find

$$\{\gamma_\alpha, \gamma_\beta\} = 2\eta_{\alpha\beta}\mathbf{1} \tag{22.1}$$

as the required relation. Note that $\mathbf{1}$ is the unit element of the algebra, and the elements γ_α are called the generators of the algebra. The generators are faithfully and irreducibly represented by 4×4 complex matrices $[\gamma_\alpha]^A{}_B$, where $A, B = 1, 2, 3, 4$ are the matrix indices. Note that unlike the case of a Lie algebra where there are an infinite number of finite dimensional representations, the Clifford algebra in 1+3 dimensional spacetime has only one finite dimensional faithful and irreducible representation. The carrier space of this representation is clearly a space of complex four component objects. We will call these objects *Dirac spinors*, in which case the matrix indices $A, B, \ldots$ introduced above will be called *Dirac spinor indices*, and the

Dirac spinors will be represented by Ψ^A. The relationship between the Dirac spinors and the Lorentz spinors will be discussed below.

The basis for the Clifford algebra may be completed by including all possible products of the generators. Note, however, that the defining relation for the Clifford algebra generators, Equation (22.1), specifies the result of the symmetric product of two of the generators. Therefore only antisymmetric products of the generators are independent elements of the Clifford algebra.

22.2.1 Complex Clifford Algebra

Note that from Equation (22.1) we have $(\gamma_0)^2 = +1$ and $(\gamma_i)^2 = -1$ so the eigenvalues of γ_0 are ± 1 and the eigenvalues of γ_i are $\pm i$. Hence the matrix γ_0 is Hermitian ($\gamma_0^\dagger = \gamma_0$) whereas the matrices γ_i are antihermitian ($\gamma_i^\dagger = -\gamma_i$). This will have important consequences when considering physical applications, and leads us to consider the complex Clifford algebra and the Dirac spinor metric.

We introduce the Dirac spinor metric by asking for a matrix γ which satisfies

$$\gamma^{-1}\gamma_\alpha^\dagger\gamma = \gamma_\alpha \tag{22.2}$$

and we find the unique solution to be $\gamma = \gamma_0 = \gamma^{-1} = \gamma^\dagger$. However, there is an important distinction between the matrix representations of γ and γ_0. In particular, the generators of the algebra have matrix components specified as $[\gamma_\alpha]^A{}_B$, whereas the Dirac spinor metric has components specified as $[\gamma]_{AB}$, and the inverse spinor metric has components specified as $[\gamma]^{AB}$. The metric may be used to raise and lower spinor indices in the usual manner. We define the *Dirac conjugate* of Dirac spinors, Ψ^A, and of arbitrary elements of the Clifford algebra, $\Xi^A{}_B$, respectively as:

$$\overline{\Psi}_A \equiv \Psi^{B\,*}\gamma_{BA}\ , \ \overline{\Xi}^A{}_B \equiv \gamma^{AC}[\Xi^\dagger]_C{}^D\gamma_{DB}\ . \tag{22.3}$$

The spinor norm and the components of the current vector:

$$\sigma \equiv \overline{\Psi}\Psi\ ,\ j_\alpha \equiv \overline{\Psi}\gamma_\alpha\Psi\ , \tag{22.4}$$

are easily shown to be real. In fact, the reality of these two expressions is the reason for defining the Dirac spinor metric as in Equation (22.2).

We will now complete the basis of the algebra by including the following elements in the set:

$$\gamma_{\alpha\beta} \equiv \frac{i}{2}[\gamma_\alpha, \gamma_\beta] \tag{22.5}$$

$$\tilde{\gamma}_\alpha \equiv -\frac{i}{3!}\epsilon_{\alpha\beta\gamma\delta}\gamma^\beta\gamma^\gamma\gamma^\delta = -i\tilde{\gamma}\gamma_\alpha = i\gamma_\alpha\tilde{\gamma} \tag{22.6}$$

$$\tilde{\gamma} \equiv -\frac{1}{4!}\epsilon_{\alpha\beta\gamma\delta}\gamma^\alpha\gamma^\beta\gamma^\gamma\gamma^\delta = -\gamma_0\gamma_1\gamma_2\gamma_3. \tag{22.7}$$

The elements $\gamma_{\alpha\beta}$ form the basis for the bivectors (planes), $\tilde{\gamma}_\alpha$ form the basis for the pseudovectors (dual to the three-volumes), and $\tilde{\gamma}$ forms a basis for the pseudoscalars (dual to the four-volumes). The choice of pseudovectors here instead of trivectors is purely a matter of convenience. The complete basis for the complex Clifford algebra may now be written as: $\{\Gamma_a\} = \{\mathbf{1}, \gamma_\alpha, \gamma_{\alpha\beta}, \tilde{\gamma}_\alpha, \tilde{\gamma}\}$, where we have again introduced Clifford indices $a = 1, 2, ..., 16$. Then an arbitrary element of the algebra may be written as: $\Xi = \Xi^a \Gamma_a$, where the Ξ^a are sixteen *complex* numbers. Note that a general member of the algebra must have complex parameters since the product of any two members will in general "mix up" where the factors of "i" appear in the basis. In other words, the factors of "i" that appear in Equations (22.5) and (22.6) do not allow these elements to be members of the real Clifford algebra basis.

One may wonder, therefore, why this particular basis has been chosen. There are several reasons, some of which will come up in later discussions. For now we mention that all of the elements of the basis defined in this way are Dirac self-conjugate and therefore all of the "bilinear covariants" formed from this basis by a single Dirac spinor are real:

$$\overline{\Gamma}_a = \gamma \Gamma_a^\dagger \gamma = \Gamma_a \ , \ \Phi_a \equiv \overline{\Psi} \Gamma_a \Psi \Rightarrow \Phi_a^* = \Phi_a. \tag{22.8}$$

Since these quantities should represent observable densities in quantum mechanics, they should be real. Of course, bilinear covariants formed from two distinct Dirac spinors will not be real in general:

$$\Omega_a \equiv \overline{\Upsilon} \Gamma_a \Psi \quad \Rightarrow \quad \Omega_a^* = \overline{\Psi} \Gamma_a \Upsilon \neq \Omega_a. \tag{22.9}$$

However, since these are physically related to transition *amplitudes*, they need not be real.

22.2.2 Automorphism Group

Consider the Lie algebra associated with the complex Clifford algebra:

$$[\Gamma_a, \Gamma_b] = i c_{ab}{}^c \Gamma_c. \tag{22.10}$$

Since all of the basis elements are Dirac self-conjugate, all of the structure factors are real: $c_{ab}{}^c = c_{ab}^*{}^c$ (the proof follows simply by taking the Dirac conjugate of Equation (22.10), and is another reason for choosing this particular complex basis for the Clifford algebra). The Lie group associated with this Lie algebra has elements:

$$\Theta = \exp i\theta^a \Gamma_a. \tag{22.11}$$

where θ^a are sixteen real group parameters. The inverse group elements are given by the Dirac conjugates:

$$\overline{\Theta} = \exp -i\theta^a \Gamma_a = \Theta^{-1}. \tag{22.12}$$

Consider the spinor transformation:

$$\Psi' = \Theta \Psi \quad , \quad \overline{\Psi}' = \overline{\Psi} \overline{\Theta}. \tag{22.13}$$

Under this transformation the spinor norm is invariant:

$$\sigma' = \overline{\Psi}' \Psi' = \overline{\Psi} \overline{\Theta} \Theta \Psi = \overline{\Psi} \Psi = \sigma. \tag{22.14}$$

In other words, under these transformations the spinor metric is preserved, just as the Minkowski metric is preserved under the Lorentz transformations. Since we may choose a basis in which $\gamma = diag(1, 1, -1, -1)$, we identify this group as $U(2, 2)$. We will show later that this group is isomorphic to the conformal group.

Consider the associated transformations of the Clifford algebra basis:

$$\Gamma_a^{'} = \Theta \Gamma_a \overline{\Theta}. \tag{22.15}$$

These transformations clearly preserve the Clifford algebra defining relation, Equation (22.1), as will as the Dirac self-conjugacy of the basis, Equation (22.8). Therefore we identify these transformations as constituting the Dirac self-conjugacy preserving automorphism group of the complex Clifford algebra. Notice that the Dirac spinors may also be considered to be elements of the carrier space of the fundamental irreducible representation of the automorphism group. Finally observe that all of the bilinear covariants (Equation (22.9)) are invariant under this group of transformations.

The spinor transformations, Equation (22.13), the Clifford algebra basis transformations, Equation (22.15), and the invariance of the bilinear covariants all follow if we assume that the automorphism transformations fundamentally act on the basis spinors of the spinor space. This observation will allow a possible generalization of General Relativity, as will be discussed in another lecture of this series. The basic idea will be to allow local choice of the spinor basis, just as we are allowed local choice of the Lorentz basis in the tangent space.

As we will show, the automorphism group is isomorphic to the conformal group, and as such contains the spinor transformations of the Poincaré and Lorentz groups as subgroups. (In fact, this may be considered to be an alternate motivation for choosing this particular basis for the complex Clifford algebra.) Therefore, since the Lorentz group is a subgroup of the automorphism group, and the Dirac spinors form the fundamental irreducible representation of the automorphism group, the Dirac spinors will be associated with *reducible* representations of the Lorentz group, just as irreducible representations of the Lorentz group are associated with reducible representations of the rotation group.

22.2.3 Lorentz Group Redux

Now it is easy to see that the bivectors of the Clifford algebra basis are related to the generators of the Lorentz group. In particular we can show that:

$$\sigma_{\alpha\beta} = \frac{1}{2}\gamma_{\alpha\beta} \tag{22.16}$$

satisfies the defining relation for the Lorentz Lie algebra. So the Lorentz transformation for a Dirac spinor is written as:

$$[S(\lambda)]^A{}_B = [\exp \frac{i}{4}\lambda^{\alpha\beta}\gamma_{\alpha\beta}]^A{}_B \tag{22.17}$$

and we recognize these transformations as forming a subgroup of the automorphism group. We should not be surprised that the Clifford bivectors generate the Lorentz group, since a Lorentz transformation is a (possibly hyperbolic) rotation in a plane. Now since the Clifford algebra basis elements are Dirac self-conjugate we also have:

$$S^{-1} = \overline{S} \tag{22.18}$$

and the Dirac spinors transform as:

$$\Psi' = S\Psi \ , \ \overline{\Psi}' = \overline{\Psi'} = \overline{\Psi}\overline{S} \Rightarrow \overline{\Psi}'\Psi' = \overline{\Psi}\Psi \tag{22.19}$$

where we have explicitly noted that the Lorentz transformations preserve the spinor metric. Of course, both of these relations (Equations (22.18) and (22.19)) are just special cases of the automorphism group results (see Equations (22.12), (22.13), and (22.14)).

Now the Lorentz transformations for the generators of the Clifford algebra, in contradistinction to the transformations of the Dirac spinors, are *not* simply automorphism transformations of these elements since the Clifford generators also carry a vector index. In particular we have:

$$\gamma_\alpha' = \Lambda_\alpha{}^\beta S \gamma_\beta \overline{S} = \gamma_\alpha. \tag{22.20}$$

Note that we have indicated that the Clifford generator does not transform. This is because the Lorentz transformation on the vector index exactly cancels the transformations on the spinor indices. In fact, one may take Equation (22.20) as the definition of the Lorentz transformation for the Dirac spinors, in which case Equation (22.16) is recovered.

The Clifford generators $\gamma_\alpha{}^A{}_B$ may be used to write the Lorentz vector in terms of Dirac spinor indices and the Lorentz transformation is then written in spinor form:

$$A^A{}_B = A^\alpha \gamma_\alpha{}^A{}_B \quad , \quad A' = SA\overline{S}. \tag{22.21}$$

We may also write an antisymmetric tensor in terms of two Dirac spinor indices by making use of the bivector and its Lorentz transformation may then also written in spinor form:

$$F^A_{\ B} = F^{\alpha\beta}\gamma_{\alpha\beta}{}^A{}_B \quad , \quad F' = SF\overline{S}. \tag{22.22}$$

In fact, we may construct a general "multivector" with two Dirac spinor indices:

$$M^A_{\ B} = [s\mathbf{1} + v^\alpha\gamma_\alpha + \frac{1}{2}t^{\alpha\beta}\gamma_{\alpha\beta} + a^\alpha\tilde{\gamma}_\alpha + p\tilde{\gamma}]^A{}_B \tag{22.23}$$

and the associated Lorentz transformation is then given by:

$$M' = SM\overline{S}. \tag{22.24}$$

Since the Dirac spinors form reducible representations of the Lorentz group, it should not be surprising that the Clifford algebra basis can incorporate more than just the vector representation of the Lorentz group.

Weyl (Chiral) Spinors

To gain insight into the nature of the Dirac spinors it is proves worthwhile to consider the following *chirality projection operators*:

$$P_R \equiv \frac{1}{2}(\mathbf{1} + i\tilde{\gamma}) \quad , \quad P_L \equiv \frac{1}{2}(\mathbf{1} - i\tilde{\gamma}). \tag{22.25}$$

It is straightforward to show that these elements of the Clifford algebra satisfy the following relations:

$$\begin{aligned} P_R P_R = P_R \quad &, \quad P_L P_L = P_L \\ P_R P_L = 0 = P_L P_R \quad &, \quad P_R + P_L = \mathbf{1} \end{aligned} \tag{22.26}$$

and therefore are projection operators as advertized. Consider the action of these projectors on the Lorentz group generators:

$$\gamma_{\alpha\beta} P_R = P_R \gamma_{\alpha\beta} = \frac{1}{2}(\gamma_{\alpha\beta} + \frac{i}{2}\epsilon_{\alpha\beta\gamma\delta}\gamma^{\gamma\delta}) \tag{22.27}$$

$$\gamma_{\alpha\beta} P_L = P_L \gamma_{\alpha\beta} = \frac{1}{2}(\gamma_{\alpha\beta} - \frac{i}{2}\epsilon_{\alpha\beta\gamma\delta}\gamma^{\gamma\delta}) \tag{22.28}$$

so that the chiral projectors acting on the bivectors produce the (anti)self-dual bivectors. In particular we find:

$$\lambda P_R = P_R \lambda = \phi_i J_i \quad , \quad \lambda P_L = P_L \lambda = \phi_i^* K_i \, . \tag{22.29}$$

Now an arbitrary element of the Dirac spinor representation of the Lorentz group may be written as:

$$\begin{aligned} S(\lambda) &= \exp i(\lambda P_R + \lambda P_L) = \exp i(\lambda P_R) \exp i(\lambda P_L) \\ &= \exp i(\lambda P_R) P_R + \exp i(\lambda P_L) P_L \end{aligned} \tag{22.30}$$

so the Lorentz transformation for the Dirac representation splits into two distinct parts. Although we did not display this split explicitly for the (anti)self-dual projection operators, it is indeed also true in that case.

Now we define the chiral (Weyl) spinors as:

$$\Psi_R = P_R \Psi \quad , \quad \Psi_L = P_L \Psi \tag{22.31}$$

and for their Lorentz transformations we find using Equations (22.29) and (22.30):

$$\Psi'_R = \exp(i\phi_i J_i)\Psi_R \quad , \quad \Psi'_L = \exp(i\phi_i^* K_i)\Psi_L. \tag{22.32}$$

Therefore, the Lorentz group representations for the chiral spinors are:

$$\Psi_R \, : \, (\frac{1}{2}, 0) \quad , \quad \Psi_L \, : \, (0, \frac{1}{2}) \tag{22.33}$$

and for the Dirac spinor we find:

$$\Psi = \Psi_R + \Psi_L \, : \, (\frac{1}{2}, 0) \oplus (0, \frac{1}{2}) \text{ complex}. \tag{22.34}$$

We can now find the Lorentz group representation structure of an arbitrary element of the Clifford algebra. Since an arbitrary element of the Clifford algebra carries two Dirac spinor indices, it must transform as the product of two Dirac spinors, so we have:

$$\begin{aligned}&((\tfrac{1}{2},0)\oplus(0,\tfrac{1}{2}))\otimes((\tfrac{1}{2},0)\oplus(0,\tfrac{1}{2}))\\&\qquad=(0,0)\oplus(\tfrac{1}{2},\tfrac{1}{2})\oplus((1,0)\oplus(0,1))\oplus(\tfrac{1}{2},\tfrac{1}{2})\oplus(0,0).\end{aligned}\tag{22.35}$$

These terms respectively correspond to the scalar, vector, bivector, pseudovector, and pseudoscalar terms that appear in Equation (22.23), and each may be either real or complex.

Majorana Spinors

Note that the representation for the Dirac spinor (Equation (22.34)) is of the form to allow a real representation, so we define the Majorana spinor to be:

$$\Upsilon\equiv\Psi+\Psi^{*}\ :\ (\frac{1}{2},0)\oplus(0,\frac{1}{2})\ \text{real}.\tag{22.36}$$

Clearly this spinor remains real under Lorentz transformations (although the form of the Lorentz transformation may generally be a bit complicated), so that just as in the case of the antisymmetric tensor field (for example), the reality of this spinor is an invariant statement. We will give an explicit demonstration of this in the following section.

We may also consider the possibility of constructing chiral Majorana spinors:

$$\Upsilon_R=P_R\Upsilon\quad,\quad\Upsilon_L=P_L\Upsilon.\tag{22.37}$$

However, just as in the case of the real antisymmetric tensor field where the (anti)self-dual parts simply encode the information in complex form, the chiral Majorana fields also contain the same information as the Majorana field but encoded in complex form. In other words, a right (left) handed Majorana spinor is just a right (left) handed Weyl spinor.

22.2.4 Poincaré Group

The Lorentz group is the set of transformations which preserve the Minkowski inner product of any two vectors, and as such they act in the tangent space at each point in spacetime. If we want to include in our set transformations which carry us from one point in spacetime to another, or from one tangent space to another, we encounter the Poincaré group. We may also consider the Poincaré group to be the entire set of coordinate transformations in flat Minkowski spacetime which preserve the infinitesimal line element:

$$ds^2=\eta_{\alpha\beta}dz^{\alpha}dz^{\beta}=ds'^2.\tag{22.38}$$

This set of coordinate transformations are clearly given by the Lorentz transformations and translations:

$$z'^{\alpha} = \Lambda^{\alpha}{}_{\beta} z^{\beta} \quad , \quad M_{\alpha\beta} = i(z_{\alpha}\partial_{\beta} - z_{\beta}\partial_{\alpha}) \tag{22.39}$$

$$z'^{\alpha} = z^{\alpha} + t^{\alpha} \quad , \quad P_{\alpha} = i\partial_{\alpha} \tag{22.40}$$

where $\Lambda^{\alpha}{}_{\beta}$ is any constant Lorentz transformation matrix, $M_{\alpha\beta}$ is the generator of the coordinate Lorentz transformations, t^{α} is any constant vector, and P_{α} is the generator of coordinate translations.

The Poincaré group is therefore generated by the Lorentz generators $M_{\alpha\beta} = \sigma_{\alpha\beta}$ and the translation generators P_{α}, and the Poincaré Lie algebra is given by the Lorentz Lie algebra and the following additional commutators:

$$[P_{\alpha}, P_{\beta}] = 0 \quad , \quad [P_{\alpha}, \sigma_{\beta\gamma}] = i(\eta_{\alpha\beta}P_{\gamma} - \eta_{\alpha\gamma}P_{\beta}). \tag{22.41}$$

We seek elements of the Clifford algebra basis which satisfy these commutation relations, and we find:

$$\sigma_{\alpha\beta} = \frac{1}{2}\gamma_{\alpha\beta} \quad , \quad P_{\alpha} = \frac{M}{2}(\gamma_{\alpha} \mp \tilde{\gamma}_{\alpha}) = MP_{R,L}\gamma_{\alpha}. \tag{22.42}$$

We have the interesting result that the translation generator is represented by either one of the chiral vectors. Also note that there is an arbitrary constant M (with the dimension of mass) which enters. This arbitrary constant reflects the fact that the Poincaré Lie algebra is homogeneous in the translation generators. Of course, it can be chosen to be zero, as is usually done, but in a later lecture we will look at the effect of considering M to be non-zero. It will turn out that the field strength tensor associated with the translation generators is the torsion, just as the field strength tensor associated with the Lorentz generators is the Riemann curvature tensor.

22.2.5 Conformal Group

The Poincaré group may be extended to the conformal group by including the set of coordinate transformations of Minkowski spacetime which preserve all angles, but not necessarily lengths:

$$ds'^2 = \eta_{\alpha\beta}dz'^{\alpha}dz'^{\beta} = \Omega(z)\eta_{\alpha\beta}dz^{\alpha}dz^{\beta} \tag{22.43}$$

and we find the additional transformations, special conformal transformations and dilatations, to be of the following form:

$$z'^{\alpha} = \frac{z^{\alpha} + z^2 c^{\alpha}}{(1 + 2z^{\alpha}c_{\alpha} + z^2c^2)} \quad , \quad K_{\alpha} = i(2z_{\alpha}z^{\beta}\partial_{\beta} - z^2\partial_{\alpha}) \tag{22.44}$$

$$z'^{\alpha} = sz^{\alpha} \quad , \quad D = iz^{\alpha}\partial_{\alpha}. \tag{22.45}$$

Here c^{α} is a constant vector parameter of the special conformal transformations, K_{α} is the generator of the special conformal transformations, s is a constant parameter of scale transformations, and D is the generator of the dilatations. The Lie algebra for the conformal group may now be constructed from the generators, and we find that the additional commutators are given by:

$$\begin{array}{c} [D,\sigma_{\alpha\beta}]=0 \quad , \quad [D,P_\alpha]=-iP_\alpha \\ [D,K_\alpha]=iK_\alpha \quad , \quad [K_\alpha,K_\beta]=0 \\ [K_\alpha,\sigma_{\beta\gamma}]=i(\eta_{\alpha\beta}K_\gamma-\eta_{\alpha\gamma}K_\beta) \\ [K_\alpha,P_\beta]=-2i(\eta_{\alpha\beta}D+\sigma_{\alpha\beta}) \end{array} \tag{22.46}$$

We seek elements of the Clifford algebra basis which satisfy these relationships and find:

$$\begin{array}{c} D=\tilde{\gamma} \quad , \quad P_\alpha=\frac{M}{2}(\gamma_\alpha-\tilde{\gamma}_\alpha)=MP_R\gamma_\alpha \\ K_\alpha=\frac{1}{2M}(\gamma_\alpha+\tilde{\gamma}_\alpha)=\frac{1}{M}P_L\gamma_\alpha \end{array} \tag{22.47}$$

Several comments are in order. First note that we have established the isomorphism between the automorphism group and the conformal group. Also note that the arbitrary constant parameter M can no longer be chosen to be zero, since it appears inversely in the definition of the special conformal generator. Also, the choice of the right handed vector to represent the translation generator and the left handed vector to represent the special conformal generator is related to the choice of dilatation generator. In particular, if we had chosen $D=-\tilde{\gamma}$ instead, then the role of the right handed and left handed vectors would have been reversed, so this choice is indeed arbitrary as expected. Finally note that the group $SU(2,2)$ is actually the covering group of the conformal group $SO(2,4)$, and therefore the Dirac spinor is related to the conformal group in exactly the same way as the Lorentz spinor is related to the Lorentz group; each one spans the carrier space of the fundamental representation of the covering group.

22.3 Summary

A Dirac spinor may be defined in a variety of equivalent ways: (1) as an element of the carrier space of the representation of the Clifford algebra of spacetime; (2) as an element of the carrier space of the fundamental representation of the Dirac spinor metric preserving automorphism group of the Clifford algebra; and (3) as an element of the carrier space of the fundamental representation of the covering group of the conformal group.

The Dirac spinor forms a complex reducible representation of the Lorentz group: $(\frac{1}{2},0)\oplus(0,\frac{1}{2})$. The irreducible Lorentz components $(\frac{1}{2},0)$ and $(0,\frac{1}{2})$ are the Weyl spinors and are obtained by a chiral decomposition of the Dirac spinor. The Lorentz representation structure of the Dirac spinor also allows for a real representation, the Majorana spinor. The relation between the Majorana spinors and the Weyl spinors is exactly analogous to the relation between a real antisymmetric tensor field and a (anti)self-dual tensor field in that the chiral=self-dual fields encode the information of the corresponding real fields in complex form. In this sense the Dirac spinor is analogous to a complex antisymmetric tensor field.

Bibliography

[1] Brauer, R. and H. Weyl, "Spinors in n dimensions" *Am. J. Math.*, **57**, 1935, 425-449.

[2] E. Cartan, *The Theory of Spinors*, M.I.T. Press, 1966.

[3] C. Chevalley, *The Algebraic Theory of Spinors*, Columbia University Press, 1954.

[4] W.K. Clifford, *Am. J. Math.* **1**, 1878, 350.

[5] *Clifford Algebas and Their Applications in Mathematical Physics*, J.S.R. Chisholm and A.K. Common eds., Reidel, 1986.

[6] *Clifford Algebas and Their Applications in Mathematical Physics*, A. Micali, R. Boudet, and J. Helmstetter eds., Kluwer, 1992.

[7] *Clifford Algebas and Their Applications in Mathematical Physics*, F. Brackx, R. Delanghe, and H. Serras eds., Kluwer, 1993.

[8] J.P. Crawford, "Clifford Algebra: Notes on the spinor metric and Lorentz, Poincaré, and conformal groups" J. Math. Phys., 32, 1991, 576-583.

[9] D. Hestenes and G. Sobczyk, *Clifford Algebra to Geometric Calculus*, Reidel, 1984.

[10] E. Majorana, *Nuovo Cimento*, **14**, 1937, 171.

[11] R. Penrose and W. Rindler, *Spinors and space-time*, Cambridge University Press, 1984.

Chapter 23

General Relativity: An Overview

James P. Crawford
Department of Physics, Penn State - Fayette,
Uniontown, PA 15401 USA

23.1 Introduction

This lecture is a brief introduction General Relativity. We begin with an outline of general coordinate covariance, tensor analysis, and Riemannian geometry. The Bianchi identities are derived and we include some comments on torsion, since torsion may arise when spinor fields are present. Classical General Relativity is then reviewed and the field equations are found from the Einstein-Hilbert action. The Palatini formalism is also discussed.

General Relativity is fundamentally based upon two principles: The Principle of General Covariance, and the Principle of Equivalence. The Principle of General Covariance is completely self evident — it states that physical reality cannot possibly depend on any particular choice of coordinate system, so that the laws of physics must be written in generally covariant form. This requires the introduction of the notions of covariant differentiation and affine connection, the purview of tensor analysis. At this stage it is not necessary to assume that the spacetime is curved, although having written the equations in generally covariant form makes them also valid in curved spacetime. Aside from this convenience, the Principle of General Covariance contains no physics, since it does not make any statement concerning the dynamics of spacetime.

It is the Principle of Equivalence that establishes the deep connection between matter and geometry — it states that at any point in spacetime we can always find a local frame of reference in which the gravitational field vanishes. Such a reference frame is called locally inertial. This implies that particles travel geodesic paths (in the absence of non-gravitational forces) and that matter induces spacetime curvature. Geodesic motion follows since we may find a sequence of inertial frames along a particle trajectory, so that the particle always obeys Newton's First Law (written relativistically, of course), and the generally covariant form of the Law of

Inertia is precisely the equation of geodesic motion. That matter induces spacetime curvature now follows since we know that the presence of matter is responsible for gravitational forces, but the gravitational forces are only a reflection of spacetime curvature, and therefore the curvature must be induced by the matter.

23.2 Tensor Analysis

23.2.1 Vectors and Tensors

Consider an arbitrary (possibly curved) region of spacetime. At any given point in this spacetime the vectors will lie in the tangent space at that point, and the tensors will lie in Cartesian products of the tangent space. Consider a set of coordinates x^μ covering this region of spacetime. The basis for the tangent space at any point may be chosen to be the set of tangent vectors $\mathbf{E}_\mu$ of the coordinate lines. For example, $\mathbf{E}_1$ at some point P, where $x^\mu(P) = (p^0, p^1, p^2, p^3)$, is any vector tangent to the curve x^1 at point P, where the coordinate curve x^1 going through point P is given by the coordinates $x^\mu = (p^0, x^1, p^2, p^3)$. Then any vector in the tangent space, or any tensor in the cartesian product of tangent spaces, may be written as:

$$\mathbf{V} = V^\mu \mathbf{E}_\mu \quad , \quad \mathbf{T} = T^{\mu_1 \cdots \mu_N} \mathbf{E}_{\mu_1} \otimes \cdots \otimes \mathbf{E}_{\mu_N} \ , \tag{23.1}$$

where the four quantities V^μ are called the contravariant components of the vector, and the 4^N quantities $T^{\mu_1 \cdots \mu_N}$ are called the contravariant components of the tensor.

Consider an arbitrary coordinate transformation $x^\mu \to x'^\mu = x'^\mu(x)$. Under such a transformation the basis vectors transform as:

$$\mathbf{E}'_\mu = \left(\frac{\partial x^\nu}{\partial x'^\mu} \right) \mathbf{E}_\nu \ , \tag{23.2}$$

so that the new basis vectors $\mathbf{E}'_\mu$ lie along the new coordinate lines. Therefore, in order for a vector to remain invariant: $\mathbf{V}' = \mathbf{V}$, its components must transform as:

$$V'^\mu = \left(\frac{\partial x'^\mu}{\partial x^\nu} \right) V^\nu \ . \tag{23.3}$$

Conversely, any set of fields which transforms in this manner constitutes the *contravariant components of a vector.* Similar expressions hold for the contravariant components of a tensor.

23.2.2 Affine Connection and Covariant Differentiation

In general, for an arbitrary coordinate system, as we move from point to point in spacetime the orientation of the basis vectors $\mathbf{E}_\mu$ will change. The change in a vector is again a vector so we may write:

$$\partial_\mu \mathbf{E}_\nu \equiv \Gamma^\rho{}_{\nu\mu} \mathbf{E}_\rho \ , \tag{23.4}$$

where the quantities $\Gamma^{\rho}{}_{\nu\mu}$ are called the *coefficients of affine connection*, or more simply the connection. Then for the derivative of an arbitrary vector we find:

$$\partial_\mu \mathbf{V} = \partial_\mu (V^\nu \mathbf{E}_\nu) = (\partial_\mu V^\nu + \Gamma^\nu{}_{\rho\mu} V^\rho)\mathbf{E}_\nu \equiv (\nabla_\mu V^\nu)\mathbf{E}_\nu \, , \tag{23.5}$$

where we have defined $\nabla_\mu V^\nu \equiv \partial_\mu V^\nu + \Gamma^\nu{}_{\rho\mu} V^\rho$ as the *covariant derivative* of the contravariant components of a vector. In a similar fashion we may easily find the covariant derivative of the contravariant components of an arbitrary tensor.

Now note that the derivative transforms as:

$$\partial'_\mu = \left(\frac{\partial}{\partial x'^\mu}\right) = \left(\frac{\partial x^\nu}{\partial x'^\mu}\right)\left(\frac{\partial}{\partial x^\nu}\right) = \left(\frac{\partial x^\nu}{\partial x'^\mu}\right)\partial_\nu \, . \tag{23.6}$$

We will refer to any set of fields which transforms in this way as constituting the *covariant components of a vector*. The covariant derivative of covariant components of a vector is then given by:

$$\nabla_\mu U_\nu = \partial_\mu U_\nu - \Gamma^\rho{}_{\nu\mu} U_\rho \, . \tag{23.7}$$

The generalization to covariant components of a tensor is straightforward, and the explicit forms of the covariant derivatives of tensors with arbitrary indices are easily obtained. Essentially we find that in addition to the ordinary derivative term, there is one additional connection term for each index, appearing with a positive sign for contravariant indices and a negative sign for covariant indices.

We want the covariant derivative to be a tensor operation so that it should transform in the same way as a covariant vector. So we have:

$$\nabla'_\mu V'^\nu = \partial'_\mu V'^\nu + \Gamma'^\nu{}_{\rho\mu} V'^\rho = \left(\frac{\partial x^\lambda}{\partial x'^\mu}\right)\left(\frac{\partial x'^\nu}{\partial x^\tau}\right)\nabla_\lambda V^\tau \, , \tag{23.8}$$

and this allows us to determine the transformation property for the affine connection:

$$\Gamma'^\nu{}_{\rho\mu} = \left(\frac{\partial x'^\nu}{\partial x^\tau}\right)\left(\frac{\partial x^\lambda}{\partial x'^\rho}\right)\left(\frac{\partial x^\sigma}{\partial x'^\mu}\right)\Gamma^\tau{}_{\lambda\sigma} + \left(\frac{\partial x'^\nu}{\partial x^\tau}\right)\left(\frac{\partial^2 x^\tau}{\partial x'^\rho \partial x'^\mu}\right) \, . \tag{23.9}$$

The inhomogeneous term in this transformation indicates that the coefficients of affine connection do not constitute the components of a tensor. Instead we see that the connection is akin to a gauge field.

23.2.3 Torsion

Note that the inhomogeneous term in Equation (23.9) only contributes to the part of the connection symmetric in the covariant indices. Therefore, the part of the connection antisymmetric in the covariant indices does indeed constitute the components of a tensor. This tensor is called the *torsion*. So we make the definitions:

$$S^\rho{}_{\mu\nu} \equiv \Gamma^\rho{}_{\mu\nu} + \Gamma^\rho{}_{\nu\mu} \, , \tag{23.10}$$

$$T^\rho{}_{\mu\nu} \equiv \Gamma^\rho{}_{\mu\nu} - \Gamma^\rho{}_{\nu\mu} \, , \tag{23.11}$$

where $S^{\rho}{}_{\mu\nu}$ is twice the symmetric part of the connection, and $T^{\rho}{}_{\mu\nu}$, the torsion, is twice the antisymmetric part of the connection. Under a general coordinate transformation we find:

$$T'^{\nu}{}_{\rho\mu} = \left(\frac{\partial x'^{\nu}}{\partial x^{\tau}}\right)\left(\frac{\partial x^{\lambda}}{\partial x'^{\rho}}\right)\left(\frac{\partial x^{\sigma}}{\partial x'^{\mu}}\right)T^{\tau}{}_{\lambda\sigma} \, . \tag{23.12}$$

We see that the torsion components do indeed transform as tensor components, as advertized. Therefore, only the symmetric part of the connection may be considered as a gauge field.

23.2.4 Parallel Transport and Curvature

Much of the geometric information of a space may be ascertained by considering the parallel transport of a vector around a closed loop. Under parallel transport along a curve $x^{\mu}(\tau)$, where τ is some parameter, the contravariant components of a vector are defined to satisfy:

$$\frac{DV^{\mu}}{D\tau} \equiv \left(\frac{dx^{\rho}}{d\tau}\right)(\nabla_{\rho}V^{\mu}) = 0 \quad \Rightarrow \quad \frac{dV^{\mu}}{d\tau} = -\Gamma^{\mu}{}_{\nu\rho}\left(\frac{dx^{\rho}}{d\tau}\right)V^{\nu} \, . \tag{23.13}$$

In other words, under parallel transport along a curve, the variation of the contravariant components of a vector is exactly and only that required to counterbalance the change in the orientation of the basis vectors, so that the vector itself remains unchanged. Note that in the case where the vector being transported is the tangent vector to the curve we find:

$$U^{\rho} \equiv \left(\frac{dx^{\rho}}{d\tau}\right) \quad , \quad \frac{d^2x^{\mu}}{d\tau^2} = -\Gamma^{\mu}{}_{\nu\rho}\left(\frac{dx^{\rho}}{d\tau}\right)\left(\frac{dx^{\nu}}{d\tau}\right) \, . \tag{23.14}$$

This is known as the geodesic equation; it defines the minimum distance between two points in a curved spacetime.

Now although it is possible to conduct any vector by parallel transport along a curve, in general it is not possible to demand that the covariant derivative of a vector field vanish everywhere. In other words, in general it is not possible to globally define a parallel vector field in arbitrary spacetimes. To see this, consider the parallel transport of a vector around an *infinitesimal closed loop* with surface element $\Delta S^{\rho\sigma}$. In this case we obtain:

$$\Delta V^{\mu} = \oint dV^{\mu} = -\frac{1}{2}\iint R^{\mu}{}_{\nu\rho\sigma}V^{\nu}dS^{\rho\sigma} = -\frac{1}{2}R^{\mu}{}_{\nu\rho\sigma}V^{\nu}\Delta S^{\rho\sigma} \, , \tag{23.15}$$

where we have used Stoke's theorem and defined the components of the Riemann curvature tensor as:

$$R^{\mu}{}_{\nu\rho\sigma} \equiv -\partial_{\rho}\Gamma^{\mu}{}_{\nu\sigma} + \partial_{\sigma}\Gamma^{\mu}{}_{\nu\rho} - \Gamma^{\mu}{}_{\kappa\rho}\Gamma^{\kappa}{}_{\nu\sigma} + \Gamma^{\mu}{}_{\kappa\sigma}\Gamma^{\kappa}{}_{\nu\rho} \, . \tag{23.16}$$

Therefore, only in spacetimes in which the Riemann curvature vanishes is it possible to find constant vector fields.

Note that we may also obtain both the Riemann curvature tensor and the torsion tensor by considering the antisymmetric combination of two successive covariant differentiations:

$$(\nabla_\rho \nabla_\sigma - \nabla_\sigma \nabla_\rho) V^\mu = -R^\mu{}_{\nu\rho\sigma} V^\nu + T^\nu{}_{\rho\sigma} \nabla_\nu V^\mu \,. \tag{23.17}$$

This expression makes manifest the fact that the Riemann curvature is indeed a tensor, and is often convenient for formal derivations. It also immediately indicates an index antisymmetry for the Riemann tensor:

$$R^\mu{}_{\nu\rho\sigma} = -R^\mu{}_{\nu\sigma\rho} \,. \tag{23.18}$$

This index antisymmetry is also apparent from Equation (23.16) and is consistent with the contraction on the surface elements in Equation (23.15). The geometric interpretation of the Riemann curvature is most transparent if we restrict our attention to a Riemann space, in which case it is easily extracted after we consider the Bianchi identities and introduce the notion of metric.

23.2.5 Bianchi Identities

The Bianchi identities are just the Jacobi identity for the covariant derivative. The Jacobi identity reads:

$$[\nabla_\mu, [\nabla_\nu, \nabla_\rho]] + [\nabla_\rho, [\nabla_\mu, \nabla_\nu]] + [\nabla_\nu, [\nabla_\rho, \nabla_\mu]] = 0 \,. \tag{23.19}$$

This expression is true regardless of what it acts upon. In particular, if we apply this expression to the contravariant components of a vector we obtain two Bianchi identities:

$$\nabla_\rho R^\lambda{}_{\kappa\mu\nu} + \nabla_\mu R^\lambda{}_{\kappa\nu\rho} + \nabla_\nu R^\lambda{}_{\kappa\rho\mu} = R^\lambda{}_{\kappa\sigma\rho} T^\sigma{}_{\mu\nu} + R^\lambda{}_{\kappa\sigma\mu} T^\sigma{}_{\nu\rho} + R^\lambda{}_{\kappa\sigma\nu} T^\sigma{}_{\rho\mu} \,, \tag{23.20}$$

$$\begin{aligned} &\nabla_\rho T^\sigma{}_{\mu\nu} + \nabla_\mu T^\sigma{}_{\nu\rho} + \nabla_\nu T^\sigma{}_{\rho\mu} \\ &\quad = R^\sigma{}_{\rho\mu\nu} + R^\sigma{}_{\mu\nu\rho} + R^\sigma{}_{\nu\rho\mu} + T^\sigma{}_{\kappa\rho} T^\kappa{}_{\mu\nu} + T^\sigma{}_{\kappa\mu} T^\kappa{}_{\nu\rho} + T^\sigma{}_{\kappa\nu} T^\kappa{}_{\rho\mu} \,. \end{aligned} \tag{23.21}$$

In the case of vanishing torsion these identities take the simple and familiar forms. We will discuss the contracted Bianchi identities after introducing the metric.

23.2.6 Metric

Affine spaces may be endowed with a metric. The metric allows the computation of the length of a vector and, more generally, the computation of the inner product of any two vectors. We assume that it is symmetric:

$$\mathbf{V} \cdot \mathbf{U} \equiv \mathbf{g}(\mathbf{V}, \mathbf{U}) = \mathbf{g}(\mathbf{U}, \mathbf{V}) \,. \tag{23.22}$$

The covariant components of the metric define the inner product of the basis vectors:

$$g_{\mu\nu} \equiv \mathbf{E}_\mu \cdot \mathbf{E}_\nu = \mathbf{g}(\mathbf{E}_\mu, \mathbf{E}_\nu) \quad \Rightarrow \quad \mathbf{V} \cdot \mathbf{U} = V^\mu U^\nu g_{\mu\nu} \,. \tag{23.23}$$

Note that the symmetry of the metric illustrated in Equation (23.22) insures that the components are symmetric: $g_{\mu\nu} = g_{\nu\mu}$. That the metric is indeed a tensor now follows from the transformation of the basis vectors:

$$g'_{\mu\nu} = \mathbf{E}'_\mu \cdot \mathbf{E}'_\nu = \left(\frac{\partial x^\rho}{\partial x'^\mu}\right)\left(\frac{\partial x^\sigma}{\partial x'^\mu}\right)\mathbf{E}_\rho \cdot \mathbf{E}_\sigma = \left(\frac{\partial x^\rho}{\partial x'^\mu}\right)\left(\frac{\partial x^\sigma}{\partial x'^\mu}\right) g_{\rho\sigma} \,. \tag{23.24}$$

We define the contravariant components of the metric tensor as constituting the inverse matrix:

$$g^{\mu\rho} g_{\rho\nu} = \delta^\mu_\nu \quad \Rightarrow \quad g'^{\mu\nu} = \left(\frac{\partial x'^\mu}{\partial x^\rho}\right)\left(\frac{\partial x'^\nu}{\partial x^\sigma}\right) g^{\rho\sigma} \,. \tag{23.25}$$

Since the identity matrix is invariant, the contravariant components transform as required for a tensor, as indicated.

At this juncture, aside from the requirements of index symmetry and invertibility, the components of the metric are completely arbitrary. In other words, the geometric structure of the space is determined by the affine connection, and the metric is a set of ten arbitrary functions which allows the definition of lengths and inner products. However, if we require that the geometry of the space be such that under parallel transport the length of vectors remains unchanged and the inner product of any two vectors remains unchanged, then we must place restrictions on the metric and the affine connection. In particular, these conditions clearly require that the covariant derivative of the metric vanish:

$$\nabla_\rho g_{\mu\nu} = 0 \,. \tag{23.26}$$

This condition (a total of 40 equations) may be used to determine the symmetric part of the connection (encompassing 40 functions) in terms of the metric and the torsion:

$$S^\rho{}_{\mu\nu} = g^{\rho\sigma}(\partial_\mu g_{\sigma\nu} + \partial_\nu g_{\sigma\mu} - \partial_\sigma g_{\mu\nu} - T_{\mu\sigma\nu} - T_{\nu\sigma\mu}) \,. \tag{23.27}$$

Equation (23.26) is known as the Riemann condition. In the case of vanishing torsion we recover the Christoffel form of the connection. Clearly not all affine spaces can be considered Riemann spaces. In particular, only those affine spaces for which a metric can be found such that the connection can be written as in Equation (23.27) may be considered to be Riemann spaces. Of course, since we will assume that the Riemann condition holds, we simply define the (symmetric part of the) connection in terms of the metric as given by Equation (23.27). Finally note that since the connection may be considered to be the potential of the curvature (see Equation (23.16)), in torsion-free Riemann spaces the metric is sometimes referred to as the "prepotential" since it serves as the "potential of the potential."

The Riemann condition also implies an additional index symmetry on the Riemann curvature tensor:

$$(\nabla_\mu \nabla_\nu - \nabla_\nu \nabla_\mu) g_{\rho\sigma} = R^\tau{}_{\rho\mu\nu} g_{\tau\sigma} + R^\tau{}_{\sigma\mu\nu} g_{\rho\tau} + T^\tau{}_{\mu\nu}(\nabla_\tau g_{\rho\sigma}) \tag{23.28}$$

$$\Rightarrow \quad R_{\rho\sigma\mu\nu} = -R_{\sigma\rho\mu\nu} \,, \tag{23.29}$$

where Equation (23.28) is formally true for any tensor of two covariant indices, and Equation (23.29) follows from the Riemann condition.

23.2.7 Contracted Bianchi Identities

We now turn to the contracted Bianchi identities. Due to the index symmetries displayed in Equations (23.18) and (23.29), there are only two independent contractions of the Riemann tensor, called the Ricci tensor and the Ricci scalar, and it also proves convenient to define the contracted torsion tensor:

$$R_{\mu\nu} \equiv R^{\rho}{}_{\mu\rho\nu} \quad , \quad R \equiv g^{\mu\nu} R_{\mu\nu} \quad , \quad T_{\mu} \equiv T^{\rho}{}_{\rho\mu} \; . \tag{23.30}$$

We may contract the Bianchi identities (Equations (23.20) and (23.21)) once without use of the metric to obtain:

$$\nabla_{\sigma} R^{\sigma}{}_{\tau\mu\nu} - \nabla_{\mu} R_{\tau\nu} + \nabla_{\nu} R_{\tau\mu} = -R_{\tau\rho} T^{\rho}{}_{\mu\nu} - R^{\sigma}{}_{\tau\rho\mu} T^{\rho}{}_{\sigma\nu} + R^{\sigma}{}_{\tau\rho\nu} T^{\rho}{}_{\sigma\mu} \, , \tag{23.31}$$

$$\nabla_{\rho} T^{\rho}{}_{\mu\nu} - \nabla_{\mu} T_{\nu} + \nabla_{\nu} T_{\mu} = R_{\mu\nu} - R_{\nu\mu} + T_{\rho} T^{\rho}{}_{\mu\nu} \; . \tag{23.32}$$

The first of these is not yet terribly transparent, but note that the second leads to the symmetry of the Ricci tensor when the torsion vanishes. Now using the metric to contract the first Bianchi identity again yields:

$$\nabla_{\mu}(R^{\mu\nu} - \frac{1}{2} g^{\mu\nu} R) = R_{\sigma\rho} T^{\rho\sigma\nu} + \frac{1}{2} R^{\sigma\mu\rho\nu} T_{\rho\mu\sigma} \, , \tag{23.33}$$

which will be useful when we consider constructing the standard version of General Relativity. Finally note that contraction of the second Bianchi identity (Equation (23.32)) is a trivial identity, so there are no other relations to be obtained in this fashion.

23.3 General Relativity

23.3.1 The Principle of Equivalence and the Einstein Equation

The Principle of Equivalence is the formal recognition of the importance of Galileo's "Leaning Tower of Pisa Experiment." In a small enough region of spacetime it is not possible to distinguish an accelerating frame of reference from a non-accelerating frame in which there is a gravitational field. Alternately, we may always find a local frame of reference in which the gravitational field vanishes.

Suppose we are in a flat spacetime ($R^{\sigma}{}_{\tau\mu\nu} = 0$), in which case we can always find a global coordinate system in which the connection vanishes ($\Gamma^{\sigma}{}_{\tau\mu} = 0$ everywhere, so we could choose a Cartesian coordinate system). Then Equation (23.14) for geodesic motion becomes:

$$\frac{d^2 x^{\mu}}{d\tau^2} = \frac{dU^{\mu}}{d\tau} = 0 \, , \tag{23.34}$$

which, if we choose τ to be the proper time, is clearly Newton's First Law of motion. In a curved spacetime ($R^{\sigma}_{\tau\mu\nu} \neq 0$) we can only find a coordinate system in which the connection vanishes at a point, in which case Equation (23.34) is only valid at that point. Such a frame is called locally inertial. Therefore, we expect the geodesic equation to govern the behavior of particles in the absence of nongravitational forces, the gravitational forces being represented by the curvature of spacetime. Also note that since the geodesic equation may be written as:

$$\frac{dU^\mu}{d\tau} = -\Gamma^\mu{}_{\rho\sigma} U^\rho U^\sigma \equiv g^\mu \tag{23.35}$$

we identify the right hand side as the four-acceleration of gravity, as indicated. So the affine connection does indeed represent the gravitational field (within factors of the four-velocity). Notice that it is only the symmetric part of the connection which contributes to the gravitational acceleration, so the dynamic behavior of a spinless point particle is independent of the torsion. Therefore, for the remainder of this section we will assume that the torsion vanishes.

Since the Principle of Equivalence leads us to treat gravitational forces as spacetime curvature, and since we know that gravitational forces arise due to the presence of matter, we are lead to consider the energy-momentum tensor $\Theta_{\mu\nu}$ as the source of curvature. Conservation of energy requires that the covariant divergence of the energy-momentum tensor vanish:

$$\nabla_\mu \Theta^{\mu\nu} = 0 \,. \tag{23.36}$$

We have seen that the affine connection plays the role of the gravitational field (Equation (23.35)), and hence the metric plays the role of the gravitational potential. Therefore we need a second order tensor differential equation for the metric which is inhomogeneous in the energy-momentum tensor. Furthermore, since the covariant divergence of the energy-momentum tensor vanishes, the tensor containing second order derivatives of the metric must also have vanishing covariant divergence. There are only two such tensors, one constructed from the Ricci tensor and the Ricci scalar (see Equation (23.33)), and the other the metric tensor itself (see Equation (23.26)). Therefore, the field equations take the form:

$$R_{\mu\nu} - \frac{1}{2} g_{\mu\nu} R + \lambda g_{\mu\nu} = -8\pi G \Theta_{\mu\nu} \,. \tag{23.37}$$

The proportionality constant for the source term has been chosen to produce the correct non-relativistic limit. The constant in the third term on the left hand side, the so called "cosmological constant," must be rather small to agree with Newtonian gravitation in the non-relativistic limit.

23.3.2 Gravitational Action

The gravitational field equations may be derived from an action principle. In general an action $\mathcal{S}$ will be written as:

$$\mathcal{S} = \int d^4x \sqrt{g} \mathfrak{L} \,. \tag{23.38}$$

Here we have defined $\mathfrak{L}$ as the Lagrangian for the system, $g \equiv -\det(g_{\mu\nu})$, and $d^4x\sqrt{g}$ is the invariant volume element. In principle, the Lagrangian may be chosen to be any real functional with mass dimension four which is invariant under all of the transformations considered for the physical theory. This insures that the action $\mathcal{S}$ is a real dimensionless scalar functional of the fields. The Lagrangian will in general be composed of several terms, each one corresponding to different fields. We may generically group these into two categories, the gravitational Lagrangian $\mathfrak{L}_{EH}$, and the matter Lagrangian $\mathfrak{L}_M$ (being composed of everything else). The field equations follow by demanding that the action be stationary under arbitrary infinitesimal variation of the fields:

$$\frac{\delta \mathcal{S}}{\delta \Xi_a(x)} = 0 \,, \tag{23.39}$$

where the $\{\Xi_a\}$ constitute a complete set of fields for the theory. Just which fields should be considered as independent is an interesting issue in gravity, but for the remainder of this lecture the metric will always be chosen to be an independent field.

We define the symmetric energy-momentum tensor to be given by:

$$\frac{\delta \mathcal{S}_M}{\delta g^{\mu\nu}(x)} \equiv \frac{1}{2}\Theta_{\mu\nu}(x) \,. \tag{23.40}$$

For example, for a scalar field we find:

$$\Theta_{\mu\nu} = \partial_\mu\phi\partial_\nu\phi - g_{\mu\nu}\mathfrak{L}_\phi = \nabla_\mu\phi\nabla_\nu\phi - g_{\mu\nu}\mathfrak{L}_\phi \,, \tag{23.41}$$

whereas for the Yang-Mills field we discover:

$$\Theta_{\mu\nu} = \frac{1}{Df^2} tr\{F_{\mu\rho}F^\rho{}_\nu + \frac{1}{4}g_{\mu\nu}F^{\rho\sigma}F_{\rho\sigma}\} \,, \tag{23.42}$$

and each expression has been written in covariant form.

The Einstein-Hilbert Lagrangian, augmented with the cosmological term, is given by:

$$\mathfrak{L}_{EH} = \frac{1}{16\pi G}(R - 2\lambda) \,, \tag{23.43}$$

where G is the Newtonian gravitational constant, λ is the cosmological constant, and R is the Ricci scalar curvature. Since the Riemann condition may be used to eliminate the affine connection in terms of the metric, we may do so and then only treat the metric as the dynamical variable. Then variation of the entire action (Einstein-Hilbert action plus matter action) with respect to the metric yields:

$$0 = \frac{1}{16\pi G}\left[R_{\mu\nu} - \frac{1}{2}g_{\mu\nu}R + \lambda g_{\mu\nu}\right] + \frac{1}{2}\Theta_{\mu\nu} \,, \tag{23.44}$$

which is clearly the Einstein equation.

Suppose, however, that we treat both the metric *and the affine connection* as independent dynamical variables, so we do not assume that the Riemann condition holds. This is known as the Palatini formulation. Then we find that stationarity of the action with respect to variation of the metric still leads to the Einstein equation. However, stationarity of the action with respect to variation of the affine connection leads to an additional field equation, and we find after a bit of algebra:

$$\Gamma^{\mu}{}_{\rho\sigma} = \frac{1}{2} g^{\mu\nu} (\partial_\rho g_{\sigma\nu} + \partial_\sigma g_{\rho\nu} - \partial_\nu g_{\rho\sigma}) \,. \tag{23.45}$$

We immediately observe that this is the Christoffel form of the affine connection. Recall that this follows from the Riemann condition in the case of vanishing torsion. This is one of the most striking and remarkable features of the Einstein-Hilbert action. We do not have to assume that the spacetime is Riemannian; treating the affine connection as an independent dynamical variable yields the Riemann condition as a field equation.

To summarize, in the original formulation of General Relativity the affine connection is assumed to be symmetric (zero torsion). The Einstein equation (Equation (23.44)) may be "written down" by asking for a tensor differential equation second order in the metric with the energy-momentum tensor as source term which yields Newton's Law of Universal Gravitation in the nonrelativistic limit, where in addition the Riemann condition (Equation (23.26)) is assumed to be valid. The Einstein equation may also be derived from the Einstein-Hilbert action. If we assume that the Riemann condition is satisfied and treat only the metric as the dynamical variable we then recover the Einstein equation; or we may treat both the metric and the (symmetric) affine connection as independent variables and so recover both the Einstein equation and the Riemann condition. In either case we obtain identical results. As we shall see, this happy state of affairs does not persist when we introduce spinor fields into the formalism.

Bibliography

[1] R. Adler, M. Bazin, and M. Schiffer, *Introduction to General Relativity*, McGraw-Hill, 1965.

[2] F.W. Hehl, P. van der Heyde, G.D. Kerlick, and J.M. Nester, "General Relativity with spin and torsion: Foundations and prospects" *Rev. Mod. Phys.*, **48**, 1976, 393-416.

[3] W. Pauli, *Theory of Relativity*, Pergamon, 1958.

[4] P.J.E. Peebles, *Principles of Physical Cosmology*, Princeton University Press, 1993.

[5] E. Schrodinger, *Space-Time Structure*, Cambridge University Press, 1950.

[6] H. Stephani, *General Relativity*, Cambridge University Press, 1982.

[7] J. Stewart, *Advanced General Relativity*, Cambridge University Press, 1991.

[8] R.M. Wald, *General Relativity*, University of Chicago Press, 1984.

[9] S. Weinberg, *Gravitation and Cosmology: Principles and Applications of the General Theory of Relativity*, Wiley, 1972.

[10] H. Weyl, *Space Time Matter*, Dover, 1952.

[11] C.M. Will, *Theory and Experiment in Gravitational Physics*, Revised Edition, Cambridge University Press, 1993.

Chapter 24

Spinors in General Relativity

James P. Crawford
Department of Physics, Penn State - Fayette,
Uniontown, PA 15401 USA

24.1 Introduction

This lecture will focus on the vierbein formalism for incorporating spinor fields in General Relativity. The vierbein formalism is fundamentally related to the notion of local Lorentz invariance in the tangent space, and may be considered regardless of the presence of spinor fields. The field equations are derived from the action and compared to those found in the standard formalism. Finally, the Clifford algebra and spinor fields are embodied in the formalism, the field equations are obtained from an action principle and compared to the previous equations. Unlike the previous cases, when spinor matter is incorporated there is an ambiguity in the resulting field equations, depending on which fields are chosen to be independent dynamical variables. At present, there are no data to guide us to the correct version.

Recall from the previous discussion of the Principle of Equivalence that we may find a sequence of local inertial frames along the trajectory of a particle. To find a sequence of local inertial frames requires spacetime dependent Lorentz transformations from one frame to the next, and therefore we expect that General Relativity may be formulated in terms of a gauge theory based on local Lorentz invariance. The vierbeins may be thought to constitute the transformation matrix between the coordinate basis vectors and a set of orthonormal basis vectors in the tangent space (a "Lorentz basis"), and as such they are often referred to as "soldering forms." Since the orthonormal basis of the tangent space may be chosen locally, we are allowed to perform local Lorentz transformations on this basis, and so we must introduce a gauge field associated with this invariance. This Lorentz gauge field is also variously known as the spin connection and the Fock-Ivanenko coefficients. The field strength tensor associated with this gauge field is found and shown to be equivalent to the Riemann curvature tensor. The Einstein-Hilbert action is written in terms of the new variables (the vierbeins and the spin connection), and the concomitant field equations are derived. If we assume that the torsion vanishes, then only the vierbein is considered to be the independent dynamical variable, and we recover the Einstein equation as its associated field equation. Alternately, if we do not assume that the torsion vanishes, then we must treat both the vierbein and the spin connection as independent dynamical variables, in which case we again recover the Einstein equation as the field equation of the vierbein, in addition to vanishing torsion as the field equation of the spin connection. Thus we obtain identical results in either case.

Finally we introduce the spinor fields into the formalism. Basically, we may think of the Clifford algebra generators as the objects which connect the spinor space to the tangent space, and as such they may also be considered to be "soldering forms." We write down the Dirac equation, substituting the covariant derivative for the ordinary derivative and inserting vierbeins where necessary, to find the equation for spinor fields in the presence of gravitational forces. We also derive the field equations from an action, and find that the generalized Dirac equation appears to differ from that obtained from naive substitution due to the possible presence of nonvanishing trace of the torsion tensor. If we assume that the torsion vanishes, then we recover the Einstein equation as the field equation for the vierbein, and the Dirac equation is in naive form. If we do not assume that the torsion vanishes, then we find that the torsion tensor is proportional to the pseudovector density of the spinor field, but the trace of the torsion tensor does indeed vanish, so we again recover the naive Dirac equation. We also obtain "almost" the Einstein equation, the only deviation appearing inside spinor matter where the torsion is nonzero. There are yet no experimental data to allow us to distinguish between these two cases.

24.2 Local Lorentz Invariance

24.2.1 Vierbeins

In the preceding lecture we have taken the basis vectors for the tangent space to be those which are tangent to the coordinate lines. However, there is nothing to prevent us from choosing a different basis. In fact, to most expeditiously introduce spinor fields and the Clifford algebra generators into the formalism, it is convenient to choose an *orthonormal* basis for the tangent space: $\mathbf{E}_\alpha$. We will refer to this basis as the Lorentz basis, and the orthonormality statement is:

$$\mathbf{E}_\alpha \cdot \mathbf{E}_\beta = \eta_{\alpha\beta} \quad , \quad \eta_{\alpha\beta} = diag(1,-1,-1,-1) \ , \tag{24.1}$$

where $\eta_{\alpha\beta}$ is the Minkowski metric. Clearly, these vectors may be written in terms of the coordinate basis vectors, so we may write:

$$\mathbf{E}_\alpha = e_\alpha{}^\mu \mathbf{E}_\mu \quad \Leftrightarrow \quad e_\alpha{}^\mu = \mathbf{E}_\alpha \cdot \mathbf{E}^\mu \tag{24.2}$$

where the $e_\alpha{}^\mu$ are called the vierbeins, and they constitute the transformation matrix between the Lorentz and coordinate bases. The inverse transformation $e^\alpha{}_\mu$ may be obtained by raising and lowering indices with the appropriate metric: $e^\alpha{}_\mu =$

$\eta^{\alpha\beta} e_\beta{}^\nu g_{\nu\mu}$, and we now find:

$$e^\alpha{}_\mu e^\beta{}_\nu \eta_{\alpha\beta} = g_{\mu\nu} \quad , \quad e_\alpha{}^\mu e_\beta{}^\nu g_{\mu\nu} = \eta_{\alpha\beta} \ . \tag{24.3}$$

The index convention will be early Greek for Lorentz basis and late Greek for coordinate basis.

Vectors may be written in terms of the Lorentz basis:

$$\mathbf{V} = V^\alpha \mathbf{E}_\alpha \quad \Rightarrow \quad V^\alpha = e^\alpha{}_\mu V^\mu \ , \ V^\mu = V^\alpha e_\alpha{}^\mu \ , \tag{24.4}$$

so the vierbeins may be used to toggle the character of an index between Lorentz and coordinate bases. This clearly generalizes to the case of tensors.

24.2.2 Local Lorentz Invariance

Since we are free to locally assign the orientation of the Lorentz basis vectors we may also perform local Lorentz transformations on these bases (this simply generates a different choice of basis vectors, and is physically equivalent to changing to a different local inertial frame of reference):

$$\mathbf{E}'_\alpha(x) = \Lambda_\alpha{}^\beta(x) \mathbf{E}_\beta(x) \ . \tag{24.5}$$

Demanding that the new basis also be orthonormal recovers the defining relation for the Lorentz group. The components of vectors and tensors (and spinors) will also transform under the action of these Lorentz transformations. We write the general transformation as:

$$\varphi'^a = [S(\lambda)]^a{}_b \varphi^b \quad \Rightarrow \quad V'^\alpha = \Lambda^\alpha{}_\beta V^\beta \tag{24.6}$$

where $[S(\lambda)]^a{}_b$ is the transformation matrix for the particular representation of the Lorentz group under consideration, and the specific case of contravariant vector components is given for convenience. More explicitly, the Lorentz transformation matrix may be written as:

$$S(\lambda) = e^{\frac{i}{2}\lambda^{\alpha\beta}\sigma_{\alpha\beta}} \quad , \quad [\sigma_{\alpha\beta}, \sigma_{\gamma\delta}] = -i(\eta_{\alpha\gamma}\sigma_{\beta\delta} - \eta_{\alpha\delta}\sigma_{\beta\gamma} - \eta_{\beta\gamma}\sigma_{\alpha\delta} + \eta_{\beta\delta}\sigma_{\alpha\gamma}) \tag{24.7}$$

where $\sigma_{\alpha\beta}$ are the matrix generators of the Lorentz group and $\lambda^{\alpha\beta}$ are the real group parameters.

24.2.3 Covariant Derivative and Spin Connection

Now since the Lorentz basis vectors may be locally assigned we must introduce new connection coefficients $\omega^{\alpha\beta}{}_\mu$ to construct the covariant derivative:

$$\nabla_\mu \varphi \equiv (\partial_\mu - \frac{i}{2}\omega^{\alpha\beta}{}_\mu \sigma_{\alpha\beta})\varphi \quad \Rightarrow \quad \nabla_\mu V^\alpha = \partial_\mu V^\alpha + \omega^\alpha{}_{\beta\mu} V^\beta, \tag{24.8}$$

and the specific case of the covariant derivative of the contravariant components of a vector is given for convenience. We may also view this connection as the set of gauge fields of the Lorentz group, and they are variously referred to as the *Fock-Ivanenko coefficients* and the *spin connection.*

Under a local Lorentz transformation the covariant derivative itself should transform as a scalar (since it does not carry a Lorentz index), from which we may obtain the transformation property of the spin connection:

$$\nabla'_\mu V'^\alpha = \Lambda^\alpha{}_\beta \nabla_\mu V^\beta \quad \Rightarrow \quad \omega'^{\alpha\beta}{}_\mu = \Lambda^\alpha{}_\gamma \Lambda^\beta{}_\delta \omega^{\gamma\delta}{}_\mu - (\partial_\mu \Lambda^\alpha{}_\gamma)\Lambda^{\beta\gamma} \,, \tag{24.9}$$

where the inhomogeneous term indicates that the spin connection is not a tensor under local Lorentz transformations, and indeed it transforms like a gauge field as expected. It is, however, a vector under general coordinate transformations and this insures that the covariant derivative itself transforms as the covariant components of a vector.

Since the basis vectors change from point to point the derivative of a vector will include this contribution:

$$\partial_\mu \mathbf{V} = (\partial_\mu V^\alpha)\mathbf{E}_\alpha + V^\alpha(\partial_\mu \mathbf{E}_\alpha) \equiv (\nabla_\mu V^\alpha)\mathbf{E}_\alpha \;\Rightarrow\; \partial_\mu \mathbf{E}_\alpha \equiv \omega^\beta{}_{\alpha\mu}\mathbf{E}_\beta \,. \tag{24.10}$$

In fact, we could have taken this equation as the definition of the spin connection and then found Equation (24.8) as a consequence.

We may obtain the equivalent of the Riemann condition for the vierbein as follows:

$$\partial_\mu e_{\alpha\nu} = \partial_\mu(\mathbf{E}_\alpha \cdot \mathbf{E}_\nu) = \omega^\beta{}_{\alpha\mu} e_{\beta\nu} + \Gamma^\rho{}_{\nu\mu} e_{\alpha\rho} \tag{24.11}$$

$$\Rightarrow \quad \nabla_\mu e_{\alpha\nu} = \partial_\mu e_{\alpha\nu} - \omega^\beta{}_{\alpha\mu} e_{\beta\nu} - \Gamma^\rho{}_{\nu\mu} e_{\alpha\rho} = 0 \,. \tag{24.12}$$

This expression (a total of 64 equations) allows us to explicitly solve for the affine connection (both the symmetric and antisymmetric parts, encompassing a total of 64 functions) in terms of the vierbein and the spin connection:

$$\Gamma^\sigma{}_{\mu\nu} = e^{\alpha\sigma}(\partial_\mu e_{\alpha\nu} - \omega^\beta{}_{\alpha\mu} e_{\beta\nu}) \tag{24.13}$$

$$\Rightarrow \quad T^\sigma{}_{\mu\nu} = e^{\alpha\sigma}(\partial_\mu e_{\alpha\nu} - \partial_\nu e_{\alpha\mu} - \omega^\beta{}_{\alpha\mu} e_{\beta\nu} + \omega^\beta{}_{\alpha\nu} e_{\beta\mu}) \,, \tag{24.14}$$

where we have explicitly solved for the torsion. In this case we treat the vierbein and the spin connection as the independent dynamical variables of the theory, with the metric recovered via Equation (24.3) and the affine connection via Equation (24.13). If we also demand that the torsion vanish (a total of 24 equations) then we may explicitly solve for the spin connection (encompassing 24 functions) in terms of the vierbeins:

$$T^\sigma_{\mu\nu} = 0 \quad \Rightarrow \quad \omega^{\alpha\beta}_{\ \ \mu} = \frac{1}{2} e_{\gamma\mu}(C^{\alpha\beta\gamma} + C^{\beta\gamma\alpha} - C^{\gamma\alpha\beta}) \,, \tag{24.15}$$

$$C^{\alpha\beta\gamma} \equiv e^{\beta\rho} e^{\gamma\sigma}(\partial_\rho e^\alpha{}_\sigma - \partial_\sigma e^\alpha{}_\rho) \,. \tag{24.16}$$

In this case only the vierbeins may be treated as dynamical variables.

24.2.4 Lorentz Field Strength Tensor

The field strength tensor $\Omega^{\alpha}{}_{\beta\mu\nu}$ for the Lorentz gauge field (the "spin curvature") may be found by considering the effect of parallel transport around a closed loop on an arbitrary vector (with components written in Lorentz basis). Alternately, it is equivalent and computationally much easier to consider:

$$(\nabla_\mu \nabla_\nu - \nabla_\nu \nabla_\mu) V^\alpha \equiv -\Omega^{\alpha}{}_{\beta\mu\nu} V^\beta + T^{\sigma}{}_{\mu\nu} (\nabla_\sigma V^\alpha) \tag{24.17}$$

$$\Rightarrow \quad \Omega^{\alpha}{}_{\beta\mu\nu} = -\partial_\mu \omega^{\alpha}{}_{\beta\nu} + \partial_\nu \omega^{\alpha}{}_{\beta\mu} - \omega^{\alpha}{}_{\gamma\mu} \omega^{\gamma}{}_{\beta\nu} + \omega^{\alpha}{}_{\gamma\nu} \omega^{\gamma}{}_{\beta\mu} \,. \tag{24.18}$$

An important relation between the Riemann curvature tensor and the Lorentz field strength tensor may be found as follows:

$$(\nabla_\mu \nabla_\nu - \nabla_\nu \nabla_\mu) e^{\alpha}{}_{\sigma} = e^{\alpha}{}_{\rho} R^{\rho}{}_{\sigma\mu\nu} - \Omega^{\alpha}{}_{\beta\mu\nu} e^{\beta}{}_{\sigma} + T^{\rho}{}_{\mu\nu} (\nabla_\rho e^{\alpha}{}_{\sigma}) \tag{24.19}$$

$$\Rightarrow R^{\alpha}{}_{\sigma\mu\nu} = \Omega^{\alpha}{}_{\sigma\mu\nu} \,, \tag{24.20}$$

where the first equation is a formal result for any tensor with one contravariant Lorentz index and one covariant coordinate index, and we have used the fact that the covariant derivative of the vierbein vanishes to obtain the second equation. This is a crucial point of this formalism, that the Lorentz field tensor is equivalent to the Riemann tensor, for so it appears that General Relativity is a gauge theory based on the Lorentz group.

24.2.5 Action and Field Equations

We are now treating the vierbein as the fundamental dynamical variable of gravitation. Furthermore, if we do not assume that the torsion vanishes, then we must also treat the spin connection as a dynamical variable. We will consider both cases. In either case, the metric and the affine connection are defined by Equations (24.3) and (24.13) respectively. So we will need to write the Lagrangian in terms of the vierbein and the spin connection. If we assume that the torsion vanishes, then the spin connection will be defined by Equations (24.15) and (24.16), and the Lagrangian will be written only in terms of the vierbein. So we write the Einstein-Hilbert Lagrangian as follows:

$$\mathfrak{L}_{EH} = \frac{1}{16\pi G} (e_{\alpha}{}^{\mu} e_{\beta}{}^{\nu} \Omega^{\alpha\beta}{}_{\mu\nu} - 2\lambda) \,, \tag{24.21}$$

where the spin curvature tensor is given in Equation (24.18).

For the case of nonvanishing torsion, the field equation arising from variation of the action with respect to the vierbein is:

$$0 = \frac{2}{16\pi G} \left[R^{\alpha}{}_{\mu} - \frac{1}{2} e^{\alpha}{}_{\mu} R + e^{\alpha}{}_{\mu} \lambda \right] + \Theta^{\alpha}{}_{\mu} \,, \tag{24.22}$$

and the field equation arising from variation with respect to the spin connection is:

$$0 = \left[T^{\mu}{}_{\alpha\beta} - (e_{\alpha}{}^{\mu} T_\beta - e_{\beta}{}^{\mu} T_\alpha) \right] \,. \tag{24.23}$$

We immediately recognize Equation (24.22) as the Einstein equation, and from Equation (24.23) we easily find that the torsion vanishes: $T^{\mu}{}_{\alpha\beta} = 0$. So if we do not assume that the torsion vanishes, we must treat the spin connection as an independent dynamical variable, and the associated field equation yields vanishing torsion as a consequence.

Finally consider the case of vanishing torsion, so that we may use Equations (24.15) and (24.16) to eliminate the spin connection in terms of the vierbein, and then treat only the vierbein as the independent dynamical variable. Variation of the action with respect to the vierbein now once again yields Equation (24.22), the Einstein equation. This result should perhaps not be surprising since vanishing torsion was found as a field equation in the previous case.

In summary, we may erect local orthonormal bases in the tangent space, and the vierbeins then constitute the transformation matrix between the coordinate bases and the orthonormal bases. Since we have the freedom to locally choose these orthonormal bases we may also perform local Lorentz transformations. Demanding invariance of the theory under the local Lorentz transformations then forces the introduction of the associated gauge field, commonly known as the spin connection. The Riemann condition then appears as vanishing covariant derivative of the vierbeins, and this allows the affine connection to be written in terms of the vierbeins and the spin connection, as well as identifying the Riemann curvature tensor with the gauge field strength tensor. Finally, we may treat both the vierbeins and the spin connection as independent dynamical variables, in which case we recover the Einstein equation and vanishing torsion as the field equations. Alternately, we may assume that the torsion vanishes and so treat only the vierbeins as independent dynamical variables, in which case we again recover the Einstein equation. However, when we consider spinor fields, this symmetry no longer persists.

24.3 Spinors in Genral Relativity

24.3.1 Dirac Equation & The Clifford Algebra

At last we come to the incorporation of spinor fields and the Clifford algebra into the formalism of General Relativity. It is the physical existence of massive spin one-half particles (leptons and quarks) which requires their introduction. In particular, in flat spacetime and in the absence of other (i.e. non-gravitational) forces, a massive spinor field satisfies the free Dirac equation:

$$(i\gamma^{\alpha}\partial_{\alpha} - m)\Psi = 0 \quad , \quad \{\gamma^{\alpha}, \gamma^{\beta}\} = 2\eta^{\alpha\beta}\mathbf{1} \, , \tag{24.24}$$

where the quantities γ^{α} satisfy the defining relation for the Clifford algebra generators, as indicated, and where the $\eta^{\alpha\beta}$ constitute the contravariant components of the Minkowski metric.

So the incorporation of the Dirac spinor fields poses two separate problems – the generalization of the ordinary derivative to the covariant derivative of the spinor field, and the generalization of the defining relation for the Clifford algebra generators to the case of curved spacetime:

$$\{\gamma^\mu(x), \gamma^\nu(x)\} = 2g^{\mu\nu}(x)\mathbf{1} . \tag{24.25}$$

Both are easily accomplished with the use of the local Lorentz invariance formalism.

The covariant derivative of the spinor field may already be found as a special case of Equation (24.8), where the generators of Lorentz transformations for the spinor field are given by the elements of the bivector basis of the Clifford algebra: $\sigma_{\alpha\beta} = \frac{1}{2}\gamma_{\alpha\beta} = \frac{i}{4}[\gamma_\alpha, \gamma_\beta]$. Explicitly, the covariant derivative of the spinor field is then written as:

$$\nabla_\mu \Psi = (\partial_\mu - \frac{i}{2}\omega^{\alpha\beta}{}_\mu \sigma_{\alpha\beta})\Psi = (\partial_\mu - \frac{i}{4}\omega^{\alpha\beta}{}_\mu \gamma_{\alpha\beta})\Psi . \tag{24.26}$$

Now note that the Clifford algebra bivectors are Dirac self-conjugate: $\overline{\sigma}_{\alpha\beta} = \sigma_{\alpha\beta}$, and hence the inverse Lorentz transformation is given by the Dirac conjugate: $S^{-1} = \overline{S}$. Therefore, under a Lorentz transformation the Clifford algebra generators are *invariant*:

$$\gamma^{\alpha\prime} = \Lambda^\alpha{}_\beta S \gamma^\beta \overline{S} = \gamma^\alpha , \tag{24.27}$$

where the Lorentz transformation induced by the spinor indices of the Clifford algebra generators is exactly canceled by the Lorentz transformation induced by the Lorentz index. So that even under a *local* Lorentz transformation the Clifford algebra generators do not acquire spacetime dependence; hence they may be *chosen* to be *constant matrices*. Another way to see this is to note that the covariant derivative of the Clifford algebra generator is just the ordinary derivative:

$$\nabla_\mu \gamma^\alpha = \partial_\mu \gamma^\alpha + \omega^\alpha{}_{\beta\mu}\gamma^\beta - \frac{i}{2}\omega^{\alpha\beta}{}_\mu[\sigma_{\alpha\beta}, \gamma^\alpha] = \partial_\mu \gamma^\alpha = 0 , \tag{24.28}$$

where the terms involving the spin connection all cancel, and the last equality follows from the assumption that γ^α are given by constant matrices. Consequently, the Clifford algebra generators written in Lorentz basis need not be considered to be independent dynamical variables, and then *all* of the spacetime dependence of the Clifford algebra generators written in coordinate basis resides in the vierbeins:

$$\gamma^\mu(x) = e_\alpha{}^\mu(x)\gamma^\alpha \quad \Rightarrow \quad \nabla_\mu \gamma_\nu(x) = 0. \tag{24.29}$$

We have also written the Riemann condition in terms of the Clifford algebra generators.

We may now write the Dirac equation in generally covariant form:

$$[i\gamma^\mu \nabla_\mu - m]\,\Psi = \left[ie_\delta{}^\mu \gamma^\delta \left(\partial_\mu - \frac{i}{4}\omega^{\alpha\beta}{}_\mu \gamma_{\alpha\beta}\right) - m\right]\Psi = 0 , \tag{24.30}$$

where we have explicitly indicated the dependence on the vierbein and the spin connection. Note that the order in which the Clifford algebra generator and the covariant derivative appear is irrelevant due to the Riemann condition (written as in Equation (24.29)), so there is no ambiguity in this regard. However, the explicit appearance of the spin connection in this equation has important consequences for the action formulation, as we shall now see.

24.3.2 Lagrangian and Field Equations

Dirac Lagrangian and Field Equation

The Lagrangian for a free Dirac particle in flat spacetime may be written in a variety of equivalent ways. We will choose the manifestly Hermitian form and state without proof that identical results follow with use of alternate forms. To obtain the generally covariant Lagrangian we substitute the covariant derivative for the ordinary derivative and obtain:

$$\mathfrak{L}_D = \overline{\Psi}\left[\frac{i}{2}(\gamma^\mu \overrightarrow{\partial}_\mu - \overleftarrow{\partial}_\mu \gamma^\mu) + \frac{1}{8}\omega^{\alpha\beta}{}_\mu \{\gamma^\mu, \gamma_{\alpha\beta}\} - m\right]\Psi\,. \tag{24.31}$$

Note that this Lagrangian contains the vierbein through $\gamma^\mu = e_\alpha{}^\mu \gamma^\alpha$, as well as the spin connection. We may now obtain the field equation by demanding stationarity of the action with respect to variations of the Dirac field, and we find:

$$0 = \left[i\gamma^\mu(\nabla_\mu + \frac{1}{2}T_\mu) - m\right]\Psi\,. \tag{24.32}$$

We now see the first surprise; this field equation is not that obtained by substitution of the covariant derivative for the ordinary derivative as in Equation (24.30). Of course, if the torsion vanishes (or if just the contracted torsion tensor vanishes) we then recover the "expected" result.

Dirac Energy-Momentum Tensor

Before proceeding to the derivation of the Einstein equation it is worthwhile considering the canonical energy-momentum tensor for a Dirac field. So we need the functional derivative of the Clifford algebra generator with respect to the metric. Clearly, the Clifford algebra generator (in coordinate basis, as it appears in the Dirac Lagrangian above) depends on the metric (since Equation (24.25) is satisfied). The required derivative is found to be:

$$\frac{\delta\gamma^\rho(y)}{\delta g^{\mu\nu}(x)} = \frac{1}{4}\left(\delta^\rho_\mu \gamma_\nu + \delta^\rho_\nu \gamma_\mu\right)\frac{\delta^4(x-y)}{\sqrt{g}}\,. \tag{24.33}$$

One may check this expression by calculating the functional derivative of Equation (24.25) with respect to the metric and obtain an identity. In fact, Equation (24.33) is obtained by asking for an expression which generates an identity from Equation (24.25) as indicated, and the displayed result is the unique solution. Note that there is an ambiguity in the computation of the energy-momentum tensor. In particular, if we assume that the torsion does not vanish, then the spin connection is an independent dynamical variable, and as such it does not depend on the metric; on the other hand, if we assume that the torsion does vanish, then the spin connection is given by Equations (24.15) and (24.16), in which case the metric implicitly appears in the spin connection through derivatives of the vierbeins, and the energy-momentum tensor gains an additional term. Therefore we find:

$$\begin{aligned}\Theta_{\mu\nu} &= -g_{\mu\nu}\mathfrak{L}_D + \frac{i}{4}\left[\overline{\Psi}\gamma_\mu(\nabla_\nu\Psi) + \overline{\Psi}\gamma_\nu(\nabla_\mu\Psi) - (\overline{\nabla_\mu\Psi})\gamma_\nu\Psi - (\overline{\nabla_\nu\Psi})\gamma_\mu\Psi\right] \\ &+\frac{1}{4}\int d^4y\left[\sqrt{g}\,\overline{\Psi}\{\gamma^\rho,\gamma_{\alpha\beta}\}\Psi\right](y)\frac{\delta\omega^{\alpha\beta}{}_\rho(y)}{\delta g^{\mu\nu}(x)}\,,\end{aligned} \tag{24.34}$$

where the last term vanishes if the spin connection is treated as independent. In fact, it turns out that this last term vanishes even if the spin connection is not treated as independent, and thus the canonical energy-momentum tensor for the Dirac field is just given by the first line of Equation (24.34).

Gravitational Field Equations

Nonvanishing Torsion We begin by considering the algebraically simpler case in which the spin connection is treated as an independent dynamical variable. Considering first the variation of the action with respect to the spin connection we find:

$$\frac{1}{16\pi G}\left[T^\mu{}_{\alpha\beta} - \left(e_\alpha{}^\mu T_\beta - e_\beta{}^\mu T_\alpha\right)\right] - \frac{1}{4}\varepsilon_{\alpha\beta\gamma\delta}e^{\gamma\mu}\overline{\Psi}\tilde{\gamma}^\delta\Psi = 0\,, \tag{24.35}$$

from which we easily obtain:

$$T_\alpha = 0 \quad , \quad T^\mu{}_{\alpha\beta} = 4\pi G\varepsilon_{\alpha\beta\gamma\delta}e^{\gamma\mu}\overline{\Psi}\tilde{\gamma}^\delta\Psi\,. \tag{24.36}$$

Now we see that although the torsion does not vanish, the contracted torsion tensor does, so that we recover the "expected" form of the Dirac equation (compare Equations (24.32) and (24.30)). Also notice that since derivatives of the torsion do not appear in Equation (24.36), the torsion does not propagate – it is only non-zero inside spinor matter.

Now demanding stationarity of the action with respect to variations of the vierbein yields:

$$\begin{aligned}0 = \frac{1}{8\pi G}\left[R^\alpha{}_\mu - \frac{1}{2}e^\alpha{}_\mu(R-2\lambda)\right] + \Theta^\alpha{}_\mu \\ -e^\alpha{}_\mu\mathfrak{L}_\Psi + \frac{i}{2}\overline{\Psi}\left[\gamma^\alpha\overrightarrow{\nabla}_\mu - \overleftarrow{\nabla}_\mu\gamma^\alpha\right]\Psi\,,\end{aligned} \tag{24.37}$$

where here $\Theta^\alpha{}_\mu$ is the energy-momentum tensor for the other (non-spinor) fields in the theory. Now contracting Equation (24.37) with $e_{\alpha\nu}$ we find that this expression is not symmetric, as it was for the previous cases considered, so we examine separately the symmetric and antisymmetric pieces:

$$\frac{1}{2}\left[R_{\mu\nu} + R_{\nu\mu} - g_{\mu\nu}(R-2\lambda)\right] = -8\pi G\Theta^{Total}_{\mu\nu}\,, \tag{24.38}$$

$$R_{\mu\nu} - R_{\nu\mu} = -i4\pi G\overline{\Psi}\left[\gamma_\mu\overrightarrow{\nabla}_\nu - \gamma_\nu\overrightarrow{\nabla}_\mu - \overleftarrow{\nabla}_\nu\gamma_\mu + \overleftarrow{\nabla}_\mu\gamma_\nu\right]\Psi\,. \tag{24.39}$$

We recognize the first expression as "almost" the Einstein equation; it is exactly the Einstein equation if the torsion vanishes since then the Ricci tensor is symmetric. In fact, the torsion does not vanish as we have just seen in Equation (24.36), but it is only nonvanishing inside spinor matter and hence the vacuum Einstein equation remains the same. It is the spherically symmetric solution of the vacuum Einstein equation (the Schwarzschild solution) which has been most extensively tested, so Equation (24.38) does not appear to be in conflict with experiment. Finally note that Equation (24.39) is not an independent relation, for it may be obtained from the contracted Bianchi identity, the equation for the spinor field (24.32), and the equation for the torsion (24.36). This is a nontrivial consistency check on these equations.

Vanishing Torsion We now complete this section with the vanishing torsion case. Here we may use Equations (24.15) and (24.16) to eliminate the spin connection in terms of the vierbein, and then treat only the vierbein as an independent variable. Variation of the entire action with respect to the vierbein now yields:

$$0 = \frac{1}{8\pi G}\left[R^{\alpha}{}_{\mu} - \frac{1}{2}e^{\alpha}{}_{\mu}(R - 2\lambda)\right] + \Theta^{\alpha}{}_{\mu} - e^{\alpha}{}_{\mu}\mathfrak{L}_{\Psi} + \frac{i}{2}\overline{\Psi}\left[\gamma^{\alpha}\overrightarrow{\nabla}_{\mu} - \overleftarrow{\nabla}_{\mu}\gamma^{\alpha}\right]\Psi + \frac{1}{4}\varepsilon^{\rho\alpha}{}_{\mu\beta}\nabla_{\rho}\left[\overline{\Psi}\tilde{\gamma}^{\beta}\Psi\right], \tag{24.40}$$

and we see that we recover Equation (24.37) but with an additional term. Now contracting with $e_{\alpha\nu}$ we again find that the expression is not symmetric, so we examine the symmetric and antisymmetric pieces separately:

$$\left[R_{\mu\nu} - \frac{1}{2}g_{\mu\nu}(R - 2\lambda)\right] = -8\pi G\Theta^{Total}_{\mu\nu}, \tag{24.41}$$

$$0 = \frac{i}{4}\overline{\Psi}\left[\gamma_{\mu}\overrightarrow{\nabla}_{\nu} - \gamma_{\nu}\overrightarrow{\nabla}_{\mu} - \overleftarrow{\nabla}_{\nu}\gamma_{\mu} + \overleftarrow{\nabla}_{\mu}\gamma_{\nu}\right]\Psi + \frac{1}{4}\varepsilon_{\mu\nu}{}^{\rho\beta}\nabla_{\rho}\left[\overline{\Psi}\tilde{\gamma}_{\beta}\Psi\right]. \tag{24.42}$$

So in this case we exactly recover the Einstein equation. Furthermore, the second equation is not an independent relation, for it follows from the Dirac equation after a bit of algebra. So unlike the previous situations we find that when spinors are included in the formalism the choice of independent variables does indeed affect the resulting field equations, although as yet there have been no experiments conducted that are capable of distinguishing between these two possibilities. Note that at the classical level there does not seem to be any theoretical justification for choosing between these cases. It remains to be seen whether the quantum version of General Relativity will address this issue, but it is reasonable to suppose that it will. In particular, it is possible that there is a *unique* self-consistent quantized version of gravity, both renormalizable and free of anomalies.

24.4 Summary and Conclusions

We have seen that the General Theory of Relativity may be conceptualized as a theory based on very simple yet profound symmetries (general coordinate covariance and local Lorentz invariance). The manner in which this occurs is quite similar but not identical to gauge theories. In particular, the local symmetry principle dictates the form of the covariant derivative. This allows the behavior of any particular field to be determined from knowledge of the appropriate connection. However, the symmetry principles do not dictate the dynamics of the connections themselves. For example, in any given spacetime of arbitrary geometry it is relatively straightforward to write down the equations for a Yang-Mills field (say), but these equations do not determine the effect of these fields on the spacetime itself. It was Einstein's brilliant and elegant observation embodied in the Principle of Equivalence which guides us to a reasonable guess. Indeed, the Einstein equation has been verified locally (in the Pound-Rebka experiment, verifying the gravitational redshift), as well as on the scale of the solar system (the perihelion shift of Newton/Kepler orbits, and the bending of light around the sun), and on the cosmic scale (gravitational lensing of distant quasars and galaxies, and the orbital spin-up of binary pulsars).

Having said all of this, it is important to keep in mind that gravity is the weakest force, and as such it is the least accurately measured of all of the fundamental forces. In fact, in all of the experimental tests performed to date, the spacetime curvature has been extremely small compared to the Planck scale: $1/l_P \sim 10^{33}/cm$. Furthermore, the problems of quantization of the gravitational field are legion, not the least of which is the lack of a deep understanding of wildly fluctuating geometry and topology at very small scales. A slightly more prosaic problem is the nonrenormalizability of the perturbative expansion of the theory when matter couplings are included. In the spinor Lagrangian, for example, the matter/graviton interaction is a derivative coupling between the spinor and the vierbein. Once again, the problems of nonrenormalizability appear at very small length scales (very high momentum transfer). It is not clear, therefore, whether the Einstein equation is truly fundamental, or an approximation that is only valid for relatively small spacetime curvature.

It is ironic that the fundamental force that was first described theoretically approximately 300 years ago is currently the least understood; whereas the other three fundamental forces, two of which were only discovered this century, have been remarkably well described in the context of gauge theory, including the classical as well as the quantum regimes. Although General Relativity clearly has many features in common with standard gauge theories, the formulations are not identical. In particular, the Einstein-Hilbert action is linear in the curvature, whereas the action of standard gauge theory is quadratic. An attempt to find a generalization of Einsteinian gravity in terms of standard gauge theory will be presented in the following lectures of this series.

Bibliography

[1] M. Carmeli, *Group Theory and General Relativity*, McGraw-Hill, 1977.

[2] J.P. Crawford, *J. Math. Phys.*, **32**, 1991, 576-583.

[3] V. Fock and D. Ivanenko, *C. R. Acad. Sci.*, **188**, 1929, 1470.

[4] F.W. Hehl, P. van der Heyde, G.D. Kerlick, and J.M. Nester, *Rev. Mod. Phys.*, **48**, 1976, 393-416.

[5] T.W.B. Kibble, *J. Math. Phys.*, **2**, 1961, 212.

[6] P.J.E. Peebles, *Principles of Physical Cosmology*, Princeton University Press, 1993.

[7] P. Peldan, *Class. Quantum Grav.*, **11**, 1994, 1087-1133.

[8] E. Schrodinger, *Sitzungsh. Akad. f. Physik*, **57**, 1929, 261.

[9] H. Stephani, *General Relativity*, Cambridge University Press, 1982.

[10] R. Utiyama, *Phys. Rev.*, **101**, 1956, 1597.

[11] S. Weinberg, *Gravitation and Cosmology: Principles and Applications of the General Theory of Relativity*, Wiley, 1972.

[12] H. Weyl, *Zeitsch. f. Physik*, **56**, 1929, 330.

Chapter 25

Hypergravity I

James P. Crawford
Department of Physics, Penn State - Fayette,
Uniontown, PA 15401 USA

25.1 Introduction

This lecture will begin to explore a generalization of General Relativity which has a closer resemblance to gauge theory than standard gravity. The generalization is obtained by demanding an additional local symmetry of the theory. In particular, just as we have the freedom to locally choose any orthonormal basis in the tangent space, we should have the freedom to choose a local orthonormal basis of the spinor space. Changing to another spinor basis induces an automorphism transformation of the Clifford algebra generators; thus we are lead to a gauge theory based on local automorphism invariance. This requires the introduction of the *Clifford connection* (the automorphism gauge field), as well as the *drehbeins* (akin to the vierbeins) as additional dynamical variables.

A symmetry principle underlying General Relativity is local Lorentz invariance in the tangent space; this invariance is related to the Principle of Equivalence. In particular, it states that we have the freedom to locally choose the orientation of the orthonormal basis vectors of the tangent space. This corresponds physically to changing from one local inertial frame to another. Thus, General Relativity also contains the gauge theory based on the Lorentz group. In this formalism, the dynamical variables are the vierbeins $e^{\alpha}{}_{\mu}$ and (possibly) the spin connection $\omega^{\alpha\beta}{}_{\mu}$, instead of the metric and (possibly) the affine connection. The vierbeins constitute the transformation matrix between the coordinate basis and the orthonormal basis, and the spin connection is the gauge field associated with the local Lorentz invariance. The Riemann condition, in this case written as $\nabla_{\rho} e^{\alpha}{}_{\mu} = 0$, then determines that the field strength tensor for the Lorentz gauge fields is identical to the Riemann curvature tensor. Note, however, that although this formalism is in many ways very much like a typical gauge theory formulation, the action is only linear in the field strength tensor, and the coupling is not dimensionless.

To incorporate spinor fields into the formalism we now must consider at each point in spacetime not only the tangent space but the spinor space as well, for although all carrier spaces of the representations of the pseudo-orthogonal group $SO(1,3)$ may be obtained as direct products of the tangent space, not all carrier spaces of the representations of the covering group $SU(2,C)$ may be obtained this way. In fact, in the following development we will consider a reducible representation of $SU(2,C)$, the Dirac *bispinor*, to be the fundamental field, though we will immediately lapse into the terminology *spinor*. One of the fundamental ingredients in the Dirac formalism is the Clifford algebra, whose defining relation is given by the familiar expression:

$$\{\gamma_\alpha, \gamma_\beta\} = 2\eta_{\alpha\beta}\mathbf{1} \ , \tag{25.1}$$

where the γ_α are the generators of the algebra, $\mathbf{1}$ is the unit element (scalar) of the algebra, and $\eta_{\alpha\beta} = diag(1,-1,-1,-1)$ is the Minkowski metric.

Under a Lorentz transformation all representations of the group transform, so that a local Lorentz transformation of the basis of the tangent space is accompanied by an associated transformation of the basis of the spinor space. This has an important consequence for the Clifford algebra generators; in particular, the transformation induced by their tangent space index is exactly canceled by the transformation induced by their spinor space indices. In other words, the Clifford algebra generators are *invariant* under the action of the Lorentz group. This implies that any particular choice of matrix representation of the Clifford algebra generators may be a global choice – the matrices need not depend on the position in spacetime; stated differently, as we move from point to point in spacetime, the orientation of both the tangent space basis and the spinor space basis will change, but their relative orientation will not change. However, the Clifford algebra generators in the coordinate basis are spacetime dependent, but all of this spacetime dependence resides in the vierbeins:

$$\gamma_\mu(x) = e^\alpha{}_\mu(x)\gamma_\alpha \quad \Rightarrow \quad \{\gamma_\mu(x), \gamma_\nu(x)\} = 2g_{\mu\nu}(x)\mathbf{1} \ , \tag{25.2}$$

which is the equivalent to Equation (25.1) but written in coordinate space basis. Now it is easy to show that $\nabla_\mu\gamma_\nu = 0$, which is the Riemann condition written in terms of the Clifford algebra generators.

I propose that the freedom of local assignment of the basis vectors of the tangent space should hold *independently* for the spinor space as well, for surely nature cannot be concerned with how we may decide to erect a spinor basis. This is equivalent to allowing the matrix representation of the Clifford algebra generators to be locally assigned. Of course, the Dirac Lagrangian is invariant under global similarity transformations of the Clifford algebra generators. However, just as the Lorentz group preserves the Minkowski metric in the tangent space, so we demand that the metric in the spinor space be preserved as well. (Preserving the metric is equivalent to demanding that the transformations preserve the orthonormality of the basis, which we require for the spinor space as well as the tangent space.) This implies that the group of allowed transformations is not the full group of similarity transformations $Gl(4,C)$, but the automorphism subgroup $U(2,2)$. In other words, a spinor metric preserving transformation of the spinor basis induces an automorphism transformation of the Clifford algebra. Thus we are lead to consider a gauge theory based on the automorphism group of the Clifford algebra. Just as the demand of local Lorentz invariance requires the introduction of the spin connection and the vierbeins, the demand of local automorphism invariance of the Clifford algebra requires the introduction of the *Clifford connection* (the automorphism gauge field) and the *drehbeins* ("spin-legs"). This also implies that the Clifford algebra generators may not be chosen globally; they are spacetime dependent: $\gamma_\alpha(x)$. In particular, and most importantly, the Riemann condition may no longer be written in terms of the Clifford algebra generators; that is, the Clifford algebra generators no longer obey the condition of parallelizability: $\nabla_\mu \gamma_\nu \neq 0$. This statement perhaps most succinctly epitomizes this generalization of General Relativity.

This lecture discusses the basics of local automorphism invariance. It includes the introduction of the covariant derivative and the drehbein fields, as well as the field strength tensor and the associated Bianchi identities. The proposed Lagrangian and concomitant field equations, as well as the recovery of Einsteinian gravity, will be discussed in the next (final) lecture.

25.2 Automorphism Invariance

25.2.1 Automorphism Group

We may choose to span the spinor space with a complete set of uninormal basis spinors: $\mathbf{S}_A$. Thus any spinor $\boldsymbol{\Psi}$ has components Ψ^A which are defined via:

$$\boldsymbol{\Psi} = \Psi^A \mathbf{S}_A \, . \tag{25.3}$$

The uninormality condition may be written as:

$$\gamma(\mathbf{S}_A, \mathbf{S}_B) \equiv \mathbf{S}_A^* \cdot \mathbf{S}_B = \gamma_{AB} \, , \tag{25.4}$$

where we have defined the spinor metric γ in an invariant manner. The spinor metric has components $\gamma_{AB} = diag(1, 1, -1, -1)$. We will refer to any such basis as a Dirac basis. More explicitly, the inner product of any two spinors is given by:

$$\gamma(\boldsymbol{\Upsilon}, \boldsymbol{\Psi}) = \boldsymbol{\Upsilon}^* \cdot \boldsymbol{\Psi} = \Upsilon^{A*} \gamma_{AB} \Psi^B = \overline{\Upsilon}_A \Psi^A = \overline{\Upsilon} \Psi \, , \tag{25.5}$$

where we have used the more familiar (indexless) notation for the last equality. After considering a few more preliminaries, we will usually not explicitly display the spinor indices.

Now suppose we change to another spinor basis:

$$\mathbf{S}'_A = \mathbf{S}_B \overline{\Theta}^B{}_A \, , \tag{25.6}$$

then in order for the spinor to remain invariant, the spinor components must transform:

$$\boldsymbol{\Psi}' = \boldsymbol{\Psi} \quad \Rightarrow \quad \Psi'^A = \Theta^A{}_B \Psi^B \, . \tag{25.7}$$

We have used the fact that the automorphism transformation is the one that preserves the spinor metric:

$$\overline{\Theta}\Theta = \Theta\overline{\Theta} = \mathbf{1} \quad \Rightarrow \quad \gamma'_{AB} = \mathbf{S}'^{*}_{A} \cdot \mathbf{S}'_{B} = \mathbf{S}^{*}_{A} \cdot \mathbf{S}_{B} = \gamma_{AB} \; . \tag{25.8}$$

Another way to see that the metric is preserved is to consider the norm of a spinor:

$$\overline{\Psi}'\Psi' = \overline{\Psi}\overline{\Theta}\Theta\Psi = \overline{\Psi}\Psi \; . \tag{25.9}$$

The spinor basis transformation also induces a transformation of the basis of the Clifford algebra:

$$\Gamma'_a = \Theta\Gamma_a\overline{\Theta} \; , \tag{25.10}$$

so that all of the physical densities constructed from the spinor are invariant:

$$\rho'_a = \overline{\Psi}'\Gamma'_a\Psi' = \overline{\Psi}\overline{\Theta}\Theta\Gamma_a\overline{\Theta}\Theta\Psi = \overline{\Psi}\Gamma_a\Psi = \rho_a \; , \tag{25.11}$$

which is expected since the spinor itself does not transform (see the first part of Equation (25.7)).

25.2.2 Covariant Derivative and Drehbeins

Since we are assuming that the spinor basis may be locally assigned, we must introduce new connection coefficients $C^{A}{}_{B\mu}$ to construct the covariant derivative of a spinor:

$$\nabla_\mu\Psi \equiv (\partial_\mu - i\omega_\mu + iC_\mu)\Psi \equiv (\partial_\mu + i\Phi_\mu)\Psi \; , \tag{25.12}$$

where we have suppressed the spinor indices and included the spin connection ω_μ since we will be maintain local Lorentz invariance of the theory. We will refer to the connection C_μ as the *Clifford connection.* Recall that the spin connection is a bivector of the Clifford algebra:

$$\omega_\mu \equiv \frac{1}{2}\omega^{\alpha\beta}{}_{\mu}\sigma_{\alpha\beta} = \frac{1}{4}\omega^{\alpha\beta}{}_{\mu}\gamma_{\alpha\beta} \; , \tag{25.13}$$

whereas the Clifford connection is an arbitrary element of the Clifford algebra:

$$C_\mu \equiv c_\mu\mathbf{1} + c^{\alpha}{}_{\mu}\gamma_\alpha + \frac{1}{2}c^{\alpha\beta}{}_{\mu}\gamma_{\alpha\beta} - d^{\alpha}{}_{\mu}\tilde{\gamma}_\alpha - d_\mu\tilde{\gamma} \; . \tag{25.14}$$

Since the combination of the spin connection and the Clifford connection appearing in Equation (25.12) occurs frequently, we have also defined the *total connection* $\Phi_\mu = C_\mu - \omega_\mu$. We will discuss the transformation properties of these connections after we introduce the drehbeins.

The local automorphism transformations imply that the Clifford algebra basis must be spacetime dependent. This is most easily implemented by introducing the *drehbein* fields $\Delta^{A}{}_{\hat{B}}(x)$, these constituting the transformation matrix between the local Clifford algebra basis and an arbitrary but global (spacetime independent) Clifford algebra basis:

$$\Delta\overline{\Delta} = \mathbf{1} \; , \; \overline{\Delta}\Delta = \hat{\mathbf{1}} \; , \tag{25.15}$$

$$\gamma^A{}_{B\alpha}(x) = \Delta^A{}_{\hat{B}}(x)\hat{\gamma}^{\hat{B}}{}_{\hat{C}\alpha}\overline{\Delta}^{\hat{C}}{}_B(x) \quad \Rightarrow \quad \Gamma_a(x) = \Delta(x)\hat{\Gamma}_a\overline{\Delta}(x) \,, \tag{25.16}$$

where the "hat" refers to the global basis. In other words, the global Clifford algebra generators $\hat{\gamma}_\alpha$, and therefore all of the elements of the global Clifford algebra basis $\hat{\Gamma}_a$, are constant (spacetime independent) matrices. Note that this does not mean that the spinor basis is constant, but that the soldering form between the tangent space Lorentz basis and the spinor basis is constant. The drehbein constraints presented in Equation (25.15) insure that the local Clifford algebra generators $\gamma_\alpha(x)$ satisfy the Clifford algebra defining relation (Equation (25.1)). We have displayed the spinor indices in the first part of Equation (25.16) for the sake of clarity; they will usually not appear explicitly. The use of "hatted" and "unhatted" indices is akin to using early Greek and late Greek letters to distinguish between the Lorentz and coordinate bases in the tangent space. Finally, just as the vierbeins may be used to transform the components of any tensor between Lorentz and coordinate bases, the drehbeins may be used to transform any object possessing spinor indices between local and global Clifford algebra bases. In fact, Equation (25.16) is already an example of this.

Under the action of the automorphism group the drehbeins transform as:

$$\Delta' = \Theta\Delta \quad , \quad \overline{\Delta}' = \overline{\Delta}\overline{\Theta} \,, \tag{25.17}$$

since we are assuming that the hatted basis is global. (Of course, we can transform the global basis by global automorphism transformations, but it is only the local transformations that affect the structure of the covariant derivative.) However, under local Lorentz transformations we have:

$$\Delta' = S\Delta\overline{\hat{S}} \quad , \quad \overline{\Delta}' = \hat{S}\overline{\Delta}\overline{S} \,, \tag{25.18}$$

since the Lorentz transformations will affect the spinor basis regardless of it being local or global. In particular, note that the hatted Lorentz transformation effects a transformation on the hatted basis, and as such the generators of this transformation are written in the hatted basis:

$$S = \Delta\hat{S}\overline{\Delta} \quad \Leftrightarrow \quad \hat{S} = \overline{\Delta}S\Delta \,. \tag{25.19}$$

Consequently we have the interesting result that the drehbeins are invariant under Lorentz transformations, since from Equations (25.15), (25.18), and (25.19) we find:

$$\Delta' = \Delta \quad , \quad \overline{\Delta}' = \overline{\Delta} \,. \tag{25.20}$$

This is similar to the result for the Clifford algebra generators, they also are invariant under Lorentz transformations.

The form of the covariant derivative of the drehbein is governed by its transformation properties. Inspection of Equations (25.17) and (25.18) yields:

$$\begin{aligned}\nabla_\mu\Delta = \partial_\mu\Delta + iC_\mu\Delta - i\omega_\mu\Delta + i\Delta\hat{\omega}_\mu &= \partial_\mu\Delta + i\Phi_\mu\Delta + i\Delta\hat{\omega}_\mu \\ &= \partial_\mu\Delta + iC_\mu\Delta \,,\end{aligned} \tag{25.21}$$

where the cancelation of the spin connection terms in the last equality is obtained with:

$$\omega_\mu = \Delta\hat{\omega}_\mu\overline{\Delta} \quad \Leftrightarrow \quad \hat{\omega}_\mu = \overline{\Delta}\omega_\mu\Delta \ . \tag{25.22}$$

Of course, this cancelation is a reflection of the invariance of the drehbein under Lorentz transformations. However, this cancelation will not occur for higher covariant differentiation, in particular when we consider the field strength tensor for the Clifford connection, since covariant derivatives do not commute with the gauge fields (connections).

For the covariant derivatives of the Clifford algebra generators we have:

$$\nabla_\mu\gamma_\alpha = \partial_\mu\gamma_\alpha - \omega^\beta{}_{\alpha\mu}\gamma_\beta + i[\Phi_\mu, \gamma_\alpha] = \partial_\mu\gamma_\alpha + i[C_\mu, \gamma_\alpha] \ , \tag{25.23}$$

where we have indicated the cancelation of all of the spin connection terms. This is a reflection of the invariance of the Clifford algebra generators under Lorentz transformations. However, the covariant derivatives of the Clifford algebra generators written in coordinate basis are given by:

$$\nabla_\mu\gamma_\nu = \partial_\mu\gamma_\nu - \Gamma^\rho{}_{\nu\mu}\gamma_\rho + i[\Phi_\mu, \gamma_\nu] \ , \tag{25.24}$$

so that no cancelation occurs. This particular covariant derivative is interesting because all of the connections are present. This is because the Clifford algebra generators written in coordinate basis transform nontrivially under the action of all of the transformations considered herein: general coordinate, local Lorentz, and local automorphism transformations. Note that we are *not* demanding that the covariant derivatives of the Clifford algebra generators vanish, as we would in standard General Relativity.

The transformation properties of the Clifford connection under the action of the local automorphism group may be obtained by considering the transformation of the covariant derivative. If we consider the covariant derivative of the drehbein we have:

$$\nabla'_\mu\Delta' = \Theta\nabla_\mu\Delta \quad \Rightarrow \quad C'_\mu = \Theta C_\mu\overline{\Theta} + i(\partial_\mu\Theta)\overline{\Theta} \ , \tag{25.25}$$

whereas if we consider the covariant derivative of the spinor we find:

$$\nabla'_\mu\Psi' = \Theta\nabla_\mu\Psi \quad \Rightarrow \quad \Phi'_\mu = \Theta\Phi_\mu\overline{\Theta} + i(\partial_\mu\Theta)\overline{\Theta} \ . \tag{25.26}$$

Now Equations (25.25) and (25.26) allow us to obtain the transformation property of the spin connection under the action of the automorphism group:

$$\omega'_\mu = \Theta\omega_\mu\overline{\Theta} \quad \Rightarrow \quad \omega'^{\alpha\beta}{}_\mu = \omega^{\alpha\beta}{}_\mu \ , \tag{25.27}$$

which is the expected result – the spin connection should not transform inhomogeneously under the action of the automorphism group, and the transformation displayed in the first part of Equation (25.27) simply reflects the transformation of the Clifford algebra basis as in Equation (25.10).

Conversely, one of the surprising aspects of this formalism is the appearance of inhomogeneous transformations of the Clifford connection under the action of the

Lorentz group. Consider the transformation of the covariant derivative of a spinor under the action of the Lorentz group:

$$\nabla'_\mu \Psi' = S\nabla_\mu \Psi \quad \Rightarrow \quad \Phi'_\mu = S\Phi_\mu \overline{S} + i(\partial_\mu S)\overline{S} . \tag{25.28}$$

Alternately, consider the transformation of the covariant derivative of the drehbein under the action of the Lorentz group:

$$\nabla'_\mu \Delta' = S(\nabla_\mu \Delta)\overline{\hat{S}} \quad \Rightarrow \quad C'_\mu = SC_\mu \overline{S} + i(\partial_\mu \Delta)\overline{\Delta} - iS(\partial_\mu \Delta)\overline{\Delta S} . \tag{25.29}$$

This transformation of the Clifford connection, involving another field of the theory (the drehbein) and its derivatives, is categorically noncanonical. However, Equations (25.28) and (25.29) do lead to the required transformation property of the spin connection:

$$\omega'_\mu = S\omega_\mu \overline{S} - i\Delta(\partial_\mu \hat{S})\overline{\hat{S}}\overline{\Delta} \quad \Rightarrow \quad \omega'^{\alpha\beta}_{\ \ \ \mu} = \Lambda^\alpha{}_\gamma \Lambda^\beta{}_\delta \omega^{\gamma\delta}{}_\mu - (\partial_\mu \Lambda^\alpha{}_\gamma)\Lambda^{\gamma\beta} . \tag{25.30}$$

Therefore, although Equation (25.29) is highly unusual, it is consistent with the transformation properties of the other connections.

25.2.3 Curvatures and Field Strength Tensors

In general, curvature tensors and field strength tensors govern behavior under parallel transport. Alternately, and more conveniently, we may consider the antisymmetric combination of two successive covariant differentiations. Recall, for example, that the spin curvature tensor $\Omega^\alpha{}_{\beta\mu\nu}$ (the Lorentz field strength tensor) may be obtained from:

$$(\nabla_\mu \nabla_\nu - \nabla_\nu \nabla_\mu)V^\alpha \equiv -\Omega^\alpha{}_{\beta\mu\nu} V^\beta + T^\sigma{}_{\mu\nu}(\nabla_\sigma V^\alpha) \tag{25.31}$$

$$\Rightarrow \quad \Omega^\alpha{}_{\beta\mu\nu} = -\partial_\mu \omega^\alpha{}_{\beta\nu} + \partial_\nu \omega^\alpha{}_{\beta\mu} - \omega^\alpha{}_{\gamma\mu}\omega^\gamma{}_{\beta\nu} + \omega^\alpha{}_{\gamma\nu}\omega^\gamma{}_{\beta\mu} . \tag{25.32}$$

Also recall that the Riemann condition implies that the spin curvature tensor is equivalent to the Riemann curvature tensor:

$$\nabla_\mu e^\alpha{}_\nu = 0 \quad \Rightarrow \quad \Omega^\alpha{}_{\beta\mu\nu} = R^\alpha{}_{\beta\mu\nu} . \tag{25.33}$$

Note that Equation (25.31) yields only the spin curvature (and the torsion) because V^α only transforms nontrivially under the action of the Lorentz group.

Now consider the parallel transport of the drehbein. We use this to define the *Clifford curvature tensor* $C_{\mu\nu}$ (the automorphism field strength tensor):

$$(\nabla_\mu \nabla_\nu - \nabla_\nu \nabla_\mu)\Delta \equiv iC_{\mu\nu}\Delta + T^\sigma{}_{\mu\nu}(\nabla_\sigma \Delta) \tag{25.34}$$

$$\begin{aligned} \Rightarrow \quad C_{\mu\nu} = \partial_\mu C_\nu - \partial_\nu C_\mu + i[C_\mu, C_\nu] \\ +i[\omega_\mu, (C_\nu + L_\nu)] - i[\omega_\nu, (C_\mu + L_\mu)] , \end{aligned} \tag{25.35}$$

where we have defined the longitudinal vector fields L_μ as follows:

$$iL_\mu \equiv (\partial_\mu \Delta)\overline{\Delta} = -\Delta(\partial_\mu \overline{\Delta}) \quad \Rightarrow \quad \partial_\mu \Gamma_a = i[L_\mu, \Gamma_a] . \tag{25.36}$$

Note that in the case of flat nondynamic spacetime ($\omega_\mu = 0$) we recover the canonical gauge theory form for the field strength tensor. To check that the Clifford curvature is indeed a tensor under local Lorentz and local automorphism transformations, we must take into account the transformation of the longitudinal vector fields, and from Equations (25.17), (25.18), and (25.36) we find:

$$\textit{Lorentz}: \quad L'_\mu = L_\mu \,, \tag{25.37}$$

$$\textit{Automorphism}: \quad L'_\mu = \Theta L_\mu \overline{\Theta} - i(\partial_\mu \Theta)\overline{\Theta} \,. \tag{25.38}$$

The first of these equations follows immediately from the fact that the drehbeins are invariant under Lorentz transformations. The second equation is very much like the inhomogeneous transformation of a gauge field, but note that the inhomogeneous term appears with the opposite sign. This suggests that we make an additional definition:

$$B_\mu \equiv C_\mu + L_\mu = -i(\nabla_\mu \Delta)\overline{\Delta} = i\Delta(\nabla_\mu \overline{\Delta}) \,, \tag{25.39}$$

and under Lorentz and automorphism transformations we find:

$$\textit{Lorentz}: \quad B'_\mu = S B_\mu \overline{S} \,, \tag{25.40}$$

$$\textit{Automorphism}: \quad B'_\mu = \Theta B_\mu \overline{\Theta} \,. \tag{25.41}$$

So this field transforms as a coordinate vector, and as a Lorentz and automorphism tensor, as is also apparent from the covariant form of Equation (25.39). This field will often reappear later when we consider the field equations. Returning to the task at hand, it is now straightforward to explicitly demonstrate that the Clifford connection transforms as follows:

$$\textit{Lorentz}: \quad C'_{\mu\nu} = S C_{\mu\nu} \overline{S} \,, \tag{25.42}$$

$$\textit{Automorphism}: \quad C'_{\mu\nu} = \Theta C_{\mu\nu} \overline{\Theta} \,. \tag{25.43}$$

We see that the Clifford curvature behaves as a tensor under all transformations considered, as required. Of course, this is guaranteed from the defining relation for the Clifford curvature, Equation (25.34), but the explicit demonstration constitutes a nontrivial check on the transformations properties of the various fields.

Now consider the parallel transport of a spinor around a closed loop. We expect this to be governed by not only the Clifford curvature (as were the drehbeins), but also by the Riemann curvature, since the spinor field suffers local Lorentz transformations as well as local automorphism transformations. Specifically we find:

$$(\nabla_\mu \nabla_\nu - \nabla_\nu \nabla_\mu)\Psi \equiv i\Phi_{\mu\nu} + T^\sigma{}_{\mu\nu}(\nabla_\sigma \Psi) \tag{25.44}$$

$$\Rightarrow \quad \Phi_{\mu\nu} = \partial_\mu \Phi_\nu - \partial_\nu \Phi_\mu + i[\Phi_\mu, \Phi_\nu] \,, \tag{25.45}$$

where we have defined the *total curvature tensor* $\Phi_{\mu\nu}$. We see that the total curvature tensor has the canonical gauge theory structure. We may also show that the total curvature is indeed the sum of the Clifford curvature and the Riemann curvature:

$$\Phi_{\mu\nu} = \Omega_{\mu\nu} + C_{\mu\nu} = \frac{1}{4}R^{\alpha\beta}{}_{\mu\nu}\gamma_{\alpha\beta} + C_{\mu\nu} , \tag{25.46}$$

as expected. We may use this result to write Equation (25.34) for the Clifford curvature as follows:

$$\begin{aligned}(\nabla_\mu\nabla_\nu - \nabla_\nu\nabla_\mu)\Delta &\equiv iC_{\mu\nu}\Delta + T^\sigma{}_{\mu\nu}(\nabla_\sigma\Delta) \\ &= i\Phi_{\mu\nu}\Delta - i\Delta\hat{\Omega}_{\mu\nu} + T^\sigma{}_{\mu\nu}(\nabla_\sigma\Delta) ,\end{aligned} \tag{25.47}$$

where the total curvature appears to the left of the drehbein (since the drehbein suffers both local Lorentz and local automorphism transformations on the left), and the Riemann curvature (written in global basis) appears on the right of the drehbein (since the drehbein suffers only local Lorentz transformations of the right).

Finally, consider parallel transport of a Clifford algebra generator. After some algebra we find that this is controlled by:

$$(\nabla_\mu\nabla_\nu - \nabla_\nu\nabla_\mu)\gamma_\alpha = i[C_{\mu\nu}, \gamma_\alpha] + T^\sigma{}_{\mu\nu}(\nabla_\sigma\gamma_\alpha) . \tag{25.48}$$

Not surprisingly, since the Clifford algebra generators are Lorentz invariant, it is only the Clifford curvature which governs their behavior under parallel transport. We now make an important observation: If we were to demand that the covariant derivatives of the Clifford algebra generators vanish, then the Clifford curvature would also vanish, aside from possible $U(1)$ pieces. Specifically we have:

$$\nabla_\mu\gamma_\alpha = 0 \Rightarrow [C_{\mu\nu}, \gamma_\alpha] = 0 \Rightarrow C_{\mu\nu} = c_{\mu\nu}\mathbf{1} . \tag{25.49}$$

Thus if we make this demand the formalism reduces to that introduced by Pagels where the total curvature is written as the sum of the Riemann curvature and the electromagnetic field strength tensor. Also note that the $U(1)$ factor does not act effectively on the Clifford algebra basis (it does not induce a transformation of the Clifford algebra basis), although it does act effectively on the spinors. Stated differently, parallelizability of the Clifford algebra generators would imply that we can always find a gauge in which the Clifford algebra basis is global. Clearly this condition is too restrictive to obtain a generalization of General Relativity as desired. Therefore, we will *not* demand parallelizability of the Clifford algebra generators.

25.2.4 Bianchi Identities

Recall that the Bianchi identities are just the Jacobi identity for the covariant derivative. The Jacobi identity reads:

$$[\nabla_\mu, [\nabla_\nu, \nabla_\rho]] + [\nabla_\rho, [\nabla_\mu, \nabla_\nu]] + [\nabla_\nu, [\nabla_\rho, \nabla_\mu]] = 0 . \tag{25.50}$$

This expression is true regardless of upon what it acts. In particular, if we apply this expression to the contravariant components of a vector in coordinate basis we obtain two familiar Bianchi identities:

$$\begin{aligned}&\nabla_\rho R^\lambda{}_{\kappa\mu\nu} + \nabla_\mu R^\lambda{}_{\kappa\nu\rho} + \nabla_\nu R^\lambda{}_{\kappa\rho\mu} \\ &\quad = R^\lambda{}_{\kappa\sigma\rho}T^\sigma{}_{\mu\nu} + R^\lambda{}_{\kappa\sigma\mu}T^\sigma{}_{\nu\rho} + R^\lambda{}_{\kappa\sigma\nu}T^\sigma{}_{\rho\mu} ,\end{aligned} \tag{25.51}$$

$$\begin{aligned}\nabla_\rho T^\sigma{}_{\mu\nu} + \nabla_\mu T^\sigma{}_{\nu\rho} + \nabla_\nu T^\sigma{}_{\rho\mu} \\ = R^\sigma{}_{\rho\mu\nu} + R^\sigma{}_{\mu\nu\rho} + R^\sigma{}_{\nu\rho\mu} \\ + T^\sigma{}_{\kappa\rho} T^\kappa{}_{\mu\nu} + T^\sigma{}_{\kappa\mu} T^\kappa{}_{\nu\rho} + T^\sigma{}_{\kappa\nu} T^\kappa{}_{\rho\mu} \,.\end{aligned} \tag{25.52}$$

If instead we apply the Jacobi identity to the drehbein field we find a Bianchi identity involving the Clifford curvature:

$$\begin{aligned}\nabla_\sigma C_{\mu\nu} + \nabla_\mu C_{\nu\sigma} + \nabla_\nu C_{\sigma\mu} \\ = C_{\rho\sigma} T^\rho{}_{\mu\nu} + C_{\rho\mu} T^\rho{}_{\nu\sigma} + C_{\rho\nu} T^\rho{}_{\sigma\mu} \\ + i[B_\sigma, \Omega_{\mu\nu}] + i[B_\mu, \Omega_{\nu\sigma}] + i[B_\nu, \Omega_{\sigma\mu}] \,,\end{aligned} \tag{25.53}$$

where we have made use of definitions appearing in Equations (25.36), (25.39), and (25.46). Note that the covariant derivative of the Clifford curvature is given by:

$$\nabla_\sigma C_{\mu\nu} = \partial_\sigma C_{\mu\nu} - \Gamma^\rho{}_{\mu\sigma} C_{\rho\nu} - \Gamma^\rho{}_{\nu\sigma} C_{\mu\rho} + i[\Phi_\sigma, C_{\mu\nu}] \,, \tag{25.54}$$

the last term arising from the spin tensor character of the Clifford curvature. This Bianchi identity exhibits noncanonical structure. Note, however, that it is a tensor equation as required, since all quantities appearing in this expression are tensors (see Equations (25.40) and (25.41) for the tensor character of B_μ). Finally, if we apply the Jacobi identity to the spinor field we find a Bianchi identity involving the total curvature:

$$\nabla_\sigma \Phi_{\mu\nu} + \nabla_\mu \Phi_{\nu\sigma} + \nabla_\nu \Phi_{\sigma\mu} = \Phi_{\rho\sigma} T^\rho{}_{\mu\nu} + \Phi_{\rho\mu} T^\rho{}_{\nu\sigma} + \Phi_{\rho\nu} T^\rho{}_{\sigma\mu} \,, \tag{25.55}$$

where the covariant derivative of the total curvature is given by:

$$\nabla_\sigma \Phi_{\mu\nu} = \partial_\sigma \Phi_{\mu\nu} - \Gamma^\rho{}_{\mu\sigma} \Phi_{\rho\nu} - \Gamma^\rho{}_{\nu\sigma} \Phi_{\mu\rho} + i[\Phi_\sigma, \Phi_{\mu\nu}] \,. \tag{25.56}$$

Note that this Bianchi identity has the canonical form. Furthermore, one may show that the three Bianchi identities given in Equations (25.51), (25.53), and (25.55) are consistent with Equation (25.46). In fact, one then sees that the "additional" terms appearing in Equation (25.53) arise from the derivatives of the Clifford algebra bivectors in Equation (25.55). These identities will be useful in our discussion of the field equations.

25.3 Discussion

Demanding local choice of the spinor space basis independent of the local choice of the tangent space basis requires the introduction of two gauge fields, the *Clifford connection* in addition to the usual spin connection, as well as two soldering forms, the *drehbein* in addition to the usual vierbein. The field strength tensor (Clifford curvature) and associated Bianchi identities for the Clifford connection are non-canonical in form. However, defining the total connection as the sum of the spin connection and the Clifford connection, we find that the total curvature tensor (the sum of the Riemann curvature and the Clifford curvature) and associated Bianchi identities do have the canonical form. Therefore, we propose to treat the total curvature as the fundamental field strength tensor for this theory. The next lecture discusses the proposed Lagrangian and field equations.

Bibliography

[1] A.O. Barut and J. McEwan, *Phys. Lett.* **B 135**, 172 (1984).

[2] J.S.R. Chisholm and R. Farwell, in *Interface of Mathematics and Particle Physics* (Oxford University Press, 1990).

[3] J.P. Crawford, *J. Math. Phys.* **31**, 1991 (1990).

[4] J.P. Crawford, *J. Math. Phys.* **32**, 576 (1991).

[5] J.P. Crawford, *J. Math. Phys.* **35**, 2701 (1994).

[6] J.P. Crawford and A.O. Barut, *Phys. Rev.* **D 27**, 2493 (1983).

[7] Z. Dongpei, *Phys. Rev.* **D 22**, 2027 (1980).

[8] L. Halpern, *First Marcel Grossmann Meeting* (North-Holland, 1977).

[9] H.G. Loos and R.P. Treat, *Phys. Lett.* **A 26**, 91 (1967).

[10] R.P. Treat, *Phys. Rev. Lett.* **12**, 407 (1964).

Chapter 26

Hypergravity II

James P. Crawford
Department of Physics, Penn State - Fayette,
Uniontown, PA 15401 USA

26.1 Introduction

This lecture completes the discussion of the basic formalism of the proposed generalization of General Relativity based on local automorphism invariance. It contains the proposed Lagrangian for this theory and the concomitant field equations. The fundamental Lagrangian is taken to be in the canonical gauge theory form (quadratic in the total curvature), and the coupling constant therefore is dimensionless. The dynamical variables of the theory are taken to be the spinor, the spin connection, the vierbeins, the Clifford connection, and the drehbeins. Internal consistency of the field equations, as well as congruency with the Bianchi identities, is demonstrated. The relationship of this theory to usual Einsteinian gravity is discussed. Since the proposed Lagrangian is quadratic in the total curvature, it does not contain the usual Einstein-Hilbert term (linear in the Ricci scalar curvature). However, the Clifford connection contains a scalar and a pseudoscalar field. When these fields are extracted from the Clifford connection, the Lagrangian is found to contain a Higgs type potential in these fields, as well as a coupling of these fields to the Ricci scalar. The minimum of the potential occurs for non-zero values of these fields and this induces a term in the Lagrangian linear in the Ricci curvature. In quantum language we would say that the automorphism gauge fields C_α obtain non-zero vacuum expectation values:

$$\langle C_\alpha \rangle = \sqrt{8\alpha_f} M_P \hat{\gamma}_\alpha \, , \tag{26.1}$$

where $\hat{\gamma}_\alpha$ are a set of global (spacetime independent) matrices satisfying the Clifford algebra defining relation, M_P is the Planck mass, and $\alpha_f = \frac{f^2}{4\pi}$ where f is the coupling constant for the automorphism gauge fields. Thus the full automorphism group is not a symmetry group of the ground state of the theory; in fact, the $U(1)$

subgroup is the only remaining automorphism symmetry of the ground state. It is important to note, however, that the Lorentz group remains a symmetry group of the ground state, since the Clifford algebra generators $\hat{\gamma}_\alpha$ are *invariant* under Lorentz transformations. Therefore, we recover Einsteinian gravity as a symmetry breaking phenomenon. We conclude with a summary and a discussion of outstanding questions and potential problems.

26.2 Lagrangian

In principle, the Lagrangian may be chosen to be any real functional with mass dimension four which is invariant under all of the transformations considered. This insures that the action, obtained by integrating the Lagrangian over the invariant volume $\det(e^\alpha{}_\mu)d^4x$, is a real dimensionless scalar functional of the fields. The Lagrangian will be composed of four parts, with terms for the gauge, spinor, and drehbein fields, as well as Lagrange multiplier terms to insure satisfaction of the drehbein constraints. We will make the standard choice for each term, thus insuring that most of the accompanying field equations will be in canonical form. However, since the total curvature tensor for the gauge fields contains the Riemann curvature tensor, we will not add a separate additional term for the Riemann curvature, and therefore we do not naively expect to recover the field equations for Einstein gravity. We will consider and compare the two distinct cases of vanishing and nonvanishing torsion. Since this lecture concerns only the classical field theory, we will not concern ourselves with Fadeev-Popov ghost fields and their associated terms in the Lagrangian.

26.2.1 Gauge Field Terms

We will use first order formalism for the gauge field terms to facilitate derivation of the field equations from the action. The second order formalism yields identical results. So we take the gauge field Lagrangian to be:

$$\mathcal{L}_\Phi = \frac{1}{16f^2} tr\{\Phi^{\mu\nu}\Phi_{\mu\nu}\} - \frac{1}{8f^2} tr\{\Phi^{\mu\nu}(\partial_\mu\Phi_\nu - \partial_\nu\Phi_\mu + i[\Phi_\mu, \Phi_\nu])\} \; , \tag{26.2}$$

where f is the dimensionless coupling constant, and all traces are over the Dirac spinor indices. Note that this Lagrangian contains a term quadratic in the Riemann curvature tensor, and a coupling term between the Riemann curvature tensor and the Clifford curvature tensor, but does not contain a term linear in the Ricci scalar, and hence does not contain the Einstein-Hilbert term.

26.2.2 Spinor Field Terms

For the spinor Lagrangian we choose the manifestly Hermitian form:

$$\begin{aligned}\mathcal{L}_\Psi &= \overline{\Psi}\left(\tfrac{i}{2}(\gamma^\mu \overrightarrow{\nabla}_\mu - \overleftarrow{\nabla}_\mu\gamma^\mu) - m\right)\Psi \\ &= \overline{\Psi}\left(\tfrac{i}{2}(\gamma^\mu \overrightarrow{\partial}_\mu - \overleftarrow{\partial}_\mu\gamma^\mu) - \tfrac{1}{2}\{\gamma^\mu, \Phi_\mu\} - m\right)\Psi \; .\end{aligned} \tag{26.3}$$

Notice the interesting interaction term between the spinor and the total connection. This term reduces to the usual gauge interaction when the gauge field and the Clifford algebra generators commute. In this case, however, the anticommutator will select only part of the total connection for interaction with the spinor field, although all of the parts of the total connection will interact "among themselves," since they all contribute to the field strength tensor. In particular, this interaction term may be written as:

$$\begin{aligned} -\tfrac{1}{2}\{\gamma^\mu, \Phi_\mu\} = &-c_\mu \gamma^\mu - c^\alpha{}_\mu e_\alpha{}^\mu \mathbf{1} \\ &+\tfrac{1}{2}\varepsilon^\mu{}_{\alpha\beta\gamma}(c^{\alpha\beta}{}_\mu - \tfrac{1}{2}\omega^{\alpha\beta}{}_\mu)\tilde{\gamma}^\gamma + \tfrac{1}{2}d^\alpha{}_\mu \varepsilon^\mu{}_{\alpha\beta\gamma}\gamma^{\beta\gamma} \,. \end{aligned} \tag{26.4}$$

We see a vector coupling exactly in the form of the electromagnetic interaction, a Yukawa (scalar) coupling to the "trace" of the vector element of the Clifford connection, the usual pseudovector coupling to the spin connection augmented with a contribution from the bivector element of the Clifford connection, and a bivector coupling to the "dual" of the pseudovector element of the Clifford connection. The pseudoscalar element of the Clifford connection does not contribute to the spinor interaction. This particular form of interaction has very important consequences for the quantized version of this theory, but this interesting line of development will not be pursued here.

26.2.3 Drehbein Terms

For the drehbein fields, we consider a natural choice to be:

$$\mathcal{L}_\Delta = \frac{M^2}{8f^2} tr\left((\nabla_\mu \overline{\Delta})(\nabla^\mu \Delta)\right) - \frac{M^4}{12f^2} tr\left(\overline{\Delta}\Delta\right) \tag{26.5}$$

where the drehbein fields are dimensionless and the parameter M has the dimension of mass. (Naively we would expect the second term to correspond to a drehbein mass, but see below.) Note that the first term in this Lagrangian is the same as used in nonlinear sigma models. Finally, to obtain the constraint equations as field equations, we introduce Lagrange multiplier fields λ and $\hat{\kappa}$, with Lagrangian given by:

$$L_{\lambda\hat{\kappa}} = tr\{\lambda(\Delta\overline{\Delta} - \mathbf{1})\} + tr\{\hat{\kappa}(\overline{\Delta}\Delta - \hat{\mathbf{1}})\} \,. \tag{26.6}$$

Note that the Lagrangian multiplier fields have mass dimension four, and reality of the Lagrangian dictates that these fields be Dirac self-conjugate.

The complete invariant Lagrangian for this theory is then the sum of Equations (26.2), (26.3), (26.5), and (26.6).

26.3 Field Equations

To obtain the field equations for this theory we must first decide which fields to treat as independent dynamical variables. If the theory is consistent, then this particular choice will have no consequential effects as long as the chosen set of independent fields is complete. We will consider the following set of fields as constituting the complete set:

$$\Phi^A{}_{B\mu\nu}\,,\;\Phi^A{}_{B\mu}\,,\;\omega^{\alpha\beta}{}_{\mu}\,,\;e_\alpha{}^\mu\,,\;\Psi^A\,,\;\overline{\Psi}_A\,,\;\Delta^A{}_{\hat{B}}\,,\;\overline{\Delta}^{\hat{A}}{}_B\,,\;\lambda^A{}_B\,,\;\hat{\kappa}^{\hat{A}}{}_{\hat{B}}\,, \tag{26.7}$$

where we have explicitly displayed all of the various indices for elucidation. The choice of considering as independent the total connection Φ_μ instead of the Clifford connection C_μ greatly facilitates the derivation of the field equations, since in this case the Lagrangian $\mathcal{L}_\Phi$ depends only on the fields $\Phi_{\mu\nu}$, Φ_μ , and $e_\alpha{}^\mu$. Had we chosen C_μ instead of Φ_μ as independent, the Lagrangian $\mathcal{L}_\Phi$ would also depend on the spin connection $\omega^{\alpha\beta}{}_\mu$ and the drehbein fields Δ and $\overline{\Delta}$, and this in turn would complicate the analysis considerably. Finally note that if we assume that the spin connection is an independent field, as indicated above, then we may not assume that the torsion vanishes. Conversely, if we assume that the torsion vanishes, then we can eliminate the spin connection; in this case the spin connection is not considered to be an independent dynamical variable, and consequently it does not generate a field equation. We will consider and compare both cases.

26.3.1 Drehbein Field Equations

Variation with respect to the Lagrange multiplier fields produces:

$$\Delta\overline{\Delta} = \mathbf{1} \quad , \quad \overline{\Delta}\Delta = \hat{\mathbf{1}}\,, \tag{26.8}$$

which are the drehbein constraint equations, as expected. These equations will often be used to simplify the subsequent field equations. Variation with respect to the drehbein fields now yields:

$$M^2\left(\nabla^\mu\nabla_\mu + \frac{2}{3}M^2 + T_\mu\nabla^\mu\right)\Delta = 8f^2\left(\lambda + \kappa + K_\mu\gamma^\mu\right)\Delta\,, \tag{26.9}$$

$$M^2\left(\nabla^\mu\nabla_\mu + \frac{2}{3}M^2 + T_\mu\nabla^\mu\right)\overline{\Delta} = 8f^2\overline{\Delta}\left(\lambda + \kappa + \gamma^\mu K_\mu\right)\,, \tag{26.10}$$

where we have defined K_μ , a Dirac self-conjugate field, as follows:

$$K_\mu \equiv \frac{i}{2}\left((\nabla_\mu\Psi)\overline{\Psi} - \Psi(\nabla_\mu\overline{\Psi})\right)\,. \tag{26.11}$$

Note that Equations (26.9) and (26.10) are conjugate pairs, as required. We may solve these equations explicitly for the Lagrange multiplier fields, leaving only one self-conjugate equation for the drehbein fields:

$$\lambda + \kappa = \frac{M^2}{8f^2}\left(\frac{2}{3}M^2 - B^\mu B_\mu\right) - \frac{1}{2}\{K_\mu,\gamma^\mu\}\,, \tag{26.12}$$

$$M^2\left(\nabla_\mu B^\mu + T_\mu B^\mu\right) = -4if^2[K_\mu,\gamma^\mu]\,, \tag{26.13}$$

where we have repeatedly used the drehbein constraint equations (26.8), and we have recalled the definition:

$$B_\mu \equiv C_\mu + L_\mu = -i(\nabla_\mu \Delta)\overline{\Delta} = i\Delta(\nabla_\mu \overline{\Delta}) . \tag{26.14}$$

As expected, the Lagrange multiplier fields may be eliminated algebraically, and they do not enter into any other field equation, although it is surprising that one of them remains completely arbitrary (since only their sum is determined). Another surprise is that what had naively appeared to be a mass term for the drehbeins in the Lagrangian, as in Equation (26.5), does not yield a mass term in the drehbein equation, although note that this term does contribute a cosmological constant when the drehbein constraints are used. More about this later.

26.3.2 Spinor Field Equations

Variation of the action with respect to the spinor fields yields:

$$0 = \left(i\gamma^\mu \nabla_\mu + \frac{i}{2}(\nabla_\mu \gamma^\mu + T_\mu \gamma^\mu) - m\right)\Psi , \tag{26.15}$$

$$0 = \overline{\Psi}\left(-i\overleftarrow{\nabla}_\mu \gamma^\mu - \frac{i}{2}(\nabla_\mu \gamma^\mu + T_\mu \gamma^\mu) - m\right) . \tag{26.16}$$

These equations are conjugate pairs, as required. Note that we recover the standard field equations for General Relativity when we set the Clifford connection as well as the covariant derivative of the Clifford algebra generator to zero. We may obtain an alternate form for these equations by using the result: $\nabla_\mu \gamma_\nu = i[B_\mu, \gamma_\nu]$, and we find:

$$\left[i\gamma^\mu\left(\partial_\mu - i\omega_\mu + \frac{1}{2}T_\mu\right) - m\right]\Psi = \left[\frac{1}{2}\{C_\mu, \gamma^\mu\} + \frac{1}{2}[L_\mu, \gamma^\mu]\right]\Psi . \tag{26.17}$$

The left-hand side is exactly the result from General Relativity; the right-hand side contains the additional interactions induced by the requirement of local automorphism invariance, the first part involving the Clifford connection and the second part containing drehbein derivative terms arising from the spacetime dependence of the Clifford algebra generators.

26.3.3 Gauge Field Equations

Variation of the action with respect to the total curvature and the total connection leads to the following:

$$\Phi_{\mu\nu} = \partial_\mu \Phi_\nu - \partial_\nu \Phi_\mu + i[\Phi_\mu, \Phi_\nu] , \tag{26.18}$$

$$\nabla_\nu \Phi^{\nu\mu} + T_\nu \Phi^{\nu\mu} + \frac{1}{2}T^\mu{}_{\rho\sigma}\Phi^{\rho\sigma} = -M^2 B^\mu + 2f^2\{\Psi\overline{\Psi}, \gamma^\mu\} , \tag{26.19}$$

where we have repeatedly used the drehbein constraint equations (26.8) as well as the definition for the field B_μ. Note that the first term on the right hand side of Equation

(26.19) is a mass term for the Clifford connection. In fact, the longitudinal part required for a massive vector field is given by the drehbein derivatives as embodied in the field $L_\mu = -i(\partial_\mu \Delta)\overline{\Delta} = i\Delta(\partial_\mu \overline{\Delta})$. This is a general aspect of automorphism gauge theory; the automorphism gauge fields gain mass without recourse to the Higgs mechanism. We will see in the next section that the mass M is on the order of the Planck mass, so these fields will have hitherto gone undetected, and yet will be expected to have important consequences for the quantum version of this theory.

Before proceeding to the derivation of the remaining field equations, we sketch a nontrivial check on the system of equations obtained thus far. We begin with the contraction of the Bianchi identity:

$$R_{\mu\nu} - R_{\nu\mu} = \nabla_\mu T_\nu - \nabla_\nu T_\mu - \nabla_\rho T^\rho{}_{\mu\nu} - T_\rho T^\rho{}_{\mu\nu} . \tag{26.20}$$

Next consider the formal identity:

$$\begin{aligned}(\nabla_\mu \nabla_\nu - \nabla_\nu \nabla_\mu)\Phi^{\rho\sigma} \\ = -R^\rho{}_{\xi\mu\nu}\Phi^{\xi\sigma} - R^\sigma{}_{\xi\mu\nu}\Phi^{\rho\xi} + T^\xi{}_{\mu\nu}(\nabla_\xi \Phi^{\rho\sigma}) + i[\Phi_{\mu\nu}, \Phi^{\rho\sigma}] .\end{aligned} \tag{26.21}$$

Upon contraction we obtain:

$$\nabla_\mu \nabla_\nu \Phi^{\mu\nu} = \frac{1}{2}\nabla_\xi (T^\xi{}_{\mu\nu}\Phi^{\mu\nu}) - \frac{1}{2}(\nabla_\mu T_\nu - \nabla_\nu T_\mu - \nabla_\xi T^\xi{}_{\mu\nu})\Phi^{\mu\nu} , \tag{26.22}$$

where we have used Equation (26.20). To reiterate, Equation (26.22) is a formal identity which depends on the Bianchi identity. Now consider the covariant divergence of Equation (26.19). From the left hand side we obtain:

$$\begin{aligned}\nabla_\mu \left(\nabla_\nu \Phi^{\nu\mu} + T_\nu \Phi^{\nu\mu} + \tfrac{1}{2}T^\mu{}_{\rho\sigma}\Phi^{\rho\sigma}\right) \\ = -\nabla_\mu \nabla_\nu \Phi^{\mu\nu} - \tfrac{1}{2}(\nabla_\mu T_\nu - \nabla_\nu T_\mu)\Phi^{\mu\nu} \\ + \tfrac{1}{2}\nabla_\xi (T^\xi{}_{\mu\nu}\Phi^{\mu\nu}) + T_\mu \nabla_\nu \Phi^{\mu\nu} .\end{aligned} \tag{26.23}$$

Whereas from the right hand side we find:

$$\begin{aligned}\nabla_\mu \left(-M^2 B^\mu + 2f^2\{\Psi\overline{\Psi}, \gamma^\mu\}\right) \\ = -T_\mu \left(-M^2 B^\mu + 2f^2\{\Psi\overline{\Psi}, \gamma^\mu\}\right) \\ = -T_\mu \left(\nabla_\nu \Phi^{\nu\mu} + T_\nu \Phi^{\nu\mu} + \tfrac{1}{2}T^\mu{}_{\rho\sigma}\Phi^{\rho\sigma}\right) ,\end{aligned} \tag{26.24}$$

where we have used Equations (26.13), (26.15), and (26.16) in obtaining the first equality in Equation (26.24), and used Equation (26.19) in obtaining the second equality. If we now set the result in Equation (26.23) equal to the result in Equation (26.24) we recover Equation (26.22). Note that this derivation involved use of the field equations for the total connection, the spinor fields, and the drehbein fields, and consequently it constitutes a complex check on the internal consistency of these equations, as well as demonstrating congruency with the Bianchi identities. Alternately, we may begin with Equation (26.19) and the formal identity (26.22) (which follows from the Bianchi identity) and derive the drehbein field equation (26.13). Finally note that all that has transpired so far is valid whether or not the torsion vanishes.

26.3.4 Gravitational Field Equations

Nonvanishing Torsion

We treat the case of nonvanishing torsion first, in which case both the spin connection and the vierbein are considered to be independent. Variation of the action with respect to the spin connection yields:

$$0 = \frac{iM^2}{8f^2} tr\left(\gamma_{\alpha\beta}(\nabla^\mu \Delta)\overline{\Delta} - \Delta(\nabla^\mu \overline{\Delta})\gamma_{\alpha\beta}\right) \Rightarrow 0 = tr(\gamma_{\alpha\beta} B^\mu) \,, \tag{26.25}$$

where we have used the drehbein constraint equations (26.8). This equation implies that the bivector part of the automorphism gauge field is pure gauge; hence all of the nontrivial bivector information in the total connection Φ_μ is contained in the spin connection. To see this clearly, suppose that Equation (26.25) is satisfied by having B_μ vanish. This is certainly over-restrictive but it suits the purposes of this argument. In this case, the entire automorphism gauge field would be pure gauge:

$$B_\mu = 0 \quad \Rightarrow \quad C_\mu = -L_\mu = i(\partial_\mu \Delta)\overline{\Delta} \quad \Rightarrow \quad C_{\mu\nu} = 0 \,, \tag{26.26}$$

where, as expected, for a pure gauge field the field strength tensor vanishes identically. Returning to the general case, observe that the trace operation in Equation (26.25) annihilates all but the bivector part of B_μ (that is, all but the bivector parts have identically vanishing trace), so that it is only the bivector part of the Clifford connection $c^{\alpha\beta}{}_\mu$ which is pure gauge. In fact, we may solve Equation (26.25) explicitly to obtain:

$$c^{\alpha\beta}{}_\mu = -\frac{1}{8} tr(L_\mu \gamma^{\alpha\beta}) = \frac{i}{8} tr\left((\partial_\mu \Delta)\overline{\Delta}\gamma^{\alpha\beta}\right) \,. \tag{26.27}$$

Now although $c^{\alpha\beta}{}_\mu$ is pure gauge, the bivector part of the Clifford curvature $C_{\mu\nu}$ does not necessarily vanish, since it will in general gain contributions from other (non-bivector) parts of the Clifford connection. Therefore, we do expect that there will remain a nontrivial coupling between the Riemann curvature and the Clifford curvature.

The field equation arising from variation of the action with respect to the vierbein is found to be:

$$\begin{aligned} 0 = -e^\alpha{}_\mu \mathcal{L} - \tfrac{1}{4f^2} tr(\Phi^{\alpha\sigma}\Phi_{\mu\sigma} + M^2 B^\alpha B_\mu) \\ + \tfrac{i}{2}\overline{\Psi}(\gamma^\alpha \overrightarrow{\nabla}_\mu - \overleftarrow{\nabla}_\mu \gamma^\alpha)\Psi \,. \end{aligned} \tag{26.28}$$

Now contracting this equation with the vierbein and extracting the symmetric and antisymmetric parts we find:

$$\begin{aligned} 0 = \tfrac{1}{4f^2} tr\left(\Phi^{\mu\sigma}\Phi_\sigma{}^\nu + \tfrac{1}{4} g^{\mu\nu}\Phi^{\rho\sigma}\Phi_{\rho\sigma}\right) + \tfrac{1}{2} tr\left(K^\mu \gamma^\nu + K^\nu \gamma^\mu\right) \\ + \tfrac{M^2}{4f^2} tr\left(B^\mu B^\nu - \tfrac{1}{2} g^{\mu\nu} B^\sigma B_\sigma\right) + \tfrac{M^4}{3f^2} g^{\mu\nu} \equiv \Theta^{\mu\nu} \,, \end{aligned} \tag{26.29}$$

$$0 = tr\left(K^\mu \gamma^\nu - K^\nu \gamma^\mu\right) \,, \tag{26.30}$$

where we have used all of the other field equations and the definitions for K_μ and B_μ to write these expressions. In Equation (26.29) we recognize the first term as the symmetric energy-momentum tensor for the gauge field, the second term as the symmetric energy-momentum tensor for the spinor field, and third term as the symmetric energy-momentum tensor for the drehbein fields. The last term is the cosmological constant term, which may be interpreted as a "background" energy-momentum density. Each of these terms has the canonical form for an energy-momentum tensor, so we have dubbed this tensor $\Theta^{\mu\nu}$ as is customary. Recall that the total curvature $\Phi_{\rho\sigma}$ contains the Riemann curvature $\Omega_{\rho\sigma}$, so although this energy-momentum tensor has canonical structure, it does include terms involving only the gravitational field, as well as cross terms involving the gravitational field and the other fields. Presumably then the pure gravitational piece should correspond to the pure gravitational energy in the system, and the cross terms should correspond to the gravitational interaction (potential) energy in the system. Then this field equation states that the total energy in the system vanishes locally, indeed an interesting result. As for Equation (26.30), it is not an independent relation, for it follows from multiplying the drehbein Equation (26.13) by the bivector $\gamma^{\mu\nu}$ and taking the trace; therefore, once the drehbein equation is satisfied, Equation (26.30) is guaranteed.

Vanishing Torsion

If we assume that the torsion vanishes, then we may eliminate the spin connection in terms of the vierbein and treat only the vierbein as independent. The field equation arising from variation of the action with respect to the vierbein then yields the following symmetric and antisymmetric equations:

$$\frac{M^2}{16f^2}\nabla_\sigma tr\left(\gamma^{\sigma\mu}B^\nu + \gamma^{\sigma\nu}B^\mu\right) = \Theta^{\mu\nu} , \tag{26.31}$$

$$\frac{M^2}{8f^2}\nabla_\sigma tr\left(\gamma^{\mu\nu}B^\sigma\right) = tr\left(K^\mu\gamma^\nu - K^\nu\gamma^\mu\right) , \tag{26.32}$$

where $\Theta^{\mu\nu}$ is defined in Equation (26.29). A quick check on these equations is to note that if the torsion is not set to zero, so that Equation (26.25) is satisfied, Equations (26.31) and (26.32) reduce to Equations (26.29) and (26.30). The additional terms in Equations (26.31) and (26.32) (the nonzero left hand sides) arise from the vierbein derivatives which appear when the spin connection is eliminated in terms of the vierbein. Also note that Equation (26.32) is not an independent relation, for it follows from multiplying the drehbein equation (26.13) by the bivector $\gamma^{\mu\nu}$ and taking the trace; therefore, once the drehbein equation is satisfied, Equation (26.32) is guaranteed. One may also show that the covariant divergence of Equation (26.31) yields an identity upon use of the field equations and the Bianchi identities.

Comparing these two approaches, we see that nonvanishing torsion allows us to solve for the bivector part of the Clifford connection $c^{\alpha\beta}{}_\mu$ in terms of derivatives of the drehbeins, whereas vanishing torsion allows us to solve for the spin connection $\omega^{\alpha\beta}{}_\mu$ in terms of derivatives of the vierbeins, so that there is a complementarity between these two cases, and in either case there are derivative couplings. However, in the case of nonvanishing torsion, the derivative couplings do not appear until the field equations are used, whereas the derivative couplings appear explicitly in the Lagrangian for the case of vanishing torsion, and therefore the quantum version of this theory may be better behaved (i.e. more likely to be renormalizable) for the case of nonvanishing torsion.

26.4 Einstein Gravity Recovered

We will now show that Einstein gravity is contained in this theory in the sense that the Einstein-Hilbert action is recovered at low energy ($E < M_P$) as a symmetry breaking phenomenon. In particular, we will extract a scalar and a pseudoscalar field from the Clifford connection, and show the gauge field and drehbein terms in the Lagrangian contain a Higgs potential in these fields. The induced vacuum expectation value for these fields then generates a term in the Lagrangian linear in the Ricci scalar, the Einstein-Hilbert form, much like the situation in Brans-Dicke theory.

We begin by modifying the Lagrangian by making use of Equation (26.18) for the total curvature, and the drehbein constraint Equations (26.8), to find:

$$\mathcal{L} = -\frac{1}{16f^2} tr\left(\Phi^{\mu\nu}\Phi_{\mu\nu}\right) + \frac{M^2}{8f^2} tr\left(B^\mu B_\mu - \frac{2}{3}M^2\right) + \mathcal{L}_\Psi \,, \tag{26.33}$$

where the spinor Lagrangian is given in Equations (26.3) and (26.4), and it is understood that the total curvature $\Phi_{\mu\nu}$ is given in Equation (26.18). Recall the expression for the Clifford connection:

$$C_\mu = c_\mu \mathbf{1} + c^\alpha{}_\mu \gamma_\alpha + \frac{1}{2} c^{\alpha\beta}{}_\mu \gamma_{\alpha\beta} - d^\alpha{}_\mu \tilde{\gamma}_\alpha - d_\mu \tilde{\gamma} \,. \tag{26.34}$$

Note that we may extract "traces" from the vector and pseudovector pieces as follows:

$$c^\alpha{}_\mu \equiv \frac{1}{4}\sigma e^\alpha{}_\mu + v^\alpha{}_\mu \,,\ \sigma \equiv c^\alpha{}_\mu e_\alpha{}^\mu \ \Leftrightarrow\ v^\alpha{}_\mu e_\alpha{}^\mu = 0 \,, \tag{26.35}$$

$$d^\alpha{}_\mu \equiv \frac{1}{4}\pi e^\alpha{}_\mu + p^\alpha{}_\mu \,,\ \pi \equiv d^\alpha{}_\mu e_\alpha{}^\mu \ \Leftrightarrow\ p^\alpha{}_\mu e_\alpha{}^\mu = 0 \,. \tag{26.36}$$

From these definitions it follows that σ is a scalar field and π is a pseudoscalar field. Note that these are the only (pseudo)scalar fields that can be extracted from the Clifford connection. Upon substitution of these definitions into the Lagrangian as written in Equation (26.33) we obtain the following expression:

$$\mathcal{L} = \frac{1}{32f^2}(\sigma^2 - \pi^2)R - \mathcal{V}(\sigma,\pi) - \frac{1}{64f^2}(\sigma^2 - \pi^2)T^\alpha{}_{\mu\nu}T_\alpha{}^{\mu\nu} + \mathcal{E} \,, \tag{26.37}$$

where we have defined the "Higgs potential" as:

$$\mathcal{V}(\sigma,\pi) \equiv \frac{3}{64f^2}(\sigma^2 - \pi^2)^2 - \frac{M^2}{8f^2}(\sigma^2 - \pi^2) + \frac{M^4}{12} \,, \tag{26.38}$$

and "everything else" (the rest of the Lagrangian) appears as $\mathcal{E}$. Notice that the Lagrangian contains a term linear in the Ricci scalar R, and a term quadratic in the torsion, although each of these terms also contains the factor $(\sigma^2 - \pi^2)$. These arise from the quadratic total curvature term in the original Lagrangian (the first term in Equation (26.33)). The first term in the Higgs potential also arises from the first term in Equation (26.33), whereas the second and third terms in the Higgs potential arise from the second term in Equation (26.33).

Now observe that the Higgs potential is minimum on the hyperbola given by:

$$(\sigma^2 - \pi^2) = \frac{4}{3}M^2 . \tag{26.39}$$

Upon substitution of the minimum value into the Lagrangian we finally obtain:

$$L = \frac{M^2}{24f^2}R - \frac{M^2}{48f^2}T^{\alpha}{}_{\mu\nu}T_{\alpha}{}^{\mu\nu} + \mathcal{E} . \tag{26.40}$$

The first term is the desired Einstein-Hilbert term. In order for the normalization to be correct we must make the identification:

$$\frac{M^2}{24f^2} = \frac{1}{16\pi G} \equiv \frac{M_P^2}{16\pi} \Rightarrow M^2 = \frac{3f^2}{2\pi}M_P^2 \equiv 6\alpha_f M_P^2 , \tag{26.41}$$

where we have made the customary definition of the Planck mass M_P , and have also defined the coupling parameter α_f in the usual manner. Then unless α_f is extremely small, the vector fields B_μ have mass M on the order of the Planck mass. Note that since the chosen normalization the "drehbein mass term" appearing in Equation (26.5) renders the value of the Higgs potential zero at its minimum, there is no cosmological constant term appearing in Equation (26.40). Clearly, for other choices of normalization this cancelation will not occur, in which case there will be a nonzero cosmological constant. Finally, we obtain a term quadratic in the torsion, which is a common choice when considering Poincaré gravity.

26.5 Discussion

Demanding local choice of the spinor space basis independent of the local choice of the tangent space basis leads to a generalization of General Relativity. The Lagrangian is chosen to be in canonical form for gauge theory and we have seen that this theory is consistent at the classical level; but canonical gauge theory form means that the Lagrangian does not contain a term linear in the Ricci scalar, and thus does not naively contain Einstein gravity. However, the theory *naturally* includes a Higgs potential which leads to a solution that does not respect the automorphism invariance, and this in turn induces a term in the Lagrangian linear in the Ricci scalar. Hence, in this formulation, Einstein gravity is recovered at low energy as a symmetry breaking phenomenon.

There are many issues that remain to be explored. Most importantly, since the theory contains terms quadratic in the Riemann curvature, this theory differs for

Einstein gravity when the curvature approaches the Planck scale. This will likely have profound implications for both the classical and quantum realms. For the classical theory, this modification will become important near spacetime singularities, possibly even eliminating them. At the quantum level, this theory becomes softer than Einstein gravity at extremely high energies; thus there is hope that this theory may be renormalizable. One of the potential problems for the quantum version is the noncompact automorphism gauge group $U(2,2)$ which could lead to ghost states and nonunitarity. However, unlike typical gauge theories, the automorphism group is precisely the group which preserves the metric of the spinor space and thus also preserves the metric of the Clifford algebra. Therefore, the symmetry itself precludes the possibility of transitions between negative and positive norm states, so that unitarity is not violated.

Another interesting issue that needs to be explored is that of the energy contained in the gravitational field. Recall that the symmetric part of the gravitational field equation (26.29) has the appearance of a canonical energy-momentum tensor, including terms for the total connection, spinor, and drehbein fields, as well as a cosmological constant. Therefore it contains a term quadratic in the Riemann curvature as well as a cross term involving the Riemann curvature and the Clifford curvature. It is tempting to claim that term quadratic in the Riemann curvature corresponds to the energy of pure gravity and the cross term is a gravitational interaction energy. Since both of these terms are proper tensors we avoid the aesthetically unappealing situation in standard gravity where the gravitational energy is represented by a pseudo-tensor.

Bibliography

[1] C. Brans and R.H. Dicke, *Phys. Rev.* **124**, 112 (1961).

[2] J.S.R. Chisholm and R. Farwell, in *Interface of Mathematics and Particle Physics* (Oxford University Press, 1990), and references therein.

[3] I. Ciufolini and J.A. Wheeler, *Gravitation and Inertia* (Princeton University Press, 1995).

[4] J.P. Crawford, *J. Math. Phys.* **31**, 1991 (1990).

[5] J.P. Crawford, *J. Math. Phys.* **32**, 576 (1991).

[6] J.P. Crawford, *J. Math. Phys.* **35**, 2701 (1994).

[7] J.P. Crawford and A.O. Barut, *Phys. Rev.* **D 27**, 2493 (1983).

[8] F. Englert and R. Brout, *Phys. Rev. Lett.* **13**, 321 (1964).

[9] G.S. Guralnik, C.R. Hagen, and T.W.B. Kibble, *Phys. Rev. Lett.* **13**, 585 (1964).

[10] P.W. Higgs, *Phys. Lett.* **12**, 132 (1964); *Phys. Rev. Lett.* **13**, 508 (1964).

[11] T.W.B. Kibble, *J. Math. Phys.* **2**, 212 (1961).

Chapter 27

Properties of Clifford Algebras for Fundamental Particles

J. S. R. Chisholm
Institute of Mathematics and Statistics,
University of Kent, Canterbury, Kent, U.K.

R. S. Farwell
University of Brighton, Brighton, East Sussex, U.K.

27.1 The basic building blocks of a gauge model: the Clifford Algebra $\mathcal{C}\ell_{p,q}$

27.1.1 Introduction

As William Kingdon Clifford commented in the abstract of his paper "On the Classification of Geometric Algebras" communicated to the London Mathematical Society on 10 March 1876, "... the system is the natural language of metrical geometry and of physics"[1]. The system which he was describing is his geometric algebra, which we now know as Clifford algebra. The paper was never finished, but was discovered following his death and published in his collected Mathematical Papers. Nevertheless the use of his algebras, not just in four dimensions, is evident today in models of physics.

In this series of three lectures, we shall review the properties of Clifford algebras of arbitrary dimension and illustrate why they are useful in developing models of the fundamental fermions, that is, quarks and leptons, and of the intermediate bosons which mediate their interaction. In doing this we shall draw out the properties of the algebras which are of particular relevance to a gauge theoretic model of the fundamental particles.

[1]W.K. Clifford, *On the Classification of Geometric Algebras* in Mathematical Papers, ed. R. Tucker, Chelsea Reprint (1968), pp 397-401.

The electromagnetic gauge theory is described elsewhere in these proceedings. Normally a gauge theory of the interactions is understood to be a Lagrangian field theory based on the direct product of space-time and some other internal symmetry space. The latter "commutes" with space-time in the sense that the space-time terms commute with the generators of the gauge symmetries in the internal space. *Spin* gauge theories have a different algebraic structure in which space-time and the "internal space" are combined within a Clifford algebraic formalism. By this we mean that the combined space is spanned by the vectors of some Clifford algebra, and elements of the algebra are used to define the Lagrangian density. This is our motivation for considering higher dimensional algebras.

If we are to describe a Lagrangian density using terms within some Clifford algebra, then we must define within the algebra the following: the imaginary unit, spinors, the hermitian conjugation operation and hence bar conjugate spinors. In investigating the definition of these in an arbitrary algebra, we shall also consider involutions of the algebra, the pseudoscalar, helicity splitting, and the generalisation of the Dirac spinor norm. In doing so we shall illustrate some of the similarities and differences between algebras of different signature.

Throughout these lectures we shall be describing the properties of *real* Clifford algebras, that is, we shall allow only linear combinations of basis elements of the algebra with *real* coefficients. The unit imaginary i may appear in the matrix representations of the algebra. However, the properties of the algebras, and the models which we build using them, are not dependent on the matrix representation used. Nevertheless, where the matrix representations are familiar to us, we shall use them to illustrate the ideas, often as a precursor to general algebraic result.

27.1.2 The Elements of the Algebra

An obvious reason for the ability of Clifford algebras to describe physics is the geometrical interpretation which can be given to the basis elements of the algebra. The Pauli algebra $\mathcal{C}\ell_{3,0}$ is chosen as an example to illustrate this geometrical interpretation. The Pauli algebra is generated by a set of three orthonormal basis vectors $\{e_i; i = 1, 2, 3\}$ satisfying

$$e_i^2 = 1, \qquad i = 1, 2, 3 \tag{27.1.1a}$$

$$e_i.e_j = 0, \qquad i \neq j. \tag{27.1.1b}$$

These equations are sufficient to describe a unique associative algebra corresponding to flat three dimensional space. The defining relations (27.1.1) of the algebra can then be written as

$$\{e_i e_j\} = 2\delta_{ij} I \qquad i, j = 1, 2, 3, \tag{27.1.2}$$

where I is the scalar of the algebra. The basis elements of the algebra and their geometrical interpretation are shown in Table 27.1.1 below. It should be noted that interpretation of the bivectors as turning operations provides an obvious motivation for the property $e_{ij}^2 = -I$, since the application twice of a turning operator through $90°$ turns a vector into minus itself.

Table 27.1.1
The basis elements of the Pauli algebra and their geometric interpretation

Basis elements	**Geometrical interpretation**
Scalar I	
Vectors $e_i, i = 1, 2, 3$	orthogonal directed line segments
$e_i^2 = I$	
Bivectors $e_{ij} = e_i e_j, i < j$	turning operators through 90° or
$e_{ij}^2 = -I$	oriented areas
Pseudoscalar $e_{123} = e_1 e_2 e_3$ $e_{123}^2 = -I$	volume

A general element of the real Pauli algebra may be written as

$$A = aI + a_i e_i + a_{ij} e_{ij} + a_{123} e_{123}, \tag{27.1.3}$$

where the coefficients $\{a, a_i, a_{ij}(i < j), a_{123}\}$ are eight real parameters and summation over the indices in each term is assumed. By using the representation of the basis vectors in terms of the 2×2 Pauli matrices,

$$e_1 = \begin{pmatrix} 0 & 1 \\ 1 & 0 \end{pmatrix} \equiv \sigma_1, e_2 = \begin{pmatrix} 0 & -i \\ i & 0 \end{pmatrix} \equiv \sigma_2, e_3 = \begin{pmatrix} 1 & 0 \\ 0 & -1 \end{pmatrix} \equiv \sigma_3 \tag{27.1.4}$$

it can be shown that the general element A is represented by

$$A = bI + b_i \sigma_i,$$

where the parameters $\{b, b_i\}$ are complex. As a 2×2 matrix A is written as

$$\begin{pmatrix} c_1 & d_1 \\ c_2 & d_2 \end{pmatrix}$$

where the entries c_i and d_i $i = 1, 2$ are complex. It is important to emphasise here the point made in the introduction that, even though the imaginary unit appears in the matrix *representation* of the algebra, the Pauli algebra described here is nevertheless over the field of the real numbers. Basis elements whose square is I $(-I)$ are represented by hermitian (anti-hermitian) matrices. This is a general property of all algebras, since the diagonal elements of the square of a hermitian matrix are non-negative.

An interesting extension of the geometric algebra associated with three dimensional space is that associated with Minkowski space-time, otherwise known as the real Dirac algebra $\mathcal{C}\ell_{1,3}$. The generalisation of the characteristic relation (27.1.2) to produce real Dirac algebra provides the relationship with the Minkowski metric of the associated geometric space:

$$\{e_i, e_j\} = 2\eta_{ij} I \qquad i, j = 0, 1, 2, 3, \tag{27.1.5}$$

where η_{ij} are the components of the Minkowski metric tensor which have matrix representation

$$\eta = \begin{pmatrix} 1 & & & \\ & -1 & & \\ & & -1 & \\ & & & -1 \end{pmatrix}. \tag{27.1.6}$$

The basis vector e_0, which squares to I, is said to be timelike, and the remaining vectors $\{e_i; i = 1, 2, 3\}$, which each square to $-I$, are said to be spacelike.

A matrix representation of the set $\{e_i\}$ can be realised by the 4×4 Dirac γ-matrices, for example. As with the Pauli matrices used to represent $\mathcal{C}\ell_{3,0}$, this representation will involve the imaginary unit, even though algebra is over the reals. A question which we shall consider later is which algebras, as distinct from their representations, have an element which behaves like the imaginary unit, allowing pairs of real coefficients to be interpreted as complex numbers. We shall show that this property is independent of the matrix representation used.

The Pauli and Dirac algebras, most familiar to us in physical models, can be extended and generalised in two ways:

- to curved spaces;
- to higher dimensions.

An example of the first generalisation is to four-dimensional curved space-time. A vector basis $\{\Gamma_\mu(x)\}$ for the algebra associated with the space may be defined by analogy with (27.1.5) as

$$\{\Gamma_\mu(x), \Gamma_\nu(x)\} = 2I g_{\mu\nu}(x) \tag{27.1.7}$$

where $g_{\mu\nu}$ are the components of the metric tensor on the space. The position dependence of the metric tensor is "inherited" from that of the basis vectors. From the relation (27.1.7), the vectors may be interpreted as the "Dirac square root" of the metric tensor and, in that sense, are more fundamental than the metric tensor. The relationship between the metrics on the curved space and on Minkowski space can be made explicit through the introduction of the vierbein field $h^i_\mu(x)$ through

$$\Gamma_\mu(x) = h^i_\mu(x) e_i \qquad \mu, i = 0, 1, 2, 3, \tag{27.1.8}$$

where the set $\{e_i\}$ are "tangent space" basis vectors satisfying (27.1.5). The vierbein fields can be used to relate the two metrics as follows:

$$g_{\mu\nu} = h^i_\mu(x) h^j_\nu(x) \eta_{ij}. \tag{27.1.9}$$

The complete formalism can be generalised to n-dimensional space of signature $p, q (p+q=n)$ and metric $g_{\mu\nu}(x)$. The equations (27.1.7)–(27.1.9) generalise exactly with the indices μ and i now both running from 1 to n and the "flat" metric η being given by

$$\eta = \begin{pmatrix} \left.\begin{matrix} 1 & & \\ & \cdots & \\ & & 1 \end{matrix}\right\} p & \\ & q \left\{\begin{matrix} -1 & & \\ & \cdots & \\ & & -1 \end{matrix}\right. \end{pmatrix}, \tag{27.1.10}$$

so that

$$\begin{aligned} \eta_{ii} &= 1 & i &= 1, 2, \ldots, p \\ \eta_{ii} &= -1 & i &= p+1, p+2, \ldots, p+q = n \\ \eta_{ij} &= 0 & i &\neq j. \end{aligned}$$

This ensures that the subset of the basis vectors $\{e_i; i = 1, 2, \ldots, p\}$ are timelike and the remaining $(n-p)$ basis vectors $\{e_i; i = p+1, p+2, \ldots, n\}$ are spacelike. In the general case, the set $\{e_i; i = 1, 2, \ldots, n\}$ forms an orthonormal vector basis of the Clifford Algebra $\mathcal{C}\ell_{p,q}$. The complete set of basis elements of $\mathcal{C}\ell_{p,q}$ is comprised of 2^n multivector elements shown in Table 27.1.2.

Table 27.1.2
The basis elements of the Clifford Algebra $\mathcal{C}\ell_{p,q}$

I	Scalar
e_i	Vectors
$e_{ij} = e_i e_j \quad i < j$	Bivectors
$e_{ijk} = e_i e_j e_k \quad i < j < k$	Trivectors
......	
$\omega = e_{123\ldots n}$	n-vector, or pseudoscalar.

We shall denote a general multivector basis element of $\mathcal{C}\ell_{p,q}$ by e_T. Thus, e_T either denotes the scalar or is a shorthand way of writing

$$e_T = e_{i_1 i_2 \ldots i_m}, \; m = 1, 2, \ldots n \quad \text{and} \quad i_1 < i_2 < \ldots < i_m \,. \tag{27.1.11}$$

A general element A of $\mathcal{C}\ell_{p,q}$ is a linear combination over the reals of the basis elements:

$$A = \sum_{T=1}^{2^n} a_T e_T, \; a_T \in \mathbf{R}\,. \tag{27.1.12}$$

We shall show that algebras of different signatures have different algebraic properties (even for the same value of n). It will emerge that some algebras with particular values of p and q are more useful than others for describing physical models.

Operations within the Algebra

tion of hermitian conjugation is well-defined for matrices, as transposition trix and complex conjugation of the elements within it. Hermitian conjugation [illegible] an operation is used in the definition of the Dirac bar conjugate spinor, the motivation for which was the construction, from a spinor and its bar conjugate, of Lorentz invariant quantities. The aim of this section is to identify an operation, within the algebra $\mathcal{C}\ell_{p,q}$, which can be seen to be equivalent to the operation of hermitian conjugation when applied to the matrix representation of the algebra.

Before considering the specific operation of "hermitian conjugation", we shall review well known operations which are involutions of the algebra[2]. The first is *reversion* of an element, which reverses the order of the basis vectors in any given multivector so that, for example, the reversal of the bivector e_{12} is

$$e_{12}^R = e_2e_1 = -e_1e_2 = -e_{12} \ .$$

In general, the reversal of e_T is given by

$$e_T^R = e_{i_1i_2\dots i_m}^R = e_{i_mi_{m-1}\dots i_1} \ .$$

Its relation to e_T depends on the number of commutations of the elements e_i, and hence on the value of m:

$$e_T^R = (-1)^{m(m-1)/2}e_{i_1i_2\dots i_m} = (-1)^{m(m-1)/2}e_T \ . \qquad (27.1.13)$$

A second operation is *inversion* in which each vector e_i is transformed into minus itself. Its effect when applied to a multivector gives

$$e_T^I = e_{i_1i_2\dots i_M}^I = (-1)^m e_{i_1i_2\dots i_m} = (-1)^m e_T \ . \qquad (27.1.14)$$

A third operation of conjugation can be defined from the combination of reversion and inversion. Conjugation applied to e_T gives

$$e_T^* = (-1)^m(-1)^{m(m-1)/2}e_{i_1i_2\dots i_m} = (-1)^{m(m+1)/2}e_T \ . \qquad (27.1.15)$$

Essentially, under each of the operations, a pure multivector of the algebra is either invariant or it transforms into minus itself. It is useful to tabulate the behaviour of each multivector under each of the operations indicating whether or not it changes sign.

---

[2] P. Lounesto, *Clifford Algebras and Spinors* in *Clifford Algebras and Their Applications in Mathematical Physics*, ed. J.S.R. Chisholm & A.K. common, Reidel, (1986), pp 25-38.

Table 27.1.3
Sign changes under the main involutions; "x" indicates no sign change; "$\surd$" indicates sign change occurs

n-vector	Reversion	Inversion	Conjugation
1	x	$\surd$	$\surd$
2	$\surd$	x	$\surd$
3	$\surd$	$\surd$	x
4	x	x	x
5	x	$\surd$	$\surd$
6	$\surd$	x	$\surd$
7	$\surd$	$\surd$	x
8	x	x	x
9	x	$\surd$	$\surd$
10	$\surd$	x	$\surd$
11	$\surd$	$\surd$	x
12	x	x	x

It is interesting to note the pattern of signs appearing in Table 27.1.3; the signs repeat on a cycle with period 4. This cyclic behaviour can be justified algebraically from the definitions (27.1.13)–(27.1.15). For example, in the case of reversion, the index appearing in the definition (27.1.13) satisfies

$$\frac{m(m-1)}{2} \quad \begin{array}{ll} \text{even,} & \text{if } m = 0, 1 \bmod(4); \\ \text{odd,} & \text{if } m = 2, 3 \bmod(4). \end{array}$$

Thus

$$e_T^R = e_T \text{ and } e_T^2 = e_T^R e_T \qquad \text{if } m = 0, 1 \bmod(4);$$

$$e_T^R = -e_T \text{ and } -e_T^2 = e_T^R e_T \qquad \text{if } m = 2, 3 \bmod(4).$$

and so

$$e_T^2 = \prod_{k=1}^{m} e_{i_k}^2 \qquad \text{if } m = 0, 1 \bmod(4), \tag{27.1.16a}$$

$$e_T^2 = -\prod_{k=1}^{m} e_{i_k}^2 \qquad \text{if } m = 2, 3 \bmod(4). \tag{27.1.16b}$$

Since each vector in the product in (27.1.16) squares to either $\pm I$, depending on whether the vector is timelike or spacelike, we can deduce that the sign of the square of the m-vector e_T depends on the value of m and the number of spacelike vectors in the multivector. Thus, denoting the number of spacelike vectors in e_T by t,

$$e_T^2 = I \quad \begin{array}{l} \text{if } m = 0, 1 \bmod(4) \text{ with } t \text{ even} \\ \text{or } m = 2, 3 \bmod(4) \text{ with } t \text{ odd} \end{array} \tag{27.1.17a}$$

and

$$e_T^2 = -I \quad \text{if } m = 0,1 \bmod(4) \text{ with } t \text{ odd} \quad \text{or } m = 2,3 \bmod(4) \text{ with } t \text{ even.} \tag{27.1.17b}$$

We are now going to consider how the operation of hermitian conjugation applied to the matrix representation of $\mathcal{C}\ell_{p,q}$ relates to these involutions. We denote the operation with a dagger superscript "†", and define[3] the hermitian conjugate of a basis multivector by

$$e_T^\dagger = e_T \quad \text{if} \quad e_T^2 = I \tag{27.1.18a}$$

$$e_T^\dagger = -e_T \quad \text{if} \quad e_T^2 = -I, \tag{27.1.18b}$$

to correspond to the properties of matrix representations. Thus, using (27.1.17),

$$e_T^\dagger = e_T \qquad m = 0,1 \bmod(4) \text{ with } t \text{ even} \quad \text{or } m = 2,3 \bmod(4) \text{ with } t \text{ odd.}$$

and

$$e_T^\dagger = -e_T \qquad m = 0,1 \bmod(4) \text{ with } t \text{ odd} \quad \text{or } m = 2,3 \bmod(4) \text{ with } t \text{ even.} \tag{27.1.19}$$

We shall now investigate whether the operation of hermitian conjugation expressed in (27.1.19) can be achieved via an inner automorphism of the algebra; that is, whether hermitian conjugation can be brought about by a similarity transformation within the algebra. To consider this, we define an element Γ in $\mathcal{C}\ell_{p,q}$ which brings about this similarity transformation for basis vectors; that is, an element such that

$$e_i^\dagger = \Gamma e_i \Gamma^{-1} \,. \tag{27.1.20}$$

From (27.1.20) we may deduce that

$$\begin{aligned} \Gamma e_T \Gamma^{-1} &= \Gamma e_{i_1} \Gamma^{-1} \Gamma e_{i_2} \Gamma^{-1} \ldots \Gamma e_{i_m} \Gamma^{-1} \\ &= e_{i_1}^\dagger e_{i_2}^\dagger \ldots e_{i_m}^\dagger \end{aligned} \tag{27.1.21a}$$

$$= \begin{cases} e_T & \text{for an even number of spacelike vectors in } e_T \\ e_T & \text{for an odd number of spacelike vectors in } e_T. \end{cases} \tag{27.1.21b}$$

The expression (27.1.21) cannot define $e_T^\dagger$ in general, since it is inconsistent with (27.1.19), and hence with the original definition (27.1.18), if $m = 2,3 \bmod(4)$. In these cases an extra change of sign is required for consistency. By considering Table 27.1.3, it is easy to see that the additional sign change for the required values of m is brought about by the operation of reversion. Thus, we must extend the definition (27.1.20) to multivectors by postulating that

$$e_T^\dagger = \Gamma e_T^R \Gamma^{-1}. \tag{27.1.22}$$

The definition (27.1.22) is consistent with (27.1.19) and (27.1.20), as required.

[3]Li Deming, private communication.

The additional sign changes have been noted by Crawford[4] for exactly the same values of m, although in a different but related context, that of imposing a reality condition on bispinor densities. Rather than use the operation of reversion to bring about the required sign changes, Crawford redefines the basis multivectors for $m = 2, 3 \bmod(4)$ to include factors of i. However, that is not in the spirit of our work here since we focus on real algebras with well-defined signatures.

From (27.1.22) the well known property of the hermitian conjugation operator can easily be deduced; namely that

$$(e_T e_S)^\dagger = e_S^\dagger e_T^\dagger, \tag{27.1.23}$$

where e_S is a second multivector basis element.

The equation (27.1.22) provides a consistent definition of the hermitian conjugation operation as an inner automorphism of the algebra. However, it is not clear that an operator Γ satisfying (27.1.22) necessarily exists. To investigate the existence question further, we consider the requirements for the operator. From (27.1.18) and (27.1.20) it follows that we require Γ to commute simultaneously with all timelike vectors and to anti-commute with all spacelike vectors; that is,

$$[\Gamma, e_i] = 0 \qquad \text{if } e_i \text{ is timelike}, \tag{27.1.24a}$$

$$\{\Gamma, e_i\} = 0 \qquad \text{if } e_i \text{ is spacelike}. \tag{27.1.24b}$$

By applying the hermitian conjugation operation to both sides of the definition (27.1.20) and then re-arranging we obtain

$$e_i^\dagger = \Gamma^\dagger e_i (\Gamma^\dagger)^{-1}. \tag{27.1.25}$$

Superficially (27.1.24) appears to suggest that consistency between (27.1.25) and (27.1.20) requires Γ to be self-conjugate. However, this is not the case since, as a result of the conditions (27.1.24), we can take Γ or $\Gamma^\dagger$ through the e_i in (27.1.20) or (27.1.25) to cancel with Γ^{-1} or $(\Gamma^\dagger)^{-1}$ respectively and in doing so bring about the change of sign required for consistency with the original definition (27.1.18). Thus, no further conditions on Γ are necessary.

A multivector basis element satisfying (27.1.24) can be found when either p is odd or q is even, $q \neq 0$. In the first case, the required element is the p-vector $e_{12\ldots p}$ which is the product of the p timelike basis vectors, and in the second case it is the q-vector $e_{p+1,p+2\ldots p+q}$ which is the product of the q spacelike vectors. If both p is even *and* q is odd, then Γ does not exist.

For algebras for which p is odd *and* q is even, $\neq 0$, a linear combination $\alpha e_{12\ldots p} + \beta e_{p+1,p+2\ldots p+q}$, for any real values of α and β, is a suitable candidate for Γ. These algebras have $p + q = n$ odd; in section 27.2.5 we shall show that this ensures that the pseudoscalar ω of the algebra commutes with all elements of the algebra. So the conjugation operator can be written in the form $ae^{b\omega}e_{12\ldots p}$, where a and b are real.

In the Dirac algebra the conjugation operator Γ exists since $p = 1$; it is equal to the single timelike vector e_0, which is γ_0 in the Dirac notation.

[4] J.P. Crawford, *J.Math.Phys.* **31**, 1991-1997 (1990).

27.2 Spinors in the Clifford Algebra $C\ell_{p,q}$

27.2.1 Minimal Left Ideals of $C\ell_{p,q}$

Ideals of the algebra $C\ell_{p,q}$ are defined using *idempotents*; these are elements P of the algebra satisfying

$$P^2 = P. \tag{27.2.1}$$

The behaviour of an idempotent when it pre– or post–multiplies any element of the algebra is equivalent to that of a projection operator. For example, in the algebra $C\ell_{3,0}$ the element $\frac{1}{2}(I + e_3)$ is an idempotent. In terms of the Pauli matrices (27.1.4) this element is represented by

$$\begin{pmatrix} 1 & 0 \\ 0 & 0 \end{pmatrix} \tag{27.2.2}$$

which when acting on another 2×2 matrix projects out the first column or first row of that matrix depending on whether it acts on the right or left respectively.

A *left ideal* of the algebra $C\ell_{p,q}$ is a subset of the algebra which is defined by the action of an idempotent P on the right of each of the elements A of $C\ell_{p,q}$. Thus, a left ideal is the set

$$\{AP; \text{ any } A \in C\ell_{p,q}\}. \tag{27.2.3}$$

A *right* ideal can be defined similarly via the action on the *left* by the idempotent P.

Continuing with the example of $C\ell_{3,0}$, we recall from section 27.1.2 that, in terms of a 2×2 matrix representation, the general element of the algebra may be written as

$$\begin{pmatrix} c_1 & d_1 \\ c_2 & d_2 \end{pmatrix},$$

where c_i and d_i are complex numbers. The idempotent (27.2.2) can be used to form a left ideal by acting as a projection operator on the right. Thus the elements of this left ideal are represented by 2×2 matrices with the second column zero,

$$\begin{pmatrix} c_1 & 0 \\ c_2 & 0 \end{pmatrix}; \tag{27.2.4}$$

that is, they depend upon two complex, or four real, parameters.

An element of a *minimal* left ideal of $C\ell_{p,q}$, has the form AP where A is any element in $C\ell_{p,q}$ and P is a *primitive* indepotent[5] A primitive idempotent has the form

$$P = \frac{1}{2}(I + p_1)\frac{1}{2}(I + p_2)\cdots\frac{1}{2}(I + p_k) \equiv \frac{1}{2^k}\prod_{i=1}^{k}(I + p_i), \tag{27.2.5}$$

where $p_i^2 = I$ and $[p_i, p_j] = 0$ for $i, j = 1, 2, \ldots, k$, and k takes its maximum value for a given algebra. The two algebraic conditions ensure that P is an idempotent. The value of k is defined uniquely for each $C\ell_{p,q}$ as

$$k = q - r_{q-p}, \tag{27.2.6}$$

[5] P. Lounesto, *Clifford Algebras and Spinors* in *Clifford Algebras and Their Applications in Mathematical Physics*, ed. J.S.R. Chisholm & A.K. Common, Reidel, (1986), pp 25-38.

where r_i is the *Radon-Hurwitz number.* Since the values of the Radon-Hurwitz number form a cycle of period 8:

$$r_{i+8} = r_i + 4, \tag{27.2.7}$$

the values of all r_i are given in Table 2.1

Table 27.2.1
The values of the Radon Hurwitz number.

i	0	1	2	3	4	5	6	7
r_i	0	1	2	2	3	3	3	3

As an example, we return again to the Pauli algebra $\mathcal{C}\ell_{3,0}$, in which

$$k = 0 - r_{-3} = 0 - (-1) = 1.$$

Thus the idempotent $\frac{1}{2}(I + e_3)$ is primitive in $\mathcal{C}\ell_{3,0}$, and any element of the Pauli algebra represented by the matrix (27.2.4) is an element of a minimal left ideal. Thus, a minimal left ideal in the Pauli algebra contains two complex parameters, which is precisely the parametric dependence of a Pauli spinor.

Having noted this feature of the minimal left ideals in the Pauli algebra, it is also instructive to consider the Dirac algebra $\mathcal{C}\ell_{1,3}$ in which the value of k is also 1. Therefore a primitive idempotent in $\mathcal{C}\ell_{1,3}$ will contain just one factor; for example,

$$P = \tfrac{1}{2}(I + e_0) \tag{27.2.8}$$

is primitive. By using this idempotent, we obtain a minimal left ideal with elements

$$A\tfrac{1}{2}(I + e_0); \quad \text{any } \; A \in \mathcal{C}\ell_{1,3}. \tag{27.2.9}$$

To get a sense of what form this minimal left ideal takes, we consider the 4×4 Dirac γ-matrix representation of the basis vectors of $\mathcal{C}\ell_{1,3}$:

$$e_0 \equiv \gamma_0 = \begin{pmatrix} I_2 & 0 \\ 0 & -I_2 \end{pmatrix} \quad e_i \equiv \gamma_i = \begin{pmatrix} 0 & \sigma_i \\ -\sigma_i & 0 \end{pmatrix} \quad i = 1, 2, 3, \tag{27.2.10}$$

where I_2 is the 2×2 identity matrix and $\sigma_i (i = 1, 2, 3)$ are the 2×2 Pauli matrices. A general element of the algebra $\mathcal{C}\ell_{1,3}$ may be written as

$$A = sI + v_k e_k + r_{ij} e_{ij} + t_{ijk} e_{ijk} + p\omega, \tag{27.2.11}$$

where the coefficients s, v_k, t_{ijk}, and p are real and $i, j, k \in \{0, 1, 2, 3\}$ with $i < j < k$. In the matrix representation (27.2.10), A can be written as

$$A = \begin{pmatrix} c_1 & -c_2^* & d_1 & -d_2^* \\ c_2 & c_1^* & d_2 & d_1^* \\ c_3 & c_4^* & d_3 & d_4^* \\ c_4 & -c_3^* & d_4 & -d_3^* \end{pmatrix} \tag{27.2.12}$$

where the entries c_i and $d_i (i = 1, 2, 3, 4)$ in the matrix are complex and are related to the coefficients in (27.2.11); the star denotes complex conjugation. In this matrix representation the elements of the minimal left ideal (27.2.9) are[6]

$$\begin{pmatrix} c_1 & -c_2^* & 0 & 0 \\ c_2 & c_1^* & 0 & 0 \\ c_3 & c_4^* & 0 & 0 \\ c_4 & -c_3^* & 0 & 0 \end{pmatrix} \tag{27.2.13}$$

Thus the elements of this minimal ideal contain eight real, or four complex, parameters, which are just sufficient to define a Dirac spinor.

We have deduced the number of parameters in certain minimal left ideals of the Pauli and Dirac algebras by considering particular matrix representations of each of the algebras. It is timely therefore to emphasise a point about matrix representations. The properties of minimal left ideals have been demonstrated using the matrix representations purely because of our familiarity with them. The conclusion about the number of parameters in the ideals is representation free. In particular, the conclusions about the number of parameters in a minimal left ideal in either the Pauli or Dirac algebras do not depend on the form of the primitive idempotent used, although we have used the particular representations (27.2.2) and (27.2.13). The number of parameters depends only on the values of n and k, as we shall see in the next section.

27.2.2 Spinors in $\mathcal{C}\ell_{p,q}$

In section 2.1 we have attempted to demonstrate, by considering particular cases, that the minimal left ideals of the Pauli and Dirac algebras contain sufficient parameters to describe Pauli and Dirac spinors respectively. It is therefore natural to define a spinor $\mathcal{C}\ell_{p,q}$ as an element of a minimal left ideal of the algebra. This will be taken as one of the fundamental principles of gauge models based on any algebra $\mathcal{C}\ell_{p,q}$.

We shall now investigate, without reverting to any matrix representation, the number of parameters in a minimal left ideal, and hence in a spinor, in $\mathcal{C}\ell_{p,q}$. A spinor is an element of $\mathcal{C}\ell_{p,q}$ which has the form

$$\left\{ A \frac{1}{2^k} \prod_{i=1}^{k} (I + p_i) \text{ any } A \in \mathcal{C}\ell_{p,q} \right\}.$$

The set of basis elements $e_T (T = 1, 2, \ldots, 2^n)$ for $\mathcal{C}\ell_{p,q}$ can be used to form a set of basis elements for the minimal left ideal, given by:

$$\left\{ e_T \frac{1}{2^k} \prod_{i=1}^{k} (1 + p_i) \right\}. \tag{27.2.14}$$

[6]This is referred to as a "mother spinor" by P. Lounesto in c. 2.

Since the 2^k terms in P form an abelian group, the set (27.2.14) can be shown to contain 2^{n-k} distinct terms. A spinor is thus a linear combination of these terms, and so contains 2^{n-k} real parameters.

The formula for the number of basis elements for a spinor space, that is, a minimal left ideal, can be illustrated using the 2×2 matrix representation (27.1.4) for the Pauli algebra $\mathcal{C}\ell_{3,0}$, and the primitive idempotent $\frac{1}{2}(I+\sigma_3)$. In terms of the Pauli matrices the eight basis elements of $\mathcal{C}\ell_{3,0}$, are

$$\{I_2, \sigma_1, \sigma_2, \sigma_3, i\sigma_1, i\sigma_2, i\sigma_3, iI_2\}.$$

By post-multiplying each of these basis elements by the primitive idempotent we obtain four distinct terms:

$$\left\{\frac{1}{2}(I+\sigma_3), \quad \sigma_1\frac{1}{2}(I+\sigma_3), \quad \sigma_2\frac{1}{2}(I+\sigma_3), \quad i\frac{1}{2}(I+\sigma_3)\right\}$$

which form a basis for the spinor space in $\mathcal{C}\ell_{3,0}$. In the matrix representation (27.1.4) the basis elements are

$$\left\{\begin{pmatrix} 1 & 0 \\ 0 & 0 \end{pmatrix}, \begin{pmatrix} 0 & 0 \\ 1 & 0 \end{pmatrix}, \begin{pmatrix} 0 & 0 \\ i & 0 \end{pmatrix}, \begin{pmatrix} i & 0 \\ 0 & 0 \end{pmatrix}\right\}.$$

A linear combination of these gives the general form (27.2.4) for the Pauli spinor.

From the formula, we deduce that in $\mathcal{C}\ell_{1,3}$ there are $2^{4-1}=8$ parameters, which agrees with the conclusion in section 2.1, where we use the Dirac matrix representation for $\mathcal{C}\ell_{1,3}$. It is interesting to note that, by contrast, for the algebra $\mathcal{C}\ell_{3,1}$ the value of k given by the formula (27.2.6) is 2 and thus the number of parameters available to define a spinor is $2^{4-2}=4$. This is insufficient for a Dirac spinor and emphasises the point that algebras with different signatures have different properties, in particular different spinor structures. By using the definition of a spinor as an element of a minimal left ideal, we understand why Dirac chose the algebra $\mathcal{C}\ell_{1,3}$, rather than $\mathcal{C}\ell_{3,1}$, in which to define his spinors.

27.2.3 Bar-Conjugate Spinors in $\mathcal{C}\ell_{p,q}$

We propose to define a bar-conjugate spinor from a given spinor using the operator Γ which brings about a hermitian conjugation of the elements in the algebra $\mathcal{C}\ell_{p,q}$ which was introduced in section 1.3. We denote a spinor by ψ, so that ψ is an element of a minimal left ideal of $\mathcal{C}\ell_{p,q}$, and then define its bar conjugate by

$$\overline{\psi} = \psi^\dagger \Gamma. \tag{27.2.15}$$

Using (27.1.22) the bar conjugate is

$$\overline{\psi} = \Gamma \psi^R. \tag{27.2.16}$$

For example, in the Dirac algebra the operator Γ is the timelike basis vector e_0, which in the Dirac matrix representation is γ_0. Using the same primitive idempotent for the Dirac algebra as in section 2.1, we can represent a spinor as any element of the Dirac algebra of the form (27.2.9). Thus, in this representation, the bar conjugate to this spinor is

$$\begin{aligned}\overline{\psi} &= \Gamma\psi^R = e_0\tfrac{1}{2}(I+e_0)A^R \\ &= \tfrac{1}{2}(I+e_0)A^R,\end{aligned} \qquad (27.2.17)$$

which is an element of a minimal right ideal of the Dirac algebra. To investigate whether (27.2.16) seems to be a reasonable definition of the bar conjugate, we shall appeal to our experience of the matrix representation of Dirac spinors.

In the Dirac matrix representation, the general element of the algebra $C\ell_{1,3}$ is given by (27.2.12). In terms of this representation the operation of reversion applied to the general element gives

$$A^R = \begin{pmatrix} c_1^* & c_2^* & -c_3^* & -c_4^* \\ -c_2 & c_1 & -c_4 & c_3 \\ -d_1^* & -d_2^* & d_3^* & d_4^* \\ -d_2 & d_1 & -d_4 & d_3 \end{pmatrix}.$$

Therefore

$$\overline{\psi} = \begin{pmatrix} c_1^* & c_2^* & -c_3^* & -c_4^* \\ -c_2 & c_1 & -c_4 & c_3 \\ 0 & 0 & 0 & 0 \\ 0 & 0 & 0 & 0 \end{pmatrix}, \qquad (27.2.18)$$

since, in the matrix representation, the operation of the primitive idempotent on the left of the element in (27.2.17) projects out the first two rows. We recall that in this representation the spinor ψ is given by (27.2.13). By using this and (27.2.18) we can evaluate the trace of the matrix which is the product of $\overline{\psi}$ and ψ as

$$\text{Trace }(\overline{\psi}\psi) = 2(|c_1|^2 + |c_2|^2 - |c_3|^2 - |c_4|^2)\,. \qquad (27.2.19)$$

In algebraic terms this evaluation is equivalent to determining the scalar part of the element $(\overline{\psi}\psi)$ of the algebra $C\ell_{1,3}$ and is denoted by $< \overline{\psi}\psi >_S$.

The real expression in brackets on the right hand side of (27.2.19) is often referred to as the "Dirac metric", or "invariant length"[7], for the spinor involving four complex parameters. The expression is exactly what we would expect from the product of the Dirac spinor and its bar conjugate, so it seems our definition (27.2.16) of the bar conjugate is correct for the Dirac algebra.

It may be considered that the derivation of (27.2.18) and (27.2.19) is a lucky accident of the choice (27.2.8) of primitive idempotent for the Dirac algebra. However, if a different primitive idempotent, say

$$\tfrac{1}{2}(I+e_{123}),$$

is used to form the minimal left ideal, so that the spinor and its bar conjugate are given respectively by

$$\begin{aligned}\psi &= A\tfrac{1}{2}(I+e_{123}) \\ \overline{\psi} &= e_0\tfrac{1}{2}(I+e_{123})A^R,\end{aligned}$$

[7] J.P. Crawford, *J.Math.Phys.* **31**, 1991-1997 (1990).

then we obtain the same form for $<\overline{\psi}\psi>_S$ as in (27.2.19).

The form (27.2.19) is also independent of the matrix representation, as we shall show in the next section by considering the scalar part of the product $(\overline{\psi}\psi)$ in $C\ell_{p,q}$.

27.2.4 The Spinor Norm in $C\ell_{p,q}$

We shall refer to the scalar part of the product $(\overline{\psi}\psi)$ as the *spinor norm.*

We noted in section 27.2.2 that a basis for a minimal left ideal $C\ell_{p,q}$ has the form

$$\left\{ e_T \frac{1}{2^k} \prod_{i=1}^{k} (I + p_i); \; T = 1, 2, \ldots, 2^{n-k} \right\} , \tag{27.2.20}$$

where $\{e_T;\ T = 1, 2, \ldots, 2^{n-k}\}$ is an appropriate subset of the basis of $C\ell_{p,q}$.

Thus, in $C\ell_{p,q}$ a spinor can be expressed as

$$\psi = \sum_{T=1}^{2^{n-k}} a_T e_T \frac{1}{2^k} \prod_{i=1}^{k} (I + p_i) \tag{27.2.21a}$$

and the operation of reversion applied to the spinor produces an expression

$$\psi^R = \sum_{T=1}^{2^{n-k}} a_T \frac{1}{2^k} \prod_{i=1}^{k} (I + p_i^R) e_T^R \, . \tag{27.2.21b}$$

Now since by definition $p_i^2 = I$, the elements p_i are self conjugate; thus from (27.1.22) we can deduce that

$$\Gamma p_i^R = p_i \Gamma$$

and the bar conjugate spinor becomes, from (27.2.16) and (27.2.21b)

$$\overline{\psi} = \sum_{T=1}^{2^{n-k}} a_T \frac{1}{2^k} \prod_{i=1}^{k} (I + p_i^R) \Gamma e_T^R \, . \tag{27.2.21c}$$

Without loss of generality, we choose a representation for the minimal left ideal in which Γ is not one of the idempotents $\{p_i\}$ defining the minimal left ideal. In this case the set $\left\{ \prod_{i=1}^{k} (I + p_i) \Gamma e_T^R \right\}$ is identical to the set $\left\{ \prod_{i=1}^{k} (I + p_i) e_T^R \right\}$, the effect of the Γ being to permute the elements of the set among themselves. Denoting the permutation operation by $\Pi(\ )$, we can write (27.2.21c) as

$$\overline{\psi} = \sum_{T=1}^{2^{n-k}} \Pi(a_T) \frac{1}{2^k} \prod_{i=1}^{k} (I + p_i^R) e_T^R \, , \tag{27.2.21d}$$

where $\Pi(a_T) = a_U$ with $T \neq U$.

The scalar part of $(\overline{\psi}\psi)$ consists only of those terms in which e_T^R multiplies e_T, that is, from the multiplication of "like" terms. Consequently we deduce that the spinor norm

$$<\overline{\psi}\psi>_S = \frac{2^k}{2^{2k}} \sum_{T=1}^{2^{n-k-1}} [\Pi(a_T) + \Pi^{-1}(a_T)] a_T e_T^R e_T \; , \qquad (27.2.22a)$$

In (27.2.22a) as compared to (27.2.21), the summation over T only runs from 1 to 2^{n-k-1}. The number of terms is halved because we have chosen to combine in pairs terms which have a common factor a_T. The product $e_T^R e_T$ is either I or $-I$ depending on T. Writing $b_T = \Pi(a_T) + \Pi^{-1}(a_T)$, any coefficient in (27.2.22a) can be expressed as a difference of two squares:

$$a_T b_T = \tfrac{1}{4}[(a_T + b_T)^2 - (a_T - b_T)^2]$$

$$-a_T b_T = \tfrac{1}{4}[(a_T - b_T)^2 - (a_T + b_T)^2]$$

and so, irrespective of the sign from $e_T^R e_T$ in (27.2.22a), we may write the spinor norm in the form

$$<\overline{\psi}\psi>_S = \frac{1}{2^k} \sum_{T=1}^{2^{n-k-1}} (A_T^2 - B_T^2) \; , \qquad (27.2.22b)$$

where A_T and B_T are defined in terms of the real coefficients a_T and b_T.

The expression (27.2.22b) demonstrates that in the algebra $C\ell_{p,q}$, the spinor norm can be represented as a sum of squares of real numbers with 2^{n-k-1} each of positive and negative signs. This formula agrees with the expression (27.2.19) which we obtained using the matrix representations for the algebra $C\ell_{1,3}$ since, when considered in terms of real numbers, it contains four of each sign.

27.2.5 Left- and Right-Handed Spinors in $C\ell_{p,q}$

To determine whether left- and right-handed spinors can be represented in $C\ell_{p,q}$ we consider the possible existence of a helicity projection operator in $C\ell_{p,q}$. For Dirac spinors the helicity projection operators are

$$h_\pm = \tfrac{1}{2}(I \pm i\gamma_5) \qquad (27.2.23)$$

where $\gamma_5 = \gamma_0\gamma_1\gamma_2\gamma_3$, the pseudoscalar in $C\ell_{1,3}$. To investigate the existence of helicity projection operators in $C\ell_{p,q}$ we shall be motivated by the form for γ_5 and thus shall consider the general properties of the pseudoscalar ω. This will also provide information on the imaginary unit i and whether it can be realised in the algebra $C\ell_{p,q}$.

The pseudoscalar element ω has different properties depending on whether the dimension of the algebra n is even or odd.

When n is even, the pseudoscalar is the product of an even number of basis vectors and hence anti-commutes with each basis vector. It commutes with all the elements of the even sub-algebra, that is, those which are a linear combination of multivector basis elements consisting of an even number of basis vectors.

The situation is quite different when n is odd since then the pseudoscalar commutes with *all* elements of the algebra $\mathcal{C}\ell_{p,q}$. In this case, taking the lead from William Clifford's paper "On the Classification of Geometric Algebra" (see section 27.1.1), we shall distinguish the cases in which the square of the pseudoscalar element is either I or $-I$. However, Clifford himself only explicitly considered in his paper algebras with signature $p = 0$.

From (27.1.16) we can deduce that

$$\begin{aligned} \omega^2 &= (-1)^q I, \quad n = 1 \bmod(4) \\ \omega^2 &= (-1)^{q+1} I, \ n = 3 \bmod(4). \end{aligned} \tag{27.2.24}$$

Thus, $\omega^2 = -I$ when

$$n = 1 \bmod(4) \text{ and } q \text{ is odd}$$

or,

$$n = 3 \bmod(4) \text{ and } q \text{ is even.}$$

In these cases, the pseudoscalar behaves like $\pm i$, and thus the algebra contains the equivalent of the imaginary unit. Indeed, in some representations, for example the Pauli matrix representation of $\mathcal{C}\ell_{3,0}$, the pseudoscalar is represented by iI. Examples of other algebras containing the equivalent of an imaginary unit are $\mathcal{C}\ell_{1,2}$, $\mathcal{C}\ell_{1,6}$ and $\mathcal{C}\ell_{1,10}$. Interestingly the Dirac algebra $\mathcal{C}\ell_{1,3}$ does not. Thus, if we restrict ourselves in $\mathcal{C}\ell_{1,3}$ to Clifford numbers which are only real combinations of the basis elements, then the helicity projection operators (27.2.23) are not contained within $\mathcal{C}\ell_{1,3}$.

From (27.2.24), we conclude that $\omega^2 = I$ when

$$n = 1 \bmod(4) \text{ and } q \text{ is even} \tag{27.2.25a}$$

or,

$$n = 3 \bmod(4) \text{ and } q \text{ is odd.} \tag{27.2.25b}$$

Then the pseudoscalar behaves like ± 1, thus accounting for its representation, like that of the scalar, by the unit matrix in some representations of these algebras. These representations are not faithful.

When (27.2.25) is satisfied, we can use the pseudoscalar to form idempotents

$$P_\pm = \tfrac{1}{2}(I \pm \omega). \tag{27.2.26}$$

The idempotents (27.2.26) can be used to split the algebra into two halves[8] consisting of two non-interacting ideals $\mathcal{C}\ell_{p,q}P_+$ and $\mathcal{C}\ell_{p,q}P_-$, since

$$P_+P_- = P_-P_+ = 0.$$

We note also that the right ideals formed using $P_\pm$ are identical to the left ideals since $P_\pm$ commutes with all the elements of the algebra. They are thus two-sided ideals.

[8] Li Deming, private communication.

It is possible to establish that, if $q \neq 0$ then $\mathcal{C}\ell_{p,q}P_{\pm}$ are both isomorphic to $\mathcal{C}\ell_{p,q-1}$. When $q = 0$ then they are both isomorphic to $\mathcal{C}\ell_{p-1,0}$. Thus, the idempotents (27.2.25) can be used to produce from $\mathcal{C}\ell_{p,q}$ two copies of either $\mathcal{C}\ell_{p,q-1}$, or when $q = 0$, $\mathcal{C}\ell_{p-1,0}$.

For example, we can show that $\mathcal{C}\ell_{0,3}$ splits into two copies of $\mathcal{C}\ell_{0,2}$. The pseudoscalar ω in $\mathcal{C}\ell_{0,3}$ is given by e_{123}. If we pre-multiply each of the basis elements by the idempotent $\frac{1}{2}(I+\omega)$ then we obtain only four distinct terms, since, for example,

$$\begin{aligned} \tfrac{1}{2}(I+\omega)e_{12} &= e_{12} - e_3 = -e_{123}e_3 - e_3 \\ &= -\tfrac{1}{2}(I+\omega)e_3. \end{aligned}$$

The four distinct terms can be written as

$$\{\tfrac{1}{2}(I+\omega), \quad \tfrac{1}{2}(I+\omega)e_1, \quad \tfrac{1}{2}(I+\omega)e_2, \quad \tfrac{1}{2}(I+\omega)e_{12}\}$$

which form a basis for $\mathcal{C}\ell_{0,2}$, with the scalar represented by $\frac{1}{2}(I+\omega)$. We are able to draw the same conclusions if we replace ω in the above analysis by $-\omega$.

If we identify $P_{\pm}$ with the helicity projection operators $h_{\pm}$, then the pseudoscalar can be said to perform a "helicity split" if (27.2.25) is satisfied. Examples of algebras where this is the case are $\mathcal{C}\ell_{0,3}$ and $\mathcal{C}\ell_{1,4}$ on which can be performed the helicity split as follows:

$$\begin{aligned} \mathcal{C}\ell_{0,3} &\rightarrow h_{+}\mathcal{C}\ell_{0,2} \oplus h_{-}\mathcal{C}\ell_{0,2} \cong \mathcal{C}\ell_{0,2} \oplus \mathcal{C}\ell_{0,2} \\ \mathcal{C}\ell_{1,4} &\rightarrow h_{+}\mathcal{C}\ell_{1,4} \oplus h_{-}\mathcal{C}\ell_{1,4} \cong \mathcal{C}\ell_{1,3} \oplus \mathcal{C}\ell_{1,3}. \end{aligned}$$

In the latter case we note that two copies of the Dirac algebra are produced.

27.3 Selecting a Higher-Dimensional Algebra for a Gauge Model

27.3.1 The Principles of the Model

As explained in section 1.1, our basic premise is that the elements of a real Clifford algebra should be used consistently to describe the fundamental fermions and their interactions. In this chapter we shall consider how we are guided towards the selection of an algebra which is appropriate for such a description.

The mathematical model used to describe the fundamental fermions and their interactions is a gauge theory, that is, a Lagrangian field theory. To take a Clifford algebraic approach to the description of a Lagrangian density, we should be able to define spinors, bar conjugate spinors, and an imaginary unit within the algebra.

Spinors are defined as minimal left ideals which exist in all algebras $C\ell_{p,q}$ irrespective of its signature. Thus we can always define a spinor. However, whether the minimal left ideal structure in a particular algebra provides a sufficient number of parameters to define an appropriate spinor remains a residual issue. We shall investigate this in more detail later in this section.

The definition (27.2.16) of a bar conjugate spinor depends on the existence of the hermitian conjugation operation Γ. The operator Γ can only be defined in $C\ell_{p,q}$ if either:

$$p \text{ is odd} \tag{27.3.1a}$$

or

$$q \text{ is even, } q \neq 0. \tag{27.3.1b}$$

The conditions under which an imaginary unit exists within the algebra are much more restrictive: from section 2.5 we require either

$$n = 1 \bmod(4) \text{ and } q \text{ odd,} \tag{27.3.2a}$$

or,

$$n = 3 \bmod(4) \text{ and } q \text{ even.} \tag{27.3.2b}$$

The condition (27.3.2a) implies that p is even and therefore (27.3.2a) directly contradicts the conditions (27.3.1) for Γ to exist. Provided that we exclude $q = 0$ in the remaining condition (27.3.2b), it is consistent with, but more restrictive than, conditions (27.3.1). Thus, we conclude that for Γ *and* an imaginary unit to exist, we require

$$n = 3 \bmod(4) \text{ and } q \text{ even, } q \neq 0. \tag{27.3.3}$$

By using the constraint (27.3.3), we can identify[9] the algebras $C\ell_{p,q}$ for $4 \leq n \leq 12$ which are suitable candidates for building our gauge model. The only allowable algebras are

$$\text{for } n = 7: \qquad C\ell_{1,6}, C\ell_{3,4}, C\ell_{5,2} \tag{27.3.4a}$$

$$\text{for } n = 11: \qquad C\ell_{1,10}, C\ell_{3,8}, C\ell_{5,6}, C\ell_{7,4}, C\ell_{9,2} \tag{27.3.4b}$$

For each of the two values of n, the allowable algebras are isomorphic. Thus, the number k of factors in a primitive idempotent is the same for all of the allowable algebras correspnding to a particular value of n. By using the formula (27.2.6) for k, we find that

$$\text{for } n = 7: \qquad k = 2 - r_1 = 3,$$

$$\text{for } n = 11: \qquad k = 6 - r_5 = 5.$$

Thus, the number 2^{n-k} of parameters in a description of spinors in the allowable seven and eleven dimensional algebras are 16 and 64, or the equivalent of two and eight Dirac spinors, respectively.

We are directed towards particular dimensional algebras on the basis of the number of different fundamental fermions we wish to describe, since this influences the number of parameters we need.

For the purposes of building a physical model it is interesting to note a property of those algebras in (27.3.4) in which $q > 3$. For any Clifford algebra in which $q > 3$, we can write

$$C\ell_{p,q} \cong C\ell_{1,3} \otimes C\ell_{p-1,q-3} \tag{27.3.5}$$

where the $\otimes$ denotes the graded direct product. In the particular case of the algebras in (27.3.4), the dimension m is

$$m = (p-1) + (q-3) = n - 4 = 3 \bmod(4),$$

since $n = 3 \bmod(4)$.

[9] Although we draw different conclusions, this follows the approach of Li Deming, *Clifford Algebraic Formulations and Quantum Theory* in *Proceedings of the International Symposium on Advanced Topics of Quantum Physics*, Ed J.Q. Liang *et al*, Science Press, (1993), pp83-89.

In section 2.5 we learnt that in an m dimensional algebra, the pseudoscalar can be used to perform a helicity split of $C\ell_{s,t}$ if, *inter alia,* $m = 3 \bmod(4)$ and t is odd. In the algebras in (27.3.4), the value of t is $q-3$ where q is even and thus t is odd. Thus the helicity split can be applied to $C\ell_{p-1,q-3}$ in (27.3.5) as follows:

$$C\ell_{p,q} \cong C\ell_{1,3} \otimes C\ell_{p-1,q-3}$$

$$h_+ \swarrow \quad \searrow h_-$$

$$C\ell_{p-1,q-4} \oplus C\ell_{p-1,q-4} \tag{27.3.6}$$

where

$$h_{\pm} = \tfrac{1}{2}(I \pm e_{56\ldots n}). \tag{27.3.7}$$

This process enables us to "separate out" space-time from the higher dimensions in gauge models based on all the algebras in (27.3.4) except those with $q < 3$. Furthermore the higher dimensions can be split into a left and right handed part using the pseudoscalar $e_{56\ldots n}$. It should also be noted that since $m = n-4$ is odd, the helicity projection operator anti-commutes with the space-time basis vectors, that is,

$$\{e_{56\ldots n}, e_i\} = 0 \quad i = 1,2,3,4.$$

27.3.2 A Model Based on $C\ell_{1,6}$

As an example, we consider what a model based on $C\ell_{1,6}$ could describe. The algebra can be separated into the graded direct product of the algebras $C\ell_{1,3}$ and $C\ell_{0,3}$ which can then be split into two copies of $C\ell_{0,2}$ using the helicity operators $h_{\pm} = \frac{1}{2}(I \pm e_{567})$:

$$C\ell_{1,6} \cong C\ell_{1,3} \otimes C\ell_{0,3}$$

$$h_+ \swarrow \quad \searrow h_-$$

$$\begin{matrix} h_+C\ell_{0,3} & \otimes & h_-C\ell_{0,3} \\ \cong C\ell_{0,2} & & \cong C\ell_{0,2} \end{matrix} \tag{27.3.8}$$

The algebra $C\ell_{1,3}$ may be associated with a curved space by introducing the vectors $\Gamma_\mu(x)$, $\mu = 1,2,3,4$, defined by (27.1.7). The vectors e_i $i = 5,6,7$ can be formally represented by $i\sigma_i$ where the σ_i are the Pauli matrices. From the definition (27.1.7) we can deduce that

$$\{\Gamma_\mu(x), e_i\} = 0, i = 5,6,7 \tag{27.3.9}$$

$$\Gamma_\mu(x)h_{\pm} = h_{\mp}\Gamma_\mu(x) \tag{27.3.10}$$

The bivectors of $C\ell_{1,6}$ contain: 6 pure space-time terms $h^j_\mu h^k_\nu e_{jk}$ which generate $SL(2,C)$; pure higher dimensional terms represented formally by $h_\pm i\sigma_i$ which generate left and right handed copies of $SU(2)$. The 12 bivectors which "mix" space-time and the higher dimensional vectors are $h^j_\mu h_\pm e_j$.

There is potential to use such an algebra to develop a model of the gravitational and electroweak interactions of the fundamental fermions using the $SL(2,C)$ and the $SU(2,C)$ bivector generators in the gauge transformations. Since each ideal in $C\ell_{1,6}$ has sufficient parameters to support 2 Dirac spinors, it is natural to use a spinor in $C\ell_{1,6}$ to represent the electroweak doublet consisting of the left handed electron ε and its neutrino ν

$$\begin{pmatrix} \varepsilon \\ \nu \end{pmatrix}_L . \tag{27.3.11}$$

The primitive idempotent used to define the spinor has the form

$$P_{+++} = \tfrac{1}{2}(I + p_1)\tfrac{1}{2}(I + p_2)\tfrac{1}{2}(I + p_3)$$

where $p_i \in C\ell_{1,6}, p_i^2 = I$ and $[p_i, p_j] = 0$. By changing the signs in the three factors comprising P_{+++} we can define seven other primitive idempotents, each one of which can be used to define an ideal in $C\ell_{1,6}$. These ideals do not interact since projection operators with different signs annihilate one another. A possible interpretation of the eight non-interacting spinors is as the eight doublets:

$$\begin{pmatrix} \varepsilon \\ \nu \end{pmatrix}_L, \begin{pmatrix} \varepsilon \\ \nu \end{pmatrix}_R, \begin{pmatrix} u_{\text{red}} \\ d_{\text{red}} \end{pmatrix}_L, \begin{pmatrix} u_{\text{red}} \\ d_{\text{red}} \end{pmatrix}_R, \begin{pmatrix} u_{\text{blue}} \\ d_{\text{blue}} \end{pmatrix}_L, \begin{pmatrix} u_{\text{blue}} \\ d_{\text{blue}} \end{pmatrix}_R, \begin{pmatrix} u_{\text{green}} \\ d_{\text{green}} \end{pmatrix}_L, \begin{pmatrix} u_{\text{green}} \\ d_{\text{green}} \end{pmatrix}_R \tag{27.3.12}$$

where the up u and down d quarks appear in each of 3 colours. The algebra $C\ell_{1,6}$ could then be used to define the electroweak and gravitational interactions of the first family of fermions, since the ideal structure would dictate that the particles in each spinor would only interact with themselves and not with the particles in another spinor.

27.3.3 A Gauge Model in $C\ell_{1,6}$

Having explored in the previous section the potential of a model based on the Clifford Algebra $C\ell_{1,6}$ to describe the electroweak and gravitational interactions of the fundamental fermions, we now consider a particular gauge model based on $C\ell_{1,6}$.

In gauge theories the particle interactions are associated with local symmetries of the spinors in the Lagrangian density. We choose a transformation of the spinors which preserves the ideal structure of the spinors ψ and bar conjugate spinors $\overline{\psi}$, defined in section 2.2. The transformations we choose are

$$\psi \to Q(x)\psi \tag{27.3.13a}$$

$$\overline{\psi} \to \overline{\psi}Q^{-1}(x) \qquad 27.13b)$$

where $Q(x)$ is an element of the algebra $C\ell_{1,6}$. The transformation (27.3.13) implies that

$$\psi\overline{\psi} \to Q(x)\psi\overline{\psi}Q^{-1}(x). \tag{27.3.14}$$

Now the product $\psi\overline{\psi}$ is not necessarily an element of an ideal of the algebra; it is an arbitrary element of $C\ell_{1,6}$. Thus, for consistency when the spinors are transformed by (27.3.13) then all the elements A in the algebra $C\ell_{1,6}$ should also be transformed according to

$$A \to Q(x)AQ^{-1}(x). \qquad (27.3.15)$$

One of the implications of the full transformations (27.3.13) and (27.3.15) is that, for example, vectors $\Gamma_\mu(x)$ $\mu = 1,2,3,4$ in the Lagrangian density will transform at the same time as the spinor. It is the transformation (27.3.15) which distinguishes spin gauge theories[10] from standard gauge theories since in the latter the vectors would be treated as constant. One of the advantages of viewing all the terms in the Lagrangian density as elements of a single Clifford algebra and thus by using the combined transformations (27.3.13) and (27.3.15) is the increased flexibility. A greater variety of gauge invariant terms can be built into the Lagrangian, since all terms of the form $\overline{\psi}A\psi$ are spin gauge invariant. Mass terms can thus be built into the Lagrangian in an invariant way.

Invariance of the Lagrangian density under the local transformations (27.3.13) and (27.3.15) is achieved by introducing a covariant derivative D_μ where

$$D_\mu = \partial_\mu - \Omega_\mu. \qquad (27.3.16)$$

The connection terms Ω_μ in D_μ transform according to

$$\Omega_\mu \to Q(x)\Omega_\mu Q^{-1}(x) - (\partial_\mu Q)Q^{-1}, \qquad (27.3.17)$$

so that

$$D_\mu\psi \to QD_\mu\psi \qquad (27.3.18)$$

and then the Lagrangian density

$$\overline{\psi}i\Gamma^\mu(D_\mu\psi) + \overline{\psi}M\psi \qquad (27.3.19)$$

is gauge invariant. The term $\overline{\psi}i\Gamma^\mu\Omega_\mu\psi$ in (27.3.19) represents the interaction of the spinor with the vector boson fields. The term $\overline{\psi}M\psi$ is the mass term for the spinor, with M being a linear combination of appropriate elements of the algebra, the real coefficients of which are the mass terms for the fundamental fermions described by the spinor ψ.

By selecting an appropriate form for the gauge transformation $Q(x)$ and hence the connection Ω_μ, the algebra $C\ell_{1,6}$ can be used to build a gauge model of the gravitational and electromagnetic interactions of the electron and its neutrino. We label the first two doublets in (27.3.12) by ψ_L and ψ_R respectively and let the spinor in the Lagrangian (27.3.19) be $\psi = \psi_L + \psi_R$. Then we define $Q(x)$ in terms of selected bivectors from $C\ell_{1,6}$, described in the previous section, as follows:

$$Q(x) = \exp[-g\theta^a(x)h_+\sigma_a - ig'\theta^4(x)(I + h_-\sigma_3) - k\theta^{ij}(x)e_{ij}] \qquad (27.3.20)$$

[10] J. S. R. Chisholm & R. S. Farwell, *Unified Spin Gauge Theories* in *Clifford Algebra and Their Applications in Mathematical Physics*, Eds J. S. R. Chisholm & A. K. Common, Reidel, (1986), pp363-370.

where $a = 1, 2, 3$ and $i, j = 1, 2, 3, 4$ $i \neq j$. The first two terms on the right hand side of (27.3.20) are the familiar $SU(2) \times U(1)$ interactions of the standard model of the electroweak interactions. The remaining term is the generator of an $SL(2, C)$ symmetry which we identify with the gravitational interactions.

The form for the connection Ω_μ follows from that of $Q(x)$ as

$$\Omega_\mu(x) = -igW_\mu^a(x)h_+\sigma_q - ig'W_\mu^4(x)(I + h_-\sigma_3) - kG_\mu^{ij}(x)e_{ij} \tag{27.3.21}$$

and contains the W vector bosons and the $SL(2, C)$ field $G_\mu^{ij}(x)$. By introducing the Weinberg mixing angle, the connection (27.3.21) may be written in terms of the electromagnetic field A_μ and the W and Z vector bosons, together with the $SL(2, C)$ field $G_\mu^{ij}(x)$. In a particular matrix representation[11] of the basis vectors in $C\ell_{1,6}$, we can "decompose" the resultant Lagrangian into four component spinors and recover exactly the electroweak interactions in the standard model together with $SL(2, C)$ interactions. However, the ability of the Lagrangian (27.3.19) with connection (27.3.21) to provide a model of the electroweak and gravitational interactions is not dependent on this particular representation being used.

The set $\{\Gamma_\mu(x)\}$ in (27.3.19) is a vector basis for $C\ell_{1,6}$ and in any representation it satisfies the identity

$$\Gamma^\mu \Gamma_\mu = 4I. \tag{27.3.22}$$

By using (27.3.22), we can factorise the mass term in the Lagrangian (27.3.19) as follows

$$\overline{\psi} IM\psi = \tfrac{1}{4}\overline{\psi}\Gamma^\mu\Gamma_\mu M\psi,$$

and hence rewrite the Lagrangian as

$$\overline{\psi} i\Gamma^\mu(D_\mu - \tfrac{1}{4}i\Gamma_\mu M)\psi. \tag{27.3.23}$$

The term in the brackets in (27.3.23) has the same form as a covariant derivative, and so we call it the "extended covariant derivative"

$$\Delta_\mu = D_\mu - \tfrac{1}{4}i\Gamma_\mu M = \partial_\mu - \Omega_\mu - \tfrac{1}{4}i\Gamma_\mu M. \tag{27.3.24}$$

It is perfectly sensible within the context of the Clifford algebra approach to define the quantity (27.3.24) since we are forming a linear combination of two algebraic expressions D_μ and $\frac{1}{4}i\Gamma_\mu M$. This approach of factorising the mass term thus applies to any spin gauge theory.

In factorising the mass term, we can recognise analogies between it and the electromagnetic interaction term. There are parallels between the electromagnetic field $A_\mu(x)$ contained in the connection Ω_μ and the term $\Gamma_\mu(x)$. We can thus consider $\Gamma_\mu(x)$ to be a field, the "frame field". Fermion mass is then interpreted as an interaction with the frame field: it is the strength of the interaction and is no longer an intrinsic property of fermions. We now consider the implications of this approach.

In the gauge model, the vector boson field strength terms may be generated in the standard way from the commutator of the covariant derivative

$$[D_\mu, D_\nu] = \lambda G_{\mu\nu} + \alpha W_{\mu\nu}^4 + \beta W_{\mu\nu}, \tag{27.3.25}$$

[11] J.S.R. Chisholm & R.S. Farwell, *J.Phys.A: Math.Gen.* **22**, 1059 (1989)

where λ, α and β are real coefficients related to the coupling constants in Ω_μ, and

$$\begin{array}{ll} G_{\mu\nu} = \partial_\mu G_\nu - \partial_\nu G_\mu - [G_\mu, G_\nu], & G_\mu = G_\mu^{ij} e_{ij} \\ W_{\mu\nu} = \partial_\mu W_\nu - \partial_\nu W_\mu - [W_\mu, W_\nu], & W_\mu = W_\mu^a h_+ \sigma_a \\ W_{\mu\nu}^4 = (I + h_- \sigma_3)(\partial_\mu W_\nu^4 - \partial_\nu W_\mu^4) & \end{array}$$

There are no cross terms mixing the fields in (27.3.25) since all generators of the gauge transformations in (27.3.20) commute with one another. The free Lagrangian for the vector bosons mediating the interactions may be generated in a standard way by using the self-commutator squared of the covariant derivative:

$$KTr\{g^{\mu\sigma} g^{\nu\tau} [D_\mu, D_\nu][D_\sigma, D_\tau]\}, \tag{27.3.26}$$

where K is taken to be an arbitrary normalisation factor.

We impose a further condition[12] that the frame field $\Gamma_\mu(x)$ be covariantly constant with respect to the "space time" covariant derivative:

$$\partial_\mu \Gamma_\nu - \Gamma_{\mu\nu}^\alpha - [G_\mu, \Gamma_\nu] = 0$$

where $\Gamma_{\mu\nu}^\alpha$ are the Christoffel symbols. This has the effect of producing a relationship between the curvature tensor $R_{\mu\nu\alpha\beta}$ and $G_{\mu\nu}\Gamma_\alpha\Gamma_\beta$, and thus, substituting for $G_{\mu\nu}$ in the Lagrangian (27.3.26), we obtain a term $R_{\mu\nu\alpha\beta}R^{\mu\nu\alpha\beta}$ in the Lagrangian. However, this term is not what is needed for a gravitational Lagrangian since it is quadratic in the curvature.

In our model it is natural to consider the consequence of replacing the covariant derivative in (27.3.25) by the extended covariant derivative to form an alternative to (27.3.26) given by

$$KTr\{g^{\mu\sigma} g^{\nu\tau} [\Delta_\mu, \Delta_\nu][\Delta_\sigma, \Delta_\tau]\}. \tag{27.3.27}$$

The terms arising by using (27.3.24) in (27.3.27) are quite astonishing.Since

$$[G_\mu^{ij} e_{ij}, \Gamma_\nu] \neq 0, \quad [W_\mu^\alpha h_+ \sigma_a, \Gamma_\nu] \neq 0, \quad [h_- \sigma_3 W_\mu^4, \Gamma_\nu] \neq 0,$$

the commutator $[\Delta_\mu, \Delta_\nu]$ contains additional terms in comparison with $[D_\mu, D_\nu]$ precisely because of the presence of the frame field term. These additional terms produce in the Lagrangian (27.3.27) the following:

- boson mass terms with the correct W/Z mass ratio and zero mass photons;
- a term proportional to the Einstein-Hilbert gravitational Lagrangian $R/16\pi G$, where R is the curvature scalar;
- a constant term.

In a quite remarkable way, the mass terms for the intermediate vector bosons *and* also the gravitational Lagrangian arise as a direct consequence of the frame field.

[12] J.S.R. Chisholm & R.S. Farwell, *Gen.Rel. and Grav.* **20**, 371 (1988).

Chapter 28

The Extended Grassmann Algebra of $\mathbf{R}^3$

Bernard Jancewicz
Institute of Theoretical Physics, University of Wrocław,
pl. Maksa Borna 9, PL-50-204 Wrocław, Poland

28.1 Introduction

When a linear space of vectors is devoid of a metric, many quantities can be introduced: multivectors, pseudo-multivectors and their duals, namely forms and pseudoforms. In $\mathbf{R}^3$, sixteen types of such quantities exist, and to each of them elementary geometric images can be ascribed by replacing traditional line segments with arrows, related to vectors. The possibility of distinguishing, for instance, vectors and linear forms is very important in theoretical physics, where formalisms are used with non-Euclidean spaces or spaces with variable metrics. This is one reason why it is important to establish the algebraic nature of physical quantities in a pre-metric space.

During the last two decades, electrodynamics has frequently been presented in the language of differential forms; see Refs [1–10]. Not all these presentations, however, took enough care in their use of pseudoforms. Only Schouten [1], Frankel [3], Ingarden and Jamiołkowski [6] and Burke [7] applied them in electrodynamics, where they called them *covariant W-p-vectors* [1], *twisted forms* [3,7] or *odd forms* [6]. One first has to consider pseudomultivectors in order to understand properly pseudoforms. This lecture gives a systematic presentation of pseudomultivectors and pseudoforms. I call these objects *directed quantities* because one can and should introduce a separate direction for each of them.

Let us start from the vector depicted as a directed segment. Its relevant features are direction and magnitude. The *direction* consists of a straight line (on which the vector lies), which I call, after Lounesto [11], an *attitude* of the vector, and an arrow on that line which we call an *orientation.* Two vectors of the same attitude are parallel. What is their magnitude? Usually, one says that magnitude is the length of the directed segment. To determine the length, however, one needs a scalar product, and here the problem starts. Is there a unique scalar product for a given vector space? The answer is "no", and its justification can be found at the beginning of Section 7. It should be obvious that a scalar product, and the norm and metric determined by it are not unique for a given vector space. Therefore, there is a need to consider vector spaces that are devoid of a metric.

One should be aware of the fact that not all of the familiar geometric notions exist in vector spaces without scalar products. Valid notions include: linearly dependent vectors, linearly independent vectors, parallel vectors and parallel planes. Undefined notions include: angles, perpendicular vectors, comparable lengths of nonparallel vectors, circles and spheres.

If there is no notion of length, what can replace the magnitude of a vector? The only possiblity is the *relative magnitude*, which is a measure for comparing parallel vectors only. If $\mathbf{a}$ and $\mathbf{b}$ are two parallel vectors, then a scalar λ exists such that $\mathbf{b} = \lambda\mathbf{a}$ and we say that $|\lambda|$ is the *magnitude of* $\mathbf{b}$ *relative to* $\mathbf{a}$. In other words, the relative magnitude of a vector parallel to $\mathbf{a}$ is its length in units of $\mathbf{a}$. Two vectors $\mathbf{b}$, $\mathbf{c}$ with the same direction and the same magnitude relative to $\mathbf{a}$ are equal: $\mathbf{b} = \mathbf{c}$.

In the case of a linear space without a scalar product, there is a need to introduce dual objects known as *linear forms* that, in a sense, replace the scalar product. The possibility of distinguishing between vectors and linear forms is very important in such parts of theoretical physics where formalisms are introduced for non-Euclidean spaces or spaces with a variable metric. A most spectacular case in which this distinction is significant is the theory of gravitation, where the metric itself becomes a dynamical quantity. This provides an additional incentive for establishing the algebraic nature of physical quantities in a space devoid of a scalar product.

The primary directed quantities are *multivectors:* vectors, bivectors, and trivectors. Just as a vector, in the process of abstraction, arises from a straight line segment with an orientation, so a *bivector* originates from a plane segment with an orientation, and a *trivector* from a solid body with an orientation. The connection of multivectors to straight lines, planes and bodies gives them the advantage of being easily depicted in illustrations. For their application in classical mechanics and in electromagnetism, see Ref. [12] and Ref. [13], respectively.

Outer forms are quantities dual to multivectors. They are called *differential forms* when they depend on their position in space. Differential forms are very popular in theoretical physics, but authors writing about them rarely use pictures to illustrate the concept to the reader; nice exceptions are Refs. [2], [7] and [14] in which outer forms are presented as specific "slicers". However, no care is put on the direction of an outer form.

Eight types of directed quantities (vectors, pseudovectors, bivectors, pseudobivectors, trivectors, pseudotrivectors, scalars and pseudoscalars) are discussed in Section 2. Their eight duals (one-forms, pseudo-one-forms, two-forms, pseudo-two-forms, three-forms, pseudo-three-forms, zero-forms and pseudo-zero-forms) are introduced in Section 3. Quantities of different kinds can be multiplied by using the outer product, or by using the contraction, which are discussed in Section 4.

Let us denote the set of all multivectors and pseudomultivectors by $\mathcal{M}$. Its linear span, endowed with the outer product, forms an algebra which we call *the extended Grassmann algebra of multivectors.* A similar extended Grassmann algebra can be introduced for the linear span of forms and pseudoforms.

Physical quantities can be assigned to each directed quantity, as we demonstrate in Section 5. All directed quantities can be replaced by vectors or pseudovectors or scalars or pseudoscalars when a metric is introduced into the vector space, as we show in Section 6.

28.2 Multivectors and pseudo-multivectors

Let us start by summarizing the relevant features of a vector:

1. *direction,* consisting of *attitude* – a straight line, and *orientation* – the sense of the line, represented by an arrow;

2. *relative magnitude* – length in units of a reference vector, defined separately for various attitudes.

The described representation of a vector can be introduced in a linear space of arbitrary dimension. The description of the next quantity, however, is specific to three-dimensional space. A *pseudovector* has an attitude and relative magnitude defined in the same way as for a vector; only the definition of orientation is different. If the line segment is given, the *orientation of a pseudovector* is taken as a ring with an arrow, surrounding the segment. The reader should agree that only two different orientations are possible for a fixed attitude. The two possible orientations, obviously, are called *opposite.* They are depicted in Figure 28.1. Summarizing, the features of a pseudovector are:

1. *direction*: *attitude* – a straight line, *orientation* – the sense of a ring surrounding the line;

2. *relative magnitude* – the length in units of a reference vector.

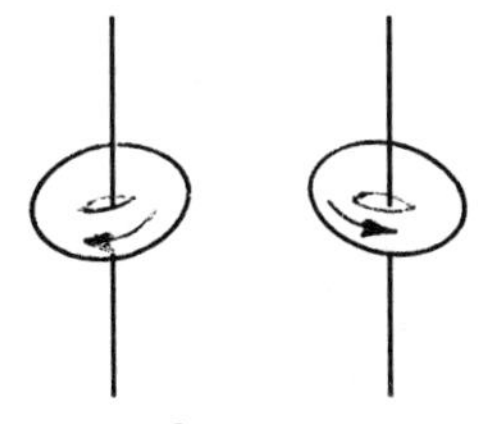

Fig. 28.1. Opposite orientations for pseudovectors with the same attitude.

The next directed quantity is a *bivector.* Its *attitude* is taken as a plane, and its *orientation,* as a circulation in the plane. It should be obvious that only two different orientations are possible for a given attitude – they are called *opposite.* The *relative magnitude* of bivectors is based on comparing areas. Thus, we can visualize bivectors as flat figures with curved arrows in their boundaries; see Figure

28.2. The shape of the figure is not important; only its area and orientation are essential. Summarizing, the features of a bivector are:

1. *direction: attitude* – a plane, *orientation* – a curved arrow in the plane;

2. *relative magnitude* – area in units of a reference figure.

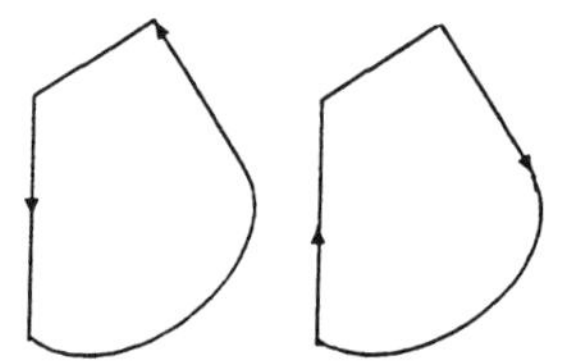

Fig. 28.2. Opposite orientations for bivectors with the same attitude.

A bivector can be obtained from two vectors **a** and **b** in the following manner: we choose one of the vectors first – let it be **a**. Then we juxtapose the origin of **b** to the tip point of **a** and draw two parallel segments to obtain a parallelogram; see Figure 28.3. The vectors **a**, **b** lying on the boundary determine the orientation of the result. In this way, we obtain the figure which represents the bivector **B**, called the *outer product of vectors* **a** and **b**. This product is denoted by a wedge:

$$\mathbf{B} = \mathbf{a} \wedge \mathbf{b} .$$

The attitude of the product is the linear span of the attitudes of the factors. Such a representation of a bivector (also called a *factorization into an outer product*) is obviously non-unique. The identity $\mathbf{a} \wedge \mathbf{b} = -\mathbf{b} \wedge \mathbf{a}$ is easily demonstrated; this means that the outer product is *anticommutative.*

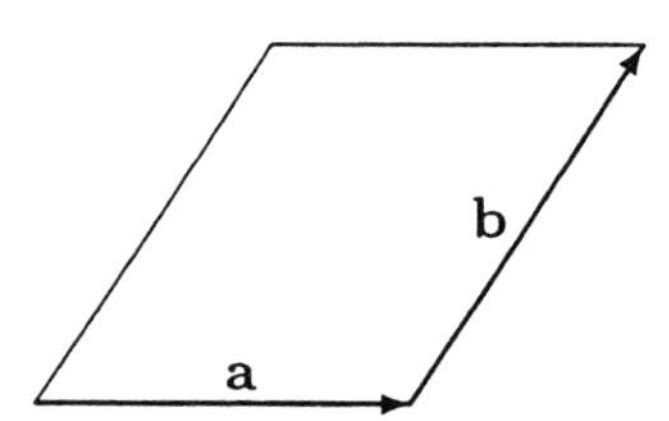

Fig. 28.3. The outer product $\mathbf{a} \wedge \mathbf{b}$ of two vectors.

Another directed quantity is the *pseudobivector*, which differs from the bivector only in the definition of the orientation. We define it as an arrow which intersects the plane of the figure. Figure 28.4 depicts two pseudobivectors with opposite orientations. Summarizing, the features of pseudobivectors are:

1. *direction: attitude* – a plane, *orientation* – an arrow not parallel to the plane;

2. *relative magnitude* – area in units of a reference figure.

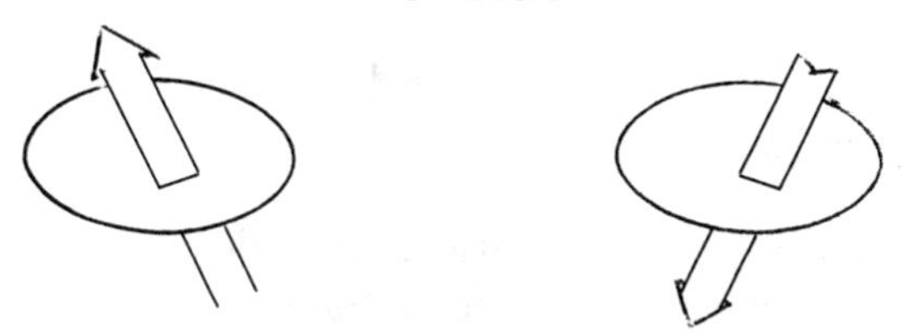

Fig. 28.4. Opposite orientations of pseudobivectors with the same attitude.

A pseudobivector $\mathbf{B}$ can be obtained from a vector $\mathbf{a}$ and a pseudovector $\mathbf{c}$ in the way illustrated in Figure 28.5. The result is called the *outer product* of $\mathbf{a}$ and $\mathbf{c}$ and is written as $\mathbf{B} = \mathbf{a} \wedge \mathbf{c}$. This product is *anticommutative*: $\mathbf{a} \wedge \mathbf{c} = -\mathbf{c} \wedge \mathbf{a}$.

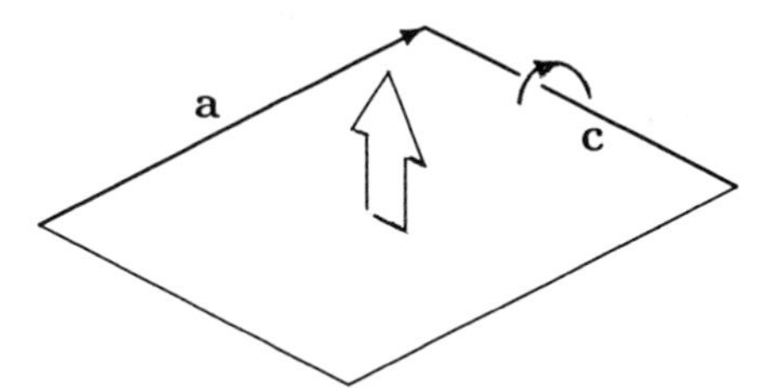

Fig. 28.5. The outer product $\mathbf{a} \wedge \mathbf{c}$ of a vector $\mathbf{a}$ with a pseudovector $\mathbf{c}$.

The missing product is that of two pseudovectors which we define now. In order to find the outer product of two pseudovectors $\mathbf{c}$ and $\mathbf{d}$, we juxtapose the segments in such a way that the ring of one segment, after passing through the junction, agrees with the orientation of the other segment. There are two possibilities of doing this; see Figure 28.6. We draw two parallel segments through free ends and obtain a parallelogram. We take this parallelogram as the outer product $\mathbf{c} \wedge \mathbf{d}$ with the orientation from $\mathbf{c}$ to $\mathbf{d}$; see the curved arrow in Figure 28.6. Attitude of the product is again the linear span of the attitudes of the factors. The other product of pseudovectors is also *anticommutative*.

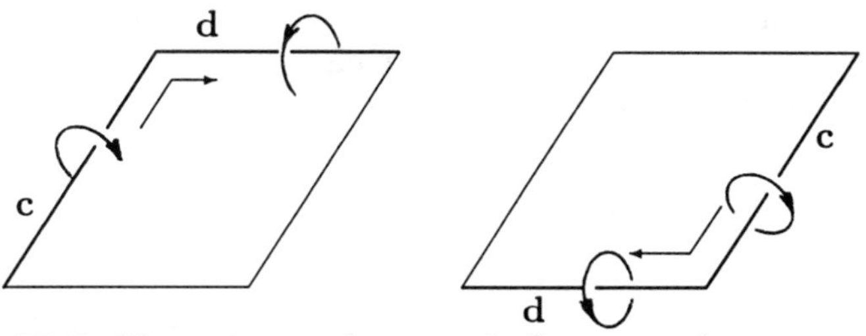

Fig. 28.6. The outer product $\mathbf{c} \wedge \mathbf{d}$ of two pseudovectors.

The examples considered above show that a given quantity differs from its corresponding pseudo-quantity only by its orientation. A dimension of the straight arrow is 1; it is natural to assume that a dimension of the curved arrow is 2. Then the following observation can be made. If the dimension of the orientation of a quantity is m, then that of the pseudo-quantity is $3 - m$. In this sense, the two quantities have orientations *complementary* to the dimension of space. The orientation of an ordinary quantity is intrinsic to its attitude (a straight arrow lies on the straight line, a curved arrow lies on the plane), whereas the orientation of a pseudo-quantity is extrinsic to its attitude (a curved arrow surrounds the straight line, a straight arrow pierces the plane).

For three-dimensional objects, the natural attitude is the volume, relative, of course, to a chosen parallelopiped. The orientation is a combination of a rotational motion with a translation; the latter cannot be parallel to the plane of rotation. The orientation is considered invariant under any rotation of both the direction of translatory motion and the plane of rotation together. Therefore, only two different orientations are possible in three-dimensional space; they are depicted in figure 7. They correspond to two kinds of screws: those with a left-handed thread (left in Figure 28.7) and those with a right-handed thread (right in Figure 28.7). The two orientations can be also visualized as spirals: left-handed and right-handed, respectively, shown in Figure 28.8.

Fig. 28.7. Two opposite three-dimensional orientations

Fig. 28.8. Two opposite three-dimensional orientations.

Now we are ready to define a *trivector* as a geometric object with the following relevant features:

1. *direction: attitude* – a 3-space, *orientation* – a handedness;
2. *relative magnitude* – volume in units of a reference parallelopiped.

Its geometric image is a solid body with a spiral or two convoluted arrows inside it, see Figure 28.9, left. If one shifts the curved arrow in the direction shown by the straight arrow, one obtains the orientation depicted on the boundary of the body, see Figure 28.9, right.

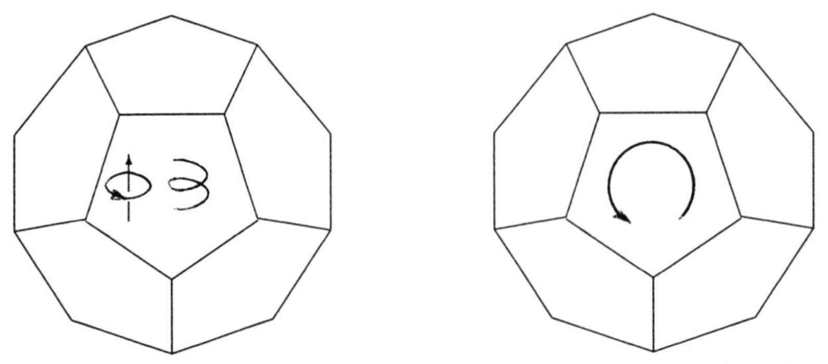

Fig. 28.9. Two ways of depicting right-hand orientation of a trivector: inside and on the boundary.

The trivector can be constructed from a vector **c** and a bivector **B** as follows. We put the tip point of **c** at the boundary of **B** and move it in a parallel fashion along this boundary to obtain an oblique cylinder with the base **B**; see Figure 28.10. In this way, all the relevant features of a trivector **T** are obtained – its orientation is the curved arrow of **B** plus straight arrow of **c** which, in the displayed case, yields the right-handed spiral. The attitude is the only possible one in the 3-dimensional space; the magnitude is naturally given as the volume of the cylinder. The presented operation ascribing **T** to the factors **c** and **B** is called the *outer product* and is denoted

$$\mathbf{T} = \mathbf{c} \wedge \mathbf{B}.$$

The attitude of the product is the span of the attitudes of the factors.

A trivector can be also treated as the outer product of three vectors after factorization $\mathbf{B} = \mathbf{a} \wedge \mathbf{b}$, namely $\mathbf{T} = \mathbf{c} \wedge (\mathbf{a} \wedge \mathbf{b})$; see Figure 28.11. The outer product of three vectors is *associative*, so one can write $\mathbf{T} = \mathbf{c} \wedge \mathbf{a} \wedge \mathbf{b}$. It is interesting to notice that the trivector can be also obtained as the outer product of a pseudovector with a pseudobivector.

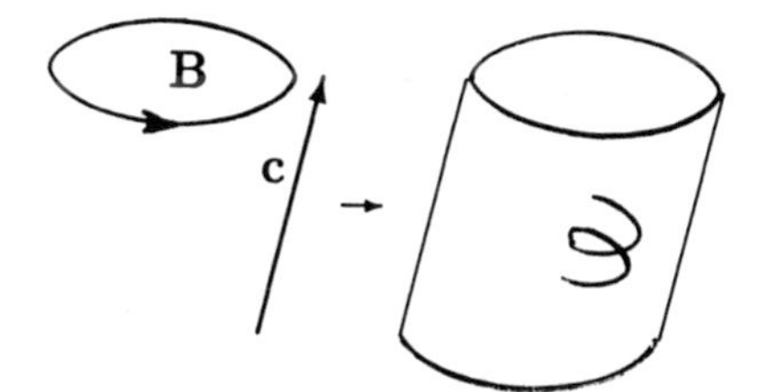

Fig. 28.10. The outer product $\mathbf{c} \wedge \mathbf{B}$.

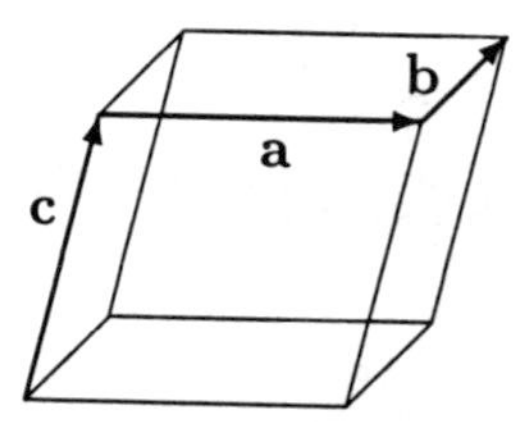

Fig. 28.11. The outer product of three vectors.

Now the next three-dimensional quantity is a *pseudotrivector*, for which we take the attitude and magnitude the same as for the trivector; only the orientation should be complementary. What is a zero-dimensional orientation? It is an orientation of the scalar, that is, sign. Hence, the two possible orientations of a pseudotrivector are *plus* and *minus* signs. In this manner, we define a *pseudotrivector* as a geometric object with the following relevant features:

1. *direction: attitude* – a 3-space, *orientation* – a sign;

2. *relative magnitude* – volume in units of a reference parallelopiped.

Geometric image of the pseudo-trivector is a solid body endowed with a sign, see Figures 28.12 and 28.13, left. One may also depict the orientation on the boundary, treated similarly to pseudovectors. The natural rule is to denote the positive orientation by the outward arrows, see Figure 28.12, right, and the negative orientation by the inward arrows, see Figure 28.13, left.

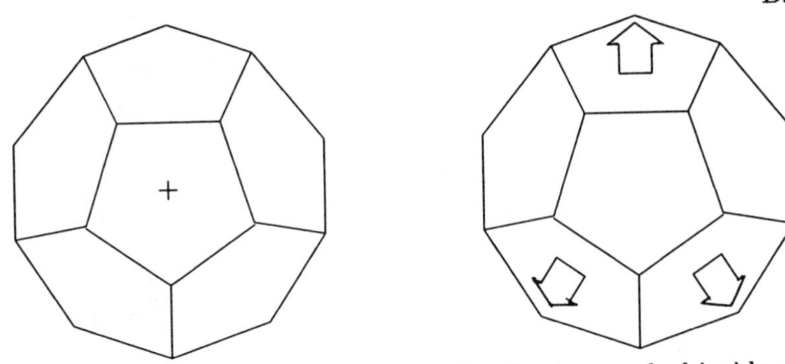

Fig. 28.12. Positive orientation of a pseudo-trivector marked inside and on the boundary.

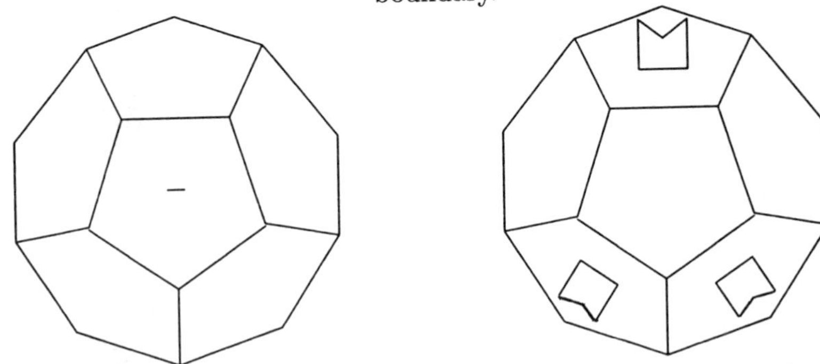

Fig. 28.13. Negative orientation of a pseudo-trivector marked inside and on the boundary.

The pseudotrivector can be represented as an outer product of a pseudobivector with a vector (Figure 28.14 shows the two possible orientations of the product) or as an outer product of a bivector with a pseudovector (the two possible results are depicted in Figure 28.15).

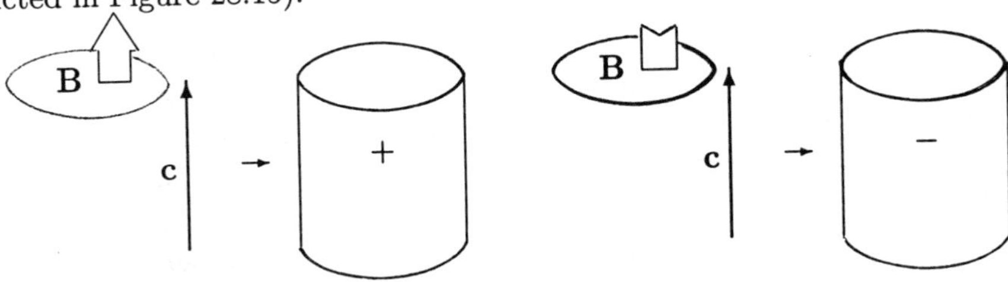

Fig. 28.14. The outer product of a vector and pseudobivector.

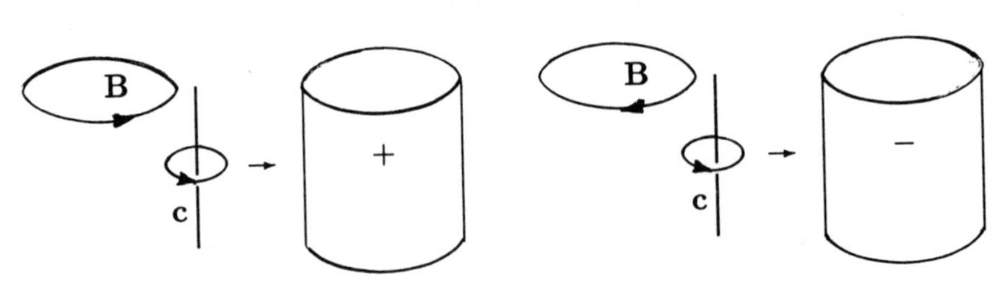

Fig. 28.15. The outer product of a pseudovector and a bivector.

By using the same reasoning for creating pseudo-quantities we form a pseudoscalar from the scalar. After recalling the features of the scalar:

1. *direction: attitude* – a point, *orientation* – a sign;

2. *magnitude* – absolute value;

we immediately list the relevant features of the *pseudoscalar:*

1. *direction: attitude* – a point, *orientation* – a handedness;

2. *magnitude* – absolute value.

The multiplication of multivectors by scalars can be treated as the outer product in which the attitude of the product is the linear span od the attitudes of the factors. In fact, linear span of a point and an arbitrary subspace is that other subspace without any alteration which fits to the fact that multiplication by scalars does not change the direction of the other factor.

As a kind of summary, we should notice from all the examples of the outer products that the attitude of the result is a span of the attitudes of the factors. A given quantity differs from its corresponding pseudo-quantity only by the orientation, which is intrinsic to the attitude for the ordinary quantities and extrinsic for the pseudo ones.

28.3 Forms and pseudoforms

The next types of directed quantities are dual to the ones previously mentioned.

The *linear form* (or *one-form* or *outer form of first order*) is a linear mapping of vectors into scalars: $\mathbf{f} : \mathbf{x} \to \mathbf{f}(\mathbf{x}) \in \mathbf{R}$. It is known from algebra that its *kernel*, i.e. the set $M = \{\mathbf{x} : \mathbf{f}(\mathbf{x}) = 0\}$ is a linear subspace of $\mathbf{R}^3$ with dimension two, so M is a plane. Thus, for each linear form, one can associate a plane M passing through the origin. One can also find other planes parallel to M, on which the form $\mathbf{f}$ takes values 1, 2, 3 and so on. In this way, we get the geometric image of a form as a family of equidistant parallel planes with an arrow joining the neighbouring planes and showing the direction of increase of the form; see Figure 28.16.

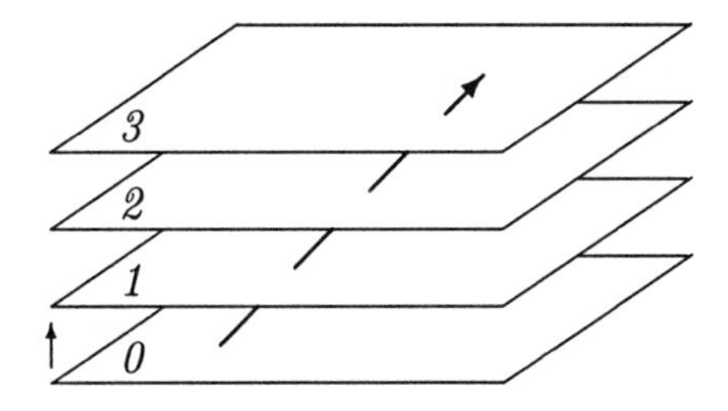

Fig. 28.16. Geometric image of a one-form.

Conversely, once we have such a family of planes, a number can be ascribed to any vector $\mathbf{x}$ as follows. If the origin of $\mathbf{x}$ lies on one of the planes, we count the number of planes pierced by $\mathbf{x}$ and this is equal to $\mathbf{f}(\mathbf{x})$; see Figure 28.16. If $\mathbf{x}$ ends on some plane, the number $\mathbf{f}(\mathbf{x})$ is integer; otherwise $\mathbf{f}(\mathbf{x})$ is not integer. If $\mathbf{x}$ intersects the planes in the direction of decreasing labels of planes, then $\mathbf{f}(\mathbf{x})$ is negative. This prescription, along with the corresponding pictures, can be found in Ref. [2].

For another form $\mathbf{f}' = 2\mathbf{f}$, which is two times $\mathbf{f}$, i.e. $\mathbf{f}'(\mathbf{x}) = 2\mathbf{f}(\mathbf{x})$ for any $\mathbf{x}$, the planes must be distributed twice as densely. In this way, we arrive at the conclusion that the (relative) *magnitude* of a form must be a linear density or the inverse of the ratio of the distance between the planes of the form $\mathbf{f}'$ and that of the form $\mathbf{f}$. In a family of parallel forms, one can use a common reference vector to measure their relative magnitudes by counting how many planes of a given form are cut by this vector. In this manner, a natural unit of the measure for the family of parallel forms is the inverse length of the reference vector.

As a matter of fact, in order to describe a linear form one does not need to draw infinitely many parallel equidistant planes. Two neighbouring ones are sufficient; see Figure 28.17. Thus we claim that the geometric image of a linear form is a *slab* (a *plane-parallel layer)* with an arrow penetrating it from one boundary to the other. Such an illustration of a linear form (called a *lamellar vector*) was considered in Ref. [15].

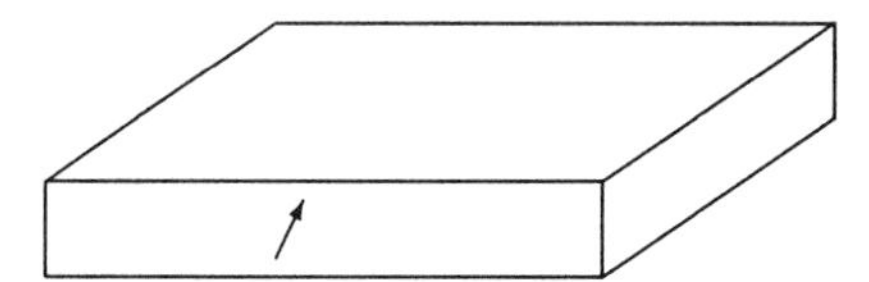

Fig. 28.17. Simplified geometric image of a one-form.

In this manner, we propose the following list of relevant features of the linear form:

1. *direction: attitude* – a plane, *orientation* – an arrow not parallel to the plane;

2. *relative magnitude* – linear density in inverse units of a reference vector.

A *pseudo-one-form* $\mathbf{f}$ is a linear mapping of pseudovectors into scalars. It can be represented geometrically by an infinite family of equidistant parallel planes with curved arrows on them, see Figure 28.18, or by a plane-parallel layer with curved arrows on its two boundaries; see Fig. 28.19. The value of $\mathbf{f}(\mathbf{x})$ is equal to the number of planes pierced by the pseudovector $\mathbf{x}$. The relevant features of a pseudo-one-form are:

1. *direction*: *attitude* – a plane, *orientation* – a curved arrow in the plane;

2. *relative magnitude* – a linear density.

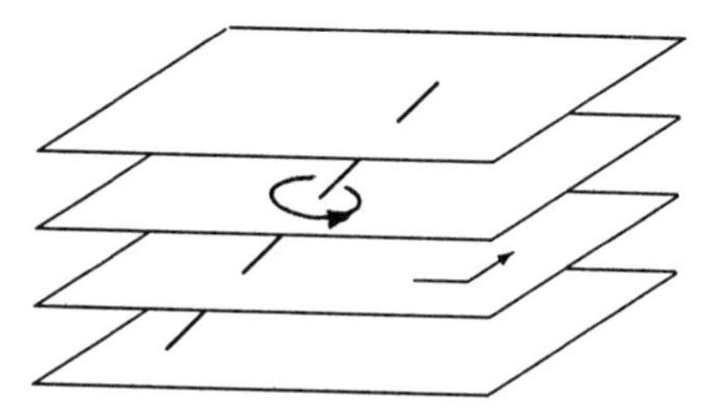

Fig. 28.18. Geometric image of a pseudo-one form.

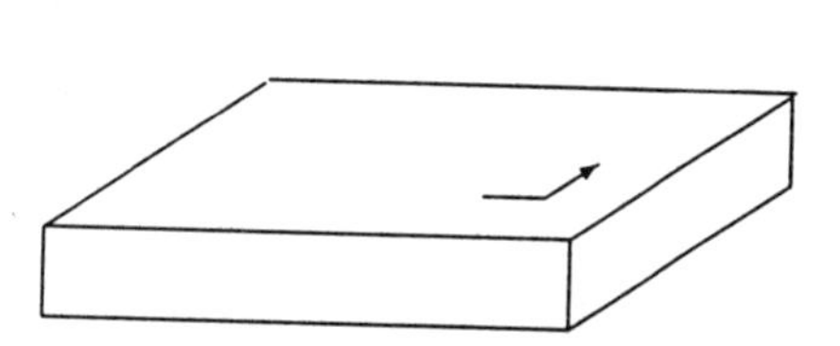

Fig. 28.19. Simplified geometric image of a pseudo-one-form.

An *outer form of second order*, or simply *two-form* is defined as a linear mapping of bivectors into scalars. It can be represented geometrically as a system of parallel prisms of equal cross-section, completely filling the space. Moreover, each prism has an arrow surrounding it on the boundary; see Figure 28.20. The value of a two-form $\mathcal{F}$ on a bivector **B** is the number of prisms intersected by the bivector, with a plus sign when the orientations of **B** and $\mathcal{F}$ are the same and a minus sign when they are opposite.

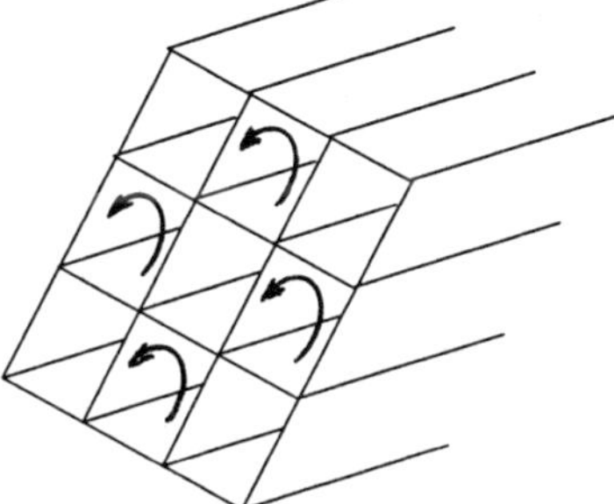

Fig. 28.20. Geometric image of a two-form.

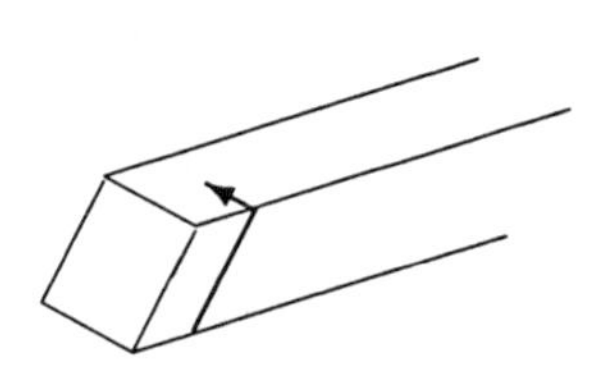

Fig. 28.21. Simplified geometric image of a two- form.

As a matter of fact, in order to determine a two-form one does not need an infinite number of prisms, one only is sufficient. Thus we claim that a geometric image of a two-form is a single infinite *bar* with an arrow surrounding it; see Figure 28.21. Its attitude is determined by the axis of the bar. Thus, the following are the relevant features of the two- form:

1. *direction: attitude* – a straight line, *orientation* – an arrow circling the line;
2. *relative magnitude* – planar density in inverse units of a reference bivector.

One may build a two-form $\mathcal{F}$ from two one-forms **f** and **g** in the way depicted in Figure 28.22. The attitude of $\mathcal{F}$ is obtained as the intersection of the attitudes (planes) of **f** and **g**, the orientation is obtained by juxtaposing the arrows of **f** and **g**. The result is called the *outer product of one-forms* **f** and **g** and is denoted as

$$\mathcal{F} = \mathbf{f} \wedge \mathbf{g} .$$

The attitude of the product is now an intersection of the attitudes of the factors. This representation of $\mathcal{F}$, also known as a *factorization into an outer product*, is not unique.

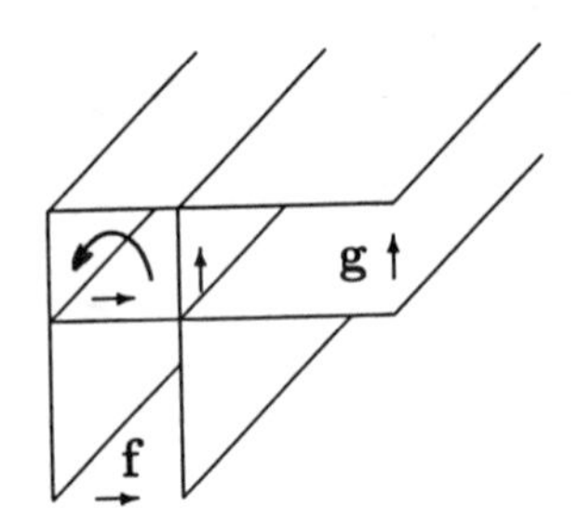

Fig. 28.22. The outer product of two one-forms.

A *pseudo-two-form* is a linear mapping of pseudobivectors into scalars. Its geometric image is a system of parallel prisms of equal cross sections, completely filling the space, each prism having an arrow along its axis, see **Fig. 28.23.** The pseudo-two-form can also be represented as an infinite bar with an arrow along its axis; see Figure 28.24. Thus, its relevant features are:

1. *direction*: *attitude* – a straight line, *orientation* – an arrow on the line;
2. *relative magnitude* – planar density.

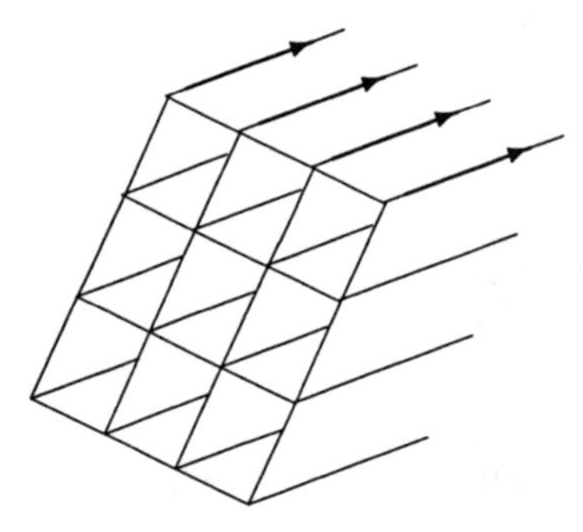

Fig. 28.23. Geometric image of a pseudo-two-form.

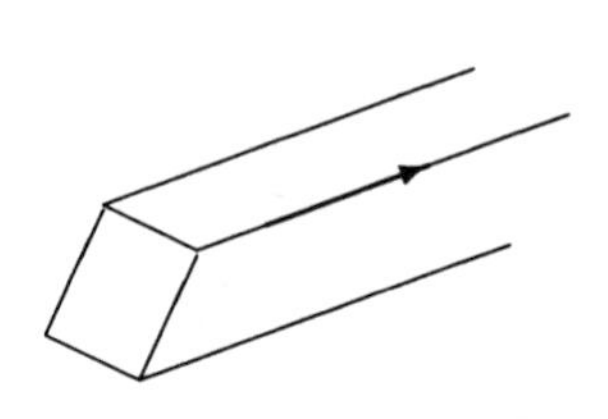

Fig. 28.24. Simplified geometric image of a pseudo-two-form.

A pseudo-two-form can be obtained from an *outer product* of a one-form **g** with a pseudo-one-form **f**, as depicted in Figure 28.25. The attitude of the product is again the intersection of the attitudes of the factors. The two last products are *anticommutative.*

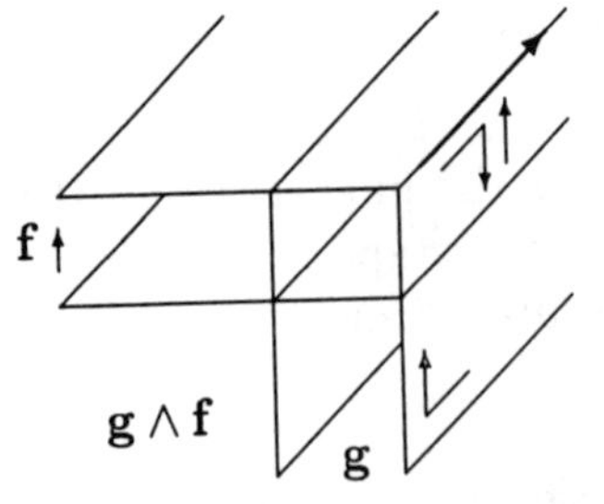

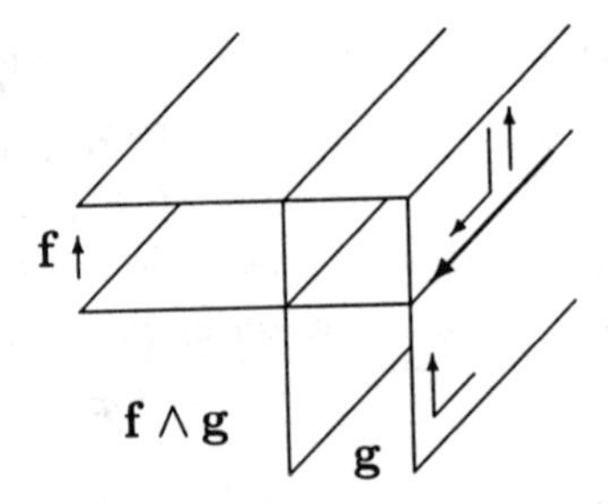

Fig. 28.25. The outer products $\mathbf{g} \wedge \mathbf{f}$ and $\mathbf{f} \wedge \mathbf{g}$ of a one-form **f** with a pseudo-one-form **g**.

An outer form of third order or simply *three-form* is a linear mapping of trivectors into scalars. It can be represented geometrically as a family of cells of equal volumes, completely filling the space. Moreover, each cell has a three- dimensional orientation – namely, a spiral in it – and all the spirals are of the same handedness; see Figure 28.26. The value of a three-form $\mathcal{R}$ on a trivector $\mathbf{T}$ is the number of cells occupied by the trivector, with the plus sign when the orientations of $\mathbf{T}$ and $\mathcal{R}$ are the same, and minus sign when they are opposite. Actually, in order to determine a three-form one does not need an infinite number of cells, one only is sufficient. Thus, we claim that the geometric image of a three-form is a single parallelopiped with a spiral inside it; see Figure 28.27.

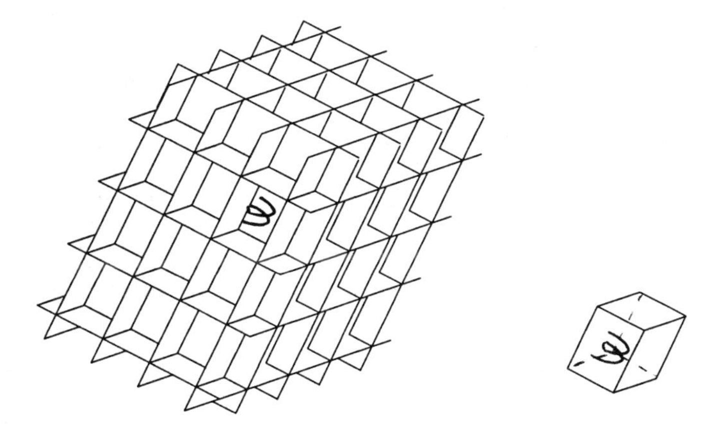

Fig. 28.26. Geometric image of a three-form.

Fig. 28.27. Simplified geometric image of a three- form.

An intruiging question is: what is an attitude of the three-form? As we may notice from cases considered, attitudes of the dual quantities are complementary to the dimension of space. Hence, we ought to assume that an attitude of the three-form must be a point. Thus, the following are the relevant features of the three-form:

1. *direction: attitude* – a point, *orientation* – a handedness;

2. *relative magnitude* – spatial density in inverse units of a reference parallelopiped.

One may build a three-form from a one-form and a two-form as the intersection of them. We depict in Figure 28.28 the operation, called the *outer product*, giving all the relevant features of the resulting trivector. The three-form can be also obtained as the outer product of a pseudo-one-form with a pseudo-two-form. In each case, the attitude of the product is the intersection of the attitudes of the factors.

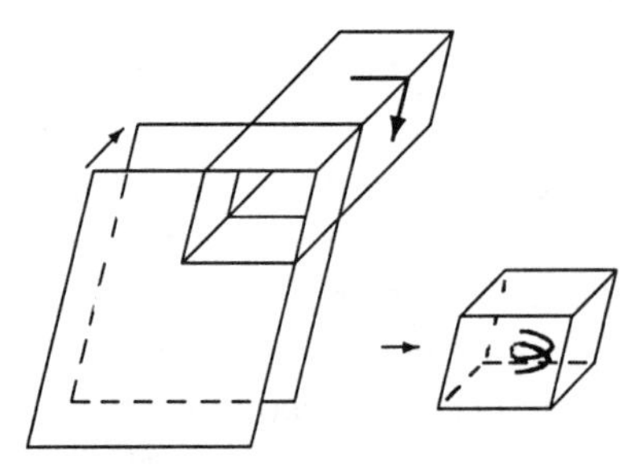

Fig. 28.28. The outer product of a one-form and a two-form.

The next directed quantity is *pseudo-three-form* for which we take the same attitude and magnitude as for the three-form. Only the orientation should be complementary; that is sign. In this manner, we define the pseudo-three-form as a geometric object with the following relevant features:

1. *direction; attitude* – a point, *orientation* – a sign;
2. *relative magnitude* – spatial density in inverse units of a reference parallelopiped.

The pseudo-three-form can be represented as an outer product of a pseudo-two-form with a one-form as shown in Figure 28.29 or as an outer product of a two-form with a pseudo-one-form as illustrated in Figure 28.30.

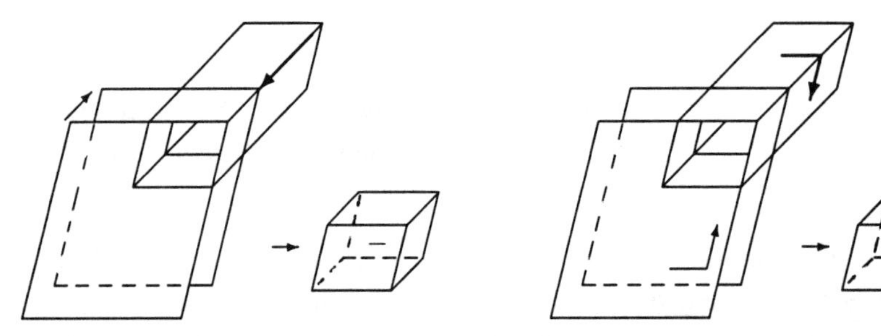

Fig. 28.29. The outer product of a one-form and a pseudo-two-form.

Fig. 28.30. The outer product of a pseudo-one-form and a two-form.

We would like to summarize directed quantities introduced in this Section. It is easy to notice that two items are lacking, namely objects dual to scalars and pseudoscalars. Obviously, the names are easy to propose for them, namely *zero-form* and *pseudo-zero-form.* The relevant features are also easily recognized: for the zero-form – 3-space, sign and absolute value; for the pseudo-zero-form – 3-space, handedness and absolute value, respectively, as the attitude, orientation and magnitude. But is there a need to introduce them? Especially in view of the widespread notion that the zero-forms are identical with scalars? My answer to this question is positive with the following justification. Multiplication of multivectors by scalars can be treated as the outer product and, in this manner, the Grassmann algebra of multivectors is obtained. Similarly, some counterpart of the scalar is needed for the outer forms to build a Grassmann algebra of them. The outer product of the forms gives the attitude of the result as the intersection of their attitudes. Hence this counterpart of the scalar has to have three-dimensional attitude which, after intersection with any other attitudes, leaves them invariant. The attitude is the only feature of zero-forms in which they differ from scalars. The same argument can be applied to pseudoscalars in order to justify the need of pseudo-zero-forms.

We have summarized the relevant features of eight ordinary directed quantities in Table 1.

Table 28.1

features	scalar	vector	bivector	trivector
attitude	point	straight line	plane	3-space
orientation	sign	arrow on	curved arrow on	handedness
magnitude	abs. value	length	area	volume
	zero-form	one-form	two-form	three-form
attitude	3-space	plane	straight line	point
orientation	sign	arrow piercing	curved arrow around	handedness
magnitude	abs. value	linear density	planar density	spatial density

We summarize the rest of the directed quantities of this Section, namely the pseudo-quantities, in Table 28.2. (They were marked by red chalk during the lectures.)

Table 28.2

features	pseudo-scalar	pseudo-vector	pseudo-bivector	pseudo-trivector
attitude	point	straight line	plane	3-space
orientation	handedness	curved arrow around	arrow piercing	sign
magnitude	abs. value	length	area	volume
	pseudo-zero-form	pseudo-one-form	pseudo-two-form	pseudo-three-form
attitude	3-space	plane	straight line	point
orientation	handedness	curved arrow on	arrow on	sign
magnitude	abs. value	linear density	planar density	spatial density

28.4 Linear spaces

For each of the directed quantities introduced above, the operations of addition and multiplication by scalars can be defined. In this manner, their sets separately form linear spaces. We proceed now to define the linear operations.

Multiplication by scalars can be defined once for all quantities. If $\mathbf{A}$ is an arbitrary directed quantity, and λ a scalar, then the quantity $\lambda\mathbf{A}$ has magnitude $|\lambda|$ times the magnitude of $\mathbf{A}$, attitude the same as $\mathbf{A}$ and orientation the same as $\mathbf{A}$ if $\lambda > 0$ or opposite if $\lambda < 0$.

Addition of pseudovectors is defined by juxtaposing the segments in such a way that the ring of one segment, after passing through the junction, agrees with the orientation of the other segment. Then, by joining the free ends by a third segment, we get the result of the addition. Figure 28.31 illustrates this procedure and also shows that the addition is commutative. Since we have two linear operations, namely addition of pseudovectors and their multiplication by scalars, we ascertain that the set of pseudovectors is a linear space.

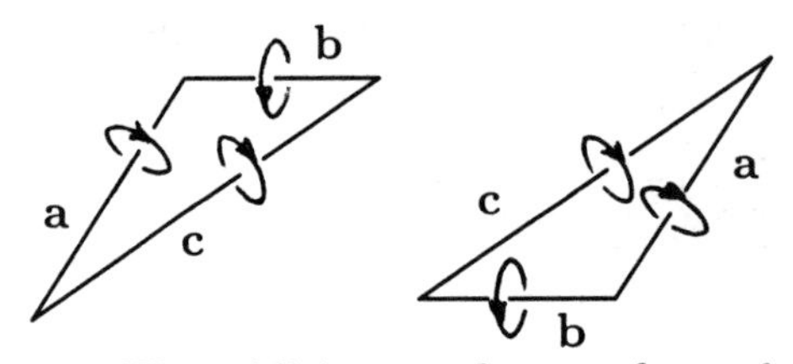

Fig. 28.31. The addition $\mathbf{a} + \mathbf{b} = \mathbf{c}$ of pseudovectors.

Given a basis $\{\mathbf{e}_1, \mathbf{e}_2, \mathbf{e}_3\}$ in the linear space of vectors, one can form a basis $\{\mathbf{e}_1^*, \mathbf{e}_2^*, \mathbf{e}_3^*\}$ in the linear space of pseudovectors as follows. The pseudovector $\mathbf{e}_1^*$ has the same attitude and length as the vector $\mathbf{e}_1$ and its orientation is the sense of rotation from $\mathbf{e}_2$ to $\mathbf{e}_3$. The definitions of $\mathbf{e}_2^*$ and $\mathbf{e}_3^*$ are similar, with cyclic permutation of the triple $\{\mathbf{e}_1, \mathbf{e}_2, \mathbf{e}_3\}$ (see Figure 28.32). A reverse order of $\mathbf{e}_2$ and $\mathbf{e}_3$ in the basis would reverse the orientation of $\mathbf{e}_1^*$, thus we notice that the order of the basis elements is essential for this definition. Since the basis introduced has three elements, we see that the linear space of pseudovectors is three- dimensional.

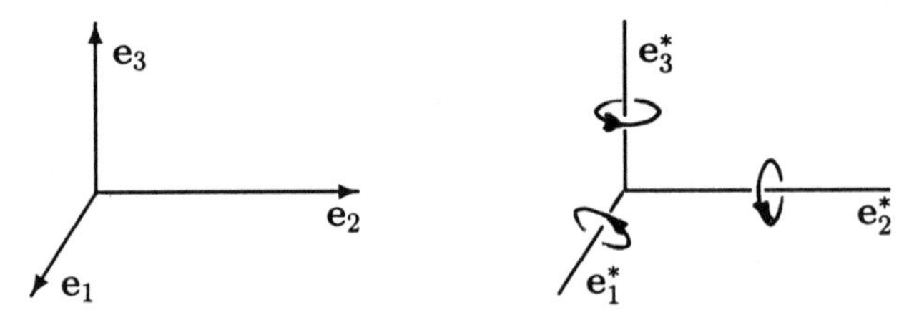

Fig. 28.32. The vector basis and the pseudovector basis determined by it.

Two bivectors $\mathbf{B}$ and $\mathbf{C}$ can be added after their factorization by a common factor $\mathbf{a}$:

$$\mathbf{B} = \mathbf{a} \wedge \mathbf{b}, \qquad \mathbf{C} = \mathbf{a} \wedge \mathbf{c}.$$

and the use of the distributive property of the wedge product over addition:

$$\mathbf{B} + \mathbf{C} = \mathbf{a} \wedge \mathbf{b} + \mathbf{a} \wedge \mathbf{c} = \mathbf{a} \wedge (\mathbf{b} + \mathbf{c}). \tag{28.1}$$

In this formula, the left-hand side is defined by the right-hand side; see Figure 28.33.

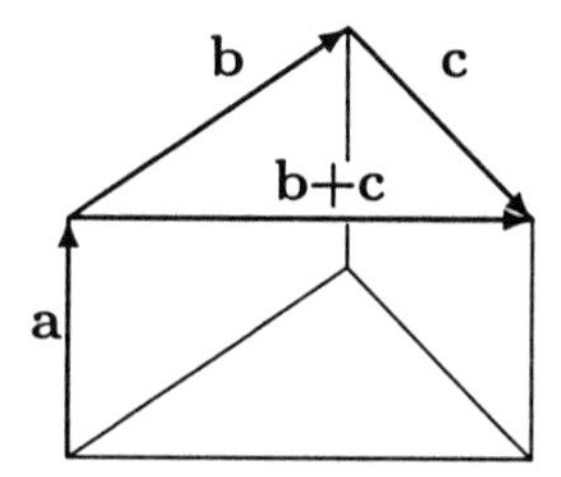

Fig. 28.33. The addition of bivectors.

For a given vector basis $\{\mathbf{e}_1, \mathbf{e}_2, \mathbf{e}_3\}$, there exists the natural bivector basis $\{\mathbf{e}_1 \wedge \mathbf{e}_2,\ \mathbf{e}_2 \wedge \mathbf{e}_3, \mathbf{e}_3 \wedge \mathbf{e}_1\}$ shown in Figure 28.34. In this manner, we see that the linear space of bivectors is three-dimensional.

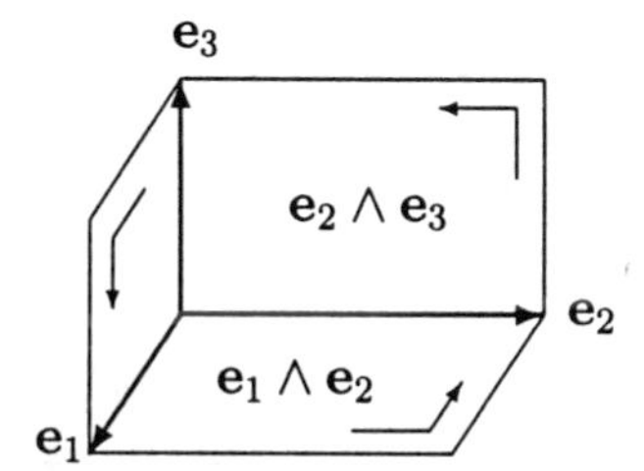

Fig. 28.34. The bivector basis.

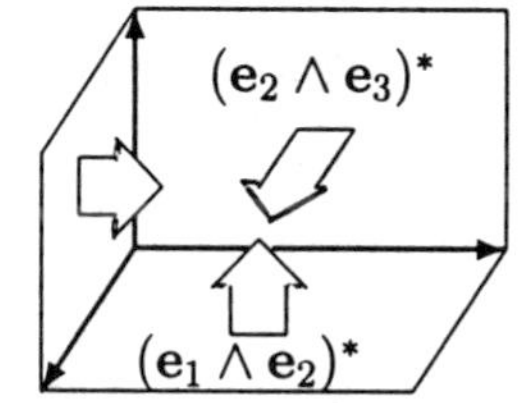

Fig. 28.35. The pseudobivector basis.

The addition of two pseudobivectors $\mathbf{B}$ and $\mathbf{C}$ is based on formula (28.1), where, however, $\mathbf{b}$ and $\mathbf{c}$ are now pseudovectors. One may introduce a natural basis $\{(\mathbf{e}_1 \wedge \mathbf{e}_2)^*,\ (\mathbf{e}_2 \wedge \mathbf{e}_3)^*,\ (\mathbf{e}_3 \wedge \mathbf{e}_1)^*\}$ for the linear space of pseudobivectors as follows. The element $(\mathbf{e}_i \wedge \mathbf{e}_j)^*$ has the attitude and magnitude of $\mathbf{e}_i \wedge \mathbf{e}_j$ but the orientation of $\mathbf{e}_k$ for $k \neq i, j$. Notice that this rule of choosing an orientation for the basic pseudobivectors is not based on the right-hand screw. It depends only on the vector basis. We depict the pseudobivector basis in Figure 28.35. In this manner, the linear space of pseudobivectors is also three-dimensional.

Making use of the outer product of vectors with pseudovectors as depicted in Figure 28.5, one is able to show the identities:

$$(\mathbf{e}_i \wedge \mathbf{e}_j)^* = \mathbf{e}_i^* \wedge \mathbf{e}_j = \mathbf{e}_i \wedge \mathbf{e}_j^*.$$

Two linear forms $\mathbf{f}$, $\mathbf{g}$ can be added by the construction depicted in Figure 28.36. One should draw a plane passing through the line of intersection of the $\mathbf{f}$,$\mathbf{g}$ planes with numbers 0,1 and through the intersection of planes 1,0 of the two summand forms, then draw a further plane parallel to it and passing through the intersection of planes 0,0. The resultant two planes, numbered 1 and 0, respectively, represent the new form $\mathbf{f} + \mathbf{g}$ when the intersections of $\mathbf{f}$ and $\mathbf{g}$ are taken to determine the planes of $\mathbf{f} + \mathbf{g}$.

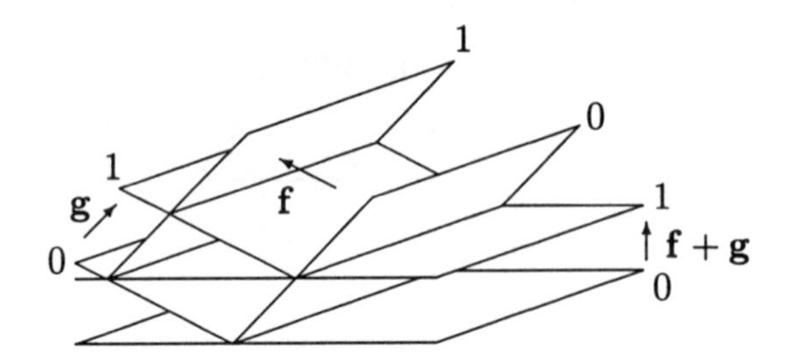

Fig. 28.36. Addition of two one-forms.

Given a basis in the linear space of vectors, one can build the so called *dual basis* $\{\mathbf{f}^1, \mathbf{f}^2, \mathbf{f}^3\}$ in the linear space of linear forms through the condition $\mathbf{f}^i(\mathbf{e}_j) = \delta^i_j$ where δ^i_j is the *Kronecker delta.* The forms $\mathbf{f}^i$ are defined geometrically as follows: the first plane of $\mathbf{f}^1$ is the plane of $(\mathbf{e}_2, \mathbf{e}_3)$, its second plane is parallel to the previous one and passes through the tip point of $\mathbf{e}_1$Thus the attitude of $\mathbf{f}^1$ is the plane of $(\mathbf{e}_2, \mathbf{e}_3)$, its orientation is the same as that of $\mathbf{e}_1$, and its relative value is one in inverse units of $\mathbf{e}_1$. The definitions of $\mathbf{f}^2$ and $\mathbf{f}^3$ are analogous. The forms $\mathbf{f}^1$, $\mathbf{f}^2$ and $\mathbf{f}^3$ are depicted in Figure 28.37.

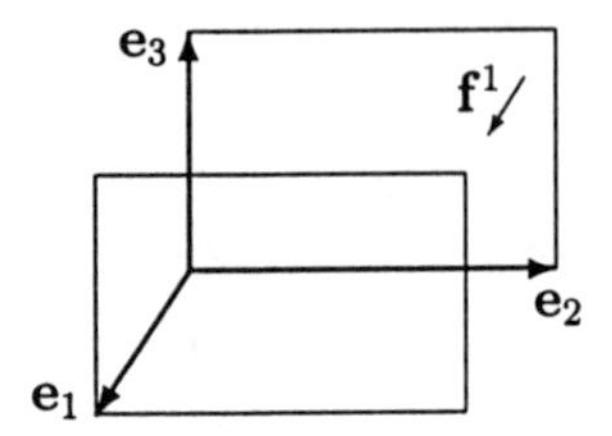

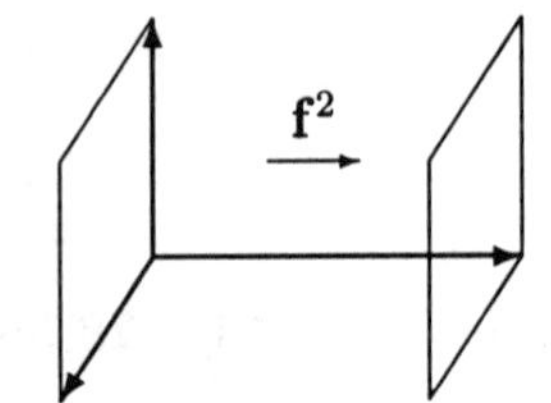

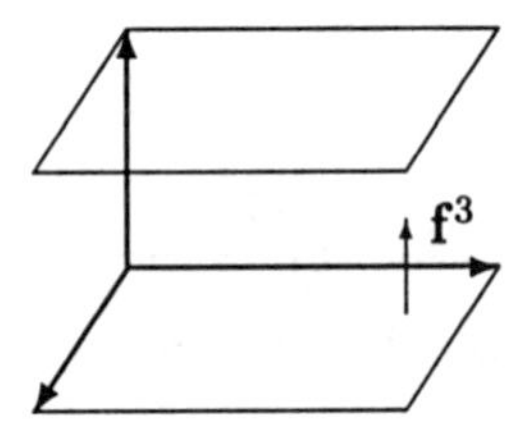

Fig. 28.37. The one-form basis.

It is worth noticing that the value of the i-th basic one-form on the radius-vector $\mathbf{r}$ is

$$\mathbf{f}^i(\mathbf{r}) = \mathbf{f}^i(x^k\mathbf{e}_k) = x^k\mathbf{f}^i(\mathbf{e}_k) = x^k\delta^i_k = x^i,$$

i.e. is equal to i-th coordinate of $\mathbf{r}$ in the vector basis $\{\mathbf{e}_j\}$.

Two pseudo-one-forms can be added according to the prescription illustrated in Figure 28.38.

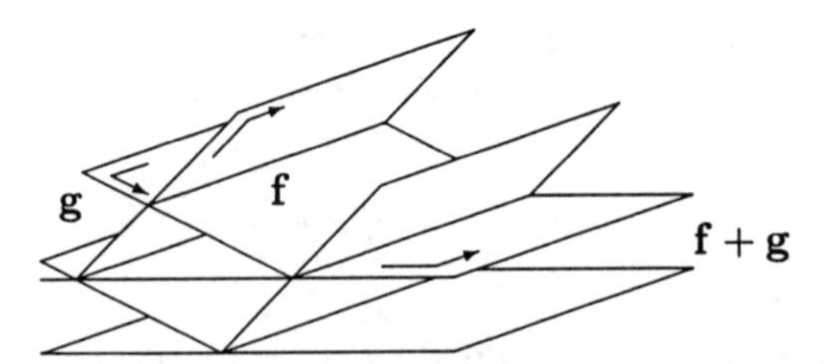

Fig. 28.38. Addition of two pseudo-one-forms.

A basis $\{\mathbf{f}^1_*, \mathbf{f}^2_*, \mathbf{f}^3_*\}$ in the linear space of pseudo-one-forms is introduced as follows: the attitude and magnitude of $\mathbf{f}^1_*$ is the same as that of $\mathbf{f}^1$, but its orientation is that of $\mathbf{e}_2 \wedge \mathbf{e}_3$, and analogously for $\mathbf{f}^2_*$ and $\mathbf{f}^3_*$ with cyclic permutation of the indices. We show this basis in Figure 28.39.

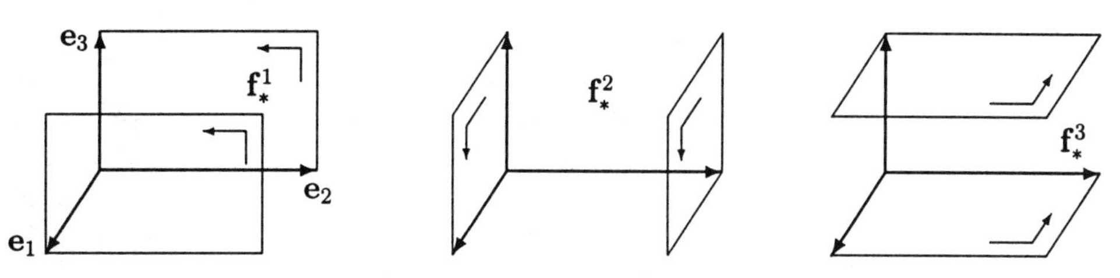

Fig. 28.39. The pseudo-one-form basis.

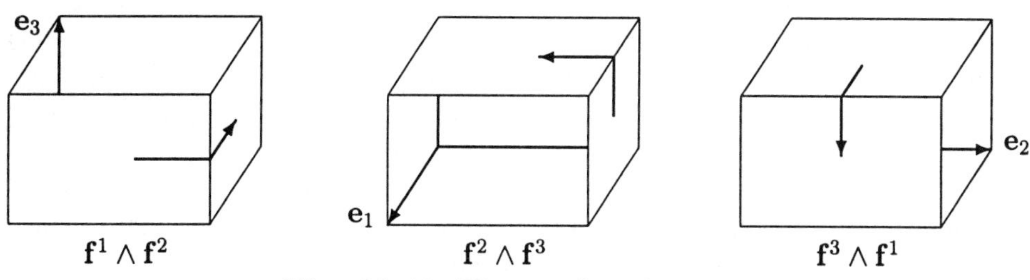

Fig. 28.40. The two-form basis.

The sets of two-forms and pseudo-two-forms are linear spaces and they again are three-dimensional. Natural bases connected with a basis in the linear space of vectors can be introduced as shown in Figures 28.40 and 28.41. The orientation of $\mathbf{f}^i \wedge \mathbf{f}^j$ is taken from the bivector $\mathbf{e}_i \wedge \mathbf{e}_j$. The orientation of $(\mathbf{f}^i \wedge \mathbf{f}^j)_*$ is the same as that of the vector of coinciding attitude, namely $\mathbf{e}_k$ with $k \neq i, k \neq j$.

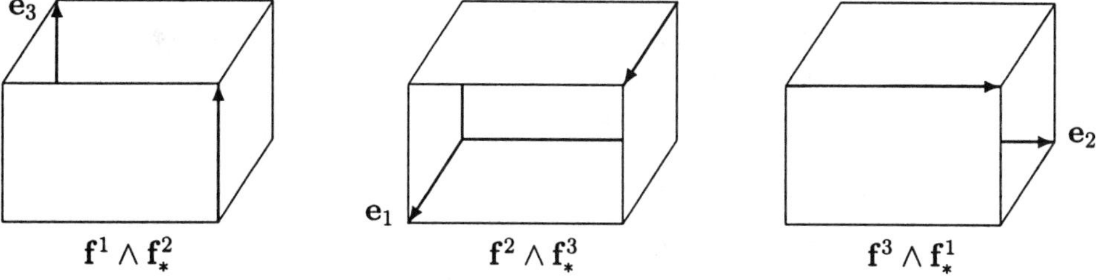

Fig. 28.41. The pseudo-two-form basis.

Making use of the outer product of one-forms with pseudo-one-forms as diplayed in Figure 28.25, one can show the identities:

$$(\mathbf{f}^i \wedge \mathbf{f}^j)_* = \mathbf{f}^i_* \wedge \mathbf{f}^j = \mathbf{f}^i \wedge \mathbf{f}^j_*.$$

We are now familiar with eight types of directed quantities connected with the three-dimensional space of vectors. Each of them separately forms a linear space in the algebraic sense (which means that elements of the same type can be added and multiplied by scalars). In each case, the linear space is three-dimensional, which implies that it can be isomorphically mapped into the space of vectors. This is the reason why physicists generally employ only vectors.

28.5 Various products

Let us denote by $\mathcal{M}$ the set of all multivectors and pseudomultivectors, including scalars and pseudoscalars. Each quantity has its grade, namely k-vector and pseudo-k-vector have grade k. The scalars and pseudoscalars have grade 0. We shall show that an outer product can be defined for each pair of quantities from $\mathcal{M}$. The grades of factors add in such a product; but, when their sum is greater than three, the product is zero.

The outer product of any quantity from $\mathcal{M}$ with the scalar is the same as the ordinary multiplication by the scalar considered in Sec. 28.4, and this product is commutative. We assume the commutivity of the outer product with a pseudoscalar and asume that this product changes a quantity into a pseudoquantity and vice versa. In this product, the magnitudes of the factors are multiplied and the directions change according to the rules displayed in Figures 28.42 – 45. Notice that, in the multiplication by pseudoscalar, the second factor has the same attitude as the product, so we may claim that they are parallel.

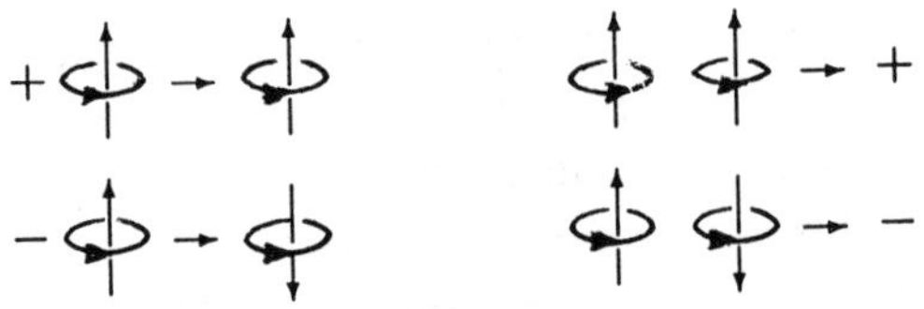

Fig. 28.42. Multiplication of scalars and pseudoscalars by pseudoscalars.

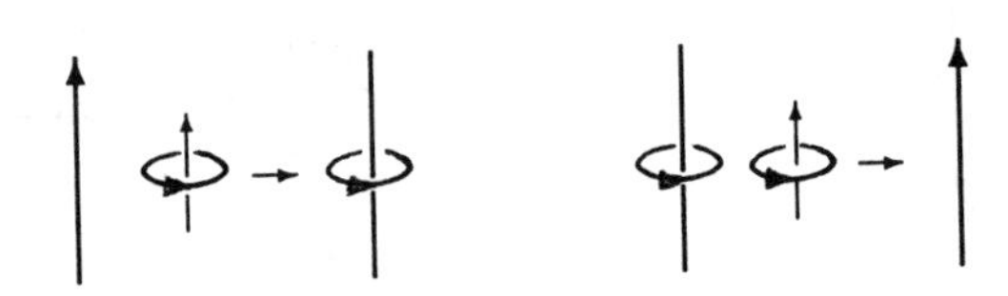

Fig. 28.43. Multiplication of vectors and pseudovectors by pseudoscalars.

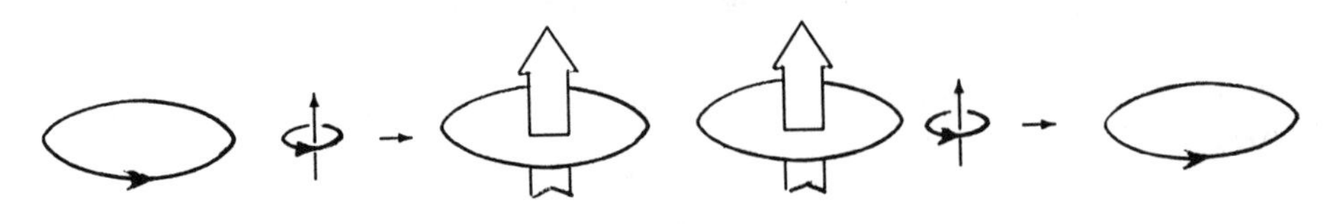

Fig. 28.44. Multiplication of bivectors and pseudobivectors by pseudoscalars.

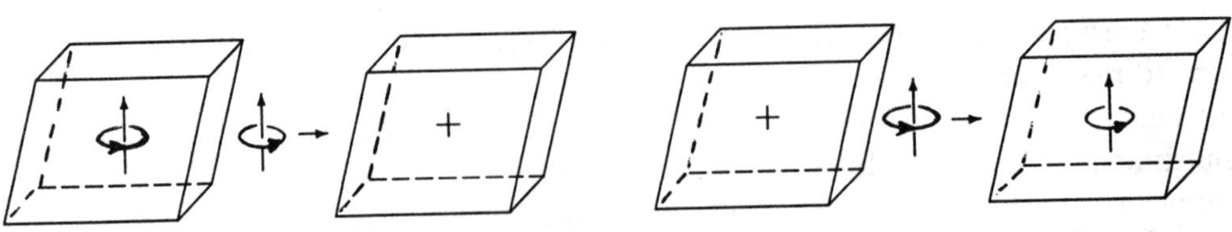

Fig. 28.45. Multiplication of trivectors and pseudotrivectors by pseudoscalars.

It is interesting to note that unit scalars and pseudoscalars form an abelian group. Their multiplication table is presented in Table 28.3, where r denotes the right-handed unit pseudoscalar, and l, the left-handed one. This group is isomorphic to the Klein four-group.

Table 28.3

1	-1	r	l
-1	1	l	r
r	l	1	-1
l	r	-1	1

The outer products of vectors with multivectors and pseudomultivectors, other than scalars and pseudoscalars, have been already defined in Figures 28.3, 28.5, 28.10 and 28.14. The only missing outer product of a vector with a trivector or pseudotrivector gives zero by the rule of adding grades.

The outer products of pseudovectors with multivectors and pseudomultivectors other than scalars and pseudoscalars are defined in Figures 28.5, 28.6, and 28.15. One can prove the homogeneity of the outer product under multiplication by a pseudoscalar μ:

$$\mu(N_1 \wedge N_2) = (\mu N_1) \wedge N_2 = N_1 \wedge (\mu N_2)$$

for $N_1, N_2 \in \mathcal{M}$.

Let $L(\mathcal{M})$ be the linear span of $\mathcal{M}$, i.e., the set of all linear combinations of various grade elements. We extend by distributivity the outer product onto $L(\mathcal{M})$. Thus we get an algebra which we could call the *extended Grassmann algebra of multivectors* in $\mathbf{R}^3$. Linear combinations of ordinary multivectors form a linear subspace $L(\mathcal{M})_+$ and a subalgebra in this algebra, namely the traditional Grassmann algebra, whereas the combinations of pseudomultivectors form only a linear subspace $L(\mathcal{M})_-$. The outer product of two pseudoquantities is an ordinary quantity; the outer product of mixed quantities (one ordinary, one pseudo-) is a pseudoquantity. In this manner, we obtain a Z_2- gradation of the algebra $L(\mathcal{M}) = L(\mathcal{M})_+ \oplus L(\mathcal{M})_-$. Pseudoquantities are here odd elements. The multiplication by two unit pseudoscalars r, l gives two natural isomorphisms between the two subspaces. We can, for instance, take the right unit r to change all pseudomultivectors into ordinary multivectors according to the prescriptions depicted at the right-hand sides of Figures 28.42–45 which can be summarized in the equality $L(\mathcal{M})_+ = rL(\mathcal{M})_-$. The extended Grassmann algebra of multivectors can be written as the direct sum

$$L(\mathcal{M}) = L(\mathcal{M})_+ \oplus rL(\mathcal{M})_+$$

in which the first summand is the ordinary Grassmann algebra of $\mathbf{R}^3$. We should point out that the choice of r is not canonical. One can equally well write l at the second term:

$$L(\mathcal{M}) = L(\mathcal{M})_+ \oplus lL(\mathcal{M})_+$$

The subset of elements of zero grade $L(\mathcal{M})_0$ contains sums of scalars and pseudoscalars of the form $x = a + br$, $a, b \in \mathbf{R}$, $r^2 = 1$. Similar extensions of the field of real numbers were considered thoroughly in mathematical literature,. We mention here, however, that only Refs. [16-20] are directly related to physics. Unfortunately, each of these papers uses its own terminology: the combinations $x = a+br,\ a,b \in \mathbf{R}$ with $r^2 = 1$ are called *double numbers* or *split complex numbers* in [16], *perplex numbers* in [17], *unipodal numbers* in [18], *duplex numbers* in [19], and *hyperbolic numbers* in [20]. The set $L(\mathcal{M})_0$ obviously forms a subalgebra and a ring but not a field because it has zero divisors. For instance, $(1+r)(1+l) = (1+r)(1-r) = 1-1 = 0$. We shall call $L(\mathcal{M})_0$ the *ring of double numbers.*

As a result of these considerations, we can say that $L(\mathcal{M})$ is the doubled Grassmann algebra over the field of reals or the Grassmann algebra over the ring of double numbers as well. The dimension of $L(\mathcal{M})$ is 16 over reals and 8 over double numbers.

Let $\mathcal{F}$ denote the set of all forms and pseudoforms. We can similarly define the outer product for all pairs of elements from $\mathcal{F}$. In this way, we obtain another algebra $L(\mathcal{F})$ which I would call the *extended Grassmann algebra of outer forms* over $\mathbf{R}^3$. Linear combinations of ordinary forms constitute a subspace $L(\mathcal{F})_+$ and a subalgebra, wehereas the combinations of pseudoforms are a subspace $L(\mathcal{F})_-$. Again, the extended Grassmann algebra of outer forms can be written as the direct sum

$$L(\mathcal{M}) = L(\mathcal{M})_+ \oplus rL(\mathcal{M})_+ \qquad \text{or} \qquad L(\mathcal{M}) = L(\mathcal{M})_+ \oplus lL(\mathcal{M})_+.$$

In this manner, multiplication by r (or equally well by l) transforms quantities from Table 28.1 into ones from Table 28.2 and vice versa.

Each pseudoquantity N can be represented as $N = rM$ for some ordinary quantity M. Using this possibility we may introduce the value of a one-form $\mathbf{f}$ on a pseudovector $\mathbf{b} = r\mathbf{a}$ as

$$\mathbf{f}(\mathbf{b}) = \mathbf{f}(r\mathbf{a}) = r\mathbf{f}(\mathbf{a}).$$

In this way, a one-form becomes a linear form on $L(\mathcal{M})_1$; that is, on the set of combinations of vectors and pseudovectors, a form linear over the ring $L(\mathcal{M})_0$. Analogous claims can be formulated about the sets $L(\mathcal{M})_2$ and $L(\mathcal{M})_3$.

Similarly, the value of a pseudo-one-form $\mathbf{g} = r\mathbf{f}$ (here r should be treated as the pseudo-zero-form) on a vector $\mathbf{a}$ is defined as

$$\mathbf{g}(\mathbf{a}) = (r\mathbf{f})(\mathbf{a}) = r(\mathbf{f}(\mathbf{a})).$$

In this manner, the set of all one-forms plus pseudo-one-forms $L(\mathcal{F})_1$ becomes a linear space over the ring $L(\mathcal{F})_0$ (called also a module over $L(\mathcal{F})_0$) and analogously for the sets $L(\mathcal{F})_2$ and $L(\mathcal{F})_3$.

Another kind of product can be introduced, namely, contractions. The *contraction* of the one-form or pseudo-one-form $\mathbf{f}$ by the vector or pseudovector $\mathbf{a}$ is $\mathbf{a} \rfloor \mathbf{f} = \mathbf{f}(\mathbf{a})$ where $\mathbf{f}(\mathbf{a})$ is the value of $\mathbf{f}$ on $\mathbf{a}$. We define, similarly, the contraction from the other side: $\mathbf{f} \lfloor \mathbf{a} = \mathbf{f}(\mathbf{a})$.

The *contraction* of higher forms and pseudo-forms with multivectors and pseudomultivectors can be defined in a manner analogous to the well known procedure. I omit the details. It is worth mentioning here, however, that the result of the contraction has an attitude parallel to the attitudes of the factors.

28.6 Physical quantities

We now present a list of physical quantities with their identification as directed objects, along with short justifications.

We assume that *electric charge* q, *electric current* I, *magnetic flux* Φ_m are scalars. Since we consider here only nonrelativistic descriptions, *time* t and *energy* E are also treated as scalars. The only pseudoscalar quantity could be *magnetic charge* m, if it exists.

The most natural vectorial quantity is the *displacement vector* $\mathbf{l}$, which is of the same nature as the *radius vector* $\mathbf{r}$ of a point in space relative to the reference point (called the *origin of a frame*). Of course, the *velocity* $\mathbf{v} = d\mathbf{r}/dt$, the derivative of $\mathbf{r}$ with respect to a scalar variable t, is also a vector. The same is true of the *acceleration* $\mathbf{a} = d\mathbf{v}/dt$, the *momentum* $\mathbf{p} = m\mathbf{v}$, the *force* $\mathbf{F} = d\mathbf{p}/dt$ and the electric dipole moment $\mathbf{d} = q\mathbf{l}$. *Magnetic dipole moment* $\mathbf{d}^* = m\mathbf{l}$ of two magnetic charges $\pm m$ separated by directed distance $\mathbf{l}$ could be treated as pseudovector quantity.

The *angular momentum* $\mathbf{L} = \mathbf{r} \wedge \mathbf{p}$ and also the *torque* $\mathbf{M} = \mathbf{r} \wedge \mathbf{F}$ are bivectors. The best physical model of a bivector is a flat electric circuit. Its magnitude is just the area encompassed by the circuit; its attitude is the plane of the circuit and orientation is given by the sense of the current. This bivector could be called a *directed area* $\mathbf{S}$ of the circuit. A connected bivectorial quantity is then the *magnetic moment* $\mathbf{m} = I\mathbf{S}$ of the circuit, where I is the current. The SI units $[\mathbf{m}] = \mathrm{Am}^2$ fit this interpretation. (Notice that the magnetic moment of the circuit is a different directed quantity than the dipole moment of two magnetic charges.)

Frankly, the identification of the force as a vector is not unique. If you consider the potential energy U as a scalar, its relation to the force in traditional language is $dU = -\mathbf{F} \cdot d\mathbf{r}$, where the dot denotes the scalar product. This means that the force is a linear map of the infinitesimal vector $d\mathbf{r}$ into the infinitesimal scalar dU. Thus, according to this observation, the force should be treated as a one-form. This in turn, through the Newton equation $\mathbf{F} = \frac{d\mathbf{p}}{dt}$, implies that the momentum $\mathbf{p}$ must also be a one-form also. This view is adopted in many modern mathematical formulations of mechanics, see e.g. [21].

Another one-form quantity occurs in the description of the plane waves. The locus of points in space with the same phase is just a plane. The family of planes with phases differing by $2\pi n$ for natural n can be viewed as the geometric image of a one-form as depicted in Figure 28.16. This one-form describes the physical quantity known as the *wave vector* $\mathbf{k}$ with magnitude $2\pi/\lambda$ (λ is the wavelength). Its physical dimension $[\mathbf{k}] = \mathrm{m}^{-1}$ fits this interpretation, because the magnitude of a one-form should be proportional to the inverse length. Thus, in my opinion, the physical quantity $\mathbf{k}$ should have a different name since, in its directed nature, it is not a vector at all. If I may create an English word, I would like to propose the name *wavity* for $\mathbf{k}$.

Another one-form is the *electric field strength* $\mathbf{E}$, since we consider it to be a linear map of the infinitesimal vector $d\mathbf{r}$ into the infinitesimal potential difference: $-dV = \mathbf{E} \cdot d\mathbf{r}$. The physical dimension in the SI system $[\mathbf{E}] = \mathrm{Vm}^{-1}$ fits this interpretation.

The *magnetic induction* $\mathbf{B}$ is an example of a two-form quantity, since it can be treated as a linear map of the directed area bivector $d\mathbf{S}$ into the magnetic flux: $d\Phi = \mathbf{B}\cdot d\mathbf{S}$. This is consistent with the SI dimension: $[\mathbf{B}] = \text{Wbm}^{-2}$.

The Stokes theorem $\int \mathbf{B}\cdot d\mathbf{S} = \oint \mathbf{A}\cdot d\mathbf{l}$, in which $\mathbf{A}$ is the so called *vector potential*, says that the product $\mathbf{A}\cdot d\mathbf{l}$ is also the magnetic flux. Thus, A is a one-form rather than a vector. We should notice here that the traditional name "vector potential" is not appropriate for the genuine directed nature of $\mathbf{A}$. Therefore, I would prefere the term *directed potential* for it.

Now consider some examples of pseudoquantities; the area $\mathbf{S}^*$ of a surface, through which a flow is measured, is the first one. The side of the surface from which a substance (mass, energy, electric charge, etc.) passes is important. Hence, the orientation of $\mathbf{S}^*$ can be marked as an arrow not parallel to the surface. This is the situation depicted in Figure 28.4. We claim that the *area of a flow* is a pseudo-bivector quantity. Accordingly, the *flux density* $\mathbf{j}$ (or the *current density* in case of the electric current flowing) has to be a pseudo-two-form quantity. It corresponds to the linear map $dI = \mathbf{j}\cdot d\mathbf{S}^*$ of the area $d\mathbf{S}^*$ into the electric current dI. The physical dimension of the current density $[\mathbf{j}] = \text{Am}^{-2}$ concords with this interpretation.

The electric induction $\mathbf{D}$ has a similar nature. We present here a prescription of its measurement quoted from Ref.[22], p. 68: "Take two identical discs each made of very thin sheet metal, and each with an isolated handle. Place one disc on top of the other, holding them by the handles, electrically discharge them, and then place them in the presence of a field. As you separate the discs, the charges induced on them (one positive, the other negative) are also separated. Now measure one of them with the aid of a Faraday cage. It turns out that for a small enough disc the charge is proportional to its area."[1] One will agree that the *disc area* $d\mathbf{S}^*$ is a pseudo- bivector quantity since its magnitude is the area, its attitude is the plane and its orientation is given by an arrow showing which disc is to be connected with the Faraday cage; see Figure 28.46. Because of the proportionality relation $dQ = \mathbf{D}\cdot d\mathbf{S}^*$, we ascertain that the *electric induction* is a linear map of the pseudo-bivectors into scalars, i.e. it is a pseudo-two-form. The SI units $[\mathbf{D}] = \text{Cm}^{-2}$ fit this interpretation. Notice that $\mathbf{D}$ is of the same directed nature as the electric current density. This is reflected in the fact that $\dot{\mathbf{D}}$ is called a *displacement current.*

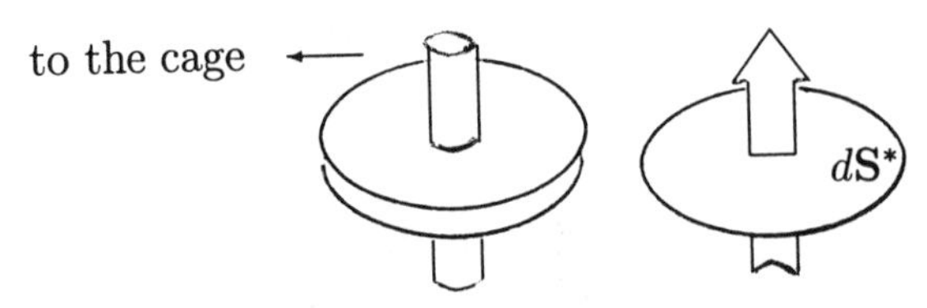

Fig. 28.46. The pseudobivector corresponding to the disc.

In another place of the same book [22], p. 347, one may find an operational definition of the *magnetic field strength*: "Take a very small wireless solenoid prepared

[1]This prescription is based on a similar one given in Ref.[23], p. 75. The same method for determination of $\mathbf{D}$ was interpreted erroneously by Levashev [15] that the pair of discs is a one-form, hence $\mathbf{D}$, as a dual quantity, is a vector.

from a superconducting material (as shown in Figure 28.47). Close the circuit in a region of space where the magnetic field vanishes. Afterwards, introduce the circuit into an arbitrary region in the field. A superconductor has the property that the magnetic flux enclosed by it is always the same; a current will be induced to compensate for this external field flux. Now measure the current dI flowing through the superconductor. It turns out to be proportional to the solenoid length: $dI = \mathbf{H} \cdot d\mathbf{l}^*$." The *solenoid length* $d\mathbf{l}^*$ in this experiment is apparently a pseudovector, hence the *magnetic field strength* $\mathbf{H}$ is a pseudo-one-form. The SI unit $[\mathbf{H}] = \text{Am}^{-1}$ confirms this interpretation.

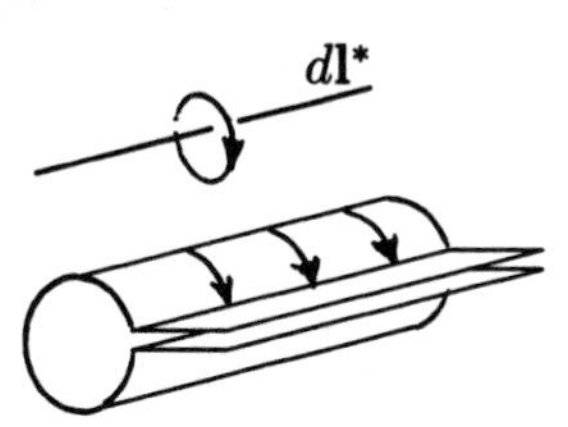

Fig. 28.47. The pseudovector corresponding to the solenoid length.

To my knowledge, the above-mentioned solenoid length is the second example of a physical quantity with a true pseudovector nature. The first one is a magnetic dipole moment $\mathbf{d}^*$ of two magnetic charges. Many other physical quantities hitherto called pseudovectors (like angular momentum, torque, magnetic moment, magnetic induction, magnetic field strength) turn out – as shown previously – to be bivectors, pseudo-one-forms or two-forms.

Two electromagnetic field quantities, namely $\mathbf{E}$ and $\mathbf{B}$, are ordinary forms, and they are called *intensity quantities* by some authors (see e.g. [6]), whereas two others, $\mathbf{D}$ and $\mathbf{H}$, are pseudo-forms, and they are called *magnitude quantities.*[2] The authors of Ref. [2] do not mention this difference.

Electric charge and energy are scalar quantities. The magnetic charge, if it existed, would be a pseudoscalar quantity. The only trivector quantity known to me is the *helicity* of a quantum particle or of a composed classical object; namely, the outer product $\mathbf{h} = \mathbf{s} \wedge \mathbf{p}$ of the internal angular momentum $\mathbf{s}$ with the linear momentum $\mathbf{p}$. A *volume* of something (a physical body or a region of space) is a pseudotrivector quantity – it is natural to take plus sign for it.[3]

As for the pseudo-three-forms, we can mention *spatial densities* of various kinds: mass density, energy density, the electric charge density. The first two occur only with the plus sign; the latter may assume two signs. The energy density of the electric field is given by the outer product of the one-form $\mathbf{E}$ with the pseudo-two-form $\mathbf{D}$

$$w_e = \frac{1}{2}\mathbf{E} \wedge \mathbf{D} = \frac{1}{2}\mathbf{D} \wedge \mathbf{E};$$

[2] Frankel [3] makes the same distinction, but calls $\mathbf{E}$, $\mathbf{B}$ *intensities* and $\mathbf{D}$,$\mathbf{H}$ *quantities* of the electromagnetic field.

[3] Most of the authors take the volume as a trivector, but there is no reason for a volume to have a handedness as its feature. Frankel [3] treats volume as a pseudoquantity, but it is pseudo-three-form rather than a pseudotrivector there.

see Figure 28.48. The energy density of the magnetic field is proportional to the outer product of the pseudo-one-form $\mathbf{H}$ with the two-form $\mathbf{B}$

$$w_m = \frac{1}{2}\mathbf{B} \wedge \mathbf{H} = \frac{1}{2}\mathbf{H} \wedge \mathbf{B};$$

see Figure 28.49. It is interesting to note that the situation depicted in Figure 28.29 can never happen with $\mathbf{E}$ and $\mathbf{D}$ because the two electric field quantities cannot have opposite orientations. The same is true for Figure 28.30 and the two magnetic field quantities. The density of the magnetic charge (if it exists) ought to be a three-form quantity because, after multiplication by the volume pseudotrivector, it should give a pseudoscalar of the magnetic charge.

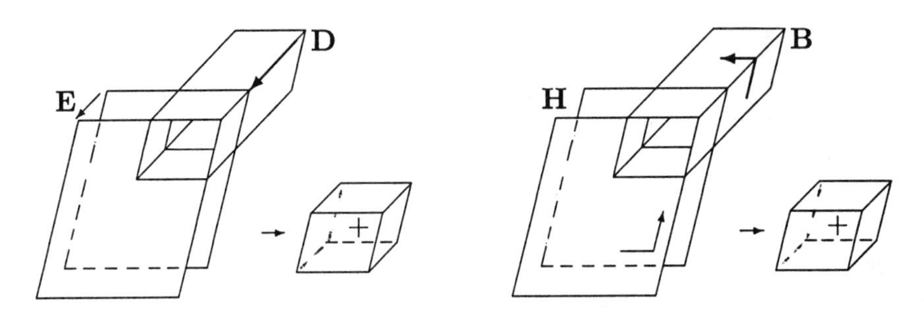

Fig. 28.48. The outer product $\mathbf{E} \wedge \mathbf{D}$. **Fig. 28.49.** The outer product $\mathbf{H} \wedge \mathbf{B}$.

28.7 Quadratic spaces

Let us review what a scalar product is in an n-dimensional vector space. Mathematically, it is a *bilinear form* that is symmetric, positive definite, and nondegenerate. When we have an arbitrary basis $\{\mathbf{e}_1, \mathbf{e}_2, \ldots, \mathbf{e}_n\}$, in our vector space, then we have, for all vectors $\mathbf{v}$, $\mathbf{w}$, the decomposition $\mathbf{v} = \sum_{i=1}^{n} v_i \mathbf{e}_i$, $\mathbf{w} = \sum_{i=1}^{n} w_i \mathbf{e}_i$, $v_i, w_i \in \mathbf{R}$. It is easy to check that the expression

$$(\mathbf{v}, \mathbf{w}) = \sum_{i=1}^{n} v_i w_i \tag{2}$$

satisfies all the demanded properties. Thus (2) can be considered as a scalar product. It yields $(\mathbf{e}_i, \mathbf{e}_j) = \delta_{ij}$ which means that vectors $\mathbf{e}_i$ form an orthonormal basis. In this manner, we see that for any basis one can choose a scalar product such that it is an orthonormal basis.

As an illustration of this procedure, choose two arbitrary linearly independent vectors in a two-dimensional linear space $\mathbf{e}_1$, $\mathbf{e}_2$, like those in Figure 28.50, and form a scalar product such that they are orthonormal.

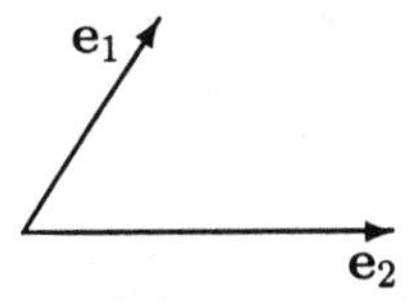

Fig. 28.50. Vectors to be orthonormal.

Such a scalar product could seem nonnatural because, as seen in Figure 28.50, the two vectors $\mathbf{e}_1$, $\mathbf{e}_2$ are "apparently" nonorthogonal and have different lengths. Nevertheless, you may be convinced that this scalar product is natural, if you draw a richer picture containing the two vectors, as in Figure 28.51, and imagine that this is a perspective view of some floor with rectangular tiles.

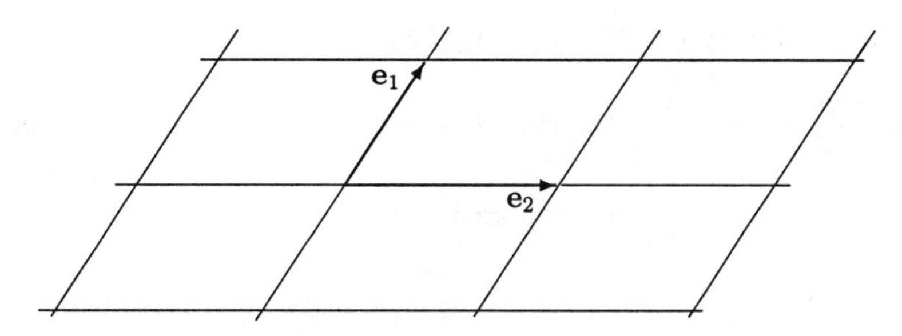

Fig. 28.51. The same vectors as a part of a tiling.

One should stress here that the scalar product formed according to prescription (2) is strongly basis dependent. For different bases, different scalar products are obtained by this prescription.

For our 3-dimensional physical space, we have no doubts about deciding which vectors are orthogonal and which nonparallel vectors have equal magnitudes. Therefore we can easily pick three unit perpendicular vectors and define a scalar product based on them according to the above prescription. But often there are reasons to introduce also another scalar products appropriate for anisotropic media.

If a *scalar product* is given for each pair of vectors, the linear space is called *Euclidean* or *quadratic.* In such a case, the *norm* is naturally given by the square root of the scalar product of a vector with itself. This is a function which ascribes to each vector a number called the *magnitude* or *length.* In contrast to the previous relative magnitude, we could call this number an *absolute magnitude.* Thus, in Euclidean linear spaces, the natural absolute magnitude is automatically given for vectors.

When we define a scalar product according to the given prescription for an arbitrary basis $\{\mathbf{e}_1, \mathbf{e}_2, \mathbf{e}_3\}$, then obviously this basis is orthonormal with respect to this scalar product. A natural question now arises: what is a sphere, i.e. a set of vectors $\mathbf{r}$ of equal magnitudes $\mid \mathbf{r} \mid = a$? One can check that this set is given as the following family of vectors

$$\mathbf{r} = a\mathbf{e}_1 \sin\theta\cos\phi + a\mathbf{e}_2 \sin\theta\sin\phi + a\mathbf{e}_3\cos\theta \tag{28.3}$$

numbered by two parameters $0 \leq \theta \leq \pi, 0 \leq \phi \leq 2\pi$. This is the surface of an ellipsoid. The parameters θ, ϕ are natural parameters for it. We shall call them *ellipsoidal (angular) coordinates* in spite of the fact that they are the spherical angular coordinates for the chosen scalar product.

The tangent vector to the coordinate line of θ is

$$\frac{\partial \mathbf{r}}{\partial \theta} = a\mathbf{e}_1 \cos\theta\cos\phi + a\mathbf{e}_2\cos\theta\sin\phi - a\mathbf{e}_3\sin\theta;$$

the tangent vector to the coordinate line of ϕ is

$$\frac{\partial \mathbf{r}}{\partial \phi} = -a\mathbf{e}_1 \sin\theta \sin\phi + a\mathbf{e}_2 \sin\theta \cos\phi.$$

It is easy to check that the two vectors are orthogonal to $\mathbf{r}$.

The tangent bivector to the surface (28.3) is

$$\mathbf{S} = \frac{\partial \mathbf{r}}{\partial \theta} \wedge \frac{\partial \mathbf{r}}{\partial \phi} = a^2 \mathbf{e}_1 \wedge \mathbf{e}_2 \sin\theta \cos\theta \cos^2\phi - a^2 \mathbf{e}_2 \wedge \mathbf{e}_1 \sin\theta \cos\theta \sin^2\phi$$

$$+a^2 \mathbf{e}_3 \wedge \mathbf{e}_1 \sin^2\theta \sin\phi - a^2 \mathbf{e}_3 \wedge \mathbf{e}_2 \sin^2\theta \cos\phi$$

$$= a^2 \sin\theta (\mathbf{e}_1 \wedge \mathbf{e}_2 \cos\theta + \mathbf{e}_2 \wedge \mathbf{e}_3 \sin\theta \cos\phi + \mathbf{e}_3 \wedge \mathbf{e}_1 \sin\theta \sin\phi).$$

As an outer product of two vectors orthogonal to $\mathbf{r}$, it is also orthogonal to $\mathbf{r}$; see Figure 28.52. This bivector can be easily transformed into a pseudobivector $\mathbf{S}^*$ just by replacing the basic bivectors by the basic pseudobivectors:

$$\mathbf{S}^* = a^2 \sin\theta [(\mathbf{e}_1 \wedge \mathbf{e}_2)^* \cos\theta + (\mathbf{e}_2 \wedge \mathbf{e}_3)^* \sin\theta \cos\phi + (\mathbf{e}_3 \wedge \mathbf{e}_1)^* \sin\theta \sin\phi]; \quad (28.4)$$

see Figure 28.53. Its orientation is outer to the ellipsoid surface which is visible after comparing expressions (28.3) and (28.4), for instance, for $\phi = 0$, $\theta = \frac{\pi}{2}$.

Fig. 28.52. The tangent bivector.

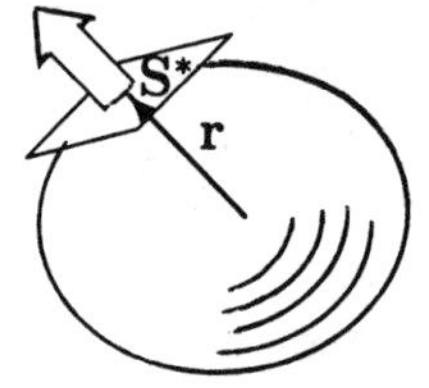

Fig. 28.53. The tangent pseudobivector.

The ellipsoid (28.3) serves to find vectors orthogonal to a given vector $\mathbf{r}$. We merely draw a plane at the tip of $\mathbf{r}$ tangent to the ellipsoid, translate this plane to the origin, and this is the plane of all vectors orthogonal to $\mathbf{r}$.

After a (nondegenerate) scalar product is established in the linear space, a natural mapping of linear forms into vectors is easy to define. Thus, for a linear form $\mathbf{f}$, there exists one and only one vector $\vec{\mathbf{f}}$ such that

$$\mathbf{f}(\mathbf{r}) = (\vec{\mathbf{f}}, \mathbf{r}) \quad \text{for each vector } \mathbf{r}$$

where $(\cdot, \cdot)$ denotes the scalar product. Of course, $\vec{\mathbf{f}}$ is perpendicular to the planes forming the attitude of $\mathbf{f}$; see Figure 28.54. (Recall that the word "perpendicular" makes sense only in the quadratic space.) Vector $\vec{\mathbf{f}}$ inherits its orientation from the form $\mathbf{f}$, but its attitude is perpendicular to that of $\mathbf{f}$. The above formula yields a prescription in which physicists define the "wavity" as a vector $\vec{\mathbf{k}}$ for a given phase function Φ of the plane wave: $\Phi(\mathbf{r}) = (\vec{\mathbf{k}}, \mathbf{r})$.

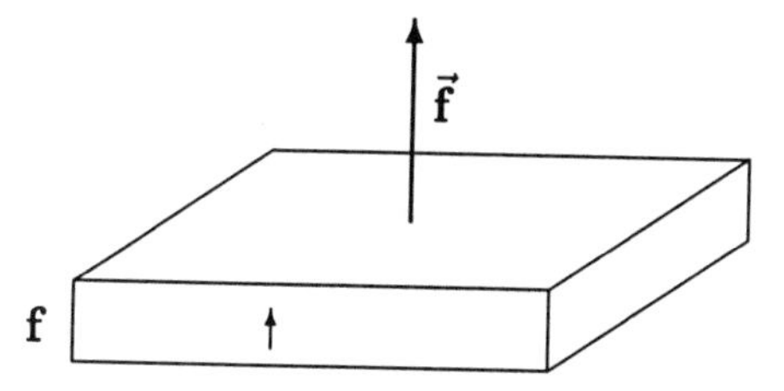

Fig. 28.54. Changing one-form f into vector $\vec{\mathbf{f}}$.

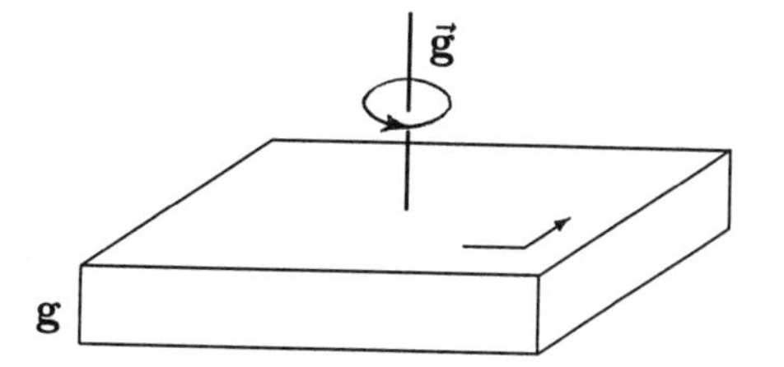

Fig. 28.55. Changing pseudo-one-form g into pseudovector $\vec{\mathbf{g}}$.

A natural mapping of pseudoforms into pseudovectors is possible to define. Namely, for a given pseudo-form **g**, there exists one and only one pseudovector $\vec{\mathbf{g}}$ such that $\mathbf{g}(\mathbf{v}) = (\vec{\mathbf{g}}, \mathbf{v})$ for each pseudovector **v**. We illustrate this prescription in Figure 28.55. The pseudovector $\vec{\mathbf{g}}$ inherits its orientation from **g**, but its attitude is perpendicular to that of **g**.

The scalar product in the linear space of vectors serves also to define a scalar product in the linear space of bivectors. Namely, for $\mathbf{B} = \mathbf{a} \wedge \mathbf{b}$, $\mathbf{C} = \mathbf{c} \wedge \mathbf{d}$ we introduce the scalar product through the determinant

$$(\mathbf{B}, \mathbf{C}) = \begin{vmatrix} (\mathbf{a}, \mathbf{c}) & (\mathbf{a}, \mathbf{d}) \\ (\mathbf{b}, \mathbf{c}) & (\mathbf{b}, \mathbf{d}) \end{vmatrix}.$$

This scalar product serves to define a magnitude $| \mathbf{B} |$ as follows $| \mathbf{B} |^2 = (\mathbf{B}, \mathbf{B})$. Analogously the scalar product $(\mathbf{v}, \mathbf{u})$ in the linear space of pseudovectors serves to define a scalar product $(\mathbf{C}, \mathbf{D})$ and a magnitude $| \mathbf{D} |$ in the linear space of pseudobivectors. Now we replace the bivectors **b** by the perpendicular pseudovectors $\vec{\mathbf{b}}$ (see Figure 28.56) and the pseudobivectors **c** by the perpendicular vectors $\vec{\mathbf{b}}$ (see Figure 28.57), preserving their magnitudes and orientations.

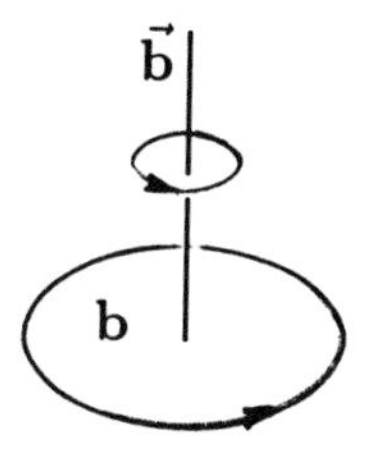

Fig. 28.56 Changing bivector b into pseudovector $\vec{\mathbf{b}}$.

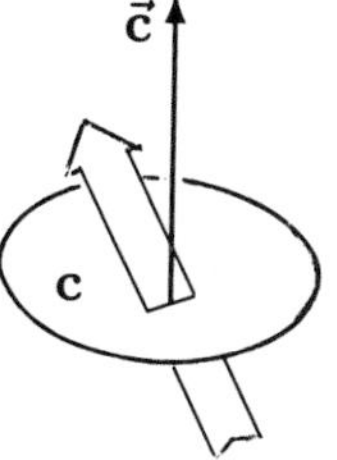

Fig. 28.57. Changing pseudobivector c into vector $\vec{\mathbf{c}}$.

Since one can define also magnitudes for two-forms, we formulate the following prescription of changing a given two-form **B** into pseudovector $\vec{\mathbf{B}}$: take the direction of **B**, ascribe it to $\vec{\mathbf{B}}$ and use the magnitude of **B** to define the length of $\vec{\mathbf{B}}$. This prescription is illustrated in Figure 28.58. Analogous prescription of transforming pseudobivectors **D** into vectors $\vec{\mathbf{D}}$ is illustrated in Figure 28.59.

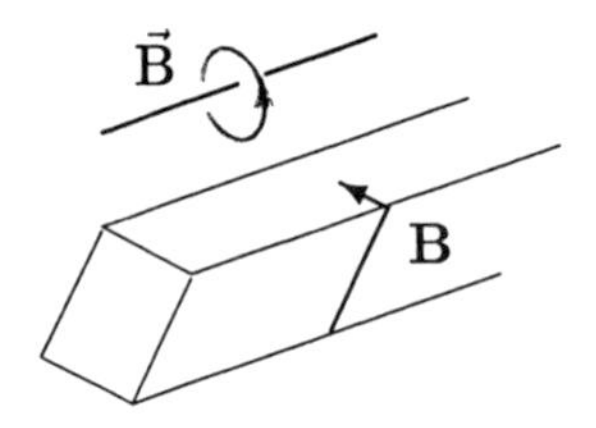

Fig. 28.58. Changing two-form $\mathbf{B}$ into pseudovector $\vec{\mathbf{B}}$.

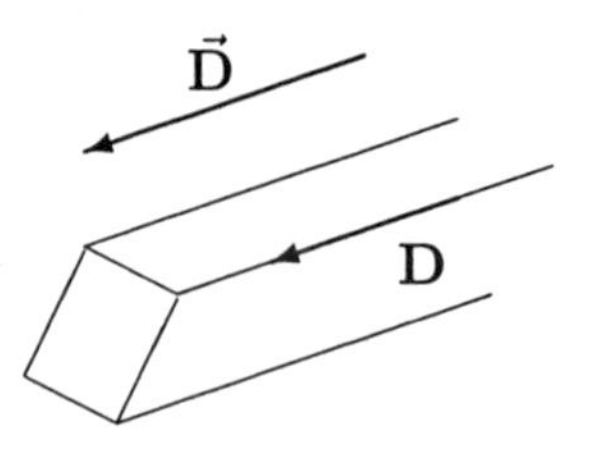

Fig. 28.59. Changing pseudo-two-form $\mathbf{D}$ into vector $\vec{\mathbf{D}}$.

The prescriptions described above allow us to replace all bivectors, pseudobivectors, forms and pseudoforms of grade one and two by vectors and pseudovectors as summarized in the Table 28.4. In this manner the number of directed quantities summarized in Table 28.1 and 28.2 is reduced from sixteen to four due to the presence of the scalar product.

Table 28.4

previous quantity	new quantity
vector	vector
bivector	pseudovector
pseudovector	pseudovector
pseudobivector	vector
form	vector
two-form	pseudovector
pseudo-form	pseudovector
pseudo-two-form	vector

After decomposition into basic quantities ($\mathbf{e}_i$'s are here basic vectors, $\mathbf{f}^j$'s are basic one-forms, and an asterisk denotes a pseudoquantity), this transition can be expressed as follows:
for bivectors,

$$\mathbf{b} = b^{12}\mathbf{e}_1 \wedge \mathbf{e}_2 + b^{23}\mathbf{e}_2 \wedge \mathbf{e}_3 + b^{31}\mathbf{e}_3 \wedge \mathbf{e}_1 \to \vec{\mathbf{b}} = b^{12}\mathbf{e}_3^* + b^{23}\mathbf{e}_1^* + b^{31}\mathbf{e}_2^*,$$

for pseudobivectors,

$$\mathbf{c} = c^{12}\mathbf{e}_1 \wedge \mathbf{e}_2^* + c^{23}\mathbf{e}_2 \wedge \mathbf{e}_3^* + c^{31}\mathbf{e}_3 \wedge \mathbf{e}_1^* \to \vec{\mathbf{c}} = c^{12}\mathbf{e}_3 + c^{23}\mathbf{e}_1 + c^{31}\mathbf{e}_2,$$

for one-forms,

$$\mathbf{E} = E_1\mathbf{f}^1 + E_2\mathbf{f}^2 + E_3\mathbf{f}^3 \quad \to \quad \vec{\mathbf{E}} = E_1\mathbf{e}_1 + E_2\mathbf{e}_2 + E_3\mathbf{e}_3,$$

for two-forms,

$$\mathbf{B} = B_{12}\mathbf{f}^1 \wedge \mathbf{f}^2 + B_{23}\mathbf{f}^2 \wedge \mathbf{f}^3 + B_{31}\mathbf{f}^3 \wedge \mathbf{f}^1 \to \vec{\mathbf{B}} = B_{12}\mathbf{e}_3^* + B_{23}\mathbf{e}_1^* + B_{31}\mathbf{e}_2^*,$$

for pseudo-one-forms,

$$\mathbf{H} = H_1\mathbf{f}_*^1 + H_2\mathbf{f}_*^2 + H_3\mathbf{f}_*^3 \quad \rightarrow \quad \vec{\mathbf{H}} = H_1\mathbf{e}_1^* + H_2\mathbf{e}_2^* + H_3\mathbf{e}_3^*$$

and for pseudo-two-forms,

$$\mathbf{D} = D_{12}\mathbf{f}^1 \wedge \mathbf{f}_*^2 + D_{23}\mathbf{f}^2 \wedge \mathbf{f}_*^3 + D_{31}\mathbf{f}^3 \wedge \mathbf{f}_*^1 \rightarrow \vec{\mathbf{D}} = D_{12}\mathbf{e}_3 + D_{23}\mathbf{e}_1 + D_{31}\mathbf{e}_2.$$

As we see, the coordinates are maintained; only the basic quantities are changed.

In a quadratic space, a natural unit of volume exists; namely, the cube with three edges formed of the orthonormal basis vectors. Thus, the absolute magnitude of a trivector or pseudotrivector is just its volume expressed in this unit. The same cube serves to ascribe magnitudes to the three-forms and pseudo-three-forms. Now a prescription can be formulated how to change dual objects into principal ones. For a three-form, take a trivector with the same orientation and equal magnitude; for a pseudo-three-form, take a pseudotrivector with the same orientation and equal magnitude. Then we are left with four quantities: scalars, pseudoscalars, trivectors and pseudotrivectors. If this is still too many, we can proceed further. We change trivectors into pseudoscalars and pseudotrivectors into scalars keeping the orientations and magnitudes. Then only two quantities remain: the familiar scalar and pseudoscalar. In the traditional presentations, however, the orientation of the pseudoscalars is assumed to be the sign. Only their distinct behaviour under the reflection of vectors is pointed out; they change sign in such a transformation.

28.8 Conclusion

We have presented a wealth of directed quantities which could replace vectors, pseudovectors, scalars, and pseudoscalars in three-dimensional space. They become necessary when the scalar product and the metric implied by it are missing in the underlying vector space. What is interesting is that almost all of them are represented by physical quantities. (We did not indicate examples of three-forms and pseudo-zero-forms.) Two outer algebras have been introduced: the extended Grassmann algebra of multivectors and pseudomultivectors, and a similar one for the forms and pseudoforms. They unite the directed quantities in appropriate algebraic structures.

The presence of a scalar product allows for the reduction of the number of directed quantities, in a first step, from sixteen to eight (through replacing the forms and pseudoforms by multivectors and pseudomultivectors) and, in a second step, to four (by leaving only scalars, pseudoscalars, vectors and pseudovectors). Pseudovectors should be depicted differently than vectors, namely with a ring surrounding the segment rather than an arrow on it. Pseudoscalars have a handedness instead of a sign.

All the notions introduced are necessary to formulate electrodynamics in a scalar-product-independent way, which we have shown elsewhere [24]. It turns out that only the principal equations of this theory can be tackled in this manner. When one seeks their solutions, that is, specific electromagnetic fields as functions of position, a scalar product is needed for writing the constitutive equations involving electric permeability and magnetic permittivity. A special scalar product can be introduced in the case of an anisotropic dielectric, for which the counterpart of the Coulomb field can be written in a very natural way. Similar considerations may be made in the case of an anisotropic (non-conducting) magnetic medium, and these lead to a natural generalization of the Biot-Savart law.

Acknowledgements

I would like to express my indebtedness to Zbigniew Oziewicz who in numerous lectures and seminars explained the significance of various geometric notions, especially of differential forms. This helped me understand matters described in the present lecture. I express my thanks to the Department of Physics at the University of Windsor where the main part of this work has been done.

Bibliography

[1] Jan Arnoldus Schouten: *Tensor Analysis for Physicists,* Dover Publ., New York 1989 (first edition: Clarendon Press, Oxford 1951).

[2] Charles Misner, Kip Thorne and John Archibald Wheeler: *Gravitation*, Freeman and Co., San Francisco 1973, Sec. 2.5.

[3] Theodore Frankel, *Gravitational Curvature. An Introduction to Einstein's Theory,* Freeman and Co., San Francisco 1979.

[4] Walther Thirring: *Course in Mathematical Physics*, vol. 2: *Classical Field Theory*, Springer Verlag, New York 1979.

[5] G.A. Deschamps: "Electromagnetics and differential forms", *Proc. IEEE* **69**(1981)676.

[6] Roman Ingarden and Andrzej Jamiołkowski: *Classical Electrodynamics*, Elsevier, Amsterdam 1985.

[7] William L. Burke: *Applied Differential Geometry*, Cambridge University Press, Cambridge 1985.

[8] D. Baldomir: "Differential forms and electromagnetism in 3-dimensional Euclidean space R^3", *IEE Proc.* **133A**(1986)139.

[9] P. Hammond and D. Baldomir: "Dual energy methods in electromagnetism using tubes and slices" *IEE Proc.* **135A**(1988)167.

[10] D. Baldomir and P. Hammond: "Global geometry of electromagnetic systems", *IEE Proc.* **140A**(1993)142.

[11] Pertti Lounesto, Risto Mikkola and Vesa Vierros: *J. Comp. Math. Sci. Teach.*, **9**(1989)93.

[12] David Hestenes: *New Foundations for Classical Mechanics*, D. Reidel, Dordrecht 1986.

[13] Bernard Jancewicz: *Multivectors and Clifford Algebra in Electrodynamics*, World Scientific, Singapore 1988.

[14] William L. Burke: *Spacetime, Geometry, Cosmology*, University Science Books, Mill Valley 1980.

[15] Anatoly E. Levashev: *Motion and Duality in Relativistic Electrodynamics* (in Russian), Publishers of the Belorussian State University, Minsk 1979.

[16] L. Sorgsepp and L Lohmus: "About nonassociativity in physics and Cayley-Graves' octonions", *Hadronic J.*, **2**(1979)1388.

[17] Paul Fjelstad: "Extending special relativity via the perplex numbers", *Am. J. Phys.*, **54**(1986)416.

[18] D. Hestenes, P. Reany and G. Sobczyk: "Unipodal algebra and roots of polynomials", *Adv. Appl. Cliff. Alg.*, **1**(1991)51.

[19] Jaime Keller: "Quaternionic, complex, duplex and real Clifford algebras", *Adv. Appl. Cliff. Alg.* **4**(1994)1.

[20] Garret Sobczyk: "The hyperbolic number plane", manuscript 1994, 12 pages.

[21] Cornelius von Westenholz: *Differential Forms in Mathematical Physics*, North-Holland, Amsterdam 1978.

[22] Jan Weyssenhoff: *Principles of classical electromagnetism and optics* (in Polish), PWN, Warsaw 1956.

[23] Arnold Sommerfeld: *Vorlesungen über Theoretische Physik.* Band III: *Elektrodynamik*, Akademie Verlag, Leipzig 1949.

[24] Bernard Jancewicz: "A variable metric electrodynamics. The Coulomb and Biot-Savart laws in anisotropic media", to be published.

Chapter 29

Geometric Algebra: Applications in Engineering

Joan Lasenby
Department of Engineering, Cambridge University,
Cambridge CB2 1PZ, UK

The lectures in this Summer School illustrate ways in which geometric algebra is simplifying certain areas of physics. The aim of this contribution is to illustrate that in some engineering fields similar simplifications are possible and that the framework of geometric algebra provides a powerful tool for the geometric manipulations we require.

In what follows we will outline the ways in which geometric algebra can be applied in the fields of **computer vision** and **robotics/mechanisms**.

It will be assumed that everyone is familiar with the basic elements of geometric algebra. Familiarity with rotors and manipulating rotors, linear algebra and simple multivector differentiation will also be assumed – these areas have been covered in other lectures in the school.[1]

29.1 Applications in Computer Vision

Problems in computer vision rely heavily on geometry: we generally observe a 3-D scene in a series of 2-D image/camera planes. We would like to infer things about the true 3-D scene structure and the object or camera motion from one or more 2-D images of the scene.

29.1.1 Projective Space and Projective Transformations

Since about the mid 1980's most of the computer vision literature discussing geometry and invariants has used the language of projective geometry. As any point on a ray from the optical centre of a camera will map to the same point in the

[1] Editor's note: see especially c. 6.

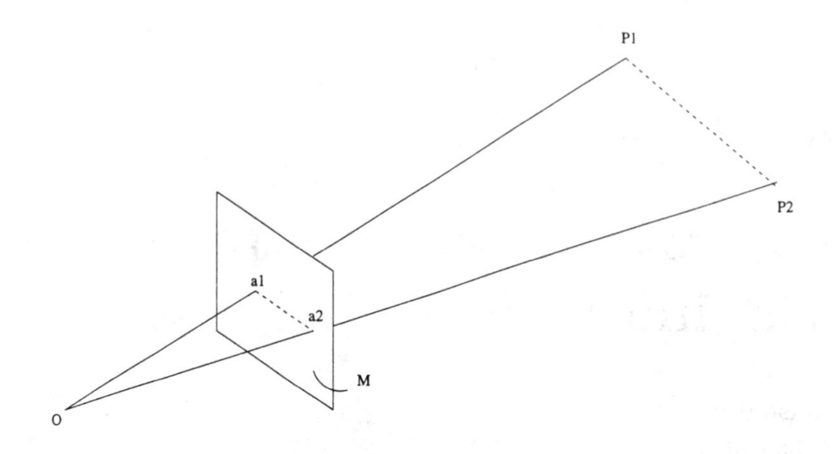

Fig. 29.1. Image geometry

camera image plane it is easy to see why a 2D view of a 3D world might well be best expressed in projective space. In classical projective geometry one defines a 3D space, $\mathcal{P}^3$, whose points are in 1-to-1 correspondence with lines through the origin in a 4D space, R^4. Similarly, k-dimensional subspaces of $\mathcal{P}^3$ are identified with $(k+1)$-dimensional subspaces of R^4. Such projective views can provide very elegant descriptions of the geometry of incidence (intersections, unions etc.), but in order to carry out any real computations one is forced to introduce some sort of basis and associated metric. From a mathematical viewpoint the projective space, $\mathcal{P}^3$, would have no metric, the basis and metric are introduced in the associated 4D space. In this 4D space a coordinate description of a projective point is conventionally brought about by using *homogeneous coordinates.* The usefulness of the projective description of space is often only realised via the introduction of such homogeneous coordinates. What we are aiming to do here is to provide a new way of looking at the problems of computer vision, using a system in which the algebra is clear and computations are completely straightforward and well-defined. Much of the conventional mathematical apparatus of projective geometry will not be needed.

Note that Hestenes and Ziegler, [*Acta Applicandae Mathematicae*,**23**, 1991] have shown how geometric algebra provides a language in which all of projective geometry can be neatly incorporated.

The Projective Split

Points in real 3D space will be represented by vectors in $\mathcal{E}^3$, a 3D space with a Euclidean metric. Since any point on a line through some origin O will be mapped to a single point in the image plane, we will find it useful to associate a point in $\mathcal{E}^3$ with a line in a 4D space, R^4. In these two distinct but related spaces we define basis vectors: $(\gamma_1, \gamma_2, \gamma_3, \gamma_4)$ in R^4 and $(\sigma_1, \sigma_2, \sigma_3)$ in $\mathcal{E}^3$. We identify R^4 and $\mathcal{E}^3$ with the geometric algebras of 4 and 3 dimensions, $\mathcal{G}_4$ and $\mathcal{G}_3$. We require that vectors, bivectors and trivectors in R^4 will represent points, lines and planes in $\mathcal{E}^3$. Suppose we choose γ_4 as a selected direction in R^4, we can then define a mapping which associates the bivectors $\gamma_i\gamma_4$, $i = 1,2,3$, in R^4 with the vectors σ_i, $i = 1,2,3$, in $\mathcal{E}^3$;

$$\sigma_1 \equiv \gamma_1\gamma_4 \quad \sigma_2 \equiv \gamma_2\gamma_4 \quad \sigma_3 \equiv \gamma_3\gamma_4. \tag{29.1}$$

To preserve the Euclidean structure of the spatial vectors $\{\sigma_i\}$ (i.e. $\sigma_i^2 = +1$) it is easy to see that we are forced to assume a non-Euclidean metric for the basis vectors in R^4. We choose to use $\gamma_4^2 = +1$, $\gamma_i^2 = -1$, $i = 1,2,3$. It is interesting to note here that this is precisely the metric structure of Minkowski spacetime. This process of associating the quantities in the higher dimensional space with quantities in the lower dimensional space is an application of what Hestenes calls the **projective split**. In the version of geometric algebra used for physics (spacetime algebra) we have the same 4D space with a Minkowski metric ($\mathcal{G}(1,3)$) and basis vectors $(\gamma_0, \gamma_1, \gamma_2, \gamma_3)$. The projective split is there called a **spacetime split** where γ_0, the time axis, is chosen as the preferred direction and performs the same function as γ_4 in R^4.

For a vector $\mathbf{X} = X_1\gamma_1 + X_2\gamma_2 + X_3\gamma_3 + X_4\gamma_4$ in R^4 the projective split is obtained by taking the geometric product of $\mathbf{X}$ and γ_4;

$$\mathbf{X}\gamma_4 = \mathbf{X}\cdot\gamma_4 + \mathbf{X}\wedge\gamma_4 = X_4\left(1 + \frac{\mathbf{X}\wedge\gamma_4}{X_4}\right) \equiv X_4(1 + \boldsymbol{x}). \tag{29.2}$$

Note that $\boldsymbol{x}$ contains terms of the form $\gamma_1\gamma_4$, $\gamma_2\gamma_4$, $\gamma_3\gamma_4$ or, via the associations in equation (29.1), terms in $\sigma_1, \sigma_2, \sigma_3$. We can therefore think of the vector $\boldsymbol{x}$ as a vector in $\mathcal{E}^3$ which is associated with the bivector $\mathbf{X}\wedge\gamma_4/X_4$ in R^4.

If we start with a vector $\boldsymbol{x} = x_1\sigma_1 + x_2\sigma_2 + x_3\sigma_3$ in $\mathcal{E}^3$, we can represent this in R^4 by the vector $\mathbf{X} = X_1\gamma_1 + X_2\gamma_2 + X_3\gamma_3 + X_4\gamma_4$ such that

$$\boldsymbol{x} = \frac{\mathbf{X}\wedge\gamma_4}{X_4} = \frac{X_1}{X_4}\gamma_1\gamma_4 + \frac{X_2}{X_4}\gamma_2\gamma_4 + \frac{X_3}{X_4}\gamma_3\gamma_4 \equiv \frac{X_1}{X_4}\sigma_1 + \frac{X_2}{X_4}\sigma_2 + \frac{X_3}{X_4}\sigma_3, \tag{29.3}$$

$\Rightarrow x_i = \frac{X_i}{X_4}$, for $i = 1,2,3$. The process of representing $\boldsymbol{x}$ in a higher dimensional space can therefore be seen to be equivalent to using **homogeneous coordinates**, $\mathbf{X}$, for $\boldsymbol{x}$. Thus, in this geometric algebra formulation we postulate distinct spaces in which we represent ordinary 3D quantities and their 4D projective counterparts, together with a well-defined way of moving between these spaces. In this way we can dispense with almost all of the machinery and conceptual construction of projective geometry, while providing all the necessary tools to carry out the desired computations. It may be worth noting that for the issues addressed here, the non-Euclidean nature of R^4 will have no effect, but the presence of a null structure (vectors which square to zero) may have interesting consequences for other problems.

Projective transformations

It is well known that there are various advantages to working in homogeneous coordinates. For example, general displacements can be expressed in terms of a single matrix and some non-linear transformations in $\mathcal{E}^3$ become linear transformations in R^4 – indeed, historically these have been the main motivations for working in homogeneous coordinates.

If a general point (x, y, z) in 3-D space is projected onto an image plane, the coordinates (x', y') in the image plane will be related to (x, y, z) via a transformation of the form:

$$x' = \frac{\alpha_1 x + \beta_1 y + \delta_1 z + \epsilon_1}{\tilde{\alpha} x + \tilde{\beta} y + \tilde{\delta} z + \tilde{\epsilon}}, \quad y' = \frac{\alpha_2 x + \beta_2 y + \delta_2 z + \epsilon_2}{\tilde{\alpha} x + \tilde{\beta} y + \tilde{\delta} z + \tilde{\epsilon}}. \tag{29.4}$$

The transformation is therefore non-linear and expressible as the ratio of two linear transformations. To make this non-linear transformation in $\mathcal{E}^3$ into a linear transformation in R^4 we define a linear function $\underline{f}_p$ mapping vectors onto vectors in R^4 such that the action of $\underline{f}_p$ on the basis vectors $\{\gamma_i\}$ is given by

$$\begin{aligned}
\underline{f}_p(\gamma_1) &= \alpha_1\gamma_1 + \alpha_2\gamma_2 + \alpha_3\gamma_3 + \tilde{\alpha}\gamma_4 \\
\underline{f}_p(\gamma_2) &= \beta_1\gamma_1 + \beta_2\gamma_2 + \beta_3\gamma_3 + \tilde{\beta}\gamma_4 \\
\underline{f}_p(\gamma_3) &= \delta_1\gamma_1 + \delta_2\gamma_2 + \delta_3\gamma_3 + \tilde{\delta}\gamma_4 \\
\underline{f}_p(\gamma_4) &= \epsilon_1\gamma_1 + \epsilon_2\gamma_2 + \epsilon_3\gamma_3 + \tilde{\epsilon}\gamma_4.
\end{aligned} \tag{29.5}$$

A general point P in $\mathcal{E}^3$ given by $\boldsymbol{x} = x\sigma_1 + y\sigma_2 + z\sigma_3$ becomes the point $\mathbf{X} = (X\gamma_1 + Y\gamma_2 + Z\gamma_3 + W\gamma_4)$ in R^4, where $x = X/W$, $y = Y/W$, $z = Z/W$. We can then see that $\underline{f}_p$ maps $\mathbf{X}$ onto $\mathbf{X}'$ where

$$\mathbf{X}' = \left[\sum_{i=1}^{3}\{(\alpha_i X + \beta_i Y + \delta_i Z + \epsilon_i W)\gamma_i\}\right] + (\tilde{\alpha} X + \tilde{\beta} Y + \tilde{\delta} Z + \tilde{\epsilon} W)\gamma_4 . \tag{29.6}$$

The vector $\boldsymbol{x}' = x'\sigma_1 + y'\sigma_2 + z'\sigma_3$ in $\mathcal{E}^3$ corresponds to $\mathbf{X}'$, where x' is given by

$$x' = \frac{\alpha_1 X + \beta_1 Y + \delta_1 Z + \epsilon_1 W}{\tilde{\alpha} X + \tilde{\beta} Y + \tilde{\delta} Z + \tilde{\epsilon} W} = \frac{\alpha_1 x + \beta_1 y + \delta_1 z + \epsilon_1}{\tilde{\alpha} x + \tilde{\beta} y + \tilde{\delta} z + \tilde{\epsilon}}. \tag{29.7}$$

Similarly we have

$$y' = \frac{\alpha_2 x + \beta_2 y + \delta_2 z + \epsilon_2}{\tilde{\alpha} x + \tilde{\beta} y + \tilde{\delta} z + \tilde{\epsilon}}, \quad z' = \frac{\alpha_3 x + \beta_3 y + \delta_3 z + \epsilon_3}{\tilde{\alpha} x + \tilde{\beta} y + \tilde{\delta} z + \tilde{\epsilon}}. \tag{29.8}$$

Note that in general we would take $\alpha_3 = f\tilde{\alpha}$, $\beta_3 = f\tilde{\beta}$ etc. so that $z' = f$ (focal length), independent of the point chosen. Via this means the non-linear transformation in $\mathcal{E}^3$ becomes a linear transformation, $\underline{f}_p$, in R^4. Use of the linear function $\underline{f}_p$ makes the invariant nature of various quantities very easy to establish.

29.1.2 Geometric Invariance in Computer Vision

In real computer vision problems a key task is to recognize objects in 3D scenes via their 2D images in the camera plane. Any one such view will differ from another due to camera calibration, viewing position, lighting conditions etc. It is therefore desirable to seek geometric properties of an object which remain invariant under such changes in the observation parameters. In what follows we will try to justify the claim that geometric algebra is the optimal framework for the theory and computation of invariants in computer vision.

Firstly we look at how to construct quantities which are invariant under projective transformations. We will try to arrive at these in a way which generalizes naturally from 1D to 2D to 3D.

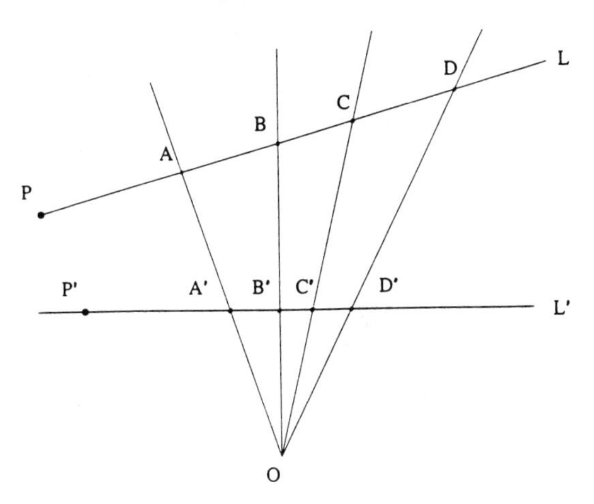

Fig. 29.2. Formation of the 1D cross-ratio

1D Cross-Ratio

The '*fundamental projective invariant*' of points on a line is the so-called **cross-ratio**, ρ, defined as

$$\rho = \frac{AC}{BC}\frac{BD}{AD} = \frac{(t_3 - t_1)(t_4 - t_2)}{(t_4 - t_1)(t_3 - t_2)},$$

where $t_1 = |PA|$, $t_2 = |PB|$, $t_3 = |PC|$, $t_4 = |PD|$ – see figure 29.2. It is fairly easy to show that for the projection through O of the collinear points A, B, C, D onto any line, ρ remains constant. For this 1D case, any point q on the line L can be written as $\boldsymbol{q} = t\sigma_1$ relative to P, where σ_1 is a unit vector in the direction of L. We then move up a dimension to a 2D space we will call R^2, with basis vectors (γ_1, γ_2), in which $\boldsymbol{q}$ is represented by the vector $\mathbf{Q}$;

$$\mathbf{Q} = T\gamma_1 + S\gamma_2$$

where, as before, we associate $\boldsymbol{q}$ with the bivector

$$\frac{\mathbf{Q}\wedge\gamma_2}{\mathbf{Q}\cdot\gamma_2} = \frac{T}{S}\gamma_1\gamma_2 \equiv \frac{T}{S}\sigma_1$$

so that $t = T/S$. When a point on line L is projected onto another line L', the distances t and t' (where $t' = |P'A'|$ etc.) are related by a projective transformation of the form

$$t' = \frac{\alpha t + \beta}{\tilde{\alpha} t + \tilde{\beta}}. \tag{29.9}$$

This non-linear transformation in $\mathcal{E}^1$ can be made into a linear transformation in R^2 by defining the linear function $\underline{f}_1$ mapping vectors onto vectors in R^2;

$$\begin{aligned} \underline{f}_1(\gamma_1) &= \alpha_1\gamma_1 + \tilde{\alpha}\gamma_2 \\ \underline{f}_1(\gamma_2) &= \beta_1\gamma_1 + \tilde{\beta}\gamma_2. \end{aligned}$$

Consider 2 vectors $\mathbf{X}_1, \mathbf{X}_2$ in R^2. Form the bivector

$$\mathcal{S}_1 = \mathbf{X}_1 \wedge \mathbf{X}_2 = \lambda_1 I_2$$

where $I_2 = \gamma_1\gamma_2$ is the pseudoscalar for R^2. We now look at how $\mathcal{S}_1$ transforms under $\underline{f}_1$:

$$\mathcal{S}_1' = \mathbf{X}_1' \wedge \mathbf{X}_2' = \underline{f}_1(\mathbf{X}_1 \wedge \mathbf{X}_2) = (\det \underline{f}_1)(\mathbf{X}_1 \wedge \mathbf{X}_2). \tag{29.10}$$

This last step follows since a linear function must map a pseudoscalar onto a multiple of itself, this multiple being the determinant of the function. Suppose that we now take 4 points on the line L whose corresponding vectors in R^2 are $\{\mathbf{X}_i\}$, $i = 1, .., 4$, and consider the ratio $\mathcal{R}_1$ of 2 wedge products;

$$\mathcal{R}_1 = \frac{\mathbf{X}_1 \wedge \mathbf{X}_2}{\mathbf{X}_3 \wedge \mathbf{X}_4}. \tag{29.11}$$

Then, under $\underline{f}_1$, $\mathcal{R}_1 \to \mathcal{R}_1'$, where

$$\mathcal{R}_1' = \frac{\mathbf{X}_1' \wedge \mathbf{X}_2'}{\mathbf{X}_3' \wedge \mathbf{X}_4'} = \frac{(\det \underline{f}_1)\mathbf{X}_1 \wedge \mathbf{X}_2}{(\det \underline{f}_1)\mathbf{X}_3 \wedge \mathbf{X}_4}. \tag{29.12}$$

$\mathcal{R}_1$ is therefore invariant under $\underline{f}_1$. However, we want to express our invariants in terms of distances on the 1D line; for this we must consider how the bivector $\mathcal{S}_1$ in R^2 projects down to $\mathcal{E}^1$.

$$\begin{aligned} \mathbf{X}_1 \wedge \mathbf{X}_2 &= (T_1\gamma_1 + S_1\gamma_2) \wedge (T_2\gamma_1 + S_2\gamma_2) \\ &= S_1 S_2 (T_1/S_1 - T_2/S_2) I_2 \\ &= S_1 S_2 (t_1 - t_2) I_2. \end{aligned} \tag{29.13}$$

In order to form a projective invariant which is independent of the choice of the arbitrary scalars S_i, we must then take *ratios* of the bivectors $\mathbf{X}_i \wedge \mathbf{X}_j$ (so that $\det \underline{f}_1$ cancels) and *multiples* of such ratios so that the S_i's cancel. More precisely, consider the following expression

$$Inv_1 = \frac{\{(\mathbf{X}_3 \wedge \mathbf{X}_1) I_2^{-1}\}\{(\mathbf{X}_4 \wedge \mathbf{X}_2) I_2^{-1}\}}{\{(\mathbf{X}_4 \wedge \mathbf{X}_1) I_2^{-1}\}\{(\mathbf{X}_3 \wedge \mathbf{X}_2) I_2^{-1}\}}.$$

Then, in terms of distances along the lines, under the projective transformation $\underline{f}_1$, Inv_1 goes to Inv_1' where

$$Inv_1' = \frac{S_3 S_1 (t_3 - t_1) S_4 S_2 (t_4 - t_2)}{S_4 S_1 (t_4 - t_1) S_3 S_2 (t_3 - t_2)} = \frac{(t_3 - t_1)(t_4 - t_2)}{(t_4 - t_1)(t_3 - t_2)}, \tag{29.14}$$

which is independent of the S_i's and is indeed the 1D classical projective invariant, the **cross-ratio**. Deriving the cross-ratio in this way enables us to easily generalize it to form invariants in higher dimensions.

2D and 3D generalizations of the Cross-Ratio

For points in a plane we again move up to a space with one higher dimension which we shall call R^3. A point P in the plane is described by the vector $\boldsymbol{x} = x\sigma_1 + y\sigma_2$ in $\mathcal{E}^2$ with an R^3 representation of $\mathbf{X} = X\gamma_1 + Y\gamma_2 + Z\gamma_3$ where $x = X/Z$ and $y = Y/Z$. As before, we can define a general projective transformation via a linear function $\underline{f}_2$ mapping vectors to vectors in R^3 such that;

$$\underline{f}_2(\gamma_i) = c_{i1}\gamma_1 + c_{i2}\gamma_2 + c_{i3}\gamma_3. \tag{29.15}$$

The procedure is then to form trivectors and take *ratios* of these trivectors (to cancel out the det $\underline{f}_2$ factors) and *multiples* of these ratios to cancel out the arbitrary Z factors. An example of an invariant is therefore given by

$$Inv_2 = \frac{\{(\mathbf{X}_5 \wedge \mathbf{X}_4 \wedge \mathbf{X}_3)I_3^{-1}\}\{(\mathbf{X}_5 \wedge \mathbf{X}_2 \wedge \mathbf{X}_1)I_3^{-1}\}}{\{(\mathbf{X}_5 \wedge \mathbf{X}_1 \wedge \mathbf{X}_3)I_3^{-1}\}\{(\mathbf{X}_5 \wedge \mathbf{X}_2 \wedge \mathbf{X}_4)I_3^{-1}\}}$$

In $\mathcal{E}^2$ we interpret this ratio as

$$Inv_2 = \frac{\{(\boldsymbol{x}_5 - \boldsymbol{x}_4)\wedge(\boldsymbol{x}_5 - \boldsymbol{x}_3)I_2^{-1}\}\{(\boldsymbol{x}_5 - \boldsymbol{x}_2)\wedge(\boldsymbol{x}_5 - \boldsymbol{x}_1)I_2^{-1}\}}{\{(\boldsymbol{x}_5 - \boldsymbol{x}_1)\wedge(\boldsymbol{x}_5 - \boldsymbol{x}_3)I_2^{-1}\}\{(\boldsymbol{x}_5 - \boldsymbol{x}_2)\wedge(\boldsymbol{x}_5 - \boldsymbol{x}_4)I_2^{-1}\}} = \frac{A_{543}A_{521}}{A_{513}A_{524}} \tag{29.16}$$

where $\frac{1}{2}A_{ijk}$ is the area of the triangle defined by the 3 vertices $\boldsymbol{x}_i, \boldsymbol{x}_j, \boldsymbol{x}_k$. This invariant is regarded as a 2D generalization of the 1D cross-ratio.

In a precisely similar manner we can form 3D invariants by taking multiples of ratios of 4-vectors in R^4, An example of such an invariant is

$$Inv_3 = \frac{\{(\mathbf{X}_1 \wedge \mathbf{X}_2 \wedge \mathbf{X}_3 \wedge \mathbf{X}_4)I_4^{-1}\}\{(\mathbf{X}_4 \wedge \mathbf{X}_5 \wedge \mathbf{X}_2 \wedge \mathbf{X}_6)I_4^{-1}\}}{\{(\mathbf{X}_1 \wedge \mathbf{X}_2 \wedge \mathbf{X}_4 \wedge \mathbf{X}_5)I_4^{-1}\}\{(\mathbf{X}_3 \wedge \mathbf{X}_4 \wedge \mathbf{X}_2 \wedge \mathbf{X}_6)I_4^{-1}\}}. \tag{29.17}$$

Using the projective split we can write each of the scalars in this expression as:

$$(\mathbf{X}_1 \wedge \mathbf{X}_2 \wedge \mathbf{X}_3 \wedge \mathbf{X}_4)I_4^{-1} \equiv W_1 W_2 W_3 W_4 \{(\boldsymbol{x}_2 - \boldsymbol{x}_1)\wedge(\boldsymbol{x}_3 - \boldsymbol{x}_1)\wedge(\boldsymbol{x}_4 - \boldsymbol{x}_1)\}I_3^{-1}. \tag{29.18}$$

We can therefore see that the invariant Inv_3 is the 3D equivalent of the 1D cross-ratio and consists of ratios of volumes;

$$Inv_3 = \frac{V_{1234}V_{4526}}{V_{1245}V_{3426}}, \tag{29.19}$$

where V_{ijkl} is the volume of the solid formed by the 4 vertices $\boldsymbol{x}_i, \boldsymbol{x}_j, \boldsymbol{x}_k, \boldsymbol{x}_l$.

Conventionally all of these invariants are well known but above we have outlined a general process for generating projective invariants in any dimension which is straightforward and simple.

29.1.3 Motion and Structure from Motion

The problem we will discuss here is easily stated and illustrates nicely how one can use geometric algebra to do things which are much more difficult in a conventional approach.

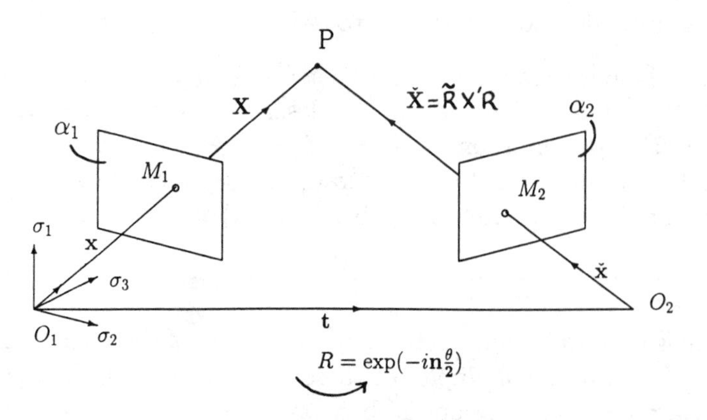

Fig. 29.3. Point in object viewed in two camera positions

Suppose we view an object in one camera, the camera then undergoes some general displacement (rotation and translation) and we have a second view of the object. Then further suppose that we can identify n point correspondences in the two images. The problem is then, from the image coordinates of the n points in the two views, we would like to reconstruct the motion of the camera and the 3-D coordinates of the points (if these are not known). We will assume in what follows that the camera calibration is known.

Figure 29.3 shows some point P observed in two camera positions. Let $\mathbf{X}$ and $\boldsymbol{x}$ be the position vectors of P and M_1 observed in position 1, and $\mathbf{X}'$ and $\boldsymbol{x}'$ be the position vectors of P and M_2 observed in the rotated frame at O_2. It is then clear that $\mathbf{X}$ and $\mathbf{X}'$ are related as follows[2]

$$\mathbf{X}' = R(\mathbf{X} - \boldsymbol{t})\tilde{R}$$

assuming $\sigma_i' = \tilde{R}\sigma_i R$. Note here that $\boldsymbol{x}$, $\mathbf{X}$ etc., are ordinary vectors in $\mathcal{E}^3$ and that we now work entirely in $\mathcal{E}^3$. Since we want the optimal estimates for the unknowns in the presence of errors on our measurements, it is sensible to adopt a least-squares approach. There are two cases we will consider; the first case is the simplest, in which the true 3-D coordinates of the n points are known in the two views. The second case, which is considerably harder, is where the range data is unknown.

Recovering motion with known range data

If the coordinates $\{\mathbf{X}_i\}$ and $\{\mathbf{X}_i'\}$ in the two views are known, then, for n matched points in the two frames we want to find the R and $\boldsymbol{t}$ which minimize

$$S = \sum_{i=1}^{n} \left[\boldsymbol{X}'_i - R(\boldsymbol{X}_i - \boldsymbol{t})\tilde{R}\right]^2. \qquad (29.20)$$

[2]Editor's note: σ_i' are the basis vectors of the rotated frame at O_2, and in figure 29.3, $\check{\mathbf{X}} = \mathbf{X} - \mathbf{t}$.

This therefore involves minimizing S w.r.t. R and $\boldsymbol{t}$. The differentiation w.r.t. $\boldsymbol{t}$ is straightforward;

$$\partial_t S = \sum_{i=1}^{n} \left[\boldsymbol{X}'_i - R(\boldsymbol{X}_i - \boldsymbol{t})\tilde{R} \right] \partial_t (R\boldsymbol{t}\tilde{R}) = 0. \qquad (29.21)$$

Which can easily be solved for $\boldsymbol{t}$ to give

$$\boldsymbol{t} = \frac{1}{n}\sum_{i=1}^{n} \left[\boldsymbol{X}_i - \tilde{R}\boldsymbol{X}'_i R \right], \qquad (29.22)$$

This can be rewritten as $\boldsymbol{t} = \bar{\boldsymbol{X}} - \tilde{R}\bar{\boldsymbol{X}}'R$ where $\bar{\boldsymbol{X}}$ and $\bar{\boldsymbol{X}}'$ are the centroids of the data points in the two views. This is a well known result, although it is often arrived at by different means.

To see how we differentiate S wrt R, we look at the following problem: *find the rotor R which 'most closely' rotates the vectors $\{\boldsymbol{u}_i\}$ into the vectors $\{\boldsymbol{v}_i\}$, $i = 1, .., n$.* i.e. we wish to find the rotor R which minimizes

$$\phi = \sum_{i=1}^{n} (\boldsymbol{v}_i - R\boldsymbol{u}_i\tilde{R})^2. \qquad (29.23)$$

Expanding ϕ gives

$$\phi = \sum_{i=1}^{n} \{(\boldsymbol{v}_i{}^2 + \boldsymbol{u}_i{}^2) - 2\langle \boldsymbol{v}_i R\boldsymbol{u}_i\tilde{R}\rangle\}. \qquad (29.24)$$

Replace $R\boldsymbol{u}_i\tilde{R}$ with $\psi\boldsymbol{u}_i\psi^{-1}$ (to avoid having to include a Lagrange multiplier to incorporate the constraint $R\tilde{R} = 1$). Then take the multivector derivative of ϕ w.r.t. ψ;

$$\partial_\psi \phi(\psi) = -2\sum_{i=1}^{n} \{\dot{\partial}_\psi \langle \dot{\psi} A_i\rangle + \dot{\partial}_\psi \langle B_i \dot{\psi}^{-1}\rangle\}, \qquad (29.25)$$

where $A_i = \boldsymbol{u}_i\psi^{-1}\boldsymbol{v}_i$ and $B_i = \boldsymbol{v}_i\psi\boldsymbol{u}_i$. The first term is easily evaluated to give A_i. To evaluate the second term we can use the result $\partial_\psi\langle M\psi^{-1}\rangle = -\psi^{-1}\mathrm{P}_\psi(M)\psi^{-1}$ for general multivectors ψ and M. Where $\mathrm{P}_\psi(M)$ projects M onto the grades of ψ.

After some manipulation this results in

$$\partial_\psi \phi(\psi) = 4\tilde{R}\sum_{i=1}^{n} \boldsymbol{v}_i \wedge (R\boldsymbol{u}_i\tilde{R}). \qquad (29.26)$$

Thus the rotor which minimizes the least-squares expression $\phi(R) = \sum_{i=1}^{n}(\boldsymbol{v}_i - R\boldsymbol{u}_i\tilde{R})^2$ must satisfy

$$\sum_{i=1}^{n} \boldsymbol{v}_i \wedge (R\boldsymbol{u}_i\tilde{R}) = 0. \qquad (29.27)$$

This is intuitively obvious – we want the R which makes $\boldsymbol{u}_i$ 'most parallel' to $\boldsymbol{v}_i$ in the average sense. To solve equation (29.27) for R we need to use the linear algebra framework of geometric algebra.

Using the definition of the vector derivative it is possible to rewrite equation (29.27) as

$$\partial_a \wedge \left[\sum_{i=1}^{n} R[(\boldsymbol{a}\cdot \boldsymbol{v}_i)\boldsymbol{u}_i]\tilde{R} \right] = 0. \tag{29.28}$$

Equation (29.28) can then be written as

$$\partial_a \wedge R\underline{f}(\boldsymbol{a})\tilde{R} = 0. \tag{29.29}$$

Where $\underline{f}$ is defined by $\underline{f}(\boldsymbol{a}) = \sum_{i=1}^{n} (\boldsymbol{a}\boldsymbol{v}_i)\boldsymbol{u}_i$. Now define another function $\underline{R}$ mapping vectors onto vectors such that $\underline{R}\boldsymbol{a} = R\boldsymbol{a}\tilde{R}$ so that

$$\partial_a \wedge \underline{R}\underline{f}(\boldsymbol{a}) = 0, \tag{29.30}$$

for any vector $\boldsymbol{a}$. This then tells us that if $\underline{G} = \underline{R}\underline{f}$ then $\underline{G}$ is symmetric. Since R is a rotor and is orthogonal, we have that $R^{-1} = \tilde{R}$ from which it follows that $\underline{R}^{-1} = \overline{R}$. This enables us to write $\underline{f}$ as

$$\underline{f}(\boldsymbol{a}) = \overline{R}\underline{G}(\boldsymbol{a}). \tag{29.31}$$

If we now perform a singular-value decomposition (SVD) on $\underline{f}$, so that $\underline{f} = \underline{U}\,\underline{S}\,\underline{V}$, we have

$$\underline{f} = \underline{U}\,\underline{V}(\overline{V}\underline{S}\,\underline{V}) \equiv \overline{R}\underline{G}. \tag{29.32}$$

$\overline{V}\underline{S}\,\underline{V}$ is obviously symmetric, as is $\underline{G}$, which tells us that the rotation $\underline{R}$ must be given by

$$\underline{R} = \overline{V}\,\overline{U}. \tag{29.33}$$

The rotation $\underline{R}$ is therefore found in terms of the SVD of the function $\underline{f}$.

To summarize: take the two sets of vectors $\{\boldsymbol{u}_i\}$ and $\{\boldsymbol{v}_i\}$ and form F, the matrix representation of $\underline{f}$ by

$$F_{\alpha\beta} \equiv \sigma_\alpha \cdot \underline{f}(\sigma_\beta) = \sum_{i=1}^{n} (\sigma_\alpha \cdot \boldsymbol{u}_i)(\sigma_\beta \cdot \boldsymbol{v}_i). \tag{29.34}$$

An SVD of F gives $F = USV^T$ and the 3×3 rotation matrix $\mathcal{R}$ is then simply given by the product $\mathcal{R} = VU^T$.

Now going back to our original question – the solution to our optimization problem can be written down directly as

$$\sum_{i=1}^{n} \boldsymbol{X}'_i \wedge R(\boldsymbol{X}_i - \boldsymbol{t})\tilde{R} = 0. \tag{29.35}$$

If we substitute our optimal value of $\boldsymbol{t}$ in this equation we have

$$\sum_{i=1}^{n} \boldsymbol{X}'_i \wedge R(\boldsymbol{X}_i - \bar{\boldsymbol{X}} - \tilde{R}\bar{\boldsymbol{X}}' R)\tilde{R} = 0. \tag{29.36}$$

Noting that the $\sum_{i=1}^{n} \boldsymbol{X}'_i \wedge \bar{\boldsymbol{X}}'$ term vanishes, this reduces to $\sum_{i=1}^{n} \boldsymbol{v}_i \wedge R\boldsymbol{u}_i\tilde{R} = 0$ with

$$\begin{aligned} \boldsymbol{u}_i &= \boldsymbol{X}_i - \overline{\boldsymbol{X}} \\ \boldsymbol{v}_i &= \boldsymbol{X}'_i. \end{aligned} \tag{29.37}$$

The rotor is then found from an SVD of the matrix F where F is defined in terms of the $\boldsymbol{u}_i$ and $\boldsymbol{v}_i$ as given in equation (29.34). There have been many methods in the literature put forward to solve the problem outlined here – and it is easy to see that although different procedures are followed (orthonormal matrices, dual quaternions etc.) the methods produce the same result. The advantage of this geometric algebra approach is that we can follow the same procedures for the more involved case of unknown range information.

Recovering motion and structure with unknown range data

Again we optimize the appropriate least squares expression simultaneously over all of the unknowns in the problem. The expression we minimize is given by

$$S = \sum_{i=1}^{n} \left[X'_{i3}\boldsymbol{x}'_i - R(X_{i3}\boldsymbol{x}_i - \boldsymbol{t})\tilde{R} \right]^2, \tag{29.38}$$

since $\mathbf{X}_i = X_{i3}\boldsymbol{x}_i$ and $\mathbf{X}'_i = X'_{i3}\boldsymbol{x}'_i$ (assuming calibrated cameras and focal lengths of 1). The $\{\boldsymbol{x}_i\}$ and the $\{\boldsymbol{x}'_i\}$ are now our observed quantities (which will suffer from possible measurement errors) and the set of unknowns in the problem increases considerably to $\{R, \boldsymbol{t}, (X_{i3}, X'_{i3}, i = 1, .., n)\}$. In order to minimize equation (29.38) we adopt the most general approach we can and differentiate S with respect to R, $\boldsymbol{t}$ and the range coordinates $\{X_{i3}, X'_{i3}\}$.

From the previous sections we know that differentiating with respect to $\boldsymbol{t}$ and R give the following equations;

$$\boldsymbol{t} = \frac{1}{n}\sum_{i=1}^{n} \left[X_{i3}\boldsymbol{x}_i - X'_{i3}\tilde{R}\boldsymbol{x}'_i R \right] \tag{29.39}$$

$$\sum_{i=1}^{n} X'_{i3}\boldsymbol{x}'_i \wedge R\left(X_{i3}\boldsymbol{x}_i - \frac{1}{n}\sum_{j=1}^{n} X_{j3}\boldsymbol{x}_j \right)\tilde{R} = 0. \tag{29.40}$$

Note now that we cannot sensibly find the 'centroid' of the data sets. Differentiation of S with respect to each of the X_{i3} and X'_{i3} is simply scalar differentiation, and solutions to the equations $\frac{\partial S}{\partial X_{i3}} = 0$ and $\frac{\partial S}{\partial X'_{i3}} = 0$ are as follows (no summation implied):

$$\begin{aligned} X_{i3} &= \frac{-(\boldsymbol{x}'_i \wedge \acute{\boldsymbol{t}})\cdot(\acute{\boldsymbol{x}}_i \wedge \boldsymbol{x}'_i)}{(\acute{\boldsymbol{x}}_i \wedge \boldsymbol{x}'_i)\cdot(\acute{\boldsymbol{x}}_i \wedge \boldsymbol{x}'_i)} && (29.41) \\ X'_{i3} &= \frac{-(\acute{\boldsymbol{x}}_i \wedge \acute{\boldsymbol{t}})\cdot(\acute{\boldsymbol{x}}_i \wedge \boldsymbol{x}'_i)}{(\acute{\boldsymbol{x}}_i \wedge \boldsymbol{x}'_i)\cdot(\acute{\boldsymbol{x}}_i \wedge \boldsymbol{x}'_i)}, && (29.42) \end{aligned}$$

for $i = 1, ..., n$, where $\acute{\boldsymbol{x}}_i = R\boldsymbol{x}_i\tilde{R}$ and $\acute{\boldsymbol{t}} = R\boldsymbol{t}\tilde{R}$. These equations can be written more concisely as $X_{i3} = -(\boldsymbol{x}'_i \wedge \acute{\boldsymbol{t}})\cdot B_i$ and $X'_{i3} = -(\acute{\boldsymbol{x}}_i \wedge \acute{\boldsymbol{t}})\cdot B_i$ where B_i is the bivector $B_i = \beta_i/(\beta_i\cdot\beta_i)$ with $\beta_i = \acute{\boldsymbol{x}}_i \wedge \boldsymbol{x}'_i$. Our task is now to find a way of simultaneously solving the equations (29.39), (29.40), (29.41), (29.42). To accomplish this solution we can develop methods of (a) finding $\boldsymbol{t}$ given R and the data, (b) finding $\{X_{i3}\}$ and $\{X'_{i3}\}$ given R, $\boldsymbol{t}$ and the data and (c) finding R given $\{X_{i3}\}$ and $\{X'_{i3}\}$ and the data.

Using equations (29.41), (29.42) we can obtain an expression for $\boldsymbol{t}$ in terms of R and the data and from this we are able to arrive at the following equations:

$$\boldsymbol{t}\cdot(\boldsymbol{x}_j - \boldsymbol{P}_j) = 0 \qquad \text{for}(j = 1, ..., n), \tag{29.43}$$

where $\boldsymbol{P}_j = \frac{1}{n}\sum_{i=1}^{n} \frac{1}{\beta_i \beta_i}\left[(\boldsymbol{A}_i\cdot\boldsymbol{x}_j)\grave{\boldsymbol{x}}'_i - (\boldsymbol{B}_i\cdot\boldsymbol{x}_j)\boldsymbol{x}_i)\right]$ and $\grave{\boldsymbol{x}}_i = \tilde{R}\boldsymbol{x}R$. If $\boldsymbol{w}_j = \boldsymbol{x}_j - \boldsymbol{P}_j$, then, since $\boldsymbol{t}$ is perpendicular to each of the $\boldsymbol{w}_j$'s we can reconstruct $\boldsymbol{t}$ by taking the cross-product of any two distinct $\boldsymbol{w}_j$'s:

$$\boldsymbol{t} = \frac{i\boldsymbol{w}_n \wedge \boldsymbol{w}_m}{|i\boldsymbol{w}_n \wedge \boldsymbol{w}_m|}, \tag{29.44}$$

for $n \neq m$. Thus, given our estimate of R and the data, we can construct an estimate for $\boldsymbol{t}$ without any estimate of the range coordinates. Given estimates of R, $\boldsymbol{t}$ and the data, $\{X_{i3}\}$ and $\{X'_{i3}\}$ can be found simply from equations (29.41), (29.42). It is clear the $\boldsymbol{t}$ and $\{\mathbf{X}_i, \mathbf{X}'_i\}$ are only determined up to a scale factor. This factor can be fixed by requiring that $\boldsymbol{t}^2 = 1$.

Since we can estimate R given the range coordinates and the data using the previously described methods, it follows that we can now construct an iterative method for solving for R, $\boldsymbol{t}$ and the range coordinates. Simulations show that the performance of such an iterative algorithm is very good, even in the presence of considerable measurement noise. As far as we are aware, this is the first such attempt at solving *directly* the relevant least squares equation simultaneously over all the unknowns in the problem – we are able to do this because of the powerful tools made available to us in the geometric algebra.

29.2 Applications in Robotics/Mechanisms

When a rigid body undergoes a general displacement this consists of a rotation plus a translation. The motion of linked rigid bodies (e.g. a robot arm) constitutes a large part of the study of mechanisms. Given some parameters governing the motion of the individual members we would generally like to be able to specify the motion of the whole mechanism in a concise way.

It is sometimes convenient to describe a general displacement by a rotation about some axis in space plus a translation parallel to that axis – such a transformation is called a *screw transformation*. When we combine several such general displacements or screw transformations we would like to know how to describe the overall displacement in terms of the individual transformations. In order to address this problem, a number of very closely related systems have been developed over the years; Clifford's

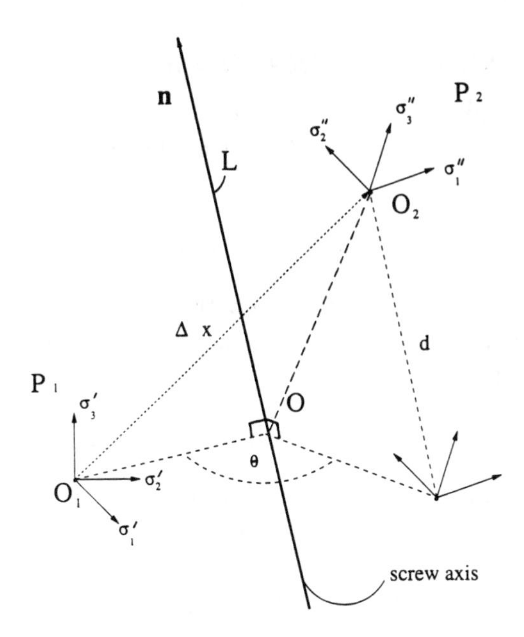

Fig. 29.4. A general displacement or screw-transformation

motor algebra, dual quaternions and Plücker coordinates being amongst those most widely used. Here we will illustrate the geometric algebra approach to this area and briefly comment on the relation of this approach to the other systems.

29.2.1 Screw Transformations

Consider a rigid body at position P_1 whose orientation is specified by a coordinate frame $(\sigma'_1, \sigma'_2, \sigma'_3)$ at O_1. The body is then rotated and translated to position P_2, such that the frame at O_1 goes to a frame $(\sigma''_1, \sigma''_2, \sigma''_3)$ at O_2. This general displacement is completely characterized by the translation $\mathbf{\Delta x} = \overrightarrow{O_1O_2}$ and the rotation R, such that $\sigma''_i = R\sigma'_i\tilde{R}$.

From figure 29.4 we can see that this general displacement can also be described by a rotation of θ about some axis L in space with direction $\boldsymbol{n}$ and a translation d parallel to that axis. This axis is called the *screw axis*.

Suppose we choose a coordinate frame $(\sigma_1, \sigma_2, \sigma_3)$ at O, where O lies on the axis L and is such that $\mathbf{x}_1 \cdot \boldsymbol{n} = 0$, where $\mathbf{x}_1 = \overrightarrow{OO_1}$. Similarly, $\mathbf{x}_2 = \overrightarrow{OO_2}$ so that $\mathbf{\Delta x} = \mathbf{x}_2 - \mathbf{x}_1$.

Let R_1 and R_2 be the rotors which specify the orientations of the body at P_1 and P_2 respectively relative to the frame at 0, then

$$R_1\sigma_i\tilde{R}_1 = \sigma'_i \quad \text{and} \quad R_2\sigma_i\tilde{R}_2 = \sigma''_i \tag{29.45}$$

Now, let R be the rotor which performs the rotation about axis L, $R = \exp(-i\frac{\theta}{2}\boldsymbol{n})$. We can then write down the equation governing the transformation from P_1 to P_2

as

$$R(R_1\sigma_i\tilde{R}_1 + \mathbf{x}_1)\tilde{R} + d\boldsymbol{n} = R_2\sigma_i\tilde{R}_2 + \mathbf{x}_2 \qquad (29.46)$$

for $i = 1, 2, 3$. The overall rotation of the axes at O_2 relative to O must result from rotating the axes at O_1, and it is therefore clear that we require $RR_1 = R_2$, which then simplifies the above equation to give

$$R\mathbf{x}_1\tilde{R} + d\boldsymbol{n} = \mathbf{x}_2. \qquad (29.47)$$

This is a very simple description of the screw transformation represented by (θ, L). Since $\mathbf{x}_1\cdot\boldsymbol{n} = 0$ we can write $\exp(-i\frac{\theta}{2}\boldsymbol{n})\mathbf{x}_1\exp(i\frac{\theta}{2}\boldsymbol{n})$ as $\exp(-i\theta\boldsymbol{n})\boldsymbol{x}_1$ so that the translation $\boldsymbol{\Delta}\mathbf{x}$ is given by

$$\boldsymbol{\Delta}\mathbf{x} = [\exp(-i\theta\boldsymbol{n}) - 1]\mathbf{x}_1 + d\boldsymbol{n} = S\mathbf{x}_1 + d\boldsymbol{n} \qquad (29.48)$$

where we have defined a new spinor (scalar plus bivector) S as $S = [\exp(-i\theta\boldsymbol{n}) - 1]$. If the inverse of S exists, we are able to write $\mathbf{x}_1$ as

$$\mathbf{x}_1 = S^{-1}\boldsymbol{\Delta}\mathbf{x} - dS^{-1}\boldsymbol{n}. \qquad (29.49)$$

Since $\mathbf{x}_1\cdot\boldsymbol{n} = 0$, taking the inner product of the above equation with $\boldsymbol{n}$ gives

$$\langle S^{-1}\boldsymbol{\Delta}\mathbf{x}\boldsymbol{n}\rangle - d\langle S^{-1}\rangle = 0 \qquad (29.50)$$

and therefore

$$d = \frac{\langle S^{-1}\boldsymbol{\Delta}\mathbf{x}\boldsymbol{n}\rangle}{\langle S^{-1}\rangle}. \qquad (29.51)$$

This set of equations enable us to obtain any of the relevant quantities in a given problem. For example, if we are given the initial position of the body and the screw parameters $(\theta, \boldsymbol{n}, d)$, we can immediately find the final position and orientation of the body from equation (29.47) and $R\sigma_i\tilde{R} = \sigma_i'$. On the other hand, if we are given the initial and final positions and orientations of the body we can obtain the rotor R from $R\sigma_i\tilde{R} = \sigma_i'$ and the translation d and the location of the screw axis in space from equations (29.51) and (29.49).

The explicit expression for the inverse, S^{-1} is

$$S^{-1} = \frac{\exp(i\theta\boldsymbol{n}) - 1}{4\sin^2\frac{\theta}{2}}. \qquad (29.52)$$

Note that there does not exist an inverse if $\theta = 0, 2\pi, ...$, but that in this case, $d\boldsymbol{n} = \boldsymbol{\Delta}\mathbf{x}$ and everything is trivially determined.

29.2.2 A simple robot arm

Complicated mechanisms can often be analysed by breaking them down into elementary sections such as the 2-bar robot arm shown in figure 29.5. Here, O_1 is fixed, O_1QO_2 is a rigid arm such that $|O_1Q| = |QO_2| = 1$ (say). If we define axes $(\sigma_1, \sigma_2, \sigma_3)$ as shown at O_1, then the arm is free to rotate (angle θ_1) about the σ_3

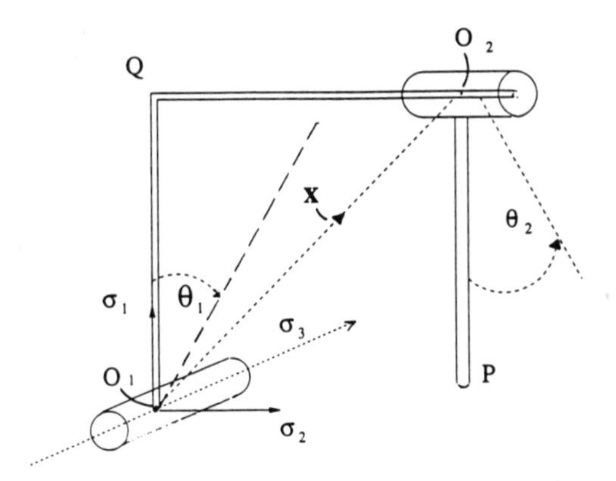

Fig. 29.5. A simple 2-bar robot arm with 2 degrees of freedom.

axis. Also, the rod O_2P can rotate (angle θ_2) about QO_2. Suppose we want an expression for the position of the end-effector P relative to O_1 for arbitrary positions of the two hinges.

The rotation of θ_1 about σ_3 is described by a rotor $R_1 = \exp(-i\frac{\theta}{2}\sigma_3)$. If $\overrightarrow{O_1O_2}=$ $\mathbf{x}_1$ initially, and $\mathbf{x}_2 =\overrightarrow{O_1O_2'}$, where O_2' is the position of O_2 after the rotation, then $\mathbf{x}_2 = R_1\mathbf{x}_1\tilde{R}_1$. Since $\mathbf{x}_1 = \sigma_1 + \sigma_2$ we can write

$$\begin{aligned}\mathbf{x}_2 &= \exp(-i\frac{\theta_1}{2}\sigma_3)(\sigma_1+\sigma_2)\exp(i\frac{\theta_1}{2}\sigma_3) &(29.53)\\ &= (\cos\theta_1 - \sin\theta_1)\sigma_1 + (\cos\theta_1+\sin\theta_1)\sigma_2 &(29.54)\end{aligned}$$

Now consider the rotation of θ_2 about σ_2' which takes P' to P'', where $\sigma_2' = R_1\sigma_2\tilde{R}_1$. If $\mathbf{x}_p'$ and $\mathbf{x}_p''$ are the position vectors of P' and P'' relative to O_2 and $R_2 = \exp(-i\frac{\theta_2}{2}\sigma_2')$ we have

$$\mathbf{x}_p'' = R_2\mathbf{x}_p'\tilde{R}_2 = \exp(-i\frac{\theta_2}{2}\sigma_2')(-R_1\sigma_1\tilde{R}_1)\exp(i\frac{\theta_2}{2}\sigma_2'). \qquad (29.55)$$

Since $\sigma_2' = R_1\sigma_2\tilde{R}_1 = -\sin\theta_1\sigma_1 + \cos\theta_1\sigma_2$ we can write $\mathbf{x}_p''$ as

$$\mathbf{x}_p'' = -\cos\theta_1\cos\theta_2\sigma_1 - \cos\theta_2\sin\theta_1\sigma_2 + \sin\theta_2\sigma_3 \qquad (29.56)$$

Relative to O_1 the final position vector of P is $\mathbf{x}_p = \mathbf{x}_2 + \mathbf{x}_p''$, from which we can explicitly write the coordinates of P as

$$\begin{aligned}\mathbf{x}_p &= [\cos\theta_1 - \sin\theta_1 - \cos\theta_1\cos\theta_2]\sigma_1 + \\ &\quad [\sin\theta_1 + cos\theta_1 - \sin\theta_1\cos\theta_2]\sigma_2 + \sin\theta_2\sigma_3. &(29.57)\end{aligned}$$

This approach is more transparent than the use of dual quaternions, Plücker coordinates etc. Although this example has been worked through step-by-step, it is possible to write down the final position of the end-effector P immediately as

$$\mathbf{x}_p = R_1\mathbf{x}_1\tilde{R}_1 + R_2\mathbf{x}_1'\tilde{R}_2. \qquad (29.58)$$

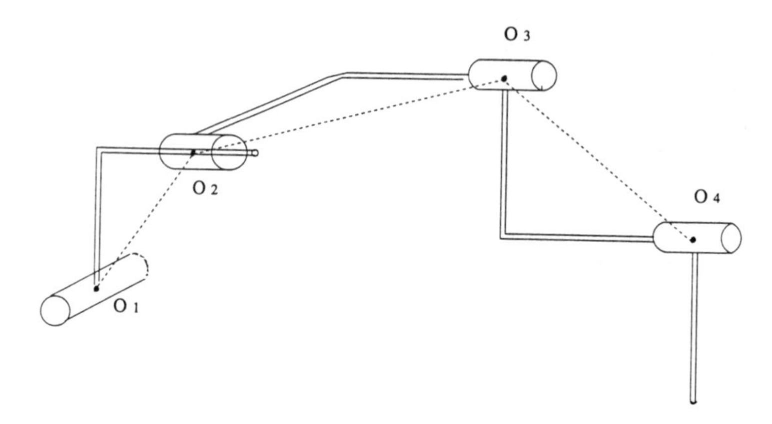

Fig. 29.6. Sketch of an n-bar robot arm.

Here, the first term gives the rotation of $\mathbf{x}_1$ by θ_1 about σ_2 ($R_1 = \exp(-i\frac{\theta_1}{2}\sigma_3)$), and the second term describes the rotation of $\mathbf{x}_1'(=\overrightarrow{O_2'P'})$ by θ_2 about σ_2' ($R_2 = \exp(-i\frac{\theta_2}{2}\sigma_2')$), where $\mathbf{x}_1' = R_1(-\sigma_1)\tilde{R}_1$ and $\sigma_2' = R_1\sigma_2\tilde{R}_1$. This form can be extended easily to an n-bar robot arm an example of which is shown in figure 29.6:

For this case the position vector $\mathbf{x}_p$ of the end-effector can be written down immediately as

$$\mathbf{x}_p = R_1\mathbf{x}_1\tilde{R}_1 + R_2\mathbf{x}_2'\tilde{R}_2 + R_3\mathbf{x}_3'\tilde{R}_3 + \ldots + R_n\mathbf{x}_n'\tilde{R}_n \tag{29.59}$$

where $\mathbf{x}_i$ denotes the initial position vector of O_{i+1} relative to O_i. If the axis through hinge i is initially $\boldsymbol{n}_i$ and R_i denotes a rotation of θ_i about the ith axis, then $R_i = \exp(-i\frac{\theta_i}{2}\boldsymbol{n}_i')$. Thus the first term tells us how to rotate $\mathbf{x}_1$ by θ_1 about $\boldsymbol{n}_1$. The second term tells us how to rotate $\mathbf{x}_2'(= R_1\mathbf{x}_2\tilde{R}_1)$ by θ_2 about $\boldsymbol{n}_2'(= R_1\boldsymbol{n}_2\tilde{R}_1)$. Similarly, the jth term will rotate $\mathbf{x}_j'$ $(= (R_{j-1}\ldots R_2R_1)\mathbf{x}_j(R_{j-1}\ldots R_2R_1)\tilde{})$ by θ_j about $\boldsymbol{n}_j'$ $(= (R_{j-1}\ldots R_2R_1)\boldsymbol{n}_j(R_{j-1}\ldots R_2R_1)\tilde{})$. With computer algebra packages such computations can be easily achieved.

When a complicated mechanism is analysed in terms of a system of linked rigid rods, one would often like to optimize its configuration for a given purpose. For example, with deployable structures such as folding satellites, one might want to find the optimal linkage positions such that the structure folds to a given size and opens smoothly to a given position satisfying various intermediate constraints. It seems likely that such an optimization problem would most efficiently be addressed using the geometric calculus framework.

29.2.3 Dual Quaternions

Dual quaternions(or biquaternions) were introduced to deal with general rigid-body displacements (rotations and translations). The aim was to be able to combine a sequence of such displacements in a straightforward way, e.g. the direct combination of two dual quaternions gives a third dual quaternion describing the overall displacement.

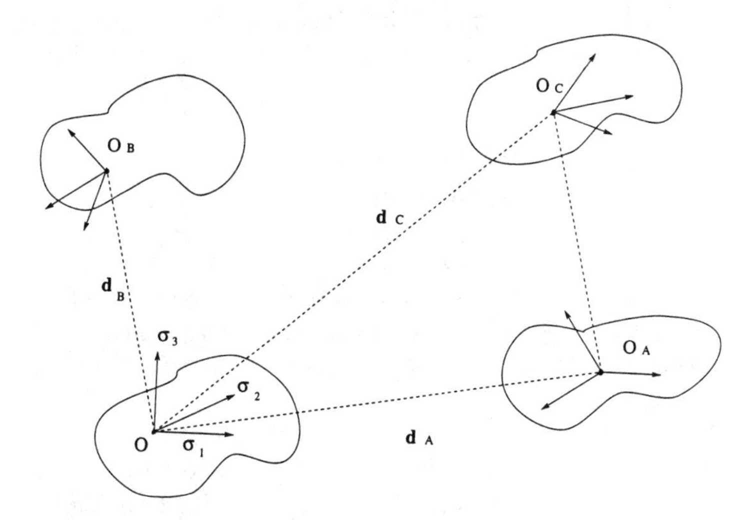

Fig. 29.7. Composition of general displacements.

Consider a rigid body referred to a frame $(\sigma_1, \sigma_2, \sigma_3)$ originally at O. This undergoes a displacement A, where A represents a rotation R_A and a translation $\boldsymbol{d}_A$, such that $O \to O_A$. A vector $\boldsymbol{x}$ then goes to $\boldsymbol{x}_A$,

$$\boldsymbol{x}_A = R_A \boldsymbol{x} \tilde{R}_A + \boldsymbol{d}_A.$$

Let R_B and $\boldsymbol{d}_B$ characterize another similar displacement B. Now suppose we perform displacement A followed by displacement B to bring the new origin to O_C where $\overrightarrow{OO_C} = \boldsymbol{d}_C$. Let the overall displacement C be described by pair $[R_C, \boldsymbol{d}_C]$. A vector $\boldsymbol{x}$ undergoing displacements A followed by B will go to $\boldsymbol{x}_C$ where

$$\boldsymbol{x}_C = R_B\{R_A \boldsymbol{x} \tilde{R}_A + \boldsymbol{d}_A\}\tilde{R}_B + \boldsymbol{d}_B \equiv R_C \boldsymbol{x} \tilde{R}_C + \boldsymbol{d}_C. \tag{29.60}$$

We therefore see that $R_C = R_B R_A$ and $\boldsymbol{d}_C = \boldsymbol{d}_B + R_B \boldsymbol{d}_A \tilde{R}_B$. If we now multiply the equation for $\boldsymbol{d}_C$ on the right by $R_C (= R_B R_A)$ we have

$$\boldsymbol{d}_C R_C = (\boldsymbol{d}_B R_B) R_A + R_B (\boldsymbol{d}_A R_A).$$

Defining $C^o = \frac{1}{2}\boldsymbol{d}_C R_C$, and similarly for A^o and B^o, reduces the above equation to

$$C^o = B^o R_A + R_B A^o.$$

Let us now introduce an operator ϵ such that $\epsilon^2 = \epsilon^3 = \ldots = 0$ and ϵ commutes with vectors. The following quantities $\mathcal{A}$ and $\mathcal{B}$ are then **dual quaternions**

$$\mathcal{A} = R_A + \epsilon A^o, \qquad \mathcal{B} = R_B + \epsilon B^o.$$

To see why this ϵ operator is introduced we consider combining the two dual quaternions $\mathcal{A}$ and $\mathcal{B}$;

$$(R_B + \epsilon B^o)(R_A + \epsilon A^o) = R_B R_A + \epsilon(R_B A^o + B^o R_A) = R_C + \epsilon C^o$$

Using ϵ we are able to combine the two dual quaternions in the required way to form the correct third dual quaternion. Note that if we are prepared to consider the way in which the displacements affect a body point $\boldsymbol{x}$ we can write down the overall displacement easily:

$$\begin{aligned} \boldsymbol{x}' &= R_A\boldsymbol{x}\tilde{R}_A + \boldsymbol{d}_A \\ \boldsymbol{x}'' &= R_B\boldsymbol{x}'\tilde{R}_B + \boldsymbol{d}_B \\ &= R_BR_A\boldsymbol{x}\tilde{R}_A\tilde{R}_B + R_B\boldsymbol{d}_A\tilde{R}_B + \boldsymbol{d}_B. \end{aligned} \tag{29.61}$$

The introduction of the ϵ operator is simply a way of encoding the fact that translations combine additively while rotations combine multiplicatively. Using ϵ one can obtain the composite transformation directly without referring to the action on some point $\boldsymbol{x}$ in the body.

Hestenes [*Neural Networks*, Vol.7, No.1, pp79-88, 1994] has also advocated the use of *hyperspinors* to describe rigid displacements – where a hyperspinor T_a represents a translation $\boldsymbol{d}_a$ and is of the form $T_a = 1+\frac{1}{2}\boldsymbol{d}_a\epsilon$, with ϵ having the properties given above. A general displacement F is then defined as $F = T_aR_a$ where

$$\boldsymbol{x} \rightarrow F(1+\boldsymbol{x}\epsilon)\tilde{F} = 1+\epsilon\{R_a\boldsymbol{x}\tilde{R}_a + \boldsymbol{d}_a\}$$

Rather than introduce the somewhat arbitrary operator ϵ, Hestenes also claims that ϵ may be interpreted geometrically as a null vector in a higher dimensional space. It is interesting to note in this context that rather than invoking higher dimensions, one may be able to interpret ϵ as a null structure in a '2-particle' system, i.e. the system consisting of 2 copies of ordinary space – in this way one would not need to create higher dimensions with no real significance, but would simply use multiple copies of the space.

29.3 Further Applications

We have discussed just two engineering areas where the advantages of using geometric algebra are obvious. There are, however, many other areas to which it might be successfully applied.

One such area is signal processing. Much of signal processing involves performing operations on data vectors or data matrices. These operations are themselves often expressed as matrix operators. In such cases one may be able to replace these procedures with equivalent linear algebra which might then lead to improved computational efficiency and facilitate optimization using the multivector and functional derivative frameworks.

Another possible application is in the field of Control Theory. Theoretical problems in Control are often dealt with in high-dimensional complex spaces. Geometric algebra is an obvious system for the analysis of such spaces and may be a useful tool for parts of control theory.

Applications of geometric algebra in wavelet theory, applied optics and image coding are also topics for future research.

Chapter 30

Projective Quadrics, Poles, Polars, and Legendre Transformations

Richard C. Pappas [1]
Department of Mathematics, Widener University
Chester, PA 19013 USA

30.1 Introduction

In a study of the solution of certain differential equations, Legendre discovered a transformation with remarkable properties, which has since been found to have applications in analysis, mechanics, and thermodynamics. (See Courant & Hilbert (1962).) In classical mechanics, for example, the Legendre transformation (LT) is used to make the transition from the Lagrangian to the Hamiltonian formalism. In modern texts, this transition is usually given a differential geometric interpretation: Lagrangian mechanics takes place on the *tangent bundle* TM of the configuration space manifold M, Hamiltonian mechanics on the *cotangent bundle* T^*M. However, there is another geometric context in which the LT may be viewed (regardless of the branch of physics in which it is applied): the LT is in fact a particular case of a construction in *projective* geometry. The symmetry, or "duality", between the new and old variables in the transformation (reflected in the duality of TM and T^*M in the differential geometric formalism) may be expressed as the projective duality between so-called "poles" and "polars" in the theory of projective quadrics.

We develop the projective interpretation of the LT here for several reasons: (1) It does not appear to be widely known, although it is mentioned by Arnold (1978); his treatment is quite different than ours, however, and he does not make the *projective* nature of the transformation very clear. (2) We speculate that it may lead to a different, if not deeper, understanding of the origin of the symplectic structure

[1]The author would like to acknowledge the financial support from the Science Division of the College of Arts and Sciences, Widener University, which made his attendence at the C.A.P. Summer School possible.

of Hamiltonian mechanics. (3) But above all, it provides a good illustration of the wide-ranging applicability of geometric algebra. The same Clifford–algebra–based formalism that has been used to describe the differential geometric and symplectic structure of Hamiltonian mechanics [Hestenes (1993); Pappas (1993)] needs no modification in order to treat the quite different geometric content of projective geometry [Hestenes & Ziegler (1991); Pappas (1966)].

The next section contains a description of the LT as it appears in mechanics texts. (We use geometric calculus, however!) Section 3 shows how the geometric algebra of $\mathbb{R}^{n+1}$ can be used to model the projective space $\mathbb{P}^n$. In section 4, we present a few definitions and prove some theorems from the projective theory of quadrics, which provides the context for the LT. The LT is constructed as the transformation of poles and polars with respect to the paraboloid $2y = z^2$ in section 5. Finally, we speculate about possible implications in section 6.

For the most part, we follow the notation and terminology of Hestenes & Sobczyk (1984). One departure from their conventions is our use of $\tilde{A}$ for the *reverse* of the multivector A and $A^\dagger$ for the *dual* of A, defined below.

30.2 The Legendre Transformation (1)

Consider a scalar–valued function $f : \mathbb{R}^n \to \mathbb{R}$, which we write as

$$f = f(z),$$

where $z = z_i\, e_i$ (summation) is a variable position vector in $\mathbb{R}^n$. Introduce a second copy of $\mathbb{R}^n$, denoted $\mathbb{R}^{n*}$, with variables $z^* = z_i^* e_i$ defined by

$$z^* = \partial_z f,$$

where ∂_z is the vector derivative. We assume that this equation is invertible for the z_i as functions of the z_i^*. Thus, we require that the "Hessian" is nonzero:

$$(\partial_1 z^* \wedge \partial_2 z^* \wedge \ldots \wedge \partial_n z^*) \cdot I_n^{-1} \neq 0,$$

where $\partial_i \equiv \partial_{z_i}$ are scalar derivatives and I_n is the unit pseudoscalar for $\mathbb{R}^n$.

The *Legendre transform* of the function f is the function $g : \mathbb{R}^{n*} \to \mathbb{R}$ given by

$$g = z \cdot z^* - f,$$

where the z_i are expressed in terms of the z_i^* on the right side, so that $g = g(z^*)$. It follows that

$$z = \partial_{z^*} g \text{ and } f = z \cdot z^* - g,$$

where the right side of the last equation has z_i^* as functions of the z_i. The transformation is thus entirely symmetrical in the "new" and the "old" variables. It is easily shown that the LT is involutive; i.e., its square is the identity.

A natural question that arises when the LT is presented this way is: "What is the motivation for the definition of the new variables $z^* = \partial_z f$?" In mechanics, this of course corresponds to defining the canonical momentum as $p = \partial_{\dot{q}} L$, and the usual motivation is that one can then write Lagrange's equation

$$\partial_t \left(\partial_{\dot{q}} L \right) - \partial_q L = 0$$

as a pair of first order equations. This provides no direct geometrical insight, however.

In terms of *differential* geometry, $\partial_{\dot{q}} L$ is (a coordinate representation of) the "fiber derivative" of the Lagrangian, which is a map from TM to T^*M [Abraham & Marsden (1978)]. But this appears more as a renaming than an explanation.

The *projective* interpretation provides a definite geometric reason for the use of the derivative, which is quite elementary: we are in fact constructing *tangent hyperplanes!* (This is mentioned in Courant & Hilbert (1962), p.32, but not in the context of projective geometry.)

30.3 Projective Spaces and Geometric Algebra

From Euclid to Hilbert, mathematicians experimented with various sets of axioms for euclidean geometry. The modern source for the axiom systems of all affine and projective geometries, both metric and nonmetric, is linear algebra. How can the algebraic structure of a vector space provide axioms for these different geometries? Roughly speaking, in purely algebraical discussions of a vector space V, the *vectors* in V are the basic elements; whereas, for geometrical purposes, one takes a different point of view. A projective space, for example, is constructed by taking the *subspaces* of V as building blocks. And an affine space consists of a set on which V acts as an additive group.

More specifically, a nonmetric projective space $\mathbb{P}^n$ of dimension n has the *lines* (one–dimensional subspaces) of an $(n+1)$–dimensional vector space V as its *points. Lines* in the projective space are identified with *planes* (two–dimensional subspaces) in V, and so forth. In general, $(k+1)$–dimensional subspaces of V are put in correspondence with k–spaces of $\mathbb{P}^n$. If V is given an inner product, one obtains an n–dimensional metric projective space. If $V = \mathbb{R}^3$, for example, and the inner product is positive-definite, one obtains the "elliptic plane". [For more details, see Busemann & Kelly (1953) and Baer (1952).]

As shown by Hestenes & Sobczyk (1984), the graded structure of the Clifford algebra $C\ell_{n+1}(V)$ provides a natural tool for characterizing the subspaces of V, and therefore, by the above identifications, the points, lines, ... , and hyperplanes of $\mathbb{P}^n$. Equivalence classes of k–blades in $C\ell_{n+1}(V)$ represent $(k-1)$–spaces in $\mathbb{P}^n$. An extensive exposition with applications may be found in Hestenes & Ziegler (1991). We assume familiarity with this paper in the next section.

30.4 Quadrics, poles, and polars

A *hypersurface of second order* in $\mathbb{P}^n$ is the set of points x satisfying

$$x \cdot x = x^2 = 0,$$

where the dot represents a nondegenerate inner product. If $n = 2$, we have a second order curve (a conic); if $n = 3$, we have a second order surface (a quadric). For brevity, we shall use the term "quadric" for the general case as well.

Suppose r and s are distinct points in $\mathbb{P}^n$ which are not on the quadric $x^2 = 0$. A third point p is on the line $r \wedge s$ if $p = r + \lambda s$ for some $\lambda \in \mathbb{R} \cup \{-\infty, +\infty\}$. If p is on the quadric $x^2 = 0$, we have $(r + \lambda s) \cdot (r + \lambda s) = 0$, that is

$$s^2\lambda^2 + 2r \cdot s\,\lambda + r^2 = 0$$

This has two roots, λ_1 and λ_2, so the line meets the quadric in general in two points. There may be only one point of intersection (when $\lambda_1 = \lambda_2$), or, in a real projective space, no points of intersection (no real roots).

Let $p_1 = r + \lambda_1 s$ and $p_2 = r + \lambda_2 s$ be the points of intersection. Let us try to choose r and s so that they are separated *harmonically* by p_1 and p_2; that is, the cross ratio $R(r, s; p_1, p_2) = -1$. But, recalling that the cross ratio can be defined as

$$R(r, s; p_1, p_2) = \frac{(r \wedge p_1)(s \wedge p_2)}{(r \wedge p_2)(s \wedge p_1)},$$

we find that

$$R(r, s; p_1, p_2) = \frac{\lambda_1}{\lambda_2},$$

so the condition for r and s to be separated harmonically by p_1 and p_2 is that the sum of the roots of the quadratic equation in λ be zero. But this will be the case if and only if

$$r \cdot s = 0.$$

Let us say that two points r and s are *conjugate* with respect to the quadric $x^2 = 0$ if they are separated harmonically by the points of intersection of $r \wedge s$ with $x^2 = 0$. Then we have proved

Theorem 1 *Two points r and s are conjugate with respect to $x^2 = 0$ iff $r \cdot s = 0$.*

If r is a point not on the quadric, then the set of all points conjugate to r satisfies

$$x \cdot r = 0$$

and is thus a *hyperplane*. It is called the *polar* of the point r, and r is called the *pole* of the hyperplane with respect to the quadric. Clearly, we have the symmetrical result:

Theorem 2 *If the polar of r contains s, then the polar of s contains r.*

More significantly, we can show that

Theorem 3 *The dual of r is the polar of r.*

Proof. Recall that the dual of a k–blade A in $C\ell_{n+1}$ is defined as the $(n+1-k)$–blade

$$A^\dagger \equiv AI_{n+1}^{-1} = A \cdot I_{n+1}^{-1}.$$

Thus the dual of a vector is a pseudovector. In the projective space $\mathbb{P}^2$, for example, the dual of a point (represented by a vector) is a line (represented by a simple bivector), as it should be.

To show that the dual of r is its polar, note that a point is in the hyperplane $r^\dagger$ iff it satisfies

$$x \wedge r^\dagger = 0.$$

But

$$\begin{aligned} x \wedge r^\dagger &= x \wedge (r \cdot I^{-1}) \\ &= (x \cdot r)I^{-1} - r \cdot (x \wedge I^{-1}) \\ &= (x \cdot r)I^{-1} \end{aligned}$$

Therefore, $x \wedge r^\dagger = 0$ iff $x \cdot r = 0$; i.e., x is in $r^\dagger$ iff x and r are conjugate. ∎

An easy consequence of Theorem 3 is the following:

Corollary 4 *The point r is incident with its polar iff it is on the quadric.*

Proof. $r \wedge r^\dagger = 0$ iff $r \cdot r = 0$. ∎

The hyperplane $x \cdot r = 0$ has another remarkable property in the case where r lies on the quadric.

If, as is natural, we designate a line as a *tangent* to a quadric if it has one point in common with or lies entirely in the quadric, then it is easy to see that the set of all tangent lines through a given point of the quadric is a hyperplane. For, as we saw above, the intersection of a line through r and s with the quadric $x^2 = 0$ is given by

$$s^2\lambda^2 + 2r \cdot s\,\lambda + r^2 = 0.$$

If r is on the quadric, $r^2 = 0$; if $s^2 = 0$ and $r \cdot s = 0$, then every point on the line is on the quadric, so the line is a tangent. If $r \cdot s = 0$ and $s^2 \neq 0$, then the quadratic has only one root ($\lambda = 0$), so there is one point of intersection, and the line is again a tangent. These are the only two cases in which a line through the point r on the quadric is a tangent, and they are both characterized by the equation

$$s \cdot r = 0.$$

Thus, *the polar of a point on the quadric is the tangent hyperplane at that point.*

Note that none of the theorems have required the specification of the inner product. This is one advantage that geometric algebra has over other approaches. However, the symmetry of the LT mentioned earlier turns out to be an expression of the pole–polar duality *with respect to a particular quadric*, so in the next section we must introduce an inner product and homogeneous coordinates. In fact, affine coordinates will make an appearance as well.

30.5 The Legendre Transformation (2)

Let $g : \mathbb{R}^{n+1} \to \mathbb{R}^{n+1}$ be a linear transformation. It determines a unique bilinear form

$$g(x, y) = x \cdot g(y),$$

and, conversely, from the bilinear form the linear transformation can be deduced:

$$g(y) = \partial_x g(x, y).$$

A quadric surface in $\mathbb{P}^n$ can be represented by the equation

$$x \cdot g(x) = 0,$$

where the dot now indicates the euclidean inner product. The equivalent bilinear form may be assumed to be symmetric without loss of generality.

Let r be a point on the quadric: $r \cdot g(r) = 0$. As shown in section 4, the tangent hyperplane at r is the polar of r with respect to the quadric, and its equation is

$$x \cdot g(r) = 0.$$

For example, suppose that we want to write the equation of the tangent line to the parabola given in homogeneous coordinates in $\mathbb{P}^2$ by

$$x_1^2 - 2x_2x_3 = 0.$$

We have

$$g(x) = x_1e_1 - x_3e_2 - x_2e_3.$$

At the point $r = r_1e_1 + r_2e_2 + r_3e_3$, the equation of the polar is therefore

$$r_1x_1 - r_3x_2 - r_2x_3 = 0.$$

Thus, *the coordinates of the pole occur as coefficients in the equation of the polar.* (This is a coordinate–dependent statement of the pole–polar duality expressed by $r^\dagger$ above.) The connection with the LT that we are seeking is clearer if we introduce affine coordinates as follows:

$$z = \frac{x_1}{x_3}, \qquad y = \frac{x_2}{x_3}, \qquad a = \frac{r_1}{r_3}, \qquad b = \frac{r_2}{r_3}.$$

The polar is then

$$y + b = az.$$

Define the LT from $\mathbb{R}^2 \to \mathbb{R}^{2*}$ by

$$y^* = z^*z - y.$$

The corresponding transformation in higher dimensions is obviously $y^* = z^* \cdot z - y$.

We can use the LT to define the transformation of a smooth function $L(z)$. Suppose that Γ is the graph

$$y = L(z).$$

Then Γ is the envelope of the set of its tangent hyperplanes. At the point $(z_0, L(z_0))$ of Γ, we find by elementary methods that the tangent hyperplane has the equation

$$y = p_0 \cdot z - H_0,$$

where

$$p_0 = \partial_z L(z_0), \qquad H_0 = p_0 \cdot z_0 - L(z_0).$$

The Legendre transform Γ^* of Γ is thus the surface in $\mathbb{R}^{n*}$ given parametrically by

$$y^* = z \cdot \partial_z L - L(z), \qquad z^* = \partial_z L$$

in terms of a parameter z. If this surface is the graph $y^* = H(z^*)$ of a single-valued function H, then H is the desired Legendre transform of L. Otherwise, we can define a *local* transform of L.

Comparison of the last set of equations with the discussion in section 2 reveals that the reason for defining the canonical momentum vector as the gradient of the Lagrangian is simply that we need the gradient to construct the equation of the tangent hyperplane to the surface specified by the Lagrangian. Clearly, *the duality (or symmetry) between the generalized velocity $\dot{q}$ and the canonical momentum p is the duality between poles and polars with respect to the paraboloid given in homogeneous coordinates by*

$$x_1^2 + x_2^2 + \ldots + x_{n-2}^2 - 2x_n x_{n-1} = 0,$$

or in affine coordinates by

$$2y = z^2,$$

where $z = \frac{x_i}{x_n} e_i$, summed from $i = 1$ to $n - 2$, and $y = \frac{x_{n-1}}{x_n}$.

30.6 Comments

We have seen that the LT used in mechanics can be given a projective interpretation in terms of the pole–polar duality with respect to a paraboloid. A straightforward investigation seems to show that, if we require the tangent hyperplane to satisfy the condition that the coordinates of the pole are the coefficients in the equation of the polar, then the quadric must be a paraboloid. In that sense, *the Legendre transformation is unique.* However, it might be interesting to examine the implications for physics of some other transformations!

One usually considers that the distinguishing geometric feature of Hamiltonian mechanics, in contrast to Lagrangian mechanics, is the symplectic structure of the phase space of the former. In the fiber-bundle approach, this feature is accounted for by the fact that the cotangent bundle has a "natural" symplectic structure, while the tangent bundle does not. However, given that the transition from the Lagrangian to the Hamiltonian formalism is made via the LT, it is intriguing to speculate that the symplectic structure may in fact have a *projective* origin. Limitations of space prevent further discussion of this point here, but it will be addressed in a future publication.

Bibliography

[1] R. Abraham and J.E. Marsden, *Foundations of Mechanics* (Benjamin/Cummings, NY, 1978)

[2] V.I. Arnold, *Mathematical Methods of Classical Mechanics* (Springer–Verlag, NY, 1978)

[3] R. Baer, *Linear Algebra and Projective Geometry* (Academic Press, NY, 1952)

[4] H. Busemann and P.J. Kelly, *Projective Geometry and Projective Metrics* (Academic Press, NY, 1953)

[5] R. Courant and D. Hilbert, *Methods of Mathematical Physics*, Vol. II (Interscience, NY, 1962)

[6] D. Hestenes, "Hamiltonian Mechanics with Geometric Calculus", in Z. Oziewicz, A. Borowicz, and B. Jancewicz (eds), *Spinors, Twistors and Clifford Algebras* (Kluwer, Dordrecht/Boston, 1993)

[7] D. Hestenes and G. Sobczyk, *Clifford Algebra to Geometric Calculus* (Reidel, Dordrecht, 1984).

[8] D. Hestenes and R. Ziegler, "Projective Geometry with Clifford Algebra," *Acta Applicandae Mathematicae* **23**, 25-63 (1991)

[9] R.C. Pappas, "A Formulation of Hamiltonian Mechanics using Geometric Calculus", in F. Brackx, R. Delanghe, and H. Serras (eds), *Clifford Algebras and their Applications in Mathematical Physics* (Kluwer, Dordrecht/Boston, 1993), pp. 251-258

[10] R.C. Pappas, "Oriented Projective Geometry with Clifford Algebra" in "Clifford Algebras with Numeric and Symbolic Computations", eds. R. Abłamowicz, P. Lounesto, and J. M. Parra, Birkhäuser, Boston, 1996,

Chapter 31

Spacetime Algebra and Line Geometry

Johannes G. Maks
Delft University of Technology
Faculty of Mathematics and Computer Science
2628 CD Delft, The Netherlands

31.1 Introduction

The hidden universal Clifford algebra structure of $M_4(R)$ is ambiguous in the sense that this matrix algebra is the universal geometric algebra belonging to each of the real four-dimensional quadratic vector spaces $R^{1,3}$ and $R^{2,2}$. As a *non*-universal Clifford algebra, however, $M_4(R)$ is a geometric algebra of the *unique* five-dimensional quadratic vector space $R^{2,3}$.

Consider the universal Clifford algebra $Cl_{1,3}$, usually referred to as spacetime algebra. Let an orthonormal basis $\{e_1, ..., e_4\}$ of Minkowski space $R^{1,3}$ be given. Then $Cl_{1,3}$ is the associative algebra which is generated by $e_1, ..., e_4$ subject to the conditions

$$\begin{cases} e_1^2 = -1, \ e_2^2 = e_3^2 = e_4^2 = 1, \\ e_i e_j + e_j e_i = 0, \ i \neq j. \end{cases}$$

The pseudoscalar $e_5 = e_1 e_2 e_3 e_4$ anti-commutes with each of the generators e_i while $e_5^2 = -1$, and so the Clifford condition $x^2 + q(x) = 0$ extends to the five-dimensional space $Span(e_1, ..., e_5)$: for each $x = x_1 e_1 + ... + x_5 e_5$ we have $x^2 + q(x) = 0$ where $q(x) = x_1^2 + x_5^2 - (x_2^2 + x_3^2 + x_4^2)$. The Clifford algebra $Cl_{1,3} \cong M_4(R)$ can thus also be thought of as being associated to the *five*-dimensional space $R^{2,3}$.

Alternatively, $M_4(R)$ has the hidden structure of $Cl_{2,2}$, which is generated by $e_1, ..., e_4$ subject to the conditions

$$\begin{cases} e_1^2 = e_2^2 = -1, \ e_3^2 = e_4^2 = 1, \\ e_i e_j + e_j e_i = 0, \ i \neq j. \end{cases}$$

In this case the pseudoscalar $e_5 = e_1e_2e_3e_4$ satisfies $e_5^2 = 1$, and so for each $x = x_1e_1 + ... + e_5e_5$ we have $x^2 + q'(x) = 0$, with $q'(x) = x_1^2 + x_2^2 - (x_3^2 + x_4^2 + x_5^2)$, so q' is *also* seen to have signature (2,3).

Note that in each of the constructions described above we could have taken $-e_5$ instead of e_5. To summarize, the algebra $M_4(R)$ can be thought of as being generated by a pentad $\{e_1, ..., e_5\}$ subject to the conditions

$$\begin{cases} e_1^2 = e_2^2 = -1, \ e_3^2 = e_4^2 = e_5^2 = 1, \\ e_ie_j + e_je_i = 0, \\ e_1e_2e_3e_4e_5 = \pm 1, \end{cases} \tag{31.1}$$

in other words, $M_4(R)$ can be thought of as a *non-universal* Clifford algebra belonging to $R^{2,3}$ (non-universal because of the two possibilities in the third condition).

The purpose of this paper is to discuss the geometrical significance of the five-dimensional space $R^{2,3}$. A fifth dimension can be used either to make spacetime into a *curved* space (embedded in the five-dimensional space) or to make spacetime into a *projective* space of dimension 4. We believe that the projective extension of spacetime is inherent in the framework of spacetime algebra. It will become clear that $R^{2,3}$, representing projective 4-space P^4 equipped with a quadric $Q^{2,3}$, actually encodes geometric information corresponding to a null polarity π or, equivalently, to a line complex C_π on the projective spinor space P^3. By passing to the space of paravectors $R \oplus R^{2,3}$ we find, more generally, that not only the complex C_π but the set of *all* lines in P^3 can be represented in the framework of spacetime algebra.

In physics the use of 5 coordinates to describe the events in spacetime is not new. Kaluza and Klein proposed such an approach in 1921 and 1926, respectively, in order to get a unified theory for electromagnetism and gravitation. A geometric interpretation of the five-dimensional Kaluza-Klein theories was given by Veblen and Hoffmann when they presented their projective theory of relativity in 1930.

Concerning the presence of line geometry in spacetime algebra, the subject of our paper, it remains to be seen whether or not this is applicable in physics.

31.2 Projective geometry

In this section we list some concepts and results from analytic projective geometry, as they will be needed later on in this paper. For details we refer the reader to a good textbook on this subject.

Recall that the real projective space P^n of dimension n is analytically defined as the set of 1-dimensional subspaces of R^{n+1}. More generally, a k-dimensional projective subspace P^k of P^n, $0 \leq k < n$, is by definition to be identified with a $(k+1)$-dimensional subspace of R^{n+1}.

It is well known that each $(k+1)$-dimensional subspace S of R^{n+1} is represented by a simple $(k+1)$-vector u in the exterior algebra ΛR^{n+1}, in the sense that $x \in S$ if and only if $u \wedge x = 0$. To be more specific, if $\{u_1, ..., u_{k+1}\}$ is a basis of S then $u = u_1 \wedge ... \wedge u_{k+1}$ and if $\{u_1', ..., u_{k+1}'\}$ is another basis of S then $u' = cu$ for some $c \in R\backslash_{\{0\}}$, where $u' = u_1' \wedge ... \wedge u_{k+1}'$.

Since the dimension of $\Lambda^{k+1}R^{n+1}$ is equal to $d = \begin{pmatrix} n+1 \\ k+1 \end{pmatrix}$, we have an embedding of $Gr(k+1, n+1)$ (the so-called Grassmannian of all $(k+1)$-dimensional subspaces of R^{n+1}, i.e., the set of all projective k-spaces in P^n) in the projective space P^{d-1}, viz. the mapping defined by

$$Span(u_1, ..., u_{k+1}) \mapsto \langle u_1 \wedge ... \wedge u_{k+1} \rangle \,,$$

where $\langle u \rangle$ denotes the set of all non-zero multiples of u. This embedding is injective, but, apart from a few trivial cases, it is not surjective : the set of *simple* $(k+1)$-vectors is, in general, a proper subset of $\Lambda^{k+1}R^{n+1}$.

In the case $(k, n) = (1, 3)$ the following is well-known.

Proposition 1 *There exists a quadratic form q of signature (3,3) on the six-dimensional space $\Lambda^2 R^4$ such that $u \in \Lambda^2 R^4$ is simple if and only if $q(u) = 0$.*

This yields the famous Klein correspondence between the set of lines in P^3 and the set of points on the Klein quadric

$$Q^{3,3} = \{\langle u \rangle \in P^5 : q(u) = 0\}.$$

To be more explicit, let be chosen a basis $\{e_1, ..., e_4\}$ of R^4. Then in terms of the so-called Plücker coordinates u_{ij} with respect to the basis $\{e_i \wedge e_j : i < j\}$ of $\Lambda^2 R^4$ the quadratic form q mentioned above is given by

$$q(u) = u_{12}u_{34} - u_{13}u_{24} + u_{14}u_{23}.$$

To recognize the signature, note that

$$q(u) = u_0^2 + u_1^2 + u_2^2 - (u_3^2 + u_4^2 + u_5^2),$$

after the change of coordinates $u_0 = \frac{1}{2}(u_{12} + u_{34})$, $u_3 = \frac{1}{2}(u_{12} - u_{34}), ..., u_2 = \frac{1}{2}(u_{14} + u_{23})$, $u_5 = \frac{1}{2}(u_{14} - u_{23})$.

Given the analytic model of P^n, a projective transformation in P^n is nothing more than a transformation which is induced by a non-singular linear transformation in R^{n+1}, i.e., the action of $a \in GL_{n+1}(R)$ on $Gr(k+1, n+1)$ is defined by

$$Span(u_1, ..., u_{k+1}) \mapsto Span(au_1, ..., au_{k+1}),$$

or, equivalently, within the framework of exterior algebra,

$$\langle u_1 \wedge ... \wedge u_{k+1} \rangle \mapsto \langle au_1 \wedge ... \wedge au_{k+1} \rangle \,.$$

Note that the inclusion relation on the collection of all Grassmannians is preserved by the $GL_{n+1}(R)$-action defined above. Since $a, b \in GL_{n+1}(R)$ represent the same projective transformation if and only if $ab^{-1} = c1$ for some $c \in R\backslash_{\{0\}}$, the group $PGL_n(R)$ of projective transformations is actually given by

$$PGL_n(R) = GL_{n+1}(R)/\{c1 : c \in R\backslash_{\{0\}}\}.$$

Let $\langle a \rangle$ denote the projective transformation represented by $a \in GL_{n+1}(R)$, in other words, let $\langle a \rangle$ denote the coset of $a \in GL_{n+1}(R)$ in the factor group $PGL_n(R)$. In the last section of this paper we shall use the following observation. A point $\langle u \rangle \in P^n$ ($u \in R^{n+1} \backslash_{\{0\}}$) is unchanged by the projective transformation $\langle a \rangle$ ($a \in GL_{n+1}(R)$) if and only if u is an eigenvector of a.

Using duality one can also consider inclusion-*reversing* transformations which induce a one-to-one correspondence between $Gr(k+1, n+1)$ and $Gr(n-k, n+1)$. Such transformations are called correlations and we conclude this preliminary section with a brief discussion of the null polarity in P^3, which is an example of a correlation of order 2 and which will play a central role in this paper.

A null polarity π on P^3 is the correlation which is induced by a non-degenerate skew bilinear form (also called a *symplectic* form) s on R^4. Before saying what is actually meant by that, we wish to remark that for each symplectic form s a symplectic basis exists such that

$$s(x, y) = x_1 y_2 - x_2 y_1 + x_3 y_4 - x_4 y_3,$$

where x_i and y_j are the coordinates of x and y with respect to that basis, in other words, in essence only one symplectic form exists (note the difference with the classification of the *symmetric* bilinear forms on R^4!), so it is justified indeed to speak of *the* null polarity on P^3.

Now, given s, the null polarity π is defined by

$$\pi : Gr(k+1, 4) \to Gr(3-k, 4), \ \pi(W) = W^{\perp}, \ k \in \{0, 1, 2\},$$

where $W^{\perp}$ denotes the orthogonal complement of W with respect to s. Note that two non-degenerate skew bilinear forms s and s' induce the same polarity if and only if $s' = cs$ for some $c \in R\backslash_{\{0\}}$. A significant property of the null polarity π is that each point $\langle u \rangle \in P^3$ is actually lying in its polar plane $\pi \langle u \rangle$ ($s(x, x) = 0$ for each $x \in R^4$) and that each line through $\langle u \rangle$ lying in the plane $\pi \langle u \rangle$ is a null line, i.e., self-polar with respect to π. More generally, the aggregate of *all* self-polar lines

$$C_\pi = \{W \in Gr(2, 4) : \pi(W) = W\}$$

is called the line complex associated to the null polarity π. The complex C_π corresponds to the intersection of the Klein quadric $Q^{3,3}$ with a non-tangent hyperplane in P^5, which is essentially a quadric in P^4 of type $Q^{2,3} = Q^{3,2}$.

Conversely, given the intersection $Q^{2,3}$ of $Q^{3,3}$ with a non-tangent hyperplane in P^5, it follows that a unique null polarity π on P^3 exists such that C_π is mapped onto $Q^{2,3}$ under the Klein correspondence (the polar plane $\pi \langle u \rangle$ of a point $\langle u \rangle$ in P^3 is precisely the join of all lines through $\langle u \rangle$ which are in C_π).

31.3 The null polarity belonging to $R^{2,3}$

Let $e_1, ..., e_5$ be five generators of $M_4(R)$ subject to the relations (31.1), where of course a particular choice for the sign in the third equation is assumed to be made.

A Clifford basis of $M_4(R)$ is then constituted by the scalar 1, the five vectors e_i and the ten bivectors e_ie_j, $i < j$.

A unique anti-automorphism $a \to \tilde{a}$ on $M_4(R)$ exists, called *reversion*, such that $\tilde{v} = v$ for each vector $v \in R^{2,3}$. Since $\tilde{b} = -b$ for each bivector $b = \sum b_{ij}e_ie_j$, the subspace of fixed elements under reversion is given by the space of paravectors $R \oplus R^{2,3}$, so $R^{2,3}$ can now be seen as the non-trivial part of $M_4(R)$ which is unchanged under reversion (at this point already it seems unnatural to single out a four-dimensional subspace of $R^{2,3}$ as the space of vectors!).

The dimension of the subspace of fixed elements being equal to six, it follows that there exists a symplectic form s on the spinor space R^4 with reversion as its adjoint anti-automorphism, i.e.,

$$s(ax, y) = s(x, \tilde{a}y) \text{ for all } x, y \in R^4 \text{ and } a \in M_4(R). \tag{31.2}$$

To make this more specific, let us assume the matrix representation

$$e_1 = \begin{pmatrix} & -1 & & \\ 1 & & & \\ & & & 1 \\ & & -1 & \end{pmatrix}, e_2 = \begin{pmatrix} & & -1 & \\ & & & -1 \\ 1 & & & \\ & 1 & & \end{pmatrix}, e_3 = \begin{pmatrix} & 1 & & \\ 1 & & & \\ & & & -1 \\ & & -1 & \end{pmatrix},$$

$$e_4 = \begin{pmatrix} & & 1 & \\ & & & 1 \\ 1 & & & \\ & 1 & & \end{pmatrix}, e_5 = \begin{pmatrix} -1 & & & \\ & 1 & & \\ & & -1 & \\ & & & 1 \end{pmatrix} (= e_1e_2e_3e_4).$$

Then we find that the skew-symmetric bilinear form $s(x, y) = x^Tuy$ on R^4 represented by the skew-symmetric matrix $u = e_1e_2$ satisfies condition (31.2). Note that in this case reversion is given by $\tilde{a} = u^{-1}a^Tu$. It is easy to verify that if we would use another representation $e_i' = p^{-1}e_ip$ for some $p \in GL_4(R)$ then $s(x, y) = x^Tu'y$ with $u' = p^Tup$ would satisfy condition (31.2).

It is obvious that if s satisfies (31.2) then so does λs for all $\lambda \in R\backslash_{\{0\}}$, and so $R^{2,3}$ singles out a null polarity on the projective spinor space P^3. Conversely, let be given a null polarity on P^3, represented by a skew-symmetric form s on R^4. Then the corresponding adjoint anti-automorphism on $M_4(R)$ has a fixed-point set of the form $R \oplus V$ with $V \cong R^{2,3}$ ($s(x, y) = x^Tu'y$ for some $u' = p^Tup$, etc.).

Consequently, there is a one-to-one correspondence between the set of null polarities on P^3 and the set of five-dimensional spaces ($\cong R^{2,3}$) which generate $M_4(R)$ as a non-universal Clifford algebra.

The general linear group $GL_4(R)$ acts transitively (as a group of inner automorphisms) on the set of five-dimensional spaces $V \cong R^{2,3}$ that generate $M_4(R)$ as a non-universal Clifford algebra. A covering of $SO(2, 3)$ is given by the isotropy subgroup

$$\{p \in GL_4(R) : p^{-1}Vp = V\}$$

of a particular $V \cong R^{2,3}$. The identity component $Spin^+(2, 3)$ is known to be isomorphic to the group $Sp_4(R)$, which is indeed the group of transformations preserving a given symplectic form.

Note that the selection of a particular Minkowski space $R^{1,3}$ lying in the five-dimensional space $R^{2,3}$ amounts to breaking the symmetry from $Spin^{+}(2,3) \cong Sp_4(R)$ to $Spin^{+}(1,3) \cong SL_2(C)$.

31.4 Line geometry in spacetime algebra

In order to give a model of the line geometry in the framework of spacetime algebra $Cl_{1,3} \cong M_4(R)$ we need to think of this algebra as a non-universal Clifford algebra belonging to $R^{2,3}$, so let us assume that such a five-dimensional generating space has been fixed. We have seen in the previous section that then also a null polarity π on P^3 is fixed and that $R^{2,3}$ and π are related by

$$s(ax,y) = s(x,\tilde{a}y); \;\; x,y \in R^4, \;\; a \in M_4(R),$$

where s is any skew-symmetric form (unique up to non-zero factor) representing π and the map $a \to \tilde{a}$ is the unique anti-automorphism on the algebra which extends the identity map on $R^{2,3}$.

In what follows the five-dimensional space $R^{2,3}$ will be employed to represent P^4, but at the same time each element of $R^{2,3}$ represents a linear transformation on the spinor space R^4, hence a projective transformation in P^3 (in the non-singular case).

Recall that each $u \in R^{2,3}$ satisfies the Clifford condition $u^2+q(u) = 0$, q denoting a quadratic form of signature (2,3), i.e., $u^2 = \lambda 1$ for some $\lambda \in R$. Let us distinguish between the cases

(1) $\lambda \neq 0$

(2) $\lambda = 0$

In the first case $u \in GL_4(R)$ with eigenvalues $\pm\sqrt{\lambda}$ (real iff $\lambda > 0$) and the dimension of each of the eigenspaces is equal to 2. The latter is necessary because $tr(u) = 0$ for each $u \in R^{2,3}$. It is obvious that the pair of eigenspaces is the same for cu for all non-zero $c \in R$. The projective transformation $\langle u \rangle$ represented by the non-isotropic $u \in R^{2,3}$ is a so-called biaxial involution, the pair of eigenspaces representing the point-wise invariant lines (axes) in P^3. Let us denote the axes of the involution $\langle u \rangle$ by $\langle u \rangle_{\pm}$. We prove

Proposition 2 *For each non-isotropic* $u \in R^{2,3}$ $\pi \langle u \rangle_{\pm} = \langle u \rangle_{\mp}$.

Proof. We prove the proposition for the cases where the axes of the involution are real (the complex cases $u^2 < 0$ are dealt with in the same way). Assume u has been normalized such that $u^2 = 1$. A point $\langle x \rangle$ in P^3 lies on the line $\langle u \rangle_{\pm}$ if and only if $ux = \pm x$. For each pair $\langle x \rangle \in \langle u \rangle_{+}$, $\langle y \rangle \in \langle u \rangle_{-}$ we have $-s(x,y) = s(x,-y) = s(ux,uy) = s(x,\tilde{u}uy) = s(x,u^2y) = s(x,y)$, and so $s(x,y) = 0$. ∎

Now consider case (2). It is easy to see that $rank(u)$=2 for each non-zero isotropic $u \in R^{2,3}$ (this of course being independent of the choice of representation). From this it follows that for each non-zero isotropic $u \in R^{2,3}$ we have $ker(u)$=$range(u) \in Gr(2,4)$. We prove

Proposition 3 *For each non-zero isotropic $u \in R^{2,3}$ $ker(u) \in C_\pi$.*

Proof. Since $ker(u) = range(u)$ we know that for each pair $x, y \in ker(u)$ a pair $x', y' \in R^4$ exists such that $x = ux'$ and $y = uy'$. Consequently, $s(x,y) = s(ux', uy') = s(x', \tilde{u}uy') = s(x', u^2y') = s(x', 0) = 0$, so $ker(u)$ is a null line with respect to the polarity π, i.e., $ker(u)$ belongs to the complex C_π. ∎

Since $ker(u') = ker(u)$ for all $u' = cu$, $c \in R\backslash_{\{0\}}$, $\langle u \rangle \to \ker(u)$ actually yields a map from the quadric

$$Q^{2,3} = \{\langle u \rangle \in P^4 : q(u) = 0\},$$

q denoting the quadratic form on $R^{2,3}$, to the line complex C_π.

The non-isotropic case (1) and the isotropic case (2) can be brought together by passing to the space of paravectors $R \oplus R^{2,3}$. Denoting a general paravector by $\underline{u} = u_0 + u$, $u_0 \in R$ and $u \in R^{2,3}$, we define

$$\hat{\underline{u}} = u_0 - u.$$

Although this map does *not* extend to an automorphism or an anti-automorphism on the non-universal Clifford algebra, a quadratic form $\underline{q}$ of signature (3,3) on $R \oplus R^{2,3}$ is defined by

$$\underline{q}(\underline{u}) = \hat{\underline{u}}\ \underline{u} (= u_0^2 - u^2 = u_0^2 + q(u)),$$

which makes $R \oplus R^{2,3}$ isomorphic to $R^{3,3}$. Recall that the set of non-zero isotropic paravectors $\underline{u} \in R^{3,3}$ yields the Klein quadric

$$Q^{3,3} = \{\langle \underline{u} \rangle \in P^5 : \underline{q}(\underline{u}) = 0\}.$$

Let be given a non-isotropic vector $u \in R^{2,3}$ with $u^2 = u_0^2$. Then the axes $\langle u \rangle_\pm$ of the involution $\langle u \rangle$ are actually given by $\langle u \rangle_- = ker(\underline{u})$ and $\langle u \rangle_+ = ker(\hat{\underline{u}})$, where of course $\underline{u} = u_0 + u$. Conversely, a paravector $\underline{u} = u_0 + u$ is singular iff $\hat{\underline{u}} = u_0 - u$ is singular iff $u^2 = u_0^2$ iff $\underline{q}(\underline{u}) = 0$. It is easy to verify that $\det(\underline{u}) = [\underline{q}(\underline{u})]^2$ for each $\underline{u} \in R \bigoplus R^{2,3}$.

A special kind of singular paravectors is given by the isotropic vectors $u \in R^{2,3}$, i.e., the singular paravectors with vanishing scalar part, for which we have seen that $ker(u) \in C_\pi$ (if $u \neq 0$). Note that a paravector $\underline{u}$ has vanishing scalar part if and only if $tr(\underline{u}) = 0$. We summarize in the following

Theorem 1 *The Klein correspondence between the* **point***-set $Q^{3,3}$ in P^5 and the set of* **lines** *in P^3 is given in spacetime algebra by the map*

$$\kappa : Q^{3,3} \to Gr(2,4), \ \ \kappa \langle \underline{u} \rangle = ker(\underline{u}),$$

where the points on $Q^{3,3}$ are represented by the non-zero singular paravectors $\underline{u} \in R \bigoplus R^{2,3} \subset M_4(R)$. Furthermore, the lines $\kappa \langle \underline{u} \rangle$ and $\kappa \langle \hat{\underline{u}} \rangle$ are polar with respect to π and $\kappa \langle \underline{u} \rangle \in C_\pi$ if and only if $tr(\underline{u}) = 0$ (if and only if $\kappa \langle \underline{u} \rangle = \kappa \langle \hat{\underline{u}} \rangle$).

Proof. The only thing we still need to check is the injectivity of the map κ, the surjectivity then being implied by the equality $\dim Q^{3,3} = \dim Gr(2,4) (= 4)$. For that purpose note that for each non-zero singular paravector $\underline{u}$ we have $rank(\underline{u}) = 2$, and so $\underline{q}(\underline{u}) = \underline{u}\ \hat{\underline{u}} = 0$ implies that $ker(\underline{u}) = range(\hat{\underline{u}})$. Given two non-zero singular $\underline{u}, \underline{v} \in R \bigoplus R^{2,3}$ with $ker(\underline{u}) = \ker(\underline{v}) = range(\hat{\underline{v}})$, we find $\underline{u}\ \hat{\underline{v}} = 0$, which is easily seen to imply $\underline{v} = c\underline{u}$ for some $c \in R \backslash_{\{0\}}$. ■

Incidentally, note that $\underline{u}^2 = 2u_0\underline{u}$ for each singular $\underline{u} \in R^{3,3}$, hence that each $\langle \underline{u} \rangle \in Q^{3,3}$ with $tr(\underline{u}) \neq 0$ is represented by a unique idempotent in the algebra ($\underline{u}^2 = 2u_0\underline{u}$ with $u_0 \neq 0$ yields the idempotent $\underline{u}' = (2u_0)^{-1}\underline{u}$ representing the same point on the Klein quadric $Q^{3,3}$). The remaining part of $Q^{3,3}$, corresponding to C_π, is represented by the set of non-zero *nilpotent* paravectors in the algebra.

Let $\underline{b}$ denote the symmetric bilinear form associated to the quadratic form $\underline{q}$ on $R \bigoplus R^{2,3}$, which is given by

$$\underline{b}(\underline{u}, \underline{v}) = \frac{1}{2}(\underline{u}\ \hat{\underline{v}} + \underline{v}\ \hat{\underline{u}}) = \frac{1}{2}(\hat{\underline{u}}\ \underline{v} + \hat{\underline{v}}\ \underline{u}),$$

for all $\underline{u}$, $\underline{v} \in R \bigoplus R^{2,3}$.

It is well-known (and easy to verify) that two points $\langle \underline{u} \rangle$ and $\langle \underline{v} \rangle$ on the Klein quadric $Q^{3,3}$ are polar to each other with respect to the polarity induced by $\underline{b}$ (i.e. $\underline{b}(\underline{u}, \underline{v}) = 0$) if and only if the corresponding lines $\kappa \langle \underline{u} \rangle$ and $\kappa \langle \underline{v} \rangle$ have non-empty intersection in P^3. Now, given such a pair of points $\langle \underline{u} \rangle, \langle \underline{v} \rangle$ on $Q^{3,3}$, it is obvious that the line $Span(\underline{u},\underline{v})$ is lying on $Q^{3,3}$ and that this line corresponds to a *pencil* (X, V) of lines in P^3, i.e., the set of all lines containing a point X and lying in a plane V. With the help of spacetime algebra the determination of (X, V) is surprisingly straightforward.

Theorem 2 *A* ***line*** *$Span(\underline{u}, \underline{v})$ on $Q^{3,3}$ corresponds to the* ***pencil*** *$(range(\hat{\underline{u}}\ \underline{v}), ker(\hat{\underline{u}}\ \underline{v}))$ in P^3.*

Proof. We prove that for each non-zero $\underline{w} \in Span(\underline{u}, \underline{v})$

$$range(\hat{\underline{u}}\ \underline{v}) \subset ker(\underline{w}) \subset ker(\hat{\underline{u}}\ \underline{v}).$$

Before doing so, let us recall that $\underline{b}(\underline{u}, \underline{v}) = 0$ is equivalent to $\hat{\underline{u}}\ \underline{v} = -\hat{\underline{v}}\ \underline{u}$. Now, let be given $\underline{w} = c\underline{u} + d\underline{v}$ with $(c, d) \neq (0, 0)$. Then the left inclusion follows from

$$\underline{w}\ \hat{\underline{u}}\ \underline{v} = c\ \underline{u}\ \hat{\underline{u}}\ \underline{v} - d\ \underline{v}\ \hat{\underline{v}}\ \underline{u} = c\underline{q}(\underline{u})\underline{v} - d\underline{q}(\underline{v})\underline{u} = 0.$$

In order to see the right inclusion, note that $\hat{\underline{u}}\ \underline{w} = d\ \hat{\underline{u}}\ \underline{v}$. If $d \neq 0$ then $\hat{\underline{u}}\ \underline{v} = \frac{1}{d}\hat{\underline{u}}\ \underline{w}$, which implies $ker(\underline{w}) \subset \ker(\hat{\underline{u}}\ \underline{v})$. If $d = 0$, then $\underline{w} = c\underline{u}$ for some $c \in R \backslash_{\{0\}}$, in which case the inclusion follows from $ker(\hat{\underline{u}}\ \underline{v}) = \ker(\hat{\underline{v}}\ \underline{u})$. Since $\hat{\underline{u}}\ \underline{v} = 0$ if and only if $\langle \underline{u} \rangle = \langle \underline{v} \rangle$ we find that for each pair of different points $\langle \underline{u} \rangle, \langle \underline{v} \rangle \in Q^{3,3}$ with $\underline{b}(\underline{u}, \underline{v}) = 0$ the rank of $\hat{\underline{u}}\ \underline{v}$ is equal to 1, hence the result. ■

Finally, we wish to consider the projective *planes* on the Klein quadric $Q^{3,3}$. Note that a plane $Span(\underline{u}, \underline{v}, \underline{w})$ in P^5 lies on $Q^{3,3}$ if and only if $\underline{b}(\underline{x}, \underline{y}) = 0$ for all $\underline{x}, \underline{y} \in Span(\underline{u}, \underline{v}, \underline{w})$. It is well-known that there are two types of planes on $Q^{3,3}$, α-planes and β-planes say, where an α-plane corresponds to a *bundle* X of lines in P^3, which is the set of all lines through a point X, and a β-plane corresponds to a *plane* V of lines in P^3, which of course means the set of all lines in a plane V.

In spacetime algebra the distinction between the α-planes and the β-planes reads as follows:

Theorem 3 *A* ***plane*** *$Span(\underline{u}, \underline{v}, \underline{w})$ on $Q^{3,3}$ is an α-plane or a β-plane with corresponding* ***bundle*** *$range(\hat{\underline{u}}\ \underline{v})$ ($=range(\hat{\underline{u}}\ \underline{w})$ $=range(\hat{\underline{v}}\ \underline{w})$) or* ***plane*** *$ker(\underline{u}\ \hat{\underline{v}}\ \underline{w})$ in P^3 if $\underline{u}\ \hat{\underline{v}}\ \underline{w} = 0$ or $\underline{u}\ \hat{\underline{v}}\ \underline{w} \neq 0$, respectively.*

Proof. From the previous theorem it follows that the lines $\kappa\,\langle \underline{v} \rangle$ and $\kappa\,\langle \underline{w} \rangle$ intersect in the point X $=range(\hat{\underline{v}}\ \underline{w})$. Now, $Span(\underline{u}, \underline{v}, \underline{w})$ is an α-plane if and only if for *each* $\langle \underline{z} \rangle$ in this plane $X \in \kappa\,\langle \underline{z} \rangle$, that is, $range(\hat{\underline{v}}\ \underline{w}) \subset ker(\underline{z})$, which is equivalent to $\underline{u}\ \hat{\underline{v}}\ \underline{w} = 0$, because $\underline{z}\ \hat{\underline{v}}\ \underline{w} = b\underline{u}\ \hat{\underline{v}}\ \underline{w}$ for each $\underline{z} = b\underline{u} + c\underline{v} + d\underline{w}$. Note that the order of $\underline{u}$, $\underline{v}$ and $\underline{w}$ in the equation $\underline{u}\ \hat{\underline{v}}\ \underline{w} = 0$ is irrelevant.

Alternatively, assume $\underline{u}\ \hat{\underline{v}}\ \underline{w} \neq 0$. Then $rank(\underline{u}\ \hat{\underline{v}}\ \underline{w}) = 1$, so that $ker(\underline{u}\ \hat{\underline{v}}\ \underline{w}) \in Gr(3,4)$. For each $\underline{z} = b\underline{u} + c\underline{v} + d\underline{w}$ we have

$$\underline{u}\ \hat{\underline{v}}\ \underline{w}\ \hat{\underline{z}} = bq(\underline{u})\underline{v}\ \hat{\underline{w}} - cq(\underline{v})\underline{u}\ \hat{\underline{w}} + dq(\underline{w})\underline{u}\ \hat{\underline{v}} = 0,$$

hence $range(\hat{\underline{z}}) \subset ker(\underline{u}\ \hat{\underline{v}}\ \underline{w})$. But $range(\hat{\underline{z}})$ $=ker(\underline{z})$, and so for *each* point $\langle \underline{z} \rangle$ in the plane $Span(\underline{u}, \underline{v}, \underline{w})$ on $Q^{3,3}$ the line $\kappa\,\langle \underline{z} \rangle$ lies in the plane V $=ker(\underline{u}\ \hat{\underline{v}}\ \underline{w})$ in P^3. ■

Chapter 32

Exploring Generalizations of Clifford Algebra

Collin C. Carbno
Saskatchewan Telecommunications
40 Dunning Crescent
Regina, Saskatchewan, Canada, S4S 3W1

Some possible extensions to standard geometrical algebras (Clifford algebra) are explored. Three type of extensions are considered, namely, Dimensions of Zero extent, Bosonic vectors, and Higher cycling vectors. The unnatural interpretations that would result from these extensions show the uniqueness and usefulness of the usual Clifford Algebra rules. Nevertheless, situations might arise in which it might be possible for the extensions to have physical significance. Most promising appears to be bosonic vectors.

32.1 Generalizations of Clifford Algebra

32.1.1 Introduction

A Clifford Algebra can be roughly described as the algebra of a vector space of 2^n dimensions $\mathrm{Cl}_{p,q}$ over the real numbers that is given by a set of orthogonal vectors $\{e_{0,}e_1, ...e_n\}$ of a base space together with a geometrical product such that

$$\begin{aligned} e_i^2 &= 1 \text{ for i =1 to p} \\ e_i^2 &= -1 \text{ for i =p+1 to n} \end{aligned}$$

and

$$e_i e_{j=-} e_j e_i \text{ with } i \neq j.$$

The different geometrical products of the set of vectors $\{e_{0,}e_1, ...e_n\}$ give rise to a basis of the Clifford Algebra. Vectors in Clifford Algebra vector space are called multi-vectors to avoid confusion with vectors $e_{1,}e_{2,}e_3, ...$of the base space. For example, $e_{1,}$ and e_1e_2 are independent multi-vectors in the Clifford Algebra sense while $5e_1e_2$ and $7e_1e_2$ are not. The vectors with square equal to one can be thought of as "unipodal numbers" and vectors with square minus one can be thought of as "imaginary" type numbers.

Addition of the Clifford Algebra elements is commutative and follows usual vector space ideas. The scalar field of the vector space is taken to be the set of real numbers. Furthermore, addition and multiplication(geometrical product) are associative, and multiplication is distributive with respect to addition. The existence of unique additive and multiplicative identities 0 and 1 are assumed ($A+0 = A$, and $1A = A$.) We note that the geometrical product is in general not commutative. Since each multiplication of vectors results in higher-grade objects, without some sort of a grade reduction operation in the geometrical product, the vector space resulting from multiplications would become of infinite extent. The square of a vector equal ± 1 nicely fulfils this requirement.

A key feature of Clifford Algebra is the concept of a grading. Scalars are elements of zero grade, vectors of 1st grade, and products of two vectors of 2nd grade. Some elements such as $5+e_1$ are of mixed grade. Every vector of the Clifford Algebra can be written as a sum of elements

$$A = < A >_0 + < A >_1 + < A >_2 + \ldots$$

where $<>_i$ denotes the part of the multi-vector of the ith grade. The grade operator becomes important in the definition of inner and outer products of multi-vectors. Objects of different grade, can be thought of as objects of different dimension. Scalars have 0-dimension, vectors have 1-dimension, and so on.

32.2 Dimensions of Zero Extent

Consider an extension of Clifford Algebra given by allowing for a third type of basis vector such that

$$\begin{aligned} e_i^2 &= 1 \text{ for i =1 to p} \\ e_i^2 &= -1 \text{ for i =p+1 to p+q} \\ e_i^2 &= 0 \text{ for i =p+q+1 to n} \end{aligned}$$

Clearly the new vectors would characterize dimensions in which vectors would have no length. The question now arises as to whether such dimensions could enter into physics. Such vectors could result in new "area" type elements such as e_1e_5 where $e_1^2 = 1$, and $e_5^2 = 0$. These new area elements suggest that there is some way in which these dimensions could enter into physical theories. Perhaps such dimensions could replace compactification dimensions. However, it appears that such dimensions cannot enter through the metric tensor, so there is some doubt about the usefulness of such an extension to the geometrical algebra.

32.3 Bosonic Vectors

The anti-commutation rule for the basis vectors reflects the fact that the geometrical algebra is created with "directed elements". In physics, the orientation of bivectors (such as e_1e_2), trivectors(such as $e_1e_2e_3$), are usually crucial to physical laws. Nevertheless, situations can arise in which non-orientable areas, volumes(and hypervolumes) might be useful. Projective geometry might be one such area of study. Consider an extension to an algebra that includes a set of vectors $\{e_0, e_1, ...e_n\}$ as before plus a new set of orthogonal vectors $\{b_0, b_1, ...b_r\}$ linearly independent of the other set of vectors such that

$$\begin{aligned} b_i^2 &= 1 \text{ for i =1 to s} \\ b_i^2 &= -1 \text{ for i =s+1 to s+t} \\ b_i^2 &= 0 \text{ for i =s+t+1 to r} \end{aligned}$$

and where

$$b_i b_j = b_j b_i$$

Now two different types of bosonic vectors can be postulated: those that commute with regular basis vectors $\{e_0, e_1, ...e_n\}$ and those that anticommute. Let us denote those bosonic vectors that anticommute with regular vectors by $\{c_0, c_1, ...c_v\}$.

Naturally one can consider Algebras in which all three types of vectors can exist. For example, for each spacial direction in the base space one could postulate that there exists three kinds of vectors, $e, b_,$ and c that are independent.

32.4 Higher Cycling Dimensions

Since in an n-dimensional space one normally does not consider objects with dimension greater than n, the usual Clifford-algebra requirement that the square of a vector be a scalar effectively cuts off the grade objects at the right dimension. However, one can imagine dimensions allowing rules such as

$$\begin{aligned} e_i^3 &= 1 \text{ or} \\ e_i^3 &= -1 \text{ or} \\ e_i^3 &= 0 \end{aligned}$$

In such a Clifford Algebra space with such dimensions one would get hypervolume elements of dimension greater than the space in which one is working. Furthermore, one would get "unnatural" area elements of the form e_ie_i. A dimension with a 3 power index could be called a tri-cycling dimension. Naturally, one could also consider dimensions have exponents of higher integer value. Also, the bosonic vectors could have higher cycling factors. Extensions having higher cycling powers might be useful in the theory of equations but appear to bear little fruit in physical theories.

32.5 Conclusion

Examining the changes brought about by these ad hoc extensions allows us to see how "natural" the standard construction of Clifford Algebra is. Of the extensions considered here, the bosonic extensions of the Clifford Algebra appear most promising and may possess possibilities for perhaps duplicating some of the ideas of superspace within the context of a Clifford Algebra type theory.

Chapter 33

Clifford Algebra Computations with Maple

Rafał Abłamowicz
Department of Mathematics, Gannon University,
Erie, PA 16541 USA

Maple Computer Algebra System provides a convenient environment for computations in the real Clifford algebra $C\ell(B)$ of an arbitrary symbolic bilinear form B on a vector space V. It is well known that the symmetric part of B determines a unique Clifford structure on $C\ell(B)$ while it is less known that the antisymmetric part of B, if present, changes the multilinear structure of $C\ell(B)$. In our computations with Maple we assume that the bilinear form B is symbolic, real, possibly degenerate, and that it may have a non-trivial antisymmetric part. Any element (multivector) of $C\ell(B)$ is represented in Maple as a multivariate Clifford polynomial in basis monomials which form a standard basis for the algebra. Multiplication of these polynomials is based on recursive application of Cartan's decomposition of Clifford product $\mathbf{x}u = \mathbf{x} \wedge u + \mathbf{x} \rfloor u$ into the wedge product part $\mathbf{x} \wedge u$ and the left contraction part $\mathbf{x} \rfloor u$ for any $\mathbf{x} \in V$ and $u \in \bigwedge V$. A package '`Clifford`' of Maple procedures described below implements, among other features, Clifford and wedge/exterior multiplications, left contraction, grade involution, reversion, conjugation, Clifford and exterior exponentiation, computation of a symbolic inverse, scalar and vector parts, matrix representations. It affords computations in division rings of quaternions and octonions including conjugation, norm, and inverse, and it allows for implementation of rotations in a 3-dimensional Euclidean space in terms of quaternions.

33.1 Introduction

The following is a demonstration of Maple V Release 3 package called '`Clifford`'. The purpose of the package is to perform algebraic computations in the Clifford algebra $C\ell(B)$ of an arbitrary symbolic bilinear form B over an n-dimensional vector space V, $1 \leq n \leq 8$. In particular, B may be degenerate, with or without an antisymmetric part, and its symmetric part could be of any signature. Entries of B as well as the dimension of V can be specified by the user, if desired, and are implemented as global Maple variables `'dim'` and `'B'`. From now on throughout this paper we will use typewriter typestyle for user-entered Maple names of procedures, variables, etc. and we will enclose them in single quotes. Notice that Maple returns these names in italics.

We can load the package as follows and see a list of available procedures:

```
> restart:with(Clifford);
```

$$[LC, builtm, c\_conjug, cbasis, cexp, cinv, clicollect, clisort, cliterms, cmul, conjugation, extract, gradeinv, init, o\_conjug, oinv, omul, onorm, ord, q\_conjug, qinv, qmul, qnorm, remove, reorder, reversion, rot3d, scalarpart, specify\_constants, vectorpart, version, wedge, wexp]$$

In order to begin one needs to specify the dimension of V. Then, procedure `'cbasis'` displays a basis for the Clifford algebra $Cl(V)$ as an ordered list. For example,

```
> dim:=4;cbasis(dim);
```

$$dim := 4$$

$$[Id, e1, e2, e3, e4, e1we2, e1we3, e1we4, e2we3, e2we4, e3we4, e1we2we3, e1we2we4, e1we3we4, e2we3we4, e1we2we3we4]$$

We will refer to these basis elements as *standard Clifford monomials* and to the whole basis as *standard Clifford basis.* Below we show how one can work with other, perhaps more suitable, bases while internal computations always will be performed by Maple in the standard basis. Notice that the identity element of the algebra is denoted by `'Id'`, a basis for V or the space of 1-vectors is $\{\mathtt{e1}, \mathtt{e2}, \mathtt{e3}, \mathtt{e4}\}$, a basis for the bivectors is $\{\mathtt{e1we2}, \mathtt{e1we3}, \mathtt{e1we4}, \mathtt{e2we3}, \mathtt{e2we4}, \mathtt{e3we4}\}$ where `e1we2` denotes $\mathbf{e}_1 \wedge \mathbf{e}_2$, etc. Thus, we have

```
> cbasis(dim,0);   #the identity element
```

$$[Id]$$

```
> cbasis(dim,1);   #1-vector basis
```

$$[e1, e2, e3, e4]$$

```
> cbasis(dim,2);   #2-vector basis
```

$$[e1we2, e1we3, e1we4, e2we3, e2we4, e3we4]$$

```
> cbasis(dim,dim); #the pseudoscalar
```

$$[e1we2we3we4]$$

```
> cbasis(dim,'even');   #basis for the even subalgebra
```

$$[\,Id, e1we2, e1we3, e1we4, e2we3, e2we4, e3we4, e1we2we3we4\,]$$

Any general multivector or cliffor is viewed here as *Clifford multivariate polynomial* in the standard basis. Clifford multiplication of these polynomials is implemented through a Maple procedure 'cmul' which has its infix form denoted by '&c'. For example, let's first define three polynomials:

```
> p1:=2*Id+e1we2; p2:=e1+e2; p3:=e3+Id;
```

$$p1 := 2\,Id + e1we2$$

$$p2 := e1 + e2$$

$$p3 := e3 + Id$$

Now, we multiply them and simplify answers using 'clicollect':

```
> p1 &c p2;
```

$$-B_{1,1}\,e2 + B_{2,1}\,e1 + 2\,e1 + B_{2,2}\,e1 - B_{1,2}\,e2 + 2\,e2$$

```
> clicollect(");
```

$$(B_{2,1} + 2 + B_{2,2})\;e1 + (-B_{1,1} - B_{1,2} + 2)\;e2$$

```
> clicollect(simplify(p1 &c p2 &c p3));
```

$$\begin{aligned}&(B_{2,1} + 2 + B_{2,2})\;e1 + (-B_{1,1} - B_{1,2} + 2)\;e2 + (B_{2,1} + 2 + B_{2,2})\;e1we3\\ &\quad + (-B_{1,1} - B_{1,2} + 2)\;e2we3\\ &\quad + (B_{2,1}\,B_{1,3} + 2\,B_{1,3} + B_{2,2}\,B_{1,3} - B_{1,1}\,B_{2,3} - B_{1,2}\,B_{2,3} + 2\,B_{2,3})\;Id\end{aligned}$$

Notice that since the bilinear form has not been specified, its entries remain unevaluated in the above expressions and the off-diagonal entries are explicitly present. It is also possible to use non-numeric indices as follows:

```
> eiwej &c ekwel;
```

$$\begin{aligned}&B_{j,k}\,eiwel + B_{j,k}\,B_{i,l}\,Id - B_{j,l}\,eiwek - B_{j,l}\,B_{i,k}\,Id + eiwejwekwel - B_{i,k}\,ejwel\\ &\quad + B_{i,l}\,ejwek\end{aligned}$$

The above computations exhibit a basic design concept of this package which is to allow computations with an arbitrary bilinear form. For example, if a user wishes to work with $C\ell_3$ then the form B should be defined as a 3×3 diagonal matrix and the Clifford product of polynomials p1 and p2 is then just a special case of the above formula for p1 &c p2:

```
> B:=linalg[diag](1,1,1):
> p1 &c p2;
```

$$e2 + 3\,e1$$

It is also possible to work with a different basis than the standard basis. (cf. [13])

For example, by adding the following 'alias' commands, one can work with a sigma basis preferred by some physicists.

```
> alias('i'*sigma[1]=e2we3,'i'*sigma[2]=-e1we3,'i'*sigma[3]=e1we2,
>             sigma[1]=e1,sigma[2]=e2,sigma[3]=e3,'i'=e1we2we3);
```

$$I, i\,\sigma_1, i\,\sigma_2, i\,\sigma_3, \sigma_1, \sigma_2, \sigma_3, i$$

```
> alias(sig1=e1,sig2=e2,sig3=e3,isig1=e2we3,isig2=-e1we3,isig3=e1we2);
```

$$I, i\,\sigma_1, i\,\sigma_2, i\,\sigma_3, \sigma_1, \sigma_2, \sigma_3, i, \mathit{sig1}, \mathit{sig2}, \mathit{sig3}, \mathit{isig1}, \mathit{isig2}, \mathit{isig3}$$

```
> alias(-'i'*sigma[2]=e1we3);
```

$$I, i\,\sigma_1, i\,\sigma_2, i\,\sigma_3, \sigma_1, \sigma_2, \sigma_3, i, \mathit{sig1}, \mathit{sig2}, \mathit{sig3}, \mathit{isig1}, \mathit{isig2}, \mathit{isig3}, -i\,\sigma_2$$

In particular, we have:

```
> sig3 &c sig1;
```

$$i\,\sigma_2$$

```
> sig1 &c sig3;
```

$$-i\,\sigma_2$$

```
> (Id + isig1 + isig2 + isig3) &c (Id + isig1 + isig2);
```

$$-\mathit{Id} + 3\,i\,\sigma_1 + i\,\sigma_2 + i\,\sigma_3$$

A Maple file Clifford.m with the 'Clifford' package is available from:

http:/www.gannon.edu/service/dept/mathdept.

33.2 Basic chores.

Basic chores of simplifying expressions involving a multitude of Clifford monomials and symbolic coefficients are performed by procedures 'cliterms', 'clisort', and 'clicollect'. Procedure 'reorder' is often used to put monomial terms in the standard order. Procedures 'scalarpart' and 'vectorpart' allow one to identify scalar and k-vector parts of the given Clifford polynomial. We will now briefly look at each procedure.

33.2.1 'cliterms'

Procedure 'cliterms' identifies Clifford monomials/indeterminates in the given multivariate Clifford polynomial. In addition to numeric coefficients, it recognizes symbolic coefficients of type 'indexed', 'constant', and 'function'. If a symbolic coefficient is incorrectly identified as a monomial, it should be first defined as a constant via the procedure 'specify_constants'.

```
> cliterms(e1+a*e2we3-Pi*e3+ab*Id);
```

$$\{\,\mathit{e2we3}, \mathit{e1}, \mathit{e3}, \mathit{Id}\,\}$$

```
> cliterms(e1-7*Id+Pi*e1we2*E*eiwej);
```

$$\{\, e1\,, e1we2, Id, eiwej\,\}$$

```
> cliterms(b*e1+e2-3*e1we3we2+e1we3we4);
```

$$\{\, e1we3we4\,, e1\,, e2, e1we3we2\,\}$$

33.2.2 'clisort'

Procedure 'clisort' sorts the given multivariate Clifford polynomial with respect to the Clifford indeterminates found in the expression by the procedure 'cliterms'.

```
> clisort(e1-e1*Pi*a+e2*Pi+B[1,2]*e1we2-e1we2*B[2,1]);
```

$$e1 - \pi\, a\, e1 + \pi\, e2 + B_{1,2}\, e1we2 - B_{2,1}\, e1we2$$

```
> clisort(ei*abc+eiwej*B[1,2]+2*ei - eiwej*B[2,5]+e1we2*Pi);
```

$$\pi\, e1we2 + B_{1,2}\, eiwej - B_{2,5}\, eiwej + 2\, ei + abc\, ei$$

33.2.3 'clicollect'

Procedure 'clicollect' collects monomial terms in a multivariate Clifford polynomial. It uses procedure 'clisort' to sort expressions and puts symbolic coefficients in front of basis monomials.

```
> clicollect(e1we2*2*a*B[1,2]+e1*abc-e1*Pi*E+a*Id-b*Id+5*Id);
```

$$(\, abc - \pi\, E\,)\, e1 + 2\, a\, B_{1,2}\, e1we2 + (\, a - b + 5\,)\, Id$$

```
> clicollect(e1*b-a*e1+Pi*e1we2-aa*e1we2+e1*n);
```

$$(\, b - a + n\,)\, e1 + (\, \pi - aa\,)\, e1we2$$

```
> clicollect(B[1,2]*e1we3-a*e1we3);
```

$$(B_{1,2} - a)\;\; e1we3$$

33.2.4 'reorder'

Procedure 'reorder' expresses an algebraic sum of elements in a Clifford algebra in terms of the standard Clifford basis, i.e., it reorders monomials using standard ordering and calculates sign of each permutation. If any one of the indices of a monomial is a letter, 'reorder' returns its argument unchanged.

```
> reorder(a*e2we1 + b*Id + 2*e5we4);
```

$$-a\, e1we2 + b\, Id - 2\, e4we5$$

```
> reorder(a*e3we2we1*Pi-2*e2we1we3*E);
```

$$-a\,\pi\;e1we2we3 + 2\,E\;e1we2we3$$

```
> reorder(eiwej);
```

$$eiwej$$

33.2.5 'scalarpart'

Procedure 'scalarpart' computes the scalar part of the given Clifford polynomial. For example:

```
> v:=a*e1 + b*e2 + c*e3;no_v:=scalarpart( v &c v); #square of the vector length
```

$$v := a\,e1 + b\,e2 + c\,e3$$

$$no\_v := a^2\,B_{1,1} + a\,b\,B_{2,1} + a\,c\,B_{3,1} + b\,a\,B_{1,2} + b^2\,B_{2,2} + b\,c\,B_{3,2} + c\,a\,B_{1,3} + c\,b\,B_{2,3} + c^2\,B_{3,3}$$

Notice that the alternating part of B gives no contribution to the norm of v since:

```
> subs({B[3,1]=-B[1,3],B[3,2]=-B[2,3],B[2,1]=-B[1,2]},no_v);
```

$$a^2\,B_{1,1} + b^2\,B_{2,2} + c^2\,B_{3,3}$$

33.2.6 'vectorpart'

Procedure 'vectorpart' computes the k-vector part of the given Clifford polynomial or multivector u where k is a nonnegative integer. When $k = 0$ the procedure returns the scalar part of u times 'Id'. For example:

```
> B:=linalg[diag](1$3):v:=a*e1+b*e2+c*e3;vectorpart(v,0);
```

$$v := a\,e1 + b\,e2 + c\,e3$$

$$0$$

```
> vectorpart(v,1);
```

$$a\,e1 + b\,e2 + c\,e3$$

```
> cliffor:=-2*Id+e1we2-2*e2+e1we2we3;
```

$$cliffor := -2\,Id + e1we2 - 2\,e2 + e1we2we3$$

```
> vectorpart(cliffor,0);  #0-vector part of the cliffor
```

$$-2\,Id$$

Notice a small difference in how the scalar part of a cliffor is returned when compared to the 0-vector part above:

```
> scalarpart(cliffor);
```

$$-2$$

33.3 Ring operations in Clifford algebra and computation of a symbolic inverse.

33.3.1 'LC': left contraction by a vector and 'wedge' multiplication.

Procedure 'LC' defines a left contraction between a vector $\mathbf{x}$ and a multivector u, i.e., vector $\mathbf{x}$ acts on the multivector u from the left. We will show some properties of contraction in the following examples.

Example 3.1.1. Contracting various multivectors.

```
> LC(b*e1,a*e2*Pi); #multicoefficients are handled properly
```

$$a\,\pi\,b\,B_{1,2}\,Id$$

```
> LC(e1,e2we4we8*Pi);
```

$$\pi\,(B_{1,2}\,e4we8 - B_{1,4}\,e2we8 + B_{1,8}\,e2we4)$$

```
> LC(2*e2-3*e1,2*e2-e1we4);
```

$$3\,B_{1,1}\,e4 - 3\,B_{1,4}\,e1 - 2\,B_{2,1}\,e4 + 2\,B_{2,4}\,e1 - 6\,B_{1,2}\,Id + 4\,B_{2,2}\,Id$$

```
> LC(ei,ejwekwel);
```

$$B_{i,j}\,ekwel - B_{i,k}\,ejwel + B_{i,l}\,ejwek$$

```
> LC(a*e2-3*e3,b*e3);
```

$$b\,(a\,B_{2,3}\,Id - 3\,B_{3,3}\,Id)$$

Example 3.1.2. In this example we verify that contraction by a vector is a *derivation* of the exterior algebra $\bigwedge \mathbb{R}^3$ and of the Clifford algebra $C\ell_3$, namely

$$\mathbf{x} \,\lrcorner\, (u \wedge v) = (\mathbf{x} \,\lrcorner\, u) \wedge v + \hat{u} \wedge (\mathbf{x} \,\lrcorner\, v) \tag{33.1}$$

where $\mathbf{x}, \mathbf{y} \in \mathbb{R}^3$ and $u, v \in \bigwedge \mathbb{R}^3$ and

$$\mathbf{x} \,\lrcorner\, (uv) = (\mathbf{x} \,\lrcorner\, u)v + \hat{u}(\mathbf{x} \,\lrcorner\, v) \tag{33.2}$$

where $\mathbf{x} \in \mathbb{R}^3$, $u, v \in C\ell_3$ and $\hat{u}$ denotes grade involuted u. (cf. [2, 3, 8]) We will define $\mathbf{x}, \mathbf{y}, u$ and v in Maple language and proceed with computation of expressions on both sides of these formulas.

```
> dim:=3:B:='B':clibasis:=cbasis(dim);
```

$$clibasis := [\,Id, e1, e2, e3, e1we2, e1we3, e2we3, e1we2we3\,]$$

```
> x:=x1*e1+x2*e2+x3*e3:
> u:=u0*Id+u1*e1+u2*e2+u3*e3+u12*e1we2+u13*e1we3+u23*e2we3+u123*e1we2we3:
> v:=v0*Id+v1*e1+v2*e2+v3*e3+v12*e1we2+v13*e1we3+v23*e2we3+v123*e1we2we3:
```

Left contraction by a vector is a derivation in the exterior algebra $\bigwedge \mathbb{R}^3$:

```
> LHS1:=simplify(LC(x,wedge(u,v))):
> RHS1:=simplify(wedge(LC(x,u),v)+wedge(gradeinv(u),LC(x,v))):
> LHS1-RHS1;
```

$$0$$

Left contraction by a vector is also a derivation in Clifford algebra $C\ell_3$:

```
> LHS2:=simplify(LC(x, u &c v)):
> RHS2:=simplify(LC(x,u) &c v + gradeinv(u) &c LC(x,v)):
> LHS2-RHS2;
```

$$0$$

Thus, for an arbitrary not necessarily symmetric bilinear form B both formulas (1) and (2) above are valid (notice that in the above computations we did not assume anything about B).

Example 3.1.3. In Clifford algebra $C\ell_3$ a symmetric *dot product* of two vectors $\mathbf{x}$ and $\mathbf{y}$ can be defined as

$$\mathbf{x} \cdot \mathbf{y} = \frac{1}{2}(\mathbf{xy} + \mathbf{yx}).$$

The following well-known formula relates the left contraction of $\mathbf{y}$ by $\mathbf{x}$ to the dot product:

$$\mathbf{x} \rfloor \mathbf{y} = \mathbf{x} \cdot \mathbf{y}.$$

We will show now that this formula is valid only if the antisymmetric part of the bilinear form B is identically zero. We proceed as follows:

```
> x:=x1*e1+x2*e2+x3*e3:y:=y1*e1+y2*e2+y3*e3:  #define two arbitrary vectors in Cl(3)
> x_dot_y:=simplify((x &c y + y &c x)/2):
> contraction_x_y:=simplify(LC(x,y)):
> collect(coeff(clicollect(contraction_x_y - x_dot_y),Id),{x1,x2,x3,x4,y1,y2,y3,y4});
```

$$\left(\left(\frac{1}{2}B_{2,1} - \frac{1}{2}B_{1,2}\right) y1 + \left(\frac{1}{2}B_{2,3} - \frac{1}{2}B_{3,2}\right) y3\right) x2 + \left(-\frac{1}{2}B_{1,3} + \frac{1}{2}B_{3,1}\right) x3\, y1$$
$$+ \left(\left(\frac{1}{2}B_{1,2} - \frac{1}{2}B_{2,1}\right) x1 + \left(-\frac{1}{2}B_{2,3} + \frac{1}{2}B_{3,2}\right) x3\right) y2$$
$$+ \left(\frac{1}{2}B_{1,3} - \frac{1}{2}B_{3,1}\right) x1\, y3$$

Notice that the above expression is zero for any two vectors $\mathbf{x}$ and $\mathbf{y}$ if and only if the antisymmetric part of B is identically 0.

Above we have already used procedure 'wedge' to calculate the wedge product in Clifford or exterior algebras. Here we mention some additional features of that procedure. Procedure 'wedge' can accept any number (≥ 2) of Clifford polynomials and its infix form is defined as an associative Maple operator '&w'. Here are some additional examples how this procedure may be used.

```
> wedge(2*e1,e2);  2*e1 &w e2;  #long form and infix form
```

$$2\, e1we2$$

$$2\, e1we2$$

```
> wedge(e1+e2,2*Id+e3+e4we3we2,e4we5);
```

$$e1we3we4we5 + e2we3we4we5 + 2\, e1we4we5 + 2\, e2we4we5$$

```
> (Id-Pi*e1we2+e4) &w (e2we3-6*e3we4we5+Id);
```

$$e2we3we4 + e2we3 + 6\,\pi\, e1we2we3we4we5 - 6\, e3we4we5 - \pi\, e1we2 + e4 + Id$$

```
> &w(2*e3+e2,e1,2*Id-e3we4);
```

$$(-e1we2 - 2\, e1we3)\, \&\text{w}\, (2\, Id - e3we4)$$

```
> eval(");
```

$$e1we2we3we4 - 4\, e1we3 - 2\, e1we2$$

33.3.2 Clifford multiplication 'cmul' or '&c'.

Procedure 'cmul' calculates the Clifford product between any two elements of the given Clifford algebra. It is based on the recursive use of Cartan's formula:

$$\mathbf{x}\, u = \text{wedge}(\mathbf{x}, u) + \text{LC}(\mathbf{x}, u) = \mathbf{x}\, \&w\, u + \text{LC}(\mathbf{x}, u) = \mathbf{x} \wedge u + \mathbf{x} \rfloor u$$

where $\mathbf{x}$ is a vector and u is any element in the algebra, wedge$(\mathbf{x}, u)$ denotes the wedge/exterior product $\mathbf{x} \wedge u$ between $\mathbf{x}$ and u, and LC$(\mathbf{x}, u)$ denotes the left contraction $\mathbf{x} \rfloor u$ of u by $\mathbf{x}$. The infix form '&c' of this multiplication is also defined. The procedure works in Clifford algebras in dimensions up to and including 8. Notice that complex coefficients are allowed permitting one to use this procedure in complexified Clifford algebras.

```
> clicollect(cmul(Id+2*e1,2*e2+e1we2we3));
```

$$-2\, B_{1,2}\, e1we3 + (2\, B_{1,3} + 4)\, e1we2 + 2\, B_{1,1}\, e2we3 + 2\, e2 + 4\, B_{1,2}\, Id + e1we2we3$$

To multiply more than two elements it is convenient to use the infix form of 'cmul', and then simplify and collect terms if necessary:

```
> clicollect(simplify(((e1+(2+I)*e3we2) &c (Id - e2)) &c (Id + e1)));
```

$$\begin{aligned}(2+I)\, e3we2 &+ (I\, B_{2,2} + 2\, B_{2,2})\, e1we3 + (-1 - 2\, B_{3,2} - I\, B_{3,2})\, e1we2 \\ &+ (-B_{1,2} - B_{2,1} + 1)\, e1 + (2\, B_{3,2} - 2\, B_{3,1} + B_{1,1} - I\, B_{3,1} + I\, B_{3,2})\, e2 \\ &+ (-2\, B_{2,2} - I\, B_{2,2} + I\, B_{2,1} + 2\, B_{2,1})\, e3\end{aligned}$$

$$+ (-I\, B_{2,2}\, B_{3,1} - B_{1,2} + 2\, B_{3,2}\, B_{2,1} - 2\, B_{2,2}\, B_{3,1} + B_{1,1} + I\, B_{3,2}\, B_{2,1})\; Id$$
$$+ (\,-2 - I\,)\; e1we2we3$$

Example 3.2.1. Suppose we want to verify whether vectors $\{\mathbf{e1}, \mathbf{e2}, \mathbf{e3}\}$ form a pseudo-orthogonal set even when the antisymmetric part of B is nonzero. We might proceed this way:

```
> vbasis:=cbasis(3,1);   #define a basis of 1-vectors
```

$$vbasis := [\, e1\,, e2\,, e3\,]$$

```
> printlevel:=2:for x in vbasis do for y in vbasis do clicollect(x &c y + y &c x) od od;
```

$$2\, B_{1,1}\, Id$$
$$(B_{1,2} + B_{2,1})\; Id$$
$$(B_{1,3} + B_{3,1})\; Id$$
$$(B_{1,2} + B_{2,1})\; Id$$
$$2\, B_{2,2}\, Id$$
$$(B_{2,3} + B_{3,2})\; Id$$
$$(B_{1,3} + B_{3,1})\; Id$$
$$(B_{2,3} + B_{3,2})\; Id$$
$$2\, B_{3,3}\, Id$$

Remembering that $B_{1,2} = -B_{2,1}$, $B_{1,3} = -B_{3,1}$, and $B_{2,3} = -B_{3,2}$, we indeed see that vectors $\{\mathbf{e1}, \mathbf{e2}, \mathbf{e3}\}$ are mutually orthogonal. A shorter way to verify the above fact is the following:

```
> 'ei &c ej + ej &c ei' = clicollect(ei &c ej + ej &c ei);
```

$$(\, ei \,\&c\, ej\,) + (\, ej \,\&c\, ei\,) = (B_{i,j} + B_{j,i})\; Id + ejwei + eiwej$$

Again, we must remember that $\mathtt{B}_{\mathtt{i},\mathtt{j}} = -\mathtt{B}_{\mathtt{j},\mathtt{i}}$ and that $\mathtt{eiwej} = -\mathtt{ejwei}$ when $i \neq j$ and $\mathtt{eiwei} = \mathtt{0}$, $i, j = 1 \ldots 3$.

Example 3.2.2. Compute and display the multiplication table of basis monomials when $\mathtt{dim} = 2$.

```
> dim:=2:monomials:=cbasis(dim):
> for x in monomials do for y in monomials do M.dim[x,y]:=clicollect(cmul(x,y)) od od:
> eval(M2);
```

table([
$$(\, e1we2, e2\,) = B_{2,2}\, e1 \,-\, B_{1,2}\, e2$$
$$(\, e2, e2\,) = B_{2,2}\, Id$$
$$(\, e1, Id\,) = e1$$
$$(\, Id, e1\,) = e1$$

$(\,e2, Id\,) = e2$
$(\,e1we2, e1we2\,) = (B_{2,1} - B_{1,2})\; e1we2 + (-B_{2,2}\, B_{1,1} + B_{2,1}\, B_{1,2})\; Id$
$(\,e2, e1\,) = -e1we2 + B_{2,1}\, Id$
$(\,Id, Id\,) = Id$
$(\,e2, e1we2\,) = -B_{2,2}\, e1 + B_{2,1}\, e2$
$(\,Id, e2\,) = e2$
$(\,e1we2, Id\,) = e1we2$
$(\,Id, e1we2\,) = e1we2$
$(\,e1, e2\,) = e1we2 + B_{1,2}\, Id$
$(\,e1we2, e1\,) = B_{2,1}\, e1 - B_{1,1}\, e2$
$(\,e1, e1\,) = B_{1,1}\, Id$
$(\,e1, e1we2\,) = -B_{1,2}\, e1 + B_{1,1}\, e2$
])

In the next subsection we use Clifford multiplication to find symbolic inverses of Clifford polynomials.

33.3.3 'cinv': symbolic inverse of a multivector.

Procedure 'cinv' calculates a symbolic inverse, if such exists, of any element u in the given Clifford algebra or in its subalgebra. The procedure call has a form cinv(u, L) where L is a list containing basis elements in the Clifford algebra or in its subalgebra. This procedure returns an error message when the elements in the list are not linearly independent and/or when no inverse exists. It is expected that the basis elements listed in L have been reordered before the procedure was called. Below we show a few examples in dimension 3 including a degenerate case.

```
> dim:=3;clibasis:=cbasis(dim);  #algebra Cl(3)
```

$$dim := 3$$

$$clibasis := [\,Id, e1, e2, e3, e1we2, e1we3, e2we3, e1we2we3\,]$$

```
> evensubalg:=cbasis(dim,'even'); #even subalgebra of Cl(3)
```

$$evensubalg := [\,Id, e1we2, e1we3, e2we3\,]$$

Example 3.3.1. In this example we will compute some inverses in $C\ell_3$.

```
> uinv:=cinv(e1,clibasis);  #uinv is the inverse of e1
```

$$uinv := \frac{e1}{B_{1,1}}$$

Let's verify that 'uninv' is the inverse of e1:

```
> 'uinv &c e1' = uinv &c e1; 'e1 &c uinv' = e1 &c uinv;
```

$$uinv \;\&c\; e1 = Id$$

$$e1 \;\&c\; uinv = Id$$

```
> u:=e1+2*e1we2; uinv:=cinv(u,clibasis);  #inverse of a mixed element
```

$$u := e1 + 2\,e1we2$$

$$uinv := 2\,\frac{(-B_{2,1} + B_{1,2})\,Id}{\%1} + \frac{e1}{\%1} + 2\,\frac{e1we2}{\%1}$$
$$\%1 := 4\,B_{2,1}\,B_{1,2} + B_{1,1} - 4\,B_{2,2}\,B_{1,1}$$

```
> 'uinv &c u' = simplify(uinv &c u); 'u &c uinv' = simplify(u &c uinv);
```

$$uinv \,\&\mathrm{c}\, u = Id$$

$$u \,\&\mathrm{c}\, uinv = Id$$

Example 3.3.2. In this example we will calculate the inverse of a bivector in $C\ell_3$. We may, of course, limit our attention to the even subalgebra of $C\ell_3$.

```
> u:=a*e1we2+b*e1we3;uinv:=cinv(u,evensubalg);
```

$$u := a\,e1we2 + b\,e1we3$$

$$uinv := \frac{(a\,B_{1,2} + b\,B_{1,3} - a\,B_{2,1} - b\,B_{3,1})\,Id}{\%1} + \frac{a\,e1we2}{\%1} + \frac{b\,e1we3}{\%1}$$
$$\%1 := -a\,B_{1,1}\,b\,B_{3,2} + a\,B_{2,1}\,B_{1,3}\,b + a^2\,B_{2,1}\,B_{1,2} + a\,b\,B_{3,1}\,B_{1,2} + B_{1,3}\,B_{3,1}\,b^2 - a^2\,B_{1,1}\,B_{2,2} - a\,B_{2,3}\,B_{1,1}\,b - b^2\,B_{3,3}\,B_{1,1}$$

```
> 'u &c uinv' = simplify(u &c uinv);'uinv &c u' = simplify(uinv &c u);
```

$$u \,\&\mathrm{c}\, uinv = Id$$

$$uinv \,\&\mathrm{c}\, u = Id$$

Example 3.3.3. In this example we will handle neutral signature and isotropic vectors whose square in a Clifford algebra is 0.

```
> dim:=4:B:=linalg[diag](1,1,-1,-1):clibasis:=cbasis(dim):
```

Notice that vector v defined below is an isotropic vector in $C\ell_{2,2}$ and, as such, it has no inverse namely:

```
> v:=e1+e3;cinv(v,clibasis); 'v &c v' = v &c v;
```

$$v := e1 + e3$$

```
Error, (in cinv) inverse of , e3+e1,  does not exist
```

$$v \,\&\mathrm{c}\, v = 0$$

Let's calculate an inverse of a parabivector:

```
> printlevel:=1:u:=(Id + e1we2)/sqrt(2);uinv:=cinv(u,clibasis);
```

$$u := \frac{1}{2}\,(\,Id + e1we2\,)\,\sqrt{2}$$

$$uinv := \frac{1}{2}\sqrt{2}\,Id - \frac{1}{2}\sqrt{2}\,e1we2$$

```
> 'u &c uinv' = simplify(u &c uinv); 'uinv &c u' = simplify(uinv &c u);
```

$$u \,\&\mathrm{c}\, uinv = Id$$

$$uinv \,\&\mathrm{c}\, u = Id$$

Example 3.3.4. In this example we consider a degenerate form B.

```
> B:=linalg[diag](1,1,0);
```

$$B := \begin{bmatrix} 1 & 0 & 0 \\ 0 & 1 & 0 \\ 0 & 0 & 0 \end{bmatrix}$$

```
> clibasis:=cbasis(3);
```

$$clibasis := [Id, e1, e2, e3, e1we2, e1we3, e2we3, e1we2we3]$$

```
> cinv(e1,clibasis);
```

$$e1$$

```
> cinv(e3,clibasis);
Error, (in cinv) inverse of , e3,  does not exist
```

```
> cinv(e1we2,clibasis);
```

$$-e1we2$$

```
> cinv(e1we3,clibasis);
Error, (in cinv) inverse of , e1we3,  does not exist
```

33.4 Clifford algebra automorphisms: grade involution, reversion and conjugation.

33.4.1 Grade involution and reversion.

Procedure 'gradeinv' is the grade involution in the Clifford algebra, i.e., it reverses signs of odd elements and leaves signs of even elements unchanged. (cf. [3, 4]) It is linear in its argument.

```
> gradeinv(2*e1we2+4*Id+e1);
```

$$-e1 + 4\,Id + 2\,e1we2$$

```
> gradeinv(2*e1we2*Pi);
```

$$2\,\pi\, e1we2$$

```
> gradeinv(5*Id);
```

$$5\,Id$$

Checking that grade involution is an automorphism of the algebra could be done this way in $C\ell(V)$ where $\dim(V) = 3$ and B is arbitrary. Verification in other dimensions can be done the same way.

```
> B:='B':dim:=3:clibasis:=cbasis(dim);
```

$$clibasis := [Id, e1, e2, e3, e1we2, e1we3, e2we3, e1we2we3]$$

```
> u:=sum(a[k]*clibasis[k],k=1..2^dim);
```

$$u := a_5\,e1we2 + a_6\,e1we3 + a_7\,e2we3 + a_1\,Id + a_2\,e1 + a_3\,e2 + a_4\,e3 + a_8\,e1we2we3$$

```
> v:=sum(b[k]*clibasis[k],k=1..2^dim);
```

$$v := b_5\,e1we2 + b_6\,e1we3 + b_7\,e2we3 + b_1\,Id + b_2\,e1 + b_3\,e2 + b_4\,e3 + b_8\,e1we2we3$$

```
> 'gradeinv(u &c v)-gradeinv(u) &c gradeinv(v)' =
>   clicollect(simplify(gradeinv(u &c v) - gradeinv(u) &c gradeinv(v)));
```

$$\mathrm{gradeinv}(u\,\&c\,v) - (\mathrm{gradeinv}(u)\,\&c\,\mathrm{gradeinv}(v)) = 0$$

Procedure 'reversion' gives reversion in the Clifford algebra and it is the algebra antiautomorphism as verified below. When the antisymmetric part of B is not zero, reversion does not preserve the gradation because it mixes grades. (cf. [1])

Example 4.1.1. Reversion mixes grades.

```
> B:='B':reversion(Id);reversion(e1);
```

$$Id$$

$$e1$$

```
> clicollect(reversion(e1we2)); #notice mixed grades in Maple's output
```

$$(B_{2,1} - B_{1,2})\,Id - e1we2$$

```
> clicollect(reversion(e1we2we3));
```

$$(-B_{3,1} + B_{1,3})\,e2 + (B_{2,1} - B_{1,2})\,e3 + (B_{3,2} - B_{2,3})\,e1 - e1we2we3$$

```
> clicollect(reversion(2*Id+3*e1+4*e1we2+5*e1we2we3));
```

$$(-5\,B_{3,1} + 5\,B_{1,3})\,e2 + (5\,B_{2,1} - 5\,B_{1,2})\,e3 + (3 + 5\,B_{3,2} - 5\,B_{2,3})\,e1 + (2 + 4\,B_{2,1} - 4\,B_{1,2})\,Id - 4\,e1we2 - 5\,e1we2we3$$

Let's verify now that at least in $C\ell(V)$ where $\dim(V) = 3$ and B is arbitrary, the operation of reversion is an algebra antiautomorphism. We define two arbitrary multivectors u and v and proceed to check that

$$\mathtt{reversion}(u \,\&c\, v) = \mathtt{reversion}(v) \,\&c\, \mathtt{reversion}(u).$$

Again, verification in other dimensions can be done the same way.

```
> B:='B':dim:=3:clibasis:=cbasis(dim);
```

$$clibasis := [\, Id, e1, e2, e3, e1we2, e1we3, e2we3, e1we2we3 \,]$$

```
> u:=sum(a[k]*clibasis[k],k=1..2^dim);
```

$$u := a_1\, Id + a_2\, e1 + a_3\, e2 + a_4\, e3 + a_5\, e1we2 + a_6\, e1we3 + a_7\, e2we3 + a_8\, e1we2we3$$

```
> v:=sum(b[k]*clibasis[k],k=1..2^dim);
```

$$v := b_1\, Id + b_2\, e1 + b_3\, e2 + b_4\, e3 + b_5\, e1we2 + b_6\, e1we3 + b_7\, e2we3 + b_8\, e1we2we3$$

```
> 'reversion(u &c v)-reversion(v) &c reversion(u)' =
> clicollect(simplify(reversion(u &c v) -  reversion(v) &c reversion(u)));
```

$$\mathrm{reversion}(\, u \,\&c\, v\,) - (\,\mathrm{reversion}(\, v\,) \,\&c\, \mathrm{reversion}(\, u\,)\,) = 0$$

33.4.2 Clifford conjugation and complex conjugation.

Procedure '`conjugation`' calculates conjugation in the Clifford algebra. It is linear in its argument. Note that '`conjugation` = `reversion` ∘ `gradeinv`' thus, like '`reversion`', '`conjugation`' does not preserve the multivector gradation when the antisymmetric part of B is non-zero.

Example 4.2.1.

```
> conjugation(e1);
```

$$-e1$$

```
> clicollect(conjugation(2*Id+3*e1+4*e1we2+5*e1we3+e1we4));
```

$$-3\, e1 - e1we4 + (5\, B_{3,1} + 4\, B_{2,1} + 2 + B_{4,1} - B_{1,4} - 5\, B_{1,3} - 4\, B_{1,2})\, Id - 4\, e1we2 - 5\, e1we3$$

```
> clicollect(conjugation(Id+e1+e1we2+e1we2we3));
```

$$(B_{3,1} - B_{1,3})\, e2 + (-B_{2,1} + B_{1,2})\, e3 + (-1 - B_{3,2} + B_{2,3})\, e1 + (1 + B_{2,1} - B_{1,2})\, Id - e1we2 + e1we2we3$$

Procedure '`c_conjug`' gives complex conjugation in complexified Clifford algebra; thus,

$$\mathtt{c\_conjug(u)} = \mathtt{c\_conjug(a + I * b)} = \mathtt{a - I * b}$$

for any elements a and b in the real Clifford algebra (here, I is the imaginary unit, $I = \sqrt{-1}$).

```
> expr:=(2+3*I)*e1we2+I*e1-(2-5*I)*e1we3;
```

$$expr := (2+3I)\,e1we2 + I\,e1 + (-2+5I)\,e1we3$$

```
> c_conjug(expr);
```

$$-I\,e1 + (2-3I)\,e1we2 + (-2-5I)\,e1we3$$

```
> c_conjug((1+I)*Id + (2*I)*e2we1-5*Id +(3-I)*e1we2we3);
```

$$(-4-I)\,Id + 2I\,e1we2 + (3+I)\,e1we2we3$$

33.5 Matrix representations.

33.5.1 Left regular representations.

Procedure 'builtm' builds a matrix for the given element u of the Clifford algebra $C\ell(B)$ in the left-regular representation with respect to a basis specified by the user. The element u is entered as the first argument and the basis, specified as the second argument, is entered as a list, e.g., builtm($u, basis$). For example, one can find the left-regular representation of the algebra on itself or one can find matrices representing spinors as elements of minimal left ideals. (cf. [2, 4, 8, 9]) However, so far only real matrices can be found with 'Clifford', i.e., complex Pauli matrices $\{\sigma_1, \sigma_2, \sigma_3\}$ representing e1, e2 and e3 in $C\ell_3$ cannot be found directly with this procedure.

Example 5.1.1. This example shows the left-regular representation of $C\ell(V)$ where $\dim(V) = 2$ and B is arbitrary. Matrices representing basis monomials will then be 4×4 real.

```
> sbasis:=cbasis(2):
> gamma0:=builtm(Id,sbasis):  #the identity matrix representing Id
> gamma1:=builtm(e1,sbasis);  #representation of e1
```

$$\gamma 1 := \begin{bmatrix} 0 & B_{1,1} & B_{1,2} & 0 \\ 1 & 0 & 0 & -B_{1,2} \\ 0 & 0 & 0 & B_{1,1} \\ 0 & 0 & 1 & 0 \end{bmatrix}$$

```
> gamma2:=builtm(e2,sbasis);  #representation of e2
```

$$\gamma 2 := \begin{bmatrix} 0 & B_{2,1} & B_{2,2} & 0 \\ 0 & 0 & 0 & -B_{2,2} \\ 1 & 0 & 0 & B_{2,1} \\ 0 & -1 & 0 & 0 \end{bmatrix}$$

```
> gamma12:=builtm(e1we2,sbasis);  #representation of e1we2
```

$$\gamma 12 := \begin{bmatrix} 0 & 0 & 0 & -B_{2,2}\,B_{1,1} + B_{2,1}\,B_{1,2} \\ 0 & B_{2,1} & B_{2,2} & 0 \\ 0 & -B_{1,1} & -B_{1,2} & 0 \\ 1 & 0 & 0 & -B_{1,2} + B_{2,1} \end{bmatrix}$$

We will verify now that $\mathtt{gamma1} \,\&{*}\, \mathtt{gamma1} = \mathtt{B}[1,1] * \mathtt{gamma0}$ where '&*' stands for matrix multiplication. In order to shorten Maple's output we will use the '`equal`' command:

```
> linalg[equal](evalm(gamma1 &* gamma1), evalm(B[1,1]*gamma0));
```

true

Likewise, $\mathtt{gamma1} \,\&{*}\, \mathtt{gamma2} + \mathtt{gamma2} \,\&{*}\, \mathtt{gamma1} = (\mathtt{B}[1,2] + \mathtt{B}[2,1]) * \mathtt{gamma0}$, which equals 0 since $B[1,2] = -B[2,1]$ and similarly for other products:

```
> linalg[equal](evalm(gamma1 &* gamma2 + gamma2 &* gamma1),
>                   evalm((B[1,2]+B[2,1])*gamma0));
```

true

Thus we have shown that $\gamma_i\gamma_j + \gamma_j\gamma_i = 2\delta_{i,j},\ i,j = 1,2$ in standard notation.[4]

Example 5.1.2. Let's now consider $C\ell(V)$ where $\dim(V) = 1$. Thus, depending whether $B[1,1]$ equals -1, 0, or 1 we have complex, dual or double numbers.

```
> fbasis:=[Id,e1];
```

$$\mathit{fbasis} := [\,\mathit{Id}, \mathit{e1}\,]$$

```
> gamma0:=builtm(Id,fbasis);
```

$$\gamma 0 := \begin{bmatrix} 1 & 0 \\ 0 & 1 \end{bmatrix}$$

```
> j:=builtm(e1,fbasis);
```

$$j := \begin{bmatrix} 0 & B_{1,1} \\ 1 & 0 \end{bmatrix}$$

```
> evalm(j &* j) = B[1,1]*'gamma0';
```

$$\begin{bmatrix} B_{1,1} & 0 \\ 0 & B_{1,1} \end{bmatrix} = B_{1,1}\,\gamma 0$$

Example 5.1.3. Consider a parabivector basis from **Example 5.1.1** above:

```
> f1:=Id;f2:=e1we2;fbasis:=['f1','f2'];
```

$$f1 := Id$$

$$f2 := e1we2$$

$$fbasis := [f1, f2]$$

```
> builtm(Id,fbasis);
```

$$\begin{bmatrix} 1 & 0 \\ 0 & 1 \end{bmatrix}$$

```
> builtm(e1we2,fbasis);
```

$$\begin{bmatrix} 0 & -B_{2,2}\,B_{1,1} + B_{2,1}\,B_{1,2} \\ 1 & -B_{1,2} + B_{2,1} \end{bmatrix}$$

```
> builtm(e1,fbasis); #the projection of e1 onto the plane spanned by {f1,f2} is null
```

$$\begin{bmatrix} 0 & 0 \\ 0 & 0 \end{bmatrix}$$

33.5.2 Spinor representations in left minimal ideals.

In this section we will use 'builtm' to find gamma matrices in irreducible representations of Clifford algebras in spinor spaces. Spinors will be defined as elements of a minimal left ideal generated by a primitive idempotent. (cf. [2, 4, 8, 9]) We begin with an example in signature $(3, 1)$.

Example 5.2.1. We proceed to find irreducible spinor representation of the Clifford algebra $C\ell_{3,1}$ in one of its left-minimal ideals generated by a primitive idempotent.

```
> dim:=4:B:=linalg[matrix](4,4,[[1,0,0,0],[0,1,0,0],[0,0,1,0],[0,0,0,-1]]):
> clibasis:=cbasis(dim);
```

$$clibasis := [Id, e1, e2, e3, e4, e1we2, e1we3, e1we4, e2we3, e2we4, e3we4, \\ e1we2we3, e1we2we4, e1we3we4, e2we3we4, e1we2we3we4]$$

```
> f1:=cmul((1/2)*(Id+e1),(1/2)*(Id+e2we4));#define 1st primitive idempotent
```

$$f1 := \frac{1}{4}\,e1we2we4 + \frac{1}{4}\,e2we4 + \frac{1}{4}\,e1 + \frac{1}{4}\,Id$$

The remaining three idempotents are:

```
> f2:=cmul((1/2)*(Id-e1),(1/2)*(Id+e2we4));
```

$$f2 := -\frac{1}{4}\, e1we2we4 + \frac{1}{4}\, e2we4 - \frac{1}{4}\, e1 + \frac{1}{4}\, Id$$

```
> f3:=cmul((1/2)*(Id+e1),(1/2)*(Id-e2we4));
```

$$f3 := -\frac{1}{4}\, e1we2we4 - \frac{1}{4}\, e2we4 + \frac{1}{4}\, e1 + \frac{1}{4}\, Id$$

```
> f4:=cmul((1/2)*(Id-e1),(1/2)*(Id-e2we4));
```

$$f4 := \frac{1}{4}\, e1we2we4 - \frac{1}{4}\, e2we4 - \frac{1}{4}\, e1 + \frac{1}{4}\, Id$$

```
> f1+f2+f3+f4;  #notice that these idempotents add up to the identity Id
```

$$Id$$

Now we will check that elements {f1, f2, f3, f4} are mutually annihilating primitive idempotents (the fact that they are primitive follows from general theory.[8, 9]

```
> [evalb(f1 = f1 &c f1), evalb(f2 = f2 &c f2), evalb(f3 = f3 &c f3),
>                          evalb(f4 = f4 &c f4)];
```

$$[\text{true}, \text{true}, \text{true}, \text{true}]$$

```
> [f1 &c f2, f1 &c f3, f1 &c f4, f2 &c f3, f2 &c f4, f3 &c f4];
```

$$[0, 0, 0, 0, 0, 0]$$

A basis {F1, F2, F3, F4} for the spinor space $S = C\ell_{3,1} f_1$ considered as a left minimal ideal of $C\ell_{3,1}$ generated by f_1 can now be found. When multiplying Maple's f1 on the left by all monomials in 'clibasis' we find the following four linearly independent basis elements for S:

```
> F1:=cmul(Id,f1);  #define first element of the spinor basis
```

$$F1 := \frac{1}{4}\, e1 + \frac{1}{4}\, e2we4 + \frac{1}{4}\, Id + \frac{1}{4}\, e1we2we4$$

```
> F2:=cmul(e2,f1);  #define second element of the spinor basis
```

$$F2 := -\frac{1}{4}\, e1we2 + \frac{1}{4}\, e4 + \frac{1}{4}\, e2 - \frac{1}{4}\, e1we4$$

```
> F3:=cmul(e3,f1);  #define third element of the spinor basis
```

$$F3 := -\frac{1}{4}\, e1we3 - \frac{1}{4}\, e2we3we4 + \frac{1}{4}\, e3 + \frac{1}{4}\, e1we2we3we4$$

```
> F4:=cmul(e2we3,f1);  #define fourth element of the spinor basis
```

$$F4 := \frac{1}{4}\, e1we2we3 - \frac{1}{4}\, e3we4 + \frac{1}{4}\, e2we3 - \frac{1}{4}\, e1we3we4$$

```
> spinbasis:=['F1','F2','F3','F4']; #define a spinor basis for S
```

$$spinbasis := [\, F1\,, F2\,, F3\,, F4\,]$$

Let's verify that $f_1\, C\ell_{3,1}\, f_1 = r * f_1$ where r is a real constant. For example, for each monomial u in '`clibasis`' defined above we calculate the product `f1&c`u`&cf1` and find out that it either equals 0 or `f1`. This could be accomplished by the following loop (Maple's output is deleted except for one product):

```
> for u in clibasis do f1 &c u &c f1 od: f1 &c e1 &c f1;
```

$$f1\ \&c\ \mathrm{e1}\ \&c\ f1 = \frac{1}{4}\, e1 + \frac{1}{4}\, e2we4 + \frac{1}{4}\, Id + \frac{1}{4}\, e1we2we4$$

Now we are ready to find with '`builtm`' gamma matrices $\{\gamma_1, \gamma_2, \gamma_3, \gamma_4\}$ representing basis vectors $\{\mathtt{e1}, \mathtt{e2}, \mathtt{e3}, \mathtt{e4}\}$ and a matrix γ_{1234} representing the pseudoscalar `e1we2we3we4`:

```
> printlevel:=1:gamma1:=builtm(e1,spinbasis);
```

$$\gamma 1 := \begin{bmatrix} 1 & 0 & 0 & 0 \\ 0 & -1 & 0 & 0 \\ 0 & 0 & -1 & 0 \\ 0 & 0 & 0 & 1 \end{bmatrix}$$

```
> gamma2:=builtm(e2,spinbasis);
```

$$\gamma 2 := \begin{bmatrix} 0 & 1 & 0 & 0 \\ 1 & 0 & 0 & 0 \\ 0 & 0 & 0 & 1 \\ 0 & 0 & 1 & 0 \end{bmatrix}$$

```
> gamma3:=builtm(e3,spinbasis);
```

$$\gamma 3 := \begin{bmatrix} 0 & 0 & 1 & 0 \\ 0 & 0 & 0 & -1 \\ 1 & 0 & 0 & 0 \\ 0 & -1 & 0 & 0 \end{bmatrix}$$

```
> gamma4:=builtm(e4,spinbasis);
```

$$\gamma 4 := \begin{bmatrix} 0 & -1 & 0 & 0 \\ 1 & 0 & 0 & 0 \\ 0 & 0 & 0 & -1 \\ 0 & 0 & 1 & 0 \end{bmatrix}$$

```
> gamma1234:=builtm(e1we2we3we4,spinbasis);
```

$$\gamma 1234 := \begin{bmatrix} 0 & 0 & -1 & 0 \\ 0 & 0 & 0 & 1 \\ 1 & 0 & 0 & 0 \\ 0 & -1 & 0 & 0 \end{bmatrix}$$

Notice that

```
> evalm(gamma1234 &* gamma1234);
```

$$\begin{bmatrix} -1 & 0 & 0 & 0 \\ 0 & -1 & 0 & 0 \\ 0 & 0 & -1 & 0 \\ 0 & 0 & 0 & -1 \end{bmatrix}$$

in agreement with the theory. (cf. [4]) Of course, we could have verified this last property just by squaring **e1we2we3we4** as follows:

```
> e1we2we3we4 &c e1we2we3we4;
```

$$-Id$$

Finally, we verify that any spinor from S decomposable over the spin basis '**spinbasis**' can be represented by a one-column matrix. (cf. [8]) First, we calculate matrices representing the basis elements {F1, F2, F3, F4} and call them {F1m, F2m, F3m, F4m} respectively:

```
> F1m:=builtm(F1,spinbasis):F2m:=builtm(F2,spinbasis):
> F3m:=builtm(F3,spinbasis):F4m:=builtm(F4,spinbasis):
```

Second, we find matrix representation for an arbitrary spinor s from S as follows:

```
> s:=psi[1]*F1m+psi[2]*F2m+psi[3]*F3m+psi[4]*F4m;
```

$$s := \psi_1\, F1m + \psi_2\, F2m + \psi_3\, F3m + \psi_4\, F4m$$

```
> s = evalm(s);
```

$$\psi_1\, F1m + \psi_2\, F2m + \psi_3\, F3m + \psi_4\, F4m = \begin{bmatrix} \psi_1 & 0 & 0 & 0 \\ \psi_2 & 0 & 0 & 0 \\ \psi_3 & 0 & 0 & 0 \\ \psi_4 & 0 & 0 & 0 \end{bmatrix}$$

33.6 Clifford and exterior exponentiations.

33.6.1 'cexp': Clifford exponentiation.

Procedure 'cexp' computes the Clifford exponential of a Clifford polynomial u up to the order specified by the second argument which is a nonnegative integer n. It returns 'Id' if $n = 0$. We will illustrate this procedure by exponentiating simple bivectors in $C\ell_3$ and $C\ell_{1,3}$.

Example 6.1.1. Exponentiation of qk. We begin by defining B to be a 3×3 identity matrix and letting u to equal qk. Then, we begin to exponentiate u and find a closed form for the series. Recall that the quaternion basis is stored under 'quatbasis[1]'.

```
> restart:with(Clifford):dim:=3:B:=linalg[diag](1$dim):
> clibasis:=cbasis(dim);evensubalg:=cbasis(dim,'even');quatbasis;
```

$$clibasis := [\,Id, e1, e2, e3, e1we2, e1we3, e2we3, e1we2we3\,]$$

$$evensubalg := [\,Id, e1we2, e1we3, e2we3\,]$$

$$[[\,Id, e3we2, e1we3, e2we1\,],\ \{\,Maple\ defined\ qi := e3we2,\ qj := e1we3,\ qk := e2we1\,\}]$$

```
> u:=qk; cexp(t*u,0);cexp(t*u,1);cexp(t*u,2);cexp(t*u,3);
> cexp(t*u,4);cexp(t*u,5);cexp(t*u,6);cexp(t*u,7);cexp(t*u,8);
```

$$u := e2we1$$

$$Id$$

$$t\, e2we1 + Id$$

$$t\, e2we1 + \left(-\frac{1}{2}t^2 + 1\right) Id$$

$$\left(-\frac{1}{6}t^3 + t\right) e2we1 + \left(-\frac{1}{2}t^2 + 1\right) Id$$

$$\left(-\frac{1}{6}t^3 + t\right) e2we1 + \left(\frac{1}{24}t^4 - \frac{1}{2}t^2 + 1\right) Id$$

$$\left(\frac{1}{120}t^5 - \frac{1}{6}t^3 + t\right) e2we1 + \left(\frac{1}{24}t^4 - \frac{1}{2}t^2 + 1\right) Id$$

$$\left(\frac{1}{120}t^5 - \frac{1}{6}t^3 + t\right) e2we1 + \left(-\frac{1}{720}t^6 + \frac{1}{24}t^4 - \frac{1}{2}t^2 + 1\right) Id$$

$$\left(-\frac{1}{5040}t^7 + \frac{1}{120}t^5 - \frac{1}{6}t^3 + t\right) e2we1 + \left(-\frac{1}{720}t^6 + \frac{1}{24}t^4 - \frac{1}{2}t^2 + 1\right) Id$$

$$\left(-\frac{1}{5040}t^7 + \frac{1}{120}t^5 - \frac{1}{6}t^3 + t\right) e2we1 + \left(\frac{1}{40320}t^8 - \frac{1}{720}t^6 + \frac{1}{24}t^4 - \frac{1}{2}t^2 + 1\right) Id$$

Notice that the above series converge to $\sin(t)$ and $\cos(t)$ respectively thus

$$\texttt{cexp}(t*\texttt{qk}) = \texttt{cexp}(t*\texttt{e2we1}) = \cos(t)*\texttt{Id}+\sin(t)*\texttt{e2we1} = \cos(t)*\texttt{Id}+\sin(t)*\texttt{qk}.$$

Example 6.1.2. Lorentz transformations in $C\ell_{1,3}$. We define B as a diagonal matrix with entries $\{1, -1, -1, -1\}$ and let u be a simple bivector `e1we2`. Then, we exponentiate u and find a closed form for the Lorentz transformation on $\mathbb{R}^{1,3}$ generated by u. Finally, we verify that the pseudo-norm of a vector v in $\mathbb{R}^{1,3}$ is preserved under the transformation

$$v \rightarrow \texttt{cexp}(t*\texttt{e1we2}) \;\&\texttt{c}\; v \;\&\texttt{c}\; \texttt{cexp}(-t*\texttt{e1we2}).$$

```
> dim:=4:B:=linalg[diag](1,-1,-1,-1):
> u:=e1we2;
> cexp(t*u,0);cexp(t*u,1);cexp(t*u,2);cexp(t*u,3);
> cexp(t*u,4);cexp(t*u,5);cexp(t*u,6);cexp(t*u,7);cexp(t*u,8);
```

$$u := e1we2$$

$$Id$$

$$Id + t\, e1we2$$

$$\left(\frac{1}{2}t^2+1\right) Id + t\, e1we2$$

$$\left(\frac{1}{2}t^2+1\right) Id + \left(\frac{1}{6}t^3+t\right) e1we2$$

$$\left(\frac{1}{24}t^4+\frac{1}{2}t^2+1\right) Id + \left(\frac{1}{6}t^3+t\right) e1we2$$

$$\left(\frac{1}{24}t^4+\frac{1}{2}t^2+1\right) Id + \left(\frac{1}{120}t^5+\frac{1}{6}t^3+t\right) e1we2$$

$$\left(\frac{1}{720}t^6+\frac{1}{24}t^4+\frac{1}{2}t^2+1\right) Id + \left(\frac{1}{120}t^5+\frac{1}{6}t^3+t\right) e1we2$$

$$\left(\frac{1}{720}t^6+\frac{1}{24}t^4+\frac{1}{2}t^2+1\right) Id + \left(\frac{1}{5040}t^7+\frac{1}{120}t^5+\frac{1}{6}t^3+t\right) e1we2$$

$$\left(\frac{1}{40320}t^8+\frac{1}{720}t^6+\frac{1}{24}t^4+\frac{1}{2}t^2+1\right) Id + \left(\frac{1}{5040}t^7+\frac{1}{120}t^5+\frac{1}{6}t^3+t\right) e1we2$$

Notice that the above series converge to $\sinh(t)$ and $\cosh(t)$ respectively thus

$$\texttt{cexp}(t*\texttt{e1we2}) = \cosh(t)*\texttt{Id}+\sinh(t)*\texttt{e1we2}$$

By replacing t with $-t$ in the above formula we have

$$\texttt{cexp}(-t*\texttt{e1we2}) = \cosh(t)*\texttt{Id}-\sinh(t)*\texttt{e1we2}$$

Consider now a vector $v \in \mathbb{R}^{1,3}$ and how it is rotated by the Lorentz transformation

$$v \rightarrow \mathtt{cexp}(t * \mathtt{e1we2}) \ \&\mathtt{c} \ v \ \&\mathtt{c} \ \mathtt{cexp}(-t * \mathtt{e1we2}).$$

At the end we will observe that the pseudo-norm of v is preserved.

```
> v:=a*e1+b*e2+c*e3+d*e4;   # a vector in R^(1,3)
```

$$v := a\, e1 + b\, e2 + c\, e3 + d\, e4$$

```
> no_v:=scalarpart(v &c v);
```

$$no\_v := -b^2 - c^2 + a^2 - d^2$$

```
> new_v:= clicollect(expand((cosh(t)*Id + sinh(t)*e1we2) &c v
>                   &c (cosh(t)*Id -sinh(t)*e1we2)));
```

$$\begin{aligned} new\_v := & \left(-2\cosh(t)\, a \sinh(t) + b\cosh(t)^2 + b\sinh(t)^2\right) e2 \\ & + \left(c\cosh(t)^2 - c\sinh(t)^2\right) e3 \\ & + \left(-2\cosh(t)\, b\sinh(t) + a\sinh(t)^2 + a\cosh(t)^2\right) e1 \\ & + \left(-d\sinh(t)^2 + d\cosh(t)^2\right) e4 \end{aligned}$$

```
> for i from 1 to 4 do c.i:=map(convert,collect(simplify(
>         convert(coeff(new_v,e.i),exp)),{a,b,c,d}),trig) od;
```

$$c1 := \cosh(2t)\, a - \sinh(2t)\, b$$

$$c2 := -\sinh(2t)\, a + \cosh(2t)\, b$$

$$c3 := c$$

$$c4 := d$$

```
> new_v:=c1*e1+c2*e2+c3*e3+c4*e4;
```

$$new\_v := (\cosh(2t)\, a - \sinh(2t)\, b)\, e1 + (-\sinh(2t)\, a + \cosh(2t)\, b)\, e2 + c\, e3 + d\, e4$$

```
> scalarpart(new_v &c new_v);
```

$$-b^2 - c^2 + a^2 - d^2$$

33.6.2 'wexp': exterior exponentiation.

Procedure 'wexp' computes the exterior exponential of a Clifford polynomial/number u up to the order specified by the second argument which is a nonnegative integer n. It returns 'Id' when $n = 0$. (cf. [4, 10])

Example 6.2.1. Let's calculate the exterior exponential of a bivector in 6 dimensions.

```
> biv:=e1 &w e2 + e3 &w e4 + e5 &w e6;  #exterior exponential of a bivector
```

$$biv := e1we2 + e3we4 + e5we6$$

```
> 'wexp(biv,0)' = wexp(biv,0);
```

$$\mathrm{wexp}(\,biv, 0\,) = Id$$

```
> 'wexp(biv,1)' = wexp(biv,1);
```

$$\mathrm{wexp}(\,biv, 1\,) = Id + e5we6 + e3we4 + e1we2$$

```
> 'wexp(biv,2)' = wexp(biv,2);
```

$$\mathrm{wexp}(\,biv, 2\,) = e3we4we5we6 + e1we2we3we4 + Id + e5we6 + e3we4 + e1we2 + e1we2we5we6$$

```
> 'wexp(biv,3)' = wexp(biv,3);
```

$$\mathrm{wexp}(\,biv, 3\,) = e3we4we5we6 + e1we2we3we4 + Id + e5we6 + e3we4 + e1we2 + e1we2we5we6 + e1we2we3we4we5we6$$

```
> 'wexp(biv,4)' = wexp(biv,4);
```

$$\mathrm{wexp}(\,biv, 4\,) = e3we4we5we6 + e1we2we3we4 + Id + e5we6 + e3we4 + e1we2 + e1we2we5we6 + e1we2we3we4we5we6$$

Example 6.2.2. Exterior exponential of a quaternion. Quaternion basis {`Id`, `qi`, `qj`, `qk`} is defined in this package as a parabivector basis in $C\ell_3$ and, more precisely, we have

$$\mathtt{qi} = \mathtt{e3we2},\ \mathtt{qj} = \mathtt{e1we3},\ \mathtt{qk} = \mathtt{e2we1}.$$

```
> specify_constants(alpha,beta,gamma);q:=alpha*qi + beta*qj + gamma*qk;
```

Maple now knows the following constant(s) : , α, FAIL, E, *Catalan*, π, false, γ, true, ∞, β

$$q := \alpha\, e3we2 + \beta\, e1we3 + \gamma\, e2we1$$

```
> wexp(q,0);
```

$$Id$$

```
> wexp(q,1);
```

$$\alpha\, e3we2 + \gamma\, e2we1 + Id + \beta\, e1we3$$

```
> wexp(q,2);  #of course result is the same as above
```

$$\alpha\, e3we2 + \gamma\, e2we1 + Id + \beta\, e1we3$$

We may use procedure 'qdisplay' to display previous answer in terms of {Id, qi, qj, qk}.

```
> qdisplay(");
```

$$Id + \gamma\, qk + \beta\, qj + \alpha\, qi$$

33.7 Quaternions and three dimensional rotations.

33.7.1 Quaternion type, conjugation, norm and inverse.

A quaternion basis is specified at the time of initialization and is stored under Maple global list 'quatbasis'. The first entry of that list is a basis list [Id, e3we2, e1we3, e2we1] while the second shows substitutions made by Maple. (cf. [8])

```
> quatbasis;
```

$$[[\,Id, e3we2, e1we3, e2we1\,],\\ \{\,Maple\ defined\ qi\ :=\ e3we2,\ qj\ :=\ e1we3,\ qk\ :=\ e2we1\,\}]$$

Thus, one may also enter quaternions in the basis {Id, qi, qj, qk} where 'qi', 'qj' and 'qk' represent basis quaternions {**i**, **j**, **k**}. Maple's output however is given in terms of {Id, e1we2, e1we3, e2we3}. Then, procedure 'qdisplay' may be used to convert Maple's display to the basis {Id, qi, qj, qk}.

```
> x:=a*Id+b*e1we2+c*e2we3+d*e1we3; #let's define a quaternion
```

$$x := a\, Id + b\, e1we2 + c\, e2we3 + d\, e1we3$$

```
> qdisplay(x); #quaternion x displayed in terms of {Id, qi, qj, qk}
```

$$a\, Id - b\, qk + d\, qj - c\, qi$$

```
> type(x,quaternion); #check type of x
```

$$\mathrm{true}$$

```
> xc:=q_conjug(x); #quaternionic conjugate is computed with 'q_conjug'
```

$$xc := -d\, e1we3 - b\, e1we2 - c\, e2we3 + a\, Id$$

```
> xnorm:=qnorm(x); #quaternionic norm is computed with 'qnorm'
```

$$xnorm := \sqrt{d^2 + b^2 + c^2 + a^2}$$

```
> xinv:=qinv(x); #quaternionic inverse 'qinv'
```

$$xinv := -\frac{d\, e1we3}{d^2 + b^2 + c^2 + a^2} - \frac{b\, e1we2}{d^2 + b^2 + c^2 + a^2} - \frac{c\, e2we3}{d^2 + b^2 + c^2 + a^2}\\ + \frac{a\, Id}{d^2 + b^2 + c^2 + a^2}$$

```
> qdisplay(xinv); #display inverse quaternion in terms of {Id, qi, qj, qk}
```

$$\frac{a\,Id}{d^2+b^2+c^2+a^2}+\frac{b\,qk}{d^2+b^2+c^2+a^2}-\frac{d\,qj}{d^2+b^2+c^2+a^2}+\frac{c\,qi}{d^2+b^2+c^2+a^2}$$

```
> xinv &q x;  x &q xinv; #'&q' is infix form of quaternionic multiplication
```

$$Id$$

$$Id$$

33.7.2 'rot3d': rotations in three dimensions.

Rotations in three dimensions are done here using quaternions. (cf. [12]) In order to see the quaternion basis used, type 'quatbasis'. Procedure 'rot3d' rotates a vector in 3-dimensional Euclidean space V using the quaternion multiplication. Any vector $v \in V$ is transformed according to the following law:

$$v \rightarrow cexp(q)\ \&c\ v\ \&c\ cinv(cexp(q))$$

where q is a pure quaternion given in the basis [e1we2, e1we3, e2we3] or [qi, qj, qk]. (cf. [7, 8]) The first entry should be a vector (or any element of the Clifford algebra $C\ell_3$) while the second element is a quaternion. Make sure to define B as a 3×3 identity matrix.

```
> specify_constants(alpha,beta,gamma); #angles specified as constants
```

Maple now knows the following constant(s) : , γ, true, α, β, ∞, E, π, FAIL, *Catalan*, false

```
> dim:=3:B:=linalg[diag](1$3):
```

Example 7.2.1. Rotations map vectors into vectors and preserve length of a vector.

```
> rot12:=cos(alpha/2)*Id+sin(alpha/2)*qk; #rotation in the 12-plane
```

$$rot12 := \cos\left(\frac{1}{2}\alpha\right) Id + \sin\left(\frac{1}{2}\alpha\right) e2we1$$

```
> v:=a*e1+b*e2+c*e3; #v is a 1-vector in Cl(3)
```

$$v := c\,e3 + a\,e1 + b\,e2$$

```
> B:=linalg[diag](1,1,1):rotated_v:=rot3d(v,rot12); #a 1-vector in Cl(3)
```

$$rotated\_v := c\,e3 + (a\cos(\alpha) - b\sin(\alpha))\,e1 + (b\cos(\alpha) + a\sin(\alpha))\,e2$$

```
> squarelength:= scalarpart(v &c v);
```

$$squarelength := c^2 + a^2 + b^2$$

```
> newsquarelength:=scalarpart(rotated_v &c rotated_v); #norm of v is preserved
```

$$newsquarelength := c^2 + a^2 + b^2$$

Example 7.2.2. Recall that by exponentiation of the pure quaternion basis {qi, qj, qk} we obtain rotations in coordinate planes. Here are the definitions and basic properties of these rotation operators. We showed in **Example 6.1.1** how to obtain these operators by using Clifford exponentiation procedure 'cexp'. Here we use already summed-up series.

```
> rot12:=cos(alpha/2)*Id+sin(alpha/2)*qk; #rotation in the 12-plane
> rot13:=cos(beta/2)*Id+sin(beta/2)*qj;   #rotation in the 13-plane
> rot23:=cos(gamma/2)*Id+sin(gamma/2)*qi; #rotation in the 23-plane
```

$$rot12 := \sin\left(\frac{1}{2}\alpha\right)\ e2we1 + \cos\left(\frac{1}{2}\alpha\right)\ Id$$

$$rot13 := \cos\left(\frac{1}{2}\beta\right)\ Id + \sin\left(\frac{1}{2}\beta\right)\ e1we3$$

$$rot23 := \cos\left(\frac{1}{2}\gamma\right)\ Id + \sin\left(\frac{1}{2}\gamma\right)\ e3we2$$

```
> rot12 &q rot12; #rotation in the 12-plane done twice
```

$$-e1we2\sin(\alpha) + Id\cos(\alpha)$$

```
> rot12 &q rot12 &q rot12; #rotation in the 12-plane done three times
```

$$-e1we2\sin\left(\frac{3}{2}\alpha\right) + Id\cos\left(\frac{3}{2}\alpha\right)$$

We find now expressions for inverse rotations in the coordinate planes:

```
> rot12inv:=map(combine,qinv(rot12),trig): #inverse of rot12
> rot13inv:=map(combine,qinv(rot13),trig): #inverse of rot13
> rot23inv:=map(combine,qinv(rot23),trig): #inverse of rot23
```

Display these inverses in the quaternion basis Id, qi, qj, qk:

```
> 'rot12inv'=qdisplay(rot12inv);'rot13inv'=qdisplay(rot13inv);'rot23'=
>                   qdisplay(rot23inv);
```

$$rot12inv = \cos\left(\frac{1}{2}\alpha\right)\ Id - \sin\left(\frac{1}{2}\alpha\right)\ qk$$

$$rot13inv = \cos\left(\frac{1}{2}\beta\right)\ Id - \sin\left(\frac{1}{2}\beta\right)\ qj$$

$$rot23 = \cos\left(\frac{1}{2}\gamma\right)\ Id - \sin\left(\frac{1}{2}\gamma\right)\ qi$$

```
> [rot12 &q rot12inv,rot13 &q rot13inv,rot23 &q rot23inv]; #verify R inverses
```

$$[Id, Id, Id]$$

```
> [rot12inv &q rot12,rot13inv &q rot13,rot23inv &q rot23]; #verify L inverses
```

$$[Id, Id, Id]$$

Example 7.2.3. Rotations preserve orthogonality. Let's rotate the basis vectors {e1, e2, e3} in the 12-plane using 'rot3d' and the rotation operator 'rot12'.

```
> e1_new:=rot3d(e1,rot12);e2_new:=rot3d(e2,rot12);e3_new:=rot3d(e3,rot12);
```

$$e1\_new := \cos(\alpha)\, e1 + \sin(\alpha)\, e2$$

$$e2\_new := -\sin(\alpha)\, e1 + \cos(\alpha)\, e2$$

$$e3\_new := e3$$

This transformation of the basis vectors can be expressed in a matrix form this way:

```
> linalg[genmatrix]([e1_new,e2_new,e3_new],[e1,e2,e3]);
```

$$\begin{bmatrix} \cos(\alpha) & \sin(\alpha) & 0 \\ -\sin(\alpha) & \cos(\alpha) & 0 \\ 0 & 0 & 1 \end{bmatrix}$$

Of course, rotations preserve orthogonality of the basis {e1, e2, e3}:

```
> map(combine,e1_new &c e2_new + e2_new &c e1_new,trig);
> map(combine,e1_new &c e3_new + e3_new &c e1_new,trig);
> map(combine,e2_new &c e3_new + e3_new &c e2_new,trig);
```

$$0$$

$$0$$

$$0$$

Example 7.2.4. Let's now see rotation of an arbitrary vector $v \in \mathbb{R}^3$:

```
> v:=a*e1+b*e2+c*e3;scalarpart(v &c v); #vector v and its square norm
```

$$v := a\, e1 + b\, e2 + c\, e3$$

$$a^2 + b^2 + c^2$$

```
> v_new:=rot3d(v,rot12); #notice that e3-direction is preserved
> no_v_new:=scalarpart( v_new &c v_new); #the vector length is preserved
```

$$v\_new := (a\cos(\alpha) - b\sin(\alpha))\, e1 + (a\sin(\alpha) + b\cos(\alpha))\, e2 + c\, e3$$

$$no\_v\_new := a^2 + b^2 + c^2$$

Likewise in the 23-plane (note that the e1 direction remains unchanged):

```
> v_new:=rot3d(v,rot23); #notice that e1-direction is preserved
> no_v_new:=scalarpart( v_new &c v_new); #the vector length is preserved
```

$$v\_new := a\, e1 + (b\cos(\gamma) - c\sin(\gamma))\, e2 + (c\cos(\gamma) + b\sin(\gamma))\, e3$$

$$no\_v\_new := a^2 + b^2 + c^2$$

We know that rotations in $\mathbb{R}^3$ do not commute:

```
> v123:=rot3d(rot3d(v,rot12),rot13): #rot12 followed by rot13, applied to v
> v132:=rot3d(rot3d(v,rot13),rot12): #rot13 followed by rot12, applied to v
> commut:=clicollect(v123 - v132);   #rotations do not commute
```

$$commut := \left(c\sin(\beta) - \frac{1}{2}b\sin(\alpha-\beta) - \frac{1}{2}b\sin(\beta+\alpha) + b\sin(\alpha) - \frac{1}{2}c\sin(\beta+\alpha) - \frac{1}{2}c\sin(\beta-\alpha)\right)e1 + \left(a\sin(\alpha) - \frac{1}{2}a\sin(\beta+\alpha) - \frac{1}{2}a\sin(\alpha-\beta) - \frac{1}{2}c\cos(\beta-\alpha) + \frac{1}{2}c\cos(\beta+\alpha)\right)e2 + \left(-\frac{1}{2}a\sin(\beta-\alpha) - \frac{1}{2}a\sin(\beta+\alpha) - \frac{1}{2}b\cos(\beta+\alpha) + \frac{1}{2}b\cos(\beta-\alpha) + a\sin(\beta)\right)e3$$

33.8 Octonions: type, conjugation, norm and inverse.

In this section we will describe a division ring of non-associative octonions. (cf. [5, 6, 7, 8, 10, 12]) Octonions are defined here as paravectors in $C\ell_{0,7}$. Octonionic type checking, conjugation, norm, multiplication and finding inverses are available in this package. In the following we will show some computations with octonions. Standard octonionic basis is defined via the initialization procedure '`init`' and is stored under a global variable '`octbasis`'.

```
> octbasis;
```

$$[Id, e1, e2, e3, e4, e5, e6, e7]$$

```
> o1:= 2*Id + 3*e1+b*e3;        #let's define an octonion
```

$$o1 := 2\,Id + 3\,e1 + b\,e3$$

```
> type(o1, octonion);           #type checking
```

true

Now we will exhibit the non-associative nature of octonions.

```
> (e1 &o e3) &o e4;
```

$$e5$$

```
> e1 &o (e3 &o e4);
```

$$-e5$$

Octonionic conjugation, norm and inverse are calculated as follows:

```
> o1:=3*Id+e1+4*e2;
```

$$o1 := e1 + 4\,e2 + 3\,Id$$

```
> o_conjug(o1);  #octonionic conjugate
```

$$-e1 - 4\,e2 + 3\,Id$$

```
> onorm(o1);   #octonionic norm
```

$$\sqrt{26}$$

```
> x:=oinv(o1);   #octonionic inverse
```

$$x := -\frac{1}{26}\,e1 - \frac{2}{13}\,e2 + \frac{3}{26}\,Id$$

```
> oinv(x) &o x;  #let's verify the inverse
```

$$Id$$

We will proceed now to verify the 'Eight-Square Identity' which says that

$$\mathtt{onorm}(u \,\&o\, v) = \mathtt{onorm}(u) * \mathtt{onorm}(v)$$

for any two octonions u and v where $*$ stands for multiplication of real numbers.

Example 8.1. 'The Eight-Square Identity'. In this example we will verify the 'Eight-Square Identity' for any two octonions u and v. (cf. [5, 7]) First, we specify 16 arbitrary constants that will serve as symbolic coefficients of these two octonions.

```
> specify_constants(u0,u1,u2,u3,u4,u5,u6,u7,v0,v1,v2,v3,v4,v5,v6,v7):
> u:=u0*Id+u1*e1+u2*e2+u3*e3+u4*e4+u5*e5+u6*e6+u7*e7;
```

$$u := u0\;Id + u1\;e1 + u2\;e2 + u3\;e3 + u4\;e4 + u5\;e5 + u6\;e6 + u7\;e7$$

```
> v:=v0*Id+v1*e1+v2*e2+v3*e3+v4*e4+v5*e5+v6*e6+v7*e7;
```

$$v := v0\;Id + v1\;e1 + v2\;e2 + v3\;e3 + v4\;e4 + v5\;e5 + v6\;e6 + v7\;e7$$

Next, we proceed to compute the left and the right sides of the identity.

```
> uv:=clicollect(u &o v);
```

$$\begin{aligned} uv := {} & (-u7\,v5 - u3\,v6 - u2\,v1 + u6\,v3 + u4\,v0 + u0\,v4 + u1\,v2 + u5\,v7)\,e4 \\ & + (u5\,v0 + u2\,v3 + u0\,v5 - u4\,v7 - u3\,v2 + u7\,v4 + u6\,v1 - u1\,v6)\,e5 \\ & + (u6\,v0 + u0\,v6 + u3\,v4 - u2\,v7 + u1\,v5 - u4\,v3 + u7\,v2 - u5\,v1)\,e6 \\ & + (u4\,v5 + u0\,v7 - u5\,v4 + u7\,v0 + u2\,v6 - u6\,v2 + u1\,v3 - u3\,v1)\,e7 \\ & + (u1\,v0 + u0\,v1 + u3\,v7 + u5\,v6 - u6\,v5 - u4\,v2 + u2\,v4 - u7\,v3)\,e1 \\ & + (u6\,v7 + u2\,v0 - u5\,v3 + u0\,v2 + u3\,v5 - u1\,v4 - u7\,v6 + u4\,v1)\,e2 \\ & + (u4\,v6 + u3\,v0 - u6\,v4 + u0\,v3 - u2\,v5 + u7\,v1 - u1\,v7 + u5\,v2)\,e3 \\ & + (-u7\,v7 - u5\,v5 - u4\,v4 + u0\,v0 - u3\,v3 - u2\,v2 - u1\,v1 - u6\,v6)\,Id \end{aligned}$$

```
> simplify(onorm(uv));
```

$$((u2^2 + u5^2 + u0^2 + u6^2 + u4^2 + u7^2 + u3^2 + u1^2)\\ (v7^2 + v4^2 + v5^2 + v2^2 + v6^2 + v0^2 + v1^2 + v3^2))^{1/2}$$

Notice that the last product is precisely `onorm(u) * onorm(v)` namely:

```
> onorm(u) * onorm(v);
```

$$\sqrt{u2^2 + u5^2 + u0^2 + u6^2 + u4^2 + u7^2 + u3^2 + u1^2}\\ \sqrt{v7^2 + v4^2 + v5^2 + v2^2 + v6^2 + v0^2 + v1^2 + v3^2}$$

Example 8.2. In this last example we verify a weak form of associativity which makes octonions an *alternating algebra.* (cf. [5]) This property of weak associativity means that the following two formulas are valid for any two octonions u and v:

$$(u \,\&o\, v) \,\&o\, v = u \,\&o\, (v \,\&o\, v) \quad \text{and} \quad v \,\&o\, (v \,\&o\, u) = (v \,\&o\, v) \,\&o\, u.$$

We will proceed as follows:

```
> vv:=clicollect(v &o v); vu:=clicollect(v &o u):
```

$$vv := 2\, v0\, v4\, e4 + 2\, v0\, v5\, e5 + 2\, v0\, v6\, e6 + 2\, v0\, v7\, e7 + 2\, v0\, v1\, e1 + 2\, v0\, v2\, e2\\ + 2\, v0\, v3\, e3 + (v0^2 - v3^2 - v2^2 - v7^2 - v4^2 - v5^2 - v1^2 - v6^2)\, Id$$

```
> simplify(uv &o v - u &o vv);    #this expression should simplify to 0
```

$$0$$

```
> simplify(v &o vu - vv &o u);  #this should also simplify to zero
```

$$0$$

A verification of *Moufang identities* (cf. [5, 6]) can be done in a manner similar to the verification of the 'Eight Square Identity' and is left as an exercise.

33.9 Working with homomorphisms of algebras.

In this final section we will show how to define in Maple algebra homomorphisms from one Clifford algebra to another. In particular, we will answer the following question posed by Z. Oziewicz in Montpellier in 1989: *Are Clifford algebras of a bilinear form more general than Clifford algebras of quadratic forms?* (cf. [11]) This question was related to a famous question posed by Albert Crumeyrolle: *What is a bivector?* (cf. [4, 9]) We will answer Oziewicz's question in dimension four.

Theorem 1. *Suppose that 'B' is a non-degenerate bilinear form in* V*,* $\dim_{\mathbf{R}}(V) = 4$*, with a non-trivial antisymmetric part and let* Q *be the quadratic form associated with the symmetric part of* B*. Then, in the category of associative algebras, there exists an isomorphism* ϕ *from the Clifford algebra* $C\ell(Q)$ *to the Clifford algebra* $C\ell(B)$ *of the bilinear form* B*. However, algebras* $C\ell(Q)$ *and* $C\ell(B)$ *are not isomorphic as Clifford algebras due to different multivector structure.*

Proof: We will proceed to prove this theorem with the package 'Clifford'. We will define a mapping ϕ from $C\ell(Q)$ to $C\ell(B)$ and prove that it is an isomorphism of algebras.

```
> with(Clifford):with(linalg):
> dim:=4: N:=2^dim:  #define dim to be 4
```

We will assume that B is nondegenerate, symbolic, and that it has a nontrivial antisymmetric part. To shorten Maple's output, we make the following symbolic assignments to the off-diagonal entries of B:

```
> B:=matrix(dim,dim,[]):
> B[1,1]:= B11: B[1,2]:= a: B[1,3]:= b: B[1,4]:= d:
> B[2,1]:=-a: B[2,2]:= B22: B[2,3]:= c: B[2,4]:= e:
> B[3,1]:=-b: B[3,2]:=-c: B[3,3]:= B33: B[3,4]:= f:
> B[4,1]:=-d: B[4,2]:=-e: B[4,3]:=-f: B[4,4]:= B44:
> 'B'=evalm(B);
```

$$B = \begin{bmatrix} B11 & a & b & d \\ -a & B22 & c & e \\ -b & -c & B33 & f \\ -d & -e & -f & B44 \end{bmatrix}$$

where a, b, c, d, e, f are totally arbitrary real numbers. We define them as Maple constants:

```
> specify_constants(a,b,c,d,e,f,B11,B22,B33,B44):
```

We know that $C\ell(Q)$ and $C\ell(B)$ have the same basis; what is different is the way basis monomials are multiplied. For example, in $C\ell(B)$ we have:

```
> clibasis:=cbasis(dim);
```

$$clibasis := [Id, e1, e2, e3, e4, e1we2, e1we3, e1we4, e2we3, e2we4, e3we4, e1we2we3, e1we2we4, e1we3we4, e2we3we4, e1we2we3we4]$$

```
> e1 &c e2;
```

$$e1we2 + a\,Id$$

```
> e1we2 &c e3we4;
```

$$e1we2we3we4 - b\, e2we4 + d\, e2we3 - e\, e1we3 - e\, b\, Id + c\, e1we4 + c\, d\, Id$$

The same multiplications done in $C\ell(Q)$ will give:

```
> zeroset:={a=0,b=0,c=0,d=0,e=0,f=0};subs(zeroset, e1 &c e2);
```

$$zeroset := \{ d = 0, b = 0, a = 0, e = 0, f = 0, c = 0 \}$$

$$e1we2$$

```
> subs(zeroset,e1we2 &c e3we4);
```

$$e1we2we3we4$$

where we have just set all off diagonal entries of B to zero. Notice that the Clifford product of `e1` and `e2` in $C\ell(B)$ is not a bivector only but that there is also a scalar part `a * Id` in that product. Thus, "what is a bivector in $C\ell(B)$"? We may verify easily that basis vectors {e1, e2} (like all other basis 1-vectors) still satisfy the following fundamental identities even though B has a non-trivial antisymmetric part:

$$\mathtt{e1}\ \&\mathrm{c}\ \mathtt{e2} + \mathtt{e2}\ \&\mathrm{c}\ \mathtt{e1} = 0,\ \mathtt{e1}\ \&\mathrm{c}\ \mathtt{e1} = \mathtt{B}[1,1] * \mathtt{Id},\ \mathtt{e2}\ \&\mathrm{c}\ \mathtt{e2} = \mathtt{B}[2,2] * \mathtt{Id}$$

```
> e1 &c e2  + e2 &c e1; e1 &c e1; e2 &c e2;
```

$$0$$

$$B11\ Id$$

$$B22\ Id$$

In the next step, we will find the 16×16 multiplication table of basic monomials of $C\ell(B)$ and store it under `MU_skew4`. To save space, we will not display these computations.

```
> MU_skew4:=matrix(N,N,[]):
> for i from 1 to N do
> for j from 1 to N do MU_skew4[i,j]:=cmul(clibasis[i],clibasis[j]) od od:
```

For example, in $C\ell(B)$ we have:

```
> 'e1we2 &c e2we4' = MU_skew4[6,10];
```

$$e1we2\ \&c\ e2we4 = -e\, e1we2 - e\, a\, Id + B22\, e1we4 + B22\, d\, Id - a\, e2we4$$

Multiplication table `MU_reg4` of $C\ell(Q)$ without the alternating part can be obtained by using the assignment of elements:

```
> MU_reg4:=matrix(N,N,[]):
> a:=0:b:=0:c:=0:d:=0:e:=0:f:=0:
> for i from 1 to N do
> for j from 1 to N do MU_reg4[i,j]:=MU_skew4[i,j]: od:od:
> a:='a':b:='b':c:='c':d:='d':e:='e':f:='f':
```

For example, in $C\ell(Q)$ we have:

```
> 'e1we2 &c e2we4' = MU_reg4[6,10];
```

$$e1we2 \&c\ e2we4 = B22\ e1we4$$

Now we will define a linear mapping ϕ between $C\ell(Q)$ and $C\ell(B)$ and proceed to prove that it is an isomorphism of algebras.

```
> phi:=proc(a:algebraic) local j,ind; global dim;options remember;
> if type(args[1],constant) then RETURN(args[1]) elif
> member(args[1],Clifford[cbasis](dim)) then do
>   ind:=Clifford[extract](args[1],'integers');
>   if ind=[] then RETURN(Id) else
>   RETURN(Clifford[clicollect](eval(&c(seq(cat(e,ind[j]),
>                                   j=1..nops(ind)))))) fi od; elif
> type(args[1],{'*','+'}) then RETURN(map(phi,args[1])) fi;
> end:
```

To see how exactly ϕ has been defined, we will show images of all 16 basis monomials of $C\ell(Q)$ contained in the list 'clibasis'. Namely,

```
> for x in clibasis do 'phi('.x.')' = phi(x) od;
```

$$phi(Id) = Id$$

$$phi(e1) = e1$$

$$phi(e2) = e2$$

$$phi(e3) = e3$$

$$phi(e4) = e4$$

$$phi(e1we2) = e1we2 + a\ Id$$

$$phi(e1we3) = e1we3 + b\ Id$$

$$phi(e1we4) = e1we4 + d\ Id$$

$$phi(e2we3) = c\ Id + e2we3$$

$$phi(e2we4) = e2we4 + e\ Id$$

$$phi(e3we4) = e3we4 + f\ Id$$

$$phi(e1we2we3) = e1we2we3 - b\ e2 + a\ e3 + c\ e1$$

$$phi(e1we2we4) = e1we2we4 - d\ e2 + e\ e1 + a\ e4$$

$$phi(e1we3we4) = e1we3we4 - d\ e3 + f\ e1 + b\ e4$$

$$phi(e2we3we4) = e2we3we4 + f\ e2 - e\ e3 + c\ e4$$

$$phi(e1we2we3we4) = -b\ e2we4 + a\ e3we4 + c\ e1we4 + f\ e1we2 - e\ e1we3 \\ + (-e\,b + a\,f + c\,d)\ Id + e1we2we3we4 + d\ e2we3$$

Then, we also have:

```
> 'phi(e1+4*e4-5*e1we2)' = phi(e1+4*e4-5*e1we2);
```

$$\phi(\, e1 + 4\, e4 - 5\, e1we2\,) = e1 + 4\, e4 - 5\, e1we2 - 5\, a\, Id$$

```
> 'phi(-2*e3we4-e2-e1we3)' = phi(-2*e3we4-e2-e1we3);
```

$$\phi(\, -2\, e3we4 - e2 - e1we3\,) = -2\, e3we4 - 2\, f\, Id - e2 - e1we3 - b\, Id$$

It is not difficult to see from the definition of ϕ that ϕ is a linear mapping (see also below).

Now we check that phi is indeed an algebra homomorphism. To do so, we need to verify the identity

$$\phi(\mathtt{x}\ \&\mathtt{c}\ \mathtt{y}) = \phi(\mathtt{x})\ \&\mathtt{c}\ \phi(\mathtt{y})$$

for any basis elements x and y in $C\ell(Q)$ and then extend it by linearity to all elements of $C\ell(Q)$.

First, we calculate the left-hand side of that identity and then we compare it with the right-hand side:

```
> leftside:=map(phi,MU_reg4):
```

For example, let's find the product of e1we2 and e2we4 in $C\ell(Q)$:

```
> u:=e1we2;v:=e2we4;  subs(zeroset,u &c v);
```

$$u := e1we2$$

$$v := e2we4$$

$$B22\ e1we4$$

and now let's find the image of e1we2 &c e2we4 under ϕ in $C\ell(B)$:

```
> clicollect(phi(subs(zeroset,u &c v)));
```

$$B22\ e1we4 + B22\ d\, Id$$

We proceed to calculate the right-hand side of the identity $\phi(\mathtt{x}\&\mathtt{c}\mathtt{y}) = \phi(\mathtt{x})\&\mathtt{c}\,\phi(\mathtt{y})$. First, we define a convenient indexing function 'ind'. This function is defined as follows:

$$(i,\, j) \xrightarrow{ind} \phi(\mathtt{clibasis}[\mathtt{i}])\ \&\mathtt{c}\ \phi(\mathtt{clibasis}[\mathtt{j}])$$

and in Maple we have:

```
> ind:=proc (i,j) option remember;
>           RETURN(cmul(phi(clibasis[i]),phi(clibasis[j]))) end:
```

Now we create a matrix called 'rightside' whose entries are calculated with the help of 'ind':

```
> rightside:=matrix(N,N,ind):
```

For example, the element of 'rightside' located in the 6th row and the 10th column is:

```
> clicollect(rightside[6,10]);
```

$$B22\ e1we4 + B22\ d\ Id$$

which is, of course, the same as

```
> 'phi(u) &c phi(v)' = clicollect(phi(u) &c phi(v));
```

$$\phi(u)\ \&c\ \phi(v) = B22\ e1we4 + B22\ d\ Id$$

On the other hand, let's recall that above we obtained

```
> 'phi(u &c v)' = clicollect(phi(subs(zeroset,u &c v)));
```

$$\phi(u\ \&c\ v) = B22\ e1we4 + B22\ d\ Id$$

Thus, for the above defined elements `u` and `v` we have verified that

$$\texttt{phi(u)}\ \&c\ \texttt{phi(v)} = \texttt{phi(u}\ \&c\ \texttt{v)}.$$

The above identity is also valid for all monomials in '`clibasis`' as it can be shown by comparing matrices '`leftside`' and '`rightside`' as follows:

```
> iszero(map(simplify,evalm(leftside-rightside)));
                              true
```

It remains to extend this verification to any two arbitrary elements u and v in $C\ell(Q)$. First, we specify arbitrary constants and define `u` and `v` (from now on long Maple outputs will not be displayed):

```
> specify_constants(u0,u1,u2,u3,u4,u12,u13,u14,u23,u24,u34,u5,u6,u7,u8,u1234):
> specify_constants(v0,v1,v2,v3,v4,v12,v13,v14,v23,v24,v34,v5,v6,v7,v8,v1234):
> u:= u0*Id+u1*e1+u2*e2+u3*e3+u4*e4+u12*e1we2+u13*e1we3+u14*e1we4+u23*e2we3+
>          u24*e2we4+u34*e3we4+u5*e1we2we3+u6*e1we2we4+u7*e1we3we4+u8*e2we3we4+
>          u1234*e1we2we3we4:
> v:= v0*Id+v1*e1+v2*e2+v3*e3+v4*e4+v12*e1we2+v13*e1we3+v14*e1we4+v23*e2we3+
>        v24*e2we4+v34*e3we4+v5*e1we2we3+v6*e1we2we4+v7*e1we3we4+v8*e2we3we4+
>        v1234*e1we2we3we4:
```

We calculate Clifford product of `u` and `v` in $C\ell(Q)$ and call it `uv` (it takes about 70 seconds of CPU time on Pentium P90 to compute and simplify that product):

```
> uv:=clicollect(subs(zeroset,simplify(cmul(u,v)))): #product uv in Cl(Q)
```

Now, we need to find the image of '`uv`' in $C\ell(B)$ under the mapping ϕ. We will call this image '`phiuv1`':

```
> phiuv1 := clicollect(simplify(phi(uv))): #image of uv under phi in Cl(B)
```

Next step is to find images of `u` and `v` under ϕ and find their product in $C\ell(B)$. We will call these images '`phiu`' and '`phiv`' respectively while the image of the product will be called '`phiuv2`'.

```
> phiu := simplify(phi(u)):phiv := simplify(phi(v)):
> phiuv2:=clicollect(simplify(cmul(phiu,phiv))):
```

Thus, we will have verified that ϕ is a homomorphism if '`phiuv1` = `phiuv2`':

```
> 'phi(u &c v) - (phi(u) &c phi(v))' =
>       collect(simplify(phiuv1-phiuv2),clibasis);
```

$$\phi(u \,\&\mathrm{c}\, v) - (\phi(u)\,\&\mathrm{c}\,\phi(v)) = 0$$

We will show now that $\ker(\phi) = \{0\}$ as follows:

```
> x:=clicollect(phiu):    #this is the image of u under phi
```

Let us assume that $\phi(\mathtt{u}) = 0$ and find u. We will show that $u = 0$. This can be done by setting up a system of 16 equations in 16 unknowns $\{\mathtt{u0}, \mathtt{u1}, \mathtt{u2}, \ldots, \mathtt{u1234}\}$ and then solving it:

```
> for i from 1 to 16 do eq.i := coeff(x,clibasis[i]) od:
> sys:={seq(eq.i,i=1..16)};  #the system of equations
```

$$\begin{aligned} sys := \{&u8, u7, u12\,a + u24\,e + u14\,d + u1234\,c\,d + u0 + u13\,b + u1234\,f\,a \\ &- u1234\,e\,b + u34\,f + u23\,c, u6, u5, u1234, u1234\,c + u14, \\ &u1 + u6\,e + u7\,f + u5\,c, u12 + u1234\,f, u24 - u1234\,b, -u1234\,e + u13, \\ &u6\,a + u7\,b + u4 + u8\,c, u5\,a - u7\,d + u3 - u8\,e, u2 - u5\,b + u8\,f - u6\,d, \\ &u1234\,d + u23, u34 + u1234\,a\} \end{aligned}$$

```
> vars:={seq(coeff(u,clibasis[i]),i=1..16)};  #the unknowns
```

$$vars := \{u8, u7, u6, u5, u34, u24, u23, u13, u14, u12, u4, u2, u3, u1, u0, u1234\}$$

```
> solve(sys,vars);
```

$$\begin{aligned} \{&u0 = 0, u3 = 0, u1 = 0, u8 = 0, u7 = 0, u6 = 0, u5 = 0, u1234 = 0, u34 = 0, \\ &u24 = 0, u23 = 0, u13 = 0, u14 = 0, u12 = 0, u4 = 0, u2 = 0\} \end{aligned}$$

Thus, if $\phi(u) = 0$ for some $u \in C\ell(Q)$ then $u = 0$ or $\ker(\phi) = \{0\}$ and ϕ is a linear isomorphism of algebras.

Finally, we check that $\phi(\alpha * \mathtt{u} + \beta * \mathtt{v}) = \alpha * \phi(\mathtt{u}) + \beta * \phi(\mathtt{v})$ for any two elements u and v of $C\ell(B)$ and some scalars α and β.

```
> specify_constants(alpha,beta):
> simplify(phi(alpha*u+beta*v) - (alpha*phi(u) + beta*phi(v)));
```

$$0$$

Thus, we have proved that $\phi : C\ell(Q) \to C\ell(B)$ is an isomorphism of associative algebras which does not preserve gradation (cf. [1]).

End of Proof.

Acknowledgements

The author thanks Joe Riel of San Diego, Francis J. Wright of London, England, and Robert Israel of British Columbia, Canada, for their valuable suggestions regarding Maple programming.

Bibliography

[1] R. Abłamowicz, and P. Lounesto, 'On Clifford algebras of a bilinear form with an antisymmetric part', in "Clifford Algebras with Numeric and Symbolic Computations", eds. R. Abłamowicz, P. Lounesto, and J. M. Parra, Birkhäuser, Boston, 1996, pages 167 – 188.

[2] R. Abłamowicz, and P. Lounesto, eds.: 1995, 'Clifford Algebras and Spinor Structures: A Special Volume Dedicated to the Memory of Albert Crumeyrolle (1919-1992)', Kluwer Academic Publishers, Dordrecht.

[3] R. Abłamowicz, P. Lounesto, and J. Maks: 1991, 'Conference Report, Second Workshop on "Clifford Algebras and Their Applications in Mathematical Physics", Université des Sciences et Techniques du Languedoc, Montpellier, France, 1989', *Found. Phys.* **21**, pp. 735–748.

[4] A. Crumeyrolle: 1990, 'Orthogonal and Symplectic Clifford Algebras: Spinor Structures', Kluwer, Dordrecht.

[5] Geoffrey M. Dixon: 1994, 'Division Algebras: Octonions, Quaternions, Complex Numbers and the Algebraic Design of Physics', Kluwer Academic Publishers, Dordrecht.

[6] F. Reese Harvey: 1990, 'Spinors and Calibrations', Academic Press, San Diego.

[7] I. L. Kantor and A. S. Solodnikov: 1980, 'Hypercomplex Numbers: An Elementary Introduction to Algebras', Springer-Verlag, New York.

[8] P. Lounesto: 1995, 'Clifford Algebras and Spinors', to be published.

[9] P. Lounesto: 1993, 'What is a bivector?', in 'Spinors, Twistors, Clifford Algebras and Quantum Deformations', Proceedings of the Second Max Born Seminar Series, Wrocław, Poland, 1992; eds. Z. Oziewicz, B. Jancewicz, and A. Borowiec Kluwer, Dordrecht, pp. 153–158.

[10] P. Lounesto, R. Mikkola, and V. Vierros: 1987, 'CLICAL User Manual', Helsinki University of Technology, Institute of Mathematics, Research Reports **A248**, Helsinki.

[11] Z. Oziewicz, and Cz. Sitarczyk: 1992,'Parallel treatment of Riemannian and symplectic Clifford algebras', in "Clifford Algebras and Their Applications in Mathematical Physics", eds. A. Micali, R. Boudet, and J. Helmstetter, Kluwer, Dordrecht.

[12] Ron Shaw: 1991, 'Symmetry', An Inaugural Lecture delivered on Tuesday, 29th January, 1991, in 'Mathematical Perspectives, Four Recent Inaugural Lectures', Hull University Press, Hull.

[13] S. Somaroo, Cambridge, England: 1995, private communication.

Index